Anatomy of Dolphins

Anatomy of Dolphins
Insights into Body Structure and Function

Bruno Cozzi
Department of Comparative Biomedicine and Food Science,
University of Padova, Padova, Italy

Stefan Huggenberger
Department II of Anatomy, University of Cologne,
Cologne, Germany

Helmut Oelschläger
Institute of Anatomy III (Dr. Senckenbergische Anatomie),
University of Frankfurt, Frankfurt, Germany

Illustrations by Massimo Demma, Uko Gorter, and Jutta Oelschläger

AMSTERDAM • BOSTON • HEIDELBERG • LONDON • NEW YORK • OXFORD • PARIS
SAN DIEGO • SAN FRANCISCO • SINGAPORE • SYDNEY • TOKYO

Academic Press is an imprint of Elsevier

ELSEVIER

Academic Press is an imprint of Elsevier
125 London Wall, London EC2Y 5AS, United Kingdom
525 B Street, Suite 1800, San Diego, CA 92101-4495, United States
50 Hampshire Street, 5th Floor, Cambridge, MA 02139, United States
The Boulevard, Langford Lane, Kidlington, Oxford OX5 1GB, United Kingdom

Notices
Knowledge and best practice in this field are constantly changing. As new research and experience broaden our understanding, changes in research methods, professional practices, or medical treatment may become necessary.

Practitioners and researchers must always rely on their own experience and knowledge in evaluating and using any information, methods, compounds, or experiments described herein. In using such information or methods they should be mindful of their own safety and the safety of others, including parties for whom they have a professional responsibility.

To the fullest extent of the law, neither the Publisher nor the authors, contributors, or editors, assume any liability for any injury and/or damage to persons or property as a matter of products liability, negligence or otherwise, or from any use or operation of any methods, products, instructions, or ideas contained in the material herein.

Library of Congress Cataloging-in-Publication Data
A catalog record for this book is available from the Library of Congress

British Library Cataloguing-in-Publication Data
A catalogue record for this book is available from the British Library

ISBN: 978-0-12-407229-9

Printed in the United States of America
Last digit is the print number: 9 8 7 6 5

For information on all Academic Press publications
visit our website at https://www.elsevier.com/

Working together
to grow libraries in
developing countries

www.elsevier.com • www.bookaid.org

Publisher: Sara Tenney
Acquisition Editor: Kristi Gomez
Editorial Project Manager: Pat Gonzalez
Production Project Manager: Lucía Pérez
Designer: Maria Inês Cruz

Typeset by Thomson Digital

Contents

About the Authors

 **Bruno Cozzi** obtained his degree in Veterinary Medicine (Dr. Med.Vet.) with honors in 1980 from the Faculty of Veterinary Medicine at the University of Milan (Italy) and his PhD degree in 1993 from the Faculty of Health Sciences at the University of Copenhagen (Denmark). In 1999 he was appointed full professor of veterinary anatomy at the University of Padova (Italy). His scientific production is focused mainly on comparative neuroendocrinology and neuroanatomy of large domestic herbivores, marine mammals, and man. In 2002 he founded the Mediterranean Marine Mammal Tissue Bank.

 **Stefan Huggenberger** studied in Cologne (Germany) and graduated with the diploma in Biology on morphological aspects of the harbor porpoise population in the Baltic Sea. Since 1999, he performed research on the echolocation system of toothed whales (PhD in Biology). Next to several scientific articles he published chapters in German textbooks about whales and dolphins illustrated with high-quality graphic work. Current scientific projects focus on the neuroanatomy and physiology of the auditory system in amphibians, rodents, and particularly toothed whales. In 2010, Dr. Huggenberger organized a marine mammal morphology workshop in Stralsund (Germany) which was the intellectual starting line of this book on the anatomy of dolphins.

 **Helmut Oelschläger** (Dr.rer.nat.) is a senior lecturer, senior scientist, and professor in human anatomy. After studying biology and chemistry in Tübingen (Germany) he did his PhD and habilitation in Frankfurt am Main. He received several scientific awards and organized two marine mammal workshops (Kyoto and Tokyo 2000, Frankfurt 2001). Prof. Oelschläger is an experienced morphologist and neurobiologist and received numerous grants. Scientific topics are: the terminal nerve, magnetic orientation in mammals, and the structure of the mammalian head and nervous system (sensory organs, brainstem, neocortex, ontogenetic development). His main focus lies on the comparative neurobiology and the evolution of whales and dolphins.

About the Illustrators

Massimo Demma is an illustrator specialized in Natural History. He collaborates with scientists to achieve a thorough understanding of the structure and morphology of their subjects, and pass on this knowledge through visual representation. He lives in Milan, Italy.

Uko Gorter was born in Arnhem, Holland, and is a natural history illustrator specialized in marine mammals. He is a fellow of several scientific societies that study whales and dolphins. His artwork appeared in several journals and books. He lives in Seattle, WA.

Jutta Oelschläger is a graduated Corel Draw artist. For over 30 years she has been a technician in photography and scientific publishing in zoology and comparative anatomy.

Foreword

Colleagues, students, veterinarians, biologists, often ask me about dolphin anatomy. I can tell them how to find the thyroid or the ear, the tonsils, or even the hippocampus. I can give them reference for specialist's papers in various journals. Quite a few of these specialist papers are written by the authors of this volume. However, there is no current dolphin anatomy textbooks where one can find nearly everything that is known in one volume. The authors are my valued colleagues and friends. In the past, I have collaborated on various projects with Dr. Helmut Oelschläger and with Dr. Stefan Huggenberger. I have kept up with Dr. Bruno Cozzi through his published writings.

I think that the information presented in this new book will be useful for beginners in the field and even for specialists focused on a narrow aspect of cetacean science. Naturalists, zoologists, marine biologists, and veterinarians will want this book on their shelf as a reference. The authors give a detailed description of each body system. The writing still remains clear enough for even a novice to understand. I especially enjoyed the numerous comparisons across several different species for a more comprehensive explanation. I also found the figures incredibly helpful. The visuals not only complement the text well, but the drawings are very clear and well-organized; it is easy to distinguish all of the detailed parts of the images. The brain diagrams are especially impressive. Kudos to Massimo Demma, Uko Gorter, and Jutta Oelschläger for their impressive contributions.

I particularly enjoyed the chapter on dolphin brains. It is relevant to work on dolphin communication, navigation, and cognition. I appreciated the fact that the authors discussed each of the brain structures from an evolutionary, comparative, and behavioral context to offer a more well-rounded description. This approach was most apparent in the brain chapter, but it was used nicely elsewhere.

Not all chapters are of the same length and depth of content. Not all figures are of very fresh or idealized specimens. Some figures come from specimens with long postmortem times. These figures will be particularly relevant to those working with the many stranding networks around the world where so many important specimens may have long postmortem times before being available for dissection. To a great extent, the number of anatomical details given in each chapter mirrors the actual status of anatomical research in dolphins. However, much of the current knowledge and information on the topic is found in obscure publications of the past, and in languages other than English, or scattered in short chapters in volumes dedicated to dolphin and whale biology.

This book is the result of decades of more or less continuous work on cetaceans. It is fortuitous that these authors were able to bring this work together. It brings back needed focus on morphology and serves as a starting point for future advances in dolphin anatomy.

Sam Ridgway

DVM, PhD, DACZM
National Marine Mammal Foundation
San Diego, CA, United States
www.nmmf.org

Introduction

The idea of writing this book came spontaneously in 2010, after a successful workshop on cetacean morphology associated with the meeting of the European Cetacean Society (ECS) was held in Stralsund (Germany) that year: we realized that the body of dolphins was (and still is) a partially uncharted territory. So we discussed, organized our notes, graphical material and references, and—after several years—here we are.

WHAT IS THIS BOOK?

We prepared this manual of functional anatomy with the idea of providing a comprehensive reference text on the body structure of the bottlenose dolphin and other closely related species. We focused on dolphins (with the other cetaceans in the background) for a number of reasons, but mostly because these relatively small-toothed whales are the commonest marine mammals and also among the most studied species. There is an ever-increasing body of literature on dolphin pathology, toxicology, behavior and other disciplines, and the place of anatomy is just at the base of all constructions. The first question should always be "How is the engine built?" and then "How does it work?" Incredibly, knowledge on dolphin physiology and pathology has progressed in many ways, but anatomy has been left behind. Dolphins are protected in many countries of the world, but the bodies of stranded animals are relatively easy to obtain for research centers and universities. So, why are there only a few studies every year dedicated to the specifics of the body of these fascinating mammals? Is anatomy a science of the past? Maybe, but there are so many questions left unanswered on the morphology of dolphins that one may wonder if there is some other factor at play behind the limited interest. Certainly anatomy has a bad reputation: a smelly, messy, old-fashioned discipline that does not have the prestigious allure of the past. Yet we are convinced that the reader of this book will find some of the descriptions of the organs and systems interesting and useful as a basis to a deeper understanding. Perhaps some of our questions and doubts, expressed here and there in the different chapters, will stimulate new research and promote innovative approaches. We hope.

Coming back to the scope of the book, our focusing on members of the dolphin family does not imply that we are downsizing the importance of beaked whales, baleen whales, and other marine mammals; quite the contrary! But dolphins are the core group from which to start. There are also practical reasons for centering our attention on this family. Veterinary medicine is expanding its field of clinical interest to marine species: dolphins are monitored for understanding their physiological and endocrine parameters in the wild, examined when wounded, and rescued whenever possible after stranding. Furthermore captive dolphins—whatever your opinion on captivity—receive first class medical care and knowledge of their body systems, and a thorough understanding of their anatomy is a fundamental requisite for correct diagnosis and treatment. Finally we mention the attention given by several research groups to the infectious diseases of dolphins, for all the implications on animal and human health. In this sense normal anatomy comes before pathology and must be the guideline for a complete postmortem and correct tissue sampling.

We three are professors of anatomy (although our background has more than one color) and obviously we think that ours is a basic science, a must. But, whatever our pride in the old glorious discipline, there is simply a cultural need to know more about the dolphin body, to understand how these species are adapted to live in such a challenging environment, how they are able to communicate, explore, and interact. And also, how they move, feed, save water, survive, and thrive. We think that this book will answer at least some of these questions and appease the reader's thirst for knowledge about these wonderful animals.

So this book could be used by professionals (biologists, veterinary medical doctors) that interact with dolphins alive or dead either during their training years, or later as a consulting manual in daily practice, at the beach, in the lab, at the side of the aquarium tank, or in the boat. But we also believe that *Anatomy of Dolphins* could be a key reference book for conservation science experts and people working in museums, marine biology departments, national agencies dealing with life in the water and interaction with fisheries. Needless to say, comparative anatomists may like the book just because it sheds light (or brings together facts) on relatively unknown species and organ systems difficult to investigate.

TERMINOLOGY AND BACKGROUND

The reader unfamiliar with anatomy may find some parts of the book more difficult to read than others, especially where some technicalities were impossible to avoid (ie, in the description of the brain, of the ear and auditory system, topography of the heart, mechanisms of water balance and so forth, including the terminology of the ribs!). However, we do hope that most of the descriptions will be easy to follow, the language common enough and the jargon reduced to a minimum (well, at least that was our intention). Footnotes are there to help, with them we hope to amuse and stimulate the interest of the curious.

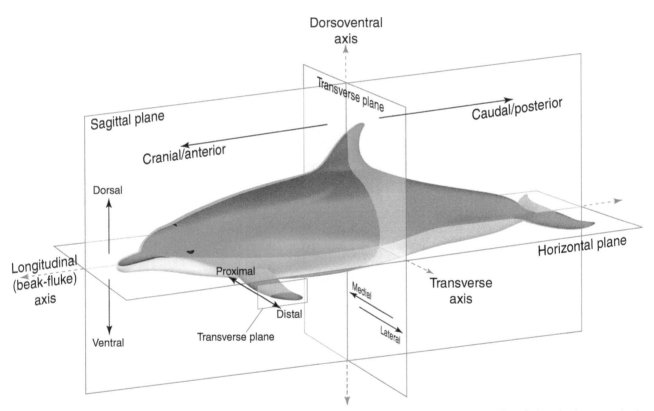

FIGURE Diagram showing the reference axis and axes of a dolphin. The horizontal plane is often referred to as frontal plane the dorsoventral axis as sagittal axis. The sagittal plane that divides the body into two equal (symmetrical) parts may be called mid-sagittal plane; planes lateral to the latter are called parasagittal. *(Artwork by Uko Gorter.)*

Readers with a medical background (PhDs, DVMs, MDs and other health professionals) will quickly recognize familiar terminology and well-known structures, even perhaps with a new shape and an unexpected functional twist. Those of you with an experience in natural sciences may find new names for parts of the body that you were accustomed to call differently (ie, the so-called "jugal" bone is in fact the zygomatic bone). This is because in this book we adopted the two internationally recognized anatomical nomenclatures, the Terminologia Anatomica (TA, human anatomical nomenclature) and the Nomina Anatomica Veterinaria (NAV, veterinary medical nomenclature). There was really no choice and we had to follow what is now the widely spoken anatomical language, essential to identify corresponding structures in the body of different mammals. The obvious and possibly more correct choice would have been to adopt only the NAV, since dolphins are animals and their morphology is in several aspects very similar to that of their land cousins, the ruminants. However, we also considered that most of you have studied or are familiar with the anatomy of man, and not with that of hoofed animals. So we have made as many references as possible to the situation of the human body. This would also simplify comprehension of some morphological aspects for readers with a different approach. The book contains many notes that just try to shed light on why a certain component of the body has a certain name (and sometimes a rather obscure name!). To make a long story short: most of the book has been written for the general reader with a minimal biological background. The only possible exception is the chapter dedicated to the brain that may appear rather complex to the lay reader. On the one hand we hope it could be stimulating and fun. On the other hand, dolphin neuroanatomy received more attention than other parts of their body, so we had more information to elaborate and discuss. In general, the length of the chapters and the details given therein represent largely the status of anatomical and functional research up to date. To this effect we added as many illustrations (macro- and microphotographs, drawings) as possible, to clarify the text as much as possible. We are also working on an anatomical atlas of dolphin anatomy (*An Interactive Atlas of the Anatomy of Dolphins, Academic Press*), a companion to the present book that we hope will be soon available to the reader.

THE CHOICE OF THE SPECIES

As we said, we centered the book on the bottlenose dolphin *Tursiops truncatus*, because it is a rather cosmopolitan cetacean, and the one most popular, and also because of the number of published studies available on the behavior and comparative psychology that concern the brain and cognitive capacities of this species. We also considered the striped dolphin *Stenella coeruleoalba* (and other members of the same genre), the short-beaked common dolphin *Delphinus delphis*, the Risso's dolphin *Grampus griseus,* and the closely related Atlantic white-sided dolphin *Lagenorhynchus acutus*, and white-beaked dolphin *Lagenorhynchus albirostris* (and other members of the same genre), for comparisons and because these species have been well studied in the wild and for the importance of their interaction with fisheries. We included—whenever possible—references to larger members of the family, like the long-finned pilot whale *Globicephala melas*, the false killer whale *Pseudorca crassidens*, and the killer whale *Orcinus orca*. In fact, we would have liked to give more details on these larger species, and especially on the killer whale. Unfortunately, their anatomy is poorly studied and largely incomplete, and we were often left without elements.

Now and then we inserted references to the harbor porpoise *Phocoena phocoena*, that is obviously not a member of the delphinid family. We reversed to the harbor porpoise where references to the corresponding body part of dolphins were missing or largely incomplete (ie, the ear region, the brachial plexus and the innervation of the flipper) and also because of the long tradition of anatomical studies (see history in Chapter 1) that made the harbor porpoise the first small-toothed whale to be relatively well known to the scientific community.

As the reader may understand, the choice of species was also influenced by the fact that we live in Europe and are more familiar with cetaceans that inhabit the Mediterranean and Black Sea, the North Sea, or the Eastern Atlantic Ocean. However most of the dolphin species that we describe here are diffuse everywhere and have therefore a worldwide importance.

A NOTE TO THE READER

We did our best to present a book as complete as possible, and to support the text with as many quotations as was reasonable (including a selection of the very old ones, sometimes the most precious for our scope). However, it is impossible to quote everything and everybody. The reader will probably find mistakes, flaws in the descriptions, or will consider that we have been too superficial in certain parts, only to be boringly detailed elsewhere. The only strategy that we may adopt is to ask you all to signal us where you do not agree with what we wrote, where you'd like to know more, or simply to suggest one or more specific and fundamental quotations that we overlooked. Dear readers, if you published a seminal paper (or are aware of somebody who did) and do not find it quoted here, let us know. If the book will have an extended life, we will fix our mistakes, with your cooperation. Thank you.

ACKNOWLEDGMENTS

This book was made possible by the vast collection of tissues, organs, and skeletons of the Mediterranean Marine Mammal Tissue Bank of the University of Padova (www.marinemammals.eu). We are all indebted to the Bank and its personnel, and especially Maristella Giurisato who found the right specimens for us, and Sandro Mazzariol, who, as a pathologist, actively contributed to the Bank and selected for us animals and tissues free of pathological conditions. A special thanks to Mattia Panin, postdoctoral fellow at the Department of Comparative Biomedicine and Food Science of the University of Padova, who coauthored Chapter 7 (Body Control), and performed many dissections with us. Another special thanks goes to Pietro Saviano (Modena), who gave us the permission to use a choice of ultrasound images of dolphin anatomy from his professional collection.

Most of the chapters have been extensively revised by external experts. These coworkers (and friends) spent their precious time in tedious revision of our confused drafts. If the final result that you are reading contains relatively few mistakes, it is mostly because they amended the text before and suggested additional articles to read to convince us of our mistakes. The errors that you may still find are due only to us, perhaps because after all we did not fully believe what our reviewers said, and/or ignored their suggestions. Here we would like to acknowledge the precious help of (in alphabetical order) Andreas Fahlman (University of Corpus Christi), Claudia Gili (Aquarium of Genova), Annamaria Grandis (University of Bologna), Michela Podestà (Museum of Natural History of Milan), Ada Rota (University of Turin). We are also indebted to Cristiano Bombardi (University of Bologna), Francesco Mascarello, Alessandro Zotti, Tommaso Banzato, Cristina Ballarin and Antonella Peruffo (University of Padova), Charlene Steinhausen (University of Cologne), Giuliano Doria and Roberto Poggi (Museum of Natural History "G. Doria", Genova). Thanks also to Giuseppe Palmisano, Michele Povinelli, and Emanuele Zanetti, of the dissecting room of the Department of Comparative Biomedicine and Food Science of the University of Padova as well as Brigitte Dengler, Maja Dinekov, Peter Hempel, and Jolanta Kozlowski from the labs in the Institute of Anatomy of the University of Cologne: they did the actual job. We would also like to thank Simone Panigada and Sabina Airoldi of the Tethys Research Institute (Milan), for sharing with us some of their beautiful images of dolphins at sea.

From the side of our tradition in cetacean research at the University of Frankfurt am Main, we would like to cordially thank the founder Milan Klima who invested decades of effort with his personnel to create unique collections of microslide series of many dolphin species which were indispensable to the investigation of the morphology and development of many organ systems, and who inspired us with his fascination for dolphins. Many people made wonderful scientific contributions to our knowledge on dolphins: Eberhard Buhl, Thomas Wanke, Thomas Holzmann, Birgit Kemp, Frank Schulmeyer, Holger Jastrow, Pia Comtesse, Michael Rauschmann, Natalie Kappesser, Lars Kossatz, Michaela Haas-Rioth, Christian Fung, Clemens Poth, Alexander Kern, Pascal Malkemper, Eliabeth van Kann, and Julian Knopf. Much of their results has been integrated into several chapters of this book. Invaluable technical assistance came from Inge Szasz-Jacobi, Irmgard Kirschenbauer, Horst L. Schneeberger, and Georg Matjasko. Without these people it would not have been possible to collect and present all the information in the book. Walter K. Schwerdtfeger and Heinz Stephan (Max- PLanck, Institute in Frankfurt am Main) are thanked for their assistance and help as to quantitative neuroanatomical research. Pieter van Bree, William F. Perrin, Oystein Froiland, Ursula Siebert, and Susanne Prahl generously collected and dedicated invaluable dolphin material.

An outstanding resource of chances and perspectives for successful research on the brain of dolphins is the famous histological collection of Giorgio Pilleri (Courgeveaux/Switzerland) on many of the dolphins we have been dealing with in our book. He spent a lifetime of intensive scientific work on dolphins and set a basis for future significant investigations on their neuroanatomy. We are grateful to Gerhard Storch and Irina Ruf, curators of the Research Institute and Natural History Museum Senckenberg in Frankfurt am Main, and Jim Mead and Charley Potter from the Natural History Museum of the Smithsonian Institution in Washington, DC, for their support to our research activities on dolphins. We kindly thank Tina Büdenbender, David Maintz, and Robert Rau (Department of Radiology, University Hospital Cologne, Germany) for their help in scanning dolphin specimens. The Dr. Senckenberg foundation in Frankfurt am Main is thanked for generous financial contributions to this book project. Jörg Stehle (Department of Anatomy, Frankfurt am Main) is thanked for his continuous support to this project over these years. We are indebted to John E. Zoeger for his important scientific input, and our personal thanks go to Sam H. Ridgway (San Diego, USA) for his enduring help throughout the last decades and friendship.

And finally the book would not be what it is without the great contribution of the three artists who illustrated it (and had the patience to bear with us): Massimo Demma (Milan), Uko Gorter (Seattle), and Jutta Oelschläger (Frankfurt am Main). The book is also theirs.

Bruno Cozzi
University of Padova, Italy

Stefan Huggenberger
University of Cologne, Germany

Helmut Oelschläger
University of Frankfurt am Main, Germany

Chapter 1

Natural History and Evolution of Dolphins: Short History of Dolphin Anatomy

EVOLUTIONARY BIOLOGY OF WHALES AND DOLPHINS

Dolphins and whales (Cetacea), in addition to manatees (Sirenia), are the only mammals that are fully adapted to life in water. The numerous adaptations to the aquatic environment represent an amazing evolution level (Laitman, 2007; Reidenberg, 2007). The monophyly of the Cetacea is well established by morphological and molecular biological characteristics. The traditional systematic classification in toothed whales (Odontoceti) and baleen whales (Mysticeti) is well secured. The ancestors of cetaceans were land mammals from the group of even-toed hoofed animals that lived about 60 million years ago. Their close relationship with the Artiodactyla is clear from molecular studies that indicate that the cetaceans are closely related to the hippos (Hippopotamidae) (Price et al., 2005). Therefore, the cetaceans are now nested within the Artiodactyla and the whole group is called accordingly Cetartiodactyla in modern textbooks.

On the morphological level, fossil finds of nearly complete skeletons revealed that the ankle joint (hock) of middle-Eozän Archaeoceti were clearly like that of even-toed ungulates since they had a talus (astragalus) of the typical double-pulley form, an autapomorphy[a] of Artiodactyla. Further apomorphies of Cetartiodactyla are the par-axis extremities and the fibroelastic type of penis with a proximal sigmoid flexure. In comparison with the hippos, cetaceans share a multiloculare (multichamber) stomach system, the nearly hairless skin, and the structure of the larynx entrance (Frey et al., 2015).

The Cetacea derived from the Mesonychidae, a group from the main line of ungulates († condylarths) in the Paleocene. From them the archaeocetes emerged in the lower Eocene, about 50 million years ago, as the first group fully adapted to the aquatic lifestyle. Due to their endothermy, it was possible for the archeocetes to populate all sea habitats regardless of the ambient temperature and also large river systems. Moreover, the fact that oxygen uptake via lungs is more effective compared with respiration by gills because of the higher oxygen content in air, this was probably one of the advantages for the evolution of these agile giant forms in the water. The sizes of cetaceans ranging from approximately 1.25 m of length and 25 kg weight for the La Plata dolphin (*Pontoporia blainvillei*) up to a length of 33.5 m and 190 t for the blue whale (*Balaenoptera musculus*), the largest animal that ever lived on earth. Other characteristics of mammals that may have a positive effect in the conquest of the aquatic environment were the completely separate two-chambered heart with an efficient circulation system, nucleus-free red blood cells, the placenta and the protected embryonic development, intensive parental care and thus greater success reproduction rate, highly social behavior repertoire and well-developed auditory organs as a basis for the development of an echolocation system. The ability for advanced hearing under water, that is, the reduced mastoid process to detach ear bones from the skull, was found in all baleen and toothed whales since the Oligocene (around 30 million years ago). Additionally, the first anatomical characteristics of echolocation were found in the earliest toothed whales of the Oligocene (Uhen, 2007). Among these characteristics are the facial fossa that houses the nasal complex and large basicranial fossae that housed the pterygoid sinuses (see Chapter 5) (Fordyce and Muizon, 2001).

---

a. An autapomorphy is a derived anatomical characteristic, that is unique to a given taxon (a taxonomic unit of related organisms), but not found in any other related group.

ODONTOCETI (TOOTHED WHALES)

Next to the baleen whales (Balaenopteroidea, Mysticeti), four monophyletic groups of toothed whales can be distinguished: Physeteroidea, Ziphioidea, Platanistoidea (river dolphins), and Delphinoidea (Huggenberger and Klima, 2015). In contrast to baleen whales, the toothed whales are characterized by the single blowhole and teeth.

1. Physeteroidea

   The Physeteroidea divide into two groups, the Physeteridae (sperm whales, one species) and the Kogiidae (dwarf sperm whales, two species) (Cagnolaro et al., 2015). Both groups are characterized by dentition only in the lower jaw.

   The giant sperm whale (*Physeter macrocephalus*) has an excessively large forehead enlarged by the nasal structures with two large fat bodies, the so-called spermaceti organ and the junk. The blowhole is asymmetrically rostrodorsal on top of the box-like nose. *P. macrocephalus* is the largest recent toothed whale species (males on average 18 m, females 12 m, up to 57 t) and its bow-shaped head is up to one third of the total length with several tons of spermaceti oil. Squid from greater depths are its main food source. It is distributed worldwide and the best known example is *Moby Dick* from H. Melville's novel.

   Moreover, there are two species of Kogiidae, *Kogia breviceps* and *Kogia sima*. Their skull is extremely asymmetrical. Their diet is squid and cuttlefish from about 200 m depth of the continental shelf. *K. breviceps*, the pygmy sperm whale, is up to 4 m and 450 kg. *K. sima* is slightly smaller (about 2.7 m and 210 kg), but otherwise very similar, so they were not distinguished until 1966.

2. Ziphioidea

   The ziphiids (beaked whales) are a group of 21 species of small- to medium-sized (4–12 m) toothed whales (Wilson and Reeder, 2005). A small dorsal fin is situated at the rear end of the body. The jaws form a long curved beak in some species. The teeth are reduced; usually there are only a pair of teeth in the lower jaw or no teeth. Above the bony nasal openings the skull bones form a roof-like projection (synvertex). Only a few species are well known. Best known is *Hyperoodon ampullatus* (Northern bottlenose whale, 7–9 m, to 8 t) who has a high bulbous forehead above its "duckbill" rostrum. One species, *Mesoplodon europaeus*, is known so far only from a few stranded remnants of dead animals and unconfirmed observations at sea.

3. River dolphins

   The artificial group of river dolphins consists of small cetacean (max. 2.6 m) characterized by long tweezers-like beaks, small dorsal fins, and large flippers. There are two groups of river dolphins (Wilson and Reeder, 2005)—Platanistoidea, endemic in South Asian river systems, and Inioidea, living in South America. In the former group there is only one species (Rice, 1998), the Ganges dolphin (*Platanista gangetica*).[b] Its head is highly movable because the cervical vertebrae are not fused and the jaws have up to 150 teeth for catching prey fish in muddy waters or sandy soil. It is almost blind as an adaptation to the turbid muddy waters of its habitat in the Indus and Ganges river systems. Males are up to 40 cm smaller than females. Their food are fish and crustaceans.

   The second group, Inioidea, is more closely related to the Delphinoidea than to the other river dolphin group (Price et al., 2005) and up to four species are discussed. Among them is the Amazon dolphin (*Inia geoffrensis*) and the La Plata dolphin (*P. blainvillei*), which have a strongly curved melon. The La Plata dolphin is the smallest cetacean, 1.25–1.70 m, 25–53 kg. It has a small dorsal fin, eyes are relatively well developed, and it has adapted to life in the coastal waters and estuaries on the east coast of South America. The Chinese river dolphin (*Lipotes vexillifer*) is regarded as extinct since 2007.

4. Delphinoidea

   The Delphinoidea can be distinguished into three groups:

   a. There are only two species belonging to the Monodontidae, the narwhale and the Beluga. These animals have spherical heads and no dorsal fin. The cervical vertebrae are not fused and thus restricted head movements are possible. They are gregarious species in the arctic and subarctic waters where they search for ground-dwelling fish, crabs, and worms as food.

      The narwhale (*Monodon monoceros*) is 3.8–5 m long (without tusks; 0.8–1.6 t). Males have one (exceptionally two) up to 3 m long tusk (left canine) protruding from the maxilla, which is the origin of the unicorn legend. The Beluga or white whale (*Delphinapterus leucas*) is slightly larger (4–6 m, 0.4–1.5 t) than the narwhale. They have a large rounded melon that can change its face obviously.

---

b. Some authors discuss a second species: *Platanista minor* (Wilson and Reeder, 2005).

b. The porpoises (Phocoenidae, six species) are a group of small toothed whales (max. 2.2 m) with a compact body and a rounded head without beak. The small and laterally flattened teeth (up to 116) are characteristic. Best known is the harbor porpoise (*Phocoena phocoena*), only 1.3–1.9 m long (40–90 kg), although it has an inconspicuous black-gray color. Its prey consists mainly of fish. For the vaquitas (*Phocoena sinus*) the situation is threatening. The number of this endemic porpoise from the Gulf of California is estimated at only 150 individuals.

c. The dolphins (Delphinidae) is the taxon with the largest number of members (approximately 34 species). They are small to medium sized toothed whales (1.25–10 m) with a prominent dorsal fin. The jaws form a beak, which is in some species dominated by a large rounded melon. The teeth are well developed, usually polyodont and homodont. Fish is the preferred prey of these fast and agile swimmers and this group is distributed from the North to the South Pole. Best known are probably the killer whale, which is the largest dolphin, and the bottlenose dolphin because it is an inhabitant of coastal regions and often kept in Delphinaria. Thus, the bottlenose dolphin is the key species in this book.

NATURAL HISTORY AND GENERAL BIOLOGY OF DOLPHINS

Dolphins are famous for their temporarily large accumulations, even with different species. For example, due to good food availability groups of these gregarious animals can count up to thousands of animals. However, groups with a consistent social structure (in dolphins called schools) usually include 10–50 animals. Within schools, there is a certain hierarchy. The primary position can be claimed by aggressive behaviors; often fought using loud sound signals. This may also lead to aggressive behaviors between groups. However, in general, dolphins use peaceful body contact to strengthen the cohesion of a group and individuals through contact swimming, petting, or nudging. The play behavior of dolphins, which is characterized by wild swimming, turns, surfing, leaps, or flipper and fluke strokes, etc., is not just limited to young animals. One can only speculate whether these playful behaviors have a communicative meaning.

Since Aristotle's era, man has puzzled about the special "mysterious" care-giving behavior (epimeletic behavior) where young, weak, sick, or injured conspecifics were given assistance in that they were pushed to the water surface so they did not drown. This behavior is so strong that sometimes other species of whales, various floating objects in the water, or even dead sharks—otherwise rivals of dolphins—are "saved" in this way. The fact that people benefit occasionally of such "rescue operations" is a logical consequence. Homer's description that Telemachus, son of Odysseus, was rescued by a dolphin, as well as many other ancient stories or fables of various primitive tribes, in which people owe their lives to dolphins, should therefore not dismissed as fictitious. Indeed, there are documented cases of people rescued by dolphins in modern times. The origin of this distinct behavior is likely the caretaking behavior of mothers who have to bring their newborn calves to the surface for breathing. For example, some porpoise species are known to have a rough back equipped with ceratin plates so that the calves can be piggybacked without slipping. Perhaps this behavior is evolutionarily old. In some dolphin fossils from the upper Miocene, a rough skin was found on their back, which could have served the same purpose (Huggenberger and Klima, 2010).

It is noteworthy that some behaviors of dolphins are actually interpreted as philanthropic traits usually not found in other mammals. These social behaviors are now used for the treatment of children with disabilities whose playful contact with dolphins in the water should bring therapeutic success.

ADAPTATIONS OF DOLPHINS TO LIFE IN WATER

In this chapter we want to compile a synthetic overview of anatomical adaptation of cetaceans in which the dolphins are nested. Because cetaceans are part of the order of even-toed ungulate (Artiodactyla), dolphins are characterized by a plethora of specialized adaptations that were accomplished rapidly during the course of cetacean evolution.

The body of dolphins has an almost perfect streamlined shape. Disturbing external attachments such as hair, ears, and hind limbs were atrophied. The head is connected to the body without a distinct neck so that it is merged to a single unit. The large boneless caudal fin (fluke) as the actual organ of locomotion is an evolutionary innovation. It is horizontally orientated and not vertical as in fishes or ichthyosaurs. Spine and trunk muscles of cetaceans are thus designed primarily for dorsoventral movements (see Locomotion). Technically, the swimming of cetaceans is an axial movement (trunk–tail as a driving member), in contrast to paraxial swimming movement (extremities as driving element as found for sea lions, penguins, etc.). The axial mode of swimming is superior paraxial energetically due to the position of the fluke at the rear end of the body. The arms are modified into flippers representing the bony bauplan of mammalian forelimbs. Moreover, a boneless, connective tissue formation represents the unpaired dorsal fin. Flippers and dorsal fin only control and stabilize the body's position while swimming.

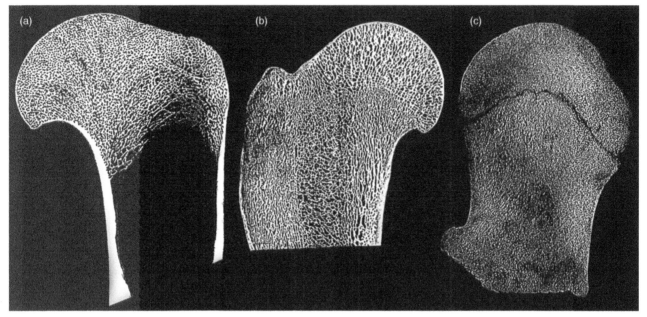

FIGURE 1.1 **Longitudinal section through the humerus of a giraffe (a), a walrus (b), and a Minke whale (c).** *(Modified from Klima, M., 1994. Anpassungen an die aquatische Lebensweise. In: Robineau, D., Duguy, R., Klima, M. (Eds.),* Meeressäuger—Wale Und Delphine 1, Handbuch Der Säuetiere Europas. *Aula, Wiesbaden, pp. 49–79)*

In general, solid compact bones are no longer necessary in cetaceans due to the buoyancy of water. The main skeletal elements are cancellous bone (Fig. 1.1), which are lighter, more flexible, and also equiped with intense blood circulation. Similar hydrostatic properties could possibly also possessed by cartilage tissue. That would explain why in dolphins have more cartilaginous skeletal elements in comparison to terrestrial mammals. Even fat is, in addition to its function for energy storage, found within the dolphin skeleton to replace solid bone tissue. All three tissues together—bone, cartilage, and fat—may thus contribute due to their densities to a positive buoyancy of the dolphin's body in water.

In comparison to terrestrial mammals, the skull of cetaceans shows three fundamental changes: (1) a far forwardly projecting rostrum, (2) shifts in the nostrils to vertex of the skull, (3) deformation and rostrocaudally shortening of the skull (see Locomotion, Head and Senses). The rostrum and mandible of the lower jaw are responsible for the typical form of the dolphin snout, especially because many species have long and narrow beaks. In the cranium, the individual skull roof bones slide over each other (teleskoping). The most shortened skull elements are the nasal and the frontal bones and the parietal is deformed. The occipital becomes the largest element of the neurocranium. It is very high and nearly perpendicular, and includes the skull to the rear. Due to these modifications, the nasal passages are nearly vertical so that the blowhole is situated at the highest point of the dolphin's head. Accordingly, it appears first at the water surface when dolphins surface. When breathing, dolphins do not need to interrupt their natural swimming movements, which is highly energy saving.

Due to the short neck region, which contributes to the streamlined body contour, the hyoid bone is very close to the sternum. The strong hyoid muscles in toothed whales are designed for quick sucking in and swallowing whole fish or squid because they cannot chew. From the skeleton of the hind limbs, only small pelvic rudiments inside the body are present in dolphins where the strong penis musculature or the vaginal muscles originate.

The holaquatic lifestyle of cetaceans places very different demands on the mammal-specific sense organs than the terrestrial habitat for terrestrial mammals (see Head and Senses, Brain, Spinal Cord, and Cranial Nerves). The sense of smell had regressed completely in dolphins. The sense of touch as well as taste are insignificant. Although the sense of sight plays only a minor role, the eyes of dolphins are relatively well developed and equipped with a spherical lens. With this spherical lens, a sharp image on the retina may be produced in low light conditions underwater as well as in bright light conditions in the air. Both eyes can be moved individually. Due to the extreme lateral position of the eyes, the visual field is probably limited.

The absolutely dominant sense of toothed whales is hearing (see Head and Senses, Brain, Spinal Cord, and Cranial Nerves). Because sound in water spreads much better than in air, the use of the sense of hearing underwater is particularly advantageous. The outer opening of the ear canal is closed and only visible as a small dot behind the eye. The middle ear has three ossicles as known from all mammals but in a somewhat modified form. The entire bony ear capsule

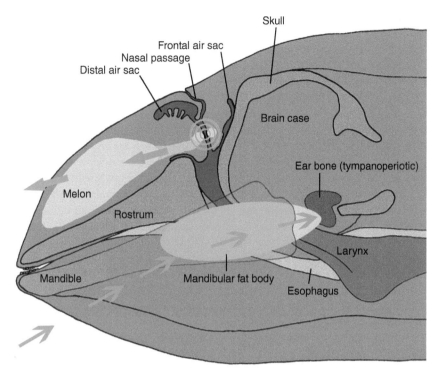

FIGURE 1.2   Schematic longitudinal (sagittal) section through the head of a toothed whale (harbor porpoise) showing its main structures involved in sound generation and hearing. The toothed whale's source of click sounds is situated in the nasal passage. Due to air flow into the distal air sac, this nasal valve system (the center of the *green circle*) is brought to vibrate. The resulting sound vibration in the wall of the nasal passage is focused by the fat body of the melon to be emitted forward into the water (*thick green arrows*). The sound conduction to the ear is a fat pad on the lower jaw (mandible, *thin green arrows*), which works similarly to the pinnae and the external auditory canals in humans and connects the ear bone acoustically to the water. This tympanoperiotic capsule contains the middle and inner ear. In toothed whales it is separated from the skull, unlike in men. So an accurate directional hearing is possible underwater.

(tympanoperioticum) is isolated acoustically from the rest of the skull by air-filled spaces, guaranteeing directional hearing under water. The ear capsules consist of very dense bone material. The audible range of dolphins includes an extensive spectrum from sonic low frequencies below 20 Hz to the ultrasonic range up to 280,000 Hz. The ultrasonic frequencies are used for their echolocation system, which is thus an active sonar system (system for sound navigation and ranging). Via echolocation, dolphins can orientate acoustically in total darkness, determine distances, scan object properties, and locate prey. The sound source is situated at the outer nasal passages. The sound is then emitted and focused through a unique fatty structure in the forehead, the so-called melon (Fig. 1.2). Then the echoed sound waves reach the middle ear through fat deposits in the lower jaw. These fat deposits thus function analogous to the pinnae and ear canals of terrestrial mammals (see Head and Senses). In addition to the ultrasonic clicking sounds used mainly for echolocation, dolphins use sonic whistles. They are not the focus but serve for communication between peers.

At least some of the occasional mass strandings of toothed whales are thought to be a result of a failure of their hearing system and their complex social system. However, the exact triggering factor for this enigmatic phenomenon is still not known. Three causes are discussed: (1) The failure of the sonar system in shallow coastal waters with a soft bottom, which dampens the sound waves; (2) the massive infestation of the ear with parasitic nematodes, leading to the lack of receiving the echo sounds, whereby only the leading animal may be affected while the whole group blindly follows; and (3) the animals follow their magnetic sense (the lines of the geomagnetic field) and rely on this navigation system even when unexpectedly land blocks their way (see Magnetosensation). Although many facts speak for the latter hypothesis, presumably mass strandings occur only when several of these factors fit together. The rescue of the stranded animals are unsuccessful often because the animals keep trying to take the wrong direction. According to historical data, the strandings are undoubtedly one of the ordinary events in the life of dolphins. Nevertheless, under natural conditions, the dolphin populations are not threatened due to strandings.

Dolphins have relatively large brains for their body mass, similar to apes (see Neurobiology and the Evolution of Dolphins). However, with the transition from a terrestrial to the aquatic lifestyle, the proportional relationships of various parts of the brain changed. With the development of an echolocation system, the acoustic pathways became the absolute domi-nant part of the brain. In parallel, dolphins have a hypertrophied cerebral cortex with a strongly developed groove pattern (gyrification) more diverse than in humans. An example of the intelligence of dolphins is reflected by their social lifestyle and by other complex behaviors such as tool use. Careful behavioral studies have shown that dolphins are undeniably "smart," which means they have high cognitive capabilities with high social skills (Gregg, 2013). However, some scientists conclude that the intelligence of dolphins is often uncritically overestimated.

A diurnally activity rhythm, caused by the regular alternation of daylight and night, is less pronounced in the cetaceans in comparison to most other mammals. Most species can be observed both day and night searching and hunting for prey. Sleeping is probably rare and there are only about 3–5 min continuous sleep interrupted by pauses of breathing. During these sleep periods, only one brain hemisphere controls the body, while the other rests.

The oxygen uptake is very effective. Dolphins are capable of exchanging about 80–90% of the air volume in the lungs with each breath (land mammals about 10–15%). However, oxygen is not stored mainly in the lungs and in the blood, but mainly in body tissues, in particular in the muscle, which is rich in myoglobin. Although the stored blood oxygen content is relatively low, the blood represents an important oxygen storage because dolphins have 2–3 times greater relative blood content per kg body weight (120–180 mL/kg) as terrestrial mammals. When diving, the oxygen consumption is reduced. Cardiac activity and blood circulation slow down and some organs can be temporarily separated from the oxygen supply.

For respiration, air is rapidly exhaled first when surfacing; what happens is a loud blow because air is blown under strong pressure through the narrow nostrils. The cooled and depressurized air may condense into a cloud of fog. After exhaling, the inhalation follows immediately. The entire respiratory exchange lasts less than a second in dolphins (Lawrence and Schevill, 1956), it can be repeated after a longer dive several times. The diving durations vary depending on the species and situation. Many dolphins usually dive only for a few minutes (see Diving: Breathing, Respiration, and the Circulatory System).

How the dolphin's body rule the pressure at great depths remains a mystery in many ways today. Since oxygen is mainly bound to body tissues, dolphins need little air in their lungs while diving. Hence, the lungs of dolphins are relatively small (see Diving: Breathing, Respiration, and the Circulatory System); their wall is stabilized up to the level of the entrance to alveoli with cartilage rings. During deep dives, the lungs collapse almost completely, similar to the bellows of an accordion and pressed to the thoracic spine and ribs. Other organs (except the middle ear) are generally rich in water and fatty liquids and thus are nearly incompressible. Therefore, they can withstand the intense pressure. Arteries with increased muscular walls form widely branched meandering systems, retia mirabilia, fill out many cavities of the body like thick cushions. These retia may control blood pressure when the water pressure is changing. Probably they also prevent the formation of gas bubbles in the blood during rapid emergence and thus reduce the risk of fatal embolism (decompression sickness).

The main diet of dolphins is fish. To catch this prey, they use their long beak-like jaws equipment with many polyodont and homodont teeth (see Feeding and the Digestive System). The teeth in the maxilla and mandible, up to 260, are usually cone shaped. The dentition appears only in a single generation. Dolphins catch their food with their beaks and swallow it as a whole.

Dolphins are known to cooperate in hunting larger schools of fish. Killer whales are famous for their strategy to encircle and concentrate fish schools near the surface so that a fluke hit debilitates a larger number of fishes to be caught easily. Moreover, killer whales have developed a hunting method in which they catch seals on the beach. For this, the whales swim with high speed so that they slide on the beach for a few meters. Additionally, killer whales may catch seals that slide down an ice floe due to larger waves made by the whales passing by at high speed. Even occasional cooperation with the people engaged in fishing operations is known where bottlenose dolphins chase fish into fishing nets.

The stomach has three sections (see Feeding and the Digestive System). First, the fundus, which is responsible for the storage and mechanical processing of food, is followed by the main body of the stomach and the pyloric stomach, both draining their digestive glands. Small intestine and colon can be distinguished from each other only histologically. The relatively large liver has no gallbladder. The kidneys show a grape-like structure with up to 10,000 lobules (renculi) in the large whales.

The testes are intraabdominal and moved far dorsally just behind the kidneys (see Urinary System, Genital Systems, and Reproduction). The fibroelastic penis is to a large extent within the abdominal cavity between the pelvic bones. It runs in an S-shaped curve into a longitudinal belly fold, which opens with a narrow elongated slit caudally of the umbilicus. The penis portion, which is located in the penis fold, is comparable to the glans (glans penis). For erection, the penis straightens be-cause of its fibroelasticity and by the blood filling the unpaired corpus cavernosus. This causes the penis, including the inner wall of the penis fold, to turn inside out. With decreasing erection, the penis is withdrawn by means of a retractor muscle.

The female reproductive organs are shifted dorsally (see Urinary System, Genital Systems, and Reproduction). The uterus has two horns and the embryo usually develops only in the left uterine horn, although the placenta extends partially into the contralateral horn. The vagina opens into an elongated fold, which lies directly in front of the anus.

There are only fragmentary information and estimates about the reproductive biology of most cetaceans. In many whales, both sexes show a seasonal reproductive cycle and the seasonal increase in activity of the testes in the males is linked to the estrous cycle of the females. Most cetaceans are promiscuous.[c]

The gestation periods are generally long in cetaceans. In dolphins, it varies depending on the species between 9 and 17 months, although it is not necessarily correlated with the body size of the animals. In general, only one young is born. Very rarely, twins are born and usually only one of them remains alive. The birth takes place tail first, as breech birth, but the head position is also possible. In any case, the newborn must be quickly brought to the surface so they can take their first breath.

The newborns are comparable with advanced precocial birds or running hooved mammals. They are fully developed and reach about one fifth the length of the mother. A correspondingly high birth weight is made possible because the birth canal can greatly expand due to the rudimentary pelvic bones. In some species, infants are nursed not only from the mother, but also sometimes from several "aunts." Since cetaceans do not have absorbent lips, the teat is taken into the mouth and the milk is injected from the mammary gland by muscle actions. The two mammary glands are on either side of a line between the umbilicus and the vaginal slit. Lactating mammae may be swollen, so that the otherwise hidden teats slightly protrude outward. The milk is extremely nutritious (fat content 10–36%, only about 5% for most land mammals) and facilitates fast growing. Nevertheless, the young dolphins are breastfed for an unusually long period, at least 4 months, and up to 8 years (bottlenose dolphin). Calving intervals range from 1 year in the smaller dolphin species to 8 years in killer whales. Sexual maturity is reached in the smallest species (common dolphin) with 6 years up to 16 years in the killer whales.

NATURAL HISTORY OF REFERENCE SPECIES

Bottlenose Dolphin

The principal reference species in this book is the (common) bottlenose dolphin (*Tursiops truncatus*) (Montagu, 1821). It is probably the most common species of dolphins. The bottlenose dolphin is commonly known from the sightings in Delphinarias and due to the fact that it is distributed worldwide, often found in costal areas of temperate to tropical waters. The central character of the popular television show *Flipper*, created by Ivan Tors, was a bottlenose dolphin. Its length is 2–4 m with an adult weight of 150–650 kg. The slightly stocky body is uniformly gray and it has a clear domed melon upon a medium-sized beak (Fig. 1.3). The prey of this dolphin is mainly fish, but also may include cephalopods and some benthic organisms. It is famous for its well-documented, distinct social life. The exact population size is unknown, but the bottlenose dolphin is generally not threatened.

Moreover, there are several coreference species in this book chosen due to their common distribution and availability in our laboratories.

Atlantic White-Sided and White-Beaked Dolphins

The two *Lagenorhynchus* species are robust and powerful, about 270 cm long and weigh c.230 kg. The back is dark, the dorsal fin is pronounced, and the belly is white (Fig. 1.4). The white-sided dolphin (*Lagenorhynchus acutus*; Gray, 1828) has a yellow-brown blaze on its caudal flanks but lacks the white rostrum, which is typical for the white-beaked dolphin (*Lagenorhynchus albirostris*; Gray, 1846). However, the rostrum is less pronounced than in the bottlenose dolphin. Both are inhabitants of the cold-temperate North Atlantic waters over the continental shelf, extending into deep oceanic waters.

Common Dolphin

The common dolphin (*Delphinus delphis*)(Linnaeus, 1758) is probably the most famous dolphin since ancient times, although it is one of the smallest dolphins (1.8–2.5 m, 70–130 kg). In contrast to the bottlenose dolphin, the common dolphin is vividly colored (Fig. 1.5): The slender body has a dark back, and a characteristic yellowish pattern on the flanks of the body and head. The long and narrow rostrum is equipped with up to 240 small pointed teeth. Its food is mainly fish and cephalopods. Common dolphins are social, often found in large schools of several thousand animals socialized with other species of the genus *Stenella* (see the section on striped and spotted dolphins). It is famous for its frequent air jumps

---

c. Mating systems with multiple partners.

Offshore adult

Coastal adult

Coastal calf

FIGURE 1.3 **Habitus of the bottlenose dolphin and spy hopping.** *(Drawings from Jefferson et al., 2015:* Marine Mammals of the World. *Elsevier; photo courtesy of J.Gonzalvo, Istituto Tethys)*

FIGURE 1.4 **Habitus of the white-beaked dolphin.** *(From Jefferson et al., 2015:* Marine Mammals of the World. *Elsevier)*

FIGURE 1.5   **Habitus of the common dolphin.** *(From Jefferson et al., 2015:* Marine Mammals of the World. *Elsevier)*

and bow riding of ships. This offshore species is distributed worldwide in temperate to tropical oceans. The stock size is not precisely known, but in some regions this species is in danger due to accidental bycatch of fishery activities.

False Killer Whale

The false killer whale (*Pseudorca crassidens*; Owen, 1846) is black with a gray throat and neck. It has a slender body with an elongated, tapered head without a marked rostrum. The dorsal fin is sickle shaped (Fig. 1.6). The average size is around 4.9 m (1200 kg). Their habitat is the open ocean of tropical and subtropical latitudes. Threads to this species cannot be defined since abundance data are unavailable. The Hawaiian population of false killer whales, which numbers around 150 individuals, is endangered.

Killer Whale

The killer whale or orca (*Orcinus orca*; Linnaeus, 1758) is the largest dolphin (6–10 m, up to 9 t). It has a characteristic black–white drawing pattern, a robust body, a tall dorsal fin (in old males up to 1.8 m), powerful rounded flippers, and a rounded head with a pointed snout (Fig. 1.7). Among all dolphins, it has the largest variety of food: from fish and cephalopods to marine birds, seals, manatees, and other cetaceans; even young whales such as gray whale and blue whale calves (thus, the name killer whale). Their distinct social behavior is best visible during their complex hunting behavior (see previous discussion). The killer whale has a worldwide distribution from polar to tropical waters. The general conservation status of this species cannot be defined because there are insufficient data.

Long-Finned Pilot Whale

The long-finned pilot whale (*Globicephala melas*; Traill, 1809) is 3.5–6 m long and weighs 1.8–3.5 t. The elongated body is black in color with a bright spot on the neck (Fig. 1.8). The dorsal fin is low and hook-shaped and it has striking long and narrow flippers. The large and spherical melon roves the whole rostrum. Its main prey is cephalopods, sometimes fish. The pilot whale is extremely social gathering in schools of 10–200 animals. Mass strandings are common worldwide due to its distribution in subpolar and temperate waters, except the North Pacific. The exact population size is unknown.

FIGURE 1.6   **Habitus of the false killer whale.** *(From Jefferson et al., 2015:* Marine Mammals of the World. *Elsevier)*

Resident killer whale

Adult female & calf

Type 2 killer whale

Adult female

FIGURE 1.7   **Habitus of the killer whale.** *(From Jefferson et al., 2015:* Marine Mammals of the World. *Elsevier)*

FIGURE 1.8   **Habitus of the long-finned pilot whale.** *(From Jefferson et al., 2015:* Marine Mammals of the World. *Elsevier)*

Risso's Dolphin

Risso's dolphins (*Grampus griseus*; G. Cuvier, 1812) are characterized by the bulbous head, which has a vertical crease frontally. Infants are dorsally gray to brown and ventrally cream-colored (Fig. 1.9). In older calves, the nonwhite areas darken, and then lighten (except for the always dark dorsal fin) in older animals. Linear white scars cover the bulk of the body. Older individuals appear mostly white. They have only two to seven pairs of teeth only in the lower jaw. The length is typically 3 m and weighs 300–500 kg so that Risso's dolphins are the largest species known as "dolphin." They are found worldwide in temperate and tropical waters, usually in deep waters. They feed almost exclusively on neritic and oceanic squid. These dolphins typically travel in groups of 10–50, but the groups may reach 300–400. There is no global population size estimate.

Striped and Spotted Dolphins (*Stenella* sp.)

The striped dolphin (*Stenella coeruleoalba*; Meyen, 1833) is slightly larger than the common dolphin (2.5 m, 155 kg). Its coloring is unique. The belly is nearly blue, white, or pink. There is a black band circling the eyes with a black band running to the flipper (Fig. 1.10). Two further dark stripes run from behind the ear. The longer band thickens along the flanks until it curves down under the belly just prior to the tail stock. Above these stripes, the dolphin's flanks are colored light

FIGURE 1.9   **Habitus of the Risso's dolphin and a Risso's dolphin surfacing.** *(Drawings from Jefferson et al., 2015:* Marine Mammals of the World. *Elsevier. Courtesy of V. Jimenez, Istituto Tethys)*

blue or gray. All body appendages are black as well as the back rostrally. The striped dolphin travels in large groups of up to thousands of individuals of different species (see the previous discussion about the common dolphin). The striped dolphin feeds on small pelagic fish and squid and is famous for its acrobatic air display. The striped dolphin inhabits temperate or tropical offshore waters worldwide and is not endangered.

The pantropical spotted dolphin (*Stenella attenuata*; Gray, 1846) varies in size and coloration according to its coastal or pelagic origin. The coastal form is larger (2.5 m, 120 kg) and more spotted. The spots are key defining characteristics in adults (Fig. 1.10), although immature individuals are generally uniformly gray colored and susceptible to confusion with the bottlenose dolphin. The spotted dolphin has a long, thin beak. The upper and lower jaws are darkly colored, separated by thin white lip strips. The ventral side is white to pale gray with a limited number of spots. The flanks are separated into three

FIGURE 1.10   **Habitus of the striped dolphins and a striped spy hopping dolphin.** *(Drawing from Jefferson et al., 2015:* Marine Mammals of the World. *Elsevier; photo courtesy of F. Bendinoni, Istituto Tethys)*

distinct color bands. This species is famous for its large, splashy leaps and bow-riding with boats. In the eastern Pacific, the dolphin is often found swimming with yellow fin tuna because the two species have similar diets of small epipelagic fish. The species may also feed on squid and crustaceans. The pantropical spotted dolphin, as its name implies, is found in high numbers across all tropical and subtropical waters around the world.

Occasionally, we refer in this book to the harbor porpoise (*P. phocoena*; Linnaeus, 1758) in cases if better original material and data of dolphins are lacking. However, we do so only if the anatomy of the porpoise does not differ notably from the dolphin anatomy. Additionally, as it is shown in the following section Research History of Dolphin Anatomy, in historical terms the harbor porpoise is the best-known toothed whale regarding its anatomy.

RESEARCH HISTORY OF DOLPHIN ANATOMY

In this section we focus on the research history of the comparative anatomy in dolphins and other Delphinoideae. For a general summarization of the history of whale research, we recommend former papers (Berta and Sumich, 1999; Bianucci and Landini, 2007; Slijper, 1979; Würsig et al., 2009).

The great Greek philosopher Aristotle (384–322 BC) defined morphology as the search for a common "bauplan" for all structures. For this search, he performed dissections in which scarce references can be found in his *Historia Anima-lium*. He differentiated between baleen and toothed whales and described whales and dolphins generating embryos, being viviparous, and producing milk. Accordingly, Aristotle was the first who separated fish from cetaceans (Romero, 2012). His detailed description means that Aristotle directly observed these animals and dissected some specimens (Bianucci and Landini, 2007), a fact that was not common during the following centuries.

The Roman writer Pliny the Elder (AD 23–79) published a book on dolphins and whales 400 years after Aristotle's time as part of his 37-volume *Naturalis Historia*. In the preface to this work, he claimed to have assembled 20,000 facts from 2,000 books written by 100 authors. Most of the works that Pliny cited have disappeared, but the information extracted has persisted (Mead and Fordyce, 2009). He illustrated the pulmonary respiration in dolphins but his anatomical descriptions were not detailed.

After Pliny the Elder, the so-called dark years followed (Bianucci and Landini, 2007) when studies on cetaceans were completely neglected for many centuries. Then, the German philosopher and bishop Albertus Magnus (~1200–1280) in

De Animalibus founded his classification of animals on the work of Aristotle. He considered whales and dolphins the most perfect marine animals because they were mammals with lungs.

During the Renaissance, the interest in natural sciences increased and several books were published. Most of the Renaissance researchers interested in marine mammals had a medical background. That is not surprising because the interest in medicine and thus human anatomy was reflected in the opportunities to dissect animals. This interest in comparative anatomy was crucial in establishing the homology between cetaceans and the terrestrial mammals. However, most of the Renaissance natural history books show figures of creatures somewhere between myth and reality (Fig. 1.11) because many naturalists of that time reproduced the written knowledge of the former centuries. In contrast, Pierre Belon (1517–1564), famous for his very early ideas on comparative anatomy, described dolphin anatomy (bottlenose, common dolphins, harbor porpoise) in detail as having a placenta, mammae, and hair on the upper lip of their fetus. He drew the fetus of a porpoise and the skull of a dolphin (Fig. 1.12). Belon wrote that apart from the presence of hind limbs, they conform to the human body plan with features such as the liver, the sternum, milk glands, lungs, the heart, the skeleton in general, the brain, genitalia (Belon, 1551). Curiously, he still classified cetaceans as fish (Cole, 1944).

Conrad Gesner (1516–1565), known as the father of natural history, presented a figure of a dolphin skull with the caption "Delphini Calvaria è libro Bellonij" that looks like the skull that Belon illustrated, however, interestingly, in a better state of preparation (Mead and Fordyce, 2009).

During Belon's activities, Guillaume Rondelet (1507–1566) described the lungs, the gut, the kidneys, and the genitals of a dolphin and compared them with the same organs in seals and cows. Moreover, he was probably the first who described

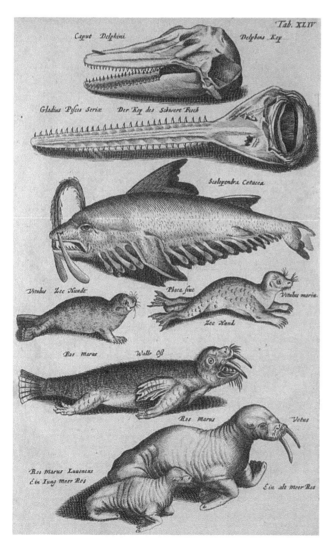
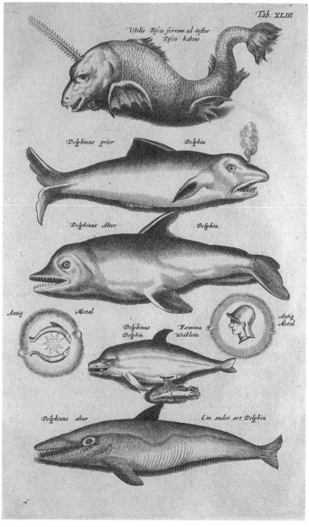

FIGURE 1.11 **In spite of the developing natural research during the Renaissance, there was still a fusion of facts and myth in natural history of sea mammals.** (*From Jonston, 1657.* Historiae Naturalis De Piscubus el Cetis, Libri V. *Frankfurt Main, Merian*).

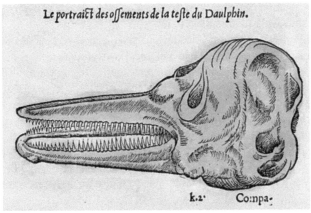

Le portraict des offements de la teste du Daulphin.

k.2·   Compa·

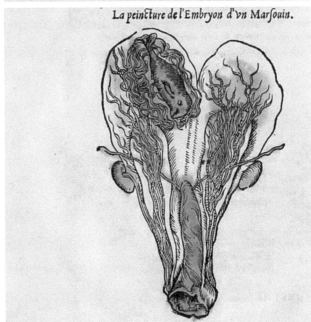

La peincture de l'Embryon d'vn Marfouin.

FIGURE 1.12 **The illustrations of Belon show these structures in some detail.** Nevertheless, Belon accounted dolphins and porpoises as fishes. *(From Belon, 1551.* L'histoire naturelle des estranges poissons marins: avec la vraie peincture & description du daulphin & de plusieurs autres de son espece. *De l'imprimerie de Regnaud Chaudiere, Paris).*

the small external auditory meatus and saw its association to the ear. Describing the dolphin's brain, Rondelet made the curious mistake that he described the transverse axis as the longitudinal axis because, obviously, he examined a brain removed from the skull (Cole, 1944). However, concerning anatomy and descriptions of the small cetaceans he dissected (probably the common dolphin and harbor porpoise), Rondelet made correlations between form, function, and environment. Similar to Belon, he compared the anatomy of dolphins to that of the pig and humans. Based on these observations, he considered marine mammals to be a type of aquatic quadruped (Romero, 2012).

During the 17th century the most significant contribution to the study of cetacean anatomy was probably given by the British anatomist Edward Tyson (1650–1708) who discovered the rete mirabilia (wonderful nets). Tyson was the first of the comparative anatomists in the modern sense and set new standards in terms of direct observation and comparative anatomy. He also established an understanding of homology not seen since Aristotle. He proved to be a very competent observer of internal anatomy and he saw comparative anatomy as a means to explain the scala naturae as proposed by Plato and Aristotle (Romero, 2012).

Tyson's description of the internal anatomy of the porpoise is remarkable, particularly of the nervous system (Tyson, 1680). In many ways he thought that the "porpess" was the transitional link between terrestrial mammals and fish (Romero, 2012). In his monograph Tyson surveyed contributions from previous authors. He corresponded with John Ray (1627–1705). Ray had also dissected a porpoise, 9 years before Tyson but was far more superficial and added very little to what other authors such as Rondelet had done.

Tyson's work *The Anatomy of a Porpess* on a harbor porpoise influenced the anatomical studies of many scholars of the 18th century. Among these, published in 1785, Alexander Monro Secundus illustraded the dissection of porpoise anatomical parts and organs with original engravings. Even more important was John Hunter's (1728–1793) work *Observations on the Structure and Oeconomy of Whales* who published in 1787 the dissection of a bottlenose dolphin (Bianucci and Landini, 2007). Hunter and others of his generation had medical training and accordingly used terms that were oriented toward human anatomy.

At about the same time as serious studies began in the West, Japanese scientists began serious work on marine mammals. That is, Otsuki began to describe the internal anatomy of cetaceans of Japan in 1808, but his manuscript remains unpublished (Würsig et al., 2009).

In the 19th century, anatomical research on cetaceans became more important thanks to an increasing interest in cetacean biology and whaling. The English explorer William Scoresby (1789–1857) published an illustrated description of porpoise anatomy. At the beginning of the 19th century, the comparative studies approved the mammalian characteristics of cetaceans, but most researchers seemed to be attracted by the specialized nasal complex of porpoises (Fig. 1.13) (Gruhl, 1911; Sibson, 1848; Von Baer, 1826).

Between 1868 and 1879, two European scientists—the Belgian zoologist Pierre-Joseph Van Beneden and the French paleontologist Paul Gervais—published an extended monograph about skeletal anatomy of fossil and extant species including detailed drawings (Fig. 1.14) (van Beneden and Gervais, 1868). Other significant European contributions on cetacean anatomy were given by the Danish naturalists Daniel Frederik Eschricht (1798–1863) and Johannes Theodore Reinhardt (1816–1882).

The anatomist William Henry Flower was probably the first to take a Darwinian approach to cetacean anatomy, pointing out that hoofed mammals share some intriguing skeletal features with whales (Flower, 1867). In parallel, Richard Owen's classical work *On the Anatomy of Vertebrates* (Owen, 1868) included some comparative aspects of dolphin anatomy.

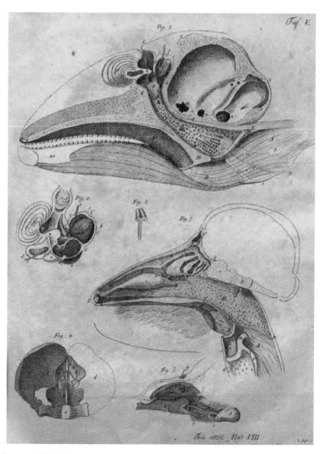

FIGURE 1.13   **Due to his comparative study Von Baer approved the blowhole of the porpoise is the true nose.** *(From Von Baer, 1826. die Nase der Cetaceen erleutert durch Untersuchung der Nase des Braunfisches Delphinus Phocoena. Isis Von Oken 19, 811–847).*

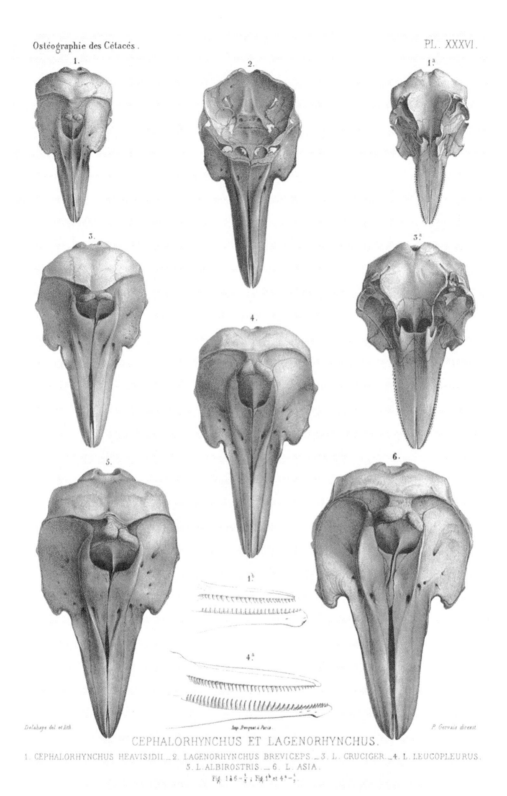

Ostéographie des Cétacés.   PL. XXXVI.

CEPHALORHYNCHUS ET LAGENORHYNCHUS.

1. CEPHALORHYNCHUS HEAVISIDII.__2. LAGENORHYNCHUS BREVICEPS __3. L. CRUCIGER.__4. L. LEUCOPLEURUS.
3. L. ALBIROSTRIS.__6. L. ASIA.

FIGURE 1.14 **The detailed illustrations on cetacean skulls and skeletons of van Beneden and Gervais are still an important reference today.** *(From van Beneden and Gervais, 1868.* Ostéographie des Cétacés Vivants et Fossiles, Comprenant la Description et l'Iconographie du Squelette et du Système Dentaire de ces Animaux: ainsi que des documentsrelatifs à leur histoire naturelle. *Arthus Bertrand, Paris).*

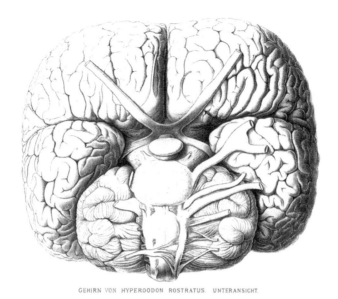

GEHIRN VON HYPEROODON ROSTRATUS. UNTERANSICHT.

FIGURE 1.15 **Detailed illustration of a toothed whale brain.** *(From Kükenthal, 1893.* Vergleichend-anatomische und entwicklungsgeschichtliche Untersuchungen an Walthieren. *Denkschr. Med. Naturwissenschftlichen Ges. Zu Jena 3, 1–447).*

In 1893, Willi Kükenthal published his monograph on the comparative anatomy and development of cetaceans (Kükenthal, 1893). This was the first work presenting a nearly comprehensive picture of the anatomy of whales and dolphins, focusing on the development of external organs, skin, and body shape as well as the comparative anatomy of the nervous system (Fig. 1.15).

The most detailed description and functional interpretation of cetacean anatomy was probably given by Georg Boenninghaus (Fig. 1.16). He published comprehensive papers on the porpoise head anatomy (Boenninghaus, 1903, 1902), which are still not surpassed today regarding the topographical aspects.

After World War II, one of the first comprehensive works on cetaceans was the book by A. G. Tomilin about Russia and adjacent countries. It was originally published in Russian in 1957 and later translated into English (1967). However, in the 20th century, one of the most famous works largely relying on whaling accumulated data consists of Everhard J. Slijper's books *Whales* and *Whales and Dolphins* (Slijper, 1979). In these comprehensive monographs, he compiled the data of many former authors as well as his own former monograph *Die Cetaceen* published first in 1936 (Slijper, 1973). The latter is a comprehensive original study on the comparative anatomy and development of the vascular, nervous, and muscular systems of whales and dolphins, based mainly on the harbor porpoise.

In 1972, a Russian book on the anatomy of whales and dolphins was published by Yablokov, Bel'kovich, and Borisov. This book reviewed extensively the literature of that time including publications written in Russian, Italian, German, Dutch, and French so that its translation is still an important source of information internationally (Yablokov et al., 1974). A further important source of collected anatomical information of that time were the three volumes of the handbook *Functional Anatomy of Marine Mammals* (Harrison, 1977, 1974, 1972). Then, in 1990 the book *The Bottlenose Dolphin* (Letherwood and Reeves, 1990) appeared with chapters on osteology, myology, and neurobiology (Pabst, 1990; Ridgway, 1990; Rommel, 1990), which are basic references for dolphin anatomy up to now. In parallel, the *CRC Handbook of Marine Mammal Medicine* appeared with a comprehensive compilation of information including medicine, surgery, pathology, physiology, anatomy, husbandry, strandings, and rehabilitation (Dierauf, 1990; Dierauf and Gulland, 2001).

Significant works focused on specific topics, such as those written by F. C. Fraser and P. E. Purves (Fraser and Purves, 1960) on cetacean hearing. This work was probably motivated by the discovery of the echolocation system of dolphins in the 1950s. In the following years Kenneth S. Norris published important concepts about the functional anatomy of the dolphin sound generation and perception system (Cranford et al., 1996; Ridgway, 1999).

Like many pioneers of the 1950s and 1960s, John C. Lilly was fascinated by the big brain of dolphins. With his book *Man and Dolphin* (Lilly, 1961), Lilly influenced the public as well as the scientific community. However, from an anatomical point of view, Lilly's work on brains was superficial. Based on a broad field of former publications and monographs (Jansen and Jansen, 1969), starting in the 1970s, Myron S. Jacobs and Peter J. Morgane published several detailed papers on the fine structure of the dolphin telencephalon (Morgane and Jacobs, 1972).

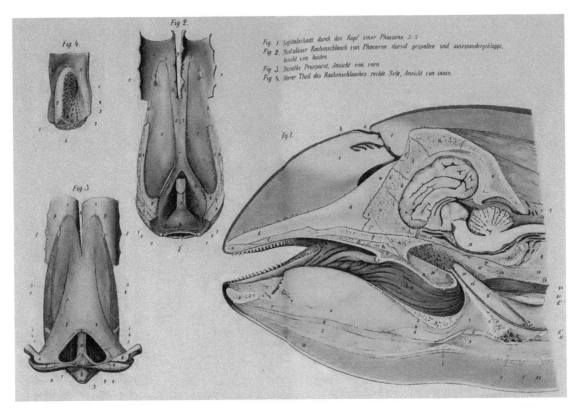

FIGURE 1.16 **The descriptions of Georg Boenninghaus were more detailed than most of the studies at the beginning of the 20th century.** *(From Boenninghaus,1902. Der Rachen von Phocaena communis Less. Eine biologische Studie. Zool. Jahrb. 17, 1–98)*

One of the pioneers of modern cetology was the US American veterinary Sam H. Ridgway, who published classical papers on the diving physiology, acoustics, and the anatomy of the nervous system of dolphins since the early 1960s. One will find these papers quoted in this book. Since that time, general references on dolphin anatomy can be found in the series *Investigations on Cetacea*, which appeared between 1969 and 1994 (25 volumes) edited by Giorgio Pilleri. He edited and published a variety of original papers and reviews on different anatomical subjects on dolphins including unique comprehensive tables of morphological data.

Up through the 1960s, cetologists studied dolphins by harpooning them. For example, the revision of the spotted dolphins (*S. attenuata* and *S. frontalis*) by William F. Perrin (Perrin et al., 1987) and the developmental studies of Milan Klima (Klima, 1999) were based in part on dolphins collected at sea in the 1960s and 1970s. Collecting animal material in the wild was the way it was done until protection of marine mammals became the norm in most countries in the 1970s. Today, dolphins are stringently protected in the wild in most places. As a result, morphological approaches on dolphins became limited to specimens from strandings and bycatch, which greatly decreased opportunities for amassing adequate series of specimens for quantitative analysis. Accordingly, new studies on marine mammals are often integrative, combining methods and ways of thinking largely gleaned from terrestrial animals and human medicine. This comparison of ideas and research techniques holds great promise for the understanding of modern marine mammalogy (Würsig et al., 2009).

REFERENCES

Belon, P., 1551. L'histoire naturelle des estranges poissons marins: avec la vraie peincture & description du daulphin & de plusieurs autres de son espece. De l'imprimerie de Regnaud Chaudiere, Paris.

Berta, A., Sumich, J.L., 1999. Marine Mammals: Evolutionary Biology. Academic Press, San Diego, CA.

Bianucci, G., Landini, W., 2007. Cetacea: an historical overview. In: Miller, D.L. (Ed.), Reproductive Biology and Phylogeny of Cetacea—Whales, Dolphins and Porpoises, Reproductive Biology and Phylogeny. Science Publishers, Enfield, pp. 1–33.

Boenninghaus, G., 1902. Der Rachen von *Phocaena communis* Less. Eine biologische Studie. Zool. Jahrb. 17, 1–98.

Boenninghaus, G., 1903. Das Ohr des Zahnwales, zugleich ein Beitrag zur Theorie der Schalleitung: Eine biologische Studie. Fischer.

Cagnolaro, L., Cozzi, B., Notarbartolo di Sciata, G., Podestà, M. (Eds.), 2015. Mammalia IV: Cetacea, Fauna d'Italia. Calderini, Bologna.

Cole, F.J., 1944. A History of Comparative Anatomy. MacMillan, London.

Cranford, T.W., Amundin, M., Norris, K.S., 1996. Functional morphology and homology in the odontocete nasal complex: implications for sound generation. J. Morphol. 228, 223–285.

Dierauf, L., 1990. CRC Handbook of Marine Mammal Medicine: Health, Disease, and Rehabilitation. CRC Press, Boca Raton.

Dierauf, L., Gulland, F.M.D., 2001. CRC Handbook of Marine Mammal Medicine: Health, Disease, and Rehabilitation, second ed. CRC Press, Boca Raton.

Flower, W.H., 1867. On the osteology of the cachalot or sperm whale (*Physeter macrocephalus*). Trans. Zool. Soc. Lond 6, 309–372.

Fordyce, R.E., de Muizon, C., 2001. Evolutionary history of the cetaceans: a review. In: Mazin, J.M., de Buffrénil, V. (Eds.), Secondary Adaptation of Tetrapods to Life in Water. Dr. Friedrich Pfeil, München, pp. 169–234.

Fraser, F.C., Purves, P.E., 1960. Hearing in cetaceans. Evolution of the accessory air sacs and the structure and function of the outer and middle ear in recent cetaceans. Bull. Br. Mus. Nat. Hist. Zool. 7, 1–140.

Frey, R., Hindrichs, H., Zachos, F.E., 2015. Artiodactyla, Paarhufer inkl. Wale. In: Westheide, W., Rieger, G. (Eds.), Spezielle Zoologie. Spektrum Akademischer Verlag, Heidelberg, pp. 575–599.

Gregg, J., 2013. Are Dolphins Really Smart?: The Mammal Behind the Myth. Oxford University Press, Oxford.

Gruhl, K., 1911. Beiträge zur anatomie und physiologie der cetaceennase. Jenaische Z. Für Naturwissenschaften 72, 367–414.

Harrison, R.J. (Ed.), 1972. Functional Anatomy of Marine Mammals. Academic Press, London.

Harrison, R.J. (Ed.), 1974. Functional Anatomy of Marine Mammals. Academic Press, London.

Harrison, R.J. (Ed.), 1977. Functional Anatomy of Marine Mammals. Academic Press, London.

Huggenberger, S., Klima, M., 2010. Cetacea, Waltiere. In: Westheide, W., Rieger, R.M. (Eds.), Spezielle Zoologie. Spektrum Akademischer Verlag, Heidelberg, pp. 658–672.

Huggenberger, S., Klima, M., 2015. Cetacea, Waltiere. In: Westheide, W., Rieger, G. (Eds.), Spezielle Zoologie. Spektrum Akademischer Verlag, Heidelberg, pp. 600–613.

Jansen, J., Jansen, J.K.S., 1969. The nervous system of Cetacea. In: Andersen, H.T. (Ed.), The Biology of Marine Mammals. Academic Press, New York, pp. 175–252.

Jefferson, T.A., Webber, M.A., Pitman, R.L., 2015. Marine Mammals of the World, second ed. Elsevier, Amsterdam.

Jonston, J., 1657. Historiae Naturalis De Piscubus el Cetis. Libri V. Merian, Frankfurt.

Klima, M., 1999. Development of the cetacean nasal skull. Adv. Anat. Embryol. Cell Biol. 149, 1–143.

Kükenthal, W., 1893. Vergleichend-anatomische und entwicklungsgeschichtliche Untersuchungen an Walthieren. Denkschr. Med. Naturwissenschftlichen Ges. Zu Jena 3, 1–447.

Laitman, J.T., 2007. Thar she blows … and dives, and feeds, and talks, and hears, and thinks: the anatomical adaptations of aquatic mammals. Anat. Rec. Adv. Integr. Anat. Evol. Biol. 290, 504–506.

Lawrence, B., Schevill, W.E., 1956. The functional anatomy of the delphinid nose. Bull. Mus. Comp. Zool. 114, 103–151.

Letherwood, S., Reeves, R.R., 1990. The Bottlenose Dolphin. Academic Press, San Diego, CA.

Lilly, J.C., 1961. Man and Dolphin. Pyramids, New York.

Mead, J.G., Fordyce, R.E., 2009. The therian skull: a lexicon with emphasis on the odontocetes. Smithson. Contrib. Zool. 627, 1–216.

Morgane, P.J., Jacobs, M.S., 1972. Comparative anatomy of the cetacean nervous system. In: Harrison, R.J. (Ed.), Functional Anatomy of Marine Mammals. Academic Press, London, pp. 117–244.

Owen, R., 1868. On the Anatomy of Vertebrates: Mammals. Longmans, Green and Company, London.

Pabst, D.A., 1990. Axial muscles and connective tissues of the bottlenose dolphin. In: Leatherwood, S., Reeves, R.R. (Eds.), The Bottlenose Dolphin. Academic Press, San Diego, CA, pp. 51–67.

Perrin, W.F., Mitchell, E.D., Mead, J.G., Caldwell, D.K., Caldwell, M.C., van Bree, P.J.H., Dawbin, W.H., 1987. Revision of the spotted dolphins Stenella spp. Mar. Mammal Sci. 3, 99–170.

Price, S.A., Bininda-Emonds, O.R.P., Gittleman, J.L., 2005. A complete phylogeny of the whales, dolphins and even-toed hoofed mammals (Cetartiodactyla). Biol. Rev. Camb. Philos. Soc. 80, 445–473.

Reidenberg, J.S., 2007. Anatomical adaptations of aquatic mammals. Anat. Rec. Adv. Integr. Anat. Evol. Biol. 290, 507–513.

Rice, D.W., 1998. Marine Mammals of the World—Systematics and Distribution. Allan Press, Lawrence.

Ridgway, S.H., 1990. The central nervous system of the bottlenose dolphin. In: Leatherwood, S., Reeves, R.R. (Eds.), The Bottlenose Dolphin. Academic Press, San Diego, CA, pp. 69–97.

Ridgway, S.H., 1999. An illustration of Norris' acoustic window. Mar. Mammal Sci. 15, 926–930.

Romero, A., 2012. When whales became mammals: the scientific journey of cetaceans from fish to mammals in the history of science. In: Romero, A., Keith, E.O. (Eds.), New Approaches to the Study of Marine Mammals. InTech, Rijeka, Croatia, pp. 3–30.

Rommel, S.A., 1990. Osteology of the bottlenose dolphin. In: Leatherwood, S., Reeves, R.R. (Eds.), The Bottlenose Dolphin. Academic Press, SanDiego, CA, pp. 29–49.

Sibson, F., 1848. On the blow-hole of the porpoise. Philos. Trans. R. Soc. Lond. 138, 117–123.

Slijper, E.J., 1973. Die Cetaceen. Asher, Amsterdam. Reprinted from Capita Zoologica Bd. VI and VII, 1936. ed.

Slijper, E.J., 1979. Whales. Cornell University Press, New York.

Tomilin, A.G., 1957. Mammals of the USSR and Adjacent Countries, vol. 9, Cetacea. (Translated by Ronen, O., 1967). Israel Program for Scientific Translations, Jerusalem, Israel, 717 pp.

Tyson, E., 1680. Phocaena or the Anatomy of a Porpess. Benj. Tooke, London.

Uhen, M.D., 2007. Evolution of marine mammals: back to the sea after 300 million years. Anat. Rec. Adv. Integr. Anat. Evol. Biol. 290, 514–522.

van Beneden, P.J., Gervais, P., 1868. Ostéographie des Cétacés Vivants et Fossiles, Comprenant la Description et l'Iconographie du Squelette et du Système Dentaire de ces Animaux: ainsi que des documentsrelatifs à leur histoire naturelle. Arthus Bertrand, Paris.

Von Baer, K.E., 1826. die Nase der Cetaceen erleutert durch Untersuchung der Nase des Braunfisches *Delphinus Phocoena*. Isis Von Oken 19, 811–847.

Wilson, D.E., Reeder, D.M., 2005. Mammal Species of the World: A Taxonomic and Geographic Reference. JHU Press, Baltimore.

Würsig, B., Perrin, W.F., Thewissen, J.G.M., 2009. History of marine mammal research. In: Perrin, W.F., Würsig, B., Thewissen, J.G.M. (Eds.), Encyclopedia of Marine Mammals. Academic Press, San Diego, CA, pp. 565–569.

Yablokov, A.V., Bel'kovich, V.M., Borisov, V.I., 1974. Whales and Dolphins (Kity i Del'finy). Joint Publications Research Service, Arlington, VA.

Chapter 2

General Appearance and Hydrodynamics (Including Skin Anatomy)

GENERAL APPEARANCE AND HYDRODYNAMICS

Dolphins are considered superior in their swimming capabilities when compared to technologies of nautical engineering. They have been recorded at speeds over 11 min/s (>21 kts), which is at the theoretical speed limits for small divers (Iosilevskii and Weihs, 2008). Turning maneuvers by dolphins are at rates as high as 450 degree/s with turn radii as low as 11–17% of body length (Fish and Rohr, 1999). Accordingly, the specialized adaptations of their morphological design for drag reduction, thrust production, energy economy, and maneuverability must be highly effective.

Dolphins have a spindle-like body form and the flukes (flippers, fin, fluke) are streamlined (Fig. 2.1) (Fish et al., 2007). Other body annexes are not present or internalized, helping to streamline the body's shape reducing drag during swimming and probably cooling. External ears (pinnae) do not develop and there is no scrotal sac. The penis is usually retracted into a skinfold. The same is true for the teats of the mammae, which are withdrawn into the mammary slits (see section Epidermis). However, pregnancy influences the swimming performance negatively (Noren et al., 2011) and also dolphin calves perform worse than expected for their body size (Noren et al., 2006).

The head of dolphins is characterized by the dorsal melon and the head of most delphinid species is additionally shaped by the beak-like rostrum. The latter increases the spindle-like appearance of the whole body. However, the shape of the head is different in some dolphins, such as the Risso's dolphin. They do not have a rostrum extending significantly ahead of the melon. The length of the rostrum seems to be correlated with food preferences: Dolphins with long rostrae mainly catch fish, whereas short rostrae are adaptations for squid feeding (suction feeding). Because the melon is essential for sound emission, both the rostrum and the melon may not be shaped to decrease the drag to an optimum level but to improve feeding and sound orientation.

The dolphin's body has large masses of musculature for powerful propulsion by the fluke. Dolphins propel themselves by vertical fluke oscillations and the fluke movement is exclusively up and down. When viewed from the side, flukes follow a sinusoidal pathway that is symmetrical about the longitudinal axis of the body and in time. However, the large moment generated about the center of gravity by the flukes produces a pitching moment in the anterior end. The pitching oscillations of the rostrum are a vertical displacement of 1–7% of body length (Fish and Rohr, 1999). Accordingly, rostrum oscillations are small compared to fluke movements, despite the distance of the rostrum from the center of gravity, which is slightly rostral of the midlength of the body. During the stroke cycles, the body does not act as a rigid beam. The pitching oscillations are nearly in phase so that the rostrum and flukes are pitched downward or upward simultaneously. The coordination of movements at the head and tail indicate that the oscillatory swimming mode of dolphins was evolutionarily derived from spinal flexion associated with the rapid asymmetrical gaits (ie, gallop, bound) used by terrestrial ancestors (Thewissen and Fish, 1997).

The propulsive musculature is composed of the longitudinal hypaxial and epaxial muscles associated with the vertebral column. The hypaxial and epaxial muscles could produce equivalent propulsive forces and movements, given the similar arrangement of the fasciculi, tendons, and muscle insertions (Arkowitz and Rommel, 1985).

Next to the propulsion musculature, it has been suggested that cetacean tendons function analogously to the elastic tendons of running mammals where collagen fibers store elastic energy generated by the stroke (Pabst, 1996a). However, mathematical models demonstrated that tendon's elastic compliance may actually increase the energy cost of swimming. Because a hydrodynamic force model indicated that the elastic elements in the body had the correct properties for near-maximum energy savings, the properties of spring-like fibers may reside in locations other than in the tendons. Accordingly, Pabst (1996b) proposed the cylinder model for swimming dolphins where the cylinder is a subdermal connective tissue sheath composed of fiber bundles from the blubber, ligaments, and muscle's tendons acting as a force

Anatomy of Dolphins. http://dx.doi.org/10.1016/B978-0-12-407229-9.00002-6

FIGURE 2.1   **The streamlined dolphin body (striped dolphin) is adapted for agile and effective locomotion in their aquatic environment.** *(Courtesy of F. Bendinoni, Tethys Research Institute)*

transmission system (Pabst, 1990). The collagen and elastin fibers in the blubber are arranged in a crossed helical geometry and the angle between the crossed fibers is greater than 60 degree. An angle of this magnitude is advantageous to stabilize a straight body position (Pabst, 1996a). However, there is variation in blubber (dermis and subcutis) compliance along the body that results from differences in the distribution and orientation of connective tissue fibers. Fibers from the blubber of the peduncle attach directly to the vertebral column and the shortest fibers were found caudally at the insertion of the fluke (Fish and Rohr, 1999). Moreover, the distal position of the sheath dorsally and ventrally far from the vertebral column would provide a large mechanical advantage for flexing the spine and the peduncle (Fish and Rohr, 1999).

Flukes are attached to the caudal vertebrae and intervertebral discs by a thick mass of collagen fibers (Felts, 1966). This attachment unites the caudal vertebrae associated with the flukes into a single resilient element. Within the fluke, the collagen fibers are arranged in horizontal, vertical, and oblique bundles (Felts, 1966). Horizontal fibers radiate out through the fluke. The fiber bundle pattern indicates an orientation appropriate for incurring high tensile stresses (Fish and Rohr, 1999). The fibrous core of the fluke is covered by a ligamentous layer that is arranged to resist tension, particularly at the edges and the lateral tips (Felts, 1966). This architecture would limit bending during the stroke cycle. The flexibility increases the efficiency compared to a rigid propulsor (Fish and Rohr, 1999).

The region of greatest stiffness of the vertebral column is at the tail base (intervertebral joints between the lumbar and caudal region). Because of the high stiffness, the lumbocaudal joints have the resistance to function as the insertion point for the powerful epaxial muscles. The joints at the fluke base are more flexible with lower stiffness, allowing control of the attack angle with small muscular input throughout the stroke (Long et al., 1997). Moreover, the vertebrae anterior to the fluke are laterally compressed, whereas the few vertebrae within the fluke are sagittally compressed. The peduncle–fluke junction is characterized by relatively large intervertebral discs and these intervertebral joints are thought to act mechanically as a low-resistance hinge (Fish and Rohr, 1999). In addition to the low stiffness of these joints, rotation is aided by the convex cranial and caudal faciae intervertebrales.

There are no muscles attaching to the dorsal fin. Accordingly, it is thought that the fin acts as a keel. In contrast, the flippers have limited mobility (see Locomotion) and are resemble hydrofoils. The flippers are moved for steering by the slow swimmers among the delphinids (eg, bottlenose dolphins). In the fastest swimmers (eg, common dolphins), there are no significant steering movements (Benke, 1993).

INTEGUMENT

The skin of dolphins is modified to fulfill different functional requirements. Next to the common functions such as protection, heat regulation, and sensory perception, it is additionally adapted to establish a minimum of frictional resistance and a maximum of body streamlining. Calculations led to the surprising finding that dolphins move faster and more effective as they should, or more accurately stated, than rigid and smooth bodies (eg, vessels of comparable size, shape, and propulsive power). This so-called Gray's paradox should be explained by the fact that the skin of cetaceans would gain excellent damping properties due to their special elastic structure. Modern studies have shown, however, that the Gray's paradox (Fish, 2006) was based on erroneous calculations, even if it gave rise to many innovations in the field of hydrodynamics.

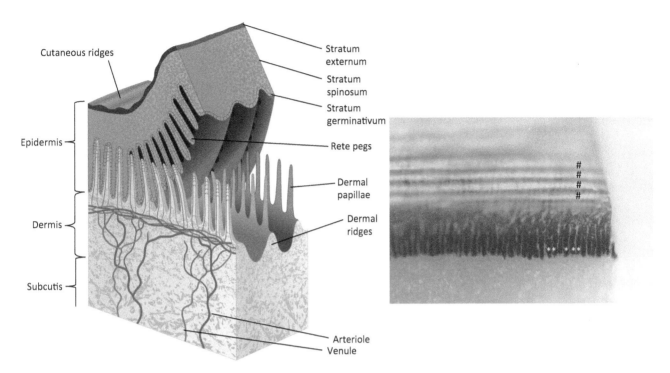

FIGURE 2.2 **Overview of epidermis and blubber (dermis and subcutis) of the dolphin (left) and cut through epidermis and dermis of the dolphin skin (right) showing the epidermal rete pegs (*) and the cutaneous ridges (#) on the surface of the skin.** *(Left, redrawn by Uko Gorter after Geraci et al., 1986. The epidermis of odontocetes: a view from within. In: Bryden, M.M., Harrison, R. (Eds.), Research on Dolphins. Clarendon Press, Oxford; and Sokolov,1982. Mammal Skin. University of California Press, Berkeley, Los Angeles, London; Right, courtesy of Sam H. Ridgway.)*

Epidermis

The structure of the skin and blubber of dolphins[a] is highly organized and complex (Figs. 2.2 and 2.3). The skin of dolphins was described to have a smooth rubbery texture without glands (except mammary glands) and hairs, except the fact that newborn dolphins possess hairs lined up in a single band on both sides on the rostrum. There are no accessory glands associated with these hair follicles. These hairs appear during the dolphin's fetal period at a total length of 16–17 cm. It is hypothesized that these curly bristle-like hairs have some tactile properties and may derive from the vibrissae of terrestrial mammals (Reidenberg and Laitman, 2009). Although dolphins lose these hairs shortly after birth (Figs. 2.4 and 2.5), they are retained as the hairless vibrissal crypts (see Passive Electroreception).

The epidermis of cetaceans is characterized by the lack of keratin (although keratin can be demonstrated histochemically; see later) and by a prominent interdigitation of epidermal rete pegs and long dermal papillae (Fig. 2.3). The thickness of the epidermis varies depending on the species, age, and location sampled between 1 and 4.5 mm in mean (Table 2.1). The ventral body wall tends to have the thickest epidermal layer. The epidermalrete pegs are uniformly long, moderately thick downward extensions which anchore the underlying dermal connective tissue. Dermal papillae interdigitate in regular fashion with the rete pegs and constitute long longitudinal dermal ridges in the cetacean skin [**see Blubber: Corium (Dermis) and Subcutis**]. The papillae extend upward one half to four-fifths of the distance to the surface and are covered by a thin epidermal layer (Table 2.1). Such a complex, serrated interface between dermis and epidermis is characteristic of the glabrous skin of terrestrial mammals (Simpson and Gardner, 1972).

The cellular strata are rather simplified compared to terrestrial mammals. A stratum granulosum and stratum lucidum are absent. Most of the epidermis is made up of the stratum spinosum.

The basal layer (stratum germinativum) is only one cell layer thick with large columnar cells and melanocytes (ratio 12:1). In the pigmented areas of the rete pegs, the melanocytes extend from the stratum germinativum into the stratum spinosum (Fig. 2.2). The melanocytes are round and clear cells with long branching processes. In the common dolphin, diffuse pigment, which is barely recognizable under light microscopy, predominates over larger granular pigment to produce

---

a. The skin (cutis) of mammals, which consists of epidermis and dermis, can be easily subdivided from the underlying subcutaneus fat layer. In dolphins, the dermis and subcutis interwine so that they constitute the blubber layer.

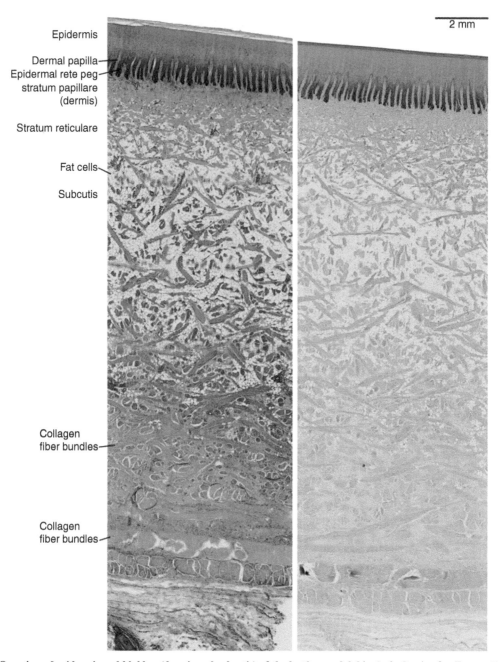

2 mm

Epidermis

Dermal papilla
Epidermal rete peg
stratum papillare
(dermis)

Stratum reticulare

Fat cells

Subcutis

Collagen
fiber bundles

Collagen
fiber bundles

FIGURE 2.3 **Overview of epidermis and blubber (dermis and subcutis) of the bottlenose dolphin.** Left, Orcein after Taenzer-Unna stain; right, HE stain.

a color. Small grains less than 5 μm are most prominent in gray skin, and black skin shows the highest densities of granules greater than 5 μm (Geraci et al., 1986). Melanin granules are numerous concentrated superficial and lateral of the nuclei. White scars, so prominent on dolphins such as the Risso's dolphin, are usually lacking the pigment.

The main columnar cells of the stratum germinativum are attached in the dermis by numerous hemidesmosomes on the basal membrane and to each other by desmosomes. The cytoplasm is typical of mammalian germinatival cells containing numerous perinuclear mitochondria, rough endoplasmic reticulum, free ribosomes, Golgi apparatus, and a woven pattern of tonofibrils. These tonofibrils connect the cytoplasm to the attachment plaques of desmosomes (Geraci et al., 1986). The oval nucleus, with its well-developed chromatin network, has one to three nucleoli. Dendritic-like cells of Langerhans, that play a central role within inflammatory and immune responses, are outlined by their melanin content in the basal portion of the epidermis (Simpson and Gardner, 1972). Other authors did not find Langerhans cells or Merkel cells in the epidermis (Geraci et al., 1986). A recent study, however, demonstrated the presence of Langerhans cells and dermal dendritic cells

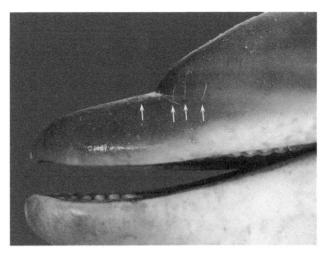

FIGURE 2.4   Vibrissal hairs (*arrows*) on the rostrum of a newborn bottlenose dolphin.

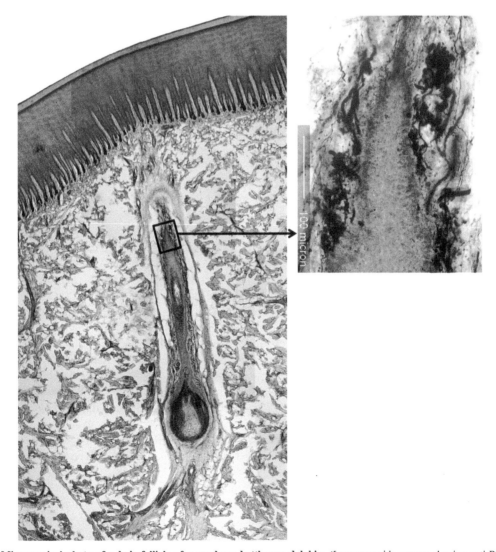

FIGURE 2.5   **Microscopical photo of a hair follicle of a newborn bottlenose dolphin.** (Imunoperoxide processed using anti-PGP9.5 antibody) *(Photographs courtesy of Dr Annamaria Grandis, University of Bologna)*

TABLE 2.1 Mean Skin Measurements

	Epidermis Thickness (mm)	Height of Dermal Papillae (mm)	Subpapillary Dermal Layer (mm)
Delphinus delphis	Mean: 1.3	0.7	0.7
	Breast: 1.3	0.7–0.8	0.3–0.6
	Back: 1.1–1.5	0.7–0.9	0.7–1
	Belly: 1.5	0.8	0.8
	Side: 1.3–1.4	0.7–0.8	0.5–1
	Peduncle: 1.3	0.7	0.4
Orcinus orca	4.5	3	2.5
Stenella coeruleoalba	1	0.5	0.5
Tursiops truncatus	Back: 2.1	1.1–1.4	
	Breast: 2.4	1.3	

Data from Sokolov (1982). Mammal Skin. University of California Press, Berkeley, Los Angeles, London.

in the striped dolphin by immunohistochemical staining for the Toll-like receptor 2 and the S-100 protein, which were considered a marker of Langerhans cells (Lauriano et al., 2014).

Above the tips of the papillae lie 35–50 rows of uniform polyhedral prickle cells with ellipsoid nuclei (Fig. 2.3). These cells are more spherical near their origin on the stratum germinativum and become flattened and elliptical toward the surface of the skin of the bottlenose dolphin. These stratum spinosum cells show fewer mitochondria, Golgi, and ribosomes but the same number and size of lipid droplets and more tonofibrils and desmosomes than the epithelial cells of the germinatival layer. Numerous lamellar bodies appear at the periphery of the cytoplasm. The large vesicular nuclei usually contain single prominent nucleoli.

Some controversy surrounds the precise nomenclature of the external layer. Although keratin can be demonstrated histochemically (Geraci et al., 1986), most authors agree that the process of cornification is incomplete. Hence, the term "stratum externum" was sometimes preferred instead of "stratum corneum." This stratum is only eight to ten cells thick and has flattened cells sometimes with deformed or destroyed nuclei, occasionally intact nuclei. Accordingly, it appears of parakeratosis known from the pouch epidermis in kangaroos (Sokolov, 1982). However, Geraci et al. (1986) have observed as many as 50 layers of cells in the stratum externum of the bottlenose dolphin. These authors suggest that far fewer than 50 cell layers are often reported, likely due to exfoliation associated with collecting and processing of the samples. The cytoplasm of these external cells contains densely packed tonofibrils, lipid droplets, and a few structures suggestive of keratohyaline granules (Geraci et al., 1986). Lamellar bodies, which are abundant in cells of the lower layer, are not apparent. Epidermal cell production of bottlenose dolphins occur at a rate 250–290 times that of humans (Palmer and Weddell, 1964).

It has been suggested that the cells of the stratum externum, because they are parakeratotic, are still living. However, the nuclei are condensed and pyknotic and the cytoplasmic organelles are sparse, suggesting that the cells are senescent. Cells of the stratum externum exfoliate intact so that sheets of epidermis can be easily collected at sea to be used for genetic fingerprinting. Moreover, the self-cleaning abilities of the dolphin's skin was recently discussed because the skin displayed a nanorough surface characterized by a pattern of nanoridge-enclosed pores (average pore size approximately $0.20 \ \mu m^2$), which is below the size of most marine biofouling organisms (Baum et al., 2002).

The mammary glands derive evolutionarily from modified sweat glands. In dolphins their teats (nipples) are positioned into short mammary slits, running longitudinal on both sides of the genital slit (Fig. 2.6). At the nipples, the dermal papillae are thin and short (Fig. 2.7). Male dolphins are known to have rudimentary nipples and slits at about the same position as females; cranial of the anal aperture. Cetacean mammary glands are elongate, narrow, and flat epithelial organs that extend in the subcutaneous connective tissue. They extend from a little posterior of the umbilicus to slightly anterior to the anus. During lactation they are swollen so that slight protrusions are visible externally. The mammary glands are divided into a large number of lobuli, all leading by a small duct into a central lactiferous duct (Fig. 2.8). The terminal duct is distended close to the nipple.

Histological criteria for lactation are well-developed alveoli, epithelium in different stages of secretion, little interalveolar connective tissue, no lencocytes, few and inactive macrophages, and the absence of corpora amylacea. These lactating glands may not necessarily have milk-filled ducts because either a calf just sucked the gland dry or glands may have lost the milk while handling the carcass (Slijper, 1966). Dolphin milk has a creamy white color. The milk contains approximately 60–77% water, 10–30% fat, and 8–11% protein. The sugar content is very low, but the protein content is about twice as much as that of terrestrial mammals.

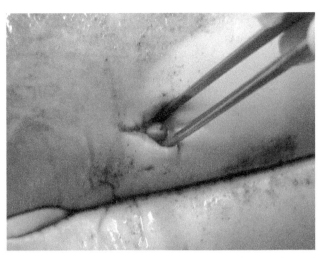

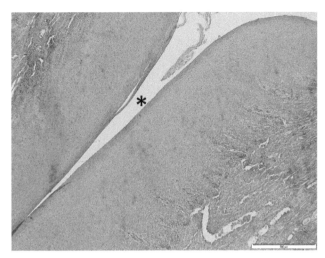

FIGURE 2.6 **Female nipple of a bottlenose dolphin (between forceps) and genital slit below.**

FIGURE 2.7 **Nipple of a bottlenose dolphin representing the lactiferous duct (*).** HE stain.

    Calves of dolphins are always suckled under water. Because the young lacks proper lips, it cannot grip the nipple firmly like the young of most terrestrial mammals. Therefore, it is believed that dolphins squirt the milk into the mouth of the calf. Accordingly, there are observations that, once the calf detaches from the nipple, the milk often continues to spout for several seconds. Thus, the milk flows under pressure, probably due to the action of the cutaneous muscles or due to contractions of the myoepithelial cells (Fig. 2.8) surrounding the alveoli (Slijper, 1966).

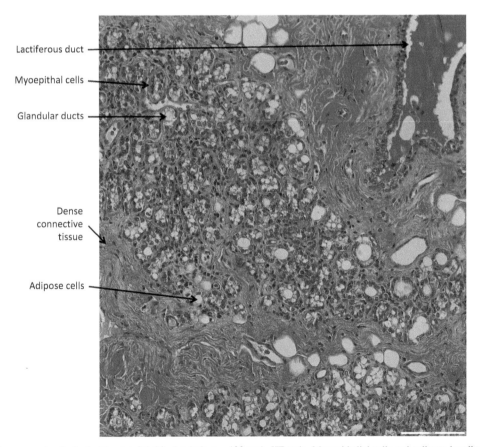

Lactiferous duct

Myoepithal cells

Glandular ducts

Dense connective tissue

Adipose cells

FIGURE 2.8 **Mammary gland of a bottlenose dolphin (scale bar = 100 μm).** HE stain. Myoepithelial cells and collagen bundles surround the epithelial cells of the glandular ducts.

Blubber: Corium (Dermis) and Subcutis

Blubber is multifunctional. It is the primary site for metabolic energy storage, provides thermal insulation, and contributes to positive buoyancy (Struntz et al., 2004). The blubber also functions to streamline the cetacean body and to form specialized locomotor structures such as the dorsal fin, propulsive fluke blades and caudal keels. Construction of the blubber structures relies not only on lipid, but also on the fiber equipment of collagen and elastin.

The thickness of the subpapillary layer of the dermis varies considerably in different species and in different parts of the body (Table 2.1). Collagen fiber bundles, 30–55 µm thick, run mostly at a flat angle to the skin or parallel to the skin and interlace at more or less right angles to a three-dimensional network (Sokolov, 1982). The dermis blends gradually with the adipose layer because fat may extend to some extent up to the epidermis border and heavy collagen bundles ramify throughout the whole subcutaneous blubber (Fig. 2.2). In the dorsal keel of the harbor porpoise considerable strong fibers run longitudinally stabilizing the locomotor apparatus caudally (Hamilton et al., 2004).

Elastin fibers in most parts of the body are numerous, running along the collagen fiber bundles and independently between the fat cells. On the border with the subcutaneous muscles, elastin fibers form intricate plexuses and even small bundles of up to 50 µm thickness (Figs. 2.2, 2.9, and 2.10).

The thickness of the subcutaneous fat tissue varies according to the animal's health status, ontogeny, geography, reproductive state, and feeding success. In dolphins, it is thickest at the dolphin's back and belly (1–2 cm in the common dolphin). Here, a stratification of blubber into outer, middle, and inner layers is visible (Montie et al., 2008) according to number and area of adipocytes, which were largest in the middle layer (Fig. 2.2). Concomitantly, in the bottlenose dolphin, the middle and inner layers had a higher percentage of microvascularity than the outer layer (McClelland et al., 2012). Although both ventrally and dorsally situated blubber can have insulator and buoyancy functions, the ventral blubber is thought to mainly serve as an energy reserve (Gómez-Campos et al., 2015). However, the blubber's fat layer is only a few millimeters (1–4 mm) at the flippers and fluke. Accordingly, the dermis is of relatively greater density in the dorsal fins, flippers, and tail. In the common dolphin, the size of the fat cells varies from 90×170 µm^2 to 120×225 µm^2 (Fig. 2.10;

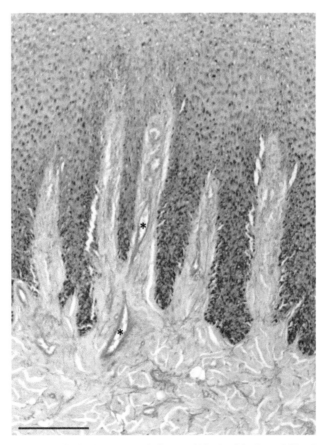

FIGURE 2.9   Dermal papillae of a bottlenose dolphin showing elastic fibers within the thin dermal fiber bundles and blood vessels (*) running in parallel to the papillae (scale bar = 100 µm). Orcein after Taenzer-Unna stain.

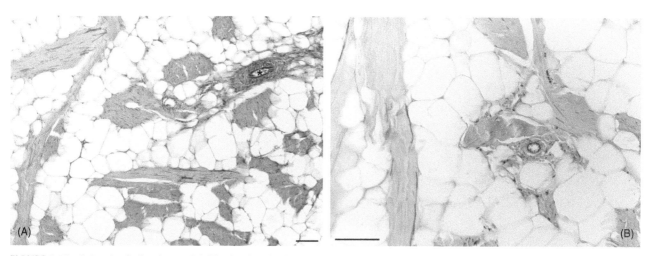

FIGURE 2.10 **Subcutis of a bottlenose dolphin showing elastic fibers within the dermal fiber bundles (scale bars = 100 μm).** Left, Orcein after Taenzer-Unna; Right, Resorcin-Fuchsin after Weigert. The blood vessel (*) on left image is heavily surrounded by elastic fibers.

Sokolov, 1982). In the deepest blubber layer, brown adipose tissue was found recently, which may play a role in regulating body temperature (Hashimoto et al., 2015).

Numerous blood vessels and nerve endings can be found in the dermis. Muscular arterioles and thin-walled venules run parallel to the surface in the superficial and deeper layers of the dermis (Figs. 2.2 and 2.9). Typically, a vertically directed thin-walled arteriole runs up the center of each dermal papilla and connects through anastomotic vessels with tributary veins on each side (Sokolov, 1982).

There is a plexus of parallel running arteries and veins in the fluke, dorsal fin, and flippers. Here, thick and muscular walled central arteries are surrounded by a plexus of veins running alongside (Fig. 2.11). These vascular arrangements

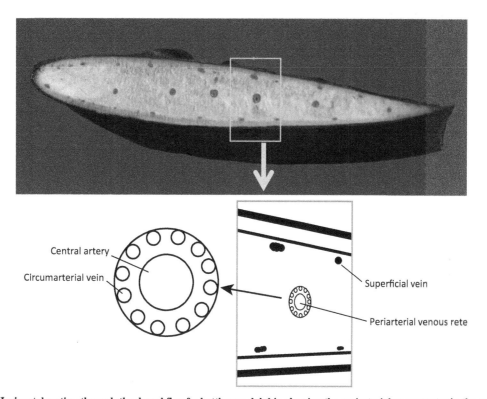

FIGURE 2.11 **Horizontal section through the dorsal fin of a bottlenose dolphin showing the periarterial venous retes in the center and superficial veins.**

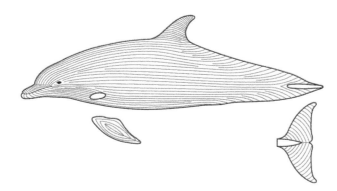

FIGURE 2.12 **Overview of dermal ridges of a dolphin.** *(Redrawn by Uko Gorter after Geraci, 1986)*

appear like a typical heat exchanger and thus it is thought that these plexus function in the body's temperature regulation. Incidentally, these vessels in the tail and flippers serve as the most useful sites for venipuncture.

The skin of dolphins is richly innervated. Numerous nerve endings resembling small, onionskin-like configurations are present in the superficial portion of the dermis, particularly in the jaw region, vulva, and perineum. In the skin of the bottlenose dolphin, potential mechanoreceptors were found in the interdigitations of the epidermis with the dermal ridges. Silver staining shows numerous nerve fibrils in the dermis, extending in the epidermis. The nerve endings are considered to be largely proprioceptive. The hair follicles on the rostrum are well innervated. Accordingly, it was suggested that the nerve endings sense low-frequency vibrations. However, today it is known that the hair follicles and thus the nerve terminals are involved in electrosensing and may serve as electroreceptors as described for the Guiana dolphin (*Sotalia guianensis*; see Passive Electroreception).

Dolphins (and other toothed whales) possess shallow, regular cutaneous ridges (Fig. 2.2) of the surface of the skin over much of their bodies (Shoemaker and Ridgway, 1991). They are usually faint and can be seen by close inspects the skin of a living animal at an oblique angle. These ridges are generally quite shallow at the surface (less than 10 to slightly over 100 μm high) and range in width from 0.4 to 1.7 mm. However, because of their faintness and because they may tend to relax or disappear from the surface of dead skin, they are difficult to see in histological cross-sections.

The association of the superficial cutaneous ridges with the underlying dermal ridges and papillae is at present unclear (Shoemaker and Ridgway, 1991). Earlier workers (Sokolov, 1955; Sokolov et al., 1968) report that dermal ridges run parallel to the long axis of the body in the harbor porpoise and the common and bottlenose dolphins (Fig. 2.12), while Purves (1963) shows dermal ridges running mainly at an oblique angle in the common dolphin and parallel to the body axis in the harbor porpoise. Stromberg (1989) also reported that dermal ridges run roughly parallel to the long axis of the body in the bottlenose dolphin, which is generally perpendicular to the direction of the surface ridges that Shoemaker and Ridgway (1991) have observed (Fig. 2.2).

Interestingly, Sokolov et al. (1968; cited after Shoemaker and Ridgway, 1991) report about 60 dermal ridges per centimeter for adult bottlenose dolphins. This value is about 3 times that of the cutaneous ridges mentioned by Shoemaker and Ridgway (1991). Accordingly, careful histological studies are needed to clarify the relationship between the subsurface dermal ridges and the cutaneous ridges on the skin surface. The hydrodynamic consequences of the ridges are still unclear and effects for drag reduction have never been demonstrated.

The dense packing of dermal papillae associated with the cutaneous ridges suggests a sensory function. The highly innervated skin has a threshold sensitivity of 10–40 mg mm^2, which is close to the most sensitive skin areas (ie, fingertips, lips, eyelids) in humans. Dolphins are sensitive to vibrations and small pressure changes. The skin, therefore could function in the detection of flow velocities and flow disruptions (Fish, 2006).

On accelerating dolphins, mobile skinfolds were observed that may result from vorticity along the body surface. The vorticity creates pressure differences, which could deform the flexible skin. Accordingly, it has been suggested that these folds may be generated from active control by muscles (Sokolov, 1982). However, passive hydrodynamically generated folds move posteriorly in a wavelike manner perpendicular to direction of the dolphin's movement. It was hypothesized that the folds were a mechanism to damp turbulence by maintaining a laminar boundary layer (Fish, 2006). Drag reduction may be possible by matching a turbulent flow with the compliant skin surface, which is stabilized by the prominent interdigitation of epidermal rete pegs and long dermal papillae. Moreover, the surface physical properties of the skin in conjunction with its high rate of sloughing (see the previous section) may help to maintain low drag characteristics by preventing fouling by encrusting organisms, such as barnacles, on the dolphin's surface (Fish, 2006).

REFERENCES

Arkowitz, R., Rommel, S., 1985. Force and bending moment of the caudal muscles in the shortfin pilot whale. Mar. Mammal Sci. 1, 203–209.

Baum, C., Meyer, W., Stelzer, R., Fleischer, L.-G., Siebers, D., 2002. Average nanorough skin surface of the pilot whale (*Globicephala melas*, Delphinidae): considerations on the self-cleaning abilities based on nanoroughness. Mar. Biol. 140, 653–657.

Benke, H., 1993. Investigations on the oseology and functional morphology of the flipper of whales and dolphins. Investigations on Cetacea. Institute of Brain Anatomy, University of Berne, Berne, pp. 9–253.

Felts, W.J.L., 1966. Some functional and structural characteristics of cetaceans flippers and flukes. In: Norris, K.S. (Ed.), Whales, Dolphins and Porpoises. University of California Press, Berkeley, Los Angeles, pp. 255–276.

Fish, F.E., 2006. The myth and reality of Gray's paradox: implication of dolphin drag reduction for technology. Bioinspir. Biomim. 1, R17.

Fish, F.E., Beneski, J.T., Ketten, D.R., 2007. Examination of the three-dimensional geometry of cetacean flukes using computed tomography scans: hydrodynamic implications. Anat. Rec. Adv. Integr. Anat. Evol. Biol. 290, 614–623.

Fish, F.E., Rohr, J.J., 1999. Review of dolphin hydrodynamics and swimming performance. Technical Report No. 1801. SSC San Diego, San Diego, CA.

Geraci, J.R., Aubin, D.J.S., Hicks, B.D., 1986. The epidermis of odontocetes: a view from within. In: Bryden, M.M., Harrison, R. (Eds.), Research on Dolphins. Clarendon Press, Oxford, pp. 3–21.

Gómez-Campos, E., Borrell, A., Correas, J., Aguilar, A., 2015. Topographical variation in lipid content and morphological structure of the blubber in the striped dolphin. Sci. Mar. 79, 189–197.

Hamilton, J.L., Dillaman, R.M., McLellan, W.A., Pabst, D.A., 2004. Structural fiber reinforcement of keel blubber in harbor porpoise (*Phocoena phocoena*). J. Morphol. 261, 105–117.

Hashimoto, O., Ohtsuki, H., Kakizaki, T., Amou, K., Sato, R., Doi, S., Kobayashi, S., Matsuda, A., Sugiyama, M., Funaba, M., Matsuishi, T., Terasawa, F., Shindo, J., Endo, H., 2015. Brown adipose tissue in cetacean blubber. PLoS One 10, e0116734.

Iosilevskii, G., Weihs, D., 2008. Speed limits on swimming of fishes and cetaceans. J. R. Soc. Interface 5, 329–338.

Lauriano, E.R., Silvestri, G., Kuciel, M., Żuwała, K., Zaccone, D., Palombieri, D., Alesci, A., Pergolizzi, S., 2014. Immunohistochemical localization of Toll-like receptor 2 in skin Langerhans' cells of striped dolphin (*Stenella coeruleoalba*). Tissue Cell 46, 113–121.

Long, Jr, J.H., Pabst, D.A., Shepherd, W.R., McLellan, W.A., 1997. Locomotor design of dolphin vertebral columns: bending mechanics and morphology of *Delphinus delphis*. J. Exp. Biol. 200, 65–81.

McClelland, S.J., Gay, M., Pabst, D.A., Dillaman, R., Westgate, A.J., Koopman, H.N., 2012. Microvascular patterns in the blubber of shallow and deep diving odontocetes. J. Morphol. 273, 932–942.

Montie, E.W., Garvin, S.R., Fair, P.A., Bossart, G.D., Mitchum, G.B., McFee, W.E., Speakman, T., Starczak, V.R., Hahn, M.E., 2008. Blubber morphology in wild bottlenose dolphins (*Tursiops truncatus*) from the southeastern United States: influence of geographic location, age class, and reproductive state. J. Morphol. 269, 496–511.

Noren, S.R., Biedenbach, G., Edwards, E.F., 2006. Ontogeny of swim performance and mechanics in bottlenose dolphins (*Tursiops truncatus*). J. Exp. Biol. 209, 4724–4731.

Noren, S.R., Redfern, J.V., Edwards, E.F., 2011. Pregnancy is a drag: hydrodynamics, kinematics and performance in pre- and post-parturition bottlenose dolphins (*Tursiops truncatus*). J. Exp. Biol. 214, 4151–4159.

Pabst, D.A., 1990. Axial muscles and connective tissues of the bottlenose dolphin. In: Leatherwood, S., Reeves, R.R. (Eds.), The Bottlenose Dolphin. Academic Press, San Diego, CA, pp. 51–67.

Pabst, D.A., 1996a. Springs in swimming animals. Am. Zool. 36, 723–735.

Pabst, D.A., 1996b. Morphology of the subdermal connective tissue sheath of dolphins: a new fibre-wound, thin-walled, pressurized cylinder model for swimming vertebrates. J. Zool. 238, 35–52.

Palmer, E., Weddell, G., 1964. The relationship between structure, innervation and function of the skin of the bottle nose dolphin (*Tursiops truncatus*). Proc. Zool. Soc. Lond. 143, 553–568.

Purves, P.E., 1963. Locomotion in Whales. Nature 197, 334–337.

Reidenberg, J.S., Laitman, J.T., 2009. Cetacean prenatal development. In: Perrin, W.F., Würsig, B., Thewissen, J.G.M. (Eds.), Encyclopedia of Marine Mammals. Academic Press, San Diego, CA, pp. 220–230.

Shoemaker, P.A., Ridgway, S.H., 1991. Cutaneous ridges in odontocetes. Mar. Mammal Sci. 7, 66–74.

Simpson, J.G., Gardner, M.B., 1972. Comparative microscopic anatomy of selected marine mammals. In: Ridgway, S.H. (Ed.), Mammals of the Sea—Biology and Medicine. Charles C Thomas, Springfield, IL, pp. 298–418.

Slijper, E.J., 1966. Functional morphology of the reproductive system in Cetacea. In: Norris, K.S. (Ed.), Whales, Dolphins and Porpoises. University of California Press, Berkeley, Los Angeles, pp. 277–319.

Sokolov, V.E., 1955. Struktura kozhnogo pokrova nekotorykh kitoobraznykh (structure of the integument of certain cetaceans). Byulleten Mosk. Otd. Biol. 60, 45–60.

Sokolov, V.E., Kuznetsov, G.V., Rodionov, V.A., 1968. Napravlenie dermal'nykh valikov v kozhe del'finov v svyazi s osobennostju obtekania tela vodoi (Direction of dermal ridges in dolphins in relation to the streamlining of their body). Byulleten Mosk. Otd. Biol. 73, 123–126.

Sokolov, V.E., 1982. Mammal Skin. University of California Press, Berkeley, Los Angeles, London.

Stromberg, M.W., 1989. Dermal-epidermal relationships in the skin of the bottlenose dolphin (*Tursiops truncatus*). Anat. Histol. Embryol. 18, 1–13.

Struntz, D.J., McLellan, W.A., Dillaman, R.M., Blum, J.E., Kucklick, J.R., Pabst, D.A., 2004. Blubber development in bottlenose dolphins (*Tursiops truncatus*). J. Morphol. 259, 7–20.

Thewissen, J.G.M., Fish, F.E., 1997. Locomotor evolution in the earliest cetaceans: functional model, modern analogues, and paleontological evidence. Paleobiology 23, 482–490.

Chapter 3

Locomotion (Including Osteology and Myology)

OSTEOLOGY AND ARTHROLOGY

Life in the water changes it all: Swimming requires different levers to face gravity in the aquatic environment. Terrestrial quadrupedal mammals move their hind limbs, act on the pelvic bones, and bend the vertebral column to lift and push forward the front limbs and the head–neck complex. Bipedal mammals who walk or jump are free to use the thoracic limbs for other tasks (from eating fruits to playing the piano). Some apes even use their upper extremities to bring themselves from one tree to another, and their lower limbs have feet able to contribute by grasping. But from the cheetah to the gibbon, all terrestrial mammals flex and extend their limbs using the ground (or a pending creeper) as a solid to push the body forward (or upward) against gravity. The liquid environment is denser, and consequently gravity lighter to contrast. Body mass grows and the skeleton changes.

Cetaceans have adapted to marine life through a long process started in the early Eocene (Uhen, 2007) and reached a degree of bodily transformation unsurpassed among other marine mammals (Reidenberg, 2007). Their entire life from birth to reproduction and senescence is spent in the water and a return to terrestrial life is not compatible with survival. Their streamlined shape is the ultimate evolutionary adaptation of complex organisms that have reverted to a liquid environment, and involves striking changes in skeletal anatomy.

The basic architecture of the cetacean skeleton (Fig. 3.1) and the structure and organization of its components is *almost* the same as in terrestrial mammals, but each element is specifically shaped for the aqueous environment and some parts—markedly the hind limbs—are missing.

Structure of Bones

The bones of dolphins (and whales) have a very high fat content, a feature useful also to increase buoyancy (Gray et al., 2007). The tightly woven osseous matrix has a moderate-to-low overall density (Felts and Spurrell, 1965, 1966; Felts, 1966), although some bones have an incredible density (see the following discussion concerning bone density). The long bones (of the arm and forearm) show no medullary cavity (Felts and Spurrell, 1965) (Fig. 3.2) and display on X-rays the famous hourglass shape, the result of a cancellous texture and reduction or even absence of compact cortices (De Buffrénil and Schoevaert, 1988).

One must keep in mind that the forelimb of dolphins bears no weight against gravity forces, and therefore the robustness of the bone does not match the same requirements found in terrestrial mammals. On the other hand, some joints (as the shoulder joint or the joints of the spine) continuously work in the ever going movement of swimming.

An exception to the general rule of moderate bone density is the tympanoperiotic bone (see the following and also Chapter 5 for a description), whose density is far higher than that of the surrounding bones of the skull, even during fetal development (Cozzi et al., 2015). This specific characteristic of the tympanoperiotic bone may be related to the need for early development of the auditory apparatus, essential for the perception of sounds in the water—and consequently for the immediate establishment of mother–calf bonding.

Skull

Skull in General

The skull of dolphins (and cetaceans in general) underwent a process of elongation, called telescopy (Miller, 1923) that gives a characteristic morphology to the bones of the face and neurocranium (see the following). Furthermore, the shape of the dolphin skull is bilaterally asymmetrical and strikingly different from that of most terrestrial mammals (Fig. 3.3) (McLeod et al., 2007). Toothed whales are the only mammals (including extinct species) in which the directional asymmetry of the skull is the normality (Howell, 1930).

Anatomy of Dolphins. http://dx.doi.org/10.1016/B978-0-12-407229-9.00003-8

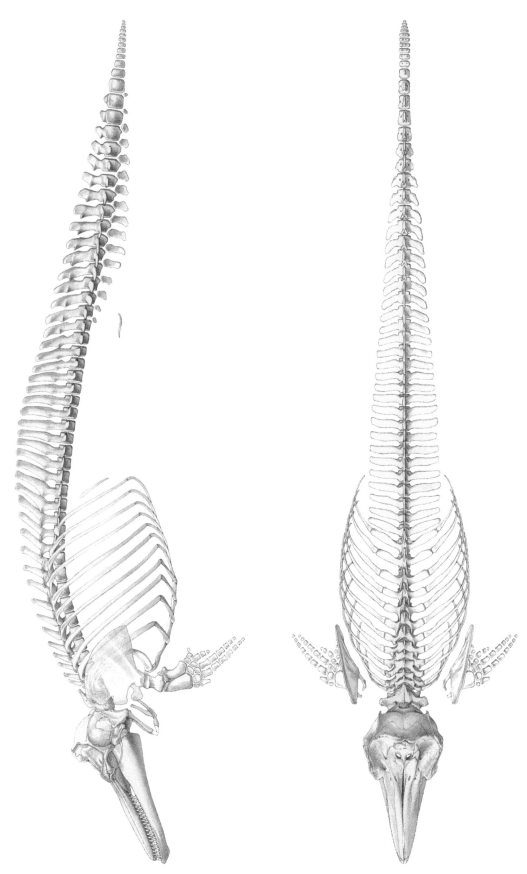

FIGURE 3.1 **Skeleton of *Tursiops truncatus*.** (*Drawings by Massimo Demma*)

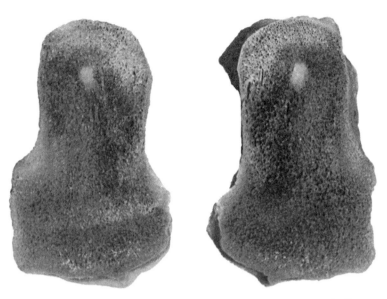

FIGURE 3.2 **Section of the humerus of a dolphin.** Note the absence of a medullary cavity and the "hourglass" structure of the cancellous bone.

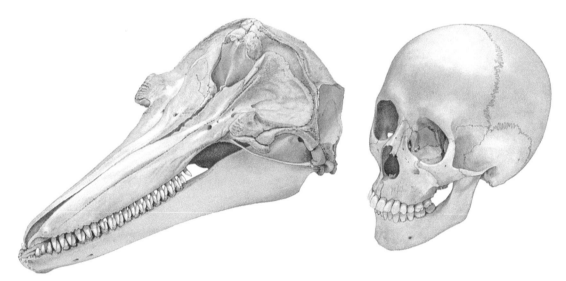

FIGURE 3.3 **Skulls of *T. truncatus* and *Homo sapiens*.** *(Drawings by Massimo Demma)*

The maxillary bones and the mandible are long and give implant to rows of uniformly cone-shaped teeth (see also Chapter 8). It is true that many terrestrial mammals have an elongated head (think of ruminants, equine, etc.), but this is generally caused by the nose acquiring more space and the mouth making room for the powerful molar teeth required for mastication. The nose of cetaceans is placed almost vertical to the maxillary bone, with small nasal bones enclosing cavities connected to the top of the head (see Chapter 5). So the elongation of the maxillary bones does not parallel a likewise expansion of the nasal cavities, and is not functional to chewing, because toothed whales and dolphins just grab and swallow their prey whole. The elongation of the skull may be due to the high number of teeth necessary for grabbing, or to the space required by the melon dorsal to the maxillary bone.

The hyoid complex is formed by the usual segments and is shaped like a dorsally concave crescent. It constitutes the basis for the attachment of the musculature of the tongue and frames the large larynx.

The skulls and mandibles of selected dolphin species are presented in the following: *T. truncatus* (Figs. 3.4 and 3.5); *Delphinus delphis* (Figs. 3.6 and 3.7); *Stenella coeruleoalba* (Figs. 3.8 and 3.9); *Grampus griseus* (Figs. 3.10 and 3.11); *Globicephala melas* (Fig. 3.12); *Orcinus orca* (Fig. 3.13); *Pseudorca crassidens* (Fig. 3.14); *Lagenorhynchus obliquidens* (Fig. 3.15).[a]

---

a. For a very detailed description of the skull of the bottlenose dolphin and other cetacean species, see Mead and Fordyce (2009). A very clear, extensive description of the bottlenose dolphin skull is also that of Rommel (1990). In the present section of the book, we choose to limit the written description of the skull bones to the essentials, and let the reader get the details through the illustrations that accompany the text.

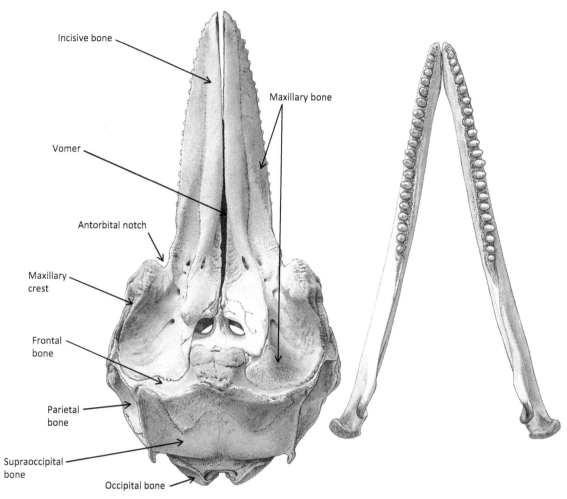

FIGURE 3.4 **Dorsal view of the skull and mandible of *T. truncatus*.** *(Drawings by Massimo Demma)*

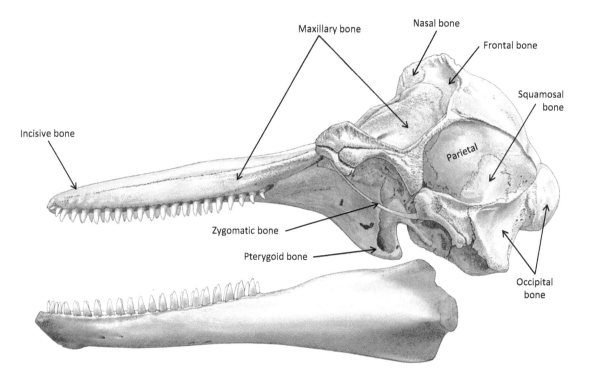

FIGURE 3.5 **Lateral view of the skull and mandible of *T. truncatus*.** *(Drawings by Massimo Demma)*

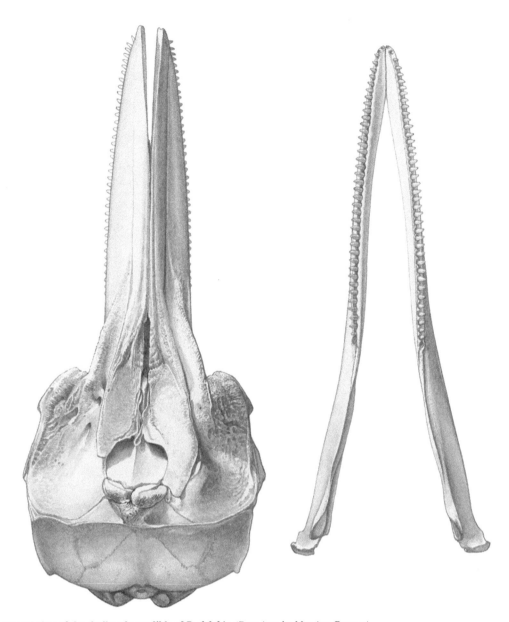

FIGURE 3.6   Dorsal view of the skull and mandible of *D. delphis*. *(Drawings by Massimo Demma)*

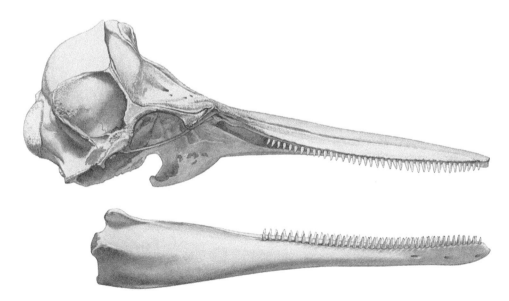

FIGURE 3.7   Lateral view of the skull and mandible of *D. delphis*. *(Drawings by Massimo Demma)*

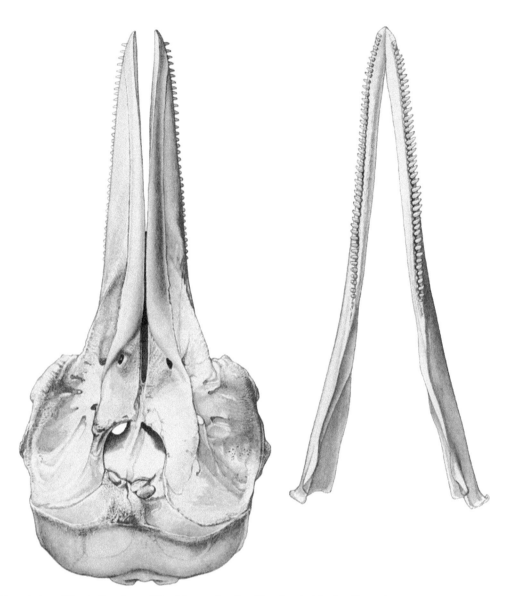

FIGURE 3.8   Dorsal view of the skull and mandible of *S. coeruleoalba*. *(Drawings by Massimo Demma)*

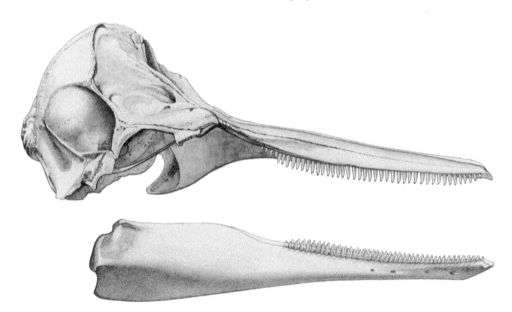

FIGURE 3.9   Lateral view of the skull and mandible of *S. coeruleoalba*. *(Drawings by Massimo Demma)*

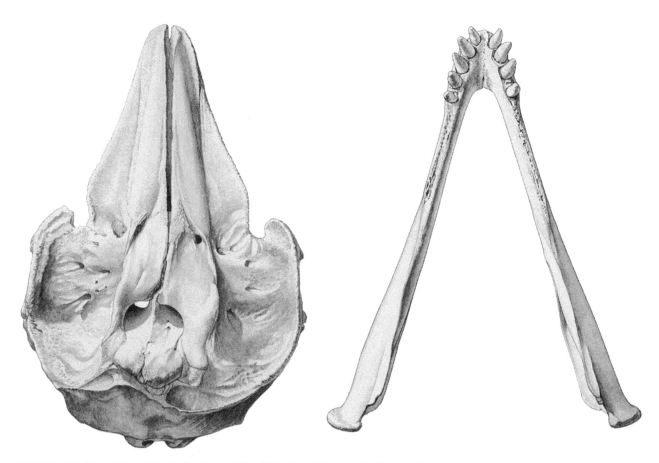

FIGURE 3.10   **Dorsal view of the skull and mandible of *G. griseus*.** *(Drawings by Massimo Demma)*

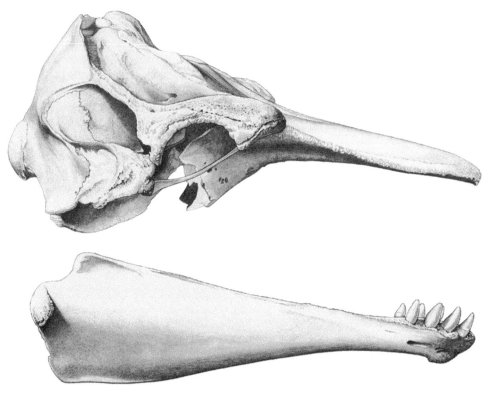

FIGURE 3.11   **Lateral view of the skull and mandible of *G. griseus*.** *(Drawings by Massimo Demma)*

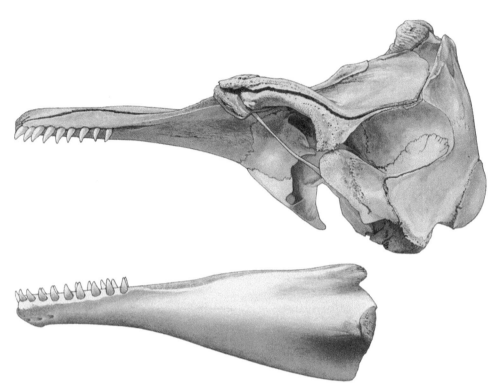

FIGURE 3.12　**Lateral view of the skull and mandible of *G. melas*.** *(Drawings by Massimo Demma)*

The bones of the skull may be described as pertinent to the face (*splanchnocranium*) or surrounding the cranial cavity (*neurocranium*). Details of single bones (including foramina) are identified in the relative figures.

Bones of the Face (Splanchnocranium)

The outline of the head of several (but not all) dolphin species presents an elongated profile with a more or less pointed beak. However, the *bony* rostrum of *all* dolphins is characterized by the very long and partially asymmetric incisive bone (frequently called premaxillary bone[b]) (Fig. 3.16) and maxillary bone (Fig. 3.17), these latter providing almost all the insertion of the upper rows of teeth.[c] The presence of longitudinal grooves along the ventral surface of the premaxillary and maxillary bones may constitute a species-specific character (Fig. 3.18).[d] The elongated and dorsally hollow vomer (Fig. 3.19) is placed along the sagittal plane and terminates in the pointed mesorostral groove at the tip of the rostrum.

The base of both the incisive and maxillary bones continues with the nasal bones that together with the ethmoid bone compose the osseous base of the nares.

The ethmoid bone (Figs. 3.20 and 3.21) is consistently different from the corresponding bone of terrestrial mammals, since odontocetes have no developed olfactory nerve and consequently there is no real cribriform plate that allows passage of minute nerve fibers from the nasal cavity into the cranial cavity. However, close inspection of the ethmoid bone of very young individuals may reveal the presence of minute foramina reminiscent of the classical cribriform lamina. A low sagittal crest constitutes the *crista galli*,[e] while the lateral part of the ethmoid bone is continue with the orbitosphenoid and the caudal part is adjacent to the presphenoid. Ventrally, the nasal passages continue with the *choanae* (Fig. 3.22), divided by the vomer into a left and a right passage, that opens in the nasopharynx. The rim of the choanae consists of the palatine (Fig. 3.23) and pterygoid bones, placed caudal to the maxillaries. The pterygoid bones (Fig. 3.24) are characterized by a large pterygoid sinus, unique to cetaceans, sustained by a pyramidal process. The complex system of air sinuses and bony

---

b. The correct term according to the Nomina Anatomica Veterinaria (NAV, 2012) is *os incisivum*, but in many descriptions of the cetacean skull as well as in the literature of comparative anatomy (Kardong, 2015) this structure is called premaxilla. Human anatomical nomenclatures (Terminologia Anatomica, 1998, or TA) do not include this term, as a separate incisive bone does not exist in man.

c. A few teeth (1–3) may be nested on the tip of the incisive bone (Rommel, 1990). However, there is no morphological feature that may justify their definition as incisive teeth.

d. The density of the bones of the face is different in dolphins and beaked whales (Cozzi et al., 2010b) and may reflect an evolutionary trait linked to diving patterns.

e. The Latin term refers to the presence of a sagittal crest in the ethmoid bone of terrestrial mammals (including man), that resembles a rooster comb.

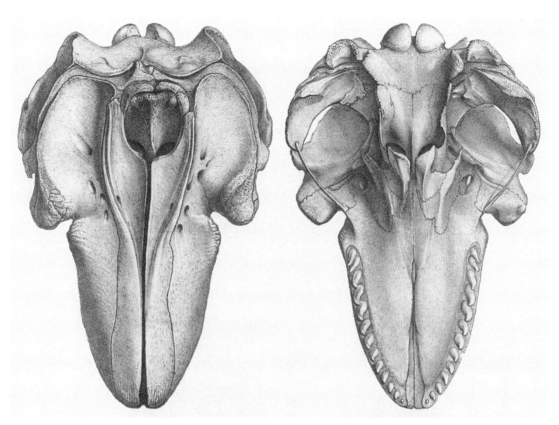

FIGURE 3.13   Skull of *O. orca*. *(From van Beneden and Gervais, 1868–1879. Ostéographie des Cétacés vivants et fossiles. Berthrand, Paris; Atlas, 64 tavv.)*

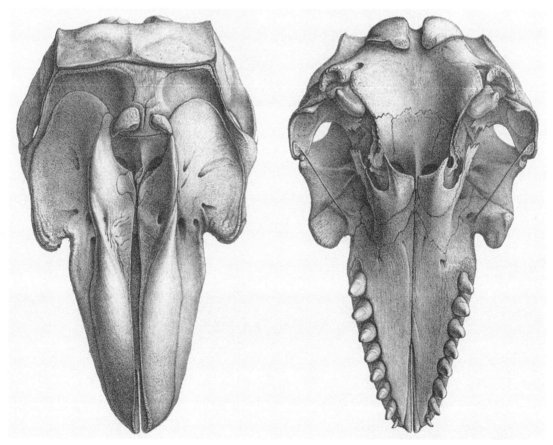

FIGURE 3.14   Skull of *P. crassidens*. *(From van Beneden and Gervais, 1868–1879. Ostéographie des Cétacés vivants et fossiles. Berthrand, Paris; Atlas, 64 tavv.)*

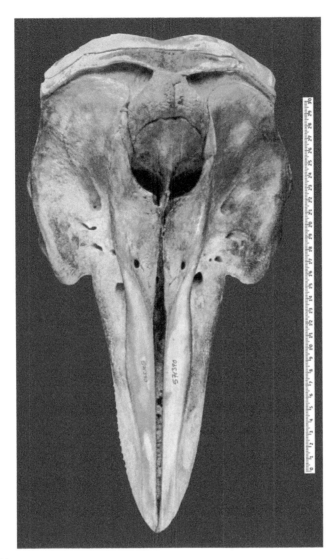

FIGURE 3.15   Skull of *L. obliquidens*.

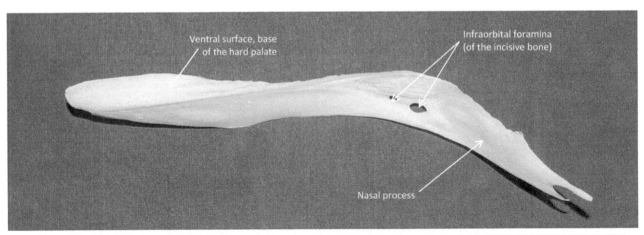

FIGURE 3.16   Os incisivum of a newborn *G. melas*.

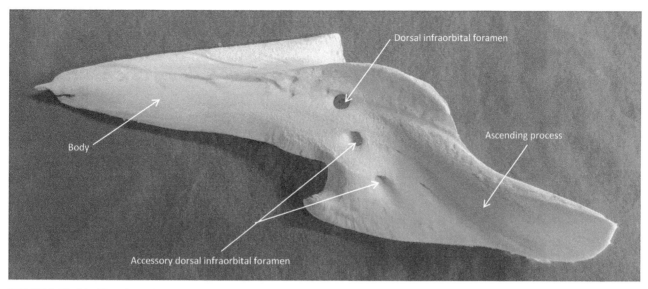

FIGURE 3.17   Maxillary bone of a newborn *G. melas*.

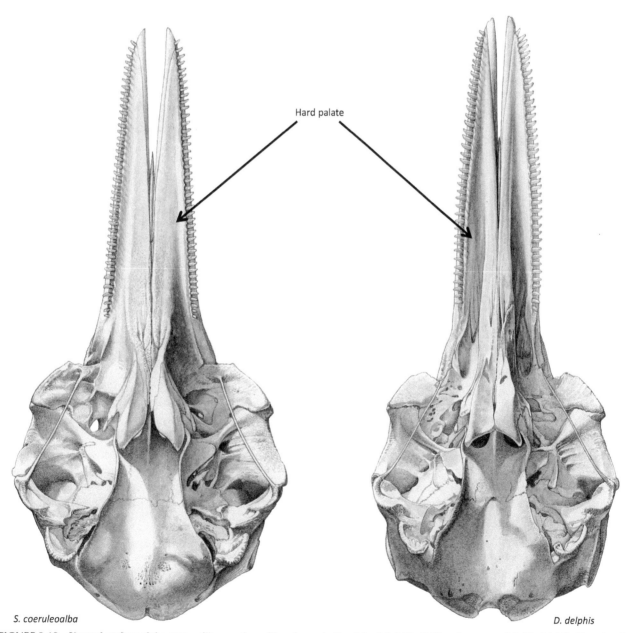

FIGURE 3.18   Ventral surface of the premaxillary and maxillary bones in the striped dolphin (left) and common dolphin (right). *(Drawings by Massimo Demma)*

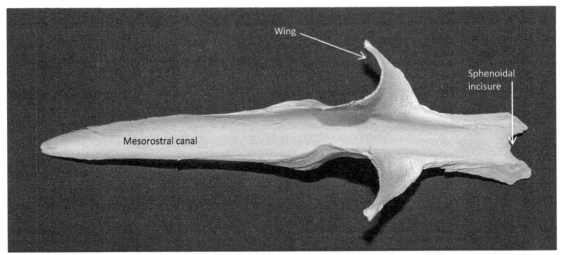

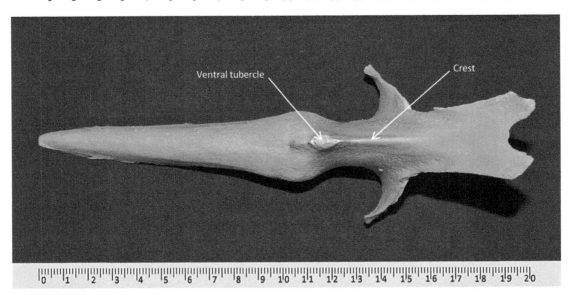

FIGURE 3.19 Dorsal (top) and ventral (bottom) view of the vomer of a newborn *G. melas*.

fossae of the basicranium constitutes an extension of the cavity of the middle ear and may have some functional analogy with the guttural pouches of the horse (see Chapter 5) (Mead and Fordyce, 2009).[f]

The orbit is made up by the zygomatic and preorbital processes of the frontal bone (Fig. 3.25), and by the lacrimal bone attached to the very thin zygomatic bone[g] (Fig. 3.26). The position of the orbit, due to the almost horizontal position of the zygomatic process of the frontal bone, is markedly separated from the rather caudal temporal fossa.

The asymmetry of the bony face is mostly evident in the caudal extremities of maxillary and nasal bones that become enlarged and bent upward over the frontal bone that constitutes the rostral wall of the neurocranium. Additionally, the midline suture of the skull is curved to the left side in the nasal region.[h] The maxillary of mammals contains the infraorbital canal in which runs the infraorbital nerve, part of the maxillary branch of the trigeminal nerve directed to the dental nerve roots (within the canal)

---

f. The guttural pouches of the horse are air-filled cavities along the course of the auditory tube and therefore in continuity with cavity of the middle ear on one side and the nasopharynx on the other. Their function in horses may be related to cooling of blood in arterial vessels directed to the brain.

g. Many articles refer to the zygomatic bone as the *jugal* bone (Kardong, 2015). However, this term is not present in either the human or the veterinary anatomical nomenclature, and—to avoid confusion—its use should perhaps be abandoned in favor of the name universally adopted in systemic anatomy.

h. For a detailed description of the external nasal passages and facial complex of *Delphinidae*, see Mead (1975). For an analysis of the asymmetry of nasal bones in dolphins and other Odontocete species see Hirose et al. (2015).

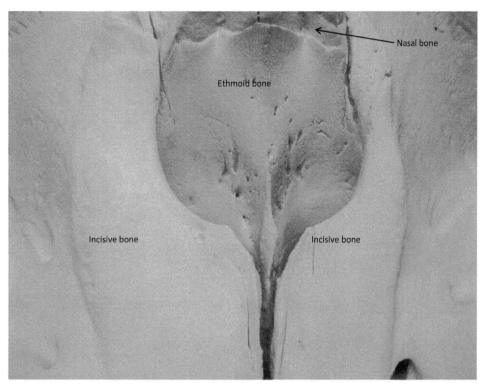

FIGURE 3.20   Frontal view of the ethmoid bone of *S. coeruleoalba* seen between the two ascending processes of the incisive bones.

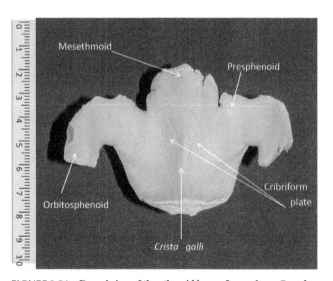

FIGURE 3.21   Dorsal view of the ethmoid bone of a newborn *G. melas*.

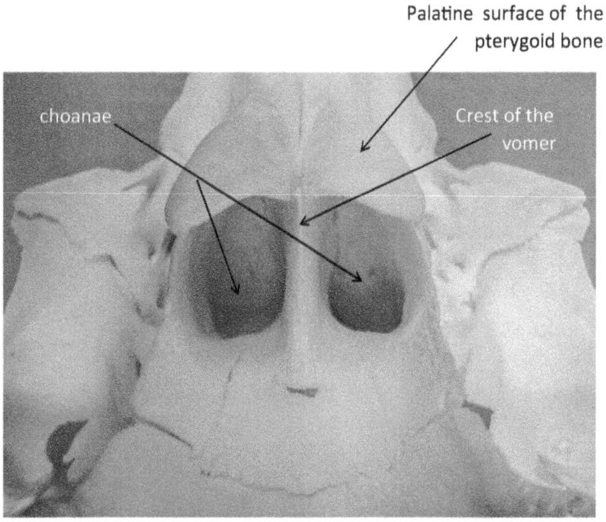

FIGURE 3.22   Ventral view of the choanae of *T. truncatus*.

and—after exiting the canal through the infraorbital foramen—to the nostrils and upper lip.[i] Interestingly, this disposition is quite different in dolphins. There is no single infraorbital foramen at the end of the canal, but a series of dorsal infraorbital foramina, carrying trigeminal braches to the lower surface of the melon. According to Rommel (1990), we should consider a dorsal and a ventral infraorbital foramen. Since the infraorbital foramen opens on the dorsal face of the proximal part of the maxillary bone, before the caudal-most upper teeth, the infraorbital nerve directed to the dental roots continues rostrally within the bone.[j] Several additional dorsal and ventral foramina may yield passage for vessels, with the ventral ones possibly directed to the palate.

---

i. In many mammals the infraorbital nerve carries somato-sensitive information from the vibrissae.
j. So, in dolphins, what is the terminal opening of the infraorbital canal is not terminal at all, and in fact is placed at the beginning of the osseous tunnel.

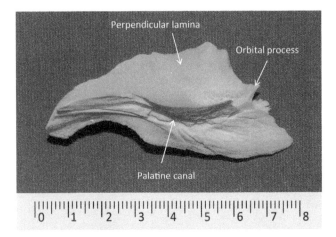

FIGURE 3.23 **Palatine bone of a newborn *G. melas*.**

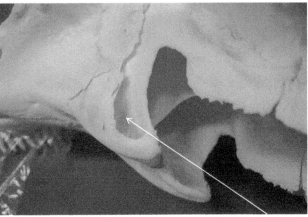

FIGURE 3.24 **Pterygoid bones and relative sinuses of *T. truncatus*.**

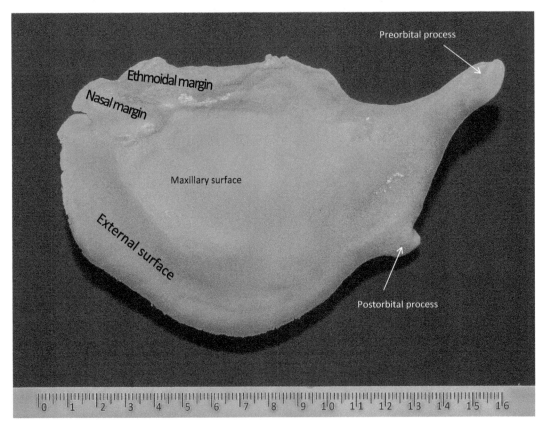

FIGURE 3.25 **Frontal bone of a newborn *G. melas*.**

Bones of the Neural Skull (Neurocranium)

The neural skull is composed of frontally flat and curved bones on the sides, and almost vertical anterior and lateral walls. The effect is that of a dome-like structure, apparently wider (and higher) than long. The external shape of the dome corresponds to the internal cranial cavity and obviously to the form of the brain. In fact, the relationship between the cerebral and cerebellar *fossae* reflects the pronounced rotation of the brain along the interinsular axis (see Chapters 5 and 6).

The frontal bone (Fig. 3.25) is partially covered by the nasal bones, the ethmoid, and the ascending processes of the maxillary bones. All together they form the *vertex,*[k] the most dorsal part of the skull, immediately followed by the nuchal crest of the occipital complex.

---

k. In the subadult skull, the maxillary bones do not reach caudally to the nuchal crest so that a large part of the frontal bone is visible in dorsal view.

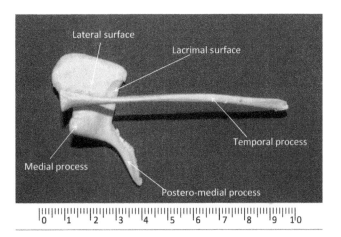

FIGURE 3.26   **Lacrimal and zygomatic bones of a newborn *G. melas*.**

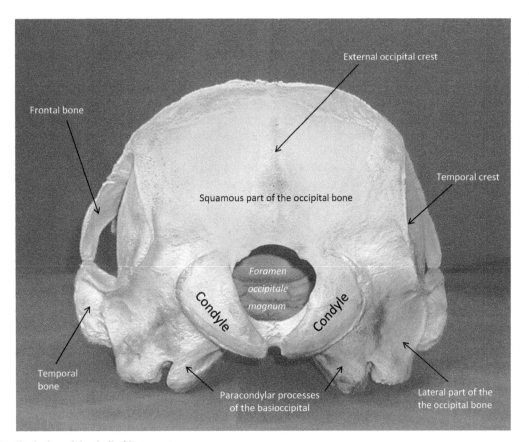

FIGURE 3.27   **Back view of the skull of *T. truncatus*.**

The parietal bones of most mammals constitute part of the lateral walls of the neurocranium and fuse along the midline. However, in dolphins there is a centrally placed interparietal bone that fuses around birth with the occipital bone[l] just posterior to the frontal bone to form the dorsal midline element of the skull (Mead and Fordyce, 2009) (Figs. 3.4, 3.6, 3.8, 3.10, 3.13–3.15). Therefore, the parietal bones proper (Figs. 3.5, 3.7, 3.9, 3.11 and 3.12) do not meet along the midline, but remain on the lateral walls of the neural skull. The occipital bone (*squama occipitalis*) forms the posterior osseous wall of the skull (Figs. 3.27 and 3.28), and allows passage of the medulla oblongata into the spinal cord at the level of the large occipital

l. In their exhaustive description of the dolphin skull, Mead and Fordyce (2009) divide the occipital bone into separate parts, called *supraoccipital*, *exoccipital*, and *basioccipital*, respectively. In this case, the *supraoccipital* bone is the part that fuses with the interparietal bone. In most mammals, the separate parts fuse around or after birth, to allow slight modification of the skull during passage of the newborn head at parturition. Since in adult individuals the separate parts fuse into a single bone, we will describe the occipital bone as a single element.

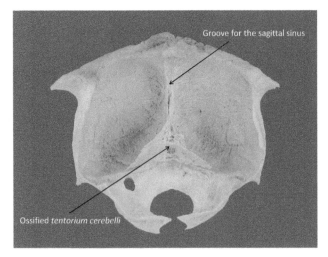

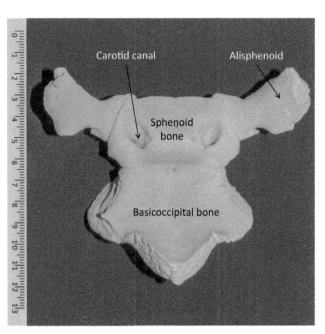

FIGURE 3.28 Internal surface of the squamous part of the occipital bone of *T. truncatus*.

FIGURE 3.29 Internal face of the sphenoid and basioccipital bones of a newborn *G. melas*.

foramen, surrounded by the occipital condyles that articulate with the atlas. The occipital bone (*pars basilaris* or *basioccipital*) constitutes also the caudal part of the base of the skull and articulates rostrally with the sphenoid bone (Figs. 3.29 and 3.30).

The ventrolateral part of the neural skull is made up by the temporal bone, composed by separate elements (*pars squamosa*, *pars petrosa*, *pars tympanica*, and *pars endotympanica*) that fuse during development. The *pars squamosa*

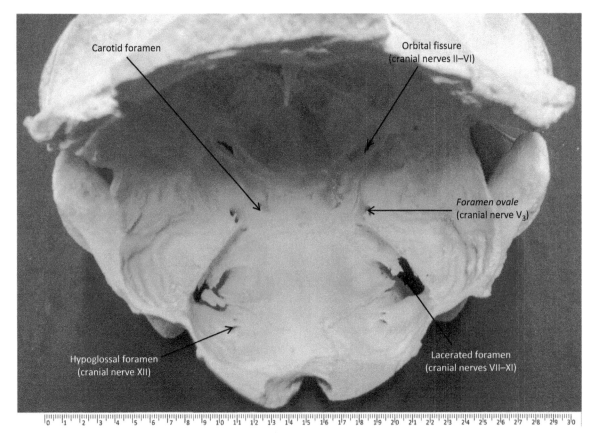

FIGURE 3.30 Base of the skull of *T. truncatus* seen from the inside after removal of the dorsal and lateral walls.

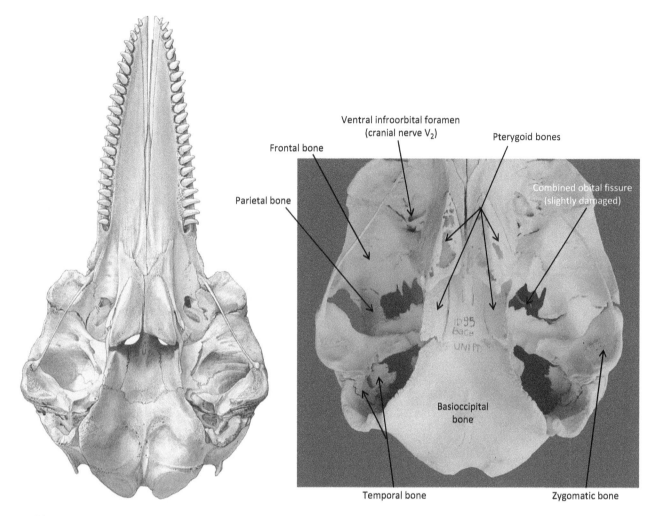

FIGURE 3.31   Ventral part of the skull of *T. truncatus* including squamosal structures. *(Drawing by Massimo Demma)*

(also called squamosal bone, Figs. 3.5, 3.7, 3.9, 3.11 and 3.12) is the large osseous plate that follows the parietal bone, and includes also the zygomatic process, the retroarticular (also called postglenoid) process, and the mandibular fossa (Fig. 3.31). The remaining parts of the temporal bone house the auditory ossicles, the tympanic bulla (Fig. 3.32) and related structures fundamental for hearing. They are described in detail in Chapter 5.

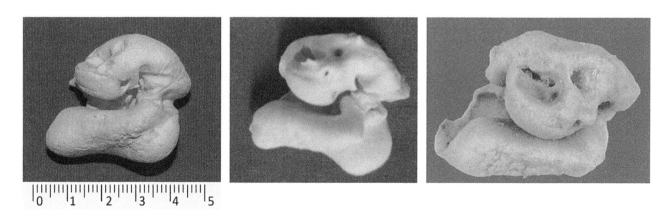

FIGURE 3.32   Tympanoperiotic bone of a newborn *G. melas* (left), adult *S. coeruleoalba* (middle) and adult *T. truncatus* (right).

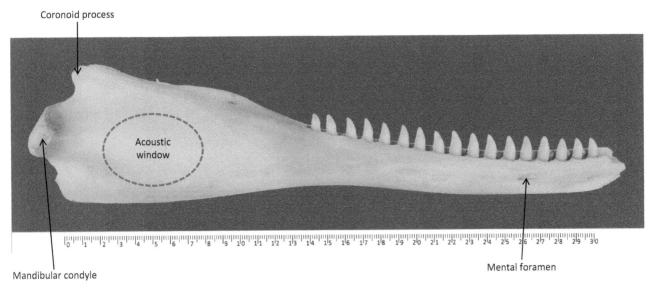

FIGURE 3.33 **Lateral face of the mandible of *T. truncatus*.**

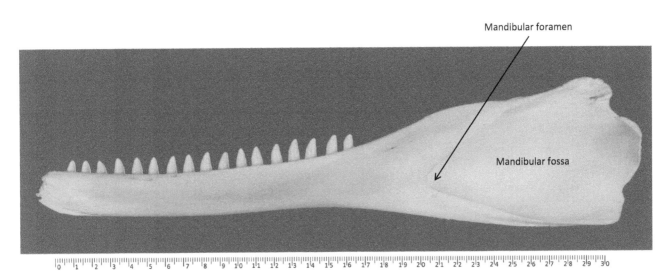

FIGURE 3.34 **Medial face of the mandible of *T. truncatus*.**

Mandible and the Hyoid Apparatus

The jaw or mandible of dolphins (Figs. 3.4–3.12, 3.33 and 3.34) is composed by two separate bones (hemimandibles) that articulate their anterior extremities with a symphysis.[m] The mandible carries the osseous alveoli that house the teeth. The mandible of dolphins consists only of the body of the mandible (*corpus mandibulae*), and is hollow. The *ramus mandibulae* (the vertical part of the mandible of terrestrial mammals) is absent. The mandibular foramen on the internal face of the mandibular body, where the mandibular branch of the trigeminal nerve enters the mandibular canal, becomes as large as the posterior margin of the bone and is filled by soft adipose tissue called intramandibular fat body (since this structure is related to sound perception, for details, see Chapter 5).

---

m. Contrarily to what happens in man and many terrestrial mammals, the symphysis between the two hemimandibles of dolphins and other odontocete does not really fully ossify to create a single jaw bone. A similar situation is found also in terrestrial Cetartiodactyla. On the contrary, the two hemimandibles remain completely distinct and separate throughout life in baleen whales.

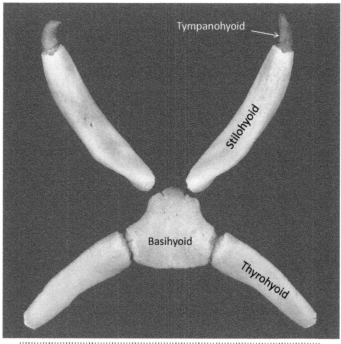

FIGURE 3.35   **Hyoid apparatus of *T. truncatus*.**

The coronoid process, that bears the insertion of the tendon of the *m. temporalis*, is low and anterior to the articular process. This latter structure (also called *condyloid* process) is the articular process of the mandible that in terrestrial mammals projects dorsally from the *ramus* of the mandible and articulates with the corresponding surface of the temporal bone (temporomandibular joint). The mandibular condyle of dolphins is placed along the posterior margin of the mandible, and therefore is not directed upward.

The temporomandibular joint of dolphins is a *syndesmosis* (Mead and Fordyce, 2009; McDonald et al., 2015) and not a *condylarthrosis* as in terrestrial mammals.[n] In dolphins, the articular surfaces of the squamous bone and condylar process of the mandible are connected by a fibrous disk. This means that there is no articular capsule with synovial fluid, and in fact, movements are different. This makes sense because cetaceans grab and suction-feed, but do not chew and thus do not require rhythmical articulate movements of the jaws.

According to the study of McDonald et al. (2015) in *G. griseus* and *P. phocoena*, the presence of adipose tissue, blood vessels, and nerves throughout the fibrous disk may suggest a role of the disk in neurological sensory functions such as echolocation.

The hyoid apparatus (Fig. 3.35) is composed by a single basihyoid bone that articulates posterolateraly with two thyrohyoid bones, and dorsally with two stylohyoid bones connected to the paraoccipital process of the skull base. The tympanohyoid cartilage[o] is placed between the stylohyoid bone and the periotic bone. Additional elements of the hyoid apparatus of other mammals (ceratohyoid and epihyoid) are transformed in dolphins into cartilage interposed between the basihyoid and the stylohyoid. According to Rommel (1990), the basihyoid and the stylohyoid are united by a synovial joint, although it is known that the ceratohyal component ossifies in some toothed whales such as the harbor porpoise (Reidenberg and Laitman, 1994). The main function of the hyoid apparatus is to assist in swallowing and, in general, in movements of the base of the tongue.

---

n. A *syndesmosis* is a joint between two bones connected by fibrous connective tissue that forms one or more ligaments. Mobility is limited depending on the shape and extension of the ligaments. On the contrary, a *condylarthrosis* is a joint in which two ellipsoidal surfaces face each other with the presence of a synovial capsule and reinforced by external ligaments. In the temporomandibular joint of several mammals (including terrestrial Cetartiodactyla and man), a meniscus is interposed between the articular surfaces.

o. According to Mead and Fordyce (2009), the tympanohyoid forms the proximal part of the styloid process of the temporal bone as seen in man, horse, and other terrestrial species.

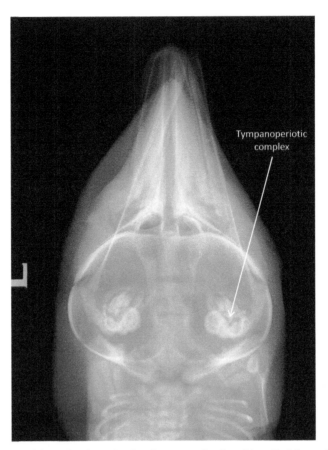

Tympanoperiotic
complex

FIGURE 3.36 **Fetus of *S. coeruleoalba* of 7 months of age showing the progressive deposition of calcium salts in the skull and first part of the vertebral column.** Note the density of the tympanoperiotic bones. *(Museum of Natural History G. Doria of Genova; specimen # 4525; X-rays courtesy of Prof Alessandro Zotti, Department of Animal Medicine, Production and Health of the University of Padova)*

Development of the Skull

In Table 9.6, we report the developmental phases (Stages) of selected dolphin species. According to Štěrba et al. (2000), during Stage 4 mesenchymal blastemas and cartilaginous tissue appear in the nasal region and elsewhere in the skull, but it is only during Stage 6 (the last embryonic stage) that it is possible to demonstrate the first lamellae of bony tissue in the maxilla and mandible. Bones of the calvaria (*neurocranium*), as well as the vomer, maxilla, *os incisivum*, and mandible start to ossify during the first fetal stage (Stage 7), while the nasal skeleton remains cartilaginous. Ossification progresses after that, and the bony face becomes progressively elongated and the head gradually assumes the shape typical of the species (Fig. 3.36). Several studies described the progressive elongation of the dolphin skull and head, in general, during fetal development and early postnatal life (for references and literature, see Rauschmann et al., 2006; Moran et al., 2011). Deposition of calcium salts is far more precocious in the tympanoperiotic complex than elsewhere in the skeleton (Cozzi et al., 2015).

Vertebral Column

The vertebral column of Cetacea is strikingly different from that of terrestrial mammals, both for the general shape and for the morphology of the individual vertebrae. The column of Cetacea is shaped like a ventrally concave bow (cyphosis), less curved than in most terrestrial mammals.

The *vertebral formula* indicates the average number of vertebrae for each sector of the column of a given species. The vertebral formulae of selected dolphin species are reported in Table 3.1.

Vertebrae are completed by epiphysary disks that fuse with the vertebral body with age. So, in young individuals, the disks are still incompletely attached since their fibrous connective tissue is not yet ossified.[p]

---

p. For a thorough analysis of the morphological variations of the vertebral column of dolphins, and the relative factors that induce them, see Viglino et al. (2014).

TABLE 3.1 Composition of the Vertebral Column in Selected Delphinidae

Species	Number of Vertebrae Cervical (% Length of Total)	Fused Vertebrae	Thoracic (%)	Lumbar (%)	Caudal (%)	Antinclinal Vertebra	Contra-antinclinal Vertebra	Total Number
T. truncatus	7 (3%)	1–2 and/or 3–5 (possibly 6–7)	12–14 (23%)	16–19 (30%)	24–27 (44%)	7 L	8 Ca	62–65
S. coeruleoalba	7	1–3 (the others are variable)	15	18–22	32–35	9–10 L		74–79
D. delphis	7 (4%)	1–2 (often complete cervical fusion)	14 (25%)	21 (32%)	31–35 (39%)	7 L	10 Ca	73 (70–75)
G. griseus	7 (2.5%)	1–6/1–7	12–13 (24.5%)	18–19 (29%)	30–31 (44%)	12–14 L	2 Ca	68–69
G. melas	7	1–5/1–6	11	12–14	28–29	7–8 L	7 Ca	58–59
O. orca	7	1–3	11–13	9–12	21–25	6 L	5 Ca	50–54
P. crassidens	7 (3%)	1–5/1–7	10–11 (17%)	9–11 (31%)	20–23 (49%)	9–10 L		47–52

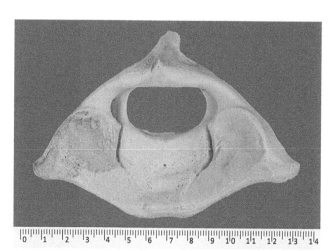

FIGURE 3.37   Rostral view of the fused atlas and epistropheum of *T. truncatus*.

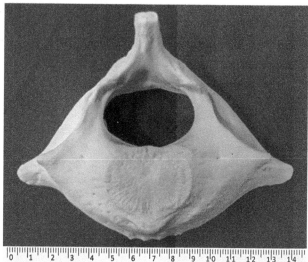

FIGURE 3.38   Caudal view of the atlas and epistropheus of *T. truncatus*.

Cervical Vertebrae

The bodies of the first cervical vertebrae are fused in smaller dolphin species (Fig. 3.37). Fusion of the cervical vertebrae involves even more elements in larger species like *O. orca* and *P. crassidens*. The fusion of the atlas (C_1) with the axis (C_2, also called *epistropheus*) gives the resulting bony complex both a relevant neural spine (*processus spinosus*), and important transverse processes (Figs. 3.38 and 3.39). Transverse processes are only partially fused in the remnant cervical spine (Figs. 3.40 and 3.41).

Thoracic Vertebrae

Thoracic (also called dorsal) vertebrae are the vertebrae connected to the ribs, and their respective numbers generally correspond (although missing fluctuating ribs may ingenerate some confusion in mounted skeletons). However, the number of thoracic vertebrae is subject to intraspecific variation.

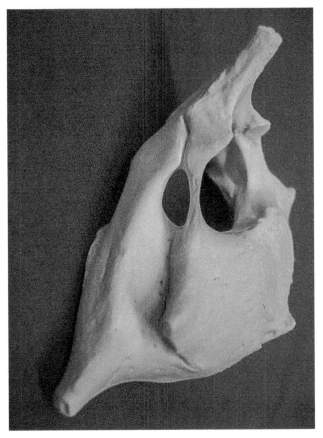

FIGURE 3.39   Latero-caudal view of the fused atlas and epistropheus of *T. truncatus*.

Thoracic vertebrae possess a short cylindrical body with evident transverse processes (with a *fovea costalis processus transversi* that articulates with the tuberculum of the ribs); articulate processes (*fovea costalis caudalis* and *fovea costalis cranialis*, that, combined in two adjacent vertebrae, constitute the articular surface for the head of the ribs); and a long neural spine (Figs. 3.42–3.44). The neural spine (*processus spinosus*) points backward in the thoracic column, until the so-called *antinclinal vertebra* (in the lumbar spine), in which the inclination changes and is reversed from backward to frontward in the vertebrae that follow.

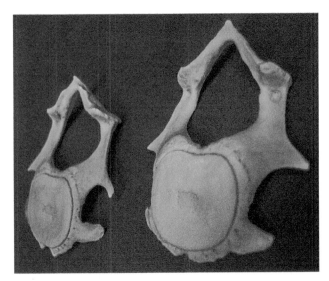

FIGURE 3.40   Antero-lateral view of C3 and C4 of *T. truncatus*.

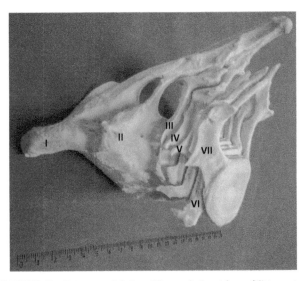

FIGURE 3.41   Latero-caudal view of the cervical vertebrae of *T. truncatus*.

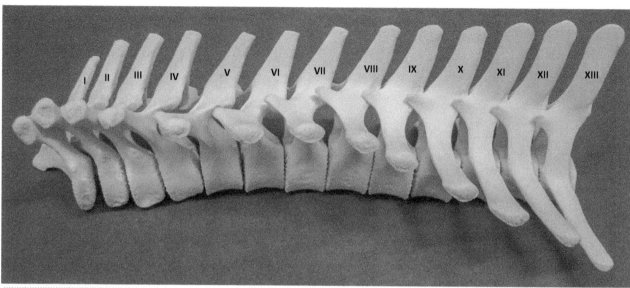

FIGURE 3.42 **Left side view of the thoracic vertebrae of *T. truncatus*.**

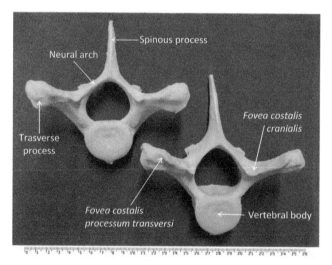

FIGURE 3.43 **Rostral side of vertebrae T3–T4 of *T. truncatus*.**

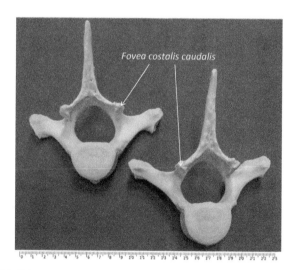

FIGURE 3.44 **Caudal side of vertebrae T3–T4 of *T. truncatus*.**

Lumbar Vertebrae

In terrestrial mammals, the lumbar tract of the vertebral column starts with the first vertebra that has no connection with the ribs and terminates before the (fused) sacral vertebrae. The first part of the definition (no more ribs attached) fits also for dolphins. However, dolphins have no fused sacral vertebrae, and all that is left of the coxal bone is the residue of the pelvis (or pelvic bones), that bears no direct relationship to the vertebral column.

Therefore, the definition of the lumbar tract of the dolphin spine is from the first vertebra without a rib to the last vertebra that shows no relationship to a ventral hemal bone. In fact, although the hemal bones (or *hemapophysis*, see the next paragraph) have no ontogenic or functional relationship with the coxal bones, the appearance of the hemal arches marks the end point of the lumbar spine.[q] A further morphological distinction between the two adjacent tracts of the column is that the bodies of lumbar vertebrae are longer than those of the caudal ones.

---

q. For an alternate classification of the spine in marine mammals, see De Smet (1977).

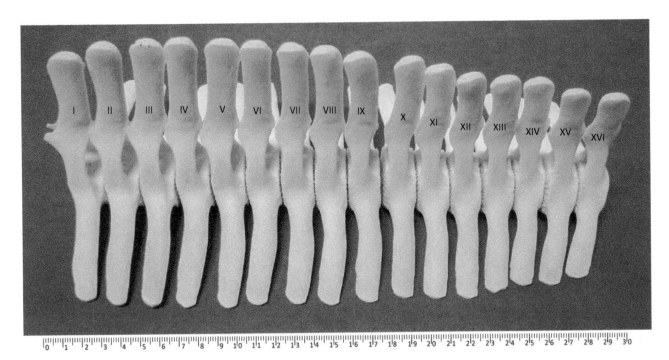

FIGURE 3.45   Side view of the lumbar vertebrae of *T. truncatus*.

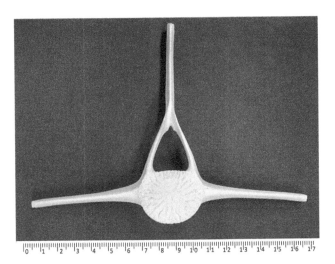

FIGURE 3.46   Rostral view of L2 of *T. truncatus*.

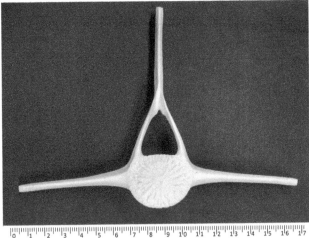

FIGURE 3.47   Caudal view of L2 of *T. truncatus*. The rostral and caudal aspects of the lumbar vertebrae are almost identical.

Like in other mammals, the lumbar vertebrae (Figs. 3.45–3.47) possess huge transverse processes, a short but broad spinous process, and a ventral keel. The *antinclinal vertebra*, the first whose spinous process is perpendicular to the body of the vertebra, corresponds to L_{7-8} in the bottlenose and common dolphin.

Caudal Vertebrae

There are no sacral vertebrae (see also below the description of the pelvic limb) and the main features that distinguish the caudal vertebrae (Figs. 3.48–3.50) from the preceding lumbar ones are the progressive disappearance of spinous and transverse processes, and the presence of a groove on the ventral surface of the body. Furthermore, a series of small Y-shaped bones, called *hemal* bones or *hemapophysis*, are connected to the ventral part of the vertebrae and form an *hemal arch* (literally meaning arch of/for blood) that protects the ventral course of the large arterial vessels directed to the flukes. The Y-shaped *hemal bones* (Fig. 3.51) attach to the caudal part of each vertebra, which has two small tubercles. The hemal

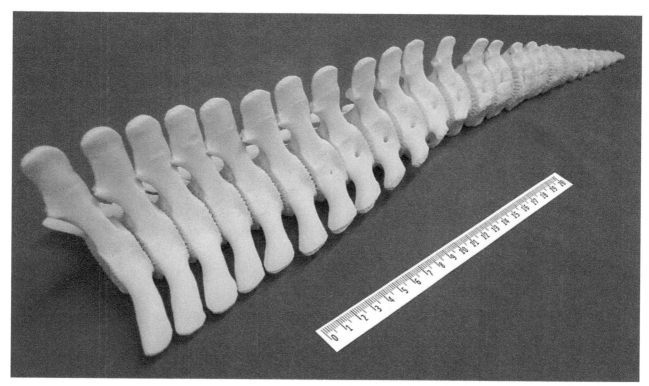

FIGURE 3.48   Antero-lateral view of vertebrae Ca1–Ca26 of *T. truncatus*.

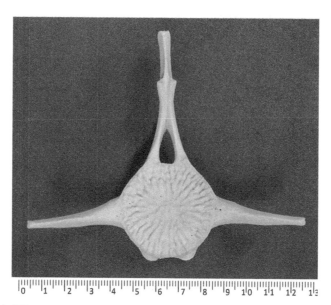

FIGURE 3.49   Rostral view of Ca4 of *T. truncatus*.

arches continue for most of the caudal sector, until the last vertebrae becomes rounded and constitutes the sustain of the flukes.

The inclination of the spinous processes changes again in the caudal spine, and the specific vertebra in which the direction varies is called *contra-antinclinal* vertebra.

Moment of Resistance and the Formula of Slijper

The fundamental studies of Slijper (1936, 1946) analyzed the dynamic morphology of the spine in several mammals and remain to date a milestone for understanding the mechanic of locomotion. Slijper proposed to adopt the formula

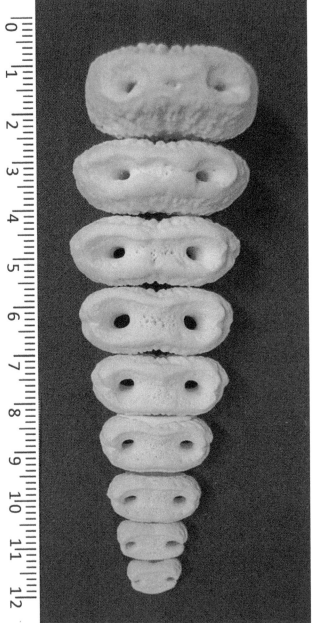

FIGURE 3.50 Dorsal view of Ca18–Ca26 of *T. truncatus*.

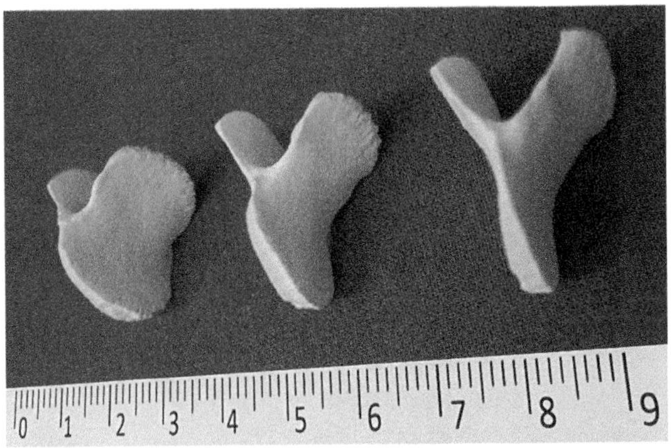

FIGURE 3.51 Antero-lateral view of hemal bones H4–H6 of *T. truncatus*.

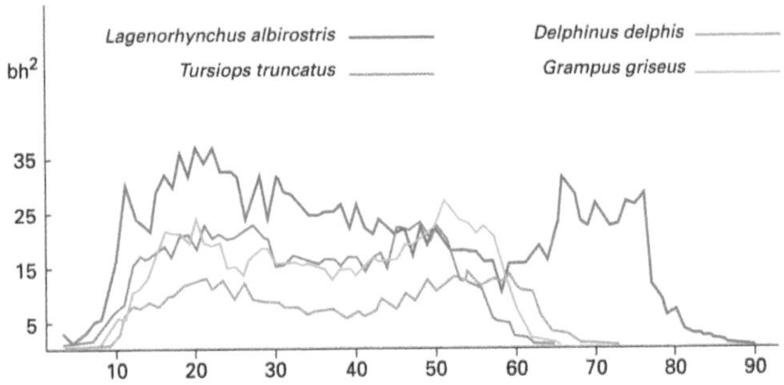

Lagenorhynchus albirostris —————— Delphinus delphis ——————
Tursiops truncatus —————— Grampus griseus ——————

FIGURE 3.52 **Moments of resistance of the vertebral column of selected dolphin species.** *(From Cozzi, B., 1981. Osservazioni sulla morfologia della colonna vertebrale nei cetacei. Atti Soc ital Sci nat Museo civ Stor nat Milano 122, 225–235.)*

$W = bh^2$ (W = moment of resistance; b = breadth of the vertebra; h = height of the vertebra[r]) to calculate the *moment of resistance* of the vertebral column and thus gain a graphic idea of the strain imposed by movements to the different sectors of the spine. The graphic representation of the formula (Fig. 3.52) emphasizes the more robust sectors of the spine and therefore indirectly points out where the higher strain forces are located and possibly also the terminal sites of powerful tendon stress. According to this formula, the different sectors of the vertebral column of smaller dolphins (up to 4 m in length) show only relative variation of the moment of resistance in the thoraco-lumbo-early caudal sectors (Cozzi, 1981), while larger species (including *G. melas*, *P. crassidens*, and *O. orca*) show a peak in the lumbar tract. Interestingly, the same part of the columns is often affected by bony pathologies, including spondylarthritis and exostosis (Cozzi et al., 2010a).

r. The height is in fact the length in mammals whose column is horizontal and not vertical.

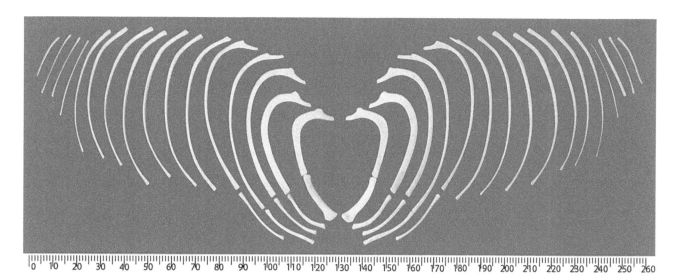

FIGURE 3.53   Ribs of *S. coeruleoalba*.

Joints of the Vertebral Column

The union of the skull with the first cervical vertebra is unique in dolphins. The occipital condyles are rather hemispherical (and not ellipsoid as in most terrestrial mammals) and therefore the occipital-axis joint allows movement/rotation along three different axes (Rommel, 1990).[s]

In fact, in all dolphins the cervical tract is rather short and the vertebrae fused in different degrees (see Table 3.1).[t] This anatomical disposition makes it impossible to twist or turn the cervical spine below C_2, with consequent increased rigidity and possible better sustain of the large skull and mandible. The extremely moveable occipital-atlas joint is therefore responsible for all movements of the dolphin head in respect to the spine.

The joints of the thoracic, lumbar, and caudal vertebrae are *symphysis*[u] and take place with the interposition of an intervertebral disk. The latter includes a denser external ring called *anulus fibrosus* and an inner core, the *nucleus polposus*. Intervertebral disks are thicker in the last thoracic, lumbar, and early caudal regions of the spine of dolphins.

Thorax

Ribs (Costae)

The mammalian ribs articulate with the thoracic vertebrae through a head and a tuberculum, and with the sternum through a direct (sternal ribs or *costae verae*; NAV, 2012) or indirect connection (asternal ribs or *costae spuriae*, connected to the sternum by a cartilage[v]). Usually one or two pairs of caudal ribs have neither direct nor indirect connection to the sternum (fluctuating ribs or *costae fluctuantes*).

The ribs of dolphins (Figs. 3.53–3.57) are generally slender and cylindrical, with the exception of the first pair that is short, robust, and wide. The first pairs of ribs are flat, then become cylindrical in section, and the most caudal rib pairs become flat again. The proximal part of the ribs in the anterior part of the rib cage (generally the initial five pairs in the bottlenose dolphin) shows a distinct head and tubercle, but the remaining (and more caudal) pairs of ribs lose the head and are attached to the vertebral column only by the tuberculum.

The anterior pairs of ribs (Table 3.2) are connected to the sternum. The first pair reaches the *manubrium* of the sternum (see the following sections), and the next four to six pairs (depending of the species) reach the *sternebrae*.

---

s. In man and most terrestrial mammals, the occipital-atlas (C_1) joint allows extension/flexion (and minimal lateral rotation) of the head, while the joint between the atlas (C_1) and the epistropheus (C_2) controls oblique (and slightly ventral) rotations along the longitudinal axis of the spine.

t. Cetaceans practically have no external neck (with the interesting exception of *Delphinapterus leucas* and some river dolphins).

u. A *symphysis* is a fibrocartilaginous joint in which two adjacent bones (in this case, two vertebrae) are united with the interposition of cartilage (in this case, the intervertebral disk).

v. Please note that the present description of the bony thorax follows standard anatomical nomenclature (TA or NAV) and therefore the terminology employed here may indicate structures quite different from what previously recognized in former descriptions (ie, the seminal paper on the bottlenose skeleton by Rommel, 1990).

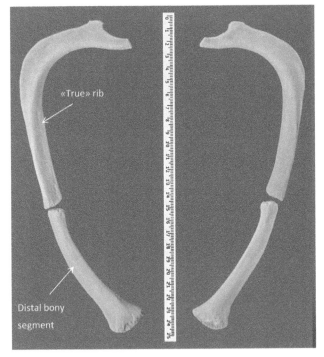

FIGURE 3.54 Caudal view of the first rib pair of *S. coeruleoalba.*

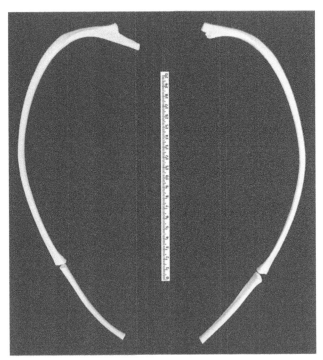

FIGURE 3.55 Rostral view of the fifth rib pair of *T. truncatus.*

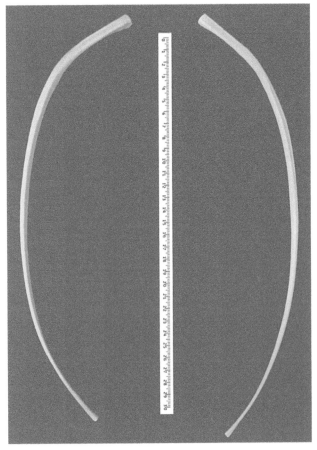

FIGURE 3.56 Rostral view of the eighth pair of ribs of *S. coeruleoalba.*

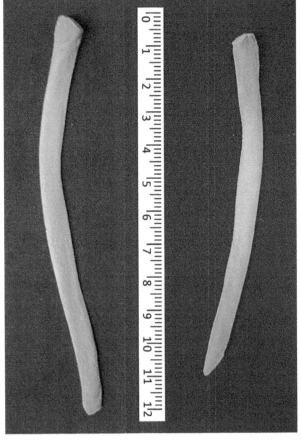

FIGURE 3.57 Caudal view of the 16th rib pair of *S. coeruleoalba.*

TABLE 3.2 Bones of the Thorax in Selected Delphinidae

Species	Number of Rib Pairs	Bicipital Ribs	Ribs With a Proximal and a Distal Bony Segment[a]		Ribs Without the Distal Bony Segment	Fluctuating[a]	Number of Sternebrae
			Distal bony segment directly connected to the sternum	Distal bony segment connected to the distal bony segment of the preceding rib			
T. truncatus[b]	12–14	5–6	5–6	3–4	4	1	3–4
S. coeruleoalba	14–16	5	4	5	4–6	1	3–4
D. delphis	13–14	4–5	4–5	2–3	5–6	1–2	4
G. griseus	12–13	6–7	4	2	5–6	(1)	3
G. melas	11	6	5	3	3	(?)	3–4
O. orca	11–13	6–7	4–5	2–3	5–6	(1)	3–4
P. crassidens	10–11	6	5	3	3	(?)	4

[a]See Footnote w for details.
[b]There is considerable variation (both in mounted specimens and in the literature) in the number of ribs and their classification in the bottlenose dolphin. This may be due to individual variations, eventual loss of bony elements during recovery of the skeleton from the carcass, or to classification under the common banner of T. truncatus of what are presently considered separate species [Tursiops aduncus, Ehrenberg, 1832 (1833); Tursiops australis, Charlton-Robb, Gershwin, Thompson, Austin, Owen & McKechnie, 2011].

The connection to the sternum in terrestrial mammals (including man) takes place with the interposition of the cartilage (*cartilago costalis*). In dolphins, the distal part of the most anterior ribs continues downward through the joint with a relatively long and thin ossified segment, that in turn reaches the sternum with the interposition of a short cartilage. Therefore, each of these ribs is thus constituted by two bony segments, the proximal one attached to the vertebral column, and the distal part attached either directly or indirectly to the sternum.[w] The distal part of the caudal-most asternal ribs may remain cartilaginous or simply fibrous, thus further increasing the complexity of the classification. The most caudal and fluctuating ribs have no bony or cartilaginous connection to the sternum and are also incompletely articulated with the vertebral column.

Since the joint between the rib and the vertebral column changes from the anterior to the posterior ribs, and the last ribs may be attached to the column only through the transverse process of the vertebrae, the total number of ribs does not always correspond to the number of thoracic vertebrae (see Table 3.2 and the description of the thoracic joints in the following paragraphs).

Sternum

The sternum (Figs. 3.58 and 3.59) is made up by separate elements (*sternebrae*) that ossify with age to constitute a single element. The first *sternebra* is called *manubrium*[x] and is wider than the others, with two evident lateral surfaces for the articulation with the first pair of ribs. On the whole, the sternum of dolphins is long and cross-shaped at the cranial

---

w. In the human and veterinary anatomical nomenclature (TA and NAV, respectively), the ribs may be classified as *sternal* or *asternal*, depending on their direct or indirect cartilaginous connection to the sternum. Ribs with absolutely no connection to the sternum are called *fluctuating* ribs. This distinction is quite arduous to make in dolphins, since the connection to the sternum is made by the distal bony segments in the more anterior pairs of ribs, and by cartilage (often lost during preparation of the skeleton) in the following pairs; the last one or two pairs generally have no junction to the sternum and their joint to the vertebral column is rather incomplete or lacking. The point is that in dolphins the more rostral rib pairs that possess a distal bony segment may be further subdivided into (group *a*) those with a direct connection to the sternum (true equivalent of the sternal ribs of TA and NAV) and (group *b*) those with a distal bony segment connected to the preceding bony segment and not directly to the sternum. The sum of these latter ribs (group *b*) with those that follow and possess only a cartilaginous connection to the preceding ribs (that is, all the ribs with an *indirect* bony or cartilaginous connection to the sternum) constitute the equivalent of the asternal ribs of TA and NAV. Finally, to calculate the number of true fluctuating ribs (no connection either bony or cartilaginous to the sternum), one should be absolutely sure of the absence of any thin strip of cartilage that may get lost during dissection. Here we considered fluctuating ribs only those that have absolutely no connection to the sternum and with an incomplete or absent vertebral joint.

x. The word *manubrium* derives from a Latin word that means handlebar, and refers to the shape of the human sternum.

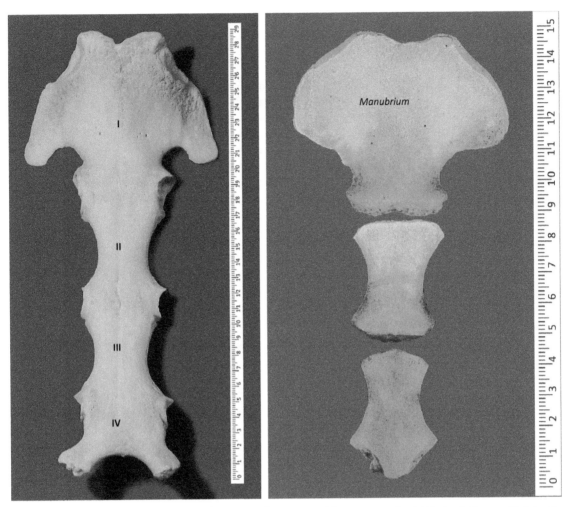

FIGURE 3.58   Ventral (left) and dorsal (right) views of the sternum of *T. truncatus*. The sternum on the right had only three ossified sternebrae.

extremity. The *manubrium* may present a central foramen that tends to close with age. In fact, the degree of ossification of the sternum is an age indicator in dolphins. The sternum of *G. melas* is rather unique for its slender shape and the persistence of central foramina in the *manubrium*, and possibly also in the following *sternebrae* (Fig. 3.60).

Joints of the Thorax

The bicipital ribs (see the previous sections) articulate with the thoracic vertebrae with synovial joints in the anterior thorax and with simple fibrous connections in the caudal thorax (Rommel, 1990). This disposition, and the shape of the articular surfaces, allows extreme rotation (see the physical manipulations of the excised bottlenose thorax, reported by Cotten et al., 2008) and consequently let the whole rib cage accommodate pressure-induced changes of the volume of the thorax during dives (Rommel, 1990).

The joints between the ribs and the sternum are placed on the articular surface of the *manubrium* (in the first pairs of ribs) or along the sides of the *sternebrae* (in the following sternal ribs). More caudal ribs have a fibrous connective connection to the sternum (fluctuating ribs have no connection).

Development of the Vertebral Column and Thorax

Developmental phases (Stages) of selected dolphin species are reported in Table 9.6 (see Chapter 9 for details). According to Štěrba et al. (2000), prochondral blastemas of the still fused vertebrae and ribs appear during Stage 3 and become more clearly delimited in Stage 4. At Stage 5, vertebrae and ribs are still cartilaginous. Hypertrophied cartilage then appears at Stage 6 in the vertebrae and ribs, and perichondral ossification of vertebral arches and ribs begins and intensifies in the later

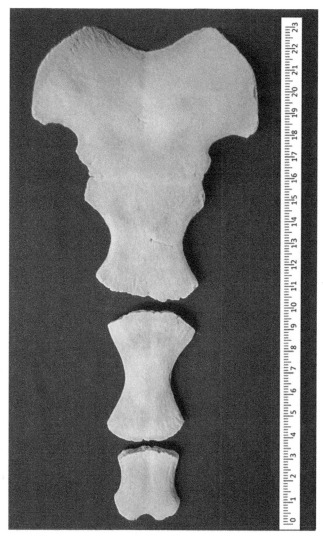

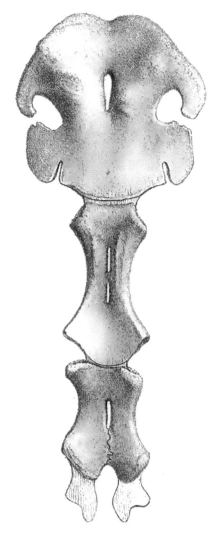

FIGURE 3.59   **Ventral view of the sternum of *S. coeruleoalba*.**

FIGURE 3.60   **Sternum of *G. melas*.** *(From van Beneden and Gervais, 1868–1879)*

stages. At Stage 9 spinous processes appear from fusion of the neural arches; endosteal ossification of neural arches and ribs and enchondral ossification of vertebral bodies are evident.

The development of the vertebral column of dolphins has been described in a series of studies. For details and references, see Ogden et al. (1981a,b). For evolutionary significance, see Buchholtz and Schur (2004) and Buchholtz (2007). For progressive development of the whole skeleton (in striped dolphins), see Ito and Miyazaki (1990).

Thoracic Limb

The scapula of the bottlenose (Figs. 3.61 and 3.62), striped, and common dolphin is flat and large, and is maintained in position by the muscles of the region. There is no spine on the lateral surface, but in some species, including *G. melas* and *G. griseus*, the external surface is almost concave and presents two or more vertical little ridges that are harder to trace in smaller species. In all dolphin there are two well-evident processes, the medial coracoid process and the lateral acromion, both pointing forward.

The absence of a well-developed spine on the external surface of the shoulder blade is possibly related to the different biomechanics of the deltoid muscle in cetaceans (see section Myology). On the whole, the shape of the scapula of dolphins (and the presence of the relevant acromion and coracoid processes) is more similar to the human than to that of quadruped mammals. However, in man the conspicuous spine separates a well-evident *supraspinous fossa* from a larger *infraspinous*

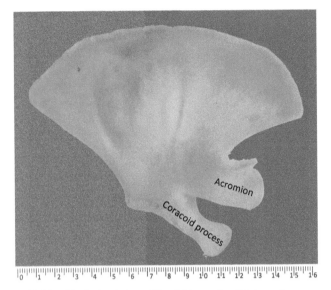

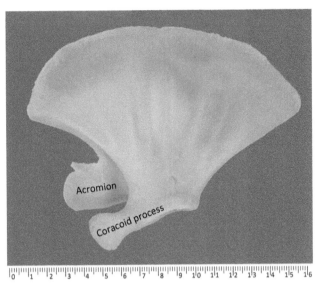

FIGURE 3.61 Lateral view of the right scapula of *T. truncatus*.

FIGURE 3.62 Medial view of the right scapula of *T. truncatus*.

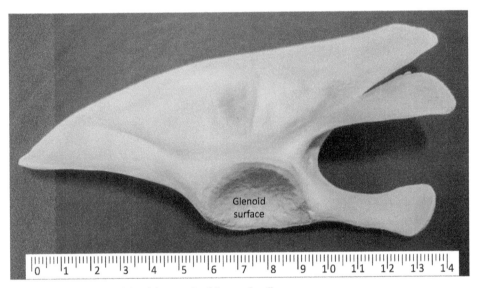

FIGURE 3.63 View of the glenoid surface of the right scapula of *S. coeruleoalba*.

fossa, while in dolphins the *supraspinous fossa* is vestigial and appears as a narrow notch on the anterior margin (Klima et al., 1980). The interosseous membrane between the coracoid process and the anterior margin of the scapula constitutes the insertion of the *m. supraspinatus* (Rommel, 1990).

The shallow and concave glenoid cavity (Fig. 3.63) faces the head of the humerus.

There is no clavicula (collar bone) in adult dolphins,[y] but all the other elements of the mammalian thoracic limb, from arm to fingers, are contained in the flipper (Figs. 3.64 and 3.65). As noted previously (see section Structure of Bones), the bones of the arm and forearm have no medullar cavity.

The humerus (Figs. 3.66–3.69) is short and thick, with a spheroid head. The greater and lesser tubercle are indistinct, but there is a single common tubercle, rotated inward, that may result from their fusion (Klima et al., 1980). The deltoid process is well developed in *O. orca*.

---

y. A residue of the clavicle maybe present in adult *Pseudorca* (Klima, 1978, 1990).

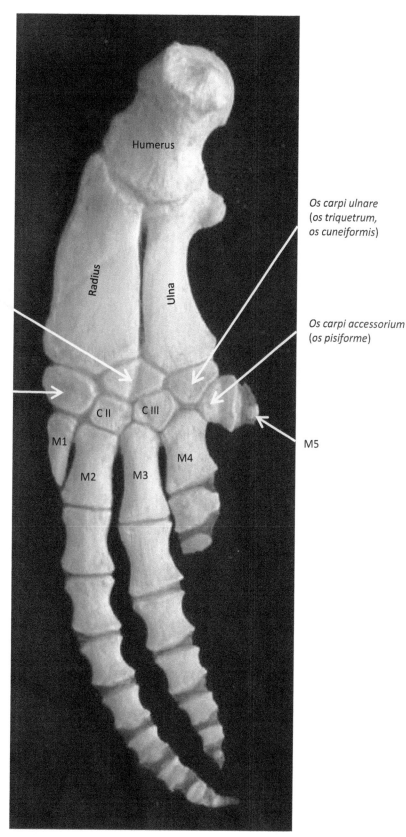

FIGURE 3.64 **Right thoracic limb of S. coeruleoalba.** *C*, carpal bone of the second row; *M*, metacarpal bone. *(Photograph courtesy of Dr. Michela Podestà, Museum of Natural History of Milan)*

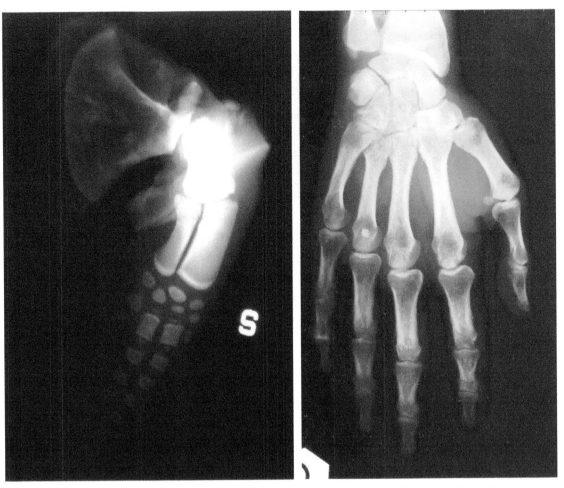

FIGURE 3.65   Radiograph of left thoracic limb of *G. griseus* and left hand of *Homo sapiens* (for indication, see Fig. 3.64).

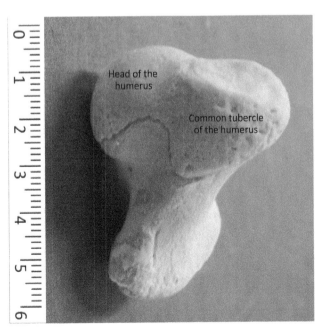

FIGURE 3.66   Anteromedial view of the left humerus of *S. coeruleoalba*.

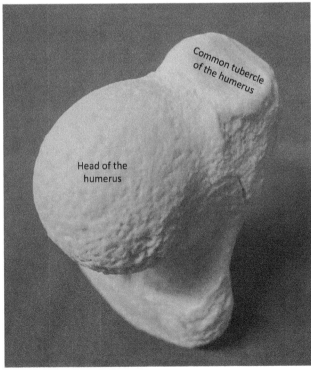

FIGURE 3.67   Dorsolateral view of the left humerus of *S. coeruleoalba*.

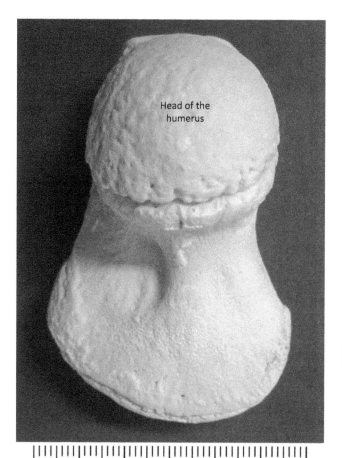

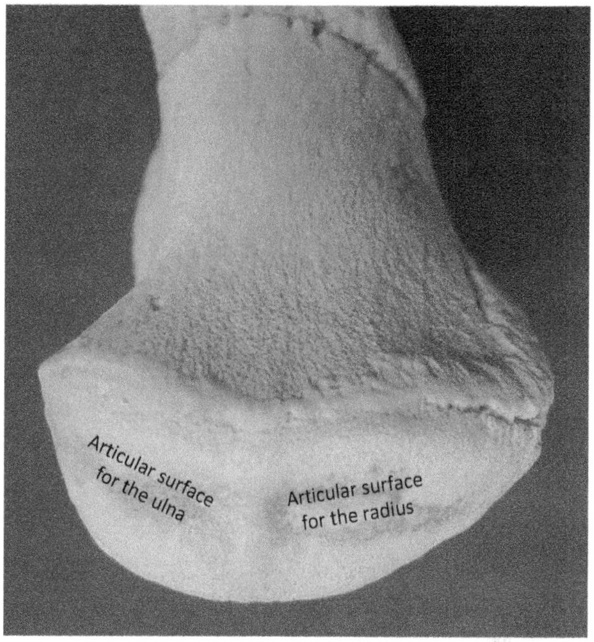

FIGURE 3.69   Ventromedial view of the distal epiphysis of the left humerus of *S. coeruleoalba*.

FIGURE 3.68   Lateral view of the right humerus of *S. coeruleoalba*.

The radius and ulna are flat (Fig. 3.70). The caudal margin of the ulna presents the olecranon proximally (Fig. 3.71) and continues distally with a curve, with a pronounced concavity corresponding to the external proximal notch of the caudal margin of the fin. The distal half of the radius of *P. crassidens* is very large (Fig. 3.72) with a convex cranial margin.

Radius and ulna articulate distally with the series of carpal bones (Figs. 3.64, 3.65, 3.70, and 3.73), followed by the long metacarpals and phalanxes.

The degree of fusion of the epiphysis of the humerus, radius, and ulna can be helpful to determine the age of the specimens in some species of dolphins (Calzada et al., 1997; Cozzi et al., 1985; DiGiancamillo et al., 1998; Podestà et al., 2001). Bone density also changes with age, and can be considered an age indicator as well (Guglielmini et al., 2002; Butti et al., 2007; Azevedo et al., 2015; Carvalho et al., 2015).

There are two rows of carpal bones (carpals I–III in the proximal row, and IV–V on the distal one). Some degree of fusion between CI and the corresponding subsequent metacarpal is possible, depending on the species and also on individual variations.

The number of fingers is five (a case of polydactyly in the bottlenose dolphin is reported by Watson et al., 1994). The thumb may be limited to the first metacarpal bone without any phalanx. The number of phalanxes of the second and third digits varies but is considerably higher than three (a phenomenon called hyperphalangy), especially in the pilot whale.[z]

Flipper formulas (includes metacarpal and phalanxes):

- *T. truncatus* I_{1-2}; II_{7-9}; III_{5-8}; IV_{2-3}; V_{1-2}
- *S. coeruleoalba* I_1; II_9; III_7; IV_4; V_1

z. For studies on the evolution of hyperphalangy in dolphins, see Cooper et al. (2007a). For a thorough analysis of the evolution of the cetacean forelimb see Sanchez and Berta, 2010.

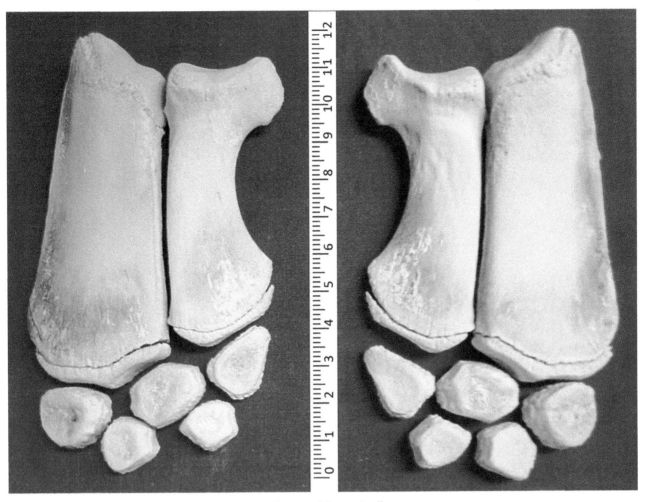

FIGURE 3.70   Lateral view of the right and left forearm and carpus of *S. coeruleoalba*.

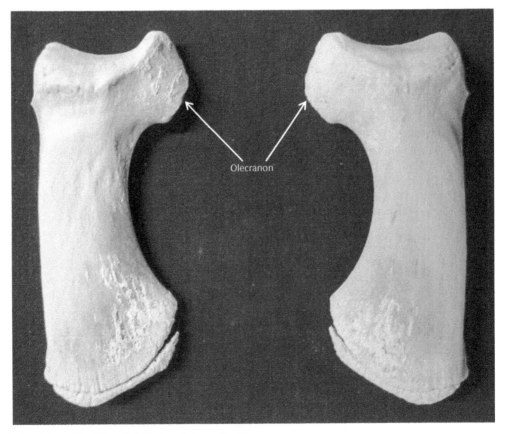

Olecranon

FIGURE 3.71   Lateral view of the left and right ulna of *S. coeruleoalba*.

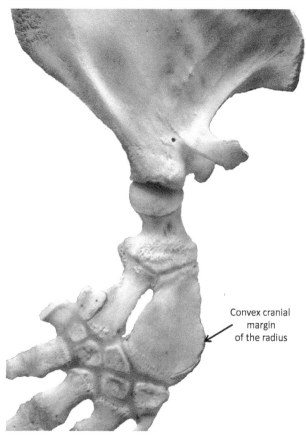

FIGURE 3.72   **Lateral view of the right forearm of *P. crassidens*.** *(Photograph courtesy of Dr. Michela Podestà, Museum of Natural History of Milan)*

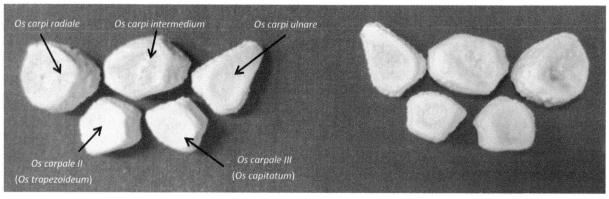

FIGURE 3.73   **Right and left carpus of *S. coeruleoalba*.**

- *D. delphis* I_{2-3}; II_{8-9}; III_{5-7}; IV_{2-4}; V_{1-2}
- *G. griseus* I_2; II_{8-10}; III_{5-8}; IV_{3-5}; V_1
- *G. melas* I_{3-4}; II_{9-14}; III_{9-11}; IV_{2-3}; V_{1-2}

Joints of the Thoracic Limb

The shoulder joint (Fig. 3.74) is a synovial joint between the concave glenoid cavity of the scapula and the spheroid head of the humerus, and is the only mobile joint of the limb. It is mostly functional in changing the inclination of the flipper during swimming.

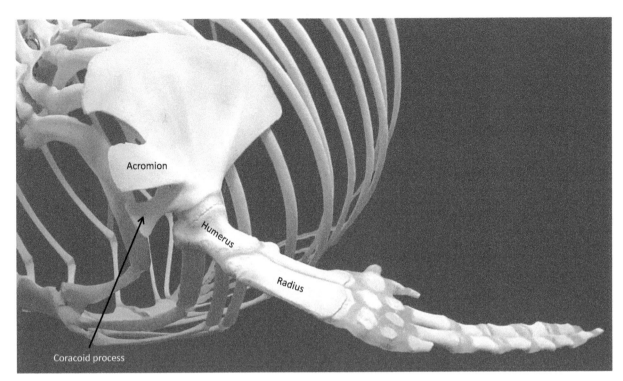

FIGURE 3.74 **Left shoulder joint of *T. truncatus*.**

According to Klima et al. (1980), the shoulder joint of dolphins (or at least of the bottlenose dolphin) is a nonspecialized spheroidea and not a real enarthrosis.[aa]

The bones of the arm, forearm, and *manus* are tightly connected by dense, rigid connective tissue and are not able to move independently.

Development of the Thoracic Limb

According to the developmental stages reported by Štěrba et al. (2000), and summarized in Table 9.6 (see Chapter 9 for details), during Stage 3 the buds of the thoracic limbs closely resemble those of other mammals. Delicate cartilaginous anlagens of the long thoracic bones (humerus, radius, and ulna) appear at Stage 5, subsequently followed by mesenchymal rays of the metacarpal bones and fingers. The fundamental changes that will turn the thoracic limb into a flipper take place during Stage 6 (the last true embryonic stage before entering the fetal phase), with changes of the axis, elongation of the hand and multiplication of the phalanxes in the second and third digit. Thin periostal bone or hypertrophied cartilage becomes evident in the diaphyses of the humerus, radius, and ulna at Stage 7, during which the future joints show initial articular cavities. Ossification begins in the scapula and clavicle and becomes evident in the humerus, radius, and ulna during Stage 9 and the following ones. The development of the thoracic limb of dolphins has been described in a series of studies (for references and general discussion, see Calzada and Aguilar, 1996; Sedmera et al., 1997a,b).

Here we emphasize that an anlage of the clavicle appears during Stage 4 in embryo of *Stenella* and other dolphins, and can be followed up to Stage 8 (Klima, 1990; Štěrba et al., 2000), after which it is no longer visible.

Pelvic Limb

The pelvic limb of dolphins is limited to a pair of vestigial bones (Fig. 3.75) embedded deeply into the ventral muscles of the abdomen, positioned dorsally to the anus and their tip is approximately aligned to the first chevron bones. The two small symmetrical residues are all that remains of the *os coxale* of land mammals, and are commonly called pelvic bones. Their only function is to allow insertion of smooth muscles related to the genital apparatus, and noticeably the

---

aa. An enarthrosis is a joint in which a segment of a sphere (here the head of the humerus) is connected to a corresponding (more or less complete) concave surface (the glenoid cavity of the scapula). A true enarthrosis (as in the primate shoulder joint) allows circumduction movements (ie, complete rotation). This is possible also because of the presence of the clavicle. In the river dolphin *Inia geoffrensis*, the shoulder joint involves also the sternum and results in a wider range of movements of the flipper (Klima et al., 1980).

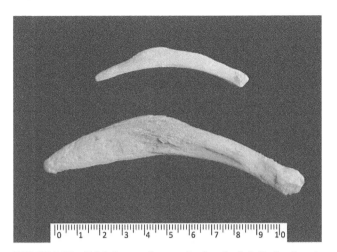

FIGURE 3.75   **Pelvic bones of young (top) and adult (bottom) male *T. truncatus*.**

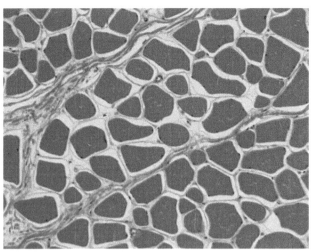

FIGURE 3.76   **Striated muscle from the *m. longissimus dorsi* of *T. truncatus*.** HE stain.

m. ischiocavernosus (in both sexes) and the *m. retractor penis* (only in males). Interestingly, pelvic bones are sexually dimorphic (Dines et al., 2014) in cetaceans.

The presence of residues of other bones of the hind limb is teratologic in postnatal dolphins.[bb]

Development of the Pelvic Limb

According to Štěrba et al. (2000) a bud of the pelvic limb appears during Stage 3, with uniform mesenchymal condensation (see Table 9.6 for definition of developmental stages in selected dolphin species). At Stage 5 the pelvic limb bud is flattened, and in cross section it is possible to distinguish anlagens not only of the pelvic cartilages, but also of the head of femur and blastema of prospective pelvic muscles. In later stages the pelvic bud turns into a skinfold with a thicker epidermis and abundant vascular plexuses, and then it disappears. Very rarely, postnatal dolphins show external knobs in the pelvic region, which seem to be homologous to the posterior limb buds of the embryo.

For developmental aspects of the disappearance of the hind limb, see Thewissen et al. (2006).

MYOLOGY

The whole body of cetaceans is adapted to life in the water, and the muscles are modified accordingly. Most of the muscular system is dedicated to locomotion (as opposed to passive sustain as in land mammals), and therefore continuously working.

If compared to terrestrial mammals, mobility of the forelimb is limited to directional control of water flow along the body surface during swimming, but has no relation to body propulsion or sustain against gravity; furthermore, as we have seen, the different elements that constitute the limb have no mobility among them (except in the shoulder joint). The musculature of the thigh, leg, and foot disappears together with the corresponding bones, and all that is left of the pelvic limb is the pelvic bones. The powerful musculature of the spine is thus the main motor for locomotion, and acts by pushing the flukes upward and downward (see Chapter 2).

The organization of the muscular system of dolphins (and cetaceans in general) hence serves hydrodynamics but also implies lack of motility in areas of the body where either reduction of the skeletal system is advanced to the extreme or muscles have lost special functions. So the mimic muscles are concentrated around the blowhole, and the face of dolphins thus appears as an expressionless mask: The general attitude of the animal might be perceived from a number of behavioral clues (body expression, acoustic signals), but, that is, not from its smile,[cc] arched eyebrows, and so forth. The same general concept applies also to movements of the forelimb.

All the muscles of dolphins contain a high quantity of myoglobin and appear very dark, almost black (Fig. 3.76). This is an obvious adaptation to facilitate oxygen storage for diving (see Chapter 4). For a comparative histological analysis of muscles in shallow and deep divers see Sierra et al., 2016.

---

bb. Tiny residues of the femur may occasionally persist in members of the family *Balaenidae*.
cc. Dolphins cannot stir, retract, or anyway move their lips. Their teeth are visible only when the jaws are open (Fig. 8.1).

Muscles of the Head and Neck

The muscles of the dolphin head include:

- mimic (facial) muscles;
- masticatory muscles;
- muscles of tongue and hyoid apparatus; and
- muscles of the larynx (discussed with the respiratory apparatus in Chapter 5).

The shortness of the neck and the modifications of the hyoid apparatus and larynx influence shape, course, and obviously functions of the relative muscles.

Facial Muscles

The facial (mimic) muscles of dolphins are markedly different from the ones of primates and land mammals in general. As previously mentioned, the facial motility and expression is reduced in cetaceans, and the muscles that move the lips, the cheeks, and the external ear disappear or follow an evolutionary transformation that makes them hard to recognize. In their accurate description of the superficial layers of the face of dolphins, Lawrence and Schevill (1965) described the *m. sphincter colli profundus*, the *m. orbicularis oris*, a *m. auricolabialis* (= *m. zygomaticus* of current nomenclature), and a vestigial *m. (levator) nasolabialis*. However, the only mimic muscles that maintain a specific topographic identity are those related to the eye (*m. orbicularis oculi*) and especially to the blowhole, which corresponds to the external nares of terrestrial mammals. In this respect, dolphins are indeed quite distinct, as the complex nasal sacs have an extraordinary importance in sound production and general pneumatic signaling. The melon (see Chapter 5) is the central structure of the forehead. It consists of a rostral compact fat cushion and caudal elongations adjacent to the soft tissue around the nasal passages and the sound-producing "monkey lip" dorsal bursae complex.

Masticatory Muscles

In mammals, these muscles open and close the jaws and help processing food through chewing. Animals that grab and tear (as carnivores) have powerful muscles that close the jaws and thus help secure the prey and tear its flesh. On the other hand, herbivore species that ingest a large quantity of fiber-rich nutriment develop potent muscles that chew. Basically, most of the muscles lift or lock the mandible, while the digastric muscle (and some muscles related to the hyoid apparatus) lowers it.

Dolphins grab but do not chew. Consequently some (but not all) of their masticatory muscles are well developed. Most evident is the *m. temporalis*, (originating in the large temporal fossa and terminating in the coronoid process of the mandible), that closes the jaws and essentially locks the mandible. On the contrary, the *m. masseter* (the key chewer of herbivore mammals) is only residual. In dolphins, the *m. digastricus* (*venter anterior*), an important agent that actively lowers the jaw and connects the skull base to the medial face of the mandible (see Chapter 5), is not subdivided into two parts.[dd] Other muscles of the region may contribute to lower the jaw (see the next paragraph).

Muscles of the Tongue and Hyoid Apparatus

The ventral floor of the mouth consists of (1) muscles proper of the tongue and (2) muscles that connect the mandible and/or the hyoid apparatus to the tongue. The former muscles are somewhat less developed in dolphins because of the reduced role of the tongue in chewing and especially in swallowing.

The muscles proper of the tongue (Figs. 3.77 and 3.78) include the *m. genioglossus*, the long cylindrical *m. styloglossus*, and the *m. hyoglossus*. Altogether they attach the tongue to the mandible and the hyoid complex. Simultaneous contractions of the *m. styloglossus* and the *m. hyoglossus* may curl the tongue into a tubular shape (Reidenberg and Laitman, 1994).

The larynx is connected to the hyoid complex and the sternum by a series of muscles that together support the movements of the laryngeal cartilages. An additional set of muscles are intrinsic to the larynx, meaning that their action concerns only the reciprocal adjustments of the separate cartilages. Their complex action is also discussed in relation to the larynx and hyoid apparatus in Chapter 5. Whereas the *mm. hyoidei* are relatively prominent as to their attachment to the skull and mandible, the posterior group of muscles that relate the hyoid apparatus to the sternum are somewhat less conspicuous in cetaceans than in terrestrial mammals.

The *m. sternohyoideus* (Fig. 3.79) is the most powerful element of this group, and its contraction may shift the whole hyoid complex (or some of its components, depending on the contraction state of the smaller intrinsic muscles) together with the

---

dd. A double-bellied *m. digastricus* has been described in *Neophocaena phocaenoides* and *Phocoenoides dalli* (Ito and Aida, 2005).

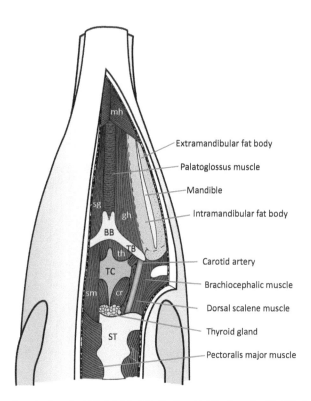

FIGURE 3.77 **Muscles of the tongue and ventral neck of _D. delphis_.** BB, Basihyoid; TB, thyreohyoid; TC, thyroid cartilage; ST, sternum; mh, m. milohyoideus; sg, m. styloglossus; gh, m. geniohyoideus; th, m. thyrohyoideus; sm, m. sternohyoideus; cr, m. cricothyroideus. *(Redrawn by Uko Gorter after Pilleri et al., 1976. Comparative anatomy of the throat of Platanista indi in reference to the sonar system. Invest. Cetacea 6, 71–88.)*

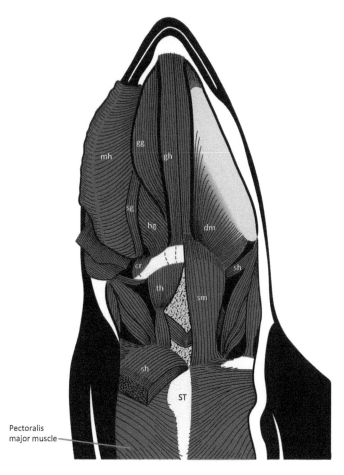

FIGURE 3.78 **Muscles of the tongue and ventral neck of near-term fetal _G. melas_.** mh, m. mylohyoideus; gg, m. genioglossus; sg, m. styloglossus; gh, m. geniohyoideus; hg, m. hyoglossus; dm, m. digastricus; th, m. thyrohyoideus; sh, m. stylohyoideus; sm, m. sternohyoideus; cr, m. cricothyroideus; ST, sternum. *(Redrawn by Uko Gorter after Reidenberg and Laitman, 1994. Anatomy of the hyoid apparatus in Odontoceti (toothed whales): specializations of their skeleton and musculature compared with those of terrestrial mammals. Anat. Rec. 240, 598–624.)*

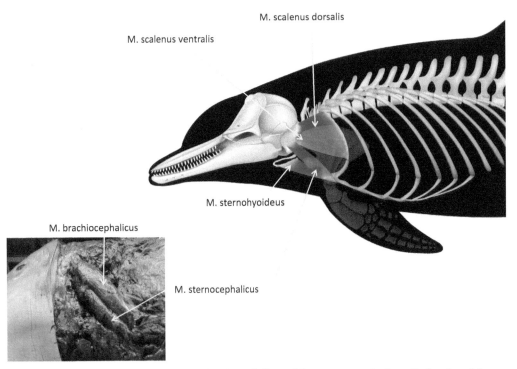

FIGURE 3.79 Dissection of the *mm. sternocephalicus*, and *brachiocephalicus* of *T. truncatus*, and schematic drawing of the same muscles and *mm. scaleni*. *(Redrawn by Uko Gorter, after Cotten et al., 2008. The gross morphology and histochemistry of respiratory muscles in bottlenose dolphins Tursiops truncatus. J. Morphol. 269, 1520–1538.)*

larynx caudally and ventrally, and even contribute to retract or lower the jaw. The *m. omohyoideus* has its origin in the lateral surface of the shoulder blade, and terminates on the thyrohyoid bone as a lateral portion of the *m. sternohyoideus* (Reidenberg and Laitman, 1994) passing deep (dorsally) to the *m. sternomastoideus* (see *m. sternocephalicus* as follows) (Klima et al., 1980).

We emphasize that the *m. sternohyoideus*—together with the *m. sternothyroideus*—is the cranial extension of the hypaxial muscles for propulsion and may thus play a minor role in locomotion. More interesting, the muscles of the hyoid apparatus are active in the process of suction feeding, by suddenly and forcefully depressing the hyoid complex and the floor of the mouth. The resulting negative pressure draws water (and prey) into the mouth (Reidenberg and Laitman, 1994; see Chapter 5).

Muscles of the Neck

The muscles of neck and ventral thorax contribute to flexing and rotating the head, while extension of the neck is limited due to the fusion of the cervical vertebrae and their relation to the skull and thorax, respectively.

The dorsal muscles of the neck mostly correspond to the rostral groups of the muscles of the spine, described later in this chapter. A precise identification of the single lesser muscles specific of the region appears difficult, also because of the tendency for fusion and thus shortening of the cervical vertebrae during prenatal development and the resulting restriction of mobility within the neck region. The lack of a movable, independent, neck improves the hydrodynamic properties of the body and, at the same time, increases the stability of the cervical region in propulsion. A series of schematic drawings of the dorsolateral muscles of the neck is presented by Pilleri et al. (1976b) for the common dolphin.

The *m. splenius* connects the cervical spine to the atlas and the back of the skull, either directly or through other ventral mm. of the neck. The *m. splenius* is easy to recognize in the common dolphin, according to Pilleri et al. (1976b); however, its precise identification is not so clear-cut in the bottlenose dolphin, at least based on a series of personal dissections.

The organization of the ventral muscles of the neck is influenced by the lack of the clavicle. These muscles, originating from the humerus and sternum and directed to the skull, are among the most important flexors and rotators of the head. In the bottlenose dolphin, the *m. sternocephalicus*[ee] (origin: sternum; attachment: latero-ventral border of the squamous part

---

ee. The *m. sternocleidomastoideus* of man does not exist as such in mammals who do not possess a clavicle. It is thus substituted by a *m. sternocephalicus* (from the sternum to the head) and a *m. brachiocephalicus* (from the humerus to the head). The *m. sternocephalicus* may terminate in different parts of the ventrolateral skull. The part connected to the mastoid process in the skull base is called *m. sternomastoideus*.

of the temporal bone) and the *m. brachiocephalicus* (*m. mastohumeralis* in Pilleri et al., 1976b) that runs from the humerus to the mandible,[ff] are well evident (Fig. 3.77). The existence of an independent *m. omotransversarius* (from the scapula the first cervical vertebrae) is debated by Pilleri et al. (1976b), at least in *D. delphis*.

In terrestrial mammals, the *mm. sternocephalicus* and *sternohyoideus* are also important in respiration because of their participation in pivoting the ribs cranially and laterally. In the bottlenose dolphin (Fig. 3.79), their contraction (together with the *m. scalenus ventralis*) can pull the sternum dorsocranially and thus expand the thoracic cavity (Cotten et al., 2008).

The *m. brachiocephalicus* may be functional to lower the mandible, and open the mouth or to flex and/or turn the whole head sideways if the mouth is closed.

The *mm. scaleni* (Fig. 3.79) travel from the ribs to the cervical vertebrae medial to the *m. brachiocephalicus*. They originate from the first pair of ribs and terminate on the transverse processes of the cervical vertebrae (*m. scalenus dorsalis*) or on the exoccipital bone (*m. scalenus ventralis* = *anterior scalenus* of Pilleri et al., 1976a). The *ventral scalenus* may lower the head (bilateral contraction) or turn it sideways (monolateral contraction) when the rib insertion is fixed. On the contrary, the *mm. scaleni* can draw the ribs cranially and thus help in respiration when the skull is fixed (Cotten et al., 2008).

Muscles of the Spine and Flukes

As we have seen in the previous pages dedicated to the skeleton, the spine acquires a special value hitherto unknown in terrestrial mammals. During locomotion the vertebral column of dolphins bends and straightens alternatively as that of a leopard or a camel, but instead of transmitting its contraction power from the hind to the front limbs, the dolphin's spine immediately produces forward thrust by moving the flukes vertically. As expected from the height and width of the vertebral processes from the middle thoracic to the early caudal regions, the epiaxial (= dorsal to the transverse processes) and hypaxial (= ventral to the transverse processes) muscle complexes are incredibly strong in dolphins. Dolphins can bend their body dorsoventrally as well as to the sides. However, the dorsal and lateral bending of the rump is somewhat limited by the thin but long and wide spinous and transverse processes.

The muscles of the back (*Musculi dorsi*) of mammals (see also NAV, 2012) are generally divided into

- muscles that connect the back to the forelimb;
- mm. that lift and bend the spine (*mm. erector spinae* and *mm. transversospinalis*);
- intrinsic mm. of the spine (*mm. interspinales* and *mm. intertransversarii*); and
- mm. of the flukes (*mm. caudae*).
- A fifth group is constituted by the hypaxial muscles, corresponding to the first group of the *mm. membri pelvini* of quadrupeds (NAV, 2012). Since their anatomical and functional relationships are along the vertebral column, we describe them here.

Muscles That Connect the Spine to the Forelimb

The first group includes muscles that in terrestrial mammals are involved in either movements of the shoulder blade or anchoring the limb to the thorax, acting together with the powerful pectoral muscles. The virtual absence of gravity in water mostly limits the scope of these muscles to the control of the shoulder. To this respect, the *m. latissimus* dorsi is reduced to a thin diagonal stripe that reaches the humerus (Fig. 3.80).

In dolphins, the whole region is covered by a thick subcutaneous muscle (*m. cutaneus trunci*) that continues to the external surface of the thorax and contributes to the smooth appearance of the body. The two main suspenders of the shoulder blade in quadruped Cetartiodactyla—*m. trapezius* and *m. rhomboideus*—originate from the spinous processes of the cervical and anterior thoracic vertebrae and associated ligaments and terminate on the spine of the scapula and the internal surface of the upper border of this bone, respectively. According to Pilleri et al. (1976b), the *m. trapezius* of the common dolphin is vestigial, and represented by a thin stripe overlying the *m. splenius*. Even the *m. rhomboideus* is vestigial according to the same authors, and only a minimal remain of the *m. rhomboideus posterior* (= *m. rhomboideus thoracis* of NAV, 2012) can be detected in the common dolphin.[gg] The relative disappearance of the *m. trapezius* and *m. rhomboideus* (accompanied by the absence of the spine in the scapula) can be related to the diminished importance of these elements in the suspension bridge mechanism of the thorax.

---

ff. When a residue of the collar bone is present (as in the tiger and to a lesser extent in smaller feline), the *m. brachiocephalicus* of terrestrial is divided into a *m. cleidobrachialis* (= *pars clavicularis m. deltoidei* of man) and a *m. cleidocephalicus*, because of the interposition of the residue of the clavicle along its course. Since this bone virtually disappears in adult dolphins, this distinction between the two parts of the muscle is impossible to make.

gg. In the same paper Pilleri et al. (1976b) state that, in contrast to *D. delphis*, the river dolphin *Platanista indi* exhibits a well-developed musculature in this region. A comparison between *Delphinidae* and *Platanistidae* is beyond our scope, but the differences known between the two Families indicate that—at least in some cetaceans—the organization of the dorsal muscles is partially similar to that of terrestrial mammals. Schulte and De Forest Smith (1918) described in detail the muscles of *Kogia breviceps*, but report also many comparative aspects of the muscles of the vertebral column, thorax, and abdomen of *Globicephala* and other dolphins with relative literature.

FIGURE 3.80   *M. latissimus dorsi* **of** *S. coeruleoalba.*

The remaining groups of muscles are dedicated to the vertebral column. They have been extensively and elegantly studied and described by Ann Pabst (1990). In her seminal paper she summarized the studies of several authors, and particularly the monographs of Slijper (1936, 1946). The following paragraphs rely heavily on Slijper (1936, 1939), Pabst (1990) and Pilleri et al. (1976b), with integration and personal observations. For a histochemical characterization of the axial musculature of the bottlenose dolphin see Bello et al., 1985.

Muscles That Lift and Bend the Spine *(*mm. erector spinae *and* mm. transversospinalis*)*

The muscles of the back of dolphins (Figs. 3.81–3.84) are enveloped in a thick subdermal sheath (SDS), formed by connective tissue derived from the tendons of the abdominal and axial muscles, and from ligaments and blubber, all interwoven by layers of collagen. The presence of blubber-derived connective tissue constitutes the main difference with the *fascia thoracolumbalis* of other mammals. The SDS provides a robust external scaffold to the whole muscular system of the region, gives rise to internal subdivisions that sheath single muscles, and increases the stiffness of the spine.

This second group of muscles of the back (*m. erector spinae* and *mm. transversospinalis*) is made up by some of the most potent muscles of the whole body. They constitute a homogeneous, red to black muscular mass attached to the vertebral column in the quadrant between the spinous and transverse processes. A thick sheath of connective tissue encompass them and further contributes to their uniform shape. The major groups of epiaxial muscles of terrestrial mammals are all present, although somewhat modified.

The essential difference between these muscles in dolphins and the corresponding ones in quadrupeds results from the strong reduction and modification of the pelvic girdle. Here the absence of a structured hip bone changes the origin (or insertion) of the long muscles and thus the mechanic of movement dramatically. Since the hind limb disappears, the long muscles of the back exert their action through robust tendons directly on the flukes.

The *m. erector spinae* as a whole is strictly connected to the SDS, partially fused with the muscles of the tail and terminates with a tendon (superficial tendon) to the flukes. It is composed of the *m. longissimus* and the *m. iliocostalis*. The *m. longissimus* originates from the temporal and occipital bones of the skull and then occupies most of the lateral lodge between the transverse processes of the vertebrae and proximal extremities of the ribs (ventrally) and the *m. multifidus* (dorsally, see the following paragraphs). The *m. iliocostalis* (Nomizo et al., 2005) lies lateral to the *longissimus* and covers the upper dorsolateral parts of the ribs.

The *m. transversospinalis* is composed by the *m. semispinalis* and the *mm. multifidi*. The *m. semispinalis* originates from the dorsolateral surface of the skull and progresses in caudal direction until the middle of the thoracic region, where

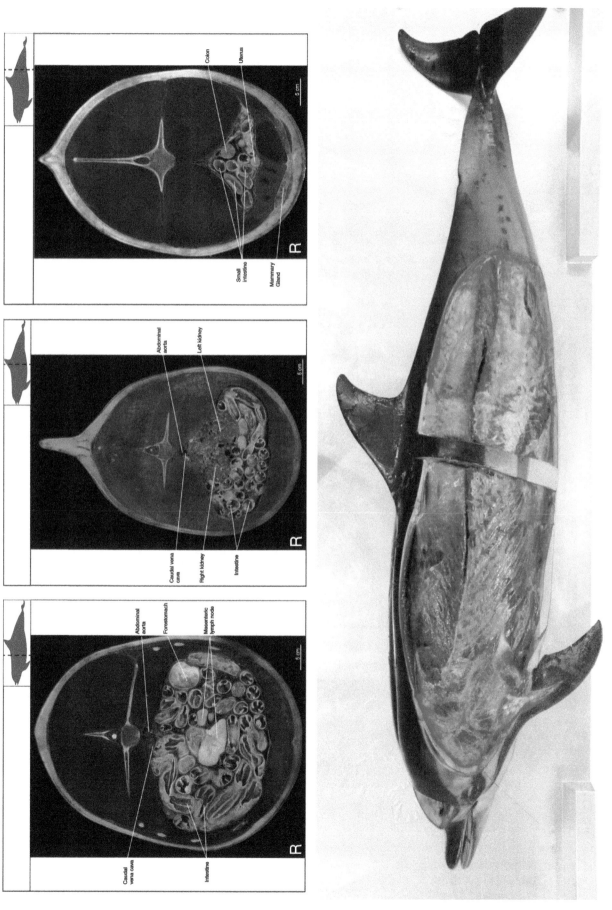

FIGURE 3.81   Muscles of the back of *T. truncatus* (transverse sections at the top) and *S. coeruleoalba* (bottom image of the whole system).

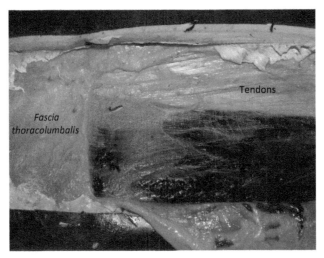

FIGURE 3.82   **Muscles of the back and relative tendons and fascia of *G. griseus*.** Note by the powerful *fascia thoracolumbalis* left in place on the left.

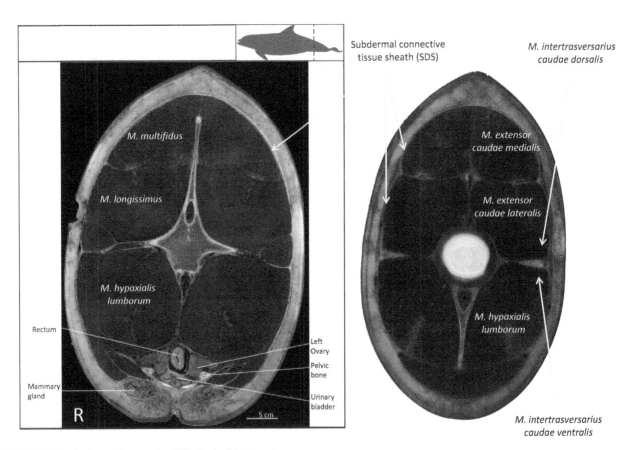

FIGURE 3.83   **Section of the muscles of the back of *T. truncatus*.**

it becomes fused with the *m. longissimus* and the *m. multifidus*. The latter muscle originates from the cervical to the caudal spine. In a transverse section of the body cut at the posterior end of the thoracic region, the dorsolateral lodge of the back (dorsal to the transverse processes) is mostly occupied by the *m. multifidus* (dorsal and medial) and the *m. longissimus* (lateral and ventral). As briefly outlined earlier, a precise distinction between the muscles of the back in dolphins is not simple, since they merge into one another and most often their terminal tendons travel together. So, for instance, the *m. transversospinalis* gives rise to smaller muscle bundles that contact adjoining vertebrae, and possibly correspond to the rotator muscles of terrestrial mammals.

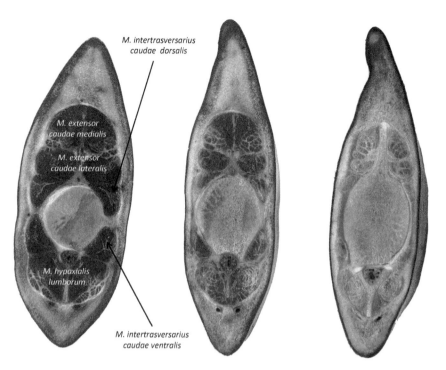

M. intertrasversarius caudae dorsalis

M. extensor caudae medialis

M. extensor caudae lateralis

M. hypaxialis lumborum

M. intertrasversarius caudae ventralis

FIGURE 3.84 Caudal section of the tail (just before the flukes) of *T. truncatus*, with termination of the muscles and fascia.

The *m. longissimus* and the *m. multifidus* terminate with tendons directed to the flukes (superficial tendon for the former; deep tendon for the latter).

Intrinsic Muscles of the Spine *(mm. interspinales and* mm. intertransversarii*)*

These relatively short but very robust muscles connect adjacent vertebrae and are particularly difficult to tell apart from the more superficial muscles that lift and bend the spine.

Muscles of the Flukes *(mm. caudae)*

Once more the lack of a conventional hip bone in dolphins changes the shape and function of the caudal part of the muscles of the back. The forces that in terrestrial mammals are applied to the coxal bones are directed to the peduncle of the tail. Therefore, this last group of column-related muscles reaches a unique level of development in cetaceans as they are mainly responsible of the propulsion of the animal.

Two powerful muscles, the *m. extensor medialis caudae* and the *m. ext. lateralis caudae*, substitute the *mm. sacrocaudales* of quadrupeds and are in direct continuation of the *mm. erector spinae* and *mm. transversospinalis* that lift and bend the back. In fact, the superficial terminal tendon of the *m. longissimus* is joined by the tendons of the *m. ext. lateralis caudae* and the deep terminal tendon of the *m. multifidus* by those of the *m. ext. medialis caudae*. Thus, the force that is applied to the flukes through the terminal tendons derives from all the muscles of the back.

The *mm. intertransversarii dorsales caudae* and the *mm. intertransversarii ventrales caudae* originate from the lumbar transverse processes and strengthen the whole caudal sector by joining the SDS and the vertebral bodies of caudal vertebrae. Perhaps they may also contribute to relative rotation (torsion) of the flukes along the longitudinal axis of the spine.

The muscles that move the flukes develop early in dolphins (Dearolf et al., 2000), and show a precocity that can be explained with the need for locomotion immediately after birth. In the process of acquiring swimming abilities, the *mm. intertransversarii dorsales caudae* limits the lateral bending of the column in neonatal dolphins (Etnier et al., 2004, 2008) and acts as a postural regulator very early in life, to become a fine-tuner of tail position in adults.

Hypaxial Muscles

The *m. ileopsoas* (composed of the *m. iliacus* and *m. psoas major*) and the *m. psoas minor* of terrestrial mammals represent a robust connection between the lower surface of the lumbar column and the hip and femur. Given the specific anatomy of dolphins, the muscles that occupy the lower lodge of the lumbar column could be named *mm. hypaxiales* (Fig. 3.85). In fact,

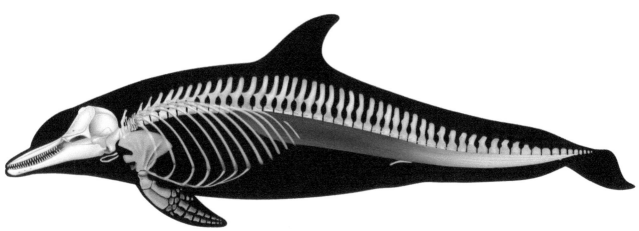

FIGURE 3.85 Schematic representation of the hypoaxial muscles of **T. truncatus.** *(Redrawn by Uko Gorter, after Cotten et al., 2008. The gross morphology and histochemistry of respiratory muscles in bottlenose dolphins Tursiops truncatus. J. Morphol. 269, 1520–1538.)*

it is possible to recognize only a conspicuous *m. hypaxialis lumborum* (Pabst, 1990), ventral to the *mm. intertransversarii ventrales caudae* and perhaps corresponding to the large *m. sacrococcygeus* (= *m. sacrocaudalis* of NAV, 2012) described by Pilleri et al. (1976b) in a young common dolphin. According to Pabst (1990), the *m. hypaxialis lumborum* of *D. delphis* may be separated into discrete constituent subunits. In a fetus of *P. crassidens*, Purves and Pilleri (1978) were able to identify traces of both the *m. psoas major* and *m. psoas minor*, but considered them bound to blend in with *m. sacrococcygeus*. In general, the flexor (hypaxial) muscles are not as powerful as the extensor (epaxial) ones (Purves and Pilleri, 1978).

Muscles of the Thorax and Abdomen

The disposition of muscles of thorax and abdomen follows the same plan as in terrestrial mammals. Most of the muscles of the thorax are active in respiration, including the external (Fig. 3.86) and internal (Fig. 3.87) intercostal muscles, the *m. transversus thoracis*, and the diaphragm. The *m. serratus ventralis caudalis* has turned into a very large, very powerful stabilizer of the thorax.

Locomotion and respiration are coupled in terrestrial mammals and the movements of the limbs and viscera during ambulation induce also a modification of the diameter of the thoracic cavity through the flattening/expansion of the diaphragm,

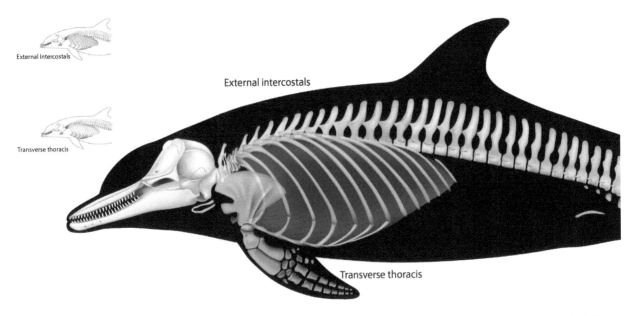

FIGURE 3.86 External intercostal and transverse thoracis muscles of **T. truncatus.** *(Redrawn by Uko Gorter, after Cotten et al., 2008. The gross morphology and histochemistry of respiratory muscles in bottlenose dolphins Tursiops truncatus. J. Morphol. 269, 1520–1538.)*

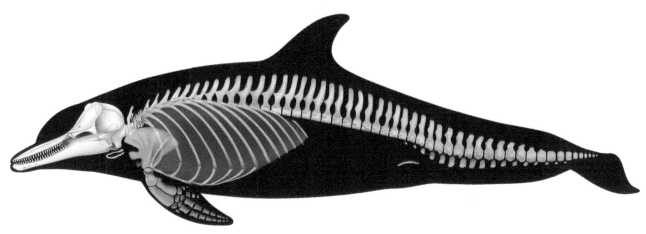

FIGURE 3.87   **Internal intercostals muscle of** *T. truncatus*. *(Redrawn by Uko Gorter, after Cotten et al., 2008. The gross morphology and histochemistry of respiratory muscles in bottlenose dolphins Tursiops truncatus. J. Morphol. 269, 1520–1538)*

and a consequent resulting change in ventilation. This is not completely true for cetaceans, who actively swim while breathholding for long periods and couple locomotion and respiration only during the surface interval (Cotten et al., 2008).

The diaphragm (Figs. 3.88 and 3.89) is greatly curved and possesses a very deep cupola that increases the intrathoracic part of the peritoneal cavity. Histochemical studies demonstrated that the diaphragm of dolphins is made up mostly by slow-twitch fibers (Dearolf, 2003) whose number greatly increases from early life to adulthood. Some of the structural and functional features necessary for respiratory muscles to achieve explosive power ventilation during rapid gas exchange at the surface require further investigations.

The ventrolateral muscles of the abdomen include the *m. rectus abdominis* (Fig. 3.90), the external and internal oblique muscles, and the *m. transversus abdominis*. Since the pelvic bones are rudimentary, the *m. rectus abdominis*

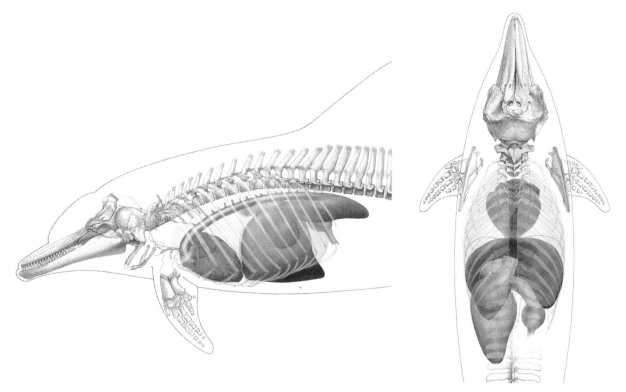

FIGURE 3.88   **Position of the diaphragm of** *T. truncatus* **seen from the side considering the position of the viscera. (heart and aorta, red; lungs, light blue; liver and stomach complex, purple and brown), and top. The outline of the diaphragm is represented by the red line.** *(Drawings by Massimo Demma).*

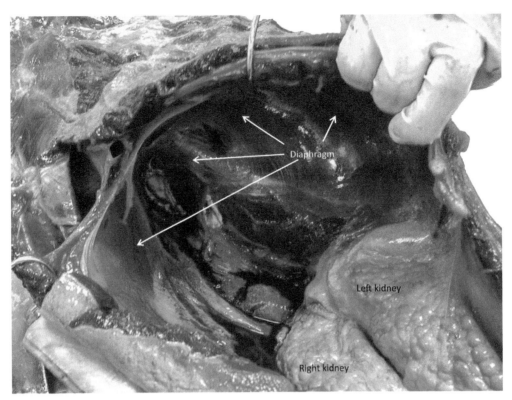

FIGURE 3.89   Diaphragm of a young *G. griseus* seen from the abdominal cavity after removal of the digestive apparatus (the kidneys remained in the extraperitoneal position).

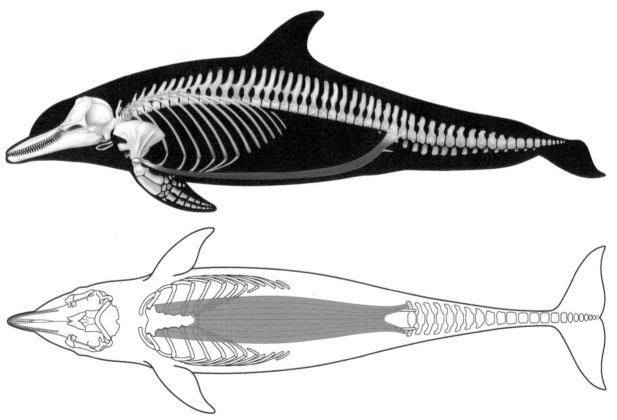

FIGURE 3.90   Schematic drawing of the *m. rectus abdominis* of *T. truncatus.* *(Redrawn by Uko Gorter, after Cotten et al., 2008. The gross morphology and histochemistry of respiratory muscles in bottlenose dolphins Tursiops truncatus. J. Morphol. 269, 1520–1538.)*

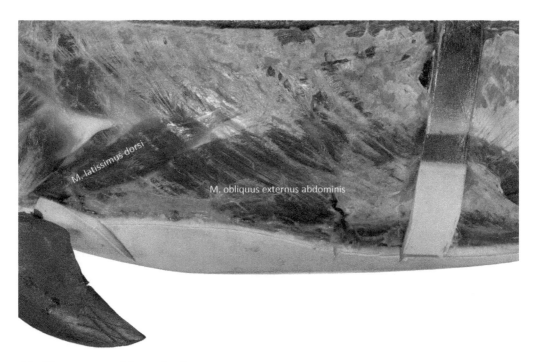

FIGURE 3.91 *M. obliquus externus* **of** *S. coeruleoalba.*

originates from the distal portions of the sternal ribs and the sternum, and terminates in part on the pelvic remains, but also continues along the connective sheath to the caudal vertebrae 2–8 (Cotten et al., 2008). Thus, this muscle is involved in respiratory movements, but contributes also to lower the tail peduncle and to the general dorsoventral flexion of the body.

The *m. obliquus externus abdominis* (Fig. 3.91) and *m. obliquus internus abdominis*, combined with *m. transversus*, constitute the ventrolateral abdominal wall and are also active in respiration and support the action of the *m. rectus abdominis*.

Summarizing, expiration at the surface (due to volume reduction of the thoracic cavity) is coupled to head flexion due to cranio-cervical muscle contraction and to downstroke action of the lumbo-pelvic muscles. Inspiration is promoted by action of the antagonistic muscles of the spine (Cotten et al., 2008). How respiration and locomotion become uncoupled during breath-holding dives at various depths remains unclear.

Muscles of the Thoracic Limb

The muscles of the forelimb are mostly active on the shoulder joint, in rotations or stabilization via stiffening. As we have seen before, in dolphins the outer surface of the shoulder blade is flat and lacks a distinct spine. In terrestrial Cetartiodactyla, the acromion is very small and the *m. deltoideus* mostly originates from the long scapular spine and, in resting position, terminates on the deltoid process of the humerus below. Therefore, its course is downward and slightly forward (almost vertical). However, the situation in dolphin is quite different (for a schematic representation see Fig. 3.92). The *m. deltoideus* of dolphins (Fig. 3.93) originates from the large acromion that projects forward, and from the anterior part of the external face of the scapula. Since the dolphin acromion is much more cranial that the deltoid tuberosity of the humerus (the whole flipper is inclined backward), the direction of the muscle is downward and backward. This disposition gives the muscle the significance of a functional paddle that may move the flipper forward against the opposite resistance of water during swimming. According to Klima et al. (1980), the tip of the coracoid process bears the insertion of the *m. pectoralis*

---

hh. The subdivision of pectoral muscles in nonprimate mammals is rather complicate. The *m. pectoralis minor* is a respiratory muscle in man, but as such it does not exist in quadrupeds. In terrestrial Cetartiodactyla (the closest relative of Cetacea), pectoral muscles are generally subdivided in two plans, with the *mm. pectoralis descendens* and *pectoralis transversus* in the superficial layer, and the *m. pectoralis profundus* (*ascendens*)] and eventually *m. subclavius* in the deeper layer. These muscles have a limited action in respiration, but are mostly functional in the muscular attachment and reciprocal movements of the thorax with the front limb. The situation in dolphins is unclear because of the difficulty to solve the homologies with other mammals.

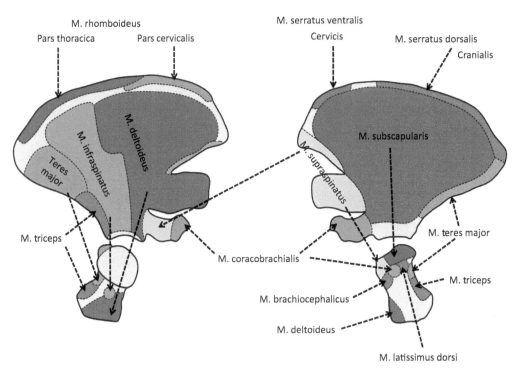

FIGURE 3.92 **Schematic drawing of the muscle insertion on the shoulder blade of *T. truncatus*.** *(Redrawn by Uko Gorter after Benke, 1993. Investigations on the osteology and functional morphology of the flipper of whales and dolphins. Investig. Cetacea 24, 9–252.)*

FIGURE 3.93 *Mm. deltoideus and other muscles of the shoulder* of *S. coeruleoalba*.

minor.[hh] A well-defined *m. coracobrachialis* (Fig. 3.94) is also present in dolphins, with an origin on the medial surface coracoid process, and the termination of the humerus.[ii] Its action further enforces the action of the deltoid as a paddle. The *m. supraspinatus* (Fig. 3.95) originates from the small *fossa supraspinata* identified on the anterior part of the scapula (see the description of the bone), from the medial surface of the large acromion and from the lateral surface of the coracoid

---

ii. The presence of the long head of *m. biceps brachii* is not clear, but it's possible at least in the false killer whale, where it may join the *m. coracobrachialis* and terminate together with the latter on the anterior convex surface of the radius.

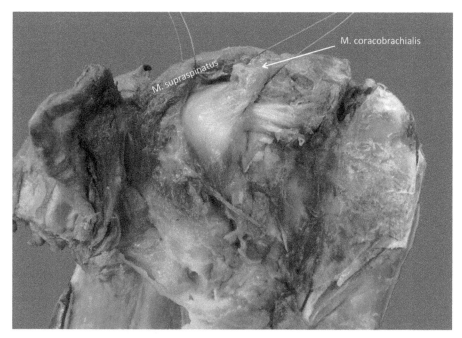

FIGURE 3.94   *M. coracobrachialis* **of a young** *T. truncatus*. *(Photograph courtesy of Dr. Annamaria Grandis, University of Bologna)*

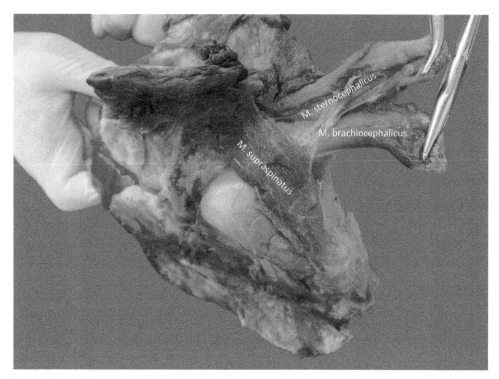

FIGURE 3.95   *M. supraspinatus* **of a young** *T. truncatus*. Pliers are placed on *m. sternocephalicus* and *m. brachiocephalicus*, respectively. *(Photograph courtesy of Dr. Annamaria Grandis, University of Bologna)*

process and terminates on the medial surface of the proximal part of the humerus, just anterior to the termination of the *m. subscapularis*. The large lateral muscle easily detected on the lateral surface of the shoulder blade represents a superimposition of the *m. deltoideus* and the large *m. infraspinatus* (Fig. 3.92). It acts on the proximo-lateral part of the short humerus to modify the angle of the flipper and help induce rotatory abduction of the arm. The *m. triceps* pulls the flipper backward, acting on the olecranon of the ulna.

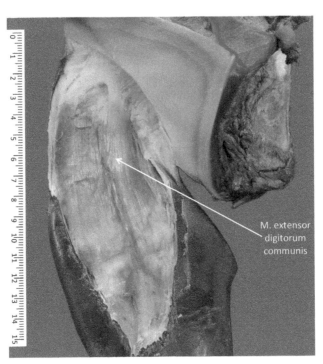

M. extensor
digitorum
communis

FIGURE 3.96 *M. extensor digitorum communis* of a young *T. truncatus*. *(Photograph courtesy of Dr. Annamaria Grandis, University of Bologna)*

The radius and ulna are not able to rotate parallel along the same axis. This is due to the fact that their reciprocal articulating surfaces are inadequate for movement and the two bones are connected by robust fibrous tissue. The skeletal elements of the wrist and hand are covered by a thick connective sheath that adds rigidity to the flipper. As a consequence, muscles of the forearm are basically useless and difficult to appreciate in dolphins, and the *m. extensor digitorum communis* can be barely identified (Fig. 3.96). According to the detailed study of the muscles and nerves of the cetacean forelimb by Cooper et al. (2007b), the reduction of the antebrachial muscles in dolphins (and specifically that of the *m. extensor digitorum communis*) is less pronounced in the killer whale. However, the same authors found no evidence of flexor muscles.

Muscles of the Pelvic Limb

The muscles relative to the pelvic limb disappeared during the evolution of cetacean together with the bones of the hind limb. Therefore, there is no trace of pelvic muscles in normal postnatal dolphins. The muscles attached to the small pelvic bones serve the genital system (see Chapter 9).

Development of the Muscular System

Developmental phases (stages) of selected dolphin species have been reported in Table 9.6 (see Chapter 9 for details). According to Štěrba et al. (2000), muscular blastemas appear during Stage 3, in which an anlage of the diaphragm also appears and separates the thoracic from the abdominal cavity.[jj] Blastemas of the muscles of the thoracic limb, and mesodermal and ectodermal layers of the muscles of the abdominal wall become evident at Stage 4. During the next stage, myoblasts of the muscles of the back, thoracic, and abdominal wall stretch toward the direction of their future anatomical position. Distinction between fleshy and tendinous parts starts to be evident at Stage 6 and progresses subsequently. Myoblast evolve into myotubes to form primary bundles during Stage 7, and the abdominal walls narrow the umbilical stalk and start to reduce the physiological umbilical hernia pushing the abdominal viscera back into the abdominal cavity (see also Development of the Digestive System in Chapter 8).

---

jj. The development of the diaphragm in the bottlenose dolphin has been described in detail by Dearolf (2003).

REFERENCES

Azevedo, C.T., Lima, J.Y., de Azevedo, R.M., Santos Neto, E.B., Tamy, W.P., de Araújo Barbosa, L., Brito, J.L., Boere, V., da Silveira, L.S., 2015. Thoracic limb bone development in *Sotalia guianensis* (Van Beneden 1864) along the coastline of Espírito Santo, Brazil. J. Mammal 96, 541–551.

Bello, M.A., Roy, R.R., Martin, T.P., Goforth, jr, H.W., Edgerton, V.R., 1985. Axial musculature in the dolphin (*Tursiops truncatus*): some architectural and histochemical characteristics. Mar. Mamm. Sci. 1, 324–336.

Benke, H., 1993. Investigations on the osteology and functional morphology of the flipper of whales and dolphins. Investig. Cetacea 24, 9–252.

Buchholtz, E.A., 2007. Modular evolution of the cetacean vertebral column. Evol. Develop. 9, 278–289.

Buchholtz, E.A., Schur, S.A., 2004. Evolution of vertebral osteology in Delphiniidae (Cetacea). Zool. J. Linn. Soc. Lon. 140, 383–401.

Butti, C., Corain, L., Cozzi, B., Podestà, M., Pirone, A., Affronte, M., Zotti, A., 2007. Age estimation in the Mediterranean bottlenose dolphin *Tursiops truncatus* (Montagu 1821) by bone density of the thoracic limb. J. Anat. 211, 639–646.

Calzada, N., Aguilar, A., 1996. Flipper development in the Mediterranean striped dolphin (*Stenella coeruleoalba*). Anat. Rec. 245, 708–714.

Calzada, N., Aguilar, A., Lockyer, C., Grau, E., 1997. Pattern of growth and physical maturity in the Western Mediterranean striped dolphin (*Stenella coeruleoalba*) (Cetacea: Odontoceti). Can. J. Zool. 75, 632–637.

Carvalho, A.P.M., Lima, J.Y., Azevedo, C.T., Botta, S., Ferreira de Queiroz, F., Sepúlveda Campos, A., de Araújo Barbosa, L., da Silveira, L.S., 2015. Ossification pattern of estuarine dolphin (*Sotalia guianensis*) forelimbs, from the coast of the state of Espírito Santo, Brazil. PLoS One 10, e0127435.

Cooper, L.N., Berta, A., Dawson, S.D., Reidenberg, J.S., 2007a. Evolution of hyperphalangy and digit reduction in the cetacean manus. Anat. Rec. 290, 654–672.

Cooper, L.N., Dawson, S.D., Reidenberg, J.S., Berta, A., 2007b. Neuromuscular anatomy and evolution of the cetacean forelimb. Anat. Rec. 290, 1121–1137.

Cotten, P.B., Piscitelli, M.A., McLellan, W.A., Rommel, S.A., Dearolf, J.L., Pabst, D.A., 2008. The gross morphology and histochemistry of respiratory muscles in bottlenose dolphins *Tursiops truncatus*. J. Morphol. 269, 1520–1538.

Cozzi, B., 1981. Osservazioni sulla morfologia della colonna vertebrale nei cetacei. Atti Soc. Ital. Sci. Nat. Mus. Civ. Stor. Nat. Milano 122, 225–235.

Cozzi, B., De Francesco, I., Cagnolaro, L., Leonardi, L., 1985. Radiological observations on the skeletal development in fetal and newborn specimens of *Delphinus delphis* L. and *Stenella coeruleoalba* (Meyen). Atti Soc. Ital. Sci. Nat. Mus. Civ. Stor. Nat. Milano 126, 120–136.

Cozzi, B., Mazzariol, S., Podestà, M., Zotti, A., 2010a. Diving adaptations of the cetacean skeleton. Open Zool. J. 2, 24–32.

Cozzi, B., Panin, M., Butti, C., Podestà, M., Zotti, A., 2010b. Bone density distribution patterns in the rostrum of delphinids and beaked whales: evidence of family-specific evolutive traits. Anat. Rec. 293, 235–242.

Cozzi, B., Podestà, M., Vaccaro, C., Poggi, R., Mazzariol, S., Huggenberger, S., Zotti, A., 2015. Precocious ossification of the tympanoperiotic bone in fetal and newborn dolphins: an evolutionary adaptation to the aquatic environment? Anat. Rec. 298, 1294–1300.

De Buffrénil, V., Schoevaert, D., 1988. On how the periosteal bone of the delphinid humerus become cancellous: ontogeny of a histological specialization. J. Morphol. 198, 149–164.

De Smet, W.M.A., 1977. The regions of the cetacean vertebral column. In: Harrison, R.J. (Ed.), Functional Anatomy of Marine Mammals, vol. 3, Academic Press, London, pp. 59–80.

Dearolf, J.L., 2003. Diaphragm muscle development in bottlenose dolphins *Tursiops truncatus*. J. Morphol. 256, 79–88.

Dearolf, J.L., McLellan, W.A., Dillaman, R.M., Frierson, D., Pabst, D.A., 2000. Precocial development of axial locomotor muscle in bottlenose dolphins (*Tursiops truncatus*). J. Morphol. 244, 203–215.

DiGiancamillo, M., Rattegni, G., Podestà, M., Cagnolaro, L., Cozzi, B., Leonardi, L., 1998. Postnatal ossification of the thoracic limb in striped dolphin (*Stenella coeruleoalba*) (Meyen, 1833) from the Mediterranean sea. Can. J. Zool. 76, 1286–1293.

Dines, J.P., Otárola-Castillo, E., Ralph, P., Alas, J., Daley, T., Smith, A.D., Dean, M.D., 2014. Sexual selection targets cetacean pelvic bones. Evolution 68, 3296–3306.

Etnier, S.A., Dearolf, J.L., McLellan, W.A., Pabst, D.A., 2004. Postural role of lateral axial muscles in developing bottlenose dolphins (*Tursiops truncatus*). Proc. R. Soc. Lond. B 271, 909–918.

Etnier, S.A., McLellan, W.A., Blum, J., Pabst, D.A., 2008. Ontogenetic changes in the structural stiffness of the tailstock of bottlenose dolphins (*Tursiops truncatus*). J. Expl. Biol. 211, 3205–3213.

Felts, W.J.L., 1966. Some functional and structural characteristics of cetacean flippers and flukes. In: Norris, K.S. (Ed.), Whales, Dolphins and Porpoises. University of California Press, Berkeley, pp. 255–276.

Felts, W.J.L., Spurrell, A., 1965. Structural orientation and density in cetacean humeri. Am. J. Anat. 116, 171–203.

Felts, W.J.L., Spurrell, F.A., 1966. Some structural and developmental characteristics of cetacean (Odontocete) radii. A study of adaptive osteogenesis. Am. J. Anat. 118, 103–134.

Gray, N.M., Kainec, K., Madar, S., Tomko, L., Wolfe, S., 2007. Sink or swim? Bone density as mechanism for buoyancy control in early cetaceans. Anat. Rec. 290, 638–653.

Guglielmini, C., Zotti, A., Bernardini, D., Pietra, M., Podestà, M., Cozzi, B., 2002. Bone density of the arm and forearm as an age indicator in specimens of stranded striped dolphins (*Stenella coeruleoalba*). Anat. Rec. 267, 225–230.

Hirose, A., Nakamura, G., Kato, H., 2015. Some aspects on an asymmetry of nasal bones in toothed whales. Mamm. Stud. 40, 101–108.

Howell, A.B., 1930. Aquatic Mammals. Their Adaptations to Life in the Water. Charles C Thomas Publisher, Springfield, IL, pp. 1–338.

Ito, H., Aida, K., 2005. The first description of the double-bellied condition of the digastrics muscle in the finless porpoise *Neophocaena phocaenoides* and Dall's porpoise *Phocoenoides dalli*. Mamm. Stud. 30, 83–87.

Ito, H., Miyazaki, N., 1990. Skeletal development of the striped dolphin (*Stenella coeruleoalba*) in Japanese waters. J. Mamm. Soc. 14, 79–96.

Kardong, K.V., 2015. Vertebrates: Comparative Anatomy, Function, Evolution, seventh ed. McGraw-Hill Education, New York, NY.

Klima, M., 1978. Comparison of early development of sternum and clavicle in striped dolphin and humpback whale. Sci. Rep. Whale Res. Inst. 30, 253–269.

Klima, M., 1990. Rudiments of the clavicle in the embryos of whales (Cetacea). Z. Säugetierkd. 55, 202–212.

Klima, M., Oelschläger, H.A., Wunsch, D., 1980. Morphology of the pectoral girdle in the Amazon dolphin *Inia geoffrensis* with special reference to the shoulder joint and movements of the flippers. Z. Säugetierkd. 45, 288–309.

Lawrence, B., Schevill, W.E., 1965. Gular musculature in delphinids. Bull. Museum Compar. Zool. Harv. Univ. 133, 1–65.

McDonald, M., Vapniarsky-Arzi, N., Verstraete, F.J.M., Staszyk, C., Leale, D.M., Woolard, K.D., Arzi, B., 2015. Characterization of the temporomandibular joint of the harbour porpoise (*Phocoena phocoena*) and Risso's dolphin (*Grampus griseus*). Arch. Oral Biol. 60, 582–592.

McLeod, C.D., Reidenberg, J., Weller, M., Santos, M.B., Herman, J., Goold, J., Pierce, G.J., 2007. Breaking symmetry: the marine environment, prey size, and the evolution of asymmetry in cetacean skulls. Anat. Rec., 539–545.

Mead, J.G., 1975. Anatomy of the external nasal passages and facial complex in the Delphinidae (Mammalia: Cetacea). Smithson. Contrib. Zool. 207, 1–72.

Mead, J.G., Fordyce, R.E., 2009. The Therian Skull. A Lexicon With Emphasis on the Odontocetes. Smithsonian Institution Scholarly Press, Washington, D.C., pp. 1–261.

Miller, Jr, G.S., 1923. The telescoping of the cetacean skull. Smithson. Misc. Coll. 76, 1–70.

Moran, M.M., Nummela, S., Thewissen, J.G.M., 2011. Development of the skull of the pantropical spotted dolphin (*Stenella attenuata*). Anat. Rec. 294, 1743–1756.

NAV, 2012. Nomina Anatomica Veterinaria—V revised edition. Published online by the Editorial Committee, Hannover (Germany), Columbia (MO, USA), Ghent (Belgium), Sapporo (Japan), pp. 1–160.

Nomizo, A., Kudoh, H., Sakai, T., 2005. Iliocostalis muscles in three mammals (dolphin, goat and human): their identification, structure and innervation. Anat. Sci. Int. 80, 212–222.

Ogden, J.A., Conlogue, G.J., Rhodin, A.G., 1981a. Roentgenographic indicators of skeletal maturity in marine mammals (Cetacea). Skeletal Radiol. 7, 119–123.

Ogden, J.A., Lee, K.E., Conlogue, G.J., Barnett, J.S., 1981b. Prenatal and postnatal development of the cervical portion of the spine in the short-finned pilot whale *Globicephala macrorhyncha*. Anat. Rec. 200, 83–94.

Pabst, D.A., 1990. Axial muscles and connective tissues of the bottlenose dolphin. In: Leatherwood, S., Reeves, R.R. (Eds.), The Bottlenose Dolphin. Academic Press, New York, pp. 51–67.

Pilleri, G., Gihr, M., Purves, P.E., Zbinden, K., Kraus, C., 1976a. Comparative anatomy of the throat of *Platanista indi* in reference to the sonar system. Invest. Cetacea 6, 71–88.

Pilleri, G., Gihr, M., Purves, P.E., Zbinden, K., Kraus, C., 1976b. On the behaviour, bioacoustics and functional morphology of the Indus river dolphin (*Platanista indi* Blyth 1859). III. Comparative study of the skin and general myology of *Platanista indi* and *Delphinus delphis* in relation to hydrodynamics and behaviour. Invest. Cetacea 6, 89–127.

Podestà, M., Rattegni, G., Leonardi, L., Cagnolaro, L., Cozzi, B., Di Giancamillo, M., 2001. Criteri per la determinazione dell'età sulla base dello sviluppo scheletrico dell'arto toracico in *Stenella coeruleoalba* (Meyen, 1833) del Mediterraneo. Atti. Soc. Ital. Sci. Nat. Mus. Civ. Stor. Nat. Milano 90, 159–162.

Purves, P.E., Pilleri, G., 1978. The functional anatomy and general biology of *Pseudorca crassidens* (Owen) with a review of the hydrodynamics and acoustics in Cetacea. Invest. Cetacea 9, 67–227.

Rauschmann, M.A., Huggenberger, S., Kossatz, L.S., Oelschläger, H.H.A., 2006. Head morphology in perinatal dolphins: a window into phylogeny and ontogeny. J. Morphol. 267, 1295–1315.

Reidenberg, J.S., 2007. Anatomical adaptations of aquatic mammals. Anat. Rec. 290, 507–513.

Reidenberg, J.S., Laitman, J.T., 1994. Anatomy of the hyoid apparatus in Odontoceti (toothed whales): specializations of their skeleton and musculature compared with those of terrestrial mammals. Anat. Rec. 240, 598–624.

Rommel, S., 1990. Osteology of the bottlenose dolphin. In: Leatherwood, S., Reeves, R.R. (Eds.), The Bottlenose Dolphin. Academic Press, San Diego, CA, pp. 29–49.

Sanchez, J.A., Berta, A., 2010. Comparative anatomy and evolution of the odontocete forelimb. Mar. Mamm. Sci. 26, 140–160.

Schulte, H.v.W., De Forest Smith, M., 1918. The external characteristics, skeletal muscle, and peripheral nerves of *Kogia breviceps* (Blainville). Bull. Am. Museum Nat. Hist. 38, 7–72.

Sedmera, D., Misek, I., Klima, M., 1997a. On the development of cetacean extremities: I Hind limb rudimentation in the spotted dolphin (*Stenella attenuata*). Eur. J. Morphol. 35, 25–30.

Sedmera, D., Misek, I., Klima, M., 1997b. On the development of cetacean extremities: II Morphogenesis and histogenesis of the flippers in the spotted dolphin (*Stenella attenuata*). Eur. J. Morphol. 35, 117–123.

Sierra, E., Fernández, A., Espinosa de los Monteros, A., Díaz-Delgado, J., Bernaldo de Quirós, Y., García-Álvarez, N., Arbelo, M., Herráez, P., 2015. Comparative histology of muscle in free ranging cetaceans: shallow versus deep diving species. Sci. Rep. 5, 15909.

Slijper, E.J., 1936. Die cetaceen. Vergleichend-anatomisch und systematisch. Capita Zool. BD VI and VII, 1–600.

Slijper, E.J., 1939. *Pseudorca crassidens* (Owen) Ein Beitrag zur vergleichenden Anatomie der Cetaceen. Zool. Mededeel. 21, 241–366.

Slijper, E.J., 1946. Comparative biological-anatomical investigations on the vertebral column and spinal musculature of mammals. Kon. Ned. Akad. Wet., Verh. (Tweede Sectie), Deel XLII, n. 5, 1–126.

Štěrba, O., Klima, M., Schildger, B., 2000. Embryology of dolphins—staging and ageing of embryos and fetuses of some cetaceans. Adv. Anat. Embryol. Cell. Biol. 157, 1–133.

Terminologia Anatomica, 1998. Prepared by the Federative Committee on Anatomical Terminology. Thieme, Stuttgart, pp. 1–292.

Thewissen, J.G.M., Cohn, M.J., Stevens, L.S., Bajpal, S., Heyning, J., Horton, Jr, W.E., 2006. Developmental basis for hind-limb loss in dolphins and origin of the cetacean bodyplan. Proc. Natl. Acad. Sci. USA 103, 8414–8418.

Uhen, M.D., 2007. Evolution of marine mammals: back to the sea after 300 million years. Anat. Rec. 290, 514–522.

van Beneden, M.P.J., Gervais, P. (1868–1879). Ostéographie des Cétacés vivants et fossiles. Berthrand, Paris; Atlas, 64 tavv.

Viglino, M., Flores, D.A., Ercoli, M.D., Álvarez, A., 2014. Patterns of morphological variation of the vertebral column in dolphins. J. Zool. 294, 267–277.

Watson, A.G., Stein, L.E., Marshall, C., Henry, G.A., 1994. Polydactyly in a bottlenose dolphin *Tursiops truncatus*. Mar. Mamm. Sci. 10, 93–100.

Chapter 4

Diving: Breathing, Respiration, and the Circulatory System

We may imagine dolphins as continuous breath-holding divers or marathon swimmers that continuously exercise. However, just by admitting this simple comparison, we immediately come out with a number of questions that has been difficult to answer: What about the energy required to move uninterruptedly in and out of the water surface? What about the oxygen stores required to perform prolonged breath-hold dives? What about heartbeat or breathing frequency, diaphragm movements, and the effects of external pressure? Some of these complex aspects of diving physiology are still partially unsolved in dolphins, and are outside the scope of the present book [for a detailed discussion of diving physiology in marine mammals see Ponganis (2015), and Castellini and Mellish (2016)]. However, we will present the structure of the respiratory and cardiovascular systems (Fig. 4.1) with an open window on the basic physiological principles of their function in the liquid environment.

BREATHING AND SWIMMING

Breathing must take place at the surface, when the blowhole is clear of the water. The blowhole, contrarily to the nostrils of terrestrial mammals, is not fixed in an open position, but air is admitted into the upper respiratory tract only when the animal voluntarily moves (= opens) the nasal plug (Fig. 4.2). The evolutionary significance of the reversal lays in the need to seal the airways during immersion and thus avoiding inflow of water (an irritation for the mucosa, an obstacle to respiration, and a potential prelude to drowning).[a] Gas is exchanged and respiratory flow-rates can be very fast (130 and 30 L s^{-1} during expiration and inspiration, respectively) and a breath completed in less than a second (Fahlman et al., 2015). Oxygen-rich air is gulped in, carbon dioxide is breathed out, sometimes very *very* rapidly (think of a school of striped dolphins swimming close to a boat, in and out of the water in seconds or fractions of seconds), even if dolphins frequently take a single breath over multiple fast surfacing events (Piscitelli et al., 2013).

Here we may find a certain similarity to humans swimming long-distance breath-stroke competitions. A difference between humans and dolphins breathing during swimming is in that humans must exhale during the underwater phase, both through the nostrils and the mouth.[b] Swimmers who exhale underwater use their expiratory muscles to do so, and expert professional swimmers increasingly use the diaphragm to optimize their performances. Humans revert their breathing to how dolphins breathe, in that there is a pause between inspiration and exhalation. Dolphins inspire then pause and then rapidly expire and then inspire. During forceful expirations, they seem to use their respiratory muscles to generate extremely high flow-rates (Fahlman, personal communication) and an impressive tidal volume of 80–90% of total lung capacity (for discussion, see Ponganis et al., 2003). Dolphins may exhale underwater,[c] but respiration through the mouth is rather uncommon (although not impossible), since the digestive tract does not normally communicate with the airways. However, dolphins can voluntarily displace the larynx when swallowing large food items. The larynx is displaced manually by veterinarians when intubating the animal for anaesthesia (Dold and Ridgway, 2007). We may also note that as a consequence dolphins possibly use their diaphragm and respiratory muscles to increase exhalation flow-rates (Fahlman et al., 2015; Cotten et al., 2008).

Things get more complicated if we consider that dolphins do not swim only at the surface, but dive to various depths to forage. Human breath-holding divers must rest at the surface for a certain amount of time between dives (as a general rule, 3–4 times the duration of the former dive) to avoid risks of decompression-sickness, recover from fatigue, allow full oxygenation of blood, and remove carbon dioxide from the blood and tissues. Cetaceans that dive to considerable depths (ie, sperm whales and beaked whales) may stop to rest at the surface for variable intervals, a feature that does not apply to

---

a. Stranded but still alive animals may show water or sand inside the blowhole, a clear indication of an agonic state.

b. Humans exhale underwater to maximize inspiration when the mouth (and the nostrils) come above the surface, but also because the outgoing bubbles underwater prevent water from coming in.

c. For example, to emit underwater and bubbles, for communication and fun.

Anatomy of Dolphins. http://dx.doi.org/10.1016/B978-0-12-407229-9.00004-X

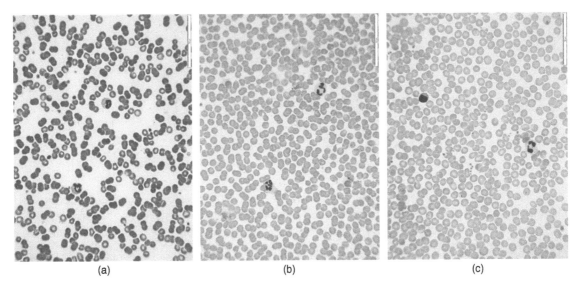

(a)	(b)	(c)

FIGURE 4.1 **Blood of *T. truncatus*, *Delphinus delphis*, and *Homo sapiens* (all scale bars = 50 μm).**

FIGURE 4.2 *Stenella coeruleoalba* **breathes at the surface.** *(Courtesy of Dr. Michela Podestà, Museum of Natural History of Milan)*

dolphins that generally do not exceed depths of 30–50 m[d] (Table 4.1). However, the surface interval of deep-diving cetaceans is not long enough by human standards, and—furthermore—is generally intermingled with other dives of intermediate depth that would be fatal for us, or simply impossible to sustain. Bottlenose dolphins have dive records of several hundred meters; Risso's dolphins and pilot whales may hunt in deep waters; killer whales have been recorded at −264 m (see Table 4.2). The question of the duration of the surface interval after long dives in these species is still under investigation.

The respiratory act—and consequently gas exchanges—must be fast and thorough to allow full oxygenation of blood at the air–blood barrier in the pulmonary alveoli. This kind of high flow "explosive ventilation" (Kooyman, 1973; Fahlman et al., 2015) allows for 75–90% of total lung volume exchange in 0.3 s, compared with 27% in 0.5 s of the galloping horse (Piscitelli et al., 2013). In dolphins, the larynx is bent at almost 90 degree to allow communication between the blowhole at the top of the head and the almost horizontal trachea, and is completely separated from the surrounding pharynx to avoid any potential leakage of water from food in transit in the alimentary tract into the respiratory system. The trachea in turn is short and very large in diameter, both characteristics that increase capacity and speed of the airflow. The mechanic of gas exchange requires a powerful effort from the respiratory muscles, including the diaphragm, whose cranial convexity is very pronounced in dolphins (see the position of the heart for reference).

The finer ramifications of the bronchial tree possess sphincters able to "shut" the respiratory bronchioles out, and thus possibly prevent reflux of the expired air into them. This is a wonderful solution to several possible inconveniences of deep dives.

---

d. Bottlenose dolphins may increase their resting/recovery time at the surface after dives longer than 2–5 min (Ridgway et al., 1969). Deep-diving species may have different and complex antioxidative mechanisms (Cantú-Medellín et al., 2011).

TABLE 4.1 Breathing Frequency at the Surface During Resting for Selected Dolphin Species

Species	Breath (min)	Observations
Tursiops truncatus	2.3[a]	Slow swimming
	5.9–12.5[b]	Heavy exercise
	2.14–2.83[c]	Complex experimental conditions
Globicephala macrorhynchus	2.5[a]	
Orcinus orca	0.9 (0.8–1.1)[a]	Resting, at the water surface

[a]Mortola, J.P., Limoges, M.-J., 2006. Resting breathing frequency in aquatic mammals: a comparative analysis with terrestrial species. Resp. Physiol. Neurobiol. 154, 500–514.
[b]Williams et al., 1993. The physiology of bottlenose dolphins (Tursiops truncatus): heart rate, metabolic rate and plasma lactate concentration during exercise. J. Expl. Biol. 179, 31–46.
[c]Karandeeva et al., 1973. Features of external respiration in the Delphinidae. In: Chapskii, K.K., Sokolov, V.E. (Eds.), Morphology and Ecology of Marine Mammals. Halsted Press; John Wiley & Sons, New York, pp. 196–212, Israel Program for Scientific Translation (Jerusalem).

TABLE 4.2 Dive Limits of Selected Dolphin Species

Species	References	Max. Diving Depth (m)	(Max.) Duration of Deep Dives
Globicephala melas	Baird et al. (2002), Ponganis et al. (2003)	−648	12 min 7 s to 15 min
G. macrorhynchus	Aguilar Soto et al. (2008)	−1019	21 min
O. orca	Baird et al. (2005)	−264	Unknown
	O'Malley Miller et al. (2010)	−254	8.5 min
T. truncatus	Ponganis et al. (2003); Ponganis (2011)	−390	8 min
Tursiops gilli		−535	
Stenella attenuata		−203	5 min
D. delphis	Ponganis et al. (2003)	−260	5 min

AN IMAGINARY DIVING SEQUENCE

Let's imagine a potential sequence of a moderately deep dive in a medium-sized dolphin (for a comparison between man and other diving mammals, see Fahlman and Schagatay, 2014). The dolphin approaches the surface, opens the nasal plug, and rapidly discharges carbon-dioxide-filled air (the diaphragm pushes forward, costal muscles bring the thorax slightly forward and turn the ribs slightly inward), immediately followed by inspiration of air (the diaphragm diminishes its convexity, the respiratory muscles enlarge the thorax by bringing the ribs backward and rotating them outside, with the general outcome of enlarging the thorax and creating a negative pressure in the pleural cavity). The nasal plug shuts, the dolphin bends its back, moves the flukes, and dives.

On the way down (Fig. 4.3), the mammalian reflex induced by the external temperature of the water maximizes peripheral vasoconstriction, potentially increasing blood volume in the thorax. The very act of diving with breath-holding and consequent stimulation of the upper respiratory receptors brings other consequences (cessation of breathing, bradycardia, lung collapse, hypoxia, hypercapnia, baroreceptor responses) that reinforce the reflex (Scholander, 1940; Ponganis et al., 2003; Terasawa et al., 2010). As the dolphin continues its descent, external hydrostatic pressure increases (one bar each 10 m[e]), thus doubling surface pressure at −10 m and increasing it fourfold at −30 m, while gas within the alveoli is compressed following Boyle's law.[f] Any diving mammal will respond to this chain of physiological events by maintaining blood flow to the heart and brain, which are the organs most intolerant to low oxygen levels (Elsner, 1999) and keep a

---

e. To simplify the discussion here, we will not make any difference between salt water (sw) and freshwater (fw), and will also use the bar unit instead of the technical Pascal or the common lay terms atmosphere (atm) or pressure/square inch (psi) (1 bar = 100 kPa = 0.986923 atm = 14.503 psi).
f. Boyle's law states that the pressure of a gas tends to decrease as the volume of a gas increases if the temperature and amount of gas remain unchanged within a closed system.

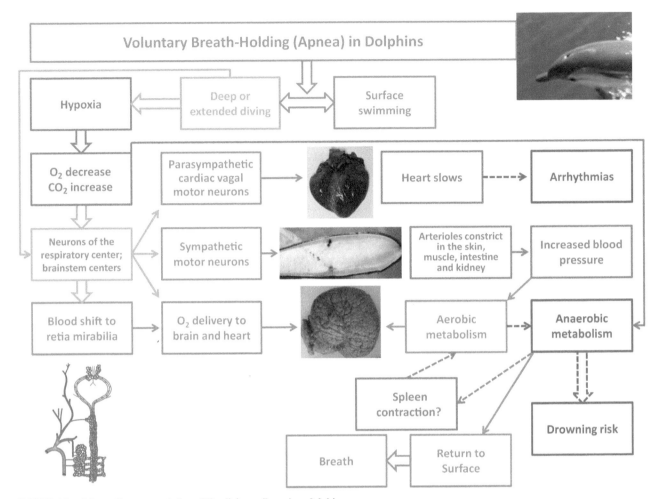

FIGURE 4.3   Schematic representation of the diving reflexes in a dolphin.

stable body temperature. After a certain period immersed in water colder than 37°C, the majority of humans will eventually feel the need to urinate, prompted by peripheral vasoconstriction and increased blood pressure. This latter effect is partially incompatible with the metabolism of the average dolphin of our example. Why so? Water is precious for the metabolism and cannot be taken directly from the surrounding sea, since the energy cost of processing salt water is superior to its metabolic value (see Chapter 9). Cetaceans extract water from their food and use specific physiological strategies to save it, with consequent reduced urinary output.

During dives, the external pressure increases and the flexible thorax recoils inward[g] to very low volumes, while the lungs gradually reduce their volume until collapse at a depth of approximately −70 m (Ridgway and Howard, 1979), or perhaps even deeper, according to CT scan experiments (Moore et al., 2011). During this phase, it is possible that part of the blood shifts into the capacious *retia mirabilia*[h] placed under the thoracic vertebral column and around the whole spinal cord. In fact, the lungs of diving mammals (including human breath-holding divers) gradually reduce their volume because of gas compression as the external pressure increases during breath-holding descent (for synthesis, see Costa, 2007). As gas is compressed and the lung volume also diminishes, the parenchyma of the lungs becomes similar to a wet squeezed sponge, in which the remaining liquid is the blood that still permeates the organs. Air is expelled from the respiratory alveoli and the finer bronchioles toward the larger bronchioles, the bronchi, and the trachea. In dolphins, return of the efflux air to the alveolar district is prevented by shutting of the bronchiolar sphincters (see the description

g. The mechanism is different in humans because our stiff chest causes negative intrathoracic pressure, which "suctions" blood into the chest. In pinnipeds, and possibly cetaceans, this negative pressure is most likely absent (Andreas Fahlman, personal communication).

h. *Rete mirabile* (pl. *retia mirabilia*) is a Latin term that means "extraordinary, wonderful net." The *retia* became the object of a famous dispute in early modern anatomy, when Vesalius (1514–64) in his *De humani corporis fabrica* (1543) proved that they are proper of the arterial supply to the brain in ruminants, but do not exist in man, contrarily to the dogma asserted by Galen (AD 131–201). For description of the *retia*, see the following.

of the lung parenchyma later in the chapter). The external pressure acts on the thoracic walls, the junctions between the ribs and their bony extensions that reach the sternum cave in, and the whole thorax recoils. The only remaining dead spaces filled with air are those of the relatively rigid airways, including the larynx, trachea, and larger bronchi, where no gas exchange takes place. Cetaceans may use this reservoir of compressed air to produce sound to locate their prey, communicate, or imply vocalize by pushing the gas into the nasal chambers. The heartbeat and the peripheral blood flow slow down (Ponganis et al., 2003), while whether the diaphragm stands still or contracts it's a matter of debate. Oxygen supply to the brain is supported also by the *retia mirabilia* (Ponganis et al., 2003), although not all the passages in this complex physiology are clear and avoidance of anaerobiosis (insufficient oxygenation of neurons) during extended dives remains somewhat difficult to explain.[i]

FOOD AS FUEL AND THE NEED FOR CONTINUOUS REFURBISHING

If we return for a moment to the initial comparison of dolphin swimming and human marathon swim athletes, we may wonder at the energy consumption required by the whole process (Williams et al., 1992; for review, see Costa and Williams, 1999; and Williams et al., 2001). Diving requires complex strategies to save energy and work on buoyancy at depth (Skrovan et al., 1999; Williams et al., 1999a,b; Fahlman et al., 2016). The limit in human performances is given by training, quality of muscle fibers, resistance, nutritional programs, and ultimately by genetics. However, human limits set the maximum duration of continuous swimming to several hours, perhaps days under special circumstances. But dolphins swim every second of their life, to move, to hunt, or simply to avoid sinking. They act as a car perennially driving along a never-ending highway: It needs gas stations. In the case of dolphins, fuel is represented by their prey and individual and pack strategies of foraging must take into account the balance between effort and gain, that is, the metabolic cost of the chase and quantity of energy the food brings. Although part of this topic will be discussed in Chapter 8, here we note that stressed, sick, or wounded animals may find the balance of *swim to hunt to swim* difficult to maintain. Hunger and weakness may be key cofactors in the stranding of sick or old individuals.

ANATOMY OF THE RESPIRATORY SYSTEM

Blowhole and Upper Airways

As we noted previously, the external nares of cetaceans have migrated to the top of the head forming an opening called the blowhole (Fig. 4.4), which is single in toothed whales and bipartite in baleen whales. In fact, the nasal passages are paired in all cetaceans, but those of dolphins fuse into a single channel on the way to the dorsal crescent-shaped opening (for an excellent review on the morphology, evolution, and functional significance of the cetacean nose, see Berta et al., 2014).

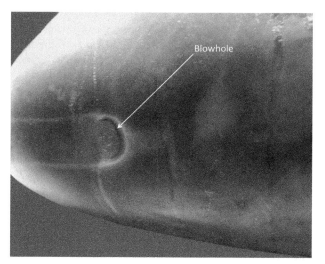

FIGURE 4.4   **Blowhole of *T. truncatus*.**

i. The aerobic dive limit is the diving duration beyond which blood lactate levels rise over resting levels, or in other words the diving duration beyond which there is a net increase in lactate production (Kooyman, 1985).

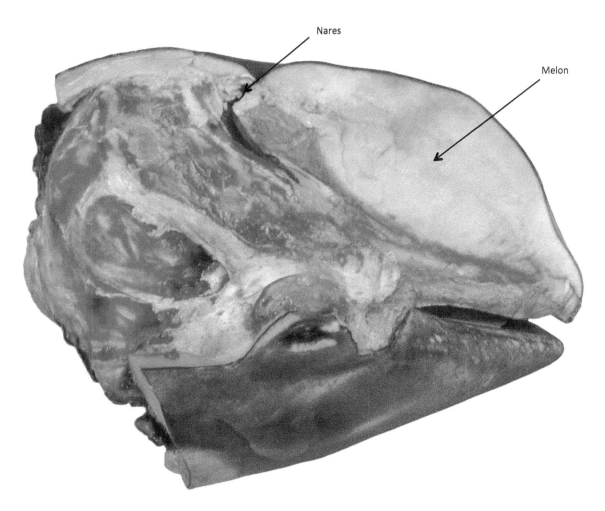

FIGURE 4.5 **Nares (and melon) of *Grampus griseus*.**

The nasal cavity is strikingly different from the human, and one must also remember that dolphins have virtually no sense of smell,[j] so there is no need for special regions of the nasal passages to carry the olfactory mucosa and its specific neurons. After the external opening of the blowhole, the dolphin nasal passageways (Fig. 4.5) open ventrally into an unpaired chamber, the vestibulum. To open the blowhole, the anterior lip (valve) of the blowhole together with the anterior wall of the vestibulum are shifted rostrally by superficial (lateral) parts of the blowhole muscles (*m. maxillonasolabialis*). This mechanism expands the vestibulum that becomes a relatively large chamber.

Below the vestibulum, the nasal passage is separated by a short soft nasal septum. At this dorsal end of the paired nasal passage there is an additional valve structure called the *monkey lips* (phonic lips), which are—together with areas of fat endowed with special acoustic properties such as the *melon*—essential for sound production (see Chapter 5). Below (exactly ventral to it) the paired nasal passage and the external nares open and close by means of a dense mass of connective tissue and muscle called the nasal plugs. The nasal plugs bulge from anterior into the lumen of each nasal passage so that the passages are only two flat transverse slits if closed. During air flow the nasal plug muscle, however, pulls the nasal plugs rostrally so that the nasal passages open as two large tubes for rapid air exchange.

The paired nasal passages (see also Chapter 5) are additionally equipped with three pairs of air sacs (Fig. 4.6) opening into the passage by slit-like horizontal openings (Cranford et al., 1996). The soft-walled nasal sacs are called:

- vestibular sacs (proximal to the blowhole);
- nasofrontal sacs, which circle the nasal passage in the horizontal plane; and
- premaxillary sacs (underneath the melon and over the incisive[k] bones).

---

j. At least in the way we conceive it, with an olfactory mucosa in the nasal cavity, olfactory nerves, and especially with olfactory bulbs, tracts and cortex in the central nervous system. For a discussion of the terminal nerve, see Chapter 5.

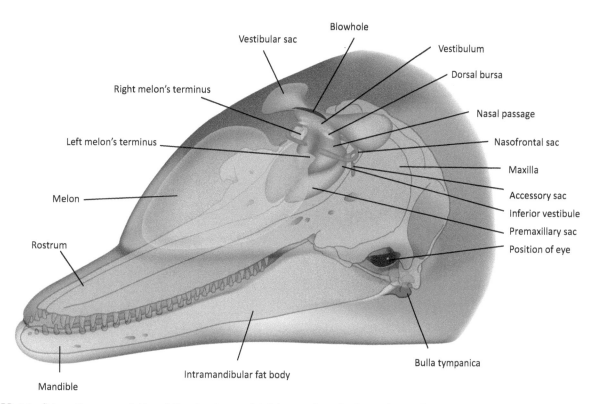

FIGURE 4.6 Schematic representation of the structures related to sound production and reception in the head of a bottlenose dolphin. *(Drawing by Uko Gorter)*

Most of the nasal sacs have no bony constraint so that they can expand depending on the quantity of air. We also note that cetaceans have no paranasal bony sinuses lined by respiratory mucosa as terrestrial mammals. Such structures would be incompatible with intensive and continuous pressure diving. However, recent evidence suggests that that fat deposits in the head and accessory air sinuses may contain extensive venous plexuses sufficient to present potential risks of diving-related trauma and pathologies (Costidis and Rommel, 2012).

Larynx

The position of the cetacean larynx is quite rostral (Figs. 4.7 and 4.8) if compared to terrestrial mammals, and corresponds to the presphenoid bone, thus rostral to the projection of the cervical vertebrae (Reidenberg and Laitman, 1987). The relatively rostral position and consequent topographical relationships are due to its shape and orientation (see the following), but also to the virtual absence of a neck in dolphins.

The larynx is composed by the epiglottal and the arytenoid cartilages that elongate to form a tube, the so-called "beak" that projects rostrally and dorsally from the pharynx floor (Figs. 4.9 and 4.10). Its proximal end rests within the internal nares. This beak-like tube has been described in several species and it has been called the "aryteno-epiglottideal" tube (see Chapter 5). This aryteno-epiglottideal tube ends dorsally with a lip so that a strong sling of the palatopharyngeus muscle connects tightly (waterproof) the laryngeal beak with the choanae.

Two additional "nasal" sacs, named pterygoid and laryngeal, are placed close to the pharynx and at the base of the larynx, respectively (Reidenberg and Laitman, 2008). The pterygoid sacs are extension of the pharyngotympanic tube and perhaps bear some evolutionary trait common to the guttural pouches of *Equidae*.[l] The laryngeal sac is indeed a complex of small diverticular blind-end sacs (Reidenberg and Laitman, 2008).

---

k. The incisive bone (*os incisivum*) of veterinary (NAV, 2012) and human nomenclature (Federative Committee on Anatomical Terminology, 1998) is often referred to as premaxillary bone in cetacean osteology. (See also Chapter 3, footnote b).

l. The equine guttural pouches are dilation of the *tuba auditiva* and are perhaps functional to cool the blood directed to the brain. (See also Chapter 3, footnote f).

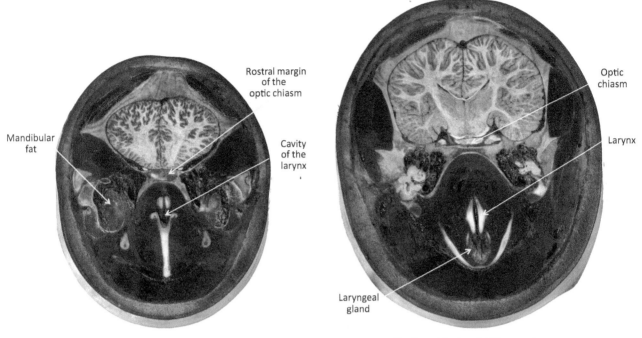

Rostral margin
of the
optic chiasm

Mandibular
fat

Cavity
of the
larynx

Optic
chiasm

Larynx

Laryngeal
gland

FIGURE 4.7   Section of the head of *T. truncatus*.

FIGURE 4.8   Section of the head of *T. truncatus*.

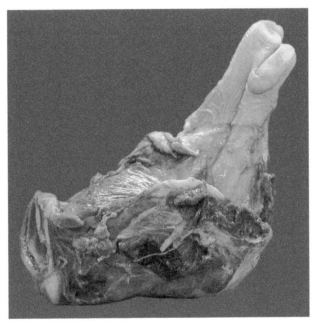

FIGURE 4.9   Right lateral view of the larynx of *T. truncatus*.

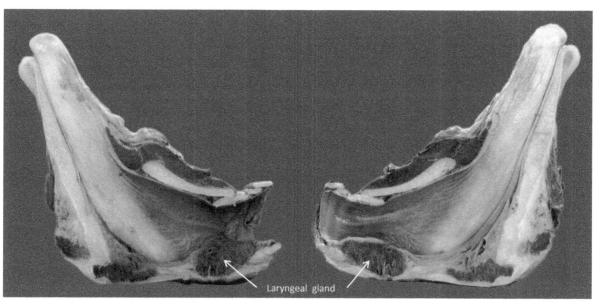

FIGURE 4.10   **Sagittal section of the larynx of *T. truncatus*.**

In the larynx, a few mucous glands can be found between the squamosal (keratinized) epithelium and the cartilage. At the base of the larynx, overlying the cricoid cartilage, there is a laryngeal gland of lympho-epithelial nature (Fig. 4.11), with several folds and crypts (Smith et al., 1999). This latter structure may be analogous of the nasopharyngeal adenoid complex of terrestrial mammals.

Trachea and Bronchi

The larynx is connected to the lungs through the trachea. The cetacean trachea is rather short, only a few cm in dolphins, but relatively large in diameter.

Once more, as elsewhere in the respiratory apparatus, the anatomical differences with the trachea of terrestrial mammals are striking. The trachea of land mammals is made up by a series of dorsally open cartilaginous rings, whose free extremities are connected by the *m. trachealis*. Consecutive rings are joined by connective tissues and the internal surface of the organ is lined by a typical respiratory mucosa with ciliated cells and sparse mucosal cells and tubule-alveolar glands. This basic architecture ensures that the trachea maintains a constantly pervious lumen and at the same time may adapt to the movements of the neck. Since cetaceans have no neck, there is no need for a mobile structure and hence the *m. trachealis* is not present. The heavy cartilaginous rings are fully circular and irregularly anastomose with the next ones, and the interposed connective tissue fills the gaps (Fig. 4.12). The mucosa is made up by the same elements as in terrestrial mammals, with a columnar ciliated epithelium and mucosal glands in the submucosa that open into the lumen, elastic fibers, and infiltrating lymphocytes.[m]

However, the most characteristic feature is the presence of large venous *lacunae* in the mucosa lining the inner surface (Cozzi et al., 2005) (Figs. 4.13–4.15).[n] The presence of large venous spaces in the lumen of the trachea is somewhat puzzling, considering that this is—together with the larynx and the initial ramification of the bronchi—the only relatively rigid dead space in the body of the dolphin. When the external pressure increases during descent, the air is probably pushed out of the lungs into the larger bronchi and trachea. The rising external pressure compresses the thorax and deforms (= flattens) the trachea (for dynamic models of tracheal compression, see Cozzi et al., 2005; Bostrom et al., 2008; Bagnoli et al., 2011; Moore et al., 2014), thus further increasing internal air pressure. During this phase, constriction of the palate-pharyngeal muscles and movements of the jaws increase pressure in the internal nasal air sacs (see Piscitelli et al., 2013, for a discussion), and manage sound emission. Tracheal compression[o] may also be functional to delay

---

m. A single report (Harrison and Fanning, 1973–74) describes the presence of goblet cells in the mucosa of the trachea of the bottlenose dolphin.

n. The venous *lacunae* were already noted by Fiebiger (1916) in a dolphin.

o. For the dramatic effects of human tracheal compression at (simulated) depth, see Lindholm and Nyrén (2005).

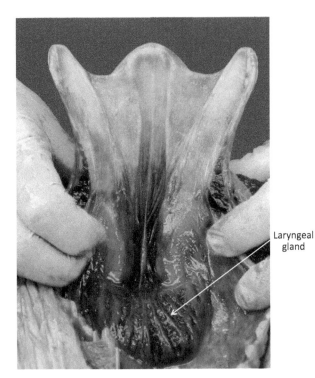

FIGURE 4.11 **Laryngeal gland of *T. truncatus*.**

alveolar collapse[p] during deep dives (Bostrom et al., 2008). When the condition is reversed during ascent, the existence of large venous lacunae may constitute a risk because expanding air may enter the blood stream. Biomechanical hypothesis (Bagnoli et al., 2011) indicate that for the trachea to resist environmental pressure and be able to return to its original shape and diameter during ascent, a certain air pressure is needed. Experimental procedures with the trachea of goats (whose structure is very similar to the human) suggest that an open-ring structure with presence of a *m. trachealis* is heavily bent in a pressure environment. Vascular *lacunae* may contribute to the restoration of the tracheal lumen during ascent by filling with blood and helping the organ to snap back to its original shape. It is also possible that the distension of the lacunae with blood reduces the lumen and therefore curtails the volume of gas in the trachea.[q] In fact, there are indications that the inner lining of the dolphin mucosa may possess the characteristics of a erectile mucosa with specific innervation and visceral ganglia (Cozzi et al., 2005).

The trachea ends with the main bronchi. As in terrestrial Cetartiodactyla, the trachea of dolphins emits a characteristic independent *bronchus trachealis* on the right side[r] before the terminal bifurcation into the two main bronchi with the typical crest in the lumen (*carina tracheae*).

The structure of the larger bronchi is the same as that of the trachea, although the lumen gradually diminishes (see the description of the respiratory structures of the finer ramifications later in this chapter).

Lungs

The lungs of Cetacea show no gross subdivision into lobes and no external segmentation (Figs. 4.16 and 4.17), except for a deep incisura of the apex, sometimes more evident in the right lung. In the thorax their base follows the pronounced curvature of the diaphragm and with a dorsal approach the lungs appear quite narrow and symmetrical (Fig. 4.18). Their costal side adapts to the shape of the ribs, and the ventrolateral extension partially covers the heart.

---

p. For a general discussion of the phenomenon of lung compression and alveolar collapse in human breath-holding divers, see Fitz-Clarke (2007a,b, 2009).
q. However, a decreased volume brings also an increased gas pressure. In this case, there should be a balance between changes in the lumen of the trachea and gas expansion during ascent, to avoid potential risks of decompression sickness given the quantity of blood in the lacunae.
r. In humans, the first bronchial ramification on the right side is called *superior lobar bronchus* (TA, 1998). The vascular relationships are the same, as this bronchus appears rostral and dorsal to the right pulmonary artery. The main difference with Cetartiodactyla is that in man the right *superior lobar bronchus* arises from the right main bronchus within the parenchyma of the lungs.

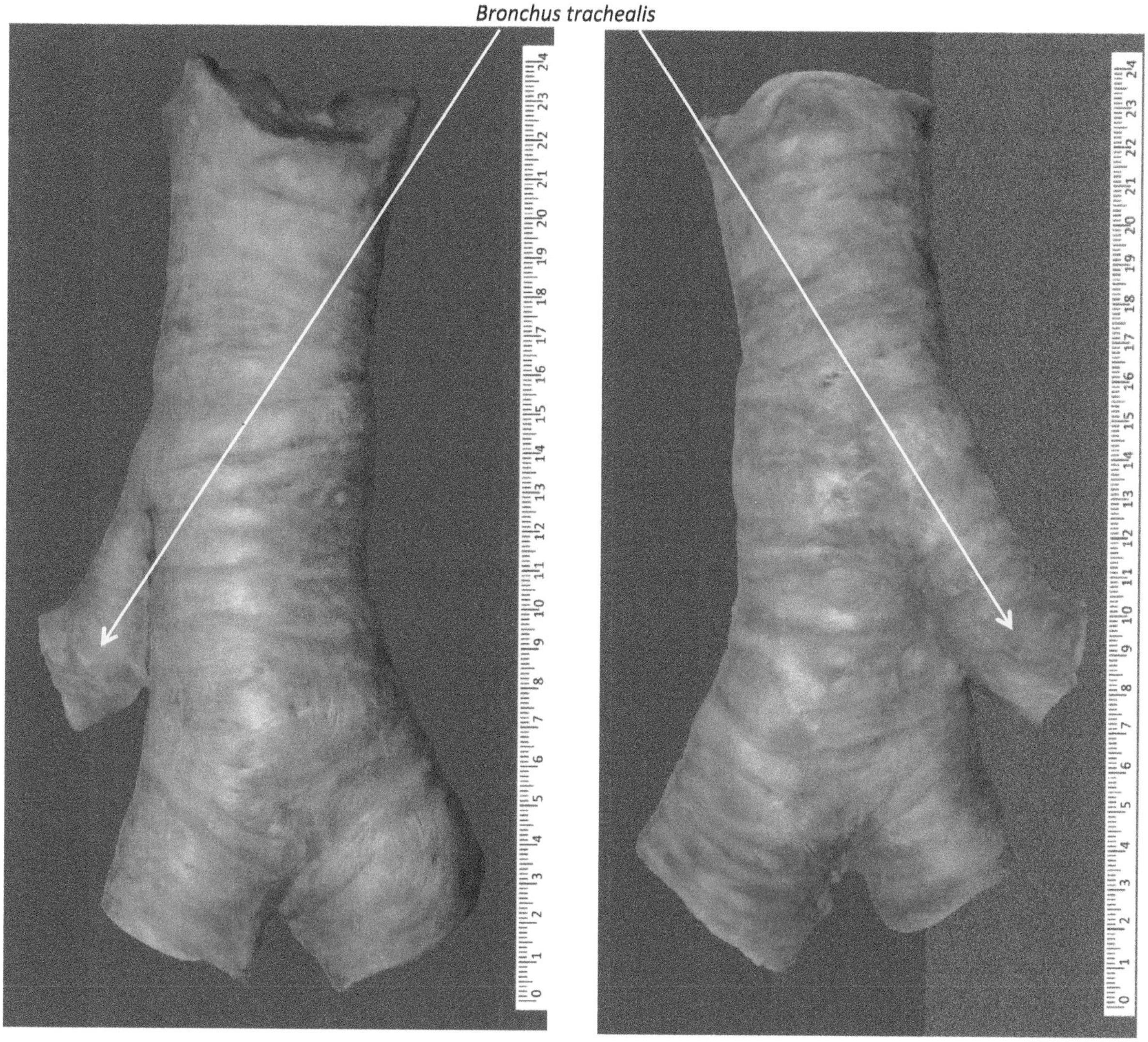

Bronchus trachealis

FIGURE 4.12   Ventral (left) and dorsal (right) view of the trachea of *T. truncatus*.

Cetaceans that normally have short-duration dives (including dolphins and porpoises) have a greater lung mass (2.7%) than terrestrial mammals (1%) (Piscitelli et al., 2013). They dive after inspiration (Ridgway, 1986, cited in Ponganis et al., 2003) and possibly their lungs act as a storage organ for oxygen at least in shallow water swimming. The intrathoracic portion of the abdominal cavity makes up only 25% of the thorax volume in the bottlenose dolphin, as opposed to 60% in the deep-diving dwarf sperm whale (Piscitelli et al., 2010): This means relatively larger body space for respiratory organs. Adaptations in the oxygen affinity of blood parallel the modifications of lung volume (see Table 4.3 for data on lung capacity), with low affinities in species where the lungs seem not to act as an oxygen store (ie, whales) and high affinities where the lungs do represent such a store. In the latter case the uptake of oxygen from the alveolar space is therefore maximized. In synthesis, cetaceans with typical long-duration dives have a small lung volume with a consequent lung collapse during the dive.[s]

Whole lung resonant frequencies of the submerged bottlenose dolphin were estimated in vivo using acoustic backscatter techniques (Finneran, 2003).

---

s. According to Ponganis et al. (2003), the respiratory (= mostly the lungs) compartment of deep divers decreases in size possibly also because less dependence on lung O_2 uptake during dives minimizes the risks of decompression sickness and nitrogen narcosis. In deep-diving marine mammals, 80–90% of body O_2 is stored in the blood and muscles: consequently a large body mass will be an advantage for deep dives (Pabst et al., 1999; Ponganis et al., 2003). For a discussion of the mechanisms that prevent decompression sickness in cetaceans, see also Fahlman et al. (2006), and Ponganis (2015).

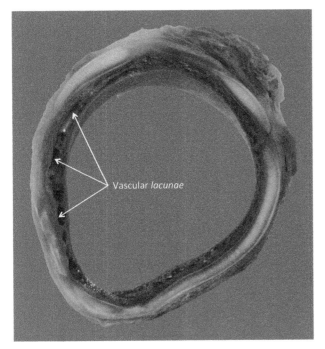

FIGURE 4.13 **Transversal section of the trachea of *T. truncatus*.**

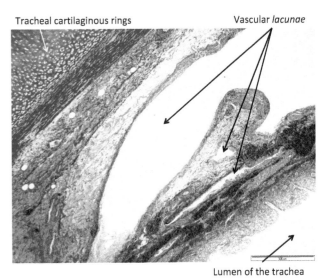

FIGURE 4.14 **Detail of structure of the vascular *lacunae* in the trachea of *T. truncatus* (scale bar = 500 μm).**

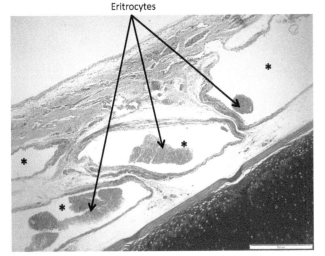

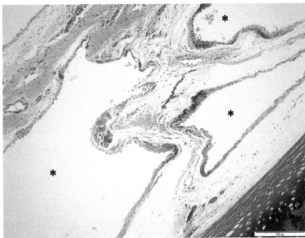

FIGURE 4.15 **Other details of the vascular *lacunae* (scale bars: left = 500 μm; right = 200 μm).** Asterisks are positioned in the lumen of the *lacunae*.

The architecture of the dolphin lungs (Ito et al., 1967; Fanning and Harrison, 1974) only partially follows the basic plan of the organ as it is known in man and land mammals, in which larger bronchi branch into smaller and smaller subdivisions with thinner cartilage. A detailed morphometric analysis of the bronchial tree of the bottlenose dolphin and other cetaceans, inclusive of the diameters of the subsequent divisions, is contained in Drabek and Kooyman (1986). Terminal and respiratory bronchioles of land mammals have no more cartilage and lead into the alveolar sacs and respiratory alveoli surrounded by thin blood vessels where gas exchanges take place. The respiratory tract (from nose to smaller bronchi) is built up to convey air in and out of the body, always maintaining an open lumen (hence, the cartilage), until the terminal parts in which the alveolar membrane that separates the gas from the blood vessels is so thin that a difference in gas partial pressure, with the aid of the surfactant factor, facilitates passage of oxygen and carbon dioxide in and out of solution.

However, this general outline is significantly different in the terminal part of the bronchial tree system of dolphins. A unique feature is the presence of myoelastic sphincters in the submucosa of bronchioles of less than 2 mm in diameter (Fig. 4.19). These sphincters are surrounded by bidirectional elastic fibers (Piscitelli et al., 2013) and act together with

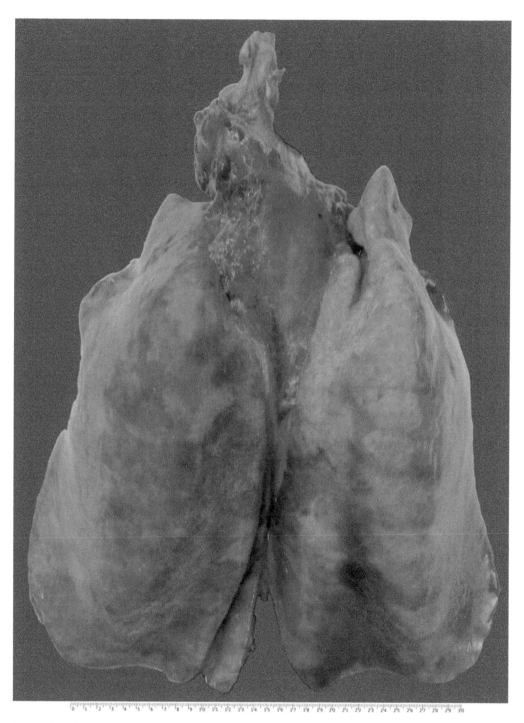

FIGURE 4.16   **Dorsal view of the lungs of a lactating *T. truncatus*.**

patches of cartilage that persist even in relatively smaller airways and bronchioles, possibly to exclude the alveolar sac complexes during the phase of lung collapse (an hypothesis first presented by Goudappel and Slijper, 1958). In smaller bronchioles, the epithelium becomes thinner and made up by cuboidal or low columnar cells and nonciliated Clara cells. In the alveoli the squamous epithelium contains type I (for gas exchange) and type II (for production of surfactant[t]) pneumocytes. Apparently brush cells or type III pneumocytes are lacking (Fanning and Harrison, 1974). Several lectins have been

---

t. The surfactant of marine mammals may have a positive effect in post-dive reinflation of the lungs (Costa, 2007).

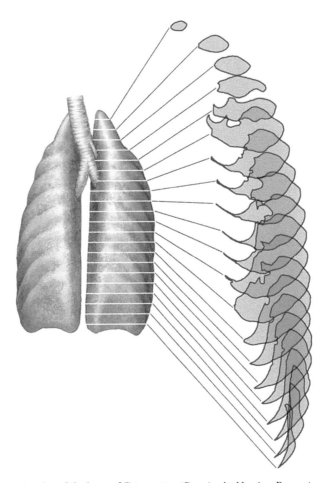

FIGURE 4.17 **Dorsal view of a reconstruction of the lungs of *T. truncatus*.** *(Drawing by Massimo Demma)*

described in the respiratory epithelium (Shimokawa et al., 2011). Lymphocytes infiltrate the epithelium everywhere except in the terminal alveolar subdivisions.

We have to keep in mind that during descent in deep dives the lungs are compressed and tend to collapse, while air is pushed into the trachea and upper divisions of the respiratory tract. Bronchiolar myoelastic sphincters, together with the persisting cartilage, may be functional in avoiding air reflux into the alveoli and thus further curtail the risk of inducing solution of gas into the blood under condition of high external pressure. On the other hand, they could shut and trap air inside the alveoli during descent, thus ensuring a minimal functional gas exchange during a dive, although at the risk of inducing nitrogen to enter the bloodstream under pressure (for discussion of the contrasting hypotheses, see Pabst et al., 1999). An additional explanation for the presence of myoelastic sphincters is that they contribute to let the alveoli collapse first, followed by the terminal airways (Costa, 2007). Basically the lungs stay collapsed also because the sphincters hinder them to inflate again, and in addition maintain air pressure within the upper respiratory tract where it is needed for sound emission. The relationship between anatomy of the cetacean lung and diving is still incompletely understood: For a thorough review of the structure, biomechanics, and diving physiology of the cetacean lungs and pleura, see Piscitelli et al. (2013).

The intense vascular network of the lung parenchyma is necessary to ensure transportation of respiratory gas in solution and the infiltration of lymphocytes guarantees immune protection to the organism. However, the high degree of vascularization is also one of the reasons why respiratory diseases are so common in marine mammals, in which repeated periods of apnea may induce proliferation of specific pathogens. The rich blood supply may also be a factor in tumor spreading, although this appears to be a relatively uncommon finding at least in stranded dolphins.

Development of the Respiratory System

Developmental phases (stages) of selected dolphin species have been reported in Table 9.6 (see Chapter 9 for details). According to Štěrba et al. (2000), during Stage 3 the laryngo-tracheal anlage branches from the ventral wall of the primitive

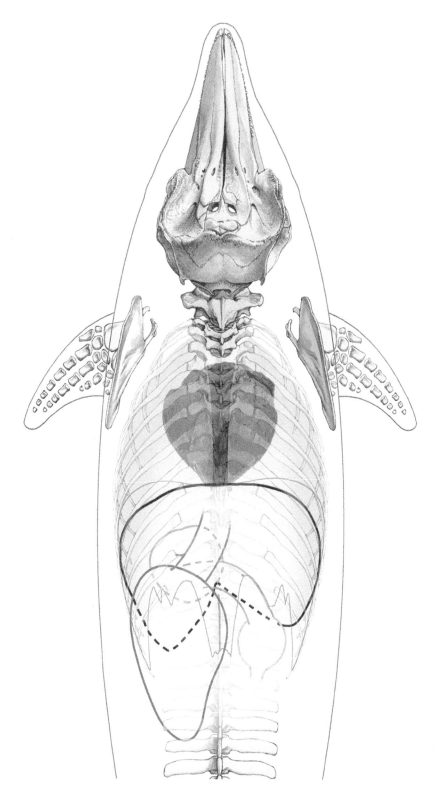

FIGURE 4.18   Position of the lungs (shaded in blue) of *T. truncatus*. *(Drawing by Massimo Demma)*

TABLE 4.3 Data on Lung Capacity of Selected Dolphin Species

Species	Body Mass (kg)	Age Class	Lung Mass	Minimum Air Volume of the Relaxed Lung (mL)	% of Estimated Total Lung Capacity
D. delphis	57–106	Immature to adult	1200–2859	302–1420	16
Lasgenorhynchus acutus	69–154	Immature to adult	2809–5850		
G. melas	305	Juvenile	7857–9930		
Grampus griseus	74	>1 year	2022	332	4

From Fahlman et al., 2011. Static inflation and deflation pressure–volume curves from excised lungs of marine mammals. J. Expl. Biol. 214, 3822–3828.

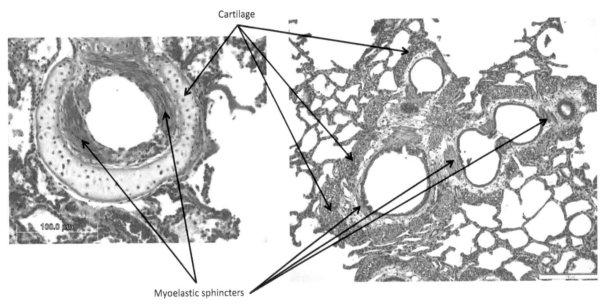

FIGURE 4.19 Myoelastic sphincters in the submucosa of bronchioles of the lungs in *S. coeruleoalba* (left) and *T. truncatus* (right) (scale bars: left = 100 μm; right = 200 μm).

esophagus and the primitive bronchi sprout as T-like branches that gradually become lobar bronchi in Stage 4. In Stage 5, the primitive lungs start their organization, when lobal bronchi divide into segmental bronchi and increase their branching. During Stage 6, the tracheal rings begin their formation, while the typical Odontocete larynx appears as such during Stage 7, and the presence of cartilage in the trachea and bronchi increases. In the following stages, the organs grow, but their organization is already similar to that of the adult.

CIRCULATORY SYSTEM

The general plan of the circulatory system of marine mammals conforms to the requirements of diving metabolism. Diving, be it to shallow waters 10–30 m deep, or to depths below 100 m, imposes changes in the outline of the circulation, and even in the vascular architecture. When the dolphin dives at increasing depths, the temperature of the surrounding environment quickly drops, pressure increases, and carbon dioxide levels rise because of extended breath holding.

The mechanisms of temperature homeostasis require several adaptations to life in the water, including the presence of blubber, and so forth. The organization of peripheral blood vessels contributes to these mechanisms and to prevent heat loss. Arterial circulation is limited in the outer parts of the body, and, moreover, every peripheral artery is surrounded by a series of small veins, useful for countercurrent heat exchange.

During the dive, flow of blood is thought to be maintained to the thorax, brain, and other vital organs, and into the huge reservoir constituted by the *retia mirabilia* (see the anatomical description later in this chapter). At the same time, when the relatively elastic ribs partially give in to external pressure, blood is thus probably diverted from the thoracic vessels into the

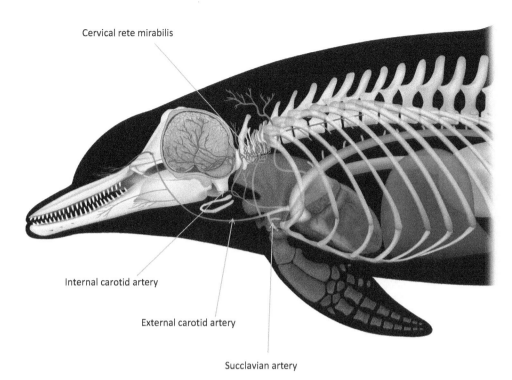

Cervical rete mirabilis

Internal carotid artery

External carotid artery

Succlavian artery

FIGURE 4.20   **General arterial plan of the anterior part of the body of *T. truncatus*.** *(Drawing by Uko Gorter)*

communicating *retia*[u] (Hui, 1975). As liquids do not compress, the presence of a sufficient quantity of blood in the thorax prevents the rib cage from crushing because of the pressure, and avoids irreversible lung collapse.[v]

Many factors contribute to the adaptation of the circulatory system of marine mammals to the marine environment. Some of these factors relate to macroscopic characteristics of the animals (blood volume of marine mammals often exceeds 12% of body weight, whereas it is 7% in man and a little more in other terrestrial mammals), or to physiological changes (such as the well-known extreme bradycardia[w]) and/or biochemical (such as the high oxygen affinity of hemoglobin and myoglobin) adaptations (Noren and Williams, 2000). Oxygen storage in the muscles of marine mammals may reach one order of magnitude more than in terrestrial species (for review, see Elsner, 1999). The slowing of the heartbeat or bradycardia, together with the general vasoconstriction, circulatory isolation of muscles, and even eventual ischemia of central organs, are all physiological mechanisms that delay depletions of O_2 stores during diving (Ponganis et al., 2003).

General Outline of the Arterial System

The general arterial plan of the dolphin body (Fig. 4.20) starts with the aorta coming out of the left ventricle and splitting into branches that will eventually give rise to the vessels of the head, thoracic girdle, thorax, abdomen, and flukes.

The pulmonary circulation includes the pulmonary artery, carrying deoxygenated blood from the right ventricle to the lungs, and the pulmonary veins, carrying oxygenated blood from the lungs to the left atrium.

The outer parts of the body, enveloped in blubber, mostly contain tiny capillaries.[x]

u. Although this hypothesis answers several questions, there is no actual proof that blood is pooled into the *retia mirabilia* during the dive because of the increased external pressure.

v. Since the dolphin chest recoils inward and does not resist pressure the way our thorax does, blood pooling is probably less conspicuous than in man. The heart does not contain gas and therefore cannot crush.

w. Dive-induced bradycardia is a general mammalian phenomenon, although more evident in marine mammals. Trained bottlenose dolphins may slow down from 100 to 12–25 beats/min during voluntary apnea at the bottom of the pool (Elsner et al., 1966). However, studies on bradycardia in freely diving dolphins (Noren et al., 2012), and the incidence of cardiac anomalies in deep-diving bottlenose dolphins (>200 m) (Williams et al., 2014), suggests that dive response is modulated by behavior and exercise and that ancestral terrestrial traits may persist in cardiac function.

x. One has to remember that in man (and in several terrestrial mammals as well), superficial veins of the limbs are essential for thermodispersion during the hot season or intense exertion. Water conducts heats away from the body approximately 25 times faster that air. So the extreme reduction of the thoracic limbs and the complete absence of the pelvic ones contribute to limit heat loss in cetaceans. On the other hand, stranded dolphins may suffer from heat stroke because of lack of the surrounding liquid environment. The blubber is such an effective insulator that dead stranded animals decompose rapidly.

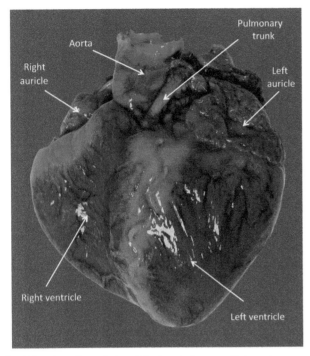

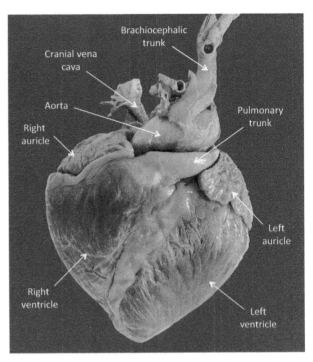

FIGURE 4.21   **Ventral view of the heart of a young *T. truncatus*.**   FIGURE 4.22   **Ventral view of the heart of an adult *T. truncatus*.**

Heart

The heart of mammals is the organ that rhythmically pumps arterial blood into the systemic arterial circle, receives venous blood from the periphery, and sends and receives blood to and from the lungs, where gas exchanges take place. This general plan turns out to require special adaptations in diving mammals, where the oxygenation is essential to sustain the long periods of diving apnea, but the heart rate may become very low during immersion. The frequency of heartbeats/minute for a bottlenose dolphin varies approximately from 60 to 130, depending on the intensity of the exercise (Williams et al., 1993). Bradycardia becomes more efficient with age, and increases markedly after maturity (Noren et al., 2004).[y]

The heart of dolphins has the general appearance of that of terrestrial mammals, with some noteworthy differences. The heart of dolphins is larger and flatter than that of most land mammals (Figs. 4.21–4.23), due to the shape of the thorax and the pressure it withstands. Terrestrial Cetartiodactyla (including bovine, pigs, camels) possess a long but narrow thorax that allows for powerful lateral insertions of muscles. The thorax of dolphins is wide and short, and the ribs are connected to the small sternum by bony appendages. This shape of the thorax is, in fact, more similar to the human anatomy (except for the virtual absence of rib cartilages in dolphins). Therefore, the heart is flat, with an oval almost circular outline, wider (but shorter) ventricles, and a slightly rotated axis. The auricles are flat and their ventral edges surpass the coronary sulcus separating atria from ventricles.

The sternal face of the heart presents the beginning of the aorta, with the brachiocephalic trunk, and the pulmonary artery. The cranial vena cava is also evident. The left ventricle here appears larger than the right. Coronary vessels are often hidden by fat.

The caudal face of the heart shows the curvature of the arch of the descending aorta with its upward branches, the pulmonary arteries (left and right), and especially the sinus of the pulmonary veins (two left vessels and two right vessels) and the entrance of the *vena cava caudalis*. The left and right ventricles appear almost even in dimension.

The weight of the heart averages 0.93% of total body weight in bottlenose dolphins, and (roughly) 0.6 % in the dolphin family in general (Slijper, 1979). So the heart of a male adult specimen of *T. truncatus* of 300 kg may weigh less than 3 kg.[z] Indications suggest that the heart of dolphins is proportionately slightly heavier than that of terrestrial mammals. The heart of *S. coeruleoalba* and *D. delphis* is slightly smaller than that of *T. truncatus*, but otherwise comparable.

The position of the dolphin heart is demarcated by its relationship to the rib cage and in this sense is markedly different from that of most terrestrial mammals (for imaging in the live bottlenose dolphin, see Ivančić et al., 2014). The heart of the bottlenose dolphin, enveloped by the thick pericardium, projects laterally on both sides on the lower third of the 1st to 4th rib (Fig. 4.24) (in man that would be the 2nd to 6th, in bovine and horse the 3rd to 6th). As a consequence, the dolphin heart may

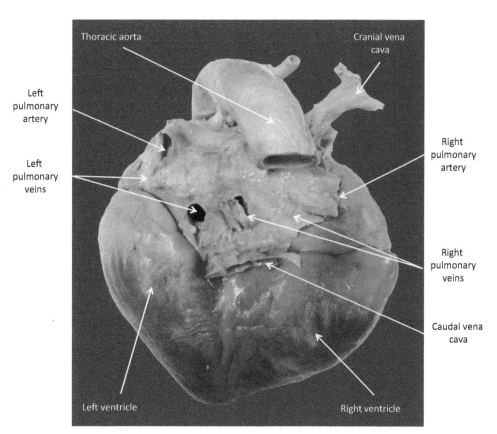

FIGURE 4.23   Dorsal view of the heart of an adult *T. truncatus*.

be reached between the 1st and the 3rd intercostal space, much more rostrally than in terrestrial species. The rostral-most limit of the diaphragm during expiration may be placed at the 3rd intercostal space, again a very rostral limit if compared to terrestrial Cetartiodactyla and primates. The flipper covers the heart region only in part, as the heart is placed with its longer dorsoventral axis turned very obliquely in the caudal sense, almost laying on the sternum (again a difference with terrestrial Cetartiodactyla in which the longer axis of the heart is almost vertical to the sternum, to which is connected by the sterno-pericardial ligament). In this way, the apex of the dolphin's heart (corresponding to the ventral tip of the left ventricle) is clear of the flipper.

Part of the former paragraph may seem somewhat obscure to readers not familiar with the basics of regional (= topographical) anatomy. To rephrase the same concepts with less technicalities, the general message is that the dolphin's heart is placed at the very start of the thorax and lays flat on the sternum and on the distal bony segments that connect the ribs to the sternum. Why so? The answer is difficult and may include a number of simultaneous causes. Possibly a key evolutionary factor is the need for a very convex—and rostral—thoracic curvature of the diaphragm during the expiratory phase, essential to a rapid and thorough gas exchange that takes place during the short surface interval typical of the dolphin's swim pattern. It may be important to note that, due to its position, the pericardium of the dolphin's heart does not have a large surface of connection to the diaphragm as in man. Therefore, the mass of the heart and the large vessels are not directly involved in the movements of the diaphragm that occur in rapid expiration and inspiration. Alternative or complementary explanations include better accommodation of the heart when the thorax is compressed by the external increase of the hydrostatic pressure in deep dives (a long, narrow, and rigid thorax with a long vertical axis of the heart—as in cows and pigs—would perhaps suffer the pressure of lateral forces on descent). One may also wonder at the position of the flipper that covers the heart only partially in its lateral projection. Here we note that the lateral projection of the heart in terrestrial quadrupeds is generally obscured by the often massive *m. triceps brachii* (in the human body, this is the muscle that helps us lift the thorax during push-ups). The *triceps* in dolphins is reduced in volume, because its function is not essential to maintenance of the limb axis during quadrupedal station, but acts mostly in moving the flipper dorsocaudally, a feat aided by the direction of water flow during swimming.

The position of the lungs in Cetaceans is dorsal to the heart, rather than latero-dorsal as in most terrestrial mammals. Therefore, the relationship of the heart of dolphins to the lungs is modified. The heart is sort of "suspended" below the lungs and turned sideways. The trachea crosses the thorax rostro-caudally dorsal to the heart. The phrenic nerves of the two sides (directed to the diaphragm) rest on the outer surface of the pericardium covering the atrial chambers.

The cetacean heart shows the usual subdivision into four chambers (Figs. 4.25 and 4.26). The atria are flat, and the venous sinus well evident. A patent *foramen ovale* (a fold of tissue shaped like a short tunnel that connects the two

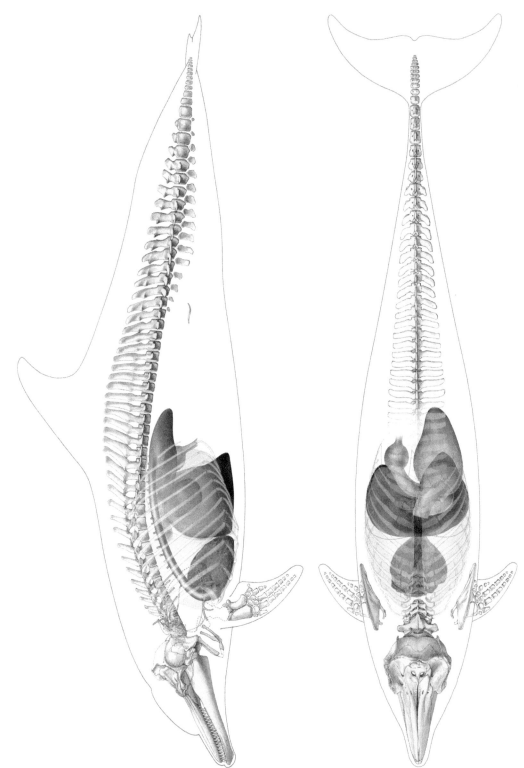

FIGURE 4.24 **Left view of the thorax of *T. truncatus* with position of the heart, left lung, and diaphragm.** (*Drawings by Massimo Demma*)

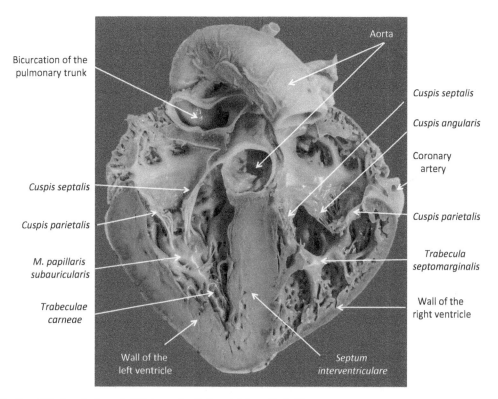

FIGURE 4.25   Section of the heart of an adult *T. truncatus* (left ventricle on the left).

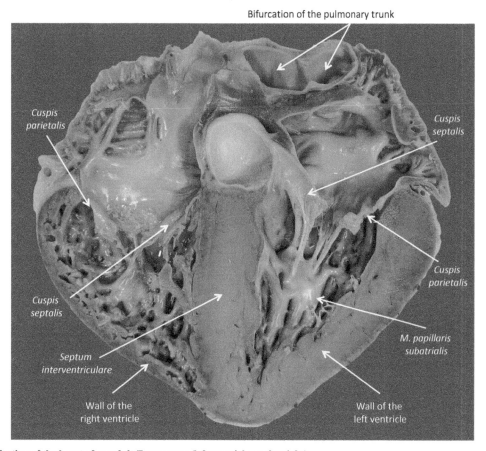

FIGURE 4.26   Section of the heart of an adult *T. truncatus* (left ventricle on the right).

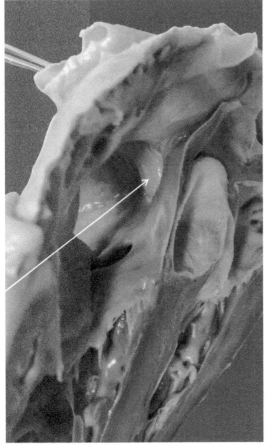

Foramen
ovale
(closed)

Ligamentum
arteriosum
(remnant of
the *ductus
arteriosus*)

FIGURE 4.27   Closed foramen ovale in an adult *T. truncatus*.

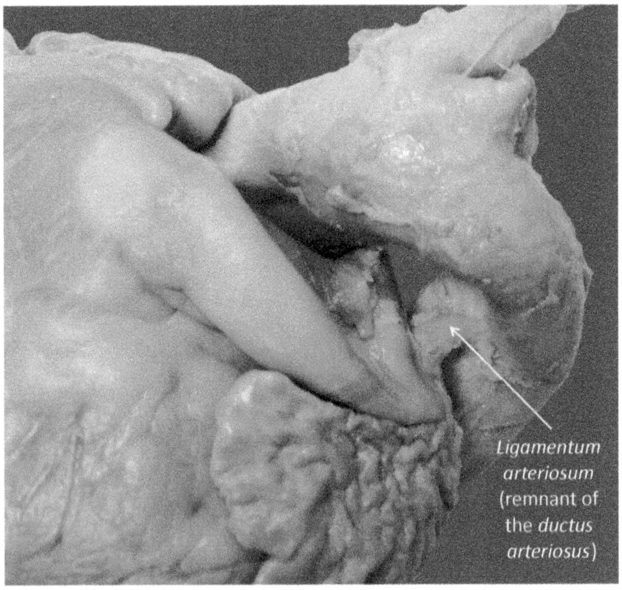

FIGURE 4.28   Remnants of the *ductus arteriosus* in an adult *T. truncatus*.

fetal atria) was observed in fetuses of several cetacean species and closes shortly after birth (Slijper, 1961; Johansen et al., 1988; Macdonald et al., 2007) (Fig. 4.27). The *doctus arteriosus* (Fig. 4.28) is also closed in adult dolphins,[aa] and therefore intermittent use of the fetal circulatory pathway[bb] to avoid potential increase in vascular resistance during diving is impossible (Slijper, 1979; Ponganis et al., 2003). Both left and right atrioventricular valves are well evident and follow the architecture of other mammals. The external wall of the left ventricle, and the interventricular septum are considerably thicker than the wall of the right ventricle. The *mm. papillares* and the *trabeculae septomarginales sinistrae*[cc] in the left ventricle, and the *trabecula septomarginales dextra* in the right ventricle are easily identified.

The cardiac muscle in dolphins is no different from that of terrestrial mammals (Fig. 4.29). The functional anatomy of the conducting system of the heart is apparently the same in dolphins and in terrestrial mammals. This essential component of the heart contraction mechanism consists of modified cardiac muscle cells (cardiac pacemaker cells) that form the sinoatrial node [of Keith-Flack, located between the right atrium and the cranial (= superior) vena cava], the atrioventricular node (of Tawara, placed at the left of the opening of the coronary sinus), and the atrioventricular bundle (bundle of His) that from the basal part of the atrial septum runs distally in the ventricular septum and divides into a left and a right branch (Purkinje fibers). The conducting system has been described in *D. delphis* by Arpino (1934), and Chiodi (1934);

aa. Once closed in postnatal life, the *doctus arteriosus* should be called *ligamentum arteriosum* (NAV, 2012).

bb. The mammalian fetus receives its oxygen supply from the maternal blood through the placenta, and the lungs do not become active until the first breath after birth. Functional circulation to the lung is bypassed by the *doctus arteriosus* that connects the aorta with the pulmonary artery, and the *foramen ovale* between the two atria also short-circuits lung circulation.

cc. The *trabeculae* are the muscle bundles that intersect the ventricles. In the human heart, the left ventricle possesses two *trabeculae*, and the right ventricle four. In the heart of the bottlenose dolphin (and at least also in the false killer whale; Slijper, 1939), the number of *trabeculae* is not easily determined because of the complex pattern of their structure and individual variation. However, on the whole, these structures are more numerous and robust in both ventricles than in man.

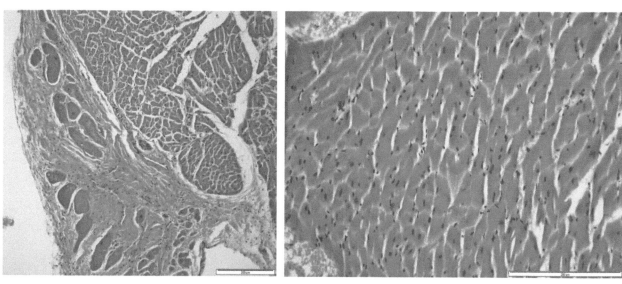

FIGURE 4.29   Cardiac muscle of *S. coeruleoalba* (scale bars = 200 μm).

in *S. coeruleoalba* by Sakata (1959); and in *Lagenorhynchus albirostris* by van Nie (1985). A recent review by Ono et al. (2009) reported the anatomy of Purkinje fibers of dolphins of unspecified species.

The coronary vasculature (and especially the arterial system) of the bottlenose dolphin has been described by Cave (1977) and can be effortlessly identified in the sectioned heart. According to Cave (1977), there is no anastomosis between the left and the right coronary artery.[dd] The drainage of the lymphatic circulation takes place in the right atrium, as in most mammals. Although no specific details have been reported about the innervation of the dolphin heart from the anatomical point of view, physiological investigations on live animals and comparative anatomical data indicate that the visceral innervation is the same as that of terrestrial mammals, including man.

Arteries

Aorta and the Retia Mirabilia

The ascending aorta arises from the left ventricle, wrapped by the thick pericardium and, after generating the coronary arteries, soon arches to form the aortic arch, with the brachiocephalic trunk, and the thoracic aorta (Fig. 4.30) that continues with the abdominal aorta. The aorta of the adult bottlenose dolphin has a diameter of approximately 3 cm at its origin, but soon decreases in size.[ee]

The subclavian and carotid arteries originate from the brachiocephalic artery (Fig. 4.31) (Slijper, 1936; De Kock, 1959). The subclavian arteries quickly decrease in size upon entering the pectoral girdle. The common carotid artery is very short (the neck is virtually absent in dolphins), and soon splits into the internal and external carotid arteries.

The intercostal and dorsal thoracic arteries, originating from the brachiocephalic trunk and the descending aorta, supply blood to the huge thoracic (Figs. 4.32 and 4.33) and spinal *retia mirabilia* (Fig. 4.34).

A *rete mirabile*[ff] is a complex structure in which a single artery branches out into a number of smaller vessels that finally reconstitute a single (or a few) larger vessel(s), a direct continuation of the artery that generated the *rete*. In some instances (as in the cranial *retia mirabilia* of ruminants), the smaller arteries derived from the generating artery are immersed in a pool (*lacuna, sinus*) of venous (cooler) blood. In ruminants the basicranial *rete mirabile*, enveloped by the dura mater,

---

dd. Anastomoses between the left and the right coronary arteries have been described in larger whales (Truex et al., 1961).

ee. Cave (1977) reported for the aorta of the bottlenose dolphin an external initial external diameter of 35 mm and an external diameter at the end of the aortic arch of 28 mm. In this latter position, the lumen was 21 mm. These values are within the normal range of the human aorta.

ff. Many even-toed Cetartiodactyla (including *Bos taurus*) have large epidural *retia mirabilia* (divided into a rostral and a caudal part) at the base of their brain. The *rete mirabile* develops on the early intracranial course of the internal carotid artery, whose proximal part does not persist in the adult. The main supply of the *rete mirabile*, in the adult terrestrial Cetartiodactyla that possess it, comes from the external carotid artery via the internal maxillary artery. The short arterial efferent that originates from the *retia mirabilia* to supply the brain is the only remaining equivalent of the (distal part of the) internal carotid artery. For a detailed description on the cerebral blood supply and *rete mirabile* in mammals, see Ask-Upmark (1935); for details on terrestrial Cetartiodactyla, see Baldwin (1964). The first modern and detailed description of the *retia mirabilia* of dolphins is that of Breschet (1836) (Fig. 4.32). (See also footnote h).

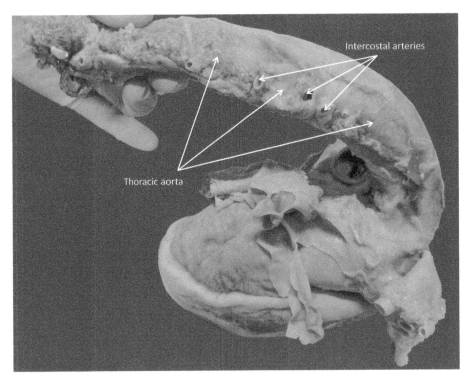

FIGURE 4.30   **Thoracic aorta of *T. truncatus*.**

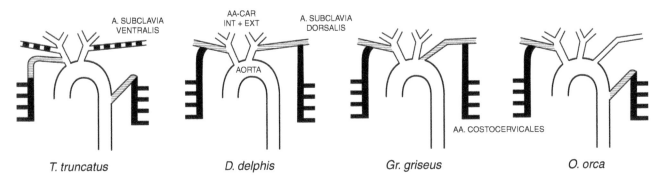

FIGURE 4.31   **Brachiocephalic trunk of selected dolphin species.** *(Modified after Slijper, 1936)*

may exert a flow- and pressure-damping effect. Thus, blood pressure in the middle cerebral artery (distal to the *rete*) is 18% lower than in the internal maxillary artery (proximal to the *rete*) (Lluch et al., 1985).

The spinal cord of dolphins is enveloped by thick extradural *retia mirabilia* that partially substitutes the epidural fat in the vertebral canal. A *rete mirabile* is also present in the dorsal wall of the thorax.

Blood Supply to the Dolphin Head

Contrary to most terrestrial mammals, including man, the internal carotid artery of dolphins remains a small vessel through postnatal life and reaches only the tympanic cavity[gg] and then the eye. Thus, what is left of the internal carotid artery of dolphins has virtually no connection with the vascular supply of the brain.[hh] Cerebral blood circulation derives from the cervical

gg. In the common porpoise *Phocoena phocoena*, the internal carotid artery reaches the floor of the skull and gives rise to an extradural *rete mirabile* connected to the others on the floor of the cranial cavity by small branches piercing the meningeal covers.

hh. In man and most mammals the three principal arteries that supply the brain are the paired internal carotid arteries and the basilar artery. All contribute to the arterial circle of the brain (formerly called *circle of Willis*) placed below the hypothalamus. From the circle originate the cerebral and cerebellar vessels. Impairment of blood flow in one of the three supplying vessels (carotids and basilar) has serious or fatal consequences in man, but not in all mammals. As an example, perennial bilateral ligation of the internal carotid arteries of the dog has no neurological or behavioral effect (Whisnant et al., 1956), since in this species the basilar artery alone is fundamental for cerebral blood flow. In ruminants, the vertebral artery alone may supply the cervical *retia mirabilia* through the occipital and maxillary collaterals.

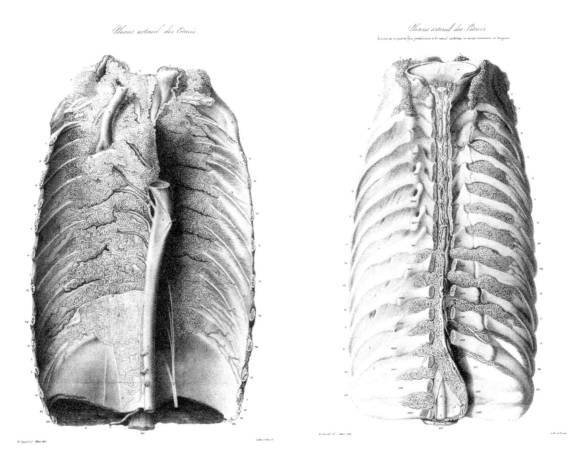

FIGURE 4.32   Ventral (left) and dorsal (right) view of the *retia mirabilia* of the harbor porpoise (Breschet, 1836, Tables II and III).

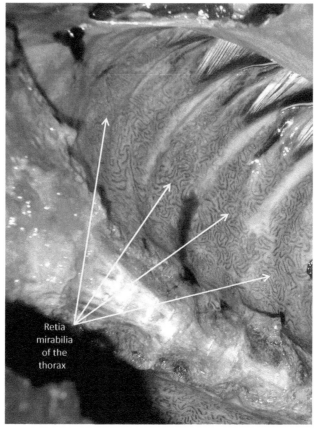

Retia
mirabilia
of the
thorax

FIGURE 4.33   *Retia mirabilia* of the thorax of *S. coeruleoalba.*

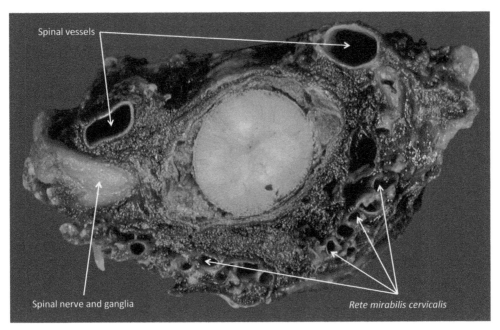

FIGURE 4.34 Section of the cervical spinal cord of *T. truncatus* surrounded by the *rete mirabilis cervicalis*.

retia mirabilia, who enters the occipital foramen with the spinal meningeal arteries and are continuous with the cervicothoracic *retia mirabilia* (Fig. 4.35) supplied by the brachiocephalic trunk and the intercostal and internal thoracic arteries (Galliano et al., 1966; McFarland et al., 1979). Homologous structures to the cerebral arteries found in man and other mammals (anterior, middle, and posterior) can be detected after careful observation, although their origin and reciprocating relationships are different in Cetacea.[ii] What is left of the internal carotid artery reaches the orbit where it contributes to the large internal ophthalmic *rete*, below and behind the eye (Galliano et al., 1966; Sommer et al., 1968; Viamonte et al., 1968; McFarland et al., 1979; Ninomiya and Yoshida, 2007). The ophthalmic *rete* is continuous with the *retia mirabilia* complex at the base of the brain.[ii]

The external carotid artery is directed to the lower parts of the head, and provides vascular supply for the face, similarly to what happens with the facial artery and derived branches in terrestrial mammals. The internal maxillary artery supplies the fibrovenous plexus of the air sacs along its course ventral and lateral to the skull.

Therefore, in dolphins, a complex vascular network is interposed between the systemic and cerebral blood flow (Nagel et al., 1968). Most of the studies on the brain circulation of dolphins date back to an era when public awareness and perception of animal existence was strikingly different from now. Only recent technology (MRI or PET) allowed new minimally invasive studies that confirmed that the main supply of blood to the brain derives from the spinal meningeal arteries (Houser et al., 2010), once considered to be nonpulsatile (Nagel et al., 1968). The MRI has also demonstrated that dolphins may show unihemispheric vasoconstriction, with consequent reduced oxygen and glucose consumption in half of the brain (Ridgway et al., 2006), a functional correlate of unihemispheric sleep. The ability to partially shut down the blood supply to half of the cerebral tissue may be important during prolonged dives, where brain metabolism may become anaerobic for short periods because of prolonged breath-holding (Ridgway et al., 1969).

Although no specific details have been reported on the innervation of the arteries of dolphin, it must be noted that the cervical *rete mirabile* has a rich innervation, probably necessary to allow its function as pulse regulator of blood flow directed to the brain, especially during dives.

Arteries of the General Circulation

The large descending aorta (Fig. 4.30) carries blood to the thorax and abdomen, letting off several intercostal and posterior branches. From the aorta, arteries reach the viscera of the thorax and abdomen as in other mammals.

ii. For example, in the narwhal *Monodon monoceros* there is no connection between the anterior cerebral arteries and the posterior compartment (Vogl and Fisher, 1981).

jj. Since in dolphins the common carotid arteries that travel in the neck have no direct bearing on the brain circulation, suppression (or slaughtering) of dolphins by severing the ventral part of the neck is particularly cruel as consciousness persists due to the spinal meningeal arteries and thoracic *retia mirabilia* that still supply the brain.

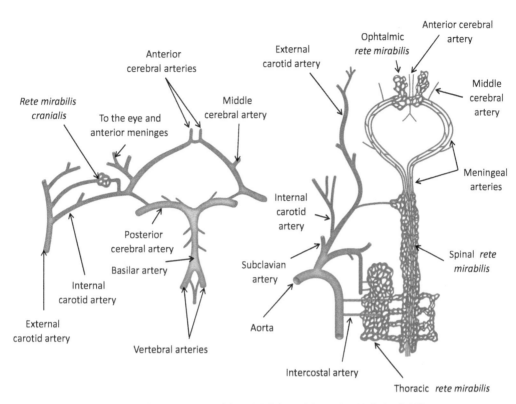

FIGURE 4.35   Blood supply to the head of nonhuman primates (left) and dolphins (right). *(After McFarland, 1979)*

However, among the main differences, we may note the presence of counter-exchange mechanisms for the testes, uterus, ovaries, and kidney. The functional anatomy of the vascular supply to the genital and urinary organs is described in Chapter 9. Here we just emphasize that the system works in such a way to supply these delicate organs with blood that has been cooled by vicinity of veins containing blood coming from the dorsal fins, and thus exposed to the heat dispersion induced by external water (25 times higher than in the air).

Another important difference with terrestrial mammals is the absence of vascular supply to the pelvic limb, represented in dolphins only by the residues of the coxal bone. So there are no external iliac arteries as in man and terrestrial quadrupeds, and a powerful and large ventral vertebral artery (protected by the hemal arches) supplies the flukes.

Peripheral Circulation

Arteries of the dorsal fin and flukes are surrounded by a spiral network of veins (Figs. 4.36 and 4.37) that act as a powerful heat-exchange tool to limit heat dispersion in the surrounding liquid environment (Elsner et al., 1974). The basic

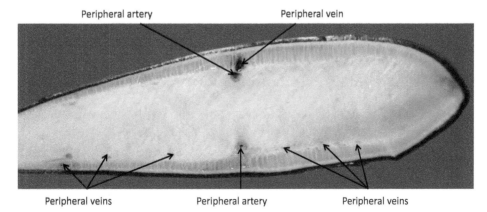

FIGURE 4.36   Peripheral counter-exchange mechanism in the dorsal fin.

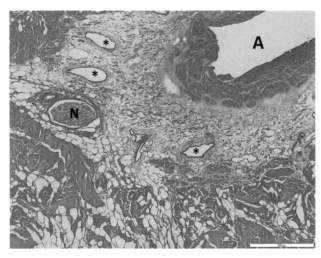

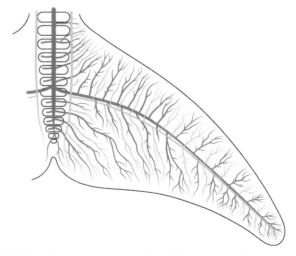

FIGURE 4.37 Peripheral artery *(A)* of the dorsal fin of *S. coeruleoalba* surrounded by small veins (*) (scale bar = 500 μm). *N*, peripheral nerve.

FIGURE 4.38 Circulatory tree of the fluke in *D. delphis*.

mechanism relies on the transfer of heat (bioheat transfer) from the warmer arterioles coming from the core of the body to the surrounding external cooler veins (also called periarterial venous retia) full of blood that is returning from the cold-exposed fins. Heat is thus maintained in the core of the body, leaving cooler blood to circulate in the relatively thin append-ages exposed to thermo-dispersion because of the water. Heat dispersion or saving can be modulated by changes in the diameter of the vessels. In fact, the dorsal fin is a spatially heterogeneous thermal surface (Meagher et al., 2002). The same mechanism applies to the flippers (Elsner et al., 1974).

The relatively regular disposition of the peripheral arterioles (Fig. 4.38) makes it simpler to draw blood for medical purposes from the flukes of trained dolphins.

Venous System

In cetaceans, as well as in terrestrial mammals, the venous system follows the arterial system. However, evolutionary adap-tation to life in the water and diving have induced a number of changes in the general architecture of the venous circulation in dolphins and other Cetacea.

A venous system of *retia mirabilia* follows the corresponding arterial structures. However, the veins of the *retia* gener-ally have no valves (so that blood can probably flow in two directions). Furthermore, these veins have no muscle layer. Arteries and veins of the *retia* are connected only by few capillaries. Veins can possibly function as shock absorbers for the arteries of the corresponding *rete mirabile*.

Veins of the General Circulation

The shape and course of both the rostral and caudal vena cava[kk] show no variation to the general mammalian plan, with the exception of the drainage of the hind limb (see section Thoracic Limbs). Cave (1977) reported a diameter of 22 mm for the lumen of the rostral vena cava, and a diameter of 28 mm for the lumen of the caudal vena cava in the bottlenose dolphin.

Veins of the Head and Thorax

The venous drainage of the head (Fig. 4.39) has been recently described by CT scans and rendering techniques (Costidis and Rommel, 2012) approach. The sinuses of the head have venous plexuses of considerable extension and complexity, constituted by repeatedly anostomotic small caliber veins. The main venous route of drainage from the pterygoid sinuses of the bottlenose dolphin is the external jugular vein (Costidis and Rommel, 2012). However, according to the same authors, in this species the linguofacial vein does not lead into the external jugular vein (as in most mammals), but directly into the brachiocephalic trunk; furthermore, the mandibular vein derives from the facial vein, and not from the maxillary. The

---

kk. The terms "cranial" and "caudal" are taken from NAV (2012). The same vessels in man are called superior and inferior, respectively (Federative Committee on Anatomical Terminology, 1998).

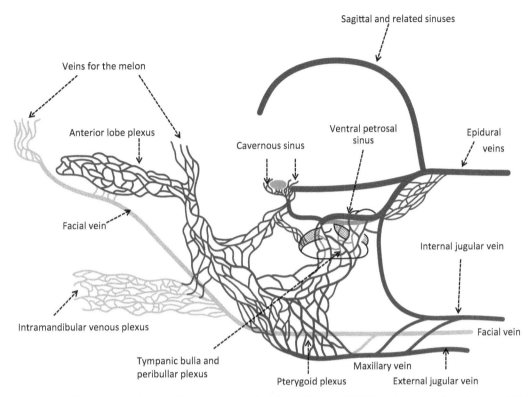

FIGURE 4.39   Deep veins of the head. *(Redrawn by Uko Gorter after Costidis and Rommel, 2012. Vascularization of air sinuses and fat bodies in the head of the bottlenose dolphin (T. truncatus): morphological implications on physiology. Front. Physiol. 3, 243.)*

facial, external jugular, and internal jugular veins drain the mandibular fat bodies and the accessory sinus system (Costidis and Rommel, 2012). The fibro-venous system of the air sacs is drained by the spinal veins through the cranial sinuses, by the large vessels that join the external jugular vein, and by tiny vessels that eventually coalesce into a single vessel that joins the mandibular vein. Therefore, it is evident that the accessory air sinuses of the head are endowed with an extraordinary venous drainage that has been related to nitrogen absorption or elimination during the dive (Fraser and Purves, 1960; Costidis and Rommel, 2012), with a physiological mechanism that still requires further examination. Possible alternate explanations for the venous plexus include the redistribution of blood to accommodate the reduction of air during descent (Fraser and Purves, 1960; Sassu and Cozzi, 2007; Costidis and Rommel, 2012).

The valveless spinal veins represent the most important route of venous drainage from the brain. They also drain the sinuses at the base of the cranium. There is always a pair of extremely well-developed large veins passing ventral to the vertebral canal, connected by costocervical and intervertebral veins to the cranial and caudal *venae cavae*, respectively.

According to Harrison and Tomlinson (1956), the presence of the system of azygos veins has not been convincingly demonstrated in cetaceans. A small vessel, embedded in the retial tissue to the right of the aorta, may represent all that is left of the right azygos vein, at least in the harbor porpoise.

Caudal Vena Cava

According to Slijper (1939), the caudal vena of *Pseudorca crassidens* is made up by two separate vessels interconnected by a mesh of smaller veins, all of them surrounding the descending aorta. These two vessels, at least in *Phocoena phocoena* (Harrison and Tomlinson, 1956), merge in the anterior part of the abdominal cavity.

The caudal vena cava of some marine mammals possess a distinct sphincter at the level of the specific foramen in the diaphragm. However, this has not been demonstrated in dolphins.[ll] A peculiar characteristic is the presence of enlarger hepatic sinuses along the course of two or three major hepatic veins (Richard and Neville, 1896), just prior to their confluence into the caudal vena cava on the dorsal margin of the liver (see also Chapter 8). The hepatic sinuses may act as a reservoir for red blood cells during diving. However, their function is still unclear.

---

ll. A sphincter (not confirmed by later studies) is, in fact, reported by Yablokov et al. (1972).

In the lateral and ventral walls of the abdominal cavity, there are *retia mirabilia*, constituted only by veins that in turn drain blood from superficial vessels coming from the tail and flukes. These venous networks direct blood flow toward the caudal vena cava, and possibly function to equalize differences in pressure between the thorax and the abdomen.

LYMPHATIC SYSTEM

Lymphatic Vessels and Lymph Nodes of the Body

There are no apparent differences between the lymphatic vessels of cetaceans and those of terrestrial mammals. Lymph nodes are numerous and—according to Slijper (1979)—comparatively larger than those observed in most terrestrial mammals. The same author also states that lymph nodes in cetaceans contribute to the breakdown of old red blood cells.

Lymphatic vessels are notoriously difficult to observe in the cadaver unless specific conditions set in (the animal dies just after a meal, lymphatic diseases swell the vessels, etc.), or specific injection methods are used. Therefore, lymphatic vessels are seldom described.

The lymph nodes are covered by a connective capsule, under which there is a subcapsular sinus. The parenchyma of the lymph node of dolphins (Figs. 4.40, 4.41) is subdivided into lobules by connective trabeculae with several blood vessels

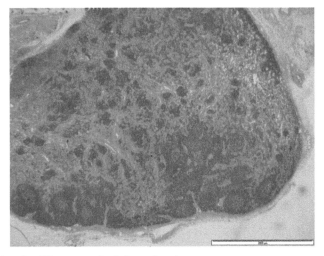

FIGURE 4.40 Deep cervical lymph node of *T. truncatus* (scale bar = 2 mm).

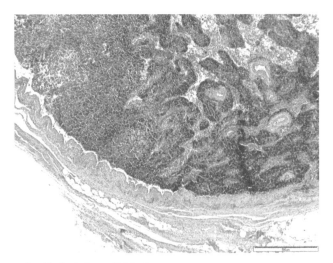

FIGURE 4.41 Lymph node of *S. coeruleoalba* (scale bar = 500 μm).

and nerves (De Oliveira e Silva et al., 2014). The lymphoid follicles (nodules) are located in the cortical region (cortex) just below the subcapsular zone, while the center of the lymph node (medulla) contains small cords of lymphocytes, with vessels, sinuses, and connective tissue. A paracortical zone is interposed between cortex and medulla (De Oliveira e Silva et al., 2014). The medullary sinuses drain the lymph from the cortical zone to the medulla. The efferent lymphatic vessels leave the hilus. The germinal zone at the center of the follicles contains lymphoblasts, surrounded by inactivated B lymphocytes at the periphery (mantle zone).

However, the structure of the cetacean lymph nodes of *D. delphis* apparently resembles that of the pig, in which peripheral sinusoids and cords surround the follicular cortex in the center of the organ (reversed structure) (Simpson and Gardner, 1972). Smooth muscle fibers have been described in the capsule and trabeculae of the visceral lymph nodes of *T. truncatus* (Cowan and Smith, 1999).

Regional Distribution of Lymph Nodes

Here follows a general summary of the distribution of lymph glands in the dolphin body, largely based on personal experience and on a recent detailed report (De Oliveira e Silva et al., 2014) (Table 4.4).

Head

Lymph nodes of the head are not easy to observe. Their diameter is approx. 1 cm. Dolphins possess well-developed tonsils at the back of the oral cavity, just beneath the laryngeal folds (laryngeal tonsil).

TABLE 4.4 Lymph Nodes of Selected Dolphin Species[a]

Lymph Node Position	NAV Name	Stenella clymene	Stenella longirostris	G. macrorhynchus	Number of Lymph Nodes
Parotid	Lymphocentrum parotideum	+	+	−	1
Mandibular	Lymphocentrum mandibulare	+	+	+	2–4
Cervical (superficial and deep)	Lymphocentrum cervicale[b] superficiale and profundum	+	+	+	1–6 in both positions
Prescapular[c]		−	−	+	1
Mediastinal	Lymphocentrum mediastinale	+	+	+	4–8
Hilar (lung)	Lymphocentrum (tracheo) bronchale	+	+	+	1–4
Pulmonary	Lymphonodi pulmonales	+	+	+	1–3
Diaphragmatic[d]	Lymphonodi phrenicoabdominalis	+	+	+	1–3
Gastric	Lymphonodi gastrici	+	+	+	1–3
Mesenteric	Lymphocentrum mesentericum (craniale and caudale)	+	+	+	15–23
Mesocolic	Lymphonodi colici	+	+	+	1–3
Hepatic hilar	Lymphonodi hepatici (portales)	+	+	+	1–2
Renal (hilar)	Lymphonodi renales	+	+	+	1–2
Pelvic[e]		+	+	+	2–5
Male and female genital	Lymphonodus ovaricus and Lymphonodus testicularis	+	+	+	2–4

[a]The original name in the article does not always correspond to a recognized topography. The corresponding NAV (2012) name has been added based on our interpretation and personal experience.
[b]Lnn. cervicales superficiales were described in every bottlenose dolphin examined by Ivančić et al. (2014).
[c]Does not exist as such in NAV (2012). Possibly part of the Lymphocentrum cervicale superficiale. The common name "prescapular" lymph node in large herbivores often identifies an enlarged and potentially pathological cervical lymph node.
[d]According to Cowan and Smith (1999), it does not exist as such in NAV (2012). In our opinion it may correspond to the Lymphonodus phrenicoabdominalis.
[e]Does not exist as such in NAV (2012). Possibly theses lymph nodes described by De Oliveira e Silva et al. (2014) correspond to the Lymphocentrum inguinofemorale and Lymphocentrum ischiadicum of terrestrial quadrupeds.
(**Source:** Modified after De Oliveira e Silva et al., 2014)

Mesenteric lymphnode

FIGURE 4.42   **Mesenteric lymph node of** *S. coeruleoalba.*

Thorax

Lymph nodes in the thorax are similar to those found in bovine. Some lymph nodes are clustered together close to termination of the tracheal tree, in the proximity of the *ilus* of the lungs (*Lnn. tracheobronchales*). A large lymph node is evident below the border of the left lung. It is difficult to ascertain whether this latter corresponds to the *marginal node of the lung* of Cowan and Smith (1999), who described it as bilaterally placed under the ventral free border of each lung, and without an NAV equivalent name. One (sometimes two) very large and long lymph node (*Lymphonodus mediastinalis caudalis*, part of the group of the *Lymphonodi pulmonales* in the caudal mediastinum) is present in the caudal mediastinum, facing the mediastinal side of both lungs. This specific large lymph node (up to 5–7 cm) is very similar to the correspondent lymph node of bovine.

Abdomen

Lymph nodes in the abdomen, sometimes very large, follow the mesentery in its attachment to the viscera (Fig. 4.42). Once more their shape, mass, and distribution closely resembles those of bovine. The mass of the mesenteric lymph nodes has been described as Aselli's *pseudopancreas* by Pilleri and Arvy (1971); however, its structure has nothing to do with the pancreas itself.

Gut-related plaques of lymphatic tissue (either diffuse or nodular) are present in the walls of the intestines throughout the postdiaphragmatic tract. Cowan and Smith (1999) wrote that a continuous sheet of lymphoid tissue, including well-organized germinal centers, is present in the *lamina propria* of the mucosa and the submucosa of the straight segment of the intestine of young animals. Lymphocytes disappear from this location in older animals, and patches were not found in extensively sampled small intestine, defined as the intestinal tract proximal to the splenic flexure (Cowan and Smith, 1999).

Although there is no appendix as in the human or rabbit, some tracts in the distal part of the intestine of young bottlenose dolphins have an appendix-like structure (Cowan and Smith, 1999).

Cetaceans have large tonsils associated to the distal part of the intestines, called anal tonsils.[mm] The anal tonsil of the bottlenose dolphin is, in fact, a complex lymphoepithelial organ placed before the external opening of the anal canal, constituted by clusters of lymphoid tissue and epithelial ducts (crypts), sometimes associated with mucous glands (Cowan and

mm. Some species of river dolphins (*Inia geoffrensis*, *Pontoporia blainvillei*) apparently lack the anal tonsil.

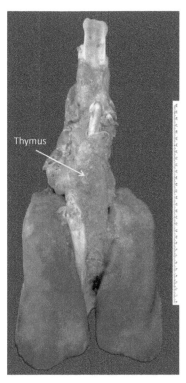

FIGURE 4.43   **The thymus of a newborn *T. truncatus*.**

Smith, 1995, 1999). The anal tonsil is active in young animals, and apparently deteriorates with age. This tonsil may play a role similar to that of the human appendix.

Lymph nodes are generally associated with the male reproductive system and especially with the female reproductive system.

Thoracic Limbs

Lymph nodes have never been observed in the modified thoracic limbs; however, they could be present, even if small.

Thymus

The thymus (Fig. 4.43) is a primary lymphoid organ that evolves with age. The thymus of dolphins is a reddish-gray lobated organ, placed on the cranial surface of the pericardium, just behind the sternum and close to the thyroid gland. The thymus may overlie the thyroid (Cowan, 1994).[nn]

The thymus consists of a cortex and a medulla (Fig. 4.44). The cortex is rich in lymphocytes, as opposed to the less cellular medulla (Rommel and Lowenstine, 2001). As in all mammals, the thymus changes with age. In newborn and younger prepubertal dolphins, the thymus enters the neck along the ventral surface of the trachea, and even embraces the heart in caudal direction. In adults, the tissue atrophies, and only fibrous remnants may be found cranially to the aortic arch. The organ undergoes degeneration of the lobules, formation of plaques of mineralized tissue, invasion by adipose tissue, and involution of Hassall's corpuscles.[oo] The thymus of an adult bottlenose dolphin may even be difficult to find, and eventually reveals a microscopic structure mostly composed of fatty lobules among which isles of thymic tissues survive, with evident Hassall's corpuscles and cysts (Cowan, 1994) (Fig. 4.44). The cysts may become filled with a colloid-like substance. The cysts may grow to be so large as to virtually substitute the thymus in location. In this case, the residual presence of islets of specific thymic tissue adjacent to the cysts may be recognized only under the microscope.

---

nn. The parathyroid glands may be found in the capsule of the thymus (Cowan, 1994).

oo. Hassall's corpuscles (or thymic corpuscles) are made up by reticular epithelial cells that have lost their autoimmune regulatory function as medullary thymic epithelial cells.

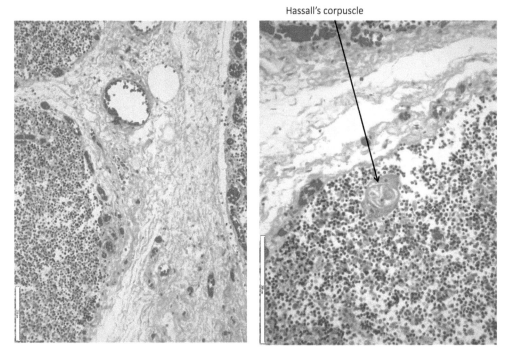

FIGURE 4.44 **Microphotographs of the thymus of *T. truncatus* (scale bars = 200 μm).** In the right image, there are evident Hassall's corpuscles.

Spleen

The spleen of dolphins is strikingly small (0.2% of body weight) if compared to that of terrestrial mammals (Slijper, 1958). It has a flat discoid shape (Figs. 4.45 and 4.46), with a dark red-bluish color. In an adult striped or common dolphin, it is no larger than 15–20 cm in diameter.

The spleen is connected to the larger curvature of the first stomach by the greater omentum and generally lies in dorsally in the left side of the abdominal cavity, but may shift elsewhere. The spleen is surrounded by a thick capsule with a fibrous exterior and a muscular interior layer. The muscle cells spread into the trabeculae.

The general architecture of the dolphin spleen (Figs. 4.47 and 4.48) generally resembles that of terrestrial mammals: From the outer connective capsule, often composed by two layers (outer fibrous and inner muscular), a series of trabeculae enter the organ. The spleen is divided into a white pulp consisting of lymphoid nodules developed at arterial terminals, evenly distributed throughout the red pulp (Cowan and Smith, 1999). Lymphoid nodules are composed of small- to medium-sized lymphocytes (Romano et al., 1993). Hematopoiesis takes place in the spleen of the bottlenose dolphins, as demonstrated by the constant presence of megakaryocytes (Cowan and Smith, 1999). According to Zwillenberg (1956, quoted by Slijper, 1979), the white pulp corresponds to 30% of the total mass of the organ in porpoises, a fact due to the larger number of lymph corpuscles.

However, some evidence indicates that the cetacean spleen is indeed somewhat different from that of terrestrial mammals. The red pulp of the spleen of *G. macrorhynchus* and *T. truncatus* is made up by an inner and an outer venous layer (Nakamine et al., 1992; Tanaka, 1994). Apparently the inner venous layer possesses an "unclear" vascular architecture and lacks the development of lymphoid-reticular tissue. Thus, since the inner layer is homologous to the intermediate zone of primitive mammals, some authors (Nakamine et al., 1992; Tanaka, 1994) have suggested that in cetaceans the vascular system of this organ has not been altered in the evolutionary process, and possibly resembles that of nonmammalian vertebrates.

The relatively small size of the spleen in cetaceans suggests that, since the mechanisms controlling blood volume and pressure during diving rely mainly on the control of blood flow through the *retia mirabilia*, a large buffer/deposit for blood like the spleen is not needed.[pp] Thus, the globular smooth-surfaced shape of the dolphin spleen is not adapted to accommodate

---

pp. The spleen of pinnipeds (Costa, 2007), and even human professional breath-holding divers (Ferrigno and Lundgren, 2003), may contract and release red blood cells during the dive, with an increase in the hematocrit value, meaning more oxygen is carried in the blood. As a result, in pinnipeds, the volume of the spleen diminishes and that of the hepatic sinuses increases.

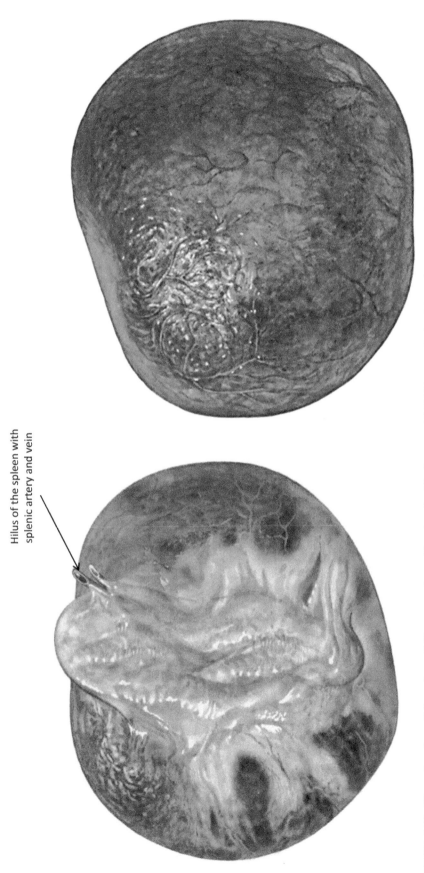

Hilus of the spleen with splenic artery and vein

FIGURE 4.46 Parietal face of the spleen of *Grampus griseus*. (*Drawings by Massimo Demma*)

FIGURE 4.45 Visceral surface of the spleen of *Grampus griseus*. (*Drawings by Massimo Demma*)

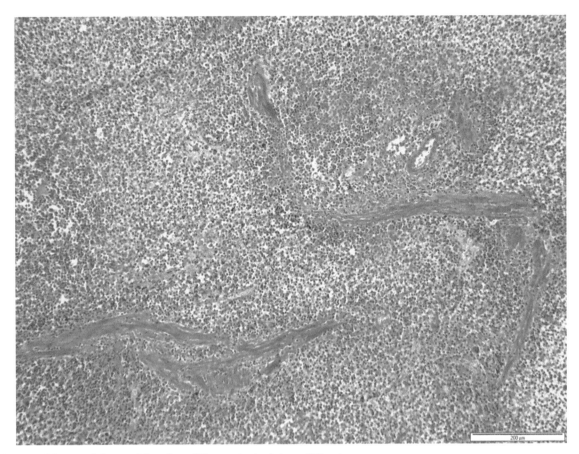

FIGURE 4.47   Microscopic image of the spleen of *T. truncatus* (scale bar = 200 μm).

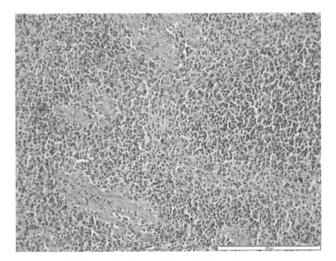

FIGURE 4.48   Microscopic image of the spleen of S. *coeruleoalba* (scale bar = 200 μm).

large changes in blood volume (Cowan and Smith, 1999). It is also possible that the large hepatic sinuses of the bottlenose and other dolphins species (Richard and Neuville, 1896) act as a reservoir of red blood cells instead of the spleen.

Accessory Spleens

There are often accessory spleens (Carvalho et al., 2009; Menezes de Oliveira e Silva et al., 2014), smaller disks of spleen tissue (Fig. 4.49), often more common in larger individuals. If present, they are distributed along the larger omentum in

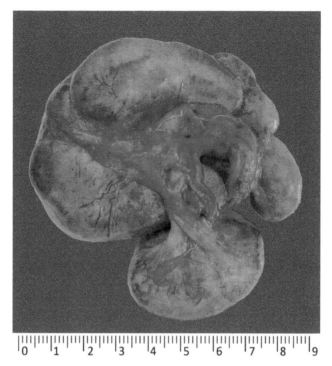

FIGURE 4.49   Accessory spleens of *T. truncatus*.

the peritoneal cavity. Accessory spleens may occur naturally in the dolphin, and may have a supplementary role in diving adaptations.

Development of the Circulatory System

Developmental phases (stages) of selected dolphin species have been reported in Table 9.6 (see Chapter 9 for details). According to Štěrba et al. (2000), during Stage 3 the heart has two atria and one ventricle and cardiac valves are not yet present. The ventricle gradually divides during Stage 4, and the four chambers are defined in Stage 5, including developing valves and arterial stems. Atrioventricular and semilunar valves are evident during Stage 6. The prenatal configuration is reached with Stage 7. In fact, there are no striking developmental differences between the human and dolphin heart, except that substantial trabecular compaction, and unification of the initially bifid ventricular apex, take place a few weeks later during the first third of the pregnancy (Sedmera et al., 2003).

REFERENCES

Aguilar Soto, N., Johnson, M.P., Madsen, P.T., Diaz, F., Domìnguez, I., Brito, A., Tyack, P., 2008. Cheetahs of the deep sea: deep foraging sprints in short-finned pilot whales off Tenerife (Canary Islands). J. Anim. Ecol. 77, 936–947.

Arpino, G., 1934. Die innervation des sinus-knotens bei *Delphinus delphis* (L.). Anat. Anz. 77, 241–252.

Ask-Upmark, E., 1935. The carotid sinus and the cerebral circulation. An anatomical, experimental, and clinical investigation, including some observations on the rete mirabile caroticum. Acta Psychiatr. Neurol. 6, 1–374.

Bagnoli, P., Cozzi, B., Zaffora, A., Acocella, F., Fumero, R., Costantino, M.L., 2011. Experimental and computational biomechanical characterization of the tracheo-bronchial tree of the bottlenose dolphin (*Tursiops truncatus*) during diving. J. Biomech. 44, 1040–1045.

Baird, R.W., Borsani, J.F., Hanson, M.B., Tyack, P.L., 2002. Diving and night-time behavior of long-finned pilot whales in the Ligurian Sea. Mar. Ecol. Prog. Ser. 237, 301–305.

Baird, R.W., Hanson, M.B., Dill, L.M., 2005. factors influencing the diving behavior of fish-eating killer whales: sex differences and diel and interannual variation in diving rates. Can. J. Zool. 83, 257–267.

Baldwin, B.A., 1964. The anatomy of the arterial supply to the cranial regions of the sheep and ox. Am. J. Anat. 115, 101–118.

Berta, A., Ekdale, E.G., Cranford, T.W., 2014. Review of the cetacean nose: form, function, and evolution. Anat. Rec. 297, 2205–2215.

Bostrom, B.L., Fahlman, A., Jones, D.R., 2008. Tracheal compression delays alveolar collapse during deep diving in marine mammals. Respir. Physiol. Neurobiol. 161, 298–305.

Breschet, G., 1836. Histoire anatomique et physiologique d'un organe de nature vasculaire découvert dans les cétacés. Bechet Jeune, Paris, pp. 1–82.

Cantú-Medellín, N., Byrd, B., Hohn, A., Vázquez-Medina, J.P., Zenteno-Savín, T., 2011. Differential antioxidant protection in tissues from marine mammals with distinct diving capacities. Shallow/short vs. deep/long divers. Comp. Biochem. Physiol. A 158, 438–443.

Carvalho, V.L., Alves Motta, M.R., Sousa Nunes-Pinheiro, N.C., Albuquerque Gomes Nogueira, T.N., Cabral Campello, C., 2009. Ocorrência de baços acessórios em boto-cinza (*Sotalia guianensis*)—aspectos histológicos. Acta Sci. Vet. 37, 177–180.

Castellini, M.A., Mellish, J.-A., 2016. Marine mammal physiology. Requisites for ocean living. CRC Press, Boca Raton, FL, pp. 1–356.

Cave, A.J.E., 1977. The coronary vasculature of the bottlenosed dolphin (*Tursiops truncatus*). Harrison, R.J. (Ed.), Functional Anatomy of Marine Mammals, vol. 3, Academic Press, New York, pp. 199–215.

Chiodi, V., 1934. Ulteriori osservazioni sull'apparato di conduzione del cuore dei mammiferi. Il nodo seno-atriale del *Delphinus delphis*. Clin. Vet. 57, 421–426.

Costa, D.P., 2007. Diving physiology of marine vertebrates. Encyclopedia of Life Sciences. John Wiley & Sons, Chichester.

Costa, D.P., Williams, T.M., 1999. Marine mammal energetics. In: Reynolds, J.E., Rommel, S.A. (Eds.), Biology of Marine Mammals. Smithsonian Institute Press, Washington, DC, pp. 176–217.

Costidis, A., Rommel, S.A., 2012. Vascularization of air sinuses and fat bodies in the head of the bottlenose dolphin (*Tursiops truncatus*): morphological implications on physiology. Front. Physiol. 3, 243.

Cotten, P.B., Piscitelli, M.A., McLellan, W.A., Rommel, S.A., Dearolf, J.L., Pabst, D.A., 2008. The gross morphology and histochemistry of respiratory muscles in bottlenose dolphins, *Tursiops truncatus*. J. Morphol. 269, 1520–1538.

Cowan, D.F., 1994. Involution and cystic transformation of the thymus in the bottlenose dolphin *Tursiops truncatus*. Vet. Pathol. 31, 648–653.

Cowan, D.F., Smith, T.L., 1995. Morphology of complex lymphoepithelial organs of the anal canal (anal tonsil) in the bottlenose dolphin, *Tursiops truncatus*. J. Morphol. 223, 263–268.

Cowan, D.F., Smith, T.L., 1999. Morphology of the lymphoid organs of the bottlenose dolphin, *Tursiops truncatus*. J. Anat. 194, 505–517.

Cozzi, B., Bagnoli, P., Acocella, F., Costantino, M.L., 2005. Structure and biomechanical properties of the trachea of the striped dolphin *Stenella coeruleoalba*: evidence of evolutionary adaptations to diving. Anat. Rec. A 284, 500–510.

Cranford, T.W., Amundin, M., Norris, K.S., 1996. Functional morphology and homology in the Odontocete nasal complex: implications for sound generation. J. Morphol. 228, 223–285.

De Kock, L.L., 1959. The arterial vessels of the neck in the pilot-whale (*Globicephala melaena* Traill) and the porpoise (*Phocaena phocaena* L.) in relation to the carotid body. Acta Anat. 36, 274–292.

De Oliveira e Silva, F.M., Guimarães, J.P., Vergara-Parente, J.E., Carvalho, V.L., De Meirelles, A.C., Marmontel, M., Ferrão, J.S., Miglino, M.A., 2014. Morphological analysis of lymph nodes in Odontocetes from north and northeast coast of Brazil. Anat. Rec. 297, 939–948.

Dold, C., Ridgway, S., 2007. Cetaceans. In: West, G., Heard, D., Caulkett, N. (Eds.), Zoo Animal and Wildlife Immobilization and Anesthesia. Wiley-Blackwell, Ames, Iowa, pp. 485–496.

Drabek, C.M., Kooyman, G.L., 1986. Bronchial morphometry of the upper conductive zones of four odontocetes cetaceans. In: Bryden, M.M., Harrison, R. (Eds.), Research on Dolphins. Clarendon Press, Oxford, pp. 109–127.

Elsner, R., 1999. Living in water. Solutions to physiological problems. In: Reynolds, III, J.E., Rommel, S.A. (Eds.), Biology of Marine Mammals. Smithsonian Institution Press, Washington, pp. 73–116.

Elsner, R., Kennedy, D.W., Burgess, K., 1966. Diving bradycardia in the trained dolphin. Nature 212, 407–408.

Elsner, R., Pirie, J., Kenney, D.D., Schemmer, S., 1974. Functional circulatory anatomy of the cetacean appendages. Harrison, R.J. (Ed.), Functional Anatomy of Marine Mammals, vol. 2, Academic Press, New York, pp. 143–159.

Fahlman, A., Schagatay, E., 2014. Man's place among the diving mammals. Hum. Evol. 29, 47–66.

Fahlman, A., Olszowska, A., Bostrom, B., Jones, D.R., 2006. Deep diving mammals: dive behavior and circulatory adjustments contribute to bends avoidance. Resp. Physiol. Neurobiol. 153, 66–77.

Fahlman, A., Loring, S.H., Ferrigno, M., Moore, C., Early, G., Niemeyer, M., Lentell, B., Wenzel, F., Joy, R., Moore, M.J., 2011. Static inflation and deflation pressure–volume curves from excised lungs of marine mammals. J. Expl. Biol. 214, 3822–3828.

Fahlman, A., Loring, S.H., Levine, G., Rocho-Levine, J., Austin, T., Brodsky, M., 2015. Lung mechanics and pulmonary function testing in cetaceans. J. Exp. Biol. 218, 2030–2038.

Fahlman, A., van der Hoop, J., Moore, M.J., Levine, G., Rocho-Levine, J., Brodskys, M., 2016. Estimating energetics in cetaceans from respiratory frequency: why we need to understand physiology. Biol. Open 5, 436–442.

Fanning, J.C., Harrison, R.J., 1974. The structure of the trachea and lungs of the South Australian bottle-nosed dolphin. Harrison, R.J. (Ed.), Functional Anatomy of Marine Mammals, vol. 2, Academic Press, New York, pp. 231–252.

Federative Committee on Anatomical Terminology, 1998. Terminologia Anatomica. Thieme, Stuttgart, pp. 1–292.

Ferrigno, M., Lundgren, C.E., 2003. Breath-hold diving. In: Brukakk, A.O., Neuman, T.S. (Eds.), Bennett and Elliott's Physiology and Medicine of Diving, fifth ed. Edinburgh, Saunders, pp. 153–180.

Fiebiger, J., 1916. Über eigentümlichkeiten im aufbau der delphinlunge und ihre physiologische bedeutung. Anat. Anz. 48, 540–565.

Finneran, J.J., 2003. Whole-lung resonance in a bottlenose dolphin (*Tursiops truncatus*) and white whale (*Delphinapterus leucas*). J. Acoust. Soc. Am. 114, 529–535.

Fitz-Clarke, J.R., 2007a. Mechanics of airway and alveolar collapse in human breath-hold diving. Resp. Physiol. Neurobiol. 159, 202–210.

Fitz-Clarke, J.R., 2007b. Computer simulation of human breath-hold diving: cardiovascular adjustments. Eur. J. Appl. Physiol. 100, 207–224.

Fitz-Clarke, J.R., 2009. Lung compression effects on gas exchange in human breath-hold diving. Resp. Physiol. Neurobiol. 165, 221–228.

Fraser, F.C., Purves, P.E., 1960. Hearing in cetaceans: evolution of the accessory air sacs and the structure and function of the outer and middle ear in recent cetaceans. Bull. Br. Mus. Nat. Hist. Zool. 7, 1–140.

Galliano, R.E., Morgane, P.J., McFarland, W.L., Nagel, E.L., Catherman, R.L., 1966. The anatomy of the cervicothoracic arterial system in the bottlenose dolphin (Tursiops truncatus) with a surgical approach suitable for guided angiography. Anat. Rec. 155, 325–338.

Goudappel, J.R., Slijper, E.J., 1958. Microscopic structure of the lungs of the bottlenose whale. Nature 182, 479.

Harrison, R.J., Fanning, J.C., 1973. Anatomical observations on the South Australian bottlenosed dolphin (Tursiops truncatus). Investig. Cetacea 5, 203–217.

Harrison, R.J., Tomlinson, J.D.W., 1956. Observations on the venous system in certain Pinnipedia and Cetacea. Proc. Zool. Soc. Lond. 126, 205–233.

Houser, D.S., Moore, P.W., Johnson, S., Lutmerding, B., Bransetter, B., Ridgway, S.H., 2010. Relationship of blood flow and metabolism to acoustic processing centers of the dolpin brain. J. Acoust. Soc. Am. 128, 1460–1466.

Hui, C.A., 1975. Thoracic collapse as affected by the retia thoracica in the dolphin. Respir. Physiol. 25, 63–70.

Ito, T., Kobayashi, K., Takahashi, Y., 1967. Histological studies on the respiratory tissue in the dolphin lung. Arch. Histol. Japan 28, 453–470.

Ivančić, M., Solano, M., Smith, C.R., 2014. Computed tomography and cross-sectional anatomy of the thorax of the live bottlenose dolphin (Tursiops truncatus). Anat. Rec. 297, 901–915.

Johansen, K., Elling, F., Paulev, F., 1988. Ductus arteriosus in pilot whales. Jap. J. Physiol. 38, 387–392.

Karandeeva, O.G., Matisheva, S.K., Shapunov, V.M., 1973. Features of external respiration in the Delphinidae. In: Chapskii, K.K., Sokolov, V.E. (Eds.), Morphology and Ecology of Marine Mammals. Halsted Press (John Wiley & Sons), New York, pp. 196–212, Israel Program for Scientific Translation (Jerusalem).

Kooyman, G.L., 1973. Respiratory adaptations in marine mammals. Am. Zool. 13, 457–468.

Kooyman, G.L., 1985. Physiology without restraint in diving mammals. Mar. Mammal Sci. 1, 166–178.

Lindholm, P., Nyrén, S., 2005. Studies on inspiratory and expiratory glossopharyngeal breathing in breath-hold divers employing magnetic resonance imaging and spirometry. Eur. J. Appl. Physiol. 94, 646–651.

Lluch, S., Diéquez, G., Garcia, A.L., Gòmez, B., 1985. Rete mirabile of goat: its flow-damping effect on cerebral circulation. Am. J. Physiol. 249, R482–R489.

Macdonald, A.A., Carr, P.A., Currie, R.J.W., 2007. Comparative anatomy of the foramen ovale in the hearts of cetaceans. J. Anat. 211, 64–77.

McFarland, W.L., Jacobs, M.S., Morgane, P.J., 1979. Blood supply to the brain of the dolphin, Tursiops truncatus, with comparative observations on special aspects of the cerebrovascular supply of other vertebrates. Neurosci. Biobehav. Rev. 3 (Suppl. 1), 1–93.

Meagher, E.M., McLellan, W.A., Westgate, A.J., Wells, R.S., Frierson, Jr, D., Pabst, D.A., 2002. The relationship between heat flow and vasculature in the dorsal fin of wild bottlenose dolphins Tursiops truncatus. J. Exp. Biol. 205, 3475–3486.

Menezes de Oliveira e Silva, F., Carvalho, V.L., Guimarães, J.P., Vergara-Parente, J.E., de Meirelles, O.A.C., Marmontel, M., Miglino, M.A., 2014. Accessory spleen in cetaceans and its relevance as a secondary lymphoid organ. Zoomorphol. 133, 343–350.

Moore, M.J., Hammar, T., Arruda, J., Cramer, S., Dennison, S., Montie, E., Fahlman, A., 2011. Hyperbaric computed tomographic measurement of lung compression in seals and dolphins. J. Expl. Biol. 214, 2390–2397.

Moore, C., Moore, M., Trumble, S., Niemeyer, M., Lentell, B., McLellan, W., Costidis, A., Fahlman, A., 2014. A comparative analysis of marine mammal tracheas. I. Expl. Biol. 217, 1154–1166.

Mortola, J.P., Limoges, M.-J., 2006. Resting breathing frequency in aquatic mammals: a comparative analysis with terrestrial species. Resp. Physiol. Neurobiol. 154, 500–514.

Nagel, E.L., Morgane, P.J., McFarlane, W.L., Galliano, R.E., 1968. Rete mirabile of dolphin: its pressure-damping effect on cerebral circulation. Science 161, 898–900.

Nakamine, H., Nagata, S., Yonezawa, M., Tanaka, Y., 1992. The whale (Odontoceti) spleen: a type of primitive mammalian spleen. Acta Anat. Japan 67, 69–82.

NAV, 2012. Nomina Anatomica Veterinaria—V revised edition. Published online by the Editorial Committee, Hannover (Germany), Columbia (MO, USA), Ghent (Belgium), Sapporo (Japan), pp. 1–160.

Ninomiya, H., Yoshida, E., 2007. Functional anatomy of the ocular circulatory system: vascular corrosion casts of the cetacean eye. Vet. Ophthalmol. 10, 231–238.

Noren, S.R., Williams, T.M., 2000. Body size and skeletal muscle myoglobin of cetaceans: adaptations for maximizing dive duration. Comp. Biochem. Physiol. A 126, 181–191.

Noren, S.R., Cuccurullo, V., Williams, T.M., 2004. The development of diving bradycardia in bottlenose dolphins (Tursiops truncatus). J. Comp. Physiol. B 174, 139–147.

Noren, S.R., Kendall, T., Cuccurullo, V., Williams, T.M., 2012. The dive response redefined: underwater behavior influences cardiac variability in freely diving dolphins. J. Expl. Biol. 215, 2735–2741.

O'Malley Miller, P.J., Shapiro, A.D., Deecke, V.B., 2010. The diving behaviour of mammal-eating killer whales (Orcinus orca): variations with ecological not physiological factors. Can. J. Zool. 88, 1103–1112.

Ono, N., Yamaguchi, T., Ishikawa, H., Arakawa, M., Takahashi, N., Saikawa, T., Shimada, T., 2009. Morphological varieties of the Purkinje fiber network in mammalian hearts, as revealed by light and electron microscopy. Arch. Histol. Cytol. 72, 139–149.

Pabst, D.A., Rommel, S.A., McLellan, W.A., 1999. The functional morphology of marine mammals. In: Reynolds, III, J.E., Rommel, S.A. (Eds.), Biology of Marine Mammals. Smithsonian Institution Press, Washington, pp. 15–72.

Pilleri, G., Arvy, L., 1971. Aselli's pseudopancreas (nodi lymphatici mesenterici) in two delphinids: Delphinus delphis and Stenella coeruleoalba. Investig. Cetacea 3, 189–193.

Piscitelli, M.A., McLellan, W.A., Rommel, S.A., Blum, J.E., Barco, S.G., Pabst, D.A., 2010. Lung size and thoracic morphology in shallow and deep-diving cetaceans. J. Morphol. 271, 654–673.

Piscitelli, M.A., Raverty, S.A., Lillie, M.A., Shadwick, R.E., 2013. A review of cetacean lung morphology and mechanics. J. Morphol. 274, 1425–1440.

Ponganis, P.I., 2015. Diving Physiology of Marine Mammals and Seabirds. Cambridge University Press, Cambridge, UK, pp. 1–334.

Ponganis, P.J., 2011. Diving mammals. Compr. Physiol. 1, 517–535.

Ponganis, P.J., Kooyman, G.L., Ridgway, S.H., 2003. Comparative diving physiology. In: Brukakk, A.O., Neuman, T.S. (Eds.), Bennett and Elliott's Physiology and Medicine of Diving, fifth ed. Saunders, Edinburgh, pp. 211–226.

Reidenberg, J.S., Laitman, J.T., 1987. Position of the larynx in Odontoceti (toothed whales). Anat. Rec. 218, 98–106.

Reidenberg, J.S., Laitman, J.T., 2008. Sisters of the sinuses: cetacean air sacs. Anat. Rec. 291, 1389–1396.

Richard, J., Neuville, H., 1896. Foie et sinus veineux intra-hepatiques du *Grampus griseus*. Bull. Mus. Hist. Nat. 2, 335–337.

Ridgway, S.H., Howard, R., 1979. Dolphin lung collapse and intramuscular circulation during free diving: evidence from nitrogen washout. Science 206, 1182–1183.

Ridgway, S.H., Scronce, B.L., Kanwisher, J., 1969. Respiration and deep diving in the bottlenose porpoise. Science 166, 1651–1654.

Ridgway, S.H., Houser, D.S., Finneran, J., Carder, D., Keogh, M., Van Bonn, W., Smith, C., Scadeng, M., Mattrey, R., Hoh, C., 2006. Functional imaging of dolphin brain metabolism and blood flow. J. Exp. Biol. 209, 2902–2910.

Romano, T.A., Felten, S.Y., Olschowka, J.A., Felten, D.L., 1993. A microscopic investigation of the lymphoid organs of the beluga, *Delphinapterus leucas*. J. Morphol. 215, 261–287.

Rommel, S.A., Lowenstine, L.J., 2001. Gross and microscopic anatomy. In: Dierauf, L.A., Gulland, F.M.D. (Eds.), CRC Handbook of Marine Mammal Medicine. second ed. CRC Press, Boca Raton, FL, pp. 129–164.

Sakata, R., 1959. Contributions to the anatomy of the cetacean heart, especially on the conducting system of the blue-white dolphin. Acta Anat. 34, 819–832.

Sassu, R., Cozzi, B., 2007. The external and middle ear of the striped dolphin *Stenella coeruleoalba* (Meyen 1833). Anat. Histol. Embryol. 36, 197–201.

Scholander, P.F., 1940. Experimental investigations on the respiratory function in diving mammals and birds. Hvalradets Skrifter 22, 1–131.

Sedmera, D., Misek, I., Klima, M., Thompson, R.P., 2003. Heart development in the spotted dolphin (*Stenella attenuata*). Anat. Rec. Part A 273, 687–699.

Shimokawa, T., Doihara, T., Makara, M., Miyawaki, K., Nabeka, H., Wakisaka, H., Kobayashi, N., Matsuda, S., 2011. Lectin histochemistry of respiratory mucosa in the Pacific white-sided dolphin. J. Vet. Med. Sci. 73, 1233–1236.

Simpson, J.G., Gardner, M.B., 1972. Comparative microscopic anatomy of selected marine mammals. In: Ridgway, S.H. (Ed.), Mammals of the Sea. Biology and Medicine. Charles C Thomas, Springfiels, IL, pp. 298–418.

Skrovan, R.C., Williams, T.M., Berry, P.S., Moore, P.W., Davis, R.W., 1999. The diving physiology of bottlenose dolphins (*Tursiops truncatus*). II. Biomechanics and changes in buoyancy at depth. J. Expl. Biol. 202, 2749–2761.

Slijper, E.J., 1936. Die cetacean. Vergleichend-anatomisch und systematisch. Ein beitrag zur vergleichenden anatomie des blutgefäss-, nerven- und muskelsystems, sowie des rumpfskelettes der säugetiere, mit studien über die theorie des aussterbens und der foetalisation. Capita Zoologica Bd VI–VII, Martin Nijhoff Publisher, pp. 1–590.

Slijper, E.J., 1939. *Pseudorca crassidens* (Owen). Ein beitrag zur vergleichenden anatomie der cetaceen. Zool. Mededeel. 21, 241–366.

Slijper, E.J., 1958. Organ weights and symmetry problems in porpoises and seals. Arch. Néerland Zool. 13, 97–113.

Slijper, E.J., 1979. Whales, second ed. Hutchinson, London, 511 p.

Slijper, J., 1961. Foramen ovale and *ductus arteriosus Botalli* in aquatic mammals. Mammalia 25, 528–570.

Smith, T.L., Turnbull, B.S., Cowan, D.F., 1999. Morphology of the complex laryngeal gland in the Atlantic bottlenose dolphin, *Tursiops truncatus*. Anat. Rec. 254, 98–106.

Sommer, L.S., McFarland, W.L., Galliano, R.E., Nagel, E.L., Morgane, P.J., 1968. Hemodynamic and coronary angiographic studies in the bottlenose dolphin (*Tursiops truncatus*). Am. J. Physiol. 215, 1498–1505.

Štěrba, O., Klima, M., Schildger, B., 2000. Embryology of dolphins—staging and ageing of embryos and fetuses of some cetaceans. Adv. Anat. Embryol. Cell Biol. 157, 1–133.

Tanaka, Y., 1994. Microscopy of vascular architecture and arterovenous communications in the spleen of two Odontocetes. J. Morphol. 221, 211–233.

Terasawa, F., Ohizumi, H., Ohshita, I., 2010. Effect of breath-hold on blood gas analysis in captive Pacific white-sided dolphins (*Lagenorhynchus obliquidens*). J. Vet. Med. Sci. 72, 1221–1224.

Truex, R.C., Nolan, F.G., Truex, Jr, R.C., Schneider, H.P., Perlmutter, H.I., 1961. Anatomy and pathology of the whale heart with special reference to the coronary circulation. Anat. Rec. 141, 325–353.

van Nie, C.J., 1985. The conducting system of the heart of the white-snouted dolphin *Lagenorhynchus albirostris* (Gray, 1846). Lutra 28, 106–112.

Viamonte, M., Morgane, P.J., Galliano, R.E., Nagel, E.L., McFarland, W.L., 1968. Angiography in the living dolphin and observations on blood supply to the brain. Am. J. Physiol. 214, 1225–1249.

Vogl, A.W., Fisher, H.D., 1981. Arterial circulation of the spinal cord and brain in the *Monodontidae* (order Cetacea). J. Morphol. 170, 171–180.

Whisnant, J.P., Millikan, C.H., Wakim, K.G., Sayre, G.P., 1956. Collateral circulation to the brain of the dog following bilateral ligation of the carotid and vertebral arteries. Am. J. Physiol. 186, 275–277.

Williams, T.M., Friedl, W.A., Fong, M.L., Yamada, R.M., Sedivy, P., Haun, J.E., 1992. Travel at low energetic cost and wave-riding bottlenose dolphins. Nature 355, 821–823.

Williams, T.M., Friedl, W.A., Haun, J.E., 1993. The physiology of bottlenose dolphins (*Tursiops truncatus*): heart rate, metabolic rate and plasma lactate concentration during exercise. J. Expl. Biol. 179, 31–46.

Williams, T.M., Haun, J.E., Friedl, W.A., 1999a. The diving physiology of bottlenose dolphins (*Tursiops truncatus*). I. Balancing the demands of exercise for energy conservation at depth. J. Expl. Biol. 202, 2739–2748.

Williams, T.M., Noren, D., Berry, P., Estes, J.A., Allison, C., Kirtland, J., 1999b. The diving physiology of bottlenose dolphins (*Tursiops truncatus*). III. Thermoregulation at depth. J. Expl. Biol. 202, 2763–2769.

Williams, T.M., Haun, J., Davis, R.W., Fuiman, L.A., Kohin, S., 2001. A killer appetite: metabolic consequences of carnivory in marine mammals. Comp. Biochem. Physiol. A 129, 785–796.

Williams, T.M., Fuiman, L.A., Kendall, T., Berry, P., Richter, B., Noren, S.R., Thometz, N., Shattock, M.J., Farrell, E., Stamper, A.M., Davis, R.W., 2014. Exercise at depth alters bradycardia and incidence of cardiac anomalies in deep-diving marine mammals. Nat. Commun. 6, 6055.

Yablokov, A.V., Bel'kovich, V.M., Borisov, V.I., 1972. The vascular system, blood and circulatory distinctions. In: Whales and Dolphins. Moscow, Translation by Joint Publication Research Service, Arlington, VA, USA, pp. 151–173.

Chapter 5

Head and Senses

DOLPHIN HEAD CHARACTERISTICS

Before going into details as to the organization of the dolphin head, it may be useful to remember its skull in comparison with that in terrestrial mammals (see Skull). In the latter, the bony tissue gives general strength to the head and resilience against forces from the outside and during jaw movements. It also houses and protects the brain and sensory organs in bony capsules, which are nearly hermetically closed (brain capsule) or more or less open to expose the sensory organs to the surroundings of an animal (nose, eyes, ears). One can find the same situation in the dolphin. However, it is evident that in the dolphin, the bony elements of the upper jaw have been extended far back to the vertex of the head, which is particularly prominent because of the relatively large size and globular shape of the brain (Fig. 5.1). In parallel, the external bony opening of the nose (bony naris) is shifted from a terminal position in other mammals to a caudal position above and behind the orbita (Figs. 5.1 and 5.2: arrows). Also the jaw apparatus has been elongated rostrally to considerable length in the delphinid species. Another important characteristic in dolphins is the comparatively long row of numerous small teeth of similar shape (homodont), which seem to act as a fishing gear for grasping and holding prey before it is swallowed whole. This latter feeding mode does not need strong biting forces and heavy jaw muscles; as a consequence, the zygomatic arch is reduced to a thin clip which helps to hold the eyeball in place.

An important result of the morphological changes within the dolphin skull is the fact that the nasal passages no longer run more or less horizontally through the anterior part of skull but instead stand upright, that is, perpendicular to the body axis (Figs. 5.1 and 5.2). This specific course of the upper respiratory tract seems to be useful for efficient surfacing to exhale and inhale very rapidly.

Head of Dolphins as an Entity

As in other mammals, the head of dolphins is complicated because different organ systems are connected to each other and are sometimes intertwined: bones, muscles, nerves, blood vessels, sensory organs, respiratory tract, alimentary tract, etc. As the topographical relations of single structures are often difficult to understand, we shall try to demonstrate and explain the principal organization of the dolphin head with conventional sections and modern imaging techniques (CT, x-ray computed tomography; MRI, magnetic resonance imaging) in sagittal and transverse planes. Such an overview on structures and/or organ systems allows an easy orientation/navigation within the head as a whole and to reliably find regions of interest, for example, for the sampling of tissue needed for histological and other purposes.

Sectional Planes Through the Dolphin Head

In this sub-chapter, the entity of the dolphin's head is illustrated and shortly described in sections through adult and subadult bottlenose dolphins and a neonate spotted dolphin (Rauschmann et al., 2006), which were done either by freezing and slicing the head with a band saw or by digital documentation (CT, MRI).

To introduce the soft anatomy of the head, it is useful to describe first nearly midsagittal sections. Here, midsagittal and parasagittal MRI scans of a subadult bottlenose dolphin (Fig. 5.3a) as well as the midsagittal section through a head of a late common dolphin fetus (Fig. 5.3b) show an overview of the head structures.

Rostrally in the forehead of the animal, there is the melon, a prominent fat body, which is involved in the emission of sound signals. The skull largely consists of the rostrum with the mesorostral cartilage and the brain case. Ventrally, the region of the jaws presents the oral cavity with the tongue and the strong gular musculature. Caudal to the melon there is the curved but nearly vertical soft nasal passage (Figs. 5.3b and 5.4). The nasal plugs, cushions of dense connective tissue and muscle, seal the upper respiratory tract at the level of the bony nostrils. The soft nasal passage opens ventrally into the bony nasal passage and then, caudoventrally, into the larynx, which is situated below the skull base. The pharyngeal muscles are rather thick below the skull base (Fig. 5.3a,b) and enclose the rostral-most part of the larynx. In the lowermost part of the

Anatomy of Dolphins. http://dx.doi.org/10.1016/B978-0-12-407229-9.00005-1

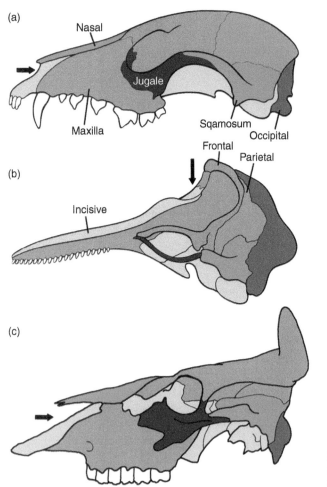

FIGURE 5.1   Comparison of skulls (lateral views) of (a) a fox (*Vulpes vulpes*), (b) a toothed whale (*Phocoena phocoena*), (c) and a cow (*Bos taurus*) brought to the same length. The *arrows* mark the entrance of the bony nasal passages. *(Modified after Huggenberger and Klima, 2015 and Nickel et al., 2003)*

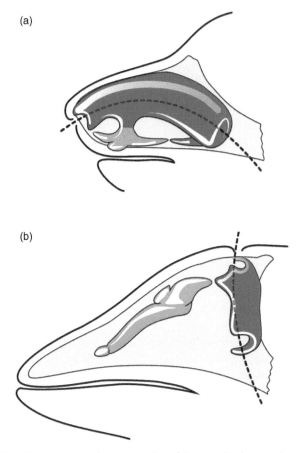

FIGURE 5.2   Schematic representation of the same developmental stage of nasal cartilaginous structures in a generalized terrestrial mammal (a) and a dolphin (b). The *dashed line* represents the route of the nasal passage. *Yellow* are the median nasal structures including the future mesorostral cartilage (Figs. 5.3, and 5.5), *blue* are the anterior nasal side wall structures, and *red* are the posterior sode wall structures including the tectum nasi. *(Modified after Klima, 1999)*

scans, the alimentary tract (pharynx) is equipped with thick muscles that attach the lower jaw and tongue to the hyoid bone. Adjacent ventrally, the pharynx is embraced by elements of the hyoid apparatus (basihyal, thyrohyal).

For a closer inspection, the above named structures are described in more detail using the transverse views. The rostral-most planes of the transverse views through the dolphin forehead Figs. 5.5a–c show dorsally the melon. The melon is braced bilaterally by sheets of the facial musculature (rostral part of the maxillonasolabial muscle); the latter can change the shape of the melon during sound emission. The skull consists largely of the base of the rostrum with the mesorostral cartilage and horizontal parts of the frontal region (maxilla, incisive bone). This bony plate bears the premaxillary sac, the ventral-most of the accessory nasal sacs, which belong to the upper respiratory tract. Ventrally, the region of the jaws presents the oral cavity with the tongue and the palatine area at the level of the angle of the gape.

Transverse views in Figs. 5.5b–g of a somewhat more caudal plane touches the orbita (not labeled). The minute bony element in cross-section is part of the zygomatic arch, which runs along the lower margin of the orbita (Fig. 5.5c). In the upper half, the section is dominated by structures of the respiratory tract; it is obvious that the nasal passage runs more or less perpendicular to the skull base. The nasal plugs and their paired muscles are prominent here. The mandibular bone is hollow; the alveolar canal is very wide and filled with fat which represents another acoustic fat body, this time involved in sound perception (intra- and extramandibular fat bodies). In this area, there is no medial wall of the mandible (see the following). In the lowermost part of the sections, the alimentary tract (pharynx) is equipped with thick muscles that attach the lower jaw and tongue to the hyoid bone.

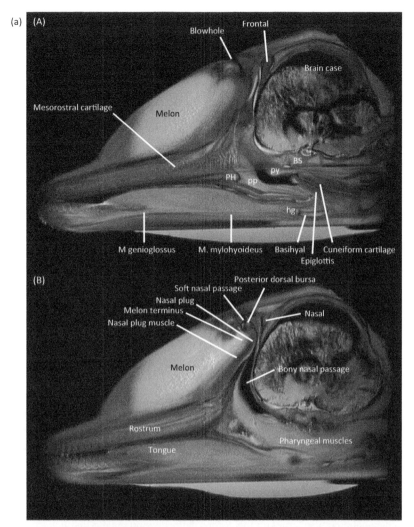

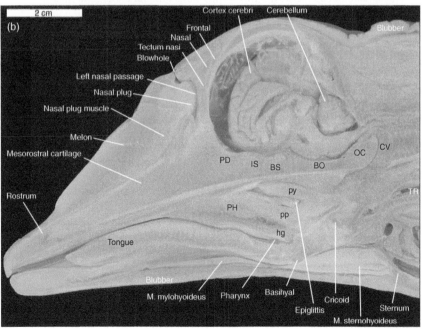

FIGURE 5.3   (a) Midsagittal (A) and parasagittal (B) MRI scans of a subadult bottlenose dolphin give a complete overview of head structures. *BS*, basi-sphenoid; *hg*, hyoglossus muscle; *PH*, pterygoid hamulus; *pp*, palatopharyngeus muscle; *py*, pterygopharyngeus muscle. (b) Nearly mediosagittal cryosection of a late fetal head of a common dolphin showing the anatomical structures of the nasal complex and the skull as well as the gular region. *BO*, basioccipital; *BS*, basisphenoid; *CV*, cervical vertebrae; *hg*, hyoglossus muscle; *IS*, intersphenoid synchrondosis; *OC*, occipital condyle; *PD*, presphenoid; *PH*, pterygoid hamulus; *pp*, palatopharyngeus muscle; *py*, pterygopharyngeus muscle; *TR*, trachea. *(Part a: Department of Radiology, University Hospital Cologne, Germany)*

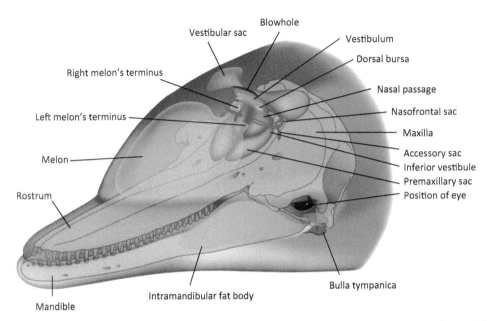

FIGURE 5.4 **Oblique view of a dolphin head showing the skull, nasal structures, and the acoustic fat body of the left mandible terminating at the tympanic bulla.** *(Artwork by Uko Gorter)*

Transverse views of Figs. 5.5g–o comprises the planes through the center of the head: The brain cavity with the large brain is wide; the eyes exhibit their round lens. Within the brain, the gray matter is hypointense (light gray) in the MRI slice, whereas the white matter is hyperintense (white) (Figs. 5.5h,k, and m). In the middle level, the pharyngeal muscle (*M. constrictor pharyngis rostralis*) is rather thick and largely obstructs the nasal passage. This muscle is essential for the dolphin; it seals the respiratory tract against incoming water. Transverse views (Figs. 5.5k–o) show sections through the center of the brain. In young animals, the cranial vault comprises a distinctly more sectional area than the rest of the head (Figs. 5.5j,k,m, and n), which is not the case in the adult (Fig. 5.5l). The brain is enclosed in the dura mater, consisting of extremely dense connective tissue. It exhibits the telencephalic hemisphere with the corpus callosum and a clear distinction of the gray and white matter of the neocortex, the diencephalon with the optic chiasm, the thalamus, and the parts of the ventricular system (Figs. 5.5l and k). The mandibular bone thins out in the direction of the temporomandibular joint (Fig. 5.5k); it releases its intramandibular fat body, which approaches in the slightly more caudal level the ear bones (tympanoperiotic complex) from laterally and ventrally. The pharyngeal sphincter muscle (*M. constrictor pharyngis rostralis*) encloses the rostral-most part of the larynx (Figs. 5.5k–m). Adjacent ventrally the pharynx is embraced by elements of the hyoid apparatus.

The sections through the caudal-most part of the head presents the posterior parts of the telencephalic hemispheres (Fig. 5.5o). Between them, the brainstem is ensheathed by rostral parts of the tentorium cerebelli. The ear bones have been uncoupled from the adjacent bony elements of the skull elements. The ceramic-like bones are white in the cryosection and the CT scan, whereas in the MRI slice they appear black. Between the periotic and tympanic bones, the tympanic (middle ear) cavity is visible. The middle ear ossicles, however, are not visible there because of their small size. The sternohyoid muscle is a strong muscle mass ventral of the larynx.

Forehead Region

The organization of the dolphin forehead is unparalleled in the animal kingdom. Its structures can be summarized in the terms "nasal complex" or "epicranial complex," that is located above the skull. They are so aberrant that, in the past, famous anatomists could not detect the functional implications of this head region (Huggenberger et al., 2009). Some of the structures do not even have a counterpart (homologon) among the terrestrial mammals, a situation not often found in morphology (Huggenberger et al., 2016). These structures are the blowhole, accessory air sacs, monkey lips, dorsal bursae, and the melon as well as the organization of the facial musculature surrounding the blowhole (Fig. 5.5a–g). Next to respiration, that is, rapid opening of the nasal passages and its water-tight closure during dives, the structures of the nasal complex act together as a pneumatically driven sound generator. Interestingly, an olfactory sensory epithelium is lacking in dolphins.

The upper respiratory tract. The forehead anatomy of dolphins has been investigated by several authors (Cranford et al., 1996; Lawrence and Schevill, 1956; Mead, 1975). Toothed whales seem to be peculiar in that they have no facial expression although their facial musculature is well developed, highly complicated, and well innervated by the facial nerve

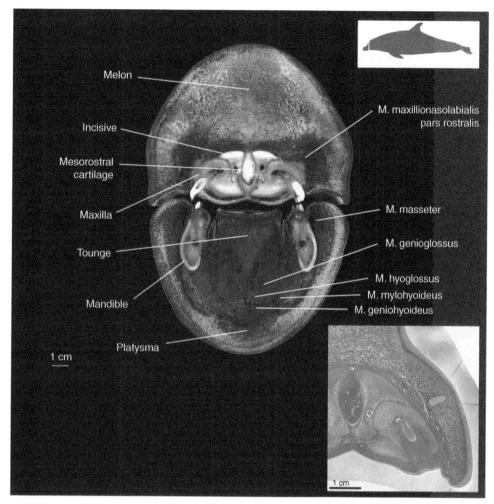

FIGURE 5.5A Transverse sawing section (caudal view) of an adult bottlenose dolphin head and a histological slide (inset: Orcein Stain) of the anterior rostrum region of a different animal.

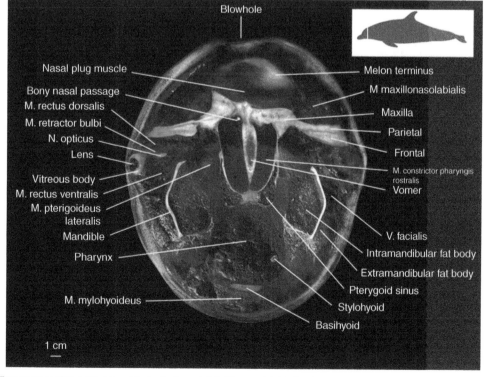

FIGURE 5.5B Transverse sawing section (caudal view) of an adult bottlenose dolphin head.

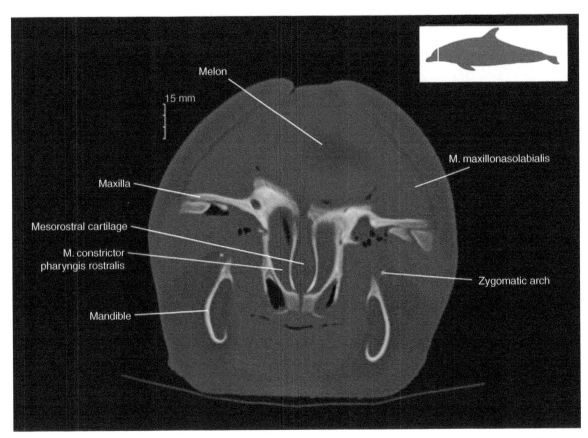

FIGURE 5.5C Transverse CT scan of a subadult bottlenose dolphin head. *(Courtesy of Department of Radiology, University Hospital Cologne, Germany)*

FIGURE 5.5D Collection of transverse sawing section, MRI scan, and CT scan of a perinatal pantropical spotted dolphin head of approximate the ▶ same level. *(Modified after Rauschmann et al., 2006. Head morphology in perinatal dolphins: a window into phylogeny and ontogeny. J. Morphol. 267, 1295–1315.)*

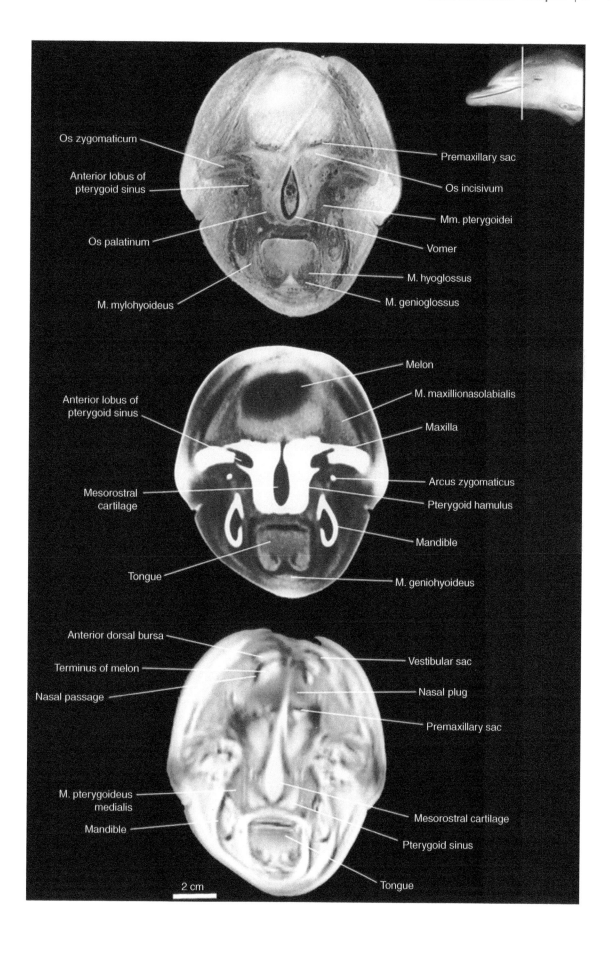

Os zygomaticum

Anterior lobus of pterygoid sinus

Os palatinum

M. mylohyoideus

Premaxillary sac

Os incisivum

Mm. pterygoidei

Vomer

M. hyoglossus

M. genioglossus

Anterior lobus of pterygoid sinus

Mesorostral cartilage

Tongue

Melon

M. maxillionasolabialis

Maxilla

Arcus zygomaticus

Pterygoid hamulus

Mandible

M. geniohyoideus

Anterior dorsal bursa

Terminus of melon

Nasal passage

M. pterygoideus medialis

Mandible

Vestibular sac

Nasal plug

Premaxillary sac

Mesorostral cartilage

Pterygoid sinus

Tongue

2 cm

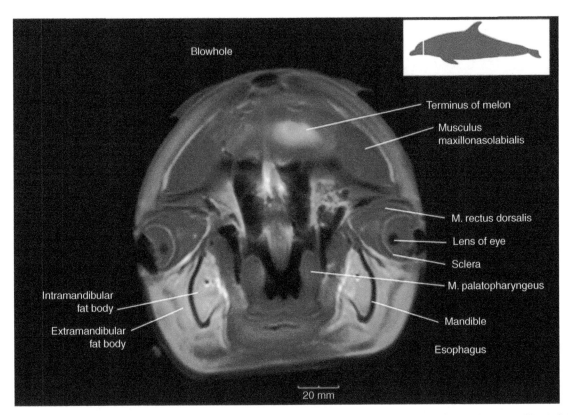

FIGURE 5.5E Transverse MRI scan of a subadult bottlenose dolphin head. *(Courtesy of Department of Radiology, University Hospital Cologne, Germany)*

FIGURE 5.5F Collection of transverse sawing section, MRI scan, and CT scan of a perinatal pantropical spotted dolphin head of approximate the ▶ same level. *(Modified after Rauschmann et al., 2006. Head morphology in perinatal dolphins: a window into phylogeny and ontogeny. J. Morphol. 267, 1295–1315.)*

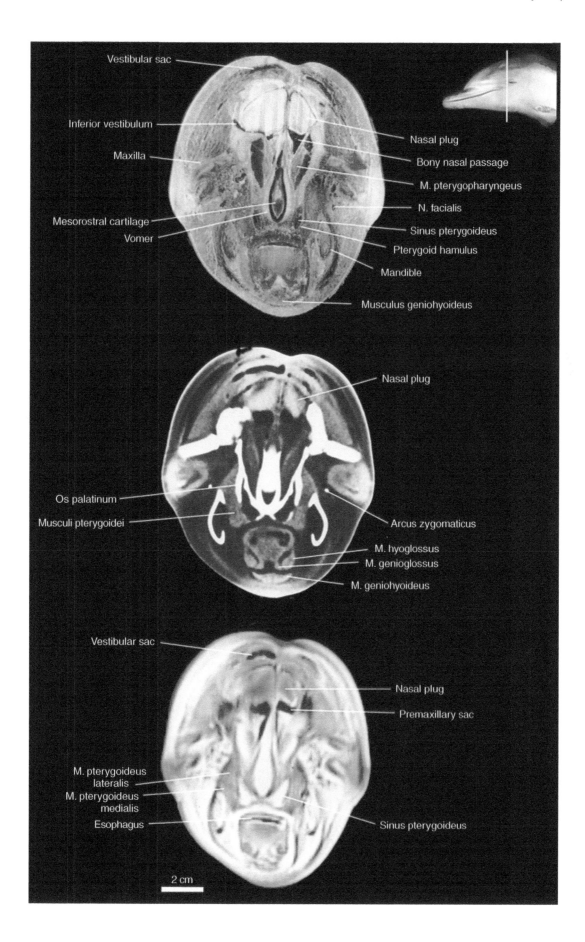

2 cm

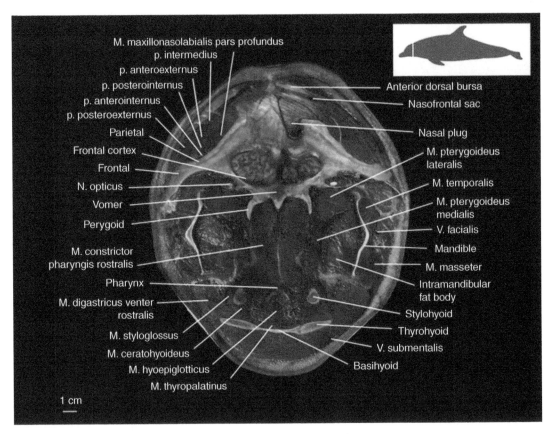

FIGURE 5.5G Transverse sawing section (caudal view) of an adult bottlenose dolphin head.

FIGURE 5.5H Collection of transverse sawing section, MRI scan, and CT scan of a perinatal pantropical spotted dolphin head of approximate the ▶ same level. *(Modified after Rauschmann et al., 2006. Head morphology in perinatal dolphins: a window into phylogeny and ontogeny. J. Morphol. 267, 1295–1315.)*

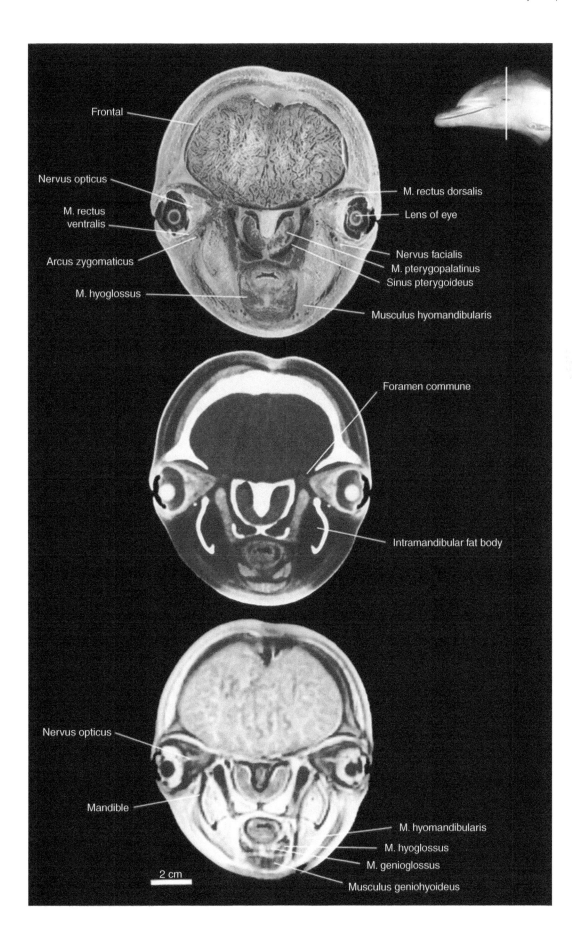

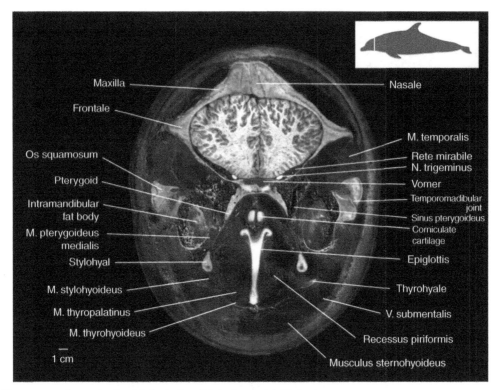

FIGURE 5.5I   Transverse sawing section (caudal view) of an adult bottlenose dolphin head.

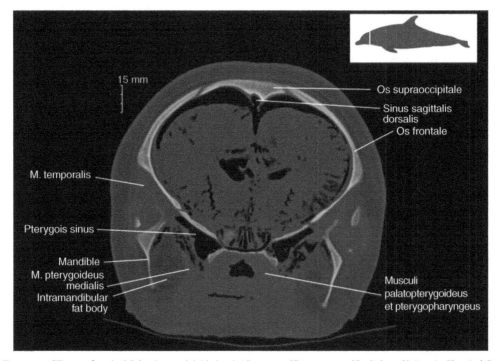

FIGURE 5.5J   Transverse CT scan of a subadult bottlenose dolphin head. *(Courtesy of Department of Radiology, University Hospital Cologne, Germany)*

FIGURE 5.5K   Collection of transverse sawing section, MRI scan, and CT scan of a perinatal pantropical spotted dolphin head of approximate the ▶ same level. *(Modified after Rauschmann et al., 2006. Head morphology in perinatal dolphins: a window into phylogeny and ontogeny. J. Morphol. 267, 1295–1315.)*

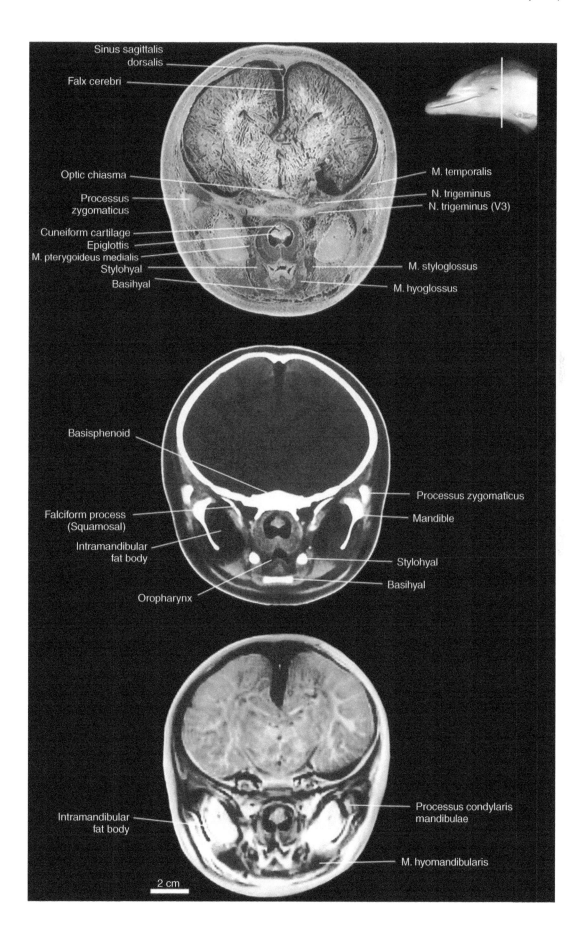

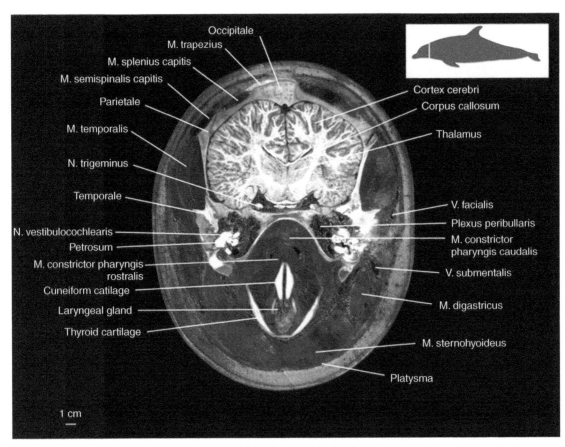

FIGURE 5.5L   Transverse sawing section (caudal view) of an adult bottlenose dolphin head.

FIGURE 5.5M   Collection of transverse sawing section, MRI scan, and CT scan of a perinatal pantropical spotted dolphin head of approximate the ▶ same level. In the sawing section, the structure of the brain is massively altered because of the freezing and thawing of the neonate animal during cryo-sectioning. *(Modified after Rauschmann et al., 2006. Head morphology in perinatal dolphins: a window into phylogeny and ontogeny. J. Morphol. 267, 1295–1315.)*

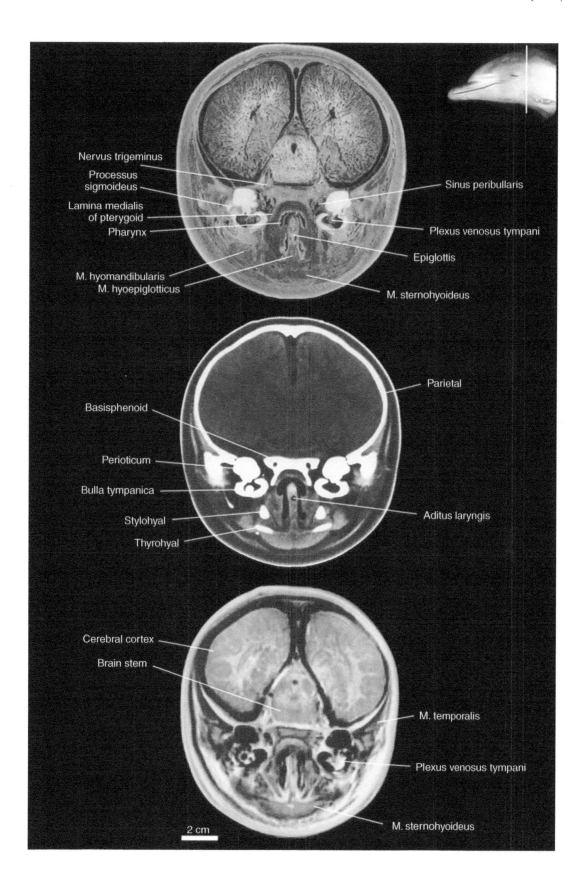

Nervus trigeminus

Processus sigmoideus

Lamina medialis of pterygoid

Pharynx

M. hyomandibularis
M. hyoepiglotticus

Sinus peribullaris

Plexus venosus tympani

Epiglottis

M. sternohyoideus

Basisphenoid

Perioticum

Bulla tympanica

Stylohyal

Thyrohyal

Parietal

Aditus laryngis

Cerebral cortex

Brain stem

M. temporalis

Plexus venosus tympani

M. sternohyoideus

2 cm

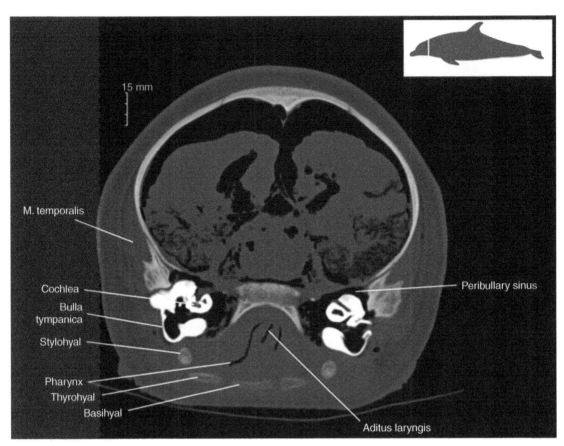

FIGURE 5.5N   Transverse CT scan of a subadult bottlenose dolphin head. *(Courtesy of Department of Radiology, University Hospital Cologne, Germany)*

FIGURE 5.5O   Collection of transverse sawing section, MRI scan, and CT scan of a perinatal pantropical spotted dolphin head of approximate the ▶
same level. *(Modified after Rauschmann et al., 2006. Head morphology in perinatal dolphins: a window into phylogeny and ontogeny. J. Morphol. 267, 1295–1315.)*

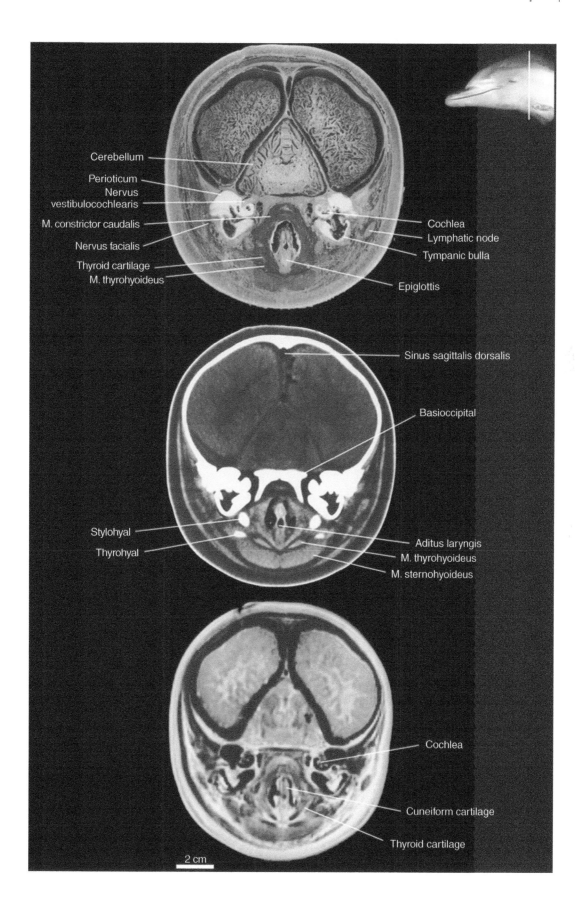

Cerebellum

Perioticum

Nervus vestibulocochlearis

M. constrictor caudalis

Nervus facialis

Thyroid cartilage
M. thyrohyoideus

Cochlea

Lymphatic node

Tympanic bulla

Epiglottis

Sinus sagittalis dorsalis

Basioccipital

Stylohyal

Thyrohyal

Aditus laryngis
M. thyrohyoideus
M. sternohyoideus

Cochlea

Cuneiform cartilage

Thyroid cartilage

2 cm

(Huggenberger et al., 2009). In these animals, it is concentrated in several sheets around the blowhole and below the melon (Fig. 5.5a and g). The nasal (epicranial) complex is seated in the so-called "facial depression" in the center of the skull. It comprises the upper respiratory tract from the single external nasal opening (blowhole) above to the paired bony nostrils below. The latter are situated in the center of this depression, immediately in front of the brain case at about half the height of the neurocranium (Fig. 5.1). The blowhole is situated dorsal and rostral to the vertex of skull (Fig. 5.3). When closed and seen from above, it resembles a transverse semicircular slit, which opens backward (see Diving: Breathing, Respiration, and the Circulatory System). The anterior lip of the blowhole can be moved significantly by hand, whereas the posterior lip is firmly attached to the skull roof. Below the blowhole, the unpaired dorsal nasal tract stands more or less vertically as a flattened transverse vestibulum (Fig. 5.4). Further ventrally, the vestibulum is divided into two nasal passages by the free edge of the soft nasal septum. Just below the division, the monkey lips (phonic lips) are situated. These monkey lips are visible as two slight, horizontal band-like thickenings in the rostral and caudal wall of each nasal passage and are characterized by delicate longitudinal (dorsoventral) wrinkles (Fig. 5.6). The epithelium of the monkey lips in the harbor porpoise comprises 70–80 layers of extremely flattened cells, that is, 4 times more layers than in the remaining epicranial air spaces (Prahl et al., 2009). This fact could not be verified in the bottlenose dolphin where the epithelium of the monkey lips were thinner than of the neighboring nasal epithelium but connected to the underlying dermis by long and complex dermal papillae (Fig. 5.6). The lips oppose each other and thus flank the air stream: a delicate horizontal ridge on the surface of the rostral monkey lip fits into the corresponding groove on the caudal lip representing a "mortise and tenon" complex. The monkey lips are based on small ellipsoid fat bodies, the dorsal bursae flanking the nasal passages rostrally and caudally (Figs. 5.4 and 5.5d). Embedded in the subepithelial connective tissue, they are part of the so-called monkey lips/dorsal bursae (MLDB) complex (Cranford et al., 1996). Apart from functional implications in phonation (see the following), this mortise and tenon complex and the underlying dorsal bursae can be regarded an additional sealing mechanism of the nasal passage below the blowhole against intruding water.

Below the monkey lips, both nasal passages are occluded by the nasal plugs, paired bodies of dense connective tissue and muscle that bulge from the anterior (rostral) wall into the lumen of the nasal passages (Figs. 5.3 and 5.5f and g). These nasal plugs fit into the apertures of the bony nostrils, a third device to close the nasal passage tightly even in dead specimens. For a tight sealing, the nasal plug is equipped with a node located dorsolaterally. This node fits from anterior into a fold formed by the diagonal membranes in the posterolateral wall at the ventral base of the soft nasal passages.

The epithelial lining of the vestibulum is dark gray, similar to the epidermis of the animal's back. At the level of the monkey lips, the color of the epithelium changes and becomes brighter and appears red in the bony nasal passages. In the rostral part of these passages, there are pinhole-like openings in the epithelium, indicating the position of mucous glands. They are supposed to serve as protection and lubrication of the nasal epithelium (Mead, 1975).

In the soft nasal passage, three pairs of air sacs (diverticula) communicate with the nasal passages by slit-like apertures (Fig. 5.4). These sacs can be taken as extensions of the associated nasal passage. The three pairs of diverticula lie in separate horizontal levels of the epicranial complex: from dorsal to ventral, the vestibular sacs (sacci dorsales) are followed by the nasofrontal sacs and the premaxillary sacs.

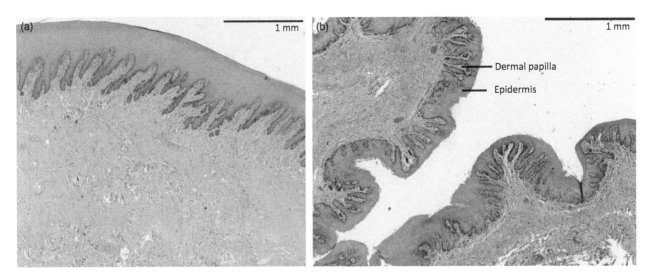

FIGURE 5.6 **Epithelia of the soft nasal passage dorsal TA (a) and at the monkey lips (b) of a bottlenose dolphin (ID196).** HE stain. Although the epidermis at the monkey lips is thinner the dermal papillae are relatively longer. For the general appearance of the skin, see Chapter 2.

TABLE 5.1 Extensions of Nasal Air Sacs in Adult Bottlenose Dolphins (Rodionov and Markov, 1992)

	Dimensions Left (cm)	Volume Left (cm^3)	Dimensions Right (cm)	Volume Right (cm^3)
Vestibular sac	65–80 long	60–65	45–60 long	50–55
Nasofrontal sac	12–13 long 1–1.5 in diameter	20–23	12–13 long 1–1.5 in diameter	20–23
Premaxillary sac	3–5 long 2.5–4 wide 1–2 high	17–30	5–7 long 4.5–6 wide 1.5–3 high	65–90

The vestibular sacs are lateral extensions of the vestibulum and reminiscent of two wings that flank the nasal passage dorsally (Fig. 5.4). Histologically, the epithelium of the vestibular sac is similar to that of the nasal passage (Fig. 5.6).

The nasofrontal sacs (sometimes called tubular sacs) are situated ventral to the vestibular sacs and on both sides encircling the nasal passages in the horizontal plane. On each side, the nasofrontal sac comprises a rostral part in front and a caudal part behind the nasal passage. The latter communicate with their nasal passage via horizontal slit-like apertures and the inferior vestibule (see the following). In this way, the posterior dorsal bursae are situated in a connective tissue lip between the soft nasal passage and the entrance into the caudal nasofrontal sac. This lip is stabilized by the blowhole ligament stretching transversely from left to right. The ligament originates from the incisive bones lateral to the bony nares.

The ventral-most and largest pair of nasal air sacs, the premaxillary sacs, rest on the caudal part of each incisive bone rostral to the bony nares. The dorsal epithelium of this sac is intensively linked to the tissue of the nasal plug via papillae of connective tissue. The premaxillary sacs expand laterocaudally around the bony nares to meet the slit-like opening of the nasofrontal sac posteriorly (Fig. 5.4). This connection, equipped with small accessory sacs, is called inferior vestibulum (Fig. 5.5f).

All nasal air sacs show an asymmetry; the right premaxillary sac is significantly larger than the left. However, the right vestibular sac is smaller than its left counterpart. Table 5.1 summarizes the dimensions of these sacs for the bottlenose dolphin.

Melon and associated connective tissue. Each pair of dorsal bursae is located in the center of connective tissue, surrounded by the nasal air sacs. The connective tissue center is permeated by muscle fiber bundles laterally and dorsocaudally (see the following). A large bulbous fat body, the melon, is situated rostral to this center and is connected via a fatty pathway (the melon's terminus) with the anterior dorsal bursa (see the following; Figs. 5.4 and 5.5e).

The melon is a symmetrical fat cushion, ovoid in shape but flattened dorsoventrally. It is supported ventrally by the maxillary and incisive bones of the rostrum (Figs. 5.3 and 5.5a–d). Therefore, in its location, the melon is responsible for the typical "face" of dolphins and their bulbous convex forehead contour. Whereas the core of the melon is nearly free of connective tissue fibers, the fiber content increases toward the periphery of the melon (Figs. 5.5 and 5.7). Ventrolaterally, these collagen fibers are interwoven with fiber bundles of the rostral facial muscles. Caudally the melon is covered by a sheet of dense collagen (tissue theca). On both sides, the ventrolateral part of the theca is continuous with the nasal plug

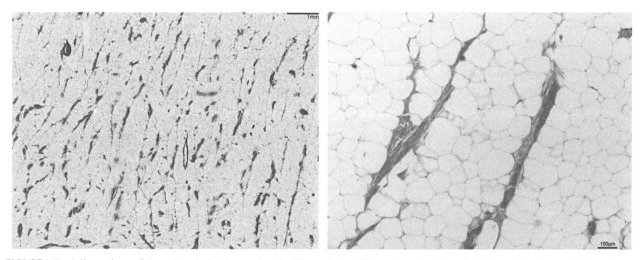

FIGURE 5.7 Adipose tissue of the melon of a bottlenose dolphin (ID196) in two different magnifications (scale bars: left = 1 mm; right = 100 μm). HE stain.

muscles, which are rich in connective tissue (see the following). Lateral and caudal to the theca, the subcutaneous connective tissue has similar properties as in the thick blubber, which covers the head except for the area above the melon and the theca, respectively. The caudal end (terminus) of the melon enters the theca. In delphinids, this terminus of the melon is bifurcated and, on the right hand side, it contacts via a short fatty pathway the right anterior dorsal bursa (Fig. 5.4).

Nasal Musculature

Superficially, the "face" of dolphins is characterized by vestigial muscles, the *m. sphincter colli profundus*, the *m. orbicularis oris*, a *m. auricolabialis* (= *m. zygomaticus*) and the *m.* (levator) *nasolabialis* (Huber, 1934; Lawrence and Schevill, 1965) (see Chapter 4). However, the only mimic muscles that maintain a specific topographic identity are those related to the eye (*m. orbicularis oculi*) and especially those of the hypertrophied nasal complex and blowhole region (epicranial complex).

The musculature of the epicranial complex in toothed whales is derived from the paired *Musculus maxillonasolabialis* in terrestrial mammals (Huber, 1934). The musculature originates from the surface of the maxillary bones and is functionally subdivided into the rostral part associated with the melon and the caudal part controlling the nasal passages and their diverticula.

The caudal part is arranged in six thin muscle layers which lie on top of each other to form a cone around the soft nasal passages, representing an onionskin-like organization (Fig. 5.5g). These bilaterally fan-shaped muscle layers arise concentrically from the maxilla and run more or less dorsomedially in the direction of the soft nasal passages and between the nasal sacs. Seen from above, the contours of the muscle sheets on each side describe semicircles around the nasal passages, starting out from the blowhole and radiating to the borders of the skull roof. From lateral to medial, the layers are named according to their position: *M. posteroexternus*, *M. intermedius*, *M. anteroexternus*, *M. posterointernus*, *M. anterointernus*, *M. profundus* (Fig. 5.5g) (Lawrence and Schevill, 1956). Moreover, the nasal plug muscle originates in front of the bony nares on the premaxilla and stretches dorsocaudally into the nasal plugs (Figs. 5.3, 5.4, and 5.5b).

The rostral part below the melon is further subdivided into a lateral and a medial portion originating both on the rostrum (Fig. 5.5a). Careful anatomical dissections, however, reveal that the rostral and caudal muscles layers are continuous. The medial rostral muscle is continuous with the profundus muscle and the origin of the lateral rostral muscle is connected to the anterointernus, posterointernus, anteroexternus, intermedius, and posteroexternus components caudally. The relative position of the medial components speaks for their identity as derivatives of the nasal portion and the topography of the lateral components for derivatives of the labial portion of the mammalian maxillonasolabialis muscle (Huggenberger et al., 2009).

Innervation

The sensory innervation of the facial structures in toothed whales is supplied by the maxillary division of the trigeminal nerve (see Genuine Peripheral Cranial Nerves) (Huber, 1934; Kern, 2012; Ridgway, 1990). The skin on the lips, around the eyes and blowhole, are heavily innervated. In the porpoise, many sensitive receptors were found in the connective tissue between the monkey lips and their adjacent dorsal bursa. They were identified as Vater-Pacini or Golgi-Mazzoni corpuscules, which react to pressure, adapt quickly, and are considered detectors of acceleration (Prahl et al., 2009). In contrast, the nasal diverticula and the melon do not appear to be heavily innervated (Mead, 1975). Free nerve endings or Merkel cells could not be found in the nasal complex. A contribution of the terminal nerve (see Cranial Nerve 0) running in parallel to the trigeminal nerve was taken into account because of topographic and functional implications but could not be substantiated yet (Buhl and Oelschläger, 1986).

Motor innervation of the nasal musculature is exclusively provided by the facial nerve (see Genuine Peripheral Cranial Nerves) (Huber, 1934; Kern, 2012; Mead, 1975; Ridgway, 1990). Quantitative analysis revealed that the bottlenose dolphin has about 7 times more axons in its facial nerve than the human (Morgane and Jacobs, 1972). In dolphins, the highly differentiated nasal musculature should have the capacity of discrete but versatile and complicated intrinsic movement patterns needed for the modulation of the nasal complex as a whole and in detail. The potential functional significance of this muscular ensemble seems to be related to its three-dimensional organization. It seems likely that single layers of the nasal muscles or parts of these layers are innervated by individual branches or axon bundles of the facial nerve and that these axons belong to specific neuron populations of the facial motor nucleus in the brainstem as is true for various components of the facial muscles in other mammals (Huggenberger et al., 2009). This kind of anatomical organization and functional implications within the epicranial complex seem to be valid for all the toothed whales investigated so far (Heyning, 1989; Huggenberger et al., 2010, 2009; Mead, 1975; Rauschmann et al., 2006). Such a detailed functional organization within the facial musculature of toothed whales and its heavy innervation, however, should allow highly complicated changes in the shape of nasal structures, including associated air spaces (as well as changes in their intrinsic air pressure). It is obvious by behavioral observations and experiments that dolphins possess a pronounced fine control of sound production, as evidenced, for example, by their proficiency in vocal mimicry (Reiss and McCowan, 1993; Ridgway et al., 2012). Thus, dolphins should be able to make "facial expressions" of considerable subtlety and variety not for visual but for auditory communication (Huggenberger et al., 2009). Variations in size and structure of facial components and a certain potential for their asymmetric development may even add to this remarkable capability a certain acoustic individuality in these animals.

Functional Aspects of the Nasal Complex

As in other mammals, the primary and main function of the nasal tract in toothed whales is respiration. Olfaction in the usual way is unlikely in toothed whales because here the olfactory bulb disappears at the beginning of the fetal period (Oelschläger and Buhl, 1985a, 1985b; see Olfaction and Sense of Taste). An olfactory sensory epithelium is lacking in postnatal animals. The second major functional aspect of the nasal apparatus in dolphins is sound generation and emission for echolocation (click sounds) and communication (whistles) (Berta et al., 2014). In this respect, the mammalian bauplan was profoundly modified and some structures in the nose are even unique to odontocetes (Huggenberger et al., 2016; Klima, 1999; Rauschmann et al., 2006).

The biomechanics of nasal respiration and phonation are difficult to separate from each other because they share the nasal passages and the facial (maxillonasolabialis) muscles. For respiration, two antagonistic muscle strands may open the dorsal part of the nasal tract, including the single blowhole and vestibulum. In front, the strong complex of the anteroexternus and anterointernus muscles pulls the rostral wall of the vestibulum and the rostral lip of the blowhole rostroventrally. Caudally, the dorsomedial-most fiber bundles of the intermedius and anteroexternus portions pull the caudal wall of the vestibulum and the caudal lip of the blowhole slightly in the caudal direction. Thus, the simultaneous action of these two antagonistic muscle groups should open the blowhole and the vestibulum, respectively. The lower (paired) section of the soft nasal tracts is opened mainly by the nasal plug muscles, which pull the nasal plugs rostrally. Correspondingly, the relaxation of the nasal plug muscles should cause the nasal plugs to slide back over the premaxillary sacs into their intranarial position. As outlined by Lawrence and Schevill (1956), the nasal portions of the anterointernus muscles may indirectly force the nasal plugs into the bony nares and, at the same time, dilate the lower nasal passages. In addition, the posteroexternus and intermedius muscles may pull the melon terminus and surrounding connective tissue caudally. This action should close the paired section of the nasal tract tightly.

The nasal structures potentially involved in sound generation exhibit the same topographical relationships in toothed whales (Huggenberger et al., 2016). According to the "unified hypothesis" for odontocete click sound generation, the "phonic lips hypothesis" (Cranford et al., 1996) implies that piston-like movements of the larynx build up positive air pressure in the area of the bony nares (see the following) (Houser et al., 2004). By means of that pressure, air quanta are driven through the lower nasal passage and into the vestibular sacs. Between the bony nares and the vestibular sacs, however, each air stream passes a pair of monkey lips and causes them to separate and then to slap together in a series of events by Bernoulli or other fluid dynamic forces. Here, the caudal monkey lip acts like a hammer that slaps against the opposing rostral epithelium of the nasal passage with the anterior monkey lip and the anterior dorsal bursa. Each clapping event causes the associated anterior monkey lip to vibrate, creating an initial sound wave of a single click that is guided via the fatty pathway and melon terminus (potential acoustic pathway) from the anterior bursa to the melon and from there into the surrounding water (Cranford et al., 1996). Within the respiratory tract, the intranarial air pressure can be generated and controlled by piston-like movements of the larynx (see the following). These two parameters (tension of the monkey lips and intranarial air pressure) should enable dolphins and porpoises to control the click repetition rate. Evidence from high-speed video endoscopy suggests that whistles are also produced at the monkey lips in the nasal passages of dolphins; in some cases, simultaneously to click generation at the contralateral monkey lips (Cranford et al., 2011).

The expanded air in the vestibular sacs should be recycled into the soft nasal passages via the contraction of the three superficial layers of the maxillonasolabialis muscles, which altogether may control the volume of these sacs. From here, the air to be recycled should be shifted down into the bony nasal passages and the throat by the retraction of the larynx (see the following).

The melon of dolphins transmits and focuses sound from the center of the nasal complex into the surrounding water and parts of the nasal diverticula and/or the skull serve as acoustic reflectors to guide the sound to the melon (McKenna et al., 2012; Cranford et al., 2013). Besides, these air sacs may help insulate the neurocranium and, above all, the ears from sound traveling caudally. The air volume and pressure in the premaxillary and posterior nasofrontal sacs may be controlled by the posterointernus muscle, which could also modulate the tension of the tissue around the posterior dorsal bursae. Moreover, the deepest layers of the maxillonasolabialis muscle (*M. profundus*) seem to regulate air flow into the inferior vestibulum and premaxillary sacs by opening and closing the laterocaudal extensions of the premaxillary sacs.

The more or less bilaterally symmetrical rostral muscles of the nasal complex presumably modulate the melon's shape by pulling its lateroventral parts in a ventral direction, thus flattening the melon on the rostrum. An elegant multitechnique investigation by Harper et al. (2008) discovered that fibers from rostral and nasal plug muscles penetrate the melon or its peripheral boundaries of connective tissue and may thus change the shape of the organ and adjust it to different phases of sound production. Furthermore, the intermedius muscle may pull the dorsal part of the melon caudally and rostral parts of the anterointernus muscle may pull the melon terminus ventrally, thus changing its height. However, it would be a matter of speculation to decide how much such a change in shape can contribute to a potential modulation of click sound focusing.

Lower Jaw

The mandible of dolphins is generally characterized by a rather simple organization. In lateral view, it has the shape of elongated triangles because there is virtually no coronoid process and the angle of the mandible is less pronounced than in most other mammals (see Skull). This simple anatomy correlates with the fact that dolphins do not chew but grasp prey items by their cone-shaped teeth to swallow it whole. Accordingly, the musculature of mastication is less well developed than in related terrestrial mammals. However, the gular musculature associated with the hyoid apparatus is very strong in dolphins because it must act, among other functions, in echolocation and locomotion to create strong negative pressure for sucking in prey prior to swallowing.

The jaws of dolphins are equipped with up to 200 uniform (homodont) teeth. The condylar processes of the mandibles and their articulation with the skull are simple and allow for only simple dorsal-ventral movements (hinge joints). The lower jaw is V-shaped in dorsal view and both dentals combined by a relatively long mandibular symphysis that ossifies usually only in old animals. The caudal tips of the V articulate with the neurocranium in the temporomandibular joints (Fig. 5.8). The temporomandibular joint lacks the typical joint cavity found in most mammals and is essentially a syndesmosis. The temporomandibular joint disks are composed of haphazardly intersecting dense connective tissue bundles separated by adipose tissue globules and small blood vessels (McDonald et al., 2015).

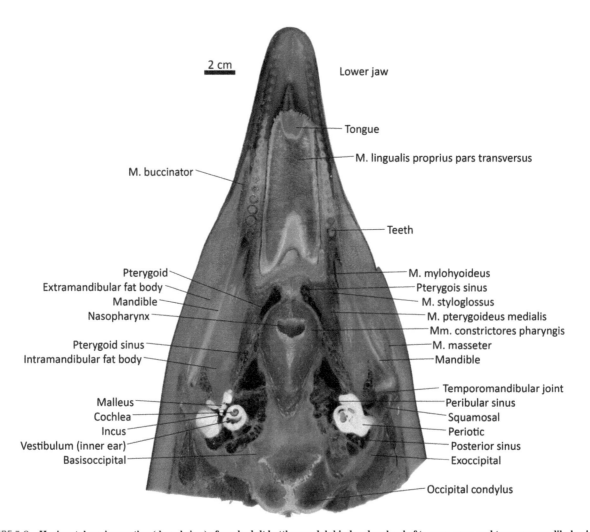

FIGURE 5.8 Horizontal sawing section (dorsal view) of a subadult bottlenose dolphin head on level of tongue, ears, and temporomandibular joint.

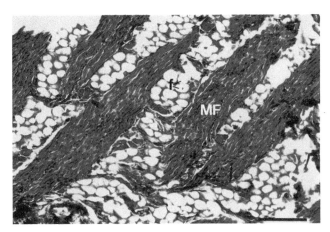

FIGURE 5.9   **Musculus masseter of a perinatal pantropical spotted dolphin (scale bar = 170 μm).** HE stain. *f*, adipose cells; *MF*, muscle fiber. *(From Rauschmann, M.A., 1992. Morphologie des Kopfes beim Schlanken Delphin Stenella attenuata mit besonderer Berücksichtigung der Hirnnerven (Inaugural-Dissertation, Fachbereich Medizin). Johann Wolfgang Goethe-Universität, Frankfurt am Main.)*

As in other toothed whales, the alveolar canal of the mandible in the bottlenose dolphin has been greatly widened in its posterior third and is filled with a fat body. As a result, the medial wall of the mandible is only membranous (periosteum) in the posterior part (Fig. 5.8). The lateral wall of the mandible in dolphins thins out to such a degree that light passes through an isolated and cleaned specimen (acoustic mandibular window) (Norris, 1968).

The morphological reconstruction of the dolphin head during evolutionary adaptation to permanent life in water involved (among other changes) a significant strengthening of certain muscle groups in the gular region (see the following). In contrast, some muscles of mastication (temporalis, masseter muscles) are relatively weak in dolphins (Fig. 5.5), obviously, they were significantly reduced secondarily (Oelschläger, 1990). Here, the pterygoid muscles (*M. pterygoideus medialis*, *M. pterygoideus lateralis*; Fig. 5.5b–i) are important for the closure of the jaws. The masseter muscle originates from the thin zygomatic arch with a short aponeurosis and converges strongly in the direction of its insertion at the lateral plane of the mandible. The light-microscopic preparation shows a situation resembling a "fatty degeneration" of the masseter muscle in correlation with a distinct reduction of the number of muscle fibers (Fig. 5.9) (Rauschmann, 1992). The temporalis muscle originates in the small fossa temporalis and attaches to the delicate coronoid process of the mandible. Thus, compared to most terrestrial mammals, both the origins and the insertions of the temporalis and masseter muscles of most dolphins are moderately developed. Killer whales are obvious exceptions to this generalization; these giant delphinids possess dominant temporalis muscles, a prerequisite for taking large prey items (Marshall, 2009).

The digastric muscle, although usually not included into the group of masticatory muscles, is one of the prominent muscles for opening the mouth because it is the most superficial hyoid muscle attaching to the mandible (Figs. 5.5g–m and 5.10) (interior of the *m. sphincter colli*). The digastric muscle originates from the lateroventral edge of the basihyal and thyrohyal. It attaches to bones the periost on the intramandibular fat pad and the ventral posterior half of the mandible.[a] This way, it opens the mouth because it lowers the mandible and the bones of the hyoid apparatus stay stable (punctum fixum).

Interior to the digastric muscle runs the mylohyoid muscle; it is located entirely anterior to the hyoid bones. This muscle originates ventrally from the mandible and stretches medioventrally. The mylohyoideus is thin but broad and both muscles meet ventrally between the mandibles (Fig. 5.5a–g). Caudal fibers attach to the lateral edge of the basihyal. Because the mylohyoid muscle spans a muscular sheet between both mandibles and the center of the hyoid (basihyoid) its action may help to open the mouth (hyoid as punctum fixum). However, if both mandible and hyoid are stabilized by actions of other muscle groups (muscles of mastication and of the hyoid; see the following), the mylohyoid muscle may elevate the whole gular region to pressurize to mouth cavity.

The geniohyoid muscle (Fig. 5.5a–g) originates as a thin, flat tendon from the anterior horn of the basihyal and some fibers originate from the rostral surface of the ceratohyal. The fleshy body of this muscle has a wide attachment on the

---

a. Although the digastric muscle was found as single-bellied in some delphinids (Reidenberg and Laitman, 1994), an additional caudal portion was described in the common dolphin (Toldt, 1905) and in porpoises (*Neophocaena phocaenoides, Phocoenoides dalli*) (Ito and Aida, 2005). This caudal portion (not shown in Fig. 5.10) originates in the area between the exoccipital and squamosal bones and attaches with a short tendon to the posterior part of the venter rostralis of the digastric muscle. However, there is presently no evidence on the persistence of a dual innervation (by the trigeminal and facial nerves, respectively) of the digastric muscles in dolphins.

medioventral surface of the mandible, caudal to the symphysis and along the anterior third of the mandible. If the hyoid is stabilized, this muscle may additionally act to open the mouth in parallel to the digastric muscle.

Tongue

Odontocete tongues are very similar to those of terrestrial mammals, being very muscular (Fig. 5.5d–g). Their musculature has large origins on the broad bones of the hyoid apparatus. This arrangement helps to pull the tongue into the throat like a piston, thereby creating enough negative pressure to draw in prey, a mechanism referred to as suction feeding or ram-suction feeding (Bloodworth and Marshall, 2005; Werth, 2007).

The hyoglossus muscle is a thin, broad muscle which originates from the lateral surface of the rostral third of the thyro-hyal bone. It extends rostrally and fuses with the ventral fibers of the genioglossus.

The styloglossus muscle originates from the ventrolateral surface of the stylohyal bone (not from the styloid process). It extends anteriorly and medially as a rather cylindrical muscle until it stretches into the tongue where it fuses with the lateral head of the hyoglossus.

The genioglossus muscle attaches to the center of the tongue and extends along its lateral surface. Some ventral fibers are continuous with those of the geniohyoid muscle (see the previous discussion). The genioglossus muscle originates at the medial surface of the mandible near the symphysis. According to its origin near the mandible's symphysis, this muscle may

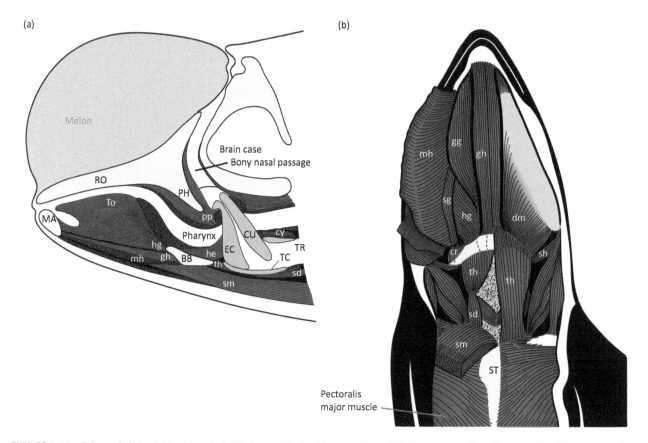

FIGURE 5.10 Schematic lateral (a) and ventral (b) views of the hyoid apparatus and the larynx as well as their associated musculature in a long finned pilot whale (modified after Reidenberg and Laitman 1994). (c) Same view as in (b) in a common dolphin (modified after Pilleri et al., 1976). (d) Schematic lateral view of the hyoid apparatus and the larynx as well as their associated musculature in the typical toothed whale (skull of *Globicephala melas*). The gular region anatomy including the hyoid apparatus of dolphins, porpoises, and most other toothed whales are alike (Werth, 2007; Huggenberger et al., 2008). Note that the hyoid region is artificially dorsoventrally elongated to show the whole laryngeal region (modified after Huggenberger et al., 2008). The free end of the stylopharyngeus muscle *(sp)* inserts onto the pharynx. The hyoid muscle is not shown. *1st*, first tracheal cartilage; *AP*, air passage; *BB*, basihyal bone; *CC*, ceratohyal cartilage; *ch*, ceratohyoid muscle; *CR*, cricoid cartilage; *cr*, cricothyroid muscle; *CU*, cuneiform cartilage; *cy*, cricoarytenoid muscle; *D*, dorsal; *dm*, digastric muscle; *EC*, epiglottic cartilage; *gh*, geniohyoid muscle; *he*, hyoepiglottic muscle; *hg*, hyoglossus muscle; *ia*, transverse arytenoid muscle; *MA*, mandible; *mh*, mylohyoid muscle; *ot*, occipitothyroid muscle; *P*, posterior; *PH*, pterygoid hamulus; *pp*, palatopharyngeus muscle; *py*, pterygopharyngeus muscle; *RO*, rostrum; *SB*, stylohyal bone; *sd*, sternothyroid muscle; *sg*, styloglossus muscle; *sh*, stylohyoid muscle; *sm*, sternohyoid muscle; *sp*, stylopharyngeus muscle (pharyngeal attachment, not shown); *ST*, sternum; *TB*, thyrohyal bone; *TC*, tympanohyal cartilage; *th*, thyrohyoid muscle; *tl*, thyropalatine muscle; *TO*, tongue; *tp*, thyropharyngeus muscle; *TR*, trachea; *ZA*, zygmatic arch. *(Artwork by Uko Gorter)*

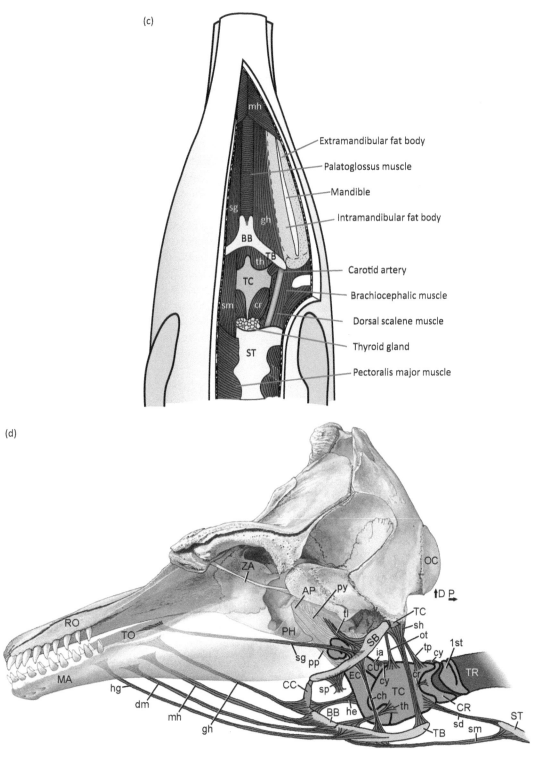

(c)

Extramandibular fat body

Palatoglossus muscle

Mandible

Intramandibular fat body

Carotid artery

Brachiocephalic muscle

Dorsal scalene muscle

Thyroid gland

Pectoralis major muscle

(d)

FIGURE 5.10 (*cont.*)

pull the tongue around rostrally and ventrally, thus open the oral cavity and oropharynx. The insertion is primarily along the ventral median raphe of the tongue, but further caudally some fibers extend dorsolaterally over the esophagus and are continuous with the palatoglossus muscle. The palatoglossus muscle surrounds the oropharynx and originates at the bony palate on both sides. This muscle may thus narrow the oropharynx.

The intrinsic muscles of the tongue are quite complex, as is typical for mammals, and stretch in all three directions, longitudinally, transversely, and vertically.

Hyoid Apparatus

In contrast to the area of the mandible described previously, the region around the nasopharynx, the esophagus, and the larynx is framed by the well-developed hyoid apparatus and the skull base. In contrast to the skull roof, the skull base of the dolphin is symmetrical and the inner bony nostrils (choanae) are centered in ventral view. Immediately in front of the choanae the paired pterygoid hamuli project ventrocaudally. Each of the choanae is bordered laterally by a crest-like extension of the pterygoid hamulus (Rommel, 1990), which bears a deep notch immediately lateral to the choanae (see Skull). However, the crest projects in a caudal direction and is built by the basioccipital bone that is in lateral contact with the periotic region. Due to these crest-like ventral projections, the skull base roofs a medioventral vault. The laryngeal passage runs through this vault where the air passage turns (by means of the larynx) from a vertical course through the bony nares to a horizontal orientation ventral to the brain case (Fig. 5.5k). Together with the ventral parts of vomer and basisphenoid, which represents the medial and caudal borders of the choanae, the pterygoid crest is the anchor of the large pterygopharyngeus muscles (*M. constrictor pharyngis rostralis*). These muscles connect the larynx tightly with the bony nares.

The hyoid apparatus of dolphins is located ventral to the skull base and C-shaped in lateral aspect so that the C of each side encompasses the larynx. The hyoid consists of three paired elements and one unpaired mediosagittal bone (see Skull). This mediosagittal bone (basihyoid) in the center has, in principle, two pairs of lateral bony wings: two ventral thyrohyal bones and two dorsal wings, the stylohyal bones. Both wing pairs point caudally to build the free ends of the C. The dorsal-most bone on each side, the stylohyal, is connected to the paroccipital process of the skull (Mead and Fordyce, 2009) by means of the tympanohyal cartilage. In the bottlenose dolphin, the cartilaginous tympanohyal may sometimes be ossified and fused to the posterior process of the periotic bone (Oelschläger, 1986). The last paired element following ventrally (ceratohyal cartilage) attaches on both sides to the ventral-most unpaired elements, the basihyal bone anteriorly and indirectly to the thyrohyal bone (Fig. 5.10) posteriorly. All of these hyoid elements articulate with their neighbors by thin pieces of cartilage (synchondroses, in old dolphins synostoses).

According to their topography and orientation, the components of the hyoid musculature may be divided into at least three different parts: a rostral, a caudal, and a dorsoventral group of muscles with different functional implications. The rostral group (running parallel in Fig. 5.10: hyoglossus, digastric, mylohyoid, geniohyoid muscles) is associated with the tongue (see the previous discussion) and pulls the hyoid apparatus forward, so that it rotates around a transverse axis through both tympanohyal cartilages (Figs. 5.5d–k and 5.10).

The caudal group, consisting of only the sternohyoid muscles, may pull the ventral part of the hyoid apparatus backward. The sternohyoid muscle is a massive muscle originating from the ventral surface of the basihyal and thyrohyal bones and inserts on the rostral margin of the sternum (Fig. 5.5m–o and 5.10). If the mouth is closed and the sternum fixed in position, contraction of this muscle may pull the ventral part of the hyoid apparatus caudally. Since the neck of the dolphin is quite reduced and the trachea very short and concealed by the sternum, so is also the *m. sternohyoideus*.

The omohyoid muscle originates on the thyrohyal as a lateral portion of the sternohyoid bone (Reidenberg and Laitman, 1994). It inserts on the lateral surface of the scapula at the acromion (Klima et al., 1980), thus passing deep (dorsal) to the sternomastoid muscle.

This movement of the sternohyoid muscle is probably antagonistic to that of the hyoglossus, digastric, mylohyoid, and geniohyoid muscles. Alternate rostrocaudal contractions of the two groups of muscles give the hyoid apparatus the appearance of a swing (Huggenberger et al., 2008). Moreover, contractions of the dorsoventral group (stylohyoid, ceratohyoid, occipitohyal muscles) should bend and lift the hyoid apparatus dorsally in direction of the skull base, supported by external laryngeal muscles (thyropalatine, thyropharyngeus, occipitothyroid; see the following).

As the dorsoventral group, there are three muscles connecting the dorsal and ventral wings of the hyoid apparatus with each other. The first (rostral-most) is the short *ceratohyoid muscle* (interhyoideus muscle; after Lawrence and Schevill, 1965). It connects the stylohyal with the thyrohyal bone. A caudal extension of the ceratohyoid muscle is the stylohyoid muscle. This muscle spans between the caudal tip of the thyrohyal bone and the tympanohyal cartilage and the adjacent areas of the skull base and the stylohyal respectively (Fig. 5.10). Further caudally is the occipitohyoid muscle, which is a small muscle in dolphins that has its origin behind the bulla in the region of the cartilaginous tip of the stylohyal (tympanohyal; see the previous discussion) and passes ventrally to the thyrohyal (not shown in the figures).

There are two "interhyoid" muscles: the small transverse hyoid muscle connects the ceratohyal cartilages near their "articulation" with the basihyal bones. The stylopharyngeus muscle originates from the medial surface of the stylohyal bone in its distal (ventral) half and on its way to the contralateral side, it stretches around the pharynx at the level of the epiglottic spout. Accordingly, this muscle is part of the *m. constrictor pharyngis rostralis*.

Larynx

With the exception of the so-called epiglottic spout, which is reminiscent of a goose beak, the cetacean larynx, in principle, is similar to that of other mammals, being composed of a cartilaginous framework held together by a number of muscles and connecting tissue. Via this epiglottic spout the larynx has an intranarial position. Its anterior end (the distal tip of the epiglottic spout) is firmly held within the choanae by a strong sphincter muscle (Fig. 5.10). The elongated intranarial larynx keeps the nasal air passages continuous with the glottis while the dolphin sucks in prey (Reidenberg and Laitman, 1994).

Because the shape of the odontocete larynx is unique among the Mammalia, various hypotheses have been put forward regarding its function. The primary task of the larynx is probably to keep the respiratory and digestive tracts separate from one another during swallowing. Moreover, during the 1980s, the idea was put forward that the larynx of toothed whales could be used in echolocation. This "larynx hypothesis" implied a similar mechanism of sound production as in other mammals (Purves and Pilleri, 1983; Reidenberg and Laitman, 1994). However, according to the recent sound generation hypotheses, whistles and echolocation clicks of toothed whales is generated in the nasal complex and the larynx and its associated musculature build up the pneumatic pressure needed for nasal sound generation (Huggenberger et al., 2016; Norris and Harvey, 1974). The mechanism of click sound generation in the nose, as described by Cranford et al. (1996), does not require a large volume of air but high pressure. Here, the larynx seems to act as a piston, which is pulled forward and upward by a strong sphincter muscle in the direction of the choanae and back by antagonistic muscles (see the following).

The thyroid cartilage is the main component of the larynx (Figs. 5.3, 5.5l, 5.10, and 5.11). Most of the extrinsic muscles, which move the complete larynx, attach to this cartilage. It is made up of right and left laminae fused at their ventral margins. Each lamina has two well-developed processes; the anteriorly directed cranial cornu and the posteriorly directed caudal cornu, the latter being the larger one. The caudal cornua of the thyroid cartilage articulate with the superior laterocaudal margins of the cricoid cartilage via small oval synovial joints. Rostrally the thyroid cartilage articulates with the epiglottic cartilage.

The cricoid cartilage (Figs. 5.10 and 5.11) does not form a complete ring as in most other mammals but it is open ventrally, forming two rather long lateral cornua, which project caudally. These cornua almost fill the caudolateral notch in each thyroid lamina. In addition to the joints formed between the cricoid and thyroid cartilages previously mentioned, there are articulations between the anterior lateral margins of the cricoid cartilage and the arytenoid cartilages.

The epiglottic cartilage in all the toothed whales examined so far is much longer than that of most mammals. It is concave medially and forms a trough with thin lateral margins. The latter projects caudally to wrap around the anterior borders of the cuneiform and arytenoid cartilages, thereby forming the "laryngeal beak" (Figs. 5.3, 5.5i, 5.10, and 5.11), which has a thick lip at its rostral end.

The arytenoid cartilages articulate with the anterior lateral margins of the cricoid cartilage by well-formed oval-shaped synovial joints. Each has a well-developed muscular process projecting laterally, but no vocal processes are evident. The arytenoid cartilages also articulate with the cuneiform cartilages.

The cuneiform cartilages are elongated blade-like processes of the arytenoid cartilage that face each other medially and lie in the trough formed by the epiglottic cartilage. Two or more small cartilages, probably representing the corniculate cartilages, attach to the caudolateral end of each cuneiform cartilage and also to the caudal two-fifths of the ventral edge of the arytenoid. The arytenoids are characterized by the attachment of an intralaryngeal "midline fold" standing in the mediosagittal plane (Reidenberg and Laitman, 1988). The fold stretches between the rostral ends of the arytenoid cartilages and the mediorostral end of the thyroid cartilage. Accordingly, it may be homologous to the vocal folds in other mammals, although some authors doubt this interpretation (Brzica et al., 2015). Lateral to this midline fold is a set of smaller folds projecting oblique along the midline fold and give way into a system of small sacculi between them (Fig. 5.11).

Relatively strong ligaments support the cartilagenous laryngeal structures (Brzica et al., 2015). There is one ligament connecting the ventral margin of the arytenoids and cuneiform cartilages with the base of the epiglottic cartilage. Moreover, the cricoarytenoid ligament connects the rostral margins of the cricoid cartilage to the arytenoid cartilages. Midiosagittally, a ligament connects the cricoid cartilage and the cuneiform cartilages. Below the cricothyroid muscle, there is the cricothyroid ligament connecting both cartilages additionally.

At the aboral end of the midline fold there is a complex lymphoepithelial gland, overlying the cricoid cartilage. It defines a heavily trabeculated area (Figs. 5.5l and 5.11). The histological appearance of the laryngeal gland is similar to the palatine, or dorsal oropharyngeal tonsils (Smith et al., 1999). Epithelial-lined folds and crypts form around aggregations of lymphocytes associated with mucosal glands. This lymphoepithelial gland may be analogous to the nasopharyngeal adenoid of terrestrial animals.

The larynx is suspended from the musculoskeletal framework of the hyoid apparatus but is connected to the hyoid itself and to the sternum by a set of three muscles: the hyoepiglottic muscle, the thyrohyoid muscle, and the sternothyroid muscle.

(a)

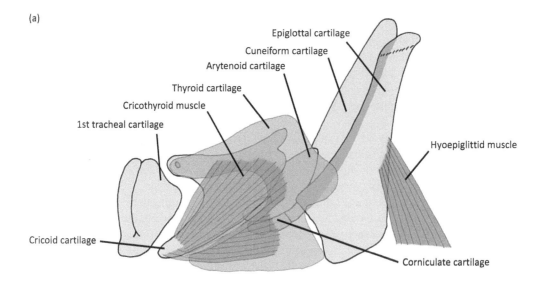

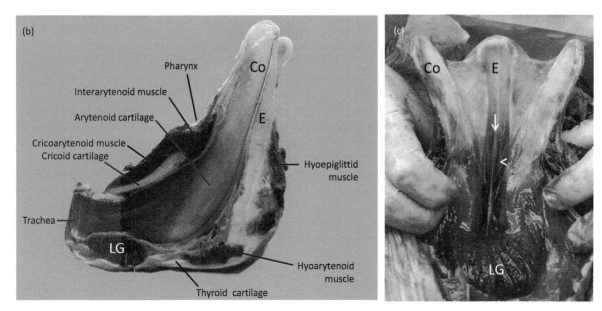

FIGURE 5.11 (a) Schematic lateral view of the larynx of a bottlenose dolphin; (b) Midsaggital section of the larynx of a bottlenose dolphin (rostral pointing right); and (c) larynx opened caudally showing its medial (*arrow*) and smaller lateral (*open arrow head*) folds and glands. *Co*, corniculate cartilage; *E*, epiglottid cartilage; *LG*, laryngeal gland. *(Part a: Green et al., 1980. Functional and descriptive anatomy of the bottlenosed dolphin nasolyngeal system with special reference to the musculature associated with sound production. In: Busnel, R.G., Fish, J.F. (Eds.), Animal Sonar Systems. Plenum Press, New York, pp. 199–238.) (Parts b and c: from Smith, T.L., Turnbull, B.S., Cowan, D.F., 1999. Morphology of the complex laryngeal gland in the Atlantic bottlenose dolphin, Tursiops truncatus. Anat. Rec. 254, 98–106.)*

The unpaired hyoepiglottic muscle originates from the middorsal basihyal and from the ventrocaudal surface of the ceratohyal and attaches to the epiglottic cartilage and more specific to the inferior half of its anterior edge (Figs. 5.3, 5.5, 5.10, and 5.11). This muscle probably functions to pull the epiglottal cartilage forward so as to enlarge the size of the anterior passageway in the laryngeal beak.

The thyrohyoid muscle arises from the rostrolateral surface of the thyroid cartilage, anterior to the attachment of the sternothyroid muscle (Fig. 5.10). It attaches to the posterior margin of the basihyal and thyrohyal bones and probably pulls the laryngeal apparatus rostrally, thus anchoring the tip of the laryngeal beak within the choanae.

The sternothyroid muscle originates from the rostroventral-most part of the sternum and inserts on the lateral surface of the thyroid cartilage, just superior to and anterior to the insertion of the cricothyroid muscle (Fig. 5.10). The sternothyroid muscle probably pulls the complete laryngeal apparatus caudally.

Furthermore, there is a set of muscles which interconnect the different cartilaginous parts of the larynx to each other: the cricothyroid muscle, the cricoarytenoid muscle, the thyroarytenoid muscle, and the transverse arytenoid muscle.

The cricothyroid muscle has its origin on the lateral posterior surface of the lateral horn of the cricoid cartilage (Fig. 5.10). It attaches to the thyroid cartilage along the inferior margin of the caudolateral part, that is, to the posterior notch and to the lateral surface of the thyroid lamina just below the posterior notch.

The cricoarytenoid muscle originates from the dorsal surface of the cricoid cartilage, extends laterad to the medial surface of the caudal horn of the thyroid cartilage and attaches to the dorsal muscular process of the arytenoid cartilage (Figs. 5.3, 5.10, and 5.11). It probably rotates the arytenoids laterally and posterodorsally.

The thyroarytenoid muscle originates from the dorsal midline and the medial surface of the thyroid cartilage (Fig. 5.11). The fibers pass dorsad to insert on the arytenoid cartilage, more precisely on the ventral surface of its muscular process. The anterior fibers pass laterally, over the posterior angle of the epiglottic cartilage. The thyroarytenoid muscles probably pull the arytenoid and cuneiform cartilages ventrad and mediad to decrease the angle between their medial surfaces. If the arytenoid and cuneiform cartilages are already approximated, the thyroarytenoid muscles probably pull the arytenoids and cuneiforms deeper into the epiglottal trough. Thus, these muscles may control the closing of the epiglottic spout.

The transverse arytenoid muscle connects the dorsal margins of the cuneiform cartilages and to a lesser degree the arytenoid cartilages (Figs. 5.10 and 5.11). Some of the posterior fibers join the anterior edge of the dorsal cricoarytenoid muscle. These muscles apparently increase the interarytenoid space between the edges of the arytenoids and thus the intercuneiform space between the anterior edges of the cuneiform cartilages. Simultaneous contraction of the cricoarytenoid muscle may, therefore, open the intercuneiform space. Rostral movement of the epiglottic (by the hyoepiglottic muscle) could then widen the connection between the nasal and the laryngeal air passages.

The larynx is suspended from the base of the cranium by a complex of muscles, the thyropalatine, the occipitothyroid, and the thyropharyngeal muscles. These muscles arise from the dorsal margin and anteromedial surface of the thyroid cartilage, and then pass dorsad and rostrad to attach to the base of the cranium. The anterior-most of these muscles is the thyropalatine (a posterior part of the palatopharyngeal muscle; see the following) (Green et al., 1980), which attaches the medial surface of the thyroid lamina and the thyroid ligament. The middle muscle, the occipitothyroid (Lawrence and Schevill, 1965), attaches to the dorsal margin of the cranial cornu of the thyroid cartilage. This muscle is not found in terrestrial mammals (Schneider, 1964). Nevertheless, in dolphins, it appears that the occipitothyroid muscle is part of the thyropharyngeus muscle that is connected to the skull base. The thyropharyngeus muscle, as part of the *m. constrictor pharyngis caudalis* (Figs. 5.3 and 5.5l), is the posterior-most muscle connecting the larynx with the skull base and attaches to the dorsal margin of the posterior cornu of the thyroid cartilage. A *m. constrictor pharyngis medialis* (hyopharyngeus muscle) was not described explicitly in dolphins and other cetaceans, although Werth (2007) showed it as part of the rostral constrictor muscle.

The *m. constrictor pharyngis rostralis* is developed as a strong sphincter muscle, a complex of the palatopharyngeus and pterygopharyngeus muscles, that suspends the epiglottic spout from the bony nares and the skull base, and is responsible for the secure intranarial position of the larynx and the reliable separation of the respiratory and digestive tracts (Figs. 5.3 and 5.5l). This sphincter muscle is reminiscent of a strong hose which arises in the bony nares and attaches to the epiglottic spout. In general, each palatopharyngeus muscle originates from the rostral and lateral surface inside the bony nasal passages, both muscles together forming the "soft palate" and surrounding the ventral part of the epiglottic spout as a loop. Accordingly, levator and tensor veli palatine muscles are not present in dolphins. The pterygoid hamulus as a posterior elongation of the bony nasal passage is extended rostrally so that it serves as an additional origin for the palatopharyngeus muscle, which largely displaces the lumen of each bony nasal passage even in dead animals. The pterygopharyngeus muscles originate from the lateral and caudal walls of the choanae as well as from the caudal area of the pterygoids and the rostral parts of the basioccipital and encircle the dorsal part of the spout. The palatopharyngeus muscle has a high (dorsorostrad) intranarial area of origin and reaches nearly as far as the dorsal bony nares. In contrast, only the lateral fibers of the pterygopharyngeus muscle originate intranarially to meet fibers of the palatopharyngeus.

According to the shape of the laryngeal sphincter muscle, it is plausible that its action should seal the larynx watertightly at the tip of the epiglottic spout when pulling the latter in the direction of the choanae. Moreover, the origin and course of the palatopharyngeus muscle within the bony nasal passage could be essential for diving maneuvers when shrinking air volumes may limit the sound generation process. Here, the shortening and thickening of the contracting muscle should replace most of the intranarial volume and thus transfer air in the direction of the nasal complex to be used for nasal sound generation (see the previous discussion) (Huggenberger et al., 2008).

Functional Aspects of the Larynx and Hyoid Complex

As described earlier, the strong gular musculature connecting the hyoid apparatus with the mandibles and the skull base may also force the larynx into the choanae. Interestingly, very strong muscles representing the rostral group (eight muscles) force both the hyoid and larynx forward and upward (protraction), whereas only two muscles seem to act as retractors (sternohyoid and sternothyroid muscles). Therefore, it is likely that the presumed laryngeal piston mechanism for sound generation is driven by the strong palatopharyngeal sphincter muscle and, in addition, by the protractors of the hyoid and the larynx (Huggenberger et al., 2008). Thus, piston-like protraction movements of the larynx, which should be closed at the tip of its epiglottis by the palatopharyngeal sphincter (*m. constrictor pharynges rostralis*), build up positive air pressure in the area of the bony nares. Due to the complex arrangement of these protractors, the larynx should be capable of subtle and fine as well as expansive and strong movements.

After each sound generation cycle, air in the nasal system can be recycled. In this step, the sternohyoid and sternothyroid muscles can retract the larynx from its intranarial position caudally and thus help to bring back air from the nasal air sacs (ie, the vestibular sacs) into the nasal passage. This retraction is all the more plausible, because, on both sides, the hyoid apparatus is "articulated" with the skull via the stylohyal and tympanohyal elements. Thus, the movements of the larynx/hyoid complex resemble that of a swing oscillating between extreme dorsorostral and ventrocaudal positions.

Beyond its high significance in sound production, the larynx/hyoid complex probably takes part in suction-like feeding behavior (Reidenberg and Laitman, 1994). In this case, the gular region and the tongue represent a "hydraulic piston," which is retracted caudally by the laryngohyoid musculature (Bloodworth and Marshall, 2007; Werth, 2007). During such negative pressure events in the oral cavity and pharynx, the full contraction of the palatopharyngeal sphincter muscle around the laryngeal spout shuts down the choanolaryngeal connection.

Of equal importance is a mechanism that prevents dolphins from choking even during rapid inspiration when surfacing, a behavior typical in these animals. Strong negative pressure in the thorax and lungs, as the prerequisite for rapid inspiration, could draw water from the alimentary tract (pharynx) into the respiratory tract. This time, however, the sphincter muscle cannot be contracted maximally around the laryngeal spout because the air passage must remain open during inhalation. However, the watertight sealing of the airway during strong negative (or positive) pressure in the respiratory tract can only be guaranteed by the active squeezing of the pharynx against the skull base, including the collapse of the alimentary tract at the level of the piriform recesses.

SENSES

Dolphins are most interesting as to sensory function because their ancestors had to adapt and readapt, respectively, their sensory organs to the physical properties of aquatic habitats. Such adaptations can generally take place at several levels, namely (1) perception of the stimulus, (2) filtering, (3) transformation into a neural impulse, and (4) projection to the central nervous system for further processing.

Mammalian vision, for example, is characterized by (1) refraction of light at the cornea, (2) control of light intensity by the pupil, (3) focusing and filtering of the incoming light by the lens and retina, respectively, (4) enhancement of light by a tapetum lucidum (night vision equipment), (5) signal processing within the different layers of the retina, and (6) projection of the resulting information via the optic nerve to nuclei in the diencephalon and mesencephalon to be processed, and from there to the visual cortex.

In hearing, incoming sound is (1) collected and filtered by the external ear, and (2) amplified in the middle ear by the lever action of the middle ear ossicles (by about 40 dB[b]; Wartzok and Ketten, 1999). (3) By its muscular equipment, the middle ear dynamic system protects the inner ear against acoustic overload, and (4) the basilar membrane in the organ of Corti of the inner ear (cochlea) selects the range of frequencies that will be sent to the brain by the auditory (cochlear) nerve. An interesting example for such acoustic processes is the need for dolphins to detect relevant signals amidst background noise (Wartzok and Ketten, 1999). This is all the more important for marine species that live in large groups of intensively communicating individuals and have to perform constant noise suppression in order to detect crucial information from nonrelevant acoustic events. In parallel, each individual has to process echolocation sounds for orientation, finding prey, and so forth.

It is evident that, seen from an evolutionary perspective, each of these structures selecting and modulating environmental stimuli has a considerable potential to play a role in adaptational processes. Therefore, it is important to be aware of these chains of structures that, single or as a whole, bring about the continuous refinement of the sensory systems needed in a specific ecological niche over time.

In the following sections, the sensory systems and organs will be presented as to their anatomical and functional specifics in comparison to the typical terrestrial mammalian conditions.

---

b. Sound loudness is usually indicated as sound pressure level (SPL); a logarithmic measure of the effective sound pressure of a sound relative to a reference value. SPL is measured in dB and defined by $SPL = 20 \log_{10} (p/p_0)$ dB (where p is the root mean square sound pressure and p_0 is the reference sound pressure, usually 1 µPa)

Olfaction and Sense of Taste

Toothed whales cannot smell. They do not have olfactory organs; all peripheral olfactory structures such as an olfactory epithelium in the nose are not present in postnatal animals. The olfactory bulbs of the telencephalon regress in early fetal periods (Fig. 5.12) (Oelschläger and Buhl, 1985a,b). The hypertrophied nasal complex of dolphins was rebuilt in the course of evolution to function first as an air pipe during very rapid lung ventilations and second for sound generation and emission. Accordingly, under the functional restrictions in their new (aquatic) habitat, the olfactory part of the mammalian nose could not be maintained (see Dolphin Head Characteristics; for brain function Olfactory System/Paleocortex).

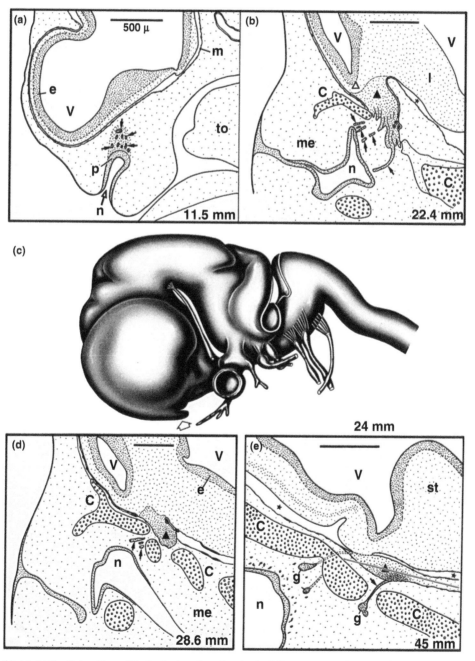

FIGURE 5.12 (a), (b), (d), (e) Sagittal sections of the forebrain and nasal region in different ontogenetic stages of the harbor porpoise. Measurements as crown-rump-lengths, calibration bars in all figures: 500 µm. (c) Brain model of a 24 mm specimen with "stadium optimum" of the olfactory bulb (*arrow*). *c*, cribriform plate; *e*, ependyma; *f*, olfactory fiber(s); *m*, primitive meninx; *v*, ventricle. *Light triangle*, primordial olfactory bulb (telencephalic component); *solid triangle*, filamentous trunk (placodal component of olfactory bulb); *, shrinkage artifacts. *(Artwork by Jutta Oelschläger)*

In comparison with terrestrial mammals, taste buds on the tongues of small toothed whales are reduced (see Chapter 8). However, behavioral and electrophysiological experiments have shown that the bottlenose dolphin, common dolphin, and harbor porpoise are able to distinguish several chemicals from sea water (Dehnhardt, 2002). Psychophysical techniques determined detection thresholds of the four primary tastes in the bottlenose dolphin (Friedl et al., 1990; Nachtigall and Hall, 1984). Due to extensive losses of genes of sweet, umami, bitter, and sour tastes (Feng et al., 2014) the thresholds for sweet, bitter, and sour (as well as salty) compounds were 2–10 times higher in the bottlenose dolphin than in the human. Although the studies reported so far show that dolphins taste chemicals dissolved in water. The functional significance of this sensory ability under natural conditions remains unclear because dolphins swallow prey items as a whole. It may be possible that the sense of taste has a function in communication as well as in orientation (Dehnhardt, 2002) (see Olfaction and Sense of Taste).

Skin Sensitivity

Somatic sensation in dolphins did not attract as much attention of the investigators as the auditory and visual senses (Supin et al., 2001). Therefore, available data in this field are restricted. Morphological studies have shown the presence of both encapsulated and free nerve endings in the dolphins' skin. The most intensive innervation was found at the head and snout, around the blowhole, and around the anus and genital slit. As another hint to the prominent head sensitivity in dolphins, the trigeminal nerve and ganglion are well developed, and the nerve is exceeded in thickness by the vestibulocochlear nerve only (see Genuine Peripheral Cranial Nerves).

The data on structural prerequisites for tactile sensitivity were complemented by physiological experiments in living dolphins. Kolchin and Bel'kovich (1973) and Ridgway and Carder (1990) used galvanic skin responses and somato-sensory evoked potentials, respectively, to study the tactile sensitivity of the bottlenose dolphin. Both studies found the areas of maximal sensitivity in the skin of the head. In the first study, a peak sensitivity was found in a zone within 2.5 cm from the blowhole and in circles of 5 cm in diameter around the eyes. Less sensitive were the snout, melon, and lower jaw. The least sensitive areas comprised the back of the animal. The second study found the peak sensitivity at the angle of the gape and decreasing sensitivity around the eyes, snout, melon, and blowhole. Both groups of authors tend to agree that the most sensitive areas of the dolphin skin may be about as sensitive as the skin of the human lips and fingers. As some kind of coincidence, Bryden and Molyneux (1986) found large numbers of encapsulated nerve endings in the region of the blowhole both in bottlenose dolphins and false killer whales (*Pseudorca crassidens*). The authors suggested that these nerve endings could be involved in the monitoring of pressure changes on the skin during surfacing of the dolphins because most receptors were found on the anterior lip of the blowhole.

Passive Electroreception

Passive electroreception is a solely aquatic sense which is mainly used for the detection of weak bioelectrical fields generated by prey and predators. For a long time, the only known mammals that possess this sense were the Monotremata (platypus and echidnas) (Czech-Damal et al., 2013). But recently, electroreception was also reported for the Guiana dolphin (*Sotalia guianensis*) (Czech-Damal et al., 2012), a delphinid species inhabiting coastal regions of northeastern South America. Psychophysical experiments revealed that the Guiana dolphin's vibrissal system perceives electric fields of at least 4.6 μV/cm. Because the perception threshold is sufficient for the detection of bioelectric fields produced by small to medium-sized fish, it is suggested that electroreception is used as a supplementary sense for close distance detection of bottom-dwelling fish species during benthic feeding and in turbid water.

Generally, whales and dolphins are born with cranial vibrissae, and baleen whales maintain this mechanosensory organ (vibrissal follicle-sinus complex) during lifetime. In contrast, in most toothed whales, the vibrissal hairs degenerate shortly after birth (Ling, 1977) and hairless vibrissal crypts remain visible in an array of 2–10 on each side on the upper jaw (Fig. 5.13). The adult Guiana dolphin possesses a row of 4–7 crypts on each side of the beak, which consist of an ampullary invagination of the epidermal integument with sizes from 4.1 to 7.1 mm in length and from 1.2 to 4.3 mm in width. The lumen is filled with a glycoprotein-based biogel, which is constantly released during desquamation of the skin. In addition, a net of shedded corneocytes and keratinous fibers is trapped in the ampulla (Fig. 5.14). The vibrissal crypts are surrounded by a dense capillary network and they are well innervated (~300 axons) in the lower two-thirds of each crypt by the infraorbital branches of the trigeminal nerve (see Cranial Nerve 5). Czech-Damal et al. (2012) hypothesize that the biogel in the lumen of the vibrissal crypt possesses a high electric conductivity and enhances the sensitivity to voltage gradients, and that intraepithelial nerve fibers which are located close to the lumen are responsible for the transduction mechanism.

FIGURE 5.13   **Vibrissal hairs and crypts (*arrows*) on the rostrum of a young bottlenose dolphin.**

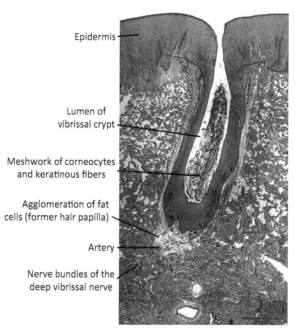

FIGURE 5.14   **Histology of the vibrissal crypts in the Guiana dolphin.** Longitudinal section through a representative vibrissal crypt stained with Masson–Goldner trichrome. *(Photo by Nicole Czech-Damal)*

Since other dolphin species also possess vibrissal crypts as adults and use a similar benthic feeding strategy as the Guiana dolphin, it is probable that electroreception is more widely distributed throughout the Delphinidae. Passive electroreception may be a valuable contribution within multisensory processing of toothed whales, which has been narrowed by the loss of olfaction (see the previous discussion) and the reduction of the vestibular system (see the following).

Magnetosensation

Magnetic orientation in dolphins and other toothed whales has been the issue for intensive discussions during the last decades (Kremers et al., 2014). There have been a number of speculations about the nature of the magnetic receptor, the most interesting one involving iron oxide crystals, usually referred to as magnetite (Fe_3O_4). Zoeger et al. (1981) reported the finding of magnetic material in the head of the common dolphin. In several specimens, magnetic particles were detected in the dura mater (pachymeninx[c]) below the vertex of the skull roof, located between the junction of the falx cerebri with the tentorium cerebelli and bone (Fig. 5.15). Scanning electron microscopy revealed small round pits with an elevated margin at the surface of a particle investigated. The authors published one figure with a fiber protruding from the center of such a trough with a diameter of about 20 µm. In one case (Fig. 5.16a and b), the process looks like a nerve fiber that was attached to a small sphere at the surface of the magnetic particle, broken off but still being located near the sphere. The potential identity of these structures as neural components to some degree is supported by their dimension: The diameter of the small processes disappearing in the magnetic material may have about 1 µm in diameter (red arrow) and the troughs about 20 µm. The broken process (red asterisk) is thicker; it may have 3- to 5-µm in diameter, the sphere itself, which also has an irregular surface, about 15 µm. The thicker part of the broken process reminds of a nerve fiber (neurite) with a more or less irregular covering by scale-like formations. The covering seems to be lacking more distally from the sphere where the smooth and thinner process seems to leave the nerve fiber. It is somewhat speculative whether the sphere could represent the cell body (soma, perikaryon) of a neuron and the process its axon covered by glia and giving off a collateral that disappears in the particle surface. Even more speculative is the possibility that this structural arrangement at the surface of the particle represents the magnetic receptor complex consisting of a nerve fiber and a recess hosting the receptor. Magnetic material was also found in the same pachymeningeal position in other toothed whales such as the bottlenose dolphin, Dall's porpoise

---

c. Pachymeninx is a synonym for the dura mater. The cavitas subduralis separates it from the leptomeninx consisting of the arachnoidea mater and pia mater.

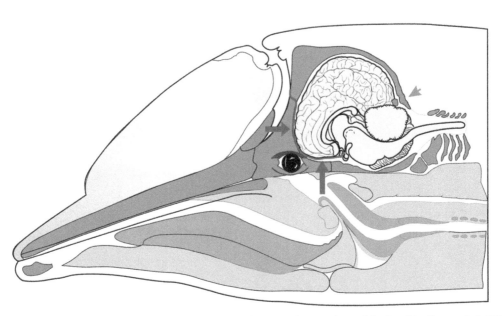

FIGURE 5.15 **Survey figure for the topography of the dolphin terminal nerve and magnetic particles found by Zoeger et al. (1981).** The vertical course of the terminal nerve is also depicted in Figure 6.44. Sagittal section of bottlenose dolphin head showing the terminal nerve, which runs around the frontal lobe of the telencephalic hemisphere between the hypothalamus of the brain, and the cribriform plate of dolphins, which has been nearly closed due to the reduction of the olfactory nerve. *Vertical arrow*, subarachnoid course of the terminal nerve; *horizontal arrow*, intradural course of the terminal nerve leading to the cribriform plate, which has leftover a few small foramina for the passage of small blood vessels and perhaps nerve fiber bundles into the pipe-like nasal passages. *(Modified from Morgane et al., 1980. The anatomy of the brain of the bottlenose dolphin (Tursiops truncatus). Brain Res Bull 5(3):1–107.)*

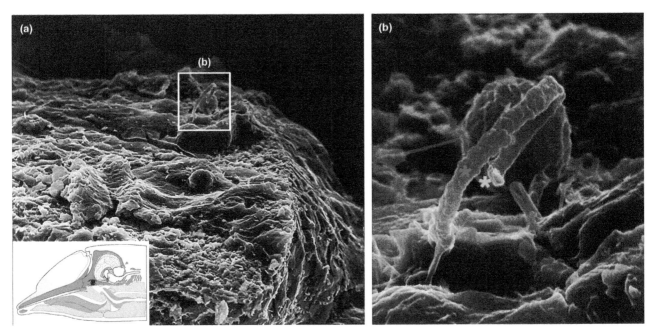

FIGURE 5.16 **Magnetosensation.** (a) Inset: Schematic midsection of dolphin head; magnetic material was found at the junction of the falx cerebri and the tentorium cerebelli close to bone (see *green arrowhead* in Fig. 5.15). Surface of magnetic particle with exotic structure. (b) Box from (a) enlarged. The process marked by the asterisks looks like a nerve fiber that was attached to a small sphere at the surface of the magnetic particle, broken off but still being located near the sphere. (For more information, see text.) *(Photographs (a) and (b) courtesy of Zoeger, taken for the study published in Science (1981) but published here for the first time)*

(*Phocoenoides dalli*), and Cuvier's beaked whale (*Ziphius cavirostris*), as well as in the humpback whale (*Megaptera novae-angliae*) (Wartzok and Ketten, 1999). The area concerned is traditionally innervated by the somatosensory part of the trigeminal nerve. However, there has so far been no proof of a coincidence of magnetite, receptor, and nerve in marine mammals.

While it is difficult to get solid data from work with magnetometers due to the risk of contamination with magnetic material from outside the animal during the preparation of the tissue samples, the experimental (behavioral) investigation of the magnetic orientation and navigation in dolphins is also cumbersome because of the dimensions of the animals and the space needed. So far there is only one study that reported the ability of bottlenose dolphins to discriminate objects on the basis of their magnetic properties, a prerequisite for magnetoreception-based navigation (Kremers et al., 2014), whereas Bauer et al. (1985) were not able to train bottlenose dolphins to respond to a magnetic stimulus in conditioning experiments. Moreover, the distribution of sighted free-ranging common dolphins also could not be related to any magnetic pattern (Hui, 1994). However, support for the assumption that dolphins have a magnetic sense comes mainly from live strandings (Kirschvink, 1990; Klinowska, 1986). It was found that there exists a correlation between magnetic field orientation, minima or gradients, and the location of live strandings of delphinids as the common dolphin, Atlantic white-sided dolphin, Atlantic spotted dolphin, striped dolphin, and long-finned pilot whale. In these areas, local lows or valleys in the magnetic fields intersected the coast or islands. In line with these findings, some strandings have been related to solar storms causing anomalies in the geomagnetic field. Accordingly, Vanselow and Ricklefs (2005) found that sperm whale strandings in the North Sea between 1712 and 2003 were related to the length of the solar cycle. In general, however, because of differences as to the consistency of earth magnetic anomalies and live strandings in different parts of the world, more data from anatomy and field work have to be awaited to get a better picture of this fascinating topic and, at present, we can only speculate about the biological significance of magnetoreception in mammals (Begall et al., 2014).

Vision: Anatomy and Optics of the Eye

During evolution, the cetacean eye has undergone many significant changes in morphology and function: It had to be readapted to the mechanical, chemical, osmotical, and optical conditions of the aquatic medium (Supin et al., 2001). Dolphin vision serves many important biological functions, like prey detection and capture, conspecific and individual identification, as well as orientation and migration. The performance of dolphins in dolphinaria suggests well-developed visual function both in water and air. The dolphin's eye is well emmetropic in water (objects are well focused without correction) but strongly myopic (nearsighted) in air.

In dolphins, the eyes are located on the sides of the head but are slightly directed forward (Fig. 5.17). Rather than being spherical as in most mammals, the dolphin eye is markedly flattened anteriorly (Figs. 5.5 and 5.17). The sclera is very thick and, together with the thickened periphery of the cornea (see the following), may serve to hold the inner globe of the eye in an elliptical shape.

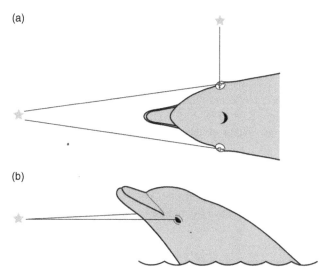

FIGURE 5.17 Characteristic positions of the dolphin body relative to visually inspected objects. (a) dorsal view (for both underwater and aerial vision) and (b) lateral view (for aerial vision). *Arrows* show directions of light rays from an object to the corresponding high-resolution area (fovea) of the retina (Fig. 5.19). *(Artwork by Uko Gorter)*

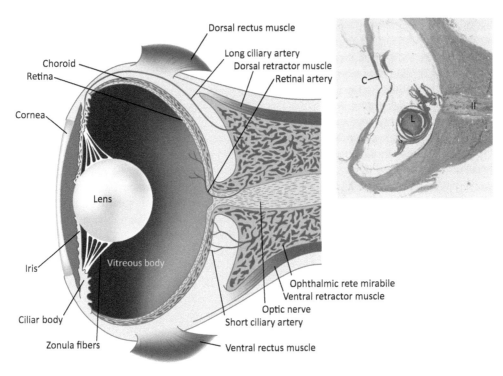

FIGURE 5.18 **Schematic transverse section trough the dolphin eye (left) and a histological longitudinal section showing the optic nerve (II), the lens (L; arificial replaced within the vitreous body), and the cornea (C).** HE stain. *(Artwork by Uko Gorter)*

In terrestrial mammals, it is usually the curved cornea which contributes most of the refractive power of the eye. This is due to the strong refraction at the air–cornea border. In dolphins, the cornea acts as a weak but divergent light-scattering lens (Mass and Supin, 2009). This divergence (distraction) of light is due to the peculiar shape of the cornea: It is thicker near the sclera's border and thinner and only slightly curved in the center (Fig. 5.18). In dolphins, refraction is largely taken over by the lens, which is no longer lens-shaped but spherical and robust as in fishes. The lens is suspended by thin and more or less diffusely arranged zonule fibers. The ciliary muscle, which in terrestrial mammals changes the shape of the lens during accomodation, is very much to completely reduced in dolphins. As a result, none or only insignificant accommodative changes of the focal distance of the lense have so far been found in dolphin eyes (Dawson, 1980). Instead, accommodation of the eye in the orbit seems to be managed by the contraction of the so-called "retractor bulbi muscle," which produces axial displacements of the eye in the orbit. When the eye is pulled back into the orbit, intraocular pressure increases, thus shifting the lens forward; when the eye is moved forward, the pressure decreases, shifting the lens backward. Such changes in intraocular pressure may be rendered possible via a large rete mirabile, which fills the orbital space behind the eyeball and surrounds the optic nerve (ophthalmic rete) (Mass and Supin, 2009).

The iris of dolphins is exceptionally resilient and highly vascularized. The dorsal part of the iris forms a so-called "operculum," which, in bright light, can extend over the central part of the lens. During strong light and thus maximal contraction of the pupil, the operculum meets the opposite border of the iris and, on both sides nasally and temporally, two separate pupil openings, a rostral and a temporal opening, form (Fig. 5.19). These openings correlate with the existence of two areas of highest neuron density of the retina (foveae), which are far off the central optic axis of the eye.[d] Particularly aquatic vision in the frontal direction through the rostral pupil opening, important for nearsighted visual control of prey during hunting, would thus have an adequate quality. This rostral (nasal) visual field is focused (represented) in the temporal fovea of the retina. Additionally, dolphins inspect under water objects situated laterally using the nasal fovea, which lies opposite to the temporal pupil opening (Fig. 5.17). The final benefit, however, for the double pinhole pupils is that they allow dolphins to have sharp vision both above and below water during strong daylight and thus a nearly panoramic visual field (Mass and Supin, 2009). The compensation in in-air myopia is due to the across position of the two pinhole-like pupil openings to the thin cornea center. The cornea center is of low curvature (Fig. 5.18) and thus has only minor additional refraction in air.

The possibility of binocular vision has been suggested for some cetaceans based on behavioral observations (Caldwell and Caldwell, 1972). Although the visual field of dolphins may overlap up to 30 degree when eyes move forward by

d. In the long-finned pilot whale a third dorsal fovea was found (Mengual et al., 2015).

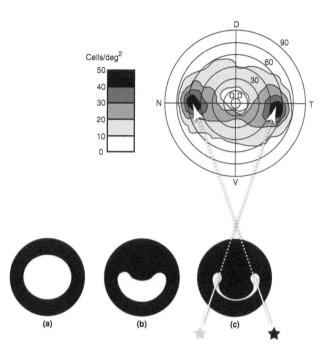

FIGURE 5.19 **The *bottom line* represents the schematic shapes of the pupil in a dolphin at various levels of illumination.** (a) *Low illumination*, nonconstricted oval pupil; (b) *moderate illumination*, partially constricted U-shaped pupil; and (c) *high illumination*, strongly constricted pupil reduced to two pinholes. The light rays (*white lines*) can pass through the pinholes to focus on the opposite fovea of the retina. The topographic distribution of ganglion cell density in the retina shows the positions of the foveae (*black areas*). Here, cell density is expressed as a number of cells per squared degree of the visual field and is shown by various colors (*yellow to black*), according to the scales. *Concentric circles* show angular coordinates on a retinal hemisphere centered on the lens. *D*, dorsal; *V*, ventral; *N*, nasal; *T*, temporal poles of the retina *(Redrawn aby Uko Gorter after Mass and Supin, 2009. Vision. In: Perrin, W.F., Wursig, B., Thewissen, J.G.M. (Eds.),* Encyclopedia of Marine Mammals. *Academic Press, San Diego, CA, pp. 1200–1211.)*

10–15 mm, other authors indicated that dolphins do not have binocular vision because uncrossed optic fibers are not described for dolphins (Ridgway, 1990; Mass and Supin, 2009).

Although dolphins have no tear glands, the Harderian glands and glands in the conjunctiva of the eyelids regularly bathe the cornea in a viscous solution, thus protecting it from the effects of sea water. Tarpley and Ridgway (1991) proposed that these special "tears" of the bottlenose dolphin, composed mainly of mucopolysaccharides, in addition to protecting the eye, these tears would help reduce frictional forces at the eye, especially when swimming at speeds where normal tears would be readily washed away (innervation of Harder's gland, see Cranial Nerve 7). The Harderian gland lies within the orbit of the dolphin eye, completely encircling the ocular globe in a belt-like fashion. In general, the mammalian Harderian gland has a poorly developed duct system. The results of histological and electron transmission microscopy, however, demonstrated that the Harderian gland of the bottlenose dolphin had characteristics of both Harderian and lacrimal glands. This gland has a multicellular composition and a well-developed, tubulo-alveolar duct system. The Harderian duct system is complex regarding its cytoarchitecture. The epithelial cells demonstrate an active transport suggested by the presence of pinocytotic vesicles (Ortiz et al., 2007).

Light Sensitivity

The dolphin eye has been modified for high light sensitivity but not high acuity (see Cranial Nerve 2). This is obvious from its complete retinal tapetalization and its large pupillary opening that can be constricted drastically under bright light conditions at the surface in and above water. The tapetalization is caused by the tapetum lucidum (tapetum fibrosum), which is a light-reflecting layer situated between the retina and the highly vascularized and thick choroid. This kind of "light amplifier" consists of an extracellular matrix of densely packed and regularly arranged collagen fibrils, which is situated behind the retinal pigment epithelium. Their external boundaries diffusely reflect incoming light. Under low light (scotopic) conditions, multiple reflections from 50 to 70 layers of fibrils results in significant back radiation of incoming light again to the retina. As to visual acuity, however, the bottlenose dolphin is well behind the human, horse, dog, and cat; it equals the harbor seal, and is followed by the goat, Indian elephant, and red deer (Peichl, 1997).

FIGURE 5.20 **Retina of a bottlenose dolphin showing the typical mammalian layering.** HE stain. *1*, Inner limiting membrane; *2*, layer of optic nerve fibers with blood vessels (*); *3*, ganglion cell layer containing the somata of ganglion cells, which are very large in dolphins; *4*, inner plexiform layer; *5*, inner nuclear layer containing the nuclei and surrounding perikarya of amacrine, bipolar, and horizontal cells; *6*, outer plexiform layer; *7*, outer nuclear layer with the cell bodies of rods and cones; *8*, outer limiting membrane; *9*, layer of rods and cones containing the outer and inner segments of photoreceptor cells (destroyed); *10*, retinal pigment epithelium (not visible here).

The organization of the retina in the dolphin is basically similar to that in terrestrial mammals but it has a number of specific cetacean features. In general, the light passes through several layers of the mammalian retina to reach the light sensors, the rods, and the cones. Rod cells function in low light conditions, while cone cells are responsible for color vision. The signal of rod and cone cells propagates first to bipolar neurons in the inner nuclear layer, which contains additionally the somata of the horizontal and amacrine interneurons (Ross and Pawlina, 2016). The bipolar cells connect to the neurons in the innermost ganglion cell layer of the eye. The axons of these ganglion cells then meet for the optic nerve.

The retina in dolphins is markedly thicker than in terrestrial mammals (370–425 µm vs. 110–240 µm) (Mass and Supin, 2009). Studies on the bottlenose and common dolphins have shown that the laminar structure of the retina in dolphins is basically similar to that in terrestrial mammals (Fig. 5.20). Cetacean retinae do contain cone receptors, but the cones seem to represent only 1–2% of the photoreceptors. The other receptor cells are rods (Peichl et al., 2001). Contrary to the majority of terrestrial mammals, which have two types of cones with different pigments providing color vision (short-wave sensitive S-opsin and middle-to-long-wave sensitive L-opsin), only L-opsin-containing cones were found in the retinae of 10 cetacean species. Several behavioral studies have supported that color vision in dolphins is poor; however, there are indications that these animals may have some mechanism of color discrimination (Madsen, 1980; Mass and Supin, 2009). This is consistent with the fact that color vision of dolphins should be blue-shifted because cones and rods of the bottlenose dolphin have a best sensitivity in the blue light area (wavelengths of the light around 525 and 488 nm, respectively).

Interestingly, the innermost (ganglionic) layer of the dolphin retina comprises a low total amount of neurons but with very large somata and thick axons. In the bottlenose dolphin, the most common diameter of ganglion somata is 20–35 µm. However, there are giant perikarya that reach up to 80 µm (Fig. 5.20), in the long-finned pilot whale up to 75 µm (mean 33.5 µm), and in the killer whale up to 100 µm (Mass and Supin, 2009; Mass et al., 2013; Mengual et al., 2015). As a whole, retinal ganglionic cells on an average seem to be larger in dolphins than in terrestrial mammals, even species as large as the bull and the elephant. Nuclei of these giant cells are also larger (>20 µm) than nearby other ganglion, bipolar, and receptor cell bodies. These large ganglionic cells in the dolphin retina may be attributed an important function in the rapid signal transmission to the brain and possibly the early recognition of visual objects (Dawson et al., 1982). Their specialized distribution in two high-density areas has been used to suggest that one of these areas, the temporal, may function for frontal vision, with the rostral (nasal) area functioning for side vision (see the previous discussion). In the bottlenose dolphin, neurons in high-density areas have slightly smaller perikarya than those in low-density areas (Mass and Supin, 2009).

The total number of ganglionic cells in the bottlenose dolphin has been reported to average at 220,000 cells per retina, 203,000 in the long-finned pilot whale, and 199,000 in the killer whale (Mengual et al., 2015). This figure is close to the range of fiber number estimations in the optic nerve of the bottlenose dolphin (147,000–185,000) (Mass et al., 2013), compared with such in the cat (250,000) and the human (1,000,000). The low density of ganglion cells in the retina of dolphins seems to correspond to the low density of axons in the optic nerve. Thus, although the diameter of the optic nerve is considerable in dolphins, their total number of optic fibers does not outnumber that in many terrestrial mammals. This is explained by the fact that (1) the axon diameters are greater than in terrestrial mammals: in a variety of dolphin species, a significant portion of the optic fibers exceeded 15 µm in diameter; (2) more than 50% of the cross-sectional area of the dolphin optic nerve are occupied by metabolically relevant extraneural space (vs. 12–20 % in terrestrial mammals). Such highly developed vegetative support and reserves could be of considerable value to a nervous system subjected to long periods of apnea.

Blood Supply of the Eye

The dolphin's eye receives its primary blood supply from the arterial ophthalmic rete located just behind the eyeball. The arteries of the rete are rather uniform in diameter (470–515 µm) and related to thermoregulation by the countercurrent heat exchange mechanism (Ninomiya and Yoshida, 2007; Ninomiya et al., 2014). The ophthalmic rete diverges from the basilar rete, which is fed by the spinal and cervical retia, a continuation of the posterior thoracic artery (see Circulatory System). The iris and the ciliary processes are supplied by iridic arteries via the major arterial circle located around the iris edge (Fig. 5.18). The position of this major arterial circle separates cetaceans from terrestrial mammals. In the latter, the major arterial circle is located between the posterior end of the iris and the anterior region of the ciliary body from which iridic arteries diverge anteriorly toward the pupillary margin. Arteries terminating in the choroid membrane run in parallel arrays so as to interdigitate the densely packed choroidal veins. The retinal vessels are of the holangiotic type (ie, the retina contains a compact plexus of blood vessels located in the major part of the light-sensitive portion of the retina) and diverge from the optic disc (blind spot) in the center of the ocular fundus (Dawson et al., 1987). Arterioles diverge from the retinal arteries as side branches, primarily at right angles. The adjoining retinal capillaries are only 3–4 µm in diameter so that red blood cells are barely able to pass through and plasma skimming is likely. These configurations (minimal diameters of blood vessels) presumably facilitate the production of a clear, sharp image without loss of light falling onto the photoreceptors (Ninomiya and Yoshida, 2007).

The vortex veins drain blood from the orbit. There are two vortex veins, located on the dorsal and ventral side of the eyeball near the iris root. They empty into the episcleral veins, following meridionally caudalward around the eyeball, run outside the rete, and then pass into the ophthalmic vein (Ninomiya and Yoshida, 2007).

Eye Movements

Dolphins have a complete set of external eye muscles, four straight and two oblique muscles. These muscles allow eye movements in both the horizontal and the vertical directions. The majority of the rectus muscle fibers insert in the eyelids and are more correctly called palpebral muscles (Pütter, 1902). In addition, they have well-developed retractor muscles (Fig. 5.18) that are innervated by the abducens nerve. They produce axial in/out movements of the eye in the orbit. Protraction of the eye by 10–15 mm occurs when bottlenose dolphins examine objects in air visually and with both eyes. So far, no systematic study on the external eye muscles (oculomotor muscles) in dolphins exists.

Ear: Sense of Balance and Hearing

The sensory part of the mammalian ear is located in the periotic or pars petrosa of the temporal bone (see the following). It consists of extremely delicate canals in a complicated spatial configuration (Figs. 5.21 and 5.22), the membranous labyrinth. The fluid within this labyrinth is called endolymph and its movements stimulate sensory cells within the membranous wall. The membranous labyrinth comprises the vestibular part with the receptor organ for balance, the cochlear part which contains sound receptors, and a duct that connects these two systems (ductus reuniens). Between the membranous labyrinth and the neighboring bone (bony labyrinth), there are fluid-filled clefts, which are connected to the subarachnoid space of the meninges by the vestibular and cochlear aqueducts. These clefts are lined by a squamous epithelium and contain perilymph, which has a composition similar to the cerebrospinal fluid.

Sense of Balance: The Vestibular System

The vestibular labyrinth of mammals comprises the utricle (utriculus), the saccule (sacculus), and the semicircular ducts (ductus semicirculares), which are filled with a fluid, the endolymph. In dolphins, the shape of the vestibular apparatus much resembles that seen in terrestrial mammals (Figs. 5.21 and 5.22). There is a sensory macula utriculi and macula sacculi within the wall of the utriculus and sacculus, respectively, and a sensory ampullary crest (crista ampullaris) within each semicircular duct (not shown). These little organs sense and conduct impulses of balance (body position, translational and rotational accelerations) via the vestibular root of the vestibulochochlear nerve (see Cranial Nerve 8) to the brainstem. The related vestibular ganglion, located within the internal acoustic meatus, receives afferent fiber bundles from the vestibular receptor (hair) cells.

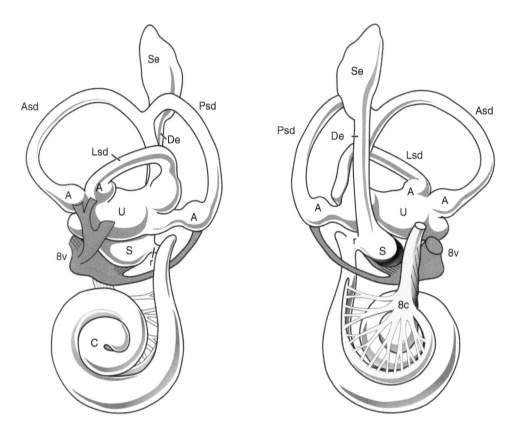

FIGURE 5.21 **Survey on the mammalian inner ear.** Lateral (left) and medial view (right) of a 30 mm human membranous labyrinth. Vestibular nerve, *red*; cochlear nerve, *yellow*. *A*, ampullae of semicircular ducts; *Asd*, anterior semicircular duct; *C*, cochlear duct; *De*, ductus endolymphaticus; *Lsd*, lateral semicircular duct; *Psd*, posterior semicircular duct; *r*, ductus reuniens; *S*, saccule; *Se*, saccus endolymphaticus; *U*, utricle; *8c*, cochlear nerve; *8v*, vestibular nerve. *Redrawn and modified from* Gray's Anatomy *(1980) Williams P.L., Warwick R., eds., 36th ed., 1578 p.*

The inner surface of the utricle and saccule is covered by a single-layered squamous epithelium (König and Liebich, 2009). In mammals, their medial wall is thickened to form the elevated and oval-shaped maculae of the utricle and the saccule (macula utriculi, macula sacculi). Within the maculae, the flattened cells of the squamous epithelium change into columnar hair cells that are embedded into supportive, nonreceptive cells and innervated by fiber bundles of the vestibular nerve. Their sensory hairs reach into a gelatinous layer on top of each macula. Adhering to the gelatinous layer are small, calcium carbonate crystals, called otoconia or otoliths. Deflections of the labyrinth cause shifts of the otoconia with respect to the sensory hairs due to gravity and initiate electrical impulses in the hair cells that are conducted to the brainstem via the labyrinth root of the vestibulocochlear nerve.

The three semicircular ducts are housed within the semicircular canals of the osseous labyrinth. Each duct arises from the utricle, forms two-thirds of a circle in a single plane to return to the utricle. The semicircular ducts stand roughly at right angles to each other. In dolphins, the anterior duct is oriented in a transverse plane of the head, the posterior in a sagittal plane, and the lateral duct in a horizontal plane. This orientation of the canals within the dolphin's head approximates that in the horse, sheep, and pig but not in the human where the canals are rotated by approximately 45 degree outward in relation to the cochlea when viewed from above (Benninghoff and Drenckhahn, 2004; Ellenberger et al., 1977; Nickel et al., 2003). Each duct has a dilatation, the ampulla, near the junction with the utricle. Within each ampulla protrudes a ridge-shaped thickening that marks the sensory ampullary crest (crista ampullaris). Similar to the situation in the maculae, the hairs (stereocilia) of the sensory cells in the crista end in a gelatinous covering (cupula). This time the acceleration of the head induces a movement of the endolymph deforming the cupula and stimulating the receptor cells, which again send an electrical impulse into the vestibular nerve.

In comparison with terrestrial mammals, the size of the vestibular system in relation to the cochlea is rather different in dolphins. Cetaceans are unique in having semicircular canals that are significantly smaller than the cochlear canal (Jansen and Jansen, 1969). The semicircular canals in dolphins and other toothed whales have smaller radii and smaller diameters than in any other mammal of comparative body size (Spoor et al., 2002). For example, whereas the utricule and saccule of a fetal narwhal are of normal size in comparison with those of other mammals, the semicircular canals are miniaturized

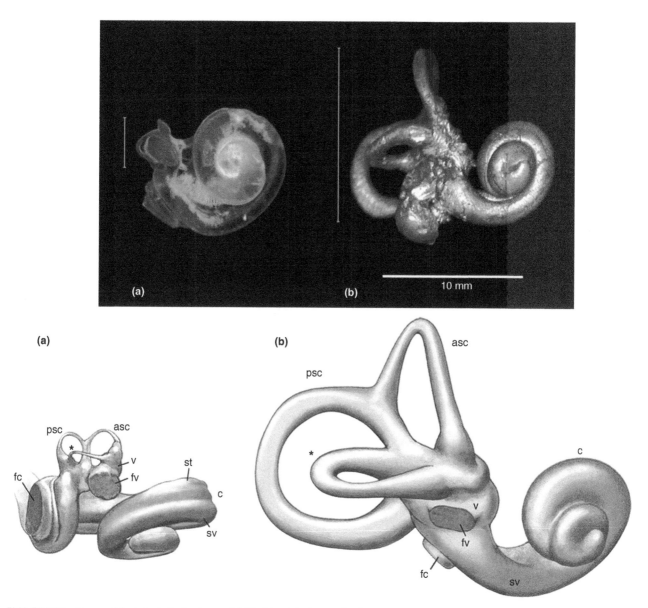

FIGURE 5.22 Osseous labyrinths in the bottlenose dolphin (a) and the human (b). Top: During careful maceration (a technique to clean bone from soft tissues), vertebrate carcasses are left to decompose until only inorganic material is retained. The cavity of the bony labyrinth can then be injected with artificial resin (left) or special alloys (right; eg, Wood's metal, an alloy from bismuth, lead, tin, and cadmium with a melting point at 70°C). After removing all bone with acid (decalcification) the original osseous labyrinth remains as a cast and can be interpreted as to its dimension and the size and shape of its components. Size reduction of the dolphin vestibular organ with respect to the human is indicated by scales (blue) representing its individual dimension in the two species. Bottom: Computer reconstructions of the bony labyrinths. In the dolphin, the cochlea (c) has been rotated during evolution by about 90° with respect to the vestibularsystem (v). The vestibulum in the dolphin is so small that the vestibular window (fv) stands out of the ventricle. *asc*, anterior semicircular canal; *fc*, cochlear foramen; *fv*, vestibular foramen; *psc*, posterior semicircular canal; *st*, scala tympani; *sv*, scala vestibuli; *v*, vestibulum; *, lateral semicircular canal. *(Endocasts on top: Courtesousy Toshiro Yamada, Tokyo, Japan).*

but not obliterated, as was reported for some adult toothed whales (Lindenlaub and Oelschläger, 2000). The same holds true for fetal pantropical spotted dolphins (Fig. 5.23). The area between the vestibule and each semicircular canal and thus their radii is smaller. The innervation of the semicircular canals is proportionately reduced. Less than 5% of the vestibulocochlear nerve is devoted to vestibular fibers, as compared to approximately 40% in other mammals (Wartzok and Ketten, 1999). No equivalent reduction of the vestibular system is known in any land mammal. In sections through the dolphin periotic, the area taken by the cochlea is about twice as large as that of the vestibular apparatus. This ratio is also seen in the early fetus of the narwhal (Comtesse-Weidner, 2007; Lindenlaub and Oelschläger, 2000) and in the fossil ancestors of dolphins and whales (Fleischer, 1978; Spoor et al., 2002). In the white whale (*Delphinapterus leucas*), the volume taken

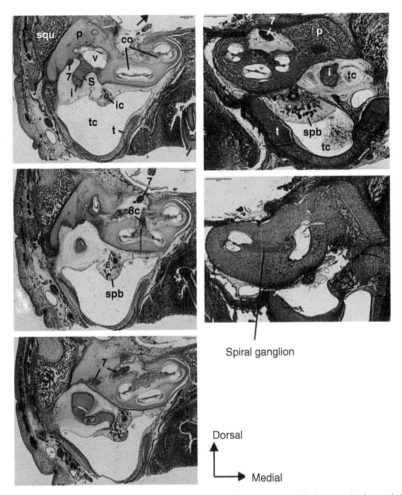

FIGURE 5.23 **Fetal pantropical spotted dolphins, transverse sections through the tympanoperiotic complex in upright position.** Azan stain. Left: tympanoperiotic complex of a 116 mm CRL fetus, chondral anlage of periotic. Right: complex of a 123 mm fetus, beginning ossification. *co*, cochlea; *CRL*, crown–rump length; *i*, incus; *ic*, internal carotid artery; *l*, larynx; *p*, periotic; *s*, stapes inoval foramen; *spb*, spongiform body (corpus spongiosum); *squ*, squamosal; *t*, tympanic element; *tc*, tympanic cavity; *v*, vestibulum; *7*, facial nerve; *8c*, cochlear nerve; *arrow*, direction of brainstem.

by the vestibular apparatus was given as 17 times smaller than that of the cochlea, compared to 1.47 times larger than the cochlea in the human (Figs. 5.21 and 5.22). That means that the cochlear part of the dolphin labyrinth is larger than that in the horse but the volume of the vestibular part is merely as large as that in the rabbit and the semicircular canals are as large as in the hamster (Claudius, 1858). In parallel, most of the vestibular nuclei in the dolphin's brainstem are also minute in comparison with terrestrial mammals (Kern et al., 2009).

A possible explanation for the reduction of the vestibular system is that the fusion of the cervical vertebrae in toothed whales resulted in limited head movements (Kandel and Hullar, 2010). In dolphins, the motility of the head is largely restricted to the atlanto-occipital joint, that is, largely to up-and-down movements in the vertical plane. Moreover, the velocity of body movements during locomotion is reduced in comparison to terrestrial mammals because of the higher density of their aquatic environment. Thus, due to both reduced head velocities and accelerations as well as slower body maneuvers, movements of the dolphin head are reduced. Such restrictions mean less input to the vestibular system and thus a reduction of related vestibular receptors. This does not mean that cetaceans do not receive acceleration and gravity cues but the neural budget for the vestibular cues is reduced (Wartzok and Ketten, 1999).

Hearing in Dolphins

The organ of hearing in dolphins is very peculiar. There are three main parts as in other mammals: the external, the middle, and the inner ear. The external ear, however, is largely rudimentary in dolphins. The middle ear is unique and highly specialized as to morphology and function, and the inner ear is well developed and comprises the typical mammalian bauplan. In the following section, we will start with a synopsis of all the relevant parts of the dolphin's ear before the background of

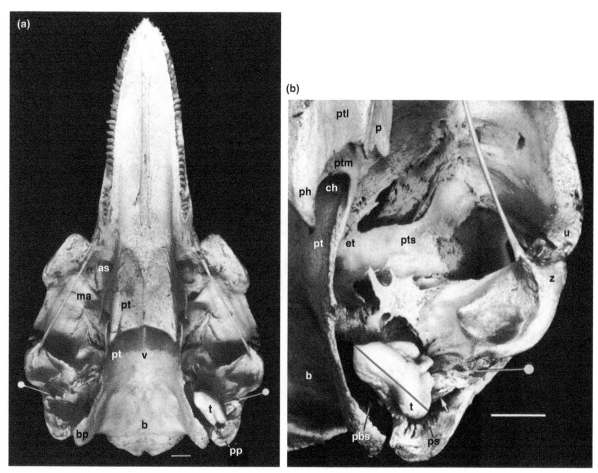

FIGURE 5.24   Basal aspect of skull (a) of an Atlantic white-sided dolphin and detail of the left otic region (scale bars = 2 cm). (a) A needle is situated in the residual external auditory meatus of the tympanic bone. (b) *Red line* marks the longitudinal axis of the tympanoperiotic complex. *as*, position of anterior lobe of pterygoid sinus; *b*, basioccipital; *bp*, lateral process of basioccipital; *ch*, choana; *et*, position of Eustachian tube; *p*, palatine; *pbs*, position of peribullary sinus; *ph*, pterygoid hamulus; *ps*, position of posterior sinus; *pt*, pterygoid; *ptl*, lateral lamina of pterygoid; *ptm*, medial lamina of pterygoid; *pts*, position of pterygoid sinus; *t*, tympanoperiotic complex; *u*, postorbital process of frontal; *v*, vomer; *z*, zygomatic process of squamosal; the *white arrow* points to the posterior processes of tympanic and periotic. *(From Oelschläger (1986). Tympanohyal bone in toothed whales and the formation of the tympano-periotic complex (Mammalia: Cetacea). J. Morphol. 188, 157–165.)*

the mammalian bauplan. With respect to specific modifications and regressions in dolphin ears, the situation in the human serves as a guideline (Figs. 5.21 and 5.22). Because hearing is the main sensory system of dolphins due to the significance of their echolocation system, we will give a detailed overview of ear morphology and function. This way a systematic presentation of the whole ear and its main parts from all aspects are given in order to help imagine how the small parts (tympanic ligament, tympanic plate, ossicles, muscles) form a continuum for sound conduction.

External Ear

Dolphins have no pinnae. With regard to the physical properties of water in comparison to air, the existence of such appendages would hamper the minimization of water resistance in order to cope with fast prey organisms like fish with their streamlined body contour. Therefore, external pinnae are never present in dolphins, and they do not even occur as anlagen[e] during ontogenesis (Štěrba et al., 2000). However, an S-shaped rudimentary external auditory meatus is found behind the temporomandibular joint (Figs. 5.24 and 5.25a); it starts at a pinhole-like external opening (porus acusticus externus) and attaches to the squamosal bone via connective tissue (needle in Fig. 5.24). Superficially, the cartilage of the meatus may be enlarged into a small rounded plate (bottlenose dolphin); deeper parts form an incomplete tube that runs aside the membranous meatus (Fig. 5.26). The latter is narrow and was reported to be obliterated in some places by cellular debris or dense cerumen (harbor porpoise; Boenninghaus, 1903) but widens before the tympanic ligament (Fig. 5.27). Its functional

---

e. Anlage is a primordium of an organ in its early stage of embryological of fetal development.

(a)

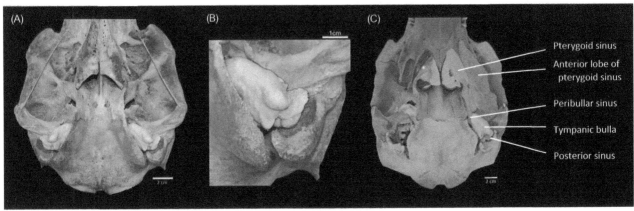

(b)

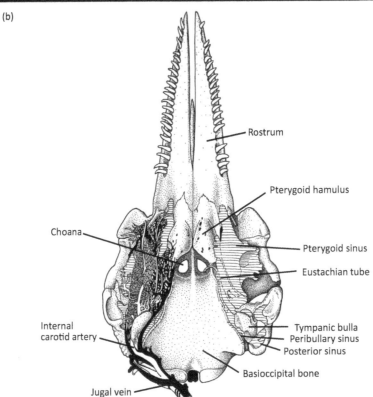

FIGURE 5.25 (a) Ventral view of skull bases of a common dolphin (A, B) and a white-beaked dolphin (C). In the common dolphin, the tympanoperiotic bones are still in place in spite of the maceration process; in the white-beaker dolphin, the corrosion preparation of the accessory air sinuses of the Eustachian tube (tuba auditiva) shows the left bulla still in place due to the rubber material. The accessory air sinuses of the Eustachian tube were segmented on the photo (transparent blue areas) according to the rubber material and after Costidis and Rommel (2012); *, the right pterygoid bone was detached so that the cavity that houses the pterygoid sinus is visible. (b) Bottlenose dolphin skull in basal aspect. Left skull base: bones of skull base with tympanoperiotic and overlying air spaces (sinuses). Right skull base: blood supply through arterial plexus from maxillary artery and a venous plexus embedded in a tough fibrous sheet of connective tissue. *(Drawing by Jutta Oelschläger, modified after Purves and Pilleri, 1983. Echolocation in Whales and Dolphins. Academic Press, London, New York.)*

implications are not clear (Fraser and Purves, 1960; Ketten, 2000). The same holds true for a set of four delicate muscles which attach to the cartilage of the external ear canal; their homologization with external ear muscles of terrestrial mammals is unclear (Boenninghaus, 1903; Huber, 1934; Lawrence and Schevill, 1965).

The inner (proximal) part of the external auditory meatus (Fig. 5.26) is equipped with glandular tissue, either compound tubuloacinar glands in a rather thick epithelium of the bottlenose dolphin or smaller (cross-sections < 0.2 mm) simple tubuloacinar glands in the common dolphin (Solntseva, 2007). The epithelial cells of the multilayered squamous epithelium seem to undergo constant proliferation and desquamation. The nuclei of the superficial layer are smaller than those in

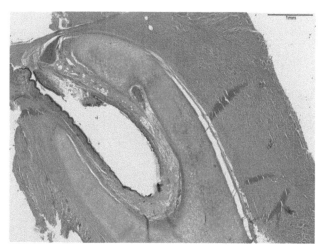

FIGURE 5.26 **Histological cross section of the external acoustic meatus of a bottlenose dolphin (ID196).** HE stain.

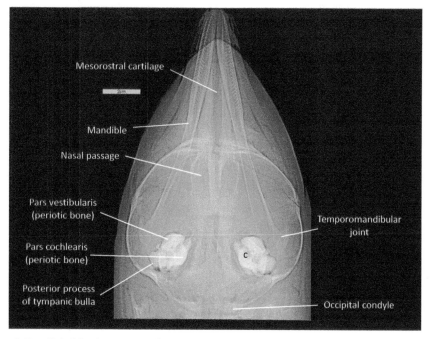

FIGURE 5.27 **High-resolution digital luminescence radiogram of a perinatal pantropical spotted dolphin head in horizontal orientation (dorsoventral view, anterior facing up)** (Rauschmann et al., 2006). *C*, cochlea.

the basal layer and pycnotic.[f] The excretory ducts open into the epithelial folds of the auditory meatus. Unlike the ducts, the secretory tubules and acini[g] are surrounded by myoepithelial cells and contain an eosinophilic secretory product with basophilic granules and degenerated nuclei (not shown). The glandular epithelium is made up of cylindrical cells and the secretion type of the auricular glands of dolphins was determined as a merocrine one with elements of a holocrine secretion (Solntseva, 2007).

There are various recent hypotheses on how sound is transmitted to the middle ear in dolphins bypassing the external auditory meatus. A possible functional counterpart of the mammalian outer ear can be regarded as an extended acoustic fat body, localized around and within each of the lower jaw (dental) bones of toothed whales (Figs. 5.4, 5.5b–k, 5.27–5.29). The fat body is supposed to guide and transmit sound directly to the tympanic bulla (tympanoperiotic complex), a medium-sized and dense shell-like bony structure which houses the air-filled middle ear cavity (see the following) (Au and Hastings, 2008;

f. Pyknosis means here a degeneration of a cell in which the nucleus shrinks in size and the chromatin condenses to a solid more or less structureless mass.
g. Acinus (Latin for berry) is a berry-shaped termination of an exocrine gland.

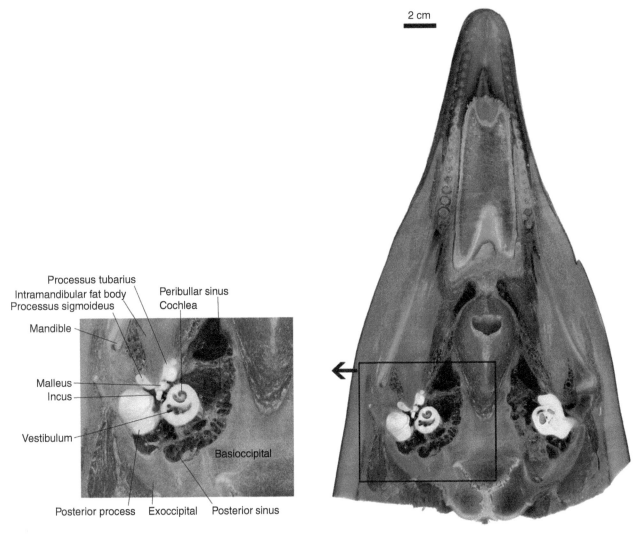

FIGURE 5.28   **Horizontal sawing section (dorsal view) of a subadult bottlenose dolphin head through the periotic bone.** (See also Fig. 5.8)

Rauschmann et al., 2006). Accordingly, the lower jaw is thought to be part of the sound-receiving system because the mandible divides the "acoustic fat body" into an internal and external component (Figs. 5.5b and 5.28). In correspondence with this, the bony lateral wall of the mandible is dense but extremely thin (pan bone[h]) and thought to act as an acoustic window in the transmission of sound (Brill et al., 1988; Norris, 1968; Ridgway, 1999). The medial bony wall of the mandible is largely lacking so that the shape of the lower jaw in toothed whales is unique. The work by Brill, Møhl, and coworkers (Brill et al., 2001; Brill and Harder, 1991; Møhl et al., 1999) clearly supports this "pan bone pathway" via the mandibular fat bodies (Figs. 5.5 and 5.28) but the mechanism for the transmission of sound across the bony wall of the mandible, that is, through the pan bone, remains largely unsettled (see the following).

Recent studies of Cranford et al. (2008a,b) suggest a pathway for sound reception in a beaked whale species (*Ziphius cavirostris*) bypassing the pan bone. According to this "gular pathway," propagated sound pressure waves may enter the head from below and between the lower jaws, passing through the soft (ie, boneless) medial wall of each posterior mandible, and proceed via the internal (medial) mandibular fat body. Both pathways, the "pan bone pathway" and the "gular pathway," are believed to transmit sound from the internal mandibular fat body to the tympanic bulla (Cranford et al., 2010; Hemilä et al., 2010, 1999; Ridgway, 1999; Ridgway and Au, 1999). It also seems likely that this fat body serves as an acoustic amplifier similar to the external auditory meatus of terrestrial mammals (Cranford et al., 2008a,b; Nummela

h. The term " pan bone" for the dolphin mandible was dubbed in the 1960s by Kenneth S. Norris. Holding the mandible toward the sun, he noticed a translucent area of very thin bone in its proximal third (Ridgway, 1999). He interpreted this area as the acoustic window for the transmission of sound on its way to the middle ear (Norris, 1968).

(a)

(b)

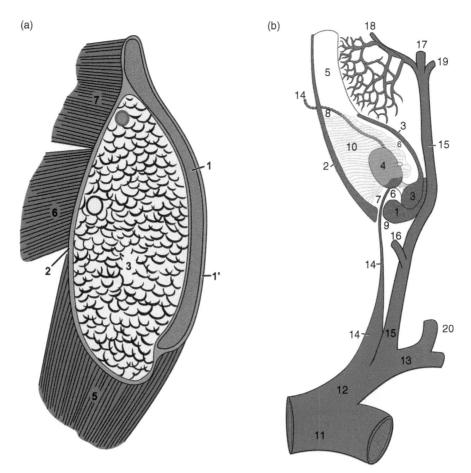

FIGURE 5.29   (a) Transverse section of the right mandibula of a toothed whale (harbor porpoise), posterior part near the lower jaw articulation: *1*, outer mandibular wall; *1'*, outer periost; *2*, inner periost; *3*, alveolar canal with mandibular fat body; *5*, mylohyoid muscle; *6*, medial pterygoid muscle; *7*, lateral pterygoid muscle (*brown*, bone; *blue*, periost; *yellow*, fat; *red*, muscles). (b) Schematic course of the right internal carotid artery of a toothed whale (harbor porpoise), ventral aspect: *1*, paroccipital process; *2*, basioccipital process; *3*, tympanic; *4*, periotic; *5*, pharyngotympanic tube; *6*, tympanic cavity; *7*, posterior entrance into peribullary sinus; *8*, anterior entrance dto.; *9*, basioccipital cleft; *10*, Corpus spongiosum; *11*, Aortic arch, anonymous artery (A. anonyma); *13*, subclavian artery; *14*, internal carotid artery; *14'*, Canalis caroticus of skull base; *15*, external carotid artery; *16*, external maxillary artery; *17*, internal maxillary artery; *18*, pterygopalatine artery (palatina descendens); *19*, deep temporal artery; *20*, occipital artery. (*Redrawn and modified by Jutta Oelschläger from Boenninghaus, 1903. Das Ohr des Zahnwales, zugleich ein Beitrag zur Theorie der Schalleitung: Eine biologische Studie. Fischer, Jena*)

et al., 2007). Cranford et al. (2010) showed that the ear ossicles may be sensitive to complex high-frequency vibrations of the tympanic bulla evoked by vibrations across the large contact area with the mandibular fat body.

Furthermore, acoustic experiments of Popov and Supin (1990a,b) suggested an additional sound reception pathway[i] further caudally and in parallel to the rudimentary external auditory meatus (lateral acoustic pathway). At first glance, this seems to contradict Norris's theory of mandibular hearing, but could also mean that two different pathways for incoming sound exist, one at the rudimentary external auditory meatus and the other via the internal mandibular fat body and that the functional implications of these pathways may change during postnatal development (Maxia et al., 2007). Recent work (Ketten, 2000, 1994; Wartzok and Ketten, 1999) seems to support this idea of an additional lateral acoustic pathway in dolphins. Based on CT and MRI data of full-term dolphin fetuses, a fatty cone-shaped structure lateral to the tympanoperiotic complex was found, probably representing this (external) lateral acoustic pathway (Rauschmann et al., 2006).

Air Sinuses The bones of the skull base adjacent to the tympanoperiotic complex seem to have receded to the same degree that accessory air sinuses of the Eustachian tube (tuba auditiva) extended. In dolphins, the middle ear cavity and auditory tube blend into a set of air-filled sinuses at the ventral surface of the skull base (Figs. 5.24 and 5.25). These sinuses are peculiar for toothed whales as to their position, structure, and volume and may be homologous to the guttural pouches in horses.

i. Recently, an additional mechanism of sound reception via the mental foramina at the tip of the mandible was proposed (Ryabov, 2010).

In dolphins, they extend between bony laminae of the pterygoid, palatine, alisphenoid, squamosal, exoccipital, and basioccipital bones. Immediately neighboring the middle ear cavity are the smaller posterior sinus, peribullar sinus, and middle sinus. The pterygoid sinus is very large in dolphins and situated lateral to the auditory tube (Oelschläger, 1986). This sinus blends rostrally in the anterior lobe of the sinuses, which extends nearly as far as the palatine bone. In parallel, but not as far rostrally, there is a hamulus lobe medially extending into the pterygoidhamulus. The central part is built by the pre- and postorbital lobes. Caudally, the pterygoid sinus reaches around the tympanic bulla forming the surface of the peribullar plexus (Costidis and Rommel, 2012). Whereas the deep walls of the sinuses are immediately connected to the adjacent bony elements, the ventral surface of the sinuses is represented by a periosteum-like tough fibrous membrane, which contains a venous plexus. The inner surface of the sinuses consists of a thin mucous membrane, which is continuous with that of the middle ear cavity and auditory tube. The mucous membrane is equipped with a laminated columnar epithelium where cilia are now and then to be seen (Boenninghaus, 1903). The fibrovenous plexus of the sinuses runs in the submucose connective tissue and is supplied by the external carotid and maxillary arteries (*A. palatina descendens*; Boenninghaus, 1903). The plexuses are drained primarily by three parent veins, namely the facial, external jugular, and internal jugular veins (Costidis and Rommel, 2012).

Apart from the middle ear ossicles, the tympanic cavity houses a venous plexus, which is embedded in delicate soft connective tissue. Interestingly, the corpus spongiosum tympanicum contains the internal carotid artery, which thins out during late fetal stages and gets obliterated functionally (Fig. 5.29b). This is a major specific indicating profound changes in the circulation of the brain (see the following). So far there is no hint that the venous plexus is supplied by the internal carotid artery, which heads for the canalis caroticus in the skull base (Costidis and Rommel, 2012).

In the harbor porpoise, the tympanic venous plexus is connected with the large jugular ganglion via a thicker fiber nerve bundle; a thin fiber bundle enters the carotid canal, whereas another bundle runs to the oval window and gives off a very thin ramus to the very small otic ganglion (Boenninghaus, 1903). The corpus spongiosum tympanicum is part of a flat convolute of blood vessels, which is connected through the tympanoperiotic fissure of the tympanic cavity to the pterygoid and peribullar sinuses (Fig. 5.34 and 5.25). Injection of the vessels of the corpus spongiosum from the jugular vein results in an increase to the double volume.

The pneumatic spaces of the sinuses mentioned may serve several functional aspects. (1) They can help to shield the TP against the dolphin's own sonar beam, which is produced in the upper respiratory tract and supporting associated structures (see Dolphin Head Characteristics). (2) They may help to isolate one ear from the other acoustically as to incoming sound, thus allowing directional hearing. (3) The sinuses can also be involved in diving mechanics. If a dolphin descends in water, the air volume in its sinuses will shrink rather quickly. The air in the pterygoid sinus may be shifted into the middle ear cavity to maintain the vibration of the ear ossicles in their "air environment." Additionally, it may be possible for the animal to transfer air from the lungs via the pharynx and the auditory (Eustachian) tube to the middle ear. In order to prevent barotrauma, blood may be shifted into the plexus venosus of the pterygoid sinus and the corpus spongiosum replacing the loss of air volume (Figs. 5.34, 5.25 and 5.28). At any rate, the air sinuses should be subject to more or less drastic volume changes. In this respect, good divers among the dolphins seem to have lost some thin bony material that still ensheathes the sinuses in pristine dolphin-like shallow-water species (La Plata dolphin, *Pontoporia blainvillei*). Thin-walled pneumatized bony cavities known in terrestrial mammals from the mastoid process of the temporal bone (mastoid cells) as well as from the surroundings of the nose (paranasal sinuses) are always lacking in dolphins.

Middle Ear

The function of the middle ear in terrestrial mammals is to match the acoustic impedance between two media, so that airborne sound is transmitted to the fluids in the inner ear. As to the fundamental characteristics of the middle ear in cetaceans, Nummela et al. (1999b) wrote:

> *The functional morphology of the ear region reflects environmental constraints. The tetrapod middle ear, at the time vertebrates first occupied land, evolved to compensate for the difference between the low characteristic acoustic impedance [resistance] of air and the much higher specific input impedance of the cochlea. Whales are mammals that are secondarily adapted to life in water. At the stage of returning to the water, tetrapods encountered the problem of impedance matching again, although a matching in the reverse direction, i.e. from a higher to a lower impedance, was now needed.*

In spite of profound changes with respect to the situation in terrestrial mammals, however, all the middle ear structures of mammals as ossicles, muscles, and nerves are present in dolphins (Fig. 5.23). The articulations between malleus and incus and between incus and stapes are functional. Both the dolphin's middle and inner ear are located within the tympanoperiotic complex. In its natural position within the animal, the complex is upright, that is, with the periotic (cochlear part) facing dorsally toward the brain, and the tympanic ventrally (Fig. 5.30). The tympanic and the periotic are the pars tympanica and the pars petrosa of the temporal bone in mammals, which do not fuse with the pars squamosa

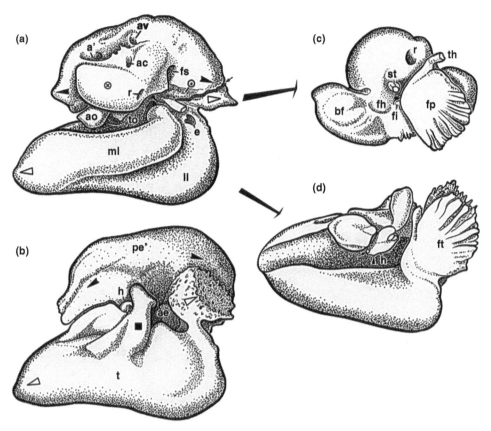

FIGURE 5.30 **Tympanoperiotic bones of bottlenose dolphin in different aspects.** (a) Medial aspect of left tympanoperiotic complex, (b) lateral aspect of right tympanoperiotic complex. (c and d) Both components of the right complex in (a) folded apart and shifted to the right to show their contact areas [*small arrow* in (a), *fp* in (c) and *ft* in (d)]. *a'*, porus acusticus internus; *ac*, external aperture of cochlear aqueduct; *ae*, bony porus acusticus externus of cetaceans; *ao*, accessory ossicle (processus tubarius); *av*, external aperture of vestibular aqueduct; *bf*, facet for processus tubarius of tympanic; *e*, elliptical foramen; *fh*, fossa for head of malleus; *fi*, fossa incudis; *fp*, articulating facet of posterior process of periotic; *fs*, fossa for stapedia muscle; *ft*, articulating facet of posterior process of tympanic; *h*, malleus; *ll*, lateral lobe of tympanic; *ml*, medial lobe of tympanic; *pe'*, periotic; *r*, fenestra cochleae; *st*, stapes; *t*, tympanic; *th*, tympanohyal; *to*, tympanic orifice; small arrow in (d), attachment site of tensor tympani muscle; small arrow head in (d), manubrium of malleus; ◄►, anterior (left) and posterior (right) process of periotic; ◁▷, anterior (left) and posterior (right) process of tympanic; ■, sigmoid process; ☉, facial canal; ⊗, cochlear part of periotic. *(Artwork by Jutta Oelschläger)*

in dolphins (see the earlier discussion). In Fig. 5.31, the ear bones are shown from different aspects in order to mediate a solid impression of their intricate three-dimensional configuration. Transverse computer tomographic (CT) sections of dolphin tympanoperiotic bones (Fig. 5.33) give good insight into the quantitative distribution of the ceramic-like bone material and inherent functional implications. The large tympanic and periotic bones are extremely different in shape (Figs. 5.31–5.33). Whereas the periotic is stout and houses the very small vestibular part and the well-developed cochlear part of the inner ear, the tympanic bulla is shell-like and U-shaped in cross-section. Both enclose the middle ear cavity, which is air filled.

From the tympanic bone, sound is conducted to the inner ear via the middle ear ossicles across the air-filled middle ear cavity (see previous discussion) (Cranford et al., 2010; Fleischer, 1978; Hemilä et al., 1999, 2010; Nummela et al., 1999a,b). During diving the hearing sensitivity of toothed whales, that is, of belugas, does not change (Au and Hastings, 2008), indicating that hydrostatic pressure at depth and thus the volume of the middle ear cavity can be adjusted by additional quanta of air or that the sound is transmitted via substrate conduction along the bony structures.

There are some peculiarities found in the middle ear of toothed whales. The tympanic bulla is thought to act as a "fourth ear ossicle" (Fleischer, 1978) that takes over the sound from the internal mandibular fat body via a large contact area (Cranford et al., 2010). Apart from this, the tympanoperiotic complex is surrounded by isolating air sinuses. The internal carotid artery is obliterated in the adult dolphin and replaced functionally by an arterial rete mirabile[j] in the area of the air

j. A rete mirabile (wonderful net) is a complex network of small arteries and sometimes accompanying veins lying close to each other. It is found in several warm-blooded vertebrates.

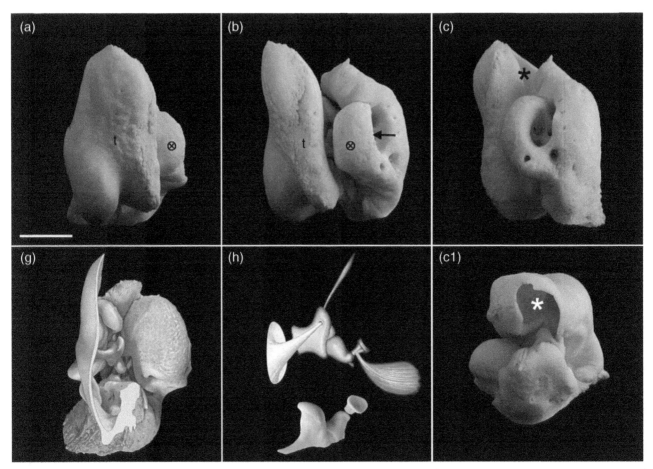

FIGURE 5.31 Systematized views of the tympanoperiotic complex and its parts (Scales = 10 mm). (a–d) Right tympanoperiotic complex of adult the white-sided dolphin. (a) Shown from ventral (same aspect of TP as in Fig. 5.24), cochlear part of the periotic (⊗) mostly hidden by the large tympanic *(t)*. (b–d) Rotated counterclockwise in each step, (b) in medial aspect, (c) slightly more rotated to show better insight into the internal acoustic opening; asterisk is in bony opening of pharyngotympanic tube, (d) lateral aspect with prominent sigmoid process (s) and external acoustic opening *(arrow)*. (c1 and c2) Seen from both ends of the tympanoperiotic complex; (c1) looking into the orifice of the pharyngotympanic tube, (c2) from the opposite end of the complex. (d) Adult bottlenose dolphin. The components of the complex are taken apart, with the result in (d1): tympanic seen from the periotic bone, (d2) periotic seen from tympanic bone. (d1') Tympanic seen from the other side, looking like an old whaler [the sigmoid process (s) as the nose, the posterior process (pp) as the beard], (d1') periotic seen from the other side, (e) neonate bottlenose dolphin. (e1 and e2) The tympanic and periotic shown as in (d1) and (d2). (e1) Tympanic with malleus and incus. (e1' and e2') Aspect on tympanic and periotic as in (d1') and (d2'), (f) neonate tympanic with ear ossicles in medial aspect. (g) Same basal aspect of tympanoperiotic complex in the adult bottlenose dolphin as for the whitesided dolphin in (a), but the medial part of the tympanic has been omitted to show the content of the tympanic cavity. (h) Ear ossicles of the bottlenose dolphin. Top: Arrangement of the ossicles with their muscles corresponding to the middle ear cavity in (g). Bottom: Isolated ossicles in their natural sequence from another dolphin. *ao*, accessory ossicle (processus tubarius); *cl*, crus longum of incus; *ct*, foramen of chorda tympani in malleus; *e*, elliptical foramen; *fp*, facet on posterior process of periotic for articulation with ft; *ft*, facet on posterior process of tympanic for articulation with fp; *hm*, head of malleus; *l*, lateral lip of tympanic seen from the rear; *ll*, lateral lobe of tympanic; *m*, medial lip of tympanic seen from the rear; *ml*, medial lobe of tympanic; *mm*, processus muscularis mallei; *p*, periotic; *pg*, processus gracilis mallei; *ppp*, posterior process of periotic; *ppt*, posterior process of tympanic; *s*, sigmoid process; *t*, tympanic bone; *tm*, tympanic membrane; *tt*, attachment site of tensor tympani muscle; wheel, cochlear part of periotic; arrow in b), porus acusticus internus; arrow in d), porus acusticus externus; asterisk in c) and d), bony opening in tympanic; asterisk in c1), tympanic cavity; blue dot in f), incus.

sinuses bypassing the middle ear cavity (Fig. 5.25b). In the fetal dolphin, the internal carotid artery is already thin but still functional (Boenninghaus, 1903). This is seen in Figs. 5.33 and 5.34 in the fetal and adult harbor porpoise.

In transverse CT sections (Fig. 5.32), the typical appearance of the two major ear bones is obvious: The periotic is massive and the tympanic hollow. Whereas the lateral wall of the tympanic cavity is continuous but thin and exposes the small external bony porus acusticus, the medial wall is discontinuous (Fig. 5.30). Here, the delicate middle part of the bony lateral wall is opposed medially by a wide gap between the cochlea and the thick inner tympanic lip (tympanoperiotic hiatus or petrotympanic fissure). Boenninghaus (1903) states that through this hiatus a remarkable body of blood vessels enters the middle ear cavity (plexus venosus caroticus; Figs. 5.33 and 5.34). This plexus may be important for pressure regulation in

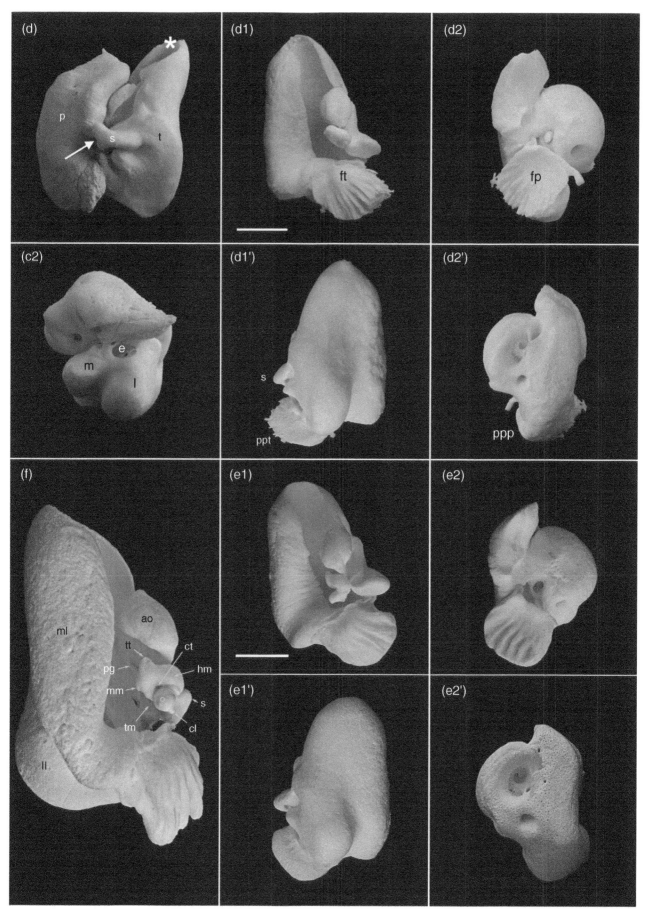

FIGURE 5.31 (*cont.*)

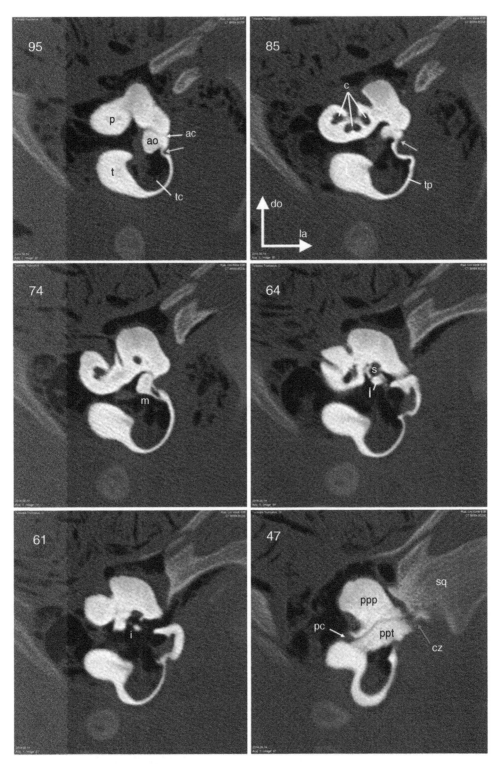

FIGURE 5.32A **Transverse CT sections of the right tympanoperiotic complex of a subadult bottlenose dolphin in relation to the tympano-periotic length axis (see 5.24) (Scales = 5.5 cm).** The *blue arrows* mark the thin and folded bony sheets connecting the tympanic plate to the processus tubarius and accessory ossicle.

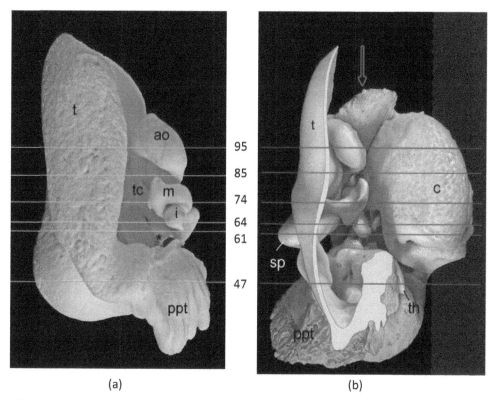

(a)   (b)

FIGURE 5.32B   (a) Tympanoperiotic complex (as in Fig. 5.31f), and (b) neonate tympanic in the dorsal aspect (as in Fig. 5.31g) to show the approximate sectional planes of the MR scans in Fig. 5.32A. *ac*, anterior contact zone between accessory ossicle and tympanic; *ao*, accessory ossicle; *c*, cochlea; *cz*, contact zone between tympanoperiotic complex and squamosal; *sq*, squamosal; *do*, dorsal; *i*, incus; *la*, lateral; *m*, malleus; *p*, periotic; *pc*, posterior contact zone; *sp*, sigmoid process; *t*, tympanic; *tc*, tympanic cavity; *th*, tympanohyal (*blue arrows*, thin and folded bony sheets connecting the tympanic plate to the processus tubarius and processus petrosus; *red hollow arrow*, entrance into tympanic cavity from pharyngotympanic tube).

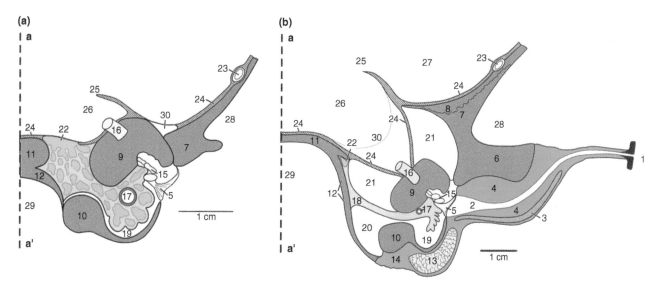

FIGURE 5.33   **Frontal section through ear region of 68 cm harbor porpoise fetus (a) and an adult (b).** *1*, external acoustic pore; *2*, external acoustic meatus; *3*, cartilage of acoustic meatus; *4*, connective tissue around acoustic meatus; *5*, tympanic membrane; *6*, zygomatic process of squamosal; *7*, squamosal bones; *8*, parietal; *9*, periotic; *10*, tympanic; *11*, basioccipital; *12*, basioccipital process; *13*, extension of mandibular fat body; *14*, thickened periost of tympanic bone; *15*, chain of middle ear ossicles; *16*, vestibulocochlear nerve; *17*, internal carotid artery (obliterated); *18*, venous plexus in corpus spongiosum tympanicum (plexus venosus caroticus); *19*, tympanic cavity; *20*, peribullary sac; *21*, Sinus pneumaticus peripetrosus; *22*, Sinus (venous) petrosus internus; *23*, spinal meningeal artery; *24*, dura mater; *25*, tentorium cerebelli; *26*, posterior cranial fossa; *27*, middle cranial fossa; *28*, temporal fossa; *29*, pharyngeal groove; *30*, contour of cerebellum; *aa'*, mediosagittal plane of skull. (*Redrawn and modified by Jutta Oelschläger from Boenninghaus, 1903. Das Ohr des Zahnwales, zugleich ein Beitrag zur Theorie der Schalleitung: Eine biologische Studie. Fischer, Jena*)

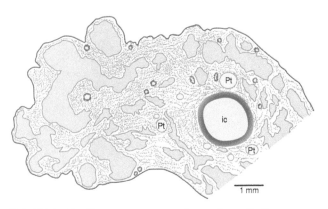

FIGURE 5.34 **Corpus cavernosum tympanicum (better: corpus spongiosum) of the harbor porpoise, a delicate vascular network embedded in a meshwork of soft connective tissue (*red structure*).** The vascular parts (*blue*) may be rather extended with respect to the internal carotid *(ic)* artery, which is still open in this specimen. *Yellow*, tympanic plexus (Pt); small arterial vessels, *red. (Redrawn and modified by Jutta Oelschläger after Boenninghaus, G., 1903. Das Ohr des Zahnwales, zugleich ein Beitrag zur Theorie der Schalleitung: Eine biologische Studie. Fischer, Jena.)*

FIGURE 5.35 **The tympanoperiotic complex and tympanic plate.** Left: basal aspect with medial part of tympanic removed (Fig. 5.31g). Right: Tympanic bone full length 34 mm (adult white-sided dolphin) taken apart and rotated by 180 degree to show the tympanic plate as a translucent ultrathin central part.

the middle ear cavity in order to prevent barotrauma during diving. At the same time, this gap is obviously essential for the function of the tympanoperiotic, that is, for the vibration patterns of the tympanic (Cranford et al., 2010; see the following).

As in terrestrial mammals, the bony porus acusticus externus of dolphins is still formed by the tympanic element (Fig. 5.30: ae). The tympanic membrane originates from the anulus tympanicus, where it seals the inner end of the rudimentary external auditory meatus (Boenninghaus, 1903). Near the anulus tympanicus the membrane forms a flattened cone, which tapers into the tympanic ligament, the tip of which attaches to the transverse part of the malleus (Figs. 5.31 and 5.35). In terrestrial mammals, the tympanic membrane firmly holds the handle of the malleus (manubrium mallei) as the beginning of the ossicular chain. In dolphins, however, the manubrium has nearly disappeared (see the following) but its position can still be determined.

Ear Ossicles Of particular interest is the position of the chain of middle ear ossicles between the two heavy bony elements of the tympanoperiotic (Figs. 5.32, 5.35–5.38). The hammer (malleus) is fused to the tympanic bone via a thin lamella (processus rostralis). At the same time, the rounded head of the malleus and its groove-like counterpart on the periotic bone seem to form a joint. Together with the rostral process and the strong malleus–incus articulation, this joint is believed to act as an amplifier for sound transmission to the stapes footplate (Fig. 5.38) (Hemilä et al., 1999). The stiffness of this system (which is also due to the thickening of the incudomallear complex) may serve as a high-pass filter for sound transmission (Hemilä et al., 2001, 2010), that is, the sound spectrum of hearing is shifted to the high-frequency components. To date, the mechanism of sound conduction to the inner ear via the ear ossicles (including the tympanic bulla) is not completely understood and so is the role of adjacent structures such as the middle ear muscles and the tympanic ligament.

The malleus is synostotically connected to the lateral lip of the tympanic bone via its rostral process (syn., processus anterior, gracilis, or goniale). This process is flat, very thin, and blends in the likewise thin "tympanic plate" in the area between the processus sigmoideus and the processus tubarius (Figs. 5.31 and 5.32). In this area, part of the round head of the malleus can be seen in lateral aspect of the tympanoperiotic (Fig. 5.31), in the corner between the anterior processes of the tympanic and the periotic and the sigmoid process of the tympanic. Seen from below, the periotic shows a shallow groove for the head of malleus, the fossa mallei (Fig. 5.30c: fh and 5.33); here, both bony surfaces seem to have a joint-like contact that may be necessary for sound conduction (Fig. 5.38: asterisk).

Dolphin ear ossicles are hyperostotic, that is, larger than in terrestrial mammals of the same body dimension. Boenninghaus (Boenninghaus, 1903) found the three ossicles together to be 5 times heavier in the harbor porpoise than in the human of about the same body mass, and 3 times heavier than in the horse. Nummela and coworkers compared the ossicles of white-beaked dolphins with those of terrestrial mammals and supplied similar data. Here, the three ossicles of dolphins were found to be about 3–4 times heavier than in the human, but about 10–12 times heavier than in the feral pig (Table 5.2). At the same time, bone density is somewhat higher in dolphins (Fleischer, 1978), showing also a "ceramic-like" consistency as the tympanoperiotic complex. In the bottlenose dolphin, ossicle density on average is 2.64 vs. 2.13 g/cm^3

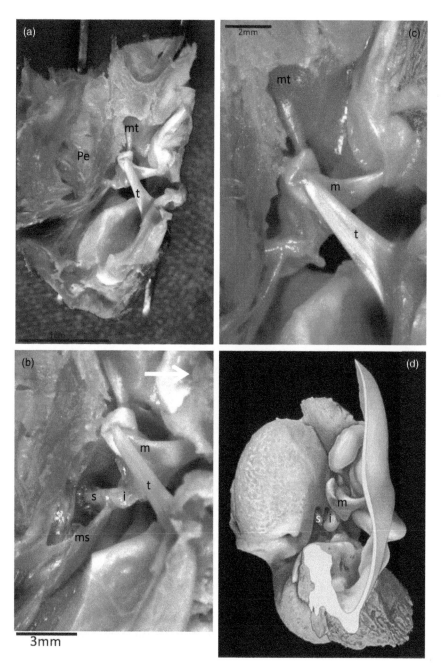

FIGURE 5.36   Decalcified ear region of a harbor porpoise (a, b, c), and dorsal view on the right middle ear cavity (lateral pointing right) with a horizontal panorama showing the chain of ossicles and their muscles (d). Right contour of specimen (a) represents the lateral surface of the tympanoperiotic complex; (b) and (c) represent magnifications of the ossicles with their muscles; (d) shows the basal view of tympanoperiotic complex and tympanic plate of a bottlenose dolphin with medial part of tympanic removed (same aspect as Fig. 5.31g). *i*, incus; *m*, malleus; *ms*, musculus stapedius; *mt*, musculus tensor tympani; *Pe*, periotic; *s*, stapes; *t*, tympanic membrane. *(Courtesy of Lars Kossatz)*

in some terrestrial mammals (Nummela et al., 1999b). The mass increase of the ossicles both through hyperostosis and a higher bone density can be interpreted as an adaptation with respect to higher stiffness in the system. In dolphins, the heavy ossicles (malleus/incus complex) seem to have lost the ability to rotate around a common axis as seen in terrestrial mammals (although the incudo-mallear and incudo-stapedial articulations are functional), obviously as a concession to the efficient conduction of ultrasound.

In dolphins, particularly the bodies of malleus and incus are increased in size, while their processes became relatively shorter or even disappeared, probably in favor of less mechanical noise in the system. As a consequence, rod-like components of the ossicles as seen in terrestrial mammals are not present here (Fleischer, 1978). This tendency to condense and,

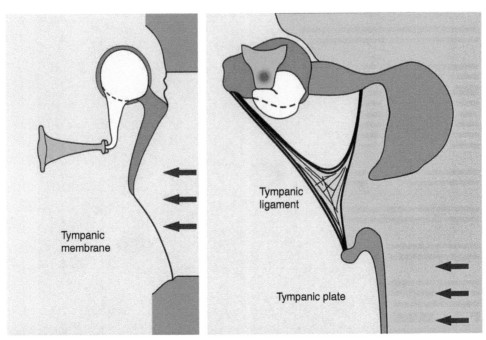

FIGURE 5.37 **Schematic comparison of two extreme types of middle ear ossicles.** Left: terrestrial low-frequency, human-like type with freely mobile ossicles; Right: ossicular chain of a dolphin with extremely compact ossicles. Manubrium of malleus appears reduced in dolphins; *red*, contour of malleus; *yellow*, incus; *green*, stapes; *blue*, air; *beige*, soft tissue; *brown*, bone; *dark arrows* represent incoming sound. *(Artwork by Jutta Oelschläger)*

at the same time, strengthen the middle ear ossicles is most obvious in delphinids. Here, the overall shape of the malleus has changed so much during the course of odontocete evolution that it is useful to compare their situation with that in terrestrial mammals. By this, Boenninghaus (1903) was able to identify the preserved residual processes of the toothed whale malleus and to reconstruct the morphological changes of the whole ossicle.

The malleus of dolphins (Figs. 5.30–5.32) is extremely stout and consists of three parts: head, manubrium, and rostral process (Mead and Fordyce, 2009). The head of malleus (caput mallei) can be identified easily by its articulation with the incus (Fig. 5.23); here, the very large malleus comprises the smaller incus with articular facets that form a right angle with each other (Nummela et al., 1999a). The neck of the malleus (collum mallei) is inconspicuous and only represented by a slight tapering of the malleus from the head in the direction of the manubrium mallei. In the bottlenose dolphin, however, the manubrium is largely reduced and much shortened, respectively. Left over is a polygonal end of the malleus, which at the one tip offers a slight protrusion with a minute round pit for the insertion of the tensor tympani muscle. The other tip is part of a low crest (manubrium mallei), which serves as an attachment site for the tympanic ligament. The rostral process of malleus (processus rostralis = processus anterior) is short and blade-like; it bears an extremely narrow sulcus for the thin chorda tympani, which enters the head of malleus and leaves between the two facets of the incudomallear joint. This rostral process—also called processus folianus, processus gracilis, or processus goniale in the literature on mammals—is confluent synostotically with the tympanic bone, that is, it can be regarded as being part of the tympanic plate to transduce vibrations from the tympanic bulla to the chain of middle ear ossicles. To remove the malleus from the middle ear, the anterior process has to be broken.

The tensor tympani muscle is well developed in dolphins (Fig. 5.36). However, in the literature, little is known about its dimensions. In the harbor porpoise, the slender and band-like muscle was depicted with a length of 7–8 mm and a width of 1–1.5 mm. Its functional implications are largely unknown because the middle ear muscles so far were not explicitly part of functional models. In general, the course of the muscle to the malleus is more or less opposite to that of the tympanic ligament and thus the malleus should be stressed somehow by the action of the tensor tympani muscle.

The incus is distinctly smaller than the malleus (Figs. 5.31, 5.32, 5.33, 5.37, and 5.38). Therefore, the facets for the articulation with the malleus comprise the whole width of the incus. The body of the incus is highly condensed and stout. The crus longum is strongly curved and the articular surfaces with the malleus and the stapes stand more or less perpendicular to each other (Mead and Fordyce, 2009). The articular surface of the incus for the stapes (lenticular process) is comparatively very small and flat. The crus breve is much more slender than the crus longum, a situation opposite to that in terrestrial mammals.

The stapes of dolphins is much less massive (hyperostotic) than malleus and incus. However, it is also distinctly more compact than in terrestrial mammals and the two crura have been largely fused with one another. As to functional implications in mammals, the stapes was reported to be tuned to a species-specific frequency range mainly by increasing or

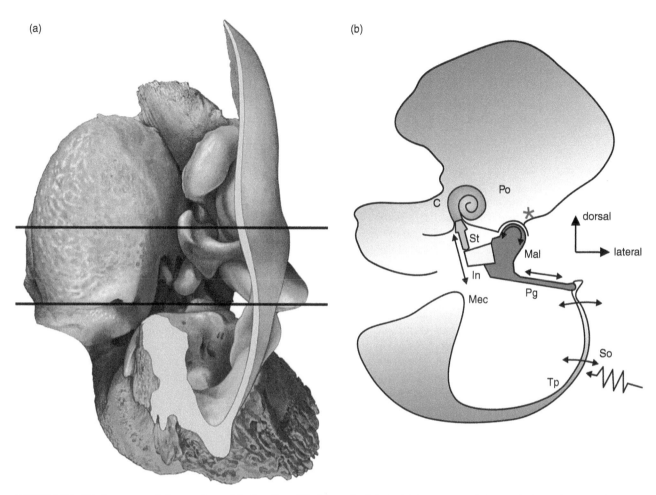

FIGURE 5.38   The tympanoperiotic complex and the function of the tympanic plate. (a) Right tympanoperiotic complex in basal aspect, with a large part of the tympanic removed, giving insight into the tympanic cavity with the middle ear ossicles. *Parallel lines*: region of interest between the articulation of the head of malleus with the periotic (scheme at the right: *double arrow* and *asterisk*) and the stapes articulating in the vestibular foramen. (b) Schematic drawing illustrating the hypothetical double lever mechanism; all the structures located between the two lines on the left brought into one plane. Incident sound *(So)* comes from the mandibular fat body to the thinnest part of the tympanic (tympanic plate, *Tp*), which acts as the first lever. The massive lower (medial) half of the tympanic bone *(T)* vibrates very little at high frequencies. The second lever is realized by the rotatory movement of the malleus-incus complex *(red-yellow)*, further increasing the particle velocity and vibration amplitude at the level of the oval window. *C*, cochlea; *In*, incus; *Mal*, malleus; *Mec*, middle ear cavity; *Pg*, processus gracilis of malleus; *Po*, periotic; *Tp*, tympanic plate; *St*, stapes, in oval window. *(Redrawn and modified by Jutta Oelschläger after Hemilä et al., 1999. A model of the odontocete middle ear. Hear. Res. 133, 82–97.)*

TABLE 5.2 Mass (g) of Ear Ossicles (Kossatz, 2006)

	Malleus	Incus	Stapes
White-beaked dolphin	0.134	0.039	0.011
Feral pig	0.009	0.005	0.001
Human	0.028	0.033	0.003

decreasing its moderate mass with respect to that of the large malleus-incus complex. Thus, toothed whales depending on ultrasound obviously possess light and short stirrups, whereas low-frequency hearing cetaceans have heavier and longer stirrups (Fleischer, 1978). Within the oval window (fenestra vestibuli), the stapes is suspended via an extremely strong annular ligament, which is thick in the axial extension (length) of the stirrup but relatively narrow with respect to the diameter of the stapes footplate (Fig. 5.23). Therefore, in dolphin cadavers, the extraction of the stapes is not possible without breaking it because of this extremely firm suspension in the vestibular foramen (Fleischer, 1982). However, in neonate bottlenose dolphins, this is still possible.

Although the pinnate stapedius muscle is small in dolphins in absolute size, it is large in comparison to that in terrestrial mammals. At the same time, it is obviously stronger in the dolphin than the tensor tympani muscle (Fig. 5.36). It attaches to the comparatively light stapes. According to these characteristics, the functional implications of the muscle are obviously similar as in terrestrial mammals (modulation of stiffness within the system, protection against noise in order to prevent damage of inner ear). As to an increase in stiffness of the system, the middle ear muscles are reported to tighten the articulations of the ossicular chain (Fleischer, 1978). Protection of the inner ear is possible by the middle ear reflex via the facial nerve, which restrains the vibrations of the stapes and thus the intensity of the sound transmitted.

Inner Ear

The sensory part of the organ of hearing is located in the wall of the membranous cochlear labyrinth and consists of the organ of Corti within the cochlear duct. The spiral canal of the cochlea is divided into three membranous ducts, which spiral together around the modiolus to the apex of the cochlea. The upper channel is the scala vestibuli, the middle the cochlear duct (also called the scala media), and the lower the scala tympani. The two scalae communicate at the apex of the cochlea (helicotrema) around the blind end of the cochlear duct. At the base of the cochlea, the scala vestibuli begins at the vestibular window, and the scala tympani at the membrane covering the cochlear window. Both scalae are lined with a single-layered epithelium and filled with perilymph.

The cochlear duct begins and ends blindly and passes up inside the spiral canal of the osseous cochlea. It is filled with endolymph and is in contact with the vestibular labyrinth via the ductus reuniens. The cochlear duct lies between the two scalae and is wedge-shaped in cross-section. Within it lies the organ of Corti so that it is immersed in endolymphatic fluid (Fig. 5.23). The walls of the cochlear duct have three distinct segments: The very thin vestibular membrane forms the roof of the cochlear duct, separating it from the scala vestibuli of the cochlea. The lateral (external) wall of the cochlear duct is formed by the spiral ligament, which is firmly adherent to the underlying periosteum of the spiral lamina. The basilar membrane forms the floor of the cochlear duct and separates it from the scala tympani.

The organ of Corti lies on the connective tissue of the basilar membrane and includes the receptor (sensory) cells (hair cells) for hearing (Fig. 5.39). Between the spirals of the scala vestibuli and scala tympani the organ of Corti runs throughout the cochlea and is covered a gel-like membrane (membrana tectoria).

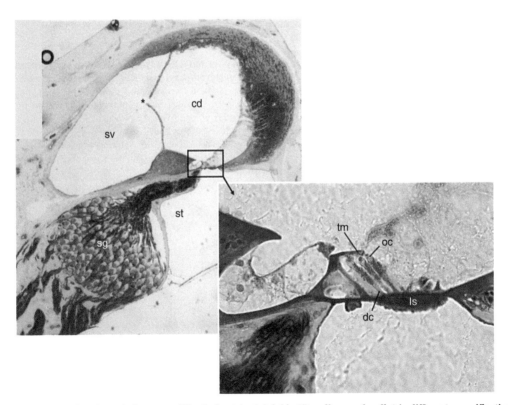

FIGURE 5.39 **Transverse section through the organ of Corti of a striped dolphin (*Stenella coeruleoalba*) in different magnifications stained with osmium tetroxide.** Note that the tectorial membrane is largely absent. *, artifact; *cd*, cochlear duct; *dc*, Deiters' supporting cell; *ls*, lamina spiralis ossea; *oc*, outer hair cell; *sg*, spiral ganglion; *st*, scala tympani; *sv*, scala vestibuli; *tm*, tectorial membrane. *(Courtesy of Maria Morell)*

Sounds are received by the external ear and provoke mechanical vibrations of the tympanic membrane, which are transmitted to the inner ear by the chain of the auditory ossicles. Since the stapes is embedded into the oval window, the perilymph of the inner ear is set in motion. Due to the incompressible nature of fluids, the movement of the perilymph is transmitted via the scala vestibuli, the helicotrema, and the scala tympani to the cochlear window. This perilymphatic motion sets the membranes of the cochlear duct in motion, which, in turn, involves the organ of Corti and the tectorial membrane, respectively. The pressure on the sensory hairs stimulates the receptor cells to send impulses to the spiral ganglion. The axons of the spiral ganglion unite to form the cochlear part of the vestibulocochlear nerve, which passes to the corresponding nuclei of the medulla oblongata. Different frequencies are analyzed due to different resonance properties of the cochlear duct membranes throughout the length of the whole cochlea. In general, high frequencies are received at the cochlear base, low frequencies near the apex.

In contrast to the middle ear, the cochlear part of the inner ear of toothed whales does not show such profound deviations from the mammalian bauplan so that its basic functional implications are likely the same as in terrestrial mammals (Ketten, 2000). As in the human, the cochlear spiral consists of nearly 2.5 turns in most dolphin species (Wever et al., 1971a, 1972). In comparison, it consists of 4 turns in the pig and of 3.5 turns in ruminants (König and Liebich, 2009). In comparison with the human (Fig. 5.22), the spatial orientation of the inner ear components and their relative size are different. Most easily, they can be compared with each other by bringing their windows (fenestrae vestibuli et cochleae) and the semicircular canals into the same orientation (Fig. 5.22). Thus, it is obvious that, in dolphins, the axis of the cochlea points ventrally (ie, perpendicular) in the skull base instead of horizontally in the human. The basal turn of the cochlea, the area of high frequency hearing, is enlarged in dolphins (Solntseva, 2007).

The cochlea of toothed whales differs in having hypertrophied cochlear duct structures, with all the components of the organ of Corti being larger or showing a higher cell density, respectively (Fig. 5.39). Most of the hypercellularity is associated with the support cells of the basilar membrane and with the stria vascularis, which plays a major role in cochlear metabolism. Auditory ganglion cell numbers are more than double those of humans and, additionally, the innervation densities in relation to basilar membrane length are two- to threefold greater than in other mammals. In the bottlenose dolphin, the total number of spiral ganglion cells was reported as 96,716 (human: 30,500), their density per mm cochlea as 2,486 (human: 950) (Wever et al., 1971b; Ketten, 1997).

Moreover, the dimensions of the basilar membrane are unique in toothed whales (Wever et al., 1971c): Although the length correlates with the animal's size, its thickness and width corresponds to high-frequency hearing. In dolphins, basilar membrane width is 35–40 µm at the base (human 120 µm) and increases up to 420 µm apically (human: 550 µm). Accordingly, basal widths of odontocetes are similar to those of bats and one third that of humans. At the same time, membrane thickness is very high, typically ranging from 25 µm at the base to 5 µm at the apex. Therefore, a typical cross-section of an odontocete basilar membrane is square at the base, becoming rectangular apically (Wartzok and Ketten, 1999). Interestingly, basilar membrane thickness to width ratios are consistent with the maximal high and low frequencies each species hears and with differences in their peak spectra. For example, harbor porpoises have a basal thickness to width ratio of 0.9 and a peak frequency around 135 kHz. The bottlenose dolphin has a thickness to width ratio of 0.7 and a peak signal of 70 kHz (Wartzok and Ketten, 1999). The terminal apical ratios are near 0.01.

A typical feature of basilar membranes in all toothed whales is their association with extensive outer bony laminae. In mammals, ossified outer spiral laminae (apart from the inner lamina spiralis ossea) are hallmarks of ultrasonic ears because they increase the stiffness of the membrane. In toothed whales, thick outer bony laminae are present throughout the basal turn and dolphins have a bony anchor for about 30% of the duct (Wartzok and Ketten, 1999).

In the organ of Corti, acoustic pressure waves enter the basilar membrane, which vibrates in a so-called "traveling wave." The shape, structure, and mechanical parameters of the membrane at a given point along its length determine the local characteristic frequency, at which the pear-shaped inner hair cells in the organ of Corti are most sensitive to a particular frequency of sound. At this specific location, the traveling wave is amplified by associated electromotile outer hair cells and the resulting glutamatergic impulses are transmitted via type 1 cells in the cochlear ganglion to the brain (see Auditory System). The outer hair cells are cylindrical and relatively small in comparison to other mammals. The small size in dolphins was interpreted as an adaptation to high-frequency hearing. Accordingly, the outer hair cells were found to be tightly anchored in the basal part of the organ of Corti by Deiters' cells, Deiters' cups, and pillar cells (Fig. 5.39). However, the total numbers of inner and outer hair cells are in the same range as in the human. The Pacific white-sided dolphin has about 3,270 inner hair cells and 12,900 outer hair cells, the bottlenose dolphin 3,450 and 13,930, and the human 3,475 and 11,500, respectively (Morell et al., 2015). Comparing these numbers with the numbers of ganglion cells (see earlier discussion), it was shown that some toothed whales, such as harbor porpoises and river dolphins, have an extremely high cochlear neuron density with a ganglion to hair cell ratio of up to 6:1. Delphinids have a ratio of approximately 5:1, echolocating bats 4:1, and humans 2.4:1 (Wartzok and Ketten, 1999). Thus, inner hair cells of toothed whales have more afferent synapses than these cells in cochleae of terrestrial mammals.

Functional Considerations of Dolphin Hearing

In the evolution of toothed whales, the skull as a whole was profoundly modified. These modifications can be interpreted mainly with its function in the echolocation system. The generation and emission of sound was shifted from the larynx to the nasal organ, which resulted in a total reconstruction of the facial skull. The nasal region is characterized by a deep facial depression (see Dolphin Head Characteristics). At the same time, the hearing organ had to be protected against high-energy (ultra) sound waves produced by the animals during echolocation (see the following). In this respect, the anterior skull roof (facial skull) was transformed by the "telescoping process" (see Skull) (Fleischer, 1978; Miller, 1923). This phenomenon is characterized by the caudal extension and elongation, respectively, of upper jaw elements (maxillae) over those of the frontal region. This way the mechanical stability and resilience of the rostrum was increased and, at the same time, the resulting stratified (multilayered) structure may improve the reflection properties of the facial skull roof for forward emission of powerful sound/ultrasound signals generated in the nasal complex. Due to these reflection properties, the skull roof may shield the ears from the high-energy vibrations of sound/ultrasound emitted by the animal ("acoustic shield"; Fleischer, 1978). The disintegration and delicateness of the zygomatic arch seems to point in the same direction by helping to dampen bone conduction to the ears.

Further, the skull was modified in such a way that the dolphin's ears were uncoupled from the skull base during evolution. By this, the two ears were also separated from each other acoustically, a prerequisite for directional hearing underwater. In completely macerated skulls, tympanoperiotic complexes are reported to fall from the carcass and can be found on the seafloor as so-called "cetoliths." To the same degree that the bones of the skull base adjacent to the periotic have receded, the anterior and posterior processes of the periotic seem to have been shortened in toothed whales (Oelschläger, 1986, 1990). In addition, also the tympanic bone was uncoupled from the skull base and united with the periotic via two connections to build the tympanoperiotic complex. The anterior contact is moderate in extension and formed by the anterior process of the periotic and the accessory ossicle (processus tubarius) on the lateral lip of the tympanic bone. The posterior contact is distinctly stronger and unites the posterior processes of the tympanic and periotic elements.

The profound modifications of dolphin head structures at the transition from the embryonic to the fetal stage reflect the adaptations of the mammalian bauplan to the requirements of a holaquatic lifestyle. However, the developmental pattern, that is, the sequence of the emerging of tissues and organs, in early prenatal life do not resemble the hypothesized evolution of the structural and functional adaptations found in the fossil record (Haddad et al., 2012). In contrast to other mammals, newborn cetaceans represent an extremely precocial state of development correlated to the fact that they have to swim and surface immediately after birth. Accordingly, the morphology of perinatal dolphins is very similar to that of the adults (Rauschmann et al., 2006) and the ear of perinatal dolphins maturates very early in development, a unique ontogenetic feature of cetaceans (Comtesse-Weidner, 2007; Cozzi et al., 2012, 2015). Thus, the maturation of sonar-relevant structures in the dolphin head resembles the course of the development of phylogenetic traits of cetacean adaptation from the terrestrial to the aquatic environment (Rauschmann et al., 2006), which is in some organ systems, such as the nose or some middle ear components, in contrast to the sequence of the emerging of tissues in early prenatal stages (Haddad et al., 2012).

In adult dolphins, the detachment of the tympanoperiotic complex from the skull is nearly complete. Only the tip of the posterior process of the tympanic has an immediate but obviously nonsynostotic[k] contact with the skull base (squamosal bone; Fig. 5.24). In the bottlenose dolphin and other delphinids, the articulating facets of the two posterior processes of the tympanoperiotic complex are particularly wide. In adult specimens, they show a characteristic pattern of parallel small plates or ridges which accurately interdigitate with each other (Fig. 5.30). The suspension of the dolphin tympanoperiotic complex ventrally and laterally, including the air sinuses, is nearly entirely taken over by a covering sheet of strong collagenous tissue (pterygoid ligament).

Coming from the outside (surrounding water), sound/ultrasound waves have to travel through the following tissues: skin, blubber, and the hollow mandibular bone with its acoustic fat body in a much widened alveolar canal (Figs. 5.28 and 5.29). In the posterior part, the lower jaw has a thin lateral wall (acoustic window, pan bone) but lacks a medial wall: Here, the caudal end of the mandibular fat body contacts the periosteum of the tympanic bone in the area of its medial lip. It is highly probable that the thinnest part in this area (tympanic plate; Figs. 5.32 and 5.35) transmits the incoming sound/ultrasound energy via the anterior process of malleus to the ossicular chain. The large malleus articulates with a much smaller incus (synchondrosis); together, they form a solid unit (Figs. 5.32a and 5.35). This malleus-incus complex seems to transfer the sound energy to the stapes and the oval window. The remaining hyperostotic part of the tympanic bone may serve as a reference mass (the mechanical framework) for this sound transmission. At the other end of the ossicular chain, the periotic seems to be the corresponding reference mass for the stapes, which articulates with the oval window. Thus, the two reference masses and other components of the middle ear region (tympanic ligament, middle ear muscles, corpus spongiosum, air sinuses) seem to guarantee the undisturbed function of the ossicular chain during perception of incoming sound.

---

k. Synostosis is the fusion of two bones. Nonsynostotic means here that the bones are not fused, although a fusion is the characteristic situation in other (terrestrial) mammals.

REFERENCES

Au, W.W.L., Hastings, M.C., 2008. Principles of Marine Bioacoustics, Modern Acoustics and Signal Processing. Springer, New York.

Bauer, G.B., Fuller, M., Perry, A., Dunn, J.R., Zoeger, J., 1985. Magnetoreception and biomineralization of magnetite in cetaceans. In: Kirschvink, J.L., Jones, D.S., MacFadden, B.J. (Eds.), Magnetite Biomineralization and Magnetoreception in Organisms. Springer, Boston, MA, pp. 489–507.

Begall, S., Burda, H., Malkemper, E.P., 2014. Magnetoreception in mammals. Advances in the Study of Behavior 46, 45–88.

Benninghoff, A., Drenckhahn, D., 2004. Anatomie: Makroskopische Anatomie, Histologie, Embryologie, Zellbiologie, 16., völligneubearb. Aufl. ed. Elsevier Urban & Fischer, München.

Berta, A., Ekdale, E.G., Cranford, T.W., 2014. Review of the cetacean nose: form, function, and evolution. Anat. Rec. 297, 2205–2215.

Bloodworth, B., Marshall, C.D., 2005. Feeding kinematics of *Kogia* and *Tursiops* (Odontoceti: Cetacea): characterization of suction and ram feeding. J. Exp. Biol. 208, 3721–3730.

Bloodworth, B.E., Marshall, C.D., 2007. A functional comparison of the hyolingual complex in pygmy and dwarf sperm whales (*Kogia breviceps* and *K. sima*), and bottlenose dolphins (*Tursiops truncatus*). J. Anat. 211, 78–91.

Boenninghaus, G., 1903. Das Ohr des Zahnwales, zugleich ein Beitrag zur Theorie der Schalleitung: Eine biologische Studie. Fischer, Jena.

Brill, R.L., Harder, P.J., 1991. The effects of attenuating returning echolocation signals at the lower jaw of a dolphin (*Tursiops truncatus*). J. Acoust. Soc. Am. 89, 2851–2857.

Brill, R.L., Sevenich, M.L., Sullivan, T.J., Sustman, J.D., Witt, R.E., 1988. Behavioral evidence for hearing through the lower jaw by an echolocating dolphin (*Tursiops truncatus*). Mar. Mamm. Sci. 4, 223–230.

Brill, R.L., Moore, P.W.B., Helweg, D.A., Dankiewicz, L.A., 2001. Investigating the Dolphin's Peripheral Hearing System: Acoustic Sensitivity about the Head and Lower Jaw. SPAWAR Systems Center, San Diego, CA.

Bryden, M.M., Molyneux, G.S., 1986. Ultrastructure of encapsulated mechanoreceptor organs in the region of the nares. In: Bryden, M.M., Harrison, R. (Eds.), Research on Dolphins. Oxford University Press, New York, pp. 99–107.

Brzica, H., Špiranec, K., Zečević, I., Lucić, H., Gomerčić, T., Ðuras, M., 2015. New aspects on the laryngeal anatomy of the bottlenose dolphin (*Tursiops truncatus*). Vet. Arh. 85, 211–226.

Buhl, E.H., Oelschläger, H.A., 1986. Ontogenetic development of the nervus terminalis in toothed whales. Evidence for its non-olfactory nature. Anat. Embryol. 173, 285–294.

Caldwell, D.K., Caldwell, M.C., 1972. The World of the Bottlenosed Dolphin. J.B. Lippincott Company, Philadelphia, New York.

Claudius F.M., 1858. Physiologische Bemerkungen über das Gehörorgan der Cetaceen und das Labyrinth der Säugethiere. Schwer, Kiel.

Comtesse-Weidner, P., 2007. Untersuchungen am Kopf des fetalen Narwals *Monodon monoceros*—Ein Atlas zur Entwicklung und funktionellen Morphologie des Sonarapparates (Dissertation Fachbereich Veterinärmedizin). Justus-Liebig-Universität, Giessen.

Costidis, A., Rommel, S.A., 2012. Vascularization of air sinuses and fat bodies in the head of the bottlenose dolphin (*Tursiops truncatus*): morphological implications on physiology. Front. Aquat. Physiol. 3, 243.

Cozzi, B., Podestà, M., Mazzariol, S., Zotti, A., 2012. Fetal and early post-natal mineralization of the tympanic bulla in fin whales may reveal a hitherto undiscovered evolutionary trait. PLoS One 7, e37110.

Cozzi, B., Podestà, M., Vaccaro, C., Poggi, R., Mazzariol, S., Huggenberger, S., Zotti, A., 2015. Precocious ossification of the tympanoperiotic bone in fetal and newborn dolphins: an evolutionary adaptation to the aquatic environment? Anat. Rec. 298, 1294–1300.

Cranford, T.W., Amundin, M., Norris, K.S., 1996. Functional morphology and homology in the odontocete nasal complex: implications for sound generation. J. Morphol. 228, 223–285.

Cranford, T.W., Krysl, P., Hildebrand, J.A., 2008a. Acoustic pathways revealed: simulated sound transmission and reception in Cuvier's beaked whale (*Ziphius cavirostris*). Bioinspir. Biomim. 3, 016001.

Cranford, T.W., Mckenna, M.F., Soldevilla, M.S., Wiggins, S.M., Goldbogen, J.A., Shadwick, R.E., Krysl, P., St. Leger, J.A., Hildebrand, J.A., 2008b. Anatomic geometry of sound transmission and reception in Cuvier's beaked whale (*Ziphius cavirostris*). Anat. Rec. 291, 353–378.

Cranford, T.W., Krysl, P., Amundin, M., 2010. A new acoustic portal into the odontocete ear and vibrational analysis of the tympanoperiotic complex. PLoS One 5, 5.131927.

Cranford, T.W., Elsberry, W.R., Van Bonn, W.G., Jeffress, J.A., Chaplin, M.S., Blackwood, D.J., Carder, D.A., Kamolnick, T., Todd, M.A., Ridgway, S.H., 2011. Observation and analysis of sonar signal generation in the bottlenose dolphin (*Tursiops truncatus*): evidence for two sonar sources. J. Exp. Mar. Biol. Ecol. 407, 81–96.

Cranford, T.W., Trijoulet, V., Smith, C.R., Krysl, P., 2013. Validation of a vibroacoustic finite element model using bottlenose dolphin simulations: the dolphin biosonar beam is focused in stages. Bioacoustics 23, 161–194.

Czech-Damal, N.U., Liebschner, A., Miersch, L., Klauer, G., Hanke, F.D., Marshall, C., Dehnhardt, G., Hanke, W., 2012. Electroreception in the Guiana dolphin (*Sotalia guianensis*). Proc. R. Soc. B 279, 663–668.

Czech-Damal, N.U., Dehnhardt, G., Manger, P., Hanke, W., 2013. Passive electroreception in aquatic mammals. J. Comp. Physiol. A 199, 555–563.

Dawson, W., 1980. The cetacean eye. In: Herman, L. (Ed.), Cetacean Behavior: Mechanisms and Functions. Wiley Interscience, New York, pp. 53–100.

Dawson, W.W., Hawthorne, M.N., Jenkins, R.L., Goldston, R.T., 1982. Giant neural systems in the inner retina and optic nerve of small whales. J. Comp. Neurol. 205, 1–7.

Dawson, W.W., Schroeder, J.P., Dawsson, J.F., 1987. The ocular fundus of two cetaceans. Mar. Mamm. Sci. 3, 1–13.

Dehnhardt, G., 2002. Sensory systems. In: Hoelzel, A.R. (Ed.), Marine Mammal Biology—An Evolutionary Approach. Blackwell Science, Oxford, pp. 116–141.

Ellenberger, W., Baum, H., Zietzschmann, O., 1977. Handbuch der vergleichenden Anatomie der Haustiere, 18. Aufl. ed. Springer, Berlin.

Feng, P., Zheng, J., Rossiter, S.J., Wang, D., Zhao, H., 2014. Massive losses of taste receptor genes in toothed and baleen whales. Genome Biol. Evol. 6 (6), 1254–1265.

Fleischer, G., 1978. Evolutionary principles of the mammalian middle ear. Adv. Anat. Embryol. Cell Biol. 55, 3–70.

Fleischer, G., 1982. Hörmechanismen bei Delphinen und Walen. HNO 30, 123–130.

Fraser, F.C., Purves, P.E., 1960. Hearing in cetaceans. Evolution of the accessory air sacs and the structure and function of the outer and middle ear in recent cetaceans. Bull. Br. Mus. Nat. Hist. Zool. 7, 1–140.

Friedl, W.A., Nachtigall, P.E., Moor, P.W.B., Chun, N.K.W., 1990. Taste reception in the Pacific bottlenose dolphin (*Tursiops truncatus* Gilli) and the California sea lion (*Zalophus californianus*). In: Thomas, J.A., Kastelein, R.A. (Eds.), Sensory Abilities of Cetaceans. Plenum Press, New York, pp. 447–451.

Green, R.F., Ridgway, S.H., Evans, W.E., 1980. Functional and descriptive anatomy of the bottlenosed dolphin nasolyngeal system with special reference to the musculature associated with sound production. In: Busnel, R.G., Fish, J.F. (Eds.), Animal Sonar Systems. Plenum Press, New York, pp. 199–238.

Haddad, D., Huggenberger, S., Haas-Rioth, M., Kossatz, L.S., Oelschläger, H.H.A., Haase, A., 2012. Magnetic resonance microscopy of prenatal dolphins (Mammalia, Odontoceti, Delphinidae)—ontogenetic and phylogenetic implications. Zool. Anz. 251, 115–130.

Harper, C.J., McLellan, W.A., Rommel, S.A., Gay, D.M., Dillaman, R.M., Pabst, D.A., 2008. Morphology of the melon and its tendinous connections to the facial muscles in bottlenose dolphins (*Tursiops truncatus*). J. Morphol. 269, 820–839.

Hemilä, S., Nummela, S., Reuter, T., 1999. A model of the odontocete middle ear. Hear. Res. 133, 82–97.

Hemilä, S., Nummela, S., Reuter, T., 2001. Modeling whale audiograms: effects of bone mass on high-frequency hearing. Hear. Res. 151, 221–226.

Hemilä, S., Nummela, S., Reuter, T., 2010. Anatomy and physics of the exceptional sensitivity of dolphin hearing (Odontoceti: Cetacea). J. Comp. Physiol. A 196, 165–179.

Heyning, J.E., 1989. Comparative facial anatomy of beaked whales (Ziphiidae) and a systematic revision among the families of extant Odontoceti. Contrib. Sci. 405, 1–64.

Houser, D.S., 2004. Structural and functional imaging of bottlenose dolphin (*Tursiops truncatus*) cranial anatomy. J. Exp. Biol. 207, 3657–3665.

Huber, E., 1934. Contribution to palaeontology IV: anatomical notes on pinnipedia and cetacea. Publ. Carnegie Inst. 447, 105–136.

Huggenberger, S., Rauschmann, M.A., Oelschläger, H.H.A., 2008. Functional morphology of the hyolaryngeal complex of the harbor porpoise (*Phocoena phocoena*): implications for its role in sound production and respiration. Anat. Rec. 291, 1262–1270.

Huggenberger, S., Rauschmann, M.A., Vogl, T.J., Oelschläger, H.H.A., 2009. Functional morphology of the nasal complex in the harbor porpoise (*Phocoena phocoena* L.). Anat. Rec. 292, 902–920.

Huggenberger, S., Vogl, T.J., Oelschläger, H.H.A., 2010. Epicranial complex of the La Plata dolphin (*Pontoporia blainvillei*): topographical and functional implications. Mar. Mamm. Sci. 26, 471–481.

Huggenberger, S., Klima, M., 2015. Cetacea, Waltiere. In: Westheide, W., Rieger, G. (Eds.), Spezielle Zoologie: Spektrum Akademischer. Verlag, Heidelberg, pp. 600–613.

Huggenberger, S., André, M., Oelschläger, H.H.A., 2016. The nose of the sperm whale: overviews of functional design, structural homologies and evolution. J. Mar. Biol. Assoc., 96 (4), 783–806.

Hui, C.A., 1994. Lack of association between magnetic patterns and the distribution of free-ranging dolphins. J. Mamm. 75, 399–405.

Ito, H., Aida, K., 2005. The first description of the double-bellied condition of the digastric muscle in the finless porpoise *Neophocaena phocaenoides* and Dall's porpoise *Phocoenoides dalli*. Mamm. Study 30, 83–87.

Jansen, J., Jansen, J.K.S., 1969. The nervous system of Cetacea. In: Andersen, H.T. (Ed.), The Biology of Marine Mammals. Academic Press, New York, pp. 175–252.

Kandel, B.M., Hullar, T.E., 2010. The relationship of head movements to semicircular canal size in cetaceans. J. Exp. Biol. 213, 1175–1181.

Kern, A., 2012. Der Neokortex der Säugetiere—Evolution und Funktion (Inaugural-Dissertation, Fachbereich Medizin). Johann Wolfgang Goethe-Universität, Frankfurt am Main.

Kern, A., Seidel, K., Oelschläger, H.H.A., 2009. The central vestibular complex in dolphins and humans: functional implications of Deiters' nucleus. Brain. Behav. Evol. 73, 102–110.

Ketten, D.R., 1994. Functional analyses of whale ears: adaptations for underwater hearing. IEEE Proc. Underw. Acoust. 1, 264–270.

Ketten, D.R., 1997. Structure and function in whale ears. Bioacoustics 8, 103–135.

Ketten, D.R., 2000. Cetacean ears. In: Au, W.W.L., Popper, A.N., Fay, R.R. (Eds.), Hearing by Whales and Dolphins, Springer Handbook of Auditory Research. Springer, New York, pp. 43–108.

Kirschvink, J.L., 1990. Geomagnetic sensitivity in cetaceans: an update with live stranding records in the United States. In: Thomas, J.A., Kastelein, R.A. (Eds.), Sensory Abilities of Cetaceans. Plenum Press, New York, pp. 639–649.

Klima, M., 1999. Development of the Cetacean Nasal Skull. Adv. Anat. Embryol. Cell Biol. 149, 1–143.

Klima, M., Oelschläger, H.H.A., Wünsch, D., 1980. Morphology of the pectoral girdle in the Amazon dolphin *Inia geoffrensis* with special reference to the shoulder joint and the movements of the flippers. Z. Für Säugetierkd. 45, 288–309.

Klinowska, M., 1986. The cetacean megnetic sense—evidence from stranding. In: Bryden, M.M., Harrison, R. (Eds.), Research on Dolphins. Oxford University Press, New York, pp. 401–432.

Kolchin, S., Bel'kovich, V., 1973. Tactile sensitivity in *Delphinus delphis*. Zool. Zhurnal 52, 620–622.

König, H.E., Liebich, H.-G., 2009. Veterinary Anatomy of Domestic Mammals. Schattauer, Stuttgart.

Kossatz, L.S., 2006. Morphologie, Anatomie und Computertomographie des Ohres bei Zahnwalen—Ein Beitrag zur Funktionalität des Mittelohres (Inaugural-Dissertation, Fachbereich Medizin). Johann Wolfgang Goethe-Universität, Frankfurt am Main.

Kremers, D., Marulanda, J.L., Hausberger, M., Lemasson, A., 2014. Behavioural evidence of magnetoreception in dolphins: detection of experimental magnetic fields. Naturwissenschaften 101, 907–911.

Lawrence, B., Schevill, W.E., 1956. The functional anatomy of the delphinid nose. Bull. Mus. Comp. Zool. 114, 103–151.

Lawrence, B., Schevill, W.E., 1965. Gular musculature in delphinids. Bull. Mus. Comp. Zool. 133, 1–63.

Lindenlaub, T., Oelschläger, H.H.A., 2000. Preliminary observations on the morphology of the inner ear of the fetal narwhal (*Monodon monoceros*). Hist. Biol. 14, 47–51.

Ling, J.K., 1977. Vibrissae of marine mammals. In: Harrison, R.J. (Ed.), Functional Anatomy of Marine Mammals. Academic Press, London, pp. 387–415.

Madsen, C.J., 1980. Social and ecological correlates of cetacean vision and visual appearance. In: Herman, L.M. (Ed.), Cetacean Behavior: Mechanisms and Functions. Wiley Interscience, New York, pp. 101–147.

Marshall, C.D., 2009. Feeding morphology. In: Perrin, W.F., Würsig, B., Thewissen, J.G.M. (Eds.), Encyclopedia of Marine Mammals. Academic Press, San Diego, CA, pp. 406–414.

Mass, A.M., Supin, A.Y., 2009. Vision. In: Perrin, W.F., Würsig, B., Thewissen, J.G.M. (Eds.), Encyclopedia of Marine Mammals. Academic Press, San Diego, CA, pp. 1200–1211.

Mass, A.M., Supin, A.Y., Abramov, A.V., Mukhametov, L.M., Rozanova, E.I., 2013. Ocular anatomy, ganglion cell distribution and retinal resolution of a killer whale (*Orcinus orca*). Brain. Behav. Evol. 81, 1–11.

Maxia, C., Scano, P., Maggiani, F., Murtas, D., Piras, F., Crnjar, R., Lai, A., Sirigu, P., 2007. A morphological and13C NMR study of the extramandibular fat bodies of the striped dolphin (*Stenella coeruleoalba*). Anat. Rec. Adv. Integr. Anat. Evol. Biol. 290, 913–919.

McDonald, M., Vapniarsky-Arzi, N., Verstraete, F.J.M., Staszyk, C., Leale, D.M., Woolard, K.D., Arzi, B., 2015. Characterization of the temporomandibular joint of the harbour porpoise (*Phocoena phocoena*) and Risso's dolphin (*Grampus griseus*). Arch. Oral Biol. 60, 582–592.

McKenna, M.F., Cranford, T.W., Berta, A., Pyenson, N.D., 2012. Morphology of the odontocete melon and its implications for acoustic function. Mar. Mamm. Sci. 28, 690–713.

Mead, J.G., 1975. Anatomy of the external nasal passages and facial complex in the Delphinidae (Mammalia, Cetacea). Smithson. Contrib. Zool. 207, 1–72.

Mead, J.G., Fordyce, R.E., 2009. The therian skull: a lexicon with emphasis on the odontocetes. Smithson. Contrib. Zool. 627, 1–216.

Mengual, R., García, M., Segovia, Y., Pertusa, J.F., 2015. Ocular morphology, topography of ganglion cell distribution and visual resolution of the pilot whale (*Globicephala melas*). Zoomorphology 134, 339–349.

Miller, G.S., 1923. The telescoping of the cetacean skull. Smithson. Misc. Collect. 76, 1–71.

Møhl, B., Au, W.W.L., Pawloski, J., Nachtigall, P.E., 1999. Dolphin hearing: relative sensitivity as a function of point of application of a contact sound source in the jaw and head region. J. Acoust. Soc. Am. 105, 3421–3424.

Morell, M., Lenoir, M., Shadwick, R.E., Jauniaux, T., Dabin, W., Begeman, L., Ferreira, M., Maestre, I., Degollada, E., Hernandez-Milian, G., Cazevieille, C., Fortuño, J.-M., Vogl, W., Puel, J.-L., André, M., 2015. Ultrastructure of the odontocete organ of Corti: scanning and transmission electron microscopy. J. Comp. Neurol. 523, 431–448.

Morgane, P.J., Jacobs, M.S., 1972. Comparative anatomy of the cetacean nervous system. In: Harrison, R.J. (Ed.), Functional Anatomy of Marine Mammals. Academic Press, London, pp. 117–244.

Morgane, P.J., Jacobs, M.S., McFarland, W.L., 1980. The anatomy of the brain of the bottlenose dolphin (*Tursiops truncatus*). Surface configurations of the telencephalon of the bottlenose dolphin with comparative anatomical observations in four other cetacean species. Brain Res. Bull. 5 (3), 1–107.

Nachtigall, P.E., Hall, R.W., 1984. Tast reception in the bottlenose dolphin. Acta Zool. Fenn. 172, 147–148.

Nickel, R., Schummer, A., Seiferle, E., 2003. Bewegungsapparat, 8., unveränd. Aufl. ed, Lehrbuch der Anatomie der Haustiere. Parey, Stuttgart.

Nickel, R., Schummer, A., Seiferle, E., 2004. Nervensystem, Sinnesorgane, endokrine Drüsen, 8., unveränd. Aufl. ed, Lehrbuch der Anatomie der Haustiere. Parey, Stuttgart.

Ninomiya, H., Yoshida, E., 2007. Functional anatomy of the ocular circulatory system: vascular corrosion casts of the cetacean eye. Vet. Ophthalmol. 10, 231–238.

Ninomiya, H., Imamura, E., Inomata, T., 2014. Comparative anatomy of the ophthalmic rete and its relationship to ocular blood flow in three species of marine mammal. Vet. Ophthalmol. 17, 100–105.

Norris, K.S., 1968. The evolution of acoustic mechanisms in odontocete cetaceans. In: Drake, E.T. (Ed.), Evolution and Environment. Yale University Press, New Haven, pp. 297–324.

Norris, K.S., Harvey, G.W., 1974. Sound transmission in the porpoise head. J. Acoust. Soc. Am. 56, 659–664.

Nummela, S., Reuter, T., Hemilä, S., Holmberg, P., Paukku, P., 1999a. The anatomy of the killer whale middle ear (*Orcinus orca*). Hear. Res. 133, 61–70.

Nummela, S., Wägar, T., Hemilä, S., Reuter, T., 1999b. Scaling of the cetacean middle ear. Hear. Res. 133, 71–81.

Nummela, S., Thewissen, J.G.M., Bajpai, S., Hussain, T., Kumar, K., 2007. Sound transmission in archaic and modern whales: anatomical adaptations for underwater hearing. Anat. Rec. Adv. Integr. Anat. Evol. Biol. 290, 716–733.

Oelschläger, H.H.A., 1986. Tympanohyal bone in toothed whales and the formation of the tympano-periotic complex (Mammalia: Cetacea). J. Morphol. 188, 157–165.

Oelschläger, H.A., 1990. Evolutionary morphology and acoustics in the dolphin skull. In: Thomas, J.A., Kastelein, R.A. (Eds.), Sensory Abilities of Cetaceans: Laboratory and Field Evidence. Plenum Press, New York, pp. 137–162.

Oelschläger, H.A., Buhl, E.H., 1985a. Occurrence of an olfactory-bulb in the early development of the harbor porpoise (*Phocoena phocoena L.*). Fortschr. Zool. 30, 695–698.

Oelschläger, H.A., Buhl, E.H., 1985b. Development and rudimentation of the peripheral olfactory system in the harbor porpoise *Phocoena phocoena* (Mammalia: Cetacea). J. Morphol. 184, 351–360.

Ortiz, G.G., Feria-Velasco, A., Tarpley, R.L., Bitzer-Quintero, O.K., Rosales-Corral, S.A., Velázquez-Brizuela, I.E., López-Navarro, O.G., Reiter, R.J., 2007. The orbital Harderian gland of the male Atlantic bottlenose dolphin (*Tursiops truncatus*): a morphological study. Anat. Histol. Embryol. J. Vet. Med. Ser. C 36, 209–214.

Peichl, L., 1997. Die Augen der Säugetiere: Unterschiedliche Blicke in die Welt. Biol. Unserer Zeit 27, 96–105.

Peichl, L., Behrmann, G., Kröger, R.H., 2001. For whales and seals the ocean is not blue: a visual pigment loss in marine mammals. Eur. J. Neurosci. 13, 1520–1528.

Popov, V.V., Supin, A.Y., 1990a. Location of an acoustic window in dolphins. Experientia 46, 53–56.

Popov, V.V., Supin, A.Y., 1990b. Localisation of the acoustic window at the dolphin's head. In: Thomas, J.A., Kastelein, R.A. (Eds.), Sensory Abilities of Cetaceans: Laboratory and Field Evidence. Plenum Press, New York, pp. 417–426.

Prahl, S., Huggenberger, S., Schliemann, H., 2009. Histological and ultrastructural aspects of the nasal complex in the harbour porpoise *Phocoena phocoena*. J. Morphol. 270, 1320–1337.

Purves, P.E., Pilleri, G., 1983. Echolocation in Whales and Dolphins. Academic Press, London, New York.

Pütter, A., 1902. Die Augen der Wassersäugetiere. Zool. Jahrb. 17, 99–402.

Rauschmann, M.A., Huggenberger, S., Kossatz, L.S., Oelschläger, H.H.A., 2006. Head morphology in perinatal dolphins: a window into phylogeny and ontogeny. J. Morphol. 267, 1295–1315.

Reidenberg, J.S., Laitman, J.T., 1988. Existence of vocal folds in the larynx of odontoceti (toothed whales). Anat. Rec. 221, 884–891.

Reidenberg, J.S., Laitman, J.T., 1994. Anatomy of the hyoid apparatus in odontoceli (toothed whales): specializations of their skeleton and musculature compared with those of terrestrial mammals. Anat. Rec. 240, 598–624.

Reiss, D., McCowan, B., 1993. Spontaneous vocal mimicry and production by bottlenose dolphins (*Tursiops truncatus*): evidence for vocal learning. J. Comp. Psychol. 107, 301–312.

Ridgway, S.H., 1990. The central nervous system of the bottlenose dolphin. In: Leatherwood, S., Reeves, R.R. (Eds.), The Bottlenose Dolphin. Academic Press, San Diego, CA, pp. 69–97.

Ridgway, S.H., 1999. An illustration of Norris' acoustic window. Mar. Mamm. Sci. 15, 926–930.

Ridgway, S.H., Au, W.W.L., 1999. Hearing and echolocation: dolphin. Elsevier's Encyclopedia of Neuroscience. Elsevier, Amsterdam, pp. 858–862.

Ridgway, S.H., Carder, D.A., 1990. Tactile sensitivity, somatosensory responses, skin vibrations, and the skin surface ridges of the bottlenose dolphin, *Tursiops truncatus*. In: Thomas, J.A., Kastelein, R.A. (Eds.), Sensory Abilities of Cetaceans: Laboratory and Field Evidence. Plenum Press, New York, pp. 163–179.

Ridgway, S., Carder, D., Jeffries, M., Todd, M., 2012. Spontaneous human speech mimicry by a cetacean. Curr. Biol. 22, R860–R861.

Rodionov, V.A., Markov, V.I., 1992. Funktional anatomy of the nasal system in the bottlenose dolphin. In: Thomas, J.A., Kastelein, R.A., Supin, A.Y. (Eds.), Marine Mammal Sensory Systems. Plenum Press, New York, pp. 147–177.

Rommel, S.A., 1990. Osteology of the bottlenose dolphin. In: Leatherwood, S., Reeves, R.R. (Eds.), The Bottlenose Dolphin. Academic Press, San Diego, CA, pp. 29–49.

Ross, M.H., Pawlina, W., 2016. Histology: a text and atlas: with correlated cell and molecular biology, seventh ed. Wolters Kluwer Health, Philadelphia.

Ryabov, V., 2010. Role of the mental foramens in dolphin hearing. Nat. Sci. 02, 646–653.

Schneider, R., 1964. Der larynx der Säugetiere. Handb. Zool. 8, 1–128.

Smith, T.L., Turnbull, B.S., Cowan, D.F., 1999. Morphology of the complex laryngeal gland in the Atlantic bottlenose dolphin, *Tursiops truncatus*. Anat. Rec. 254, 98–106.

Solntseva, G.N., 2007. Morphology of the Auditory and Vestibular Organs in Mammals, with Emphasis on Marine Species. Pensoft Pub/Brill Academic, Sofia, Bulgaria/Leiden, The Netherlands.

Spoor, F., Bajpai, S., Hussain, S.T., Kumar, K., Thewissen, J.G.M., 2002. Vestibular evidence for the evolution of aquatic behaviour in early cetaceans. Nature 417, 163–166.

Štěrba, O., Klima, M., Schildger, B., 2000. Embryology of dolphins—staging and ageing of embryos and fetuses of some cetaceans. Adv. Anat. Embryol. Cell Biol. 157, 1–133.

Supin, A.Y., Popov, V.V., Mass, A.M., 2001. The Sensory Physiology of Aquatic Mammals. Kluwer Academic Publishers, Boston.

Tarpley, R.J., Ridgway, S.H., 1991. Orbital gland structure and secretions in the Atlantic bottlenose dolphin (*Tursiops truncatus*). J. Morphol. 207, 173–184.

Toldt, C., 1905. Der Winkelfortsatz des Unterkiefersbeim Menschen und bei den Säugetieren und die Beziehung der Kaumuskelnzudemselben (II. Teil). Sitzungsberichte Kais. Akad.Wisssenschaften—Math.-Naturwissenschaftliche Kl. Abt. III 114, 315–476.

Vanselow, K.H., Ricklefs, K., 2005. Are solar activity and sperm whale *Physeter macrocephalus* strandings around the North Sea related? J. Sea Res. 53, 319–327.

Wartzok, D., Ketten, D.R., 1999. Marine mammal sensory systems. In: Reynolds, III, J.E., Rommel, S.A. (Eds.), Biology of Marine Mammals. Smithsonian Institution Press, Washington, DC, pp. 117–175.

Werth, A.J., 2007. Adaptations of the cetacean hyolingual apparatus for aquatic feeding and thermoregulation. Anat. Rec. Adv. Integr. Anat. Evol. Biol. 290, 546–568.

Wever, E.G., McCormick, J.G., Palin, J., Ridgway, S.H., 1971a. The cochlea of the dolphin, *Tursiops truncatus*: general morphology. Proc. Natl. Acad. Sci. USA 68 (10), 2381–2385.

Wever, E.G., McCormick, J.G., Palin, J., Ridgway, S.H., 1971b. The cochlea of the dolphin, *Tursiopstruncatus*: hair cells and ganglion cells. Proc. Natl. Acad. Sci. USA 68 (12), 2908–2912.

Wever, E.G., McCormick, J.G., Palin, J., Ridgway, S.H., 1971c. The cochlea of the dolphin, *Tursiops truncatus*: the basilar membrane. Proc. Natl. Acad. Sci. USA 68 (11), 2708–2711.

Wever, E.G., McCormick, J.G., Palin, J., Ridgway, S.H., 1972. Cochlear structure in the dolphin, *Lagenorhynchus obliquidens*. Proc. Natl. Acad. Sci. USA 69 (3), 657–661.

Zoeger, J., Dunn, J.R., Fuller, M., 1981. Magnetic material in the head of the common Pacific dolphin. Science 213, 892–894.

Chapter 6

Brain, Spinal Cord, and Cranial Nerves

CENTRAL NERVOUS SYSTEM: GENERAL ASPECTS

As in other vertebrates, the brain of mammals is designed as a concentration of nervous tissue at the anterior end of the body and spinal cord, respectively. It receives afferent biological information coming from the environment of an animal and returns an efferent response serving survival. Such a concentration of nervous tissue is plausible because in a moving, bilaterally symmetrical body, the rostral end is confronted first with influences from the surroundings. As a consequence, the head of the animal is equipped with a set of sensory organs, which help in orientation, the detection of prey, and many other aspects as, for example, the communication with conspecifics (for an overview see Fig. 6.1). For adequate reactions of an animal within its environment, the brain provides interfaces between the sensory structures and motor units that execute locomotion, vocalization, etc. Such a "universal" interface, the "Reticular formation" (see section later in this chapter), is located within the brainstem of mammals and responsible for all kinds of interaction between the outside and inside of an animal, and for the control of all tissues in the locomotory apparatus and the internal organs.

With respect to their biological design, dolphins are typical representatives of the mammalia; in principle, they have the same genetic equipment anlagen and bauplan[a] (Fig. 6.2). However, they experienced profound changes during evolution from a tetrapod terrestrial life to a swimming aquatic existence. Such adaptations are found in all body structures and organ systems; they are more or less obvious to the investigator. Most fascinating, in this respect, is the structure of the dolphin brain where the centers devoted to specific functional implications have "responded" to these changes in the physical parameters from one environment (land) to the other (water). In the following, we will repeatedly encounter allometric[b] phenomena and size correlations between brain structures that help elucidate the life and functional implications of the whole animal.

As in all other mammals, the dolphin brain encompasses five main subdivisions from the forehead of the animal to the foramen magnum (Fig. 6.3). Each of these brain parts is specialized, that is, has associated sensory organs devoted to specific functions (nose, eyes, ears). All these brain parts and structures are highly interconnected with each other and with the spinal cord; together they represent a unique and powerful network for successful locomotion, orientation, feeding, and so forth. The size of a brain structure is one of the parameters for the assessment of its functional importance within the concert of the sensory perceptions and motor reactions. Another important phenomenon in the brain is the hierarchy of structures within functional systems as, for example, in the auditory system. In this respect, morphological units may be extremely large as, for example, the neocortex (see the following), which can be regarded as superior to the rest of the brain.

Fundamental Organization

The early ontogenetic development of the anterior part of the neural tube is characterized by the appearance of two brain vesicles, which comprise the future prosencephalon and the rhombencephalon (Figs. 6.3 and 6.4). Further differentiation leads then to five consecutive parts: Thus the prosencephalon or forebrain divides into the telencephalon (1), which mainly comprises the telencephalic hemispheres and the diencephalon (2). The rhombencephalon consists of the mesencephalon (3), the metencephalon (4; pons and cerebellum), and the myelencephalon (5; syn. medulla oblongata). Here, we use this morphological classification because of fundamental differences in the organization of the prosencephalon and the rhombencephalon. In the embryo, the neural tube on both sides consists of two plates, which are separated by the sulcus limitans,

---

a. Anlagen are primordial structures in their ontogenetic or phylogenetic development; in the field of comparative anatomy such structures are called homologous in different species if they show a similar architecture and corresponding topographical relationships with other structures. Bauplan is a basic structural and functional pattern of animals that belong to a systematic group or taxon as a natural entity.

b. Allometric phenomena refer to quantitative relationships between two parameters, (eg, between brain mass and body mass). Such relationships are established for developing as well as for adult mammals of different body size.

Anatomy of Dolphins. http://dx.doi.org/10.1016/B978-0-12-407229-9.00006-3

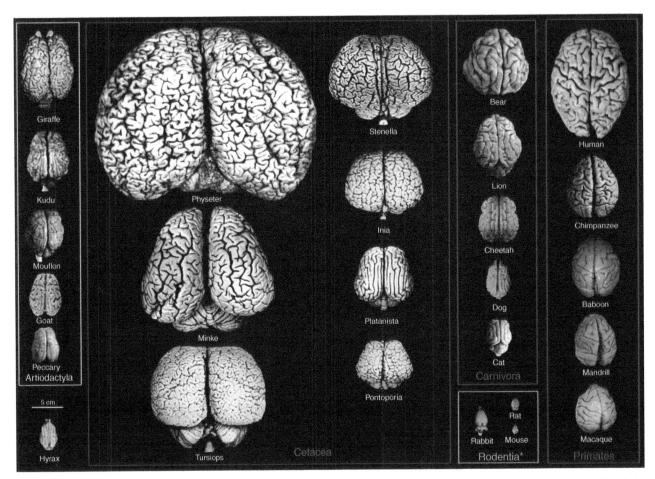

FIGURE 6.1 **Photographs of the brains of different species and their masses in grams.** Hyracoidea: Rock hyrax (*Procavia capensis*, 12 g). Artiodactyls: giraffe (*Giraffa camelopardalis*, 700 g), kudu (*Strepsiceros strepsiceros*, 166 g, mouflon (*Ovis orientalis*, 118 g), ibex (goat, *Capra pyrenaica*, 115 g), peccary (*Tayassu pecari*, 41 g). Lagomorph: rabbit (*Oryctolagus cuniculus*, 5 g). Rodents: rat (*Rattus rattus*, 2.6 g), mouse (*Mus musculus*, 0.5 g). Odontoceti: sperm whale (*Physeter macrocephalus*, adult), bottlenose dolphin (*Tursiops truncatus*, ca. 1500 g), striped dolphin (*Stenella coeruleoalba*, 1230 g), Amazon river dolphin (*Inia geoffrensis*, 460 g), South Asian river dolphin (*Platanista gangetica*, 292 g), Franciscana (*Pontoporia blainvillei*, 230 g). Mysticeti: minke whale (Balaenoptera acuto-rostrata, 2590 g). Carnivores: bear (*Ursus arctos*, 289 g), lion (*Panthera leo*, 165 g), cheetah (*Acinonyx jubatus*, 119 g), dog (*Canis familiaris*, 95 g), cat (*Felis catus*, 32 g). Primates: human (*Homo sapiens*, 1176 g), chimpanzee (*Pan troglodytes*, 273 g), baboon (*Papio cynocephalus*, 151 g), mandrill (*Mandrillus sphinx*, 123 g), macaque (*Macaca tonkeana*, 110 g). *(Modified after Brauer and Schober, 1976. Catalogue of Mammalian Brains. VEB Gustav Fischer Verlag, Jena; and de Felipe et al. 2007. Specializations of the cortical microstructure of humans. In: Kaas, J.H., Preuss, T.M. (Eds.), Evolution of the Nervous Systems. A Comprehensive Reference. Elsevier Academic Press, Amsterdam.)*

that is, the basal plate ventrally and the alar plate dorsally (Fig. 6.5). During prenatal ontogenesis, the plates on each side give rise to four longitudinal zones or columns (1–4, from medial to lateral: red, light violet, green, blue) consisting of masses of neuronal somata. In principle, these columns extend throughout most of the neural tube; they consist of specific neural tissues, which later differentiate into peculiar neurobiological systems (cf. Fig. 6.47, and 6.48). At the transition from the spinal cord to the brainstem (medulla oblongata, myelencephalon), the plates begin to swing to the sides like opening a book. Thus, the general and special somatic sensory (afferent) columns of both sides (Fig. 6.5: in blue) are now running laterally at the sides of the rhomboid fossa (which represents the bottom of the open fourth ventricle). This fundamental design of the spinal nuclear columns proceeds rostrally in the brainstem up to the caudal border of the diencephalon (Fig. 6.6) (Romer and Parsons, 1977, 1991). The nuclear columns in the brain are the first part of the so-called 'tegmentum' ontogenetically which spans the distance of the rhombencephalon between the diencephalon and the spinal cord. The longitudinal organization, however, is increasingly modified in the rostral direction due to the differential growth of neighboring structures and particularly the development of strong fiber tracts. By this, the nuclear columns are divided into single nuclei of the respective physiological quality (eg, somatic efferent (motor); Fig. 6.5). During fetal ontogenesis, the space between and around the columnar nuclei is more and more widened by a diffuse network of neurons connecting the nuclei with each other and with the longitudinal ascending and descending fiber systems. This dense network of neurons is called "Reticular Formation" and represents the second part of the tegmentum (see later in this chapter).

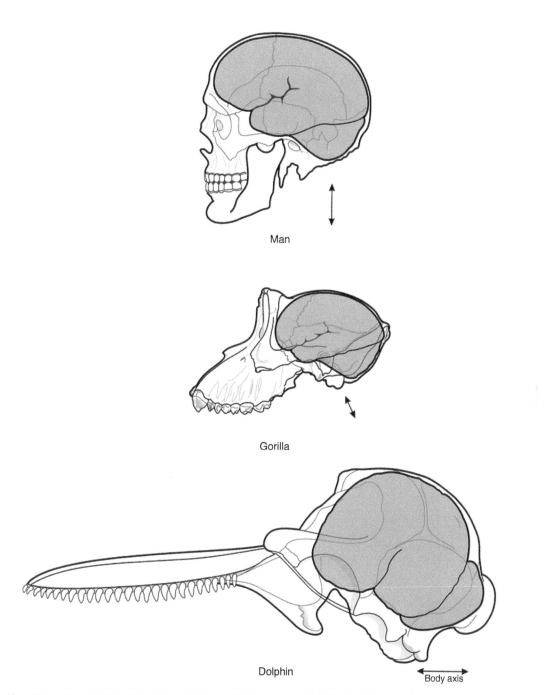

Man

Gorilla

Dolphin

Body axis

FIGURE 6.2   Size relationship of skull and brain (gray) in the gorilla, human, and dolphin. *Double arrows* indicate the orientation of the body axis. *(Redrawn by Jutta Oelschläger from Morgane et al., 1980. The anatomy of the brain of the bottlenose dolphin* (Tursiops truncatus). *Surface configurations of the telencephalon of the bottlenose dolphin with comparative anatomical observations in four other cetacean species. Brain Res. Bull. 5, 1–107.)*

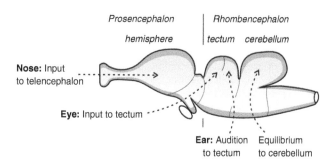

Prosencephalon   *Rhombencephalon*

hemisphere   *tectum   cerebellum*

Nose: Input to telencephalon

Eye: Input to tectum

Ear: Audition to tectum   Equilibrium to cerebellum

FIGURE 6.3   Schematic representation of the brain with its main parts and their relations to the sensory organs. *(Redrawn and modified by Jutta Oelschläger from Romer and Parsons, 1986. The Vertebrate Body, sixth ed. Saunders College Publishing, Philadelphia.)*

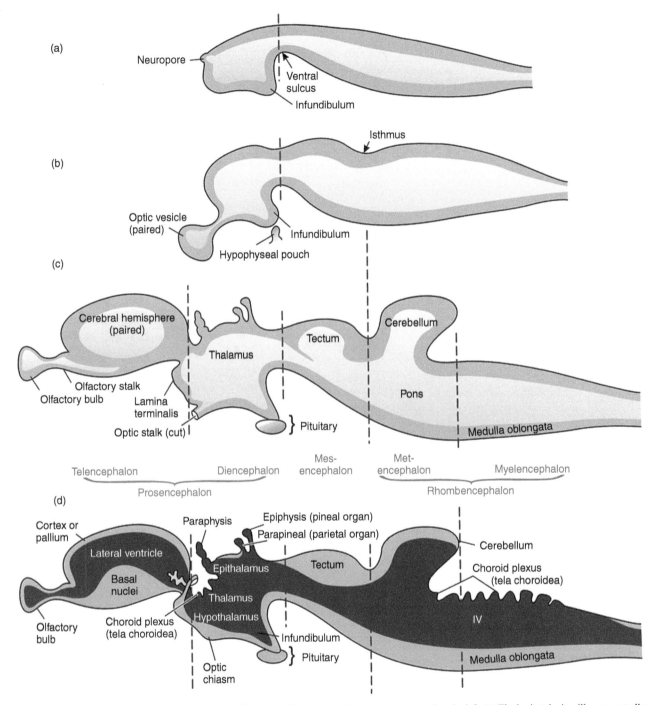

FIGURE 6.4 **The development of principal brain divisions and structures.** Lateral aspect, rostral to the left. (a) The brain tube is still open rostrally (anterior neuropore). Two divisions are obvious, the prosencephalon and the rhombencephalon. (b) The optic vesicle is obvious; the rhombencephalon is still dominant. (c) The prosencephalon is divided into the telencephalon (left hemisphere) and the diencephalon (optic vesicle omitted at optic stalk). The telencenphalic hemisphere protrudes rostrally as the olfactory bulb anlage. In the diencephalon, the two parts of the hypophysis (neuro-, adenohypophysis) are colocalized beneath the hypothalamus. In the rhombencephalon, primordia of the tectum and cerebellum are obvious. Main divisions of the brain in red. (d) Sagittal section through the brain and parasagittal section through the right telencephalic hemisphere. The thickness of the brain wall is rather different; it is advanced in the basal telencephalon (basal nuclei), optic chiasm, tectum, cerebellum, and the basal wall of the rhombencephalon. It is minimal in the area of the choroid plexus (lateral, third and fourth ventricles). Abbreviations: IV, fourth ventricle. *(Redrawn and modified by Jutta Oelschläger from Romer and Parsons, 1986. The Vertebrate Body, sixth ed. Saunders College Publishing, Philadelphia.)*

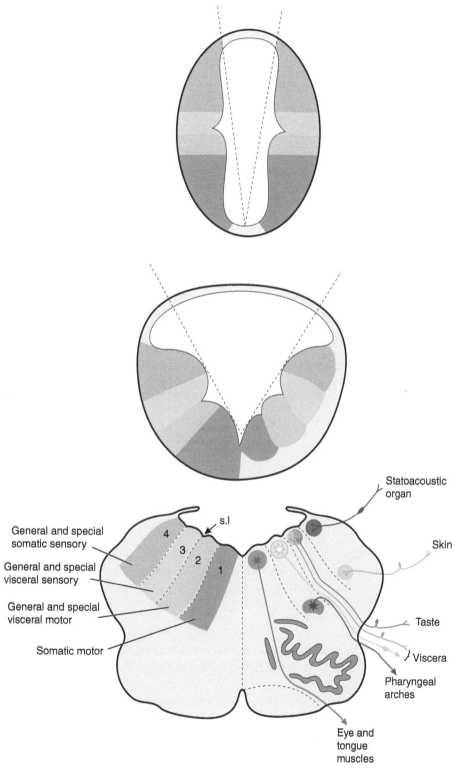

FIGURE 6.5   **Schematic cross-section through the brainstem during ontogenesis.** Top: brain tube with the basal and alar plates, divided by the Sulcus limitans (s.l.). Middle and bottom: expansion of the central canal and formation of the fourth ventricle with unfolding of the nuclear columns. Bottom left: the different nuclear columns and their physiological quality. Right: the origin of the components (qualities) of the cranial nerves and their respective periphery. *(Artwork by Jutta Oelschläger).* For more details, see Fig. 6.47.

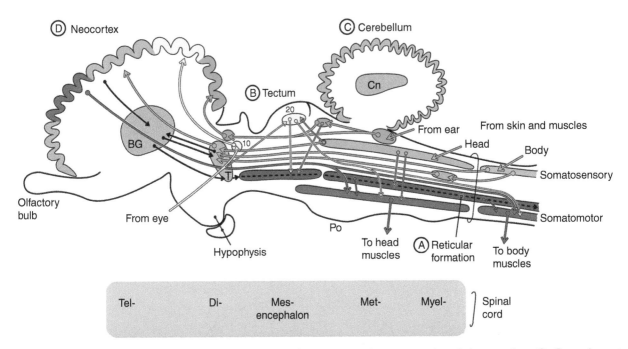

FIGURE 6.6 Scheme of mammalian brain organization and circuitry. Sagittal section; structures brought into one plane. Ⓐ–Ⓓ superior centers of sensory (and motor) modalities. Ⓐ, reticular formation (*brown* labeling); Ⓑ, tectum; Ⓒ, cerebellum; Ⓓ, telencephalon (neocortex). Only part of the physiological qualities are shown (cf. Figs. 6.5 and 6.47). Somatosensory structures in *blue*, somatomotor structures in *red*, visual system in *yellow*, auditory system in *green*. The figure shows where the input from the sensory organs and the afferent peripheral nervous system enter the brain and how the information is distributed via projections to unimodal nuclei and to multimodal superior centers of integration and coordination (A–D). In this ensemble, the mesencephalic tectum is reduced to a minor reflex center, and the corpus striatum (BG) is relatively unimportant; most sensory impulses are are directed "upward" to the cerebral cortex from where a direct motor pathway (pyramidal tract) extends to the motor centers of brainstem and spinal cord; Po, pons. *(Redrawn and modified by Jutta Oelschläger from Romer and Parsons, 1986. The Vertebrate Body, sixth ed. Saunders College Publishing, Philadelphia.)*

THE DOLPHIN BRAIN AS AN ENTITY

Sectional Planes and Three-Dimensional Architecture

Comparison with data in the literature (Igarashi and Kamiya, 1972; Morgane et al., 1980; Nickel et al., 2004) show that, in spite of many specializations, the general appearance of the dolphin brain (lateral aspect) is rather similar to that in hoofed animals (cattle). However, the brain is distinctly shortened along the beak–fluke axis and wider than long (Figs. 6.7 and 6.8). The maximal width of the dolphin brain corresponds to the (intertemporal) distance between the lateral surfaces of the prominent temporal lobes. All the brain flexures (mesencephalic, pontine, cervical flexure) are more pronounced in the dolphin. The cervical flexure seems to have been shifted caudally into the spinal cord, presumably in the course of the shortening of the neck region (see Vertebral Column). The telencephalic hemisphere is extraordinarily high and somewhat rotated in a rostroventral position, the brainstem short and thick, and the myelencephalon and cervical spinal cord arch ventrally around the very large cerebellum.

Dolphins not only possess large brains with respect to other mammals; their brain also shows a rather similar structural complexity. To date, however, there is limited knowledge as to the neurophysiological capacity of their brain in comparison with that of other mammals and primates including the human. Hoofed animals (pig, cattle, sheep) as their next relatives among the terrestrial mammals so far are also scarcely investigated. Concluding from the existing literature, dolphins seem to be at least as capable regarding the capacities of neuronal computation as other mammals with a large brain for a similar body size, but some caution is appropriate with respect to the human. Morphological studies of the dolphin brain as a whole have revealed that there are no fundamental differences with respect to brains of terrestrial mammals because the development of all these animals follows a common bauplan. This implies that, for example, the brainstem of dolphins contains the same set of structures (nuclei, fiber tracts), although there may be considerable differences as to their size, even in brains of the same dimension. This variability in the size of brain structures is obviously one important parameter in the diversity of mammalian brain function, apart from their connectivity with each other.

As was shown for the head as an entity (see Head and Senses), the mammalian brain with its complexity needs to be visualized by means of parallel sections made in different planes in order to reconstruct the three-dimensional architecture within our own brain and/or with technical procedures. In the following first step, the external appearance of the dolphin brain will be given in comparison with that of the well-known human brain (Figs. 6.7 and 6.8). In a second step, the gross structure of the dolphin brain will be presented and described along a series of MRI scans in four transverse (coronal) planes in a rostrocaudal sequence (Fig. 6.9). These MRI slices can serve as a reference tool for the detection and interpretation of important structural and topographical details throughout the whole brain. Accordingly, the MRI scans are labeled for major characteristics. In these figures, concentrations of cell bodies of neurons (perikarya), forming the gray substance in the cortex and the nuclei in the brain, are shown in light (hyperintense) encoding. In contrast, fiber tracts consisting of bundles of efferent neural processes (axons) generally appear dark (hypointense).

In the coronal scans (Fig. 6.9), it is obvious that most of the brain volume is contributed by the endbrain (telencephalon; te), more precisely of the telencephalic neocortex and the white matter later it. In consequence, the neocortex also shapes the outer surface of the brain and only leaves the brainstem and the surface of the cerebellum uncovered. The neocortex (gray substance) is extensively folded (gyri and sulci), much more than in the human (Figs. 6.7 and 6.8), and is only about half as thick as in the human. Other characteristics of the dolphin brain are the strong development of the temporal lobe and the distinct insular formation, which is hidden in the telencephalic wall. Long fiber tracts impress as dark streaks that originate or terminate in their respective cortical areas. Descending (efferent) fibers, for example, that originate in the cortical gray (cortical plate) and run deep in the white matter, join like tributaries and constitute a broad stream of fiber masses converging in the direction of the mesencephalon and pons. Inversely, ascending (afferent) fibers are switched in the large thalamus, proceed in the dark strands of the white matter, and divert in the direction of their target cortices.

The rostral-most MR section (plane A in Fig. 6.9) exhibits from top to bottom a few important landmarks: the corpus callosum (cc), which interconnects the two very large telencephalic hemispheres, is extremely thin in dolphins. In contrast, the posterior commissure (pc) is well developed (see Section "Commissures" later in this chapter). The ventricular system is rendered black and seen as ovoid or elongate hollow spaces (i, ii, lateral ventricles; iii, third ventricle). The diencephalon is rather large and extends between the two telencephalic hemispheres. Here, the central part around the posterior commissure blends laterally into the metathalamus consisting of the light (hyperintense) medial geniculate body and the dark (hypointense) lateral geniculate body (with fine oblique texture). Ventrally, the diencephalon extends into the hypothalamus with the well-developed optic chiasm (oc). As the first (rostral-most) part of the midbrain (mesencephalon), the elliptic nucleus (e) is obvious here. The latter is rather large in dolphins with respect to other mammalia and consists of two components: the interstitial nucleus of Cajal and the nucleus of Darkschewitsch (see the following). In these animals, the elliptic nucleus is obviously involved in the transmission of auditory projections to premotor and motor nuclei with respect to locomotion and to phonation in communication and sonar orientation. Together with other nuclei and fiber tracts throughout the brainstem and with the cerebellum, the elliptic nucleus seems to be responsible for mass movements of the body trunk (audiomotor navigation) (Oelschläger, 2008).

The second section (plane B in Fig. 6.9) runs slightly behind the middle of the brain. With respect to plane A, major changes in the shape and topography of structures regard the lower middle part of the section. The didactic center of the midbrain is the cerebral aqueduct (*asterisk*) and the surrounding central gray, which appears hyperintense. Previously the latter is the tectum with the rostral (superior) colliculi belonging to the visual system. Later is the tegmentum, a superior integration system within the brainstem. The lowermost level of the midbrain is represented by the crura cerebri, which contain the long descending (efferent) fiber tracts. At the bottom of the section, there is a first cut of the pons, which belongs to the metencephalon.

The third section (plane C in Fig. 6.9) runs through the caudal (inferior) colliculi and other parts of the auditory system (superior olive, lateral lemniscus), the tegmentum, and the middle of the pons. Compared with the superior colliculi in plane B, the inferior colliculi are several times larger in volume. This observation stands for the overwhelming significance of the auditory system in dolphins as a permanent source of information due to their echolocation system. On both sides of the brainstem, the small rostral-most parts of the cerebellar hemispheres appear between the pons and the telencephalon.

In the fourth section (plane D in Fig. 6.9), the situation changes dramatically in comparison to the planes described earlier. Here, the brain is characterized by the presence of the large cerebellum, which attains the size of that in the human. In contrast to the moderately developed vermis, the cerebellar hemispheres are very large. Together with the brainstem, they form a triangle. On both sides between the cerebellar hemispheres and the brainstem, the thick vestibulocochlear nerves and the much thinner facial nerves are obvious. The eighth nerve projects to the associated ventral cochlear nucleus and further to the trapezoid body and nucleus, as well as to the superior olive; all these structures belong to the ascending auditory pathway.

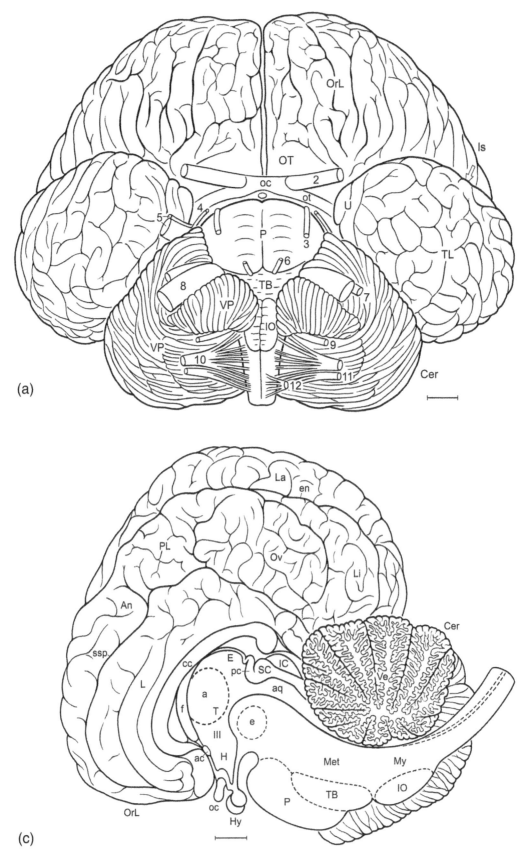

FIGURE 6.7 External aspects and mediosagittal view of bottlenose dolphin brains (a–d) and human brains (e–h). (a and e) basal aspect, (b and f) dorsal aspect, (c and g) mediosagittal aspect, (d and h) lateral aspect. All brains are of about the same size, but the drawings of the dolphin brains come from different specimens. Apart from some superficial commonalities, which indicate a more or less similar macroscopical level of developmental, there are profound differences between the two species. This refers to the general shape of brain (wider than long in the dolphin), the flexures in the brainstem, the surface configurations of the neocortex, and the diameters of the cranial nerves. In dorsal aspect, the cerebellum is seen in the dolphin but not in the human. Abbreviations: *Arrow*, pointing into sylvian cleft; a, interthalamic adhesion; ac, anterior commissure; An, anterior lobule of limbic lobe; aq, cerebral aqueduct; cc, corpus callosum; CG, cingulate gyrus; Ch, cerebral hemisphere; Cer, cerebellum; cr, cruciate sulcus; e, elliptic nucleus; E, epithalamus; ec, ectolateral; en, entolateral sulcus; ES, ectosylvian gyrus; f, fornix; fp, frontal pole; H, hypothalamus; Hy, hypophysis; IC, inferior colliculus; IO, inferior olive; L, limbic lobe; La, lateral gyrus; Li, lingual lobule; ls, lateral sulcus; MB, mamillary body; Met, metencephalon; My, myelencephalon; OB, olfactory bulb; oc, optic chiasm; OL, olfactory lobe; op, occipital pole; OrL, orbital lobe; ot, optic tract; OT, olfactory tubercle;

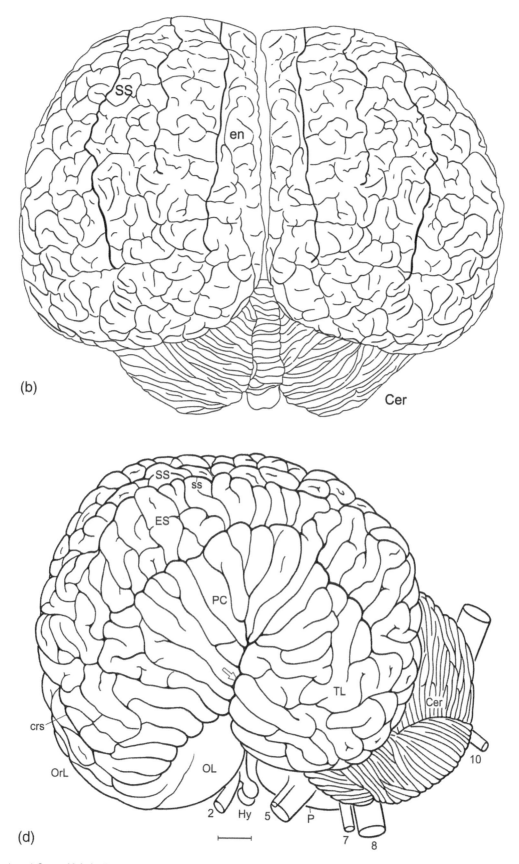

FIGURE 6.7 (*cont.*) Ov, oval lobule; P, pons; pc, posterior commissure; PC, perisylvian cortex; PL, paralimbic lobe; SC, spinal canal (in g); SC, superior colliculus (in c); SP, spinal cord; ss, suprasylvian sulcus; SS, suprasylvian gyrus; ssp, suprasplenial (limbic) sulcus; T, thalamus; TB, trapezoid body; TL, temporal lobe; U, uncus; Ve, vermis; VP, ventral paraflocculus; 2–12, cranial nerves: 2, optic nerve; 3, oculomotor nerve; 4, trochlear nerve; 5, trigeminal nerve; 6, abducens nerve; 7, facial nerve; 8, vestibulocochlear nerve; 9, glossopharyngeus nerve; 10, vagus nerve; 11, accessory nerve; 12, hypoglossus nerve; III, third ventricle; IV, fourth ventricle. [*Redrawn and modified by Jutta Oelschläger; a, c, and d, from Langworthy, 1931. A description of the central nervous system of the porpoise* (Tursiops truncatus). *J. Comp. Neurol. 54, 437–499; b, from Brauer and Schober, 1976. Catalogue of Mammalian Brains. VEB Gustav Fischer Verlag, Jena; Pilleri and Gihr, 1970. The central nervous system of the mysticete and odentocete whales. In: Pilleri, G. (Ed.), Investigations on Cetacea. Institute of Brain Anatomy, Bern, pp. 89–128; Morgane and Jacobs, 1972. Comparative anatomy of the cetacean nervous system. In: Harrison, R.J. (Ed.), Functional Anatomy of Marine Mammals. Academic Press, London, pp. 117–244.); e–h, from Köpf-Maier, P. (Ed.), 2000. Wolf-Heidegger's Atlas of Human Anatomy, fifth ed. Karger, Basel; New York.]*

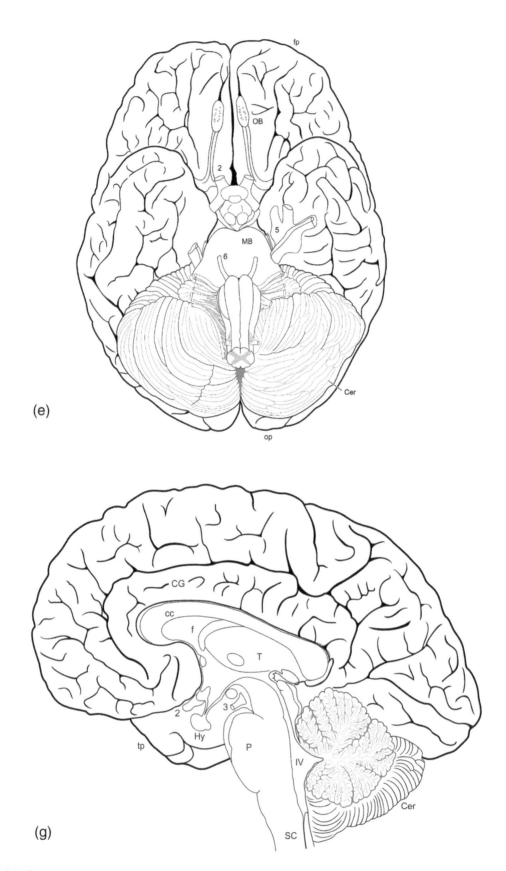

(e)

(g)

FIGURE 6.7 *(cont.)*

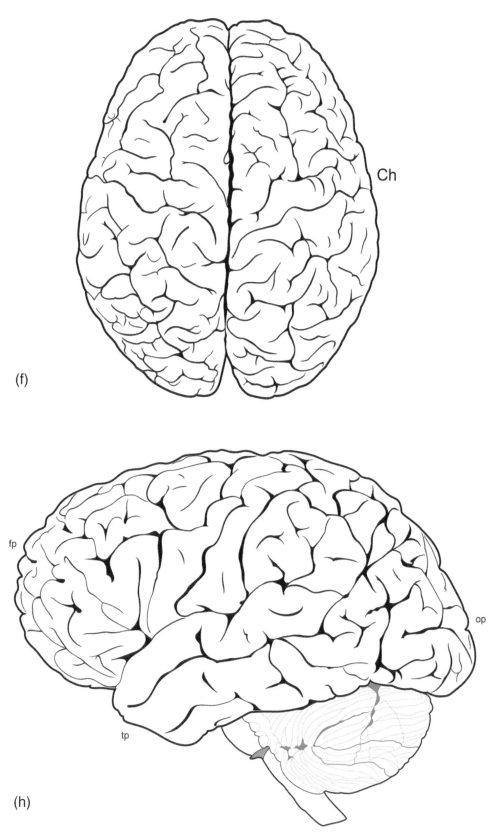

(f)

(h)

FIGURE 6.7   (*cont.*)

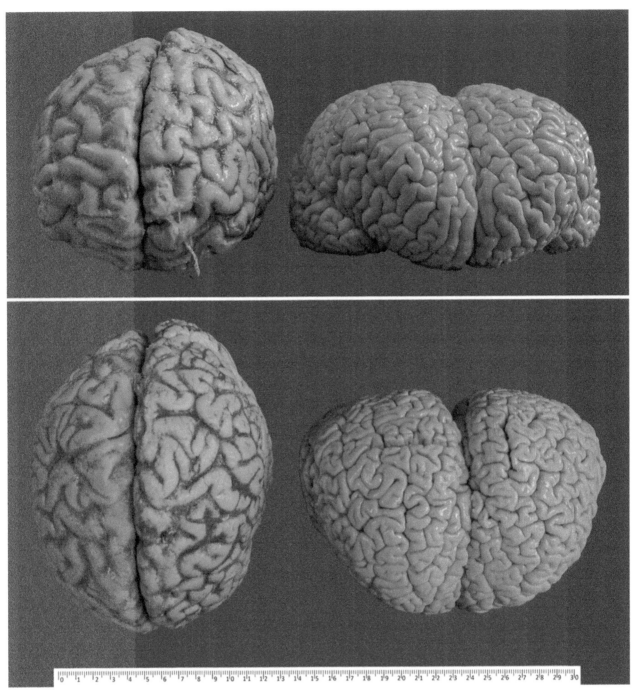

FIGURE 6.8   External aspects of a human brain (a) and a bottlenose dolphin brain (b) in rostral (upper row) and dorsal views. The general shape of brain, wider than long in the dolphin, is clearly visible. For further information on surface configurations of the neocortex, see Fig. 6.7. Note that, in dorsal aspect, the cerebellum is not seen artificially in the dolphin.

Blood Supply to the Brain

The arterial part of the dolphin vascular system is characterized by the existence of so-called retia mirabilia (wonderful nets; see Chapter 4), which are localized in different places and consist of a special network of communicating arterial blood vessels. Dolphins possess two types of arterial retia, two caudal ones in the dorsal thorax (posterior mediastinum), lateral to the vertebral column, and a bipartite one in the center of the skull base, on both sides of the sella turcica, the hypophyseal fossa (sinus cavernosus), and the optic chiasm (Fig. 6.10). The rostral rete has extensions on each side

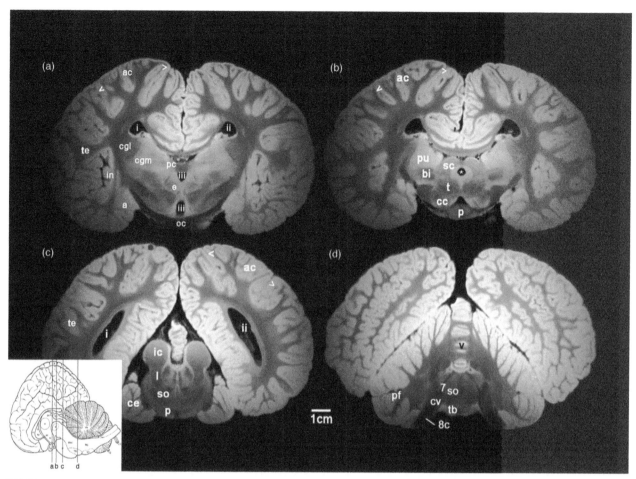

FIGURE 6.9   Common dolphin brain. Coronal MRI scans. (a) Level of the diencephalon. (b) Anterior mesencephalon with rostral first cut of the pons (p). (c) Posterior mesencephalon with rostral extremities of the cerebellum (ce). (d) Anterior medulla oblongata with seventh and eighth nerves (7, 8c), and anterior part of cerebellum (pf, v). In the rear part of the brain (parietal–temporal region) the extreme density of the cortical folding is obvious. Structures of the auditory system are in italics. Abbreviations: i–iii, ventricles; a, amygdala; ac, auditory cortex; bi, brachium of inferior colliculus; cc, crus cerebri; cv, ventral cochlear nucleus; e, elliptic nucleus; ic, inferior (caudal) collicle; in, insula; l, nucleus and fiber tract of lateral lemniscus; cgl, cgm, lateral and medial geniculate bodies; oc, optic chiasm; p, pons; pc, posterior commissure; pf, paraflocculus; pu, pulvinar; sc, superior (rostral) collicle; so, superior olive; t, thalamus; tb, nucleus of trapezoid body; te, telencephalon; v, vermis. *Asterisk* in b: cerebral aqueduct. *(Modified from Oelschläger et al., 2008. Morphology and evolutionary biology of the dolphin* (Delphinus *sp.) brain—MR imaging and conventional histology. Brain Behav. Evol. 71, 68–86.)*

along the optic nerve in the direction of the orbita. These ophthalmic retia are probably involved in the accommodation of the dolphin eye triggered by the retractor bulbi muscle (see Section "Visual System", and Chapter 5, Section "Vision: Anatomy and Optics of the Eye"). The two types of wonderful nets (caudal, rostral) are connected with each other via an interposed spinal rete (Fig. 6.10D). Rostrally, the latter transforms into two large spinal meningeal arteries (no basilar artery, no circle of Willis) which enter the cranial vault via the foramen magnum and split into many smaller and smallest arteries/arterioles that communicate with each other across the midline and thus constitute the impressive rostral (cranial) rete. All these vessels are primarily meningeal (epidural) arteries, but, interestingly, give rise to typical cerebral arteries (Fig. 6.10).

The rostral wonderful net in dolphins to a certain degree reminds of the situation in some terrestrial mammals. Among the artiodactyl domestic animals (ruminants), which are regarded closely related to cetaceans phylogenetically (see Chapter 1), the adult cattle seems to be rather similar to dolphins with respect to the rostral rete, although less specialized. In adult cattle, this cranial rete is supplied by branches of the maxillary artery. At birth, the cattle's internal carotid artery seems to be nearly completely atrophied as was reported for dolphins. Thus, the brain of cattle is nearly exclusively supplied by the external carotid artery which feeds in the carotid *rete mirabile*, and the latter gives rise to the major cerebral vessels (anterior, middle cerebral arteries, superior cerebellar artery; McFarland et al., 1979; König and Liebich, 2009). In dolphins, the internal carotid and vertebral arteries are not involved in the blood supply of the brain; instead, the brain totally depends

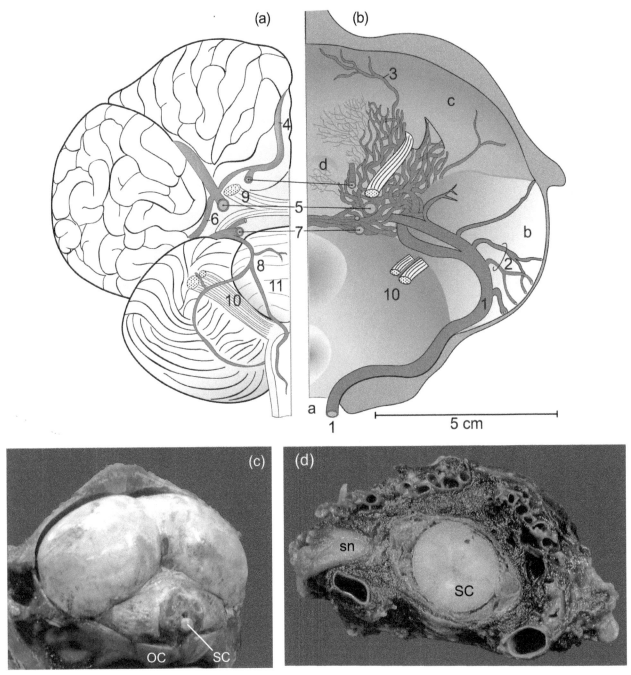

FIGURE 6.10 Arterial blood flow in the cranial vault of a dolphin- like toothed whale (harbor porpoise, a, b). Right half of head. The brain in (a) has been rotated out of the skull base in (b) to the left after cutting cranial nerves and arteries. The photo in (c) showns the intact dura mater covering the brain within the skull of a Risso's dolphin (caudolateral view) and the photo in (d) a transverse cut through the spinal cord of a bottlenose dolphin surrounded by the spinal rete mirabile. Abbreviations: 1, spinal meningeal artery; 2, posterior and medial cerebral meningeal arteries; 3, anterior cerebral meningeal artery; 4, anterior cerebral artery; 5, middle cerebral artery; 6, posterior cerebral artery; 7, superior cerebellar artery; 8, inferior cerebellar artery; 9, optic nerve; 10, vestibulocochlear nerve and facial nerve; 11, pons; a, foramen magnum; b, middle cranial fossa; c, anterior cranial fossa; d, sella turcica; OC, occipital condyles; SC, spinal cord; sn, spinal nerve. *(a, and b, redrawn and modified by Jutta Oelschläger from Boenninghaus, 1903. Das Ohr des Zahnwales, zugleich ein Beitrag zur Theorie der Schalleitung: Eine biologische Studie. Fischer.)*

on the thoraco-spinal rete, which is supplied by intercostal and posterior thoracic arteries. In contrast, other mammals have two sources of arterial supply, an anterior component (internal carotid or at least branches from the external carotid) and a posterior component (vertebral arteries). In parallel to the arterial supply in dolphins, the venous system of the cranial vault is also substantially drained via the spinal canal (Boenninghaus, 1903).

Although wonderful nets in dolphins have been known for a long time, Morgane and Jacobs as well as other members of their group (cf. Morgane and Jacobs, 1972, and citations therein) were the first to systematically investigate the spino-vascular supply in dolphins using modern casting and angiographic methods. This was necessary as a basis for functional analysis of the remarkably developed retia mirabilia in these animals. Already Tyson (1680) had noted the masses of vascular tissue lying dorsally behind the pleura epithelium within the rib cage of a dolphin. But still today the retia are not fully understood, also because their closest relatives, the ruminant artiodactyls, are typical terrestrial mammals and because there are no obvious similarities between the two groups as to behavior and ecological adaptations that would explain the correspondence between artiodactyls and dolphins as to the rostral retia within the braincase.

In dolphins, the thoracic retia are supplied segmentally from the intercostal arteries given off by the descending aorta, and they communicate through the intervertebral foramina with intravertebral retia opening into the two spinal meningeal arteries mentioned previously (Fig. 6.10b; Boenninghaus, 1903). The latter enter the foramen magnum as strong dorsal longitudinal vessels to form the principal, if not the only, blood supply to the brain (Viamonte et al., 1968). Of equal interest is the fact that the internal carotid artery was shown to be nonfunctional in the adult harbor porpoise and pilot whale, and there are neither vertebral arteries nor a basilar artery. Instead, the two thick spinal meningeal vessels take a most unusual route around the temporal lobes of the brain where they give off a series of separate arteries in the middle cranial fossa. They then curve medially to the area of the pituitary gland (sella turcica; hypophysis) and split into a large number of branches, which constitute an impressive cranial meningeal arterial rete. After Boenninghaus (1903), this central rete mirabile seems to give rise to all of the remaining arteries within the cranial vault recalling the rostral/anterior, middle/medial, and caudal/posterior cerebral arteries as well as the cerebellar vessels. It is, however, unclear to what degree these dolphin arteries are homologous to those in terrestrial mammals (Geisler and Luo, 1998; McFarland et al., 1979; Viamonte et al., 1968). On the other hand, the existence of the retia mirabilia previously mentioned are obviously common to all toothed whales (Boenninghaus, 1903). In adult dolphins and other toothed whales, there are impressions of the spinal meningeal arteries in the bony middle cranial fossa. In a late fetal (68 cm) specimen of the harbor porpoise, the ontogenesis of this peculiar arterial system is still obvious (Boenninghaus, 1903): Here, the spinal meningeal artery is still relatively thin with respect to the internal carotid artery (see Section "Ear: Sense of Balance and Hearing").

Modern investigations of the retia connected to the brain have confirmed the morphological descriptions of older papers and added some new results concerning microcirculation. The retia show up in angiograms, and histological examination revealed that small arteries give rise to a vast number of thin-walled sinusoids (Blix et al., 2013). The whole system seems to have a large cross-sectional area, which reduces blood perfusion to a very slow speed as is known from capillary loops all over the body, here facilitating the exchange of respiratory gases and nutrients. Blix et al. (2013) reported the vessels of the retia to be embedded in deposits of fat but could not confirm an intense autonomic innervation of the arteries reported by earlier authors (Nagel et al., 1968); they speculate that the special conditions within the retia mirabilia may hold for a transvascular exchange between blood and the surrounding adipose tissue with respect to nitrogen. The latter is known to be six times more soluble in fat than in water; in human divers with nitrogen super-saturated blood, nitrogen bubbles may form during depressurization and cause severe diver's sickness (also called decompression sickness, the bends or Caisson disease) with problematic symptoms all over the body. By means of the thoracic retia mirabilia, nitrogen bubbles in the blood and tissues should be avoided or at least not affect the brain and spinal cord because of being caught within the retia mirabilia. However, repeated diving cycles may bring the respiratory system to its limits. Ridgway and coworkers have suggested that the retia might play some role in the dolphin's resistance to Caisson disease while making many repeated dives (Ridgway and Harrison, 1986; Ridgway and Howard, 1979). Large retia placed dorsally in the thorax could well be able to trap bubbles occuring during such dives. In addition, it was shown (Vogl et al., 1981; Vogl and Fisher, 1982) that the thoracospinal rete, with respect to the brain as a target, has much more afferent than efferent connections. Temporary blockage of a large part of the rete would presumably bear only little risk for the blood supply of the brain and spinal cord. The fact that all deep-diving mammals have such retia mirabilia is also in favor of the assumption that they are involved in the prevention of Caisson disease. In this respect, a shift of blood volume into the rete of the thoracic cavity could replace shrinking air quanta of the lungs during diving. Also, there seem to be some indications that the retia may help to avoid or dampen the pressure-pulse in the cerebral arteries and thus help provide a continuous and steady flow of blood to the brain that is unaffected by any fluctuations in heart rate during diving (bradycardia) (Morgane and Jacobs, 1972). Although prolonged breathholding is essential for deep-diving, other structural prerequisites seem to be necessary to enable a complete collapse of the lung while air is shifted to the upper respiratory tract. This theoretically prevents nitrogen from going into the blood and protects the animal from Caisson disease (see Section "Diving"). In this respect, a bubble-trapping mechanism like the rete mirabile could serve as an additional mechanism of security.

Ventricular System

The ventricular system of mammals contains the cerebrospinal fluid in four extensions, the ventricles (Fig. 6.11). There is one in each telencephalic hemisphere (lateral ventricles), one in the diencephalon (third ventricle), and one in the metencephalon/myelencephalon (fourth ventricle). The two lateral ventricles communicate with the third ventricle via the two interventricular foramina (ivf). The third and fourth ventricles are connected via the cerebral aqueduct in the midbrain. In generalized (terrestrial) mammals, each lateral ventricle has a thin rostral extension that reaches into the olfactory bulb and is called "olfactory ventricle" or "olfactory recess" (Fig. 6.11b,d). The olfactory bulbs and tracts are parts of the central nervous system, which is completely reduced in adult toothed whales (see related Sections in Chapters 5, 6, and 10).

Caudally, the ventricular system has three outlets into the subarachnoid space: They are openings in the wall of the fourth ventricle, two paired ones at the end of each lateral recess, and one in the roof. After leaving the ventricular system, the cerebrospinal fluid runs in the subarachnoid space, suspending the brain, and is drained via special formations of the arachnoid mater (arachnoid granulations) into the dural venous system of the dorsal sagittal sinus.

In dolphins, the ventricular system largely corresponds with the situation in terrestrial mammals (Fig. 6.11a,b). However, there are some interesting modifications that can be correlated with peculiar morphological and functional specializations of their brain. As is the case with the brain as a whole, the ventricular system of the dolphin is short but wide between the temporal syn. inferior horns (ih), in comparison with that in the sheep (dorsal aspect; Fig. 6.11c,d). The lateral ventricles (1 and 2) are more strictly semicircular than in ungulate and carnivorous mammals, in correlation with the strong rotation of the telencephalic hemispheres (Fig. 6.11a,b). An olfactory recess is constantly lacking, concomitant to the loss of the anterior part of the olfactory system in the postnatal animal (see Section "Olfactory System" later in this chapter). At the same time, the rostral syn. anterior horn (ah) is short, presumably in correlation with a hypoplastic frontal lobe of the hemisphere. An occipital (posterior) horn, seen in primates and the human, is absent as is an occipital lobe of the hemisphere (Morgane and Jacobs, 1972). The central part of the lateral ventricle is well developed as is the inferior horn. The third ventricle is comparatively high but short and largely displaced by the extended fusion of the two thalami (interthalamic adhesion; cf. Fig. 6.12a). In the bottlenose dolphin, there are well-developed preoptic and infundibular recesses. However, only rarely a pineal body is reported in adult animals, whereas dolphin embryos show a regular anlage[a] (Buhl and Oelschläger, 1988; Oelschläger and Buhl, 1985). The cerebral aqueduct (Fig. 6.9b: *asterisk*) is tubular beneath the superior colliculi. In the posterior part, it is nearly collapsed and extends laterally beneath the very large inferior colliculi. The fourth ventricle of the dolphin (Fig. 6.11) does not show any conspicuous specializations; it opens in the central canal caudally and in the subarachnoid space bilaterally and dorsally. In its floor, Morgane and Jacobs (1972) reported the facial eminence where the facial nerve courses around the nucleus of the abducens nerve but did not find protrusions of the dorsal motor nucleus of the vagus nerve and of the nucleus of the hypoglossal nerve. However, Langworthy (1931) figured the nucleus gracilis at the posterior end of the ventricle.

Choroid plexuses, which consist of villus-like composits of the embryonal brain wall (ependyma) and the pia mater with its choroidal vessels, produce the cerebrospinal fluid. They are found in the left and right lateral ventricles (central part and temporal horn) as well as in the third ventricle (roof) and the fourth ventricle (roof and lateral recesses) but not in the cerebral aqueduct (McFarland et al., 1969). The cerebrospinal fluid flows along the ventricular system and exits in the fourth ventricle via the medial and lateral apertures (foramina Luschkae and Magendie, respectively) into the subarachnoid space. Here it helps suspend and protect the brain within the cranial vault and it passes over into the dural venous system (dorsal sagittal sinus).

Main Divisions of the Dolphin Brain

The brain of dolphins is very large for their body size (see Chapter 10, Section "Specific and Allometric Phenomena in Toothed Whales"). Much rounded and similar to a boxing glove, it shapes the dolphin skull in a way known from the human. At the beginning of development, the dolphin brain forms as in other mammals. In late embryonal and fetal stages, the tube-like brain changes its shape and becomes shorter along the longitudinal axis, much wider and thus more compact (Figs. 6.2, 6.3, 6.7, and 6.8). This is paralleled by the so-called "telescoping process" of the skull (Miller, 1923). The brainstem becomes stout and its flexures are different from those in other mammals (Buhl and Oelschläger, 1988). It is dominated by a very large cerebellum, which seems to have a certain size correlation with the telencephalon (Figs. 6.7 and 6.12).

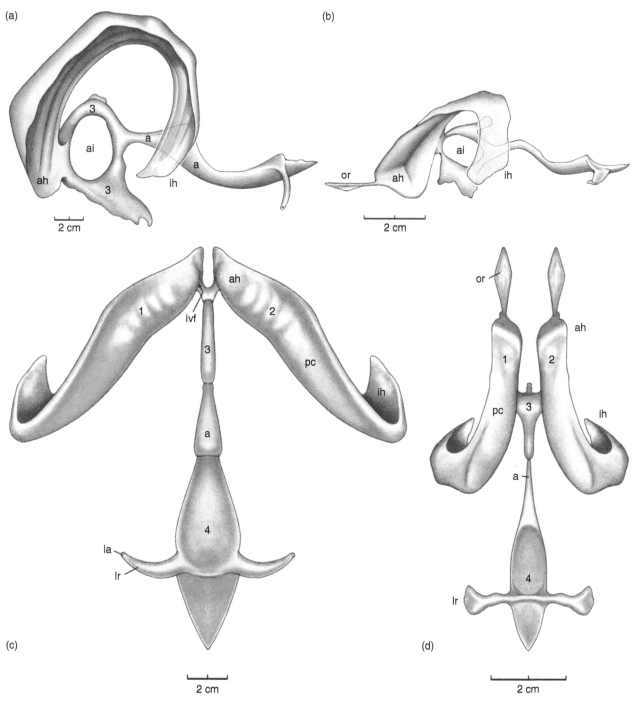

FIGURE 6.11   Ventricular system in dolphins (a, c) and the sheep (b, d). Both ventricular systems nicely correspond with the shape of the relevant brain, that is, in the dolphin, the system is very broad and about as wide as long. The sheep ventricular system is slender and its length is about twice the width. The relative shortness of the dolphin ventricles is an immediate hint for the so-called telescoping of the neurocranium and brain in dolphins. Also there is no olfactory ventricle in dolphins in correlation to the reduction of the anterior olfactory system. Interestingly, the cerebral aqueduct is very high in the dolphin, obviously in relation to the strong development of the inferior colliculi in the midbrain tectum. Abbreviations: 1, 2, lateral ventricles; 3, third ventricle; 4, fourth ventricle; a, cerebral aqueduct; ah, anterior horn; ai, adhaesio interthalamica (interthalamic adhesion); ih, inferior horn; ivf, interventricular foramen; la, lateral aperture of fourth ventricle; lr, lateral recess; or, olfactory recess; pc, pars centralis. *(Redrawn by Massimo Demma from McFarland et al., 1969. Ventricular system of the brain of the dolphin,* Tursiops truncatus, *with comparative anatomical observations and relations to brain specializations. J. Comp. Neurol. 135, 275–368.)*

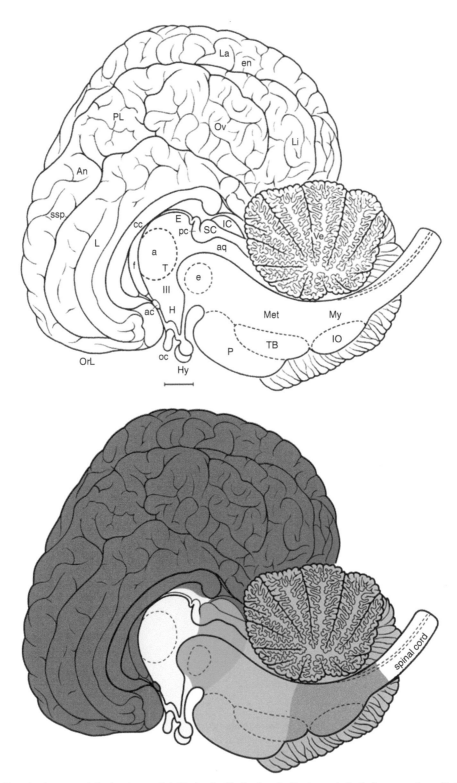

FIGURE 6.12 **Mediosagittal aspect of the bottlenose dolphin brain (Scale: 1 cm).** (Top) morphological presentation with labeling of structures. (Bottom) same drawing with the main parts of the brain in color. *Red*, telencephalon; *yellow*, diencephalon; *green*, mesencephalon; *blue*, metenephalon (cerebellum/pons); lilac, myelenephalon. Abbreviations: a, interthalamic adhesion; ac, anterior commissure; An, anterior lobule; aq, cerebral aqueduct; cc, corpus callosum; e, elliptic nucleus; E, epithalamus; en, entolateral sulcus; f, fornix; H, hypothalamus; Hy, hypophysis; IC, inferior colliculus; IO, inferior olive; L, limbic lobe; La, lateral gyrus; Li, lingual lobule; Met, metencephalon; My, myelencephalon; oc, optic chiasm; OrL, orbital lobe; Ov, oval lobule; P, pons; pc, posterior commissure; PL, paralimbic lobe; SC, superior colliculus; ssp, splenial (limbic) sulcus; T, thalamus; TB, trapezoid body; Ve, vermis; III, third ventricle. *(From Oelschläger and Oelschläger, 2009. Brain. In: Perrin, W.F., Würsig, B., Thewissen, J.G.M. (Eds.), Encyclopedia of Marine Mammals. Academic Press, San Diego, CA, pp. 134–149.)*

In adult dolphins, the telencephalic hemispheres tower over the rest of the brain. With respect to hoofed animals, the telencephalic hemisphere of dolphins seems to have been rotated rostrally and ventrally on the skull base; thus, the corpus callosum, interconnecting both telencephalic hemispheres, stands at an angle of about 55° against the base of the brain (Fig. 6.12).

TELENCEPHALON

The Cortices

Whereas in many groups of mammals, there is a general trend to increase the size of the brain as a whole (with respect to the size of the body as a whole; *encephalization*); some subgroups impress by a considerable increase of the telencephalon as a whole (*telencephalization*) and of its cortex formations (*corticalization*) (Manger, 2006). One part of the telencephalic cortex, the neocortex, was particularly successful during evolution as to size increase (*neocorticalization*). As a convergent development, this is seen in two distantly related mammalian groups, the primates and the cetaceans (Figs. 6.2 and 6.7). Here, all the phenomena just mentioned have been driven to a maximum (Boddy et al., 2012; Manger, 2006; Marino et al., 2004; Montgomery et al., 2013) so that the dolphin brain seems to rival the human brain in several quantitative and qualitative respects. Both brains exhibit an overwhelming dominance of the neocortex over the two other cortices, the paleocortex and the archicortex. The paleocortex seems to be much reduced in dolphins, in correlation with the loss of the anterior olfactory system (see Section "Olfactory System" later in this chapter).

The archicortex (see Section "Limbic System") shows characteristics of a multisensory system superior to the somatosensory, auditory, and the visual cortices, but also serving vegetative functions, memory formation and retrieval, and emotion as well as the capacity of orientation and navigation. It is the core unit of the limbic system. In dolphins, the hippocampus is the main component of the archicortex but it is also much smaller than in terrestrial mammals, including primates and the human. This is paralleled by the small size of other components of the limbic system (Papez circuit) in dolphins: The fornix as the efferent fiber tract of the hippocampus is thin, and the mamillary body and anterior nuclei of the thalamus are small. The accessory secondary cortical areas within the limbic system (periarchicortex), however, seem to be well developed. They are intermediate between the archicortex and the neocortex and comprise the much pronounced gyrus cinguli above the corpus callosum as well as the gyrus parahippocampalis with the entorhinal cortex on the medial temporal lobe of the telencephalic hemisphere (limbic lobe) (Morgane et al., 1980).

In comparison with those in the human brain, the paleocortex and archicortex of dolphins are relatively small and located at the anterior basal surface of the telencephalic hemisphere (paleocortex) or lie at the mediobasal margin of the temporal lobe (archicortex). The surface of the neocortex reaches a maximum in marine dolphins as does its folding index (gyrification). The volume ratio of the neocortex (grey matter) in the whole brain, however, is smaller than in the human because the dolphin's neocortex is much thinner. We will see later what that means with respect to the potential computing power of a brain and the respective neurobiological adaptations of the species.

Neocortex

Surface Configurations

Within the dolphin telencephalic hemisphere, the rostroventral part is presumed to be equivalent to the frontal lobe; part of this formation is called "orbital lobe" by Morgane et al. (1980; Figs. 6.7 and 6.13). The long middle part far dorsally along the vertex of the hemisphere may correspond to the parietal lobe; most of its areal extension is not to be seen in lateral aspect. At the dorsal (occipital) contour of the hemisphere, it blends smoothly in the occipital/temporal lobe complex. As in hoofed animals, an occipital lobe is not obvious morphologically but a functional equivalent was found with electrophysiological and histological methods (see the following). The temporal lobe is obviously very large and comprises much of the lateral surface of the hemisphere (Fig. 6.7).

In principle, the telencephalic hemisphere of dolphins may be classified in two different ways. The first method tries to use the lobar formations of terrestrial mammals. The second approach applies the nomenclature of Morgane et al. (1980).

Using the nomenclature of Morgane et al. (1980), it seems that the gyral (fissural) pattern of the dolphin neocortex (Figs. 6.7 and 6.12–6.15) is more or less similar to the situation in ungulates and carnivores. Nevertheless, the homologization[c] of neocortical areas between dolphins and other mammals following topographical criteria of the cortical relief

---

c. Two structures in different species are called homologous in the field of comparative anatomy if they are of a similar architecture and situated in the same relative position to neighboring structures.

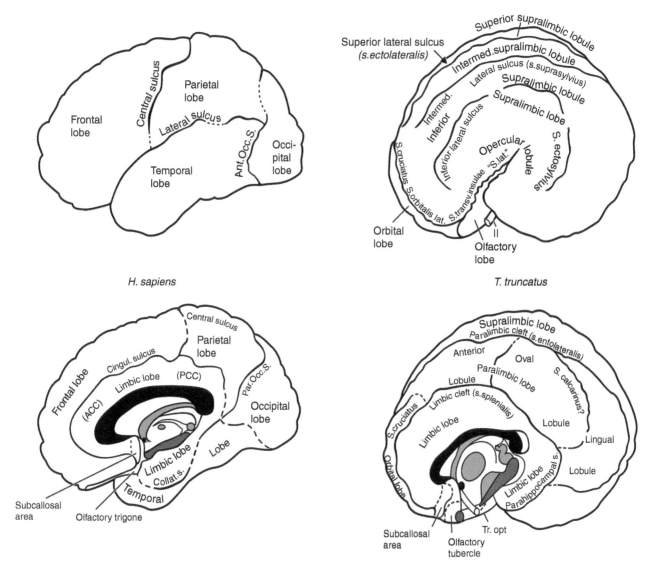

FIGURE 6.13 **Left and right cerebral hemispheres of the human in comparison with those of the bottlenose dolphin (Scale = 1 cm).** Top, left hemispheres in lateral aspect; bottom, right hemispheres in mediosagittal aspect. There are two terminologies as to the labeling of the cortical surface configurations (see text). *(Redrawn and modified by Jutta Oelschläger after Morgane et al., 1980. The anatomy of the brain of the bottlenose dolphin* (Tursiops truncatus). *Surface configurations of the telencephalon of the bottlenose dolphin with comparative anatomical observations in four other cetacean species. Brain Res. Bull. 5, 1–107.)*

is difficult or impossible. An exception to this may be the primary projection areas as far as they have been detected with electrophysiological methods and histological analysis (see the following). Several major longitudinal fissures run concentrically around the sylvian cleft (lateral sulcus) at different distances over the lateral and dorsal surface of the hemisphere. In Fig. 6.14, these sulci are labeled as the ectosylvian (es), suprasylvian (ss), lateral (la), and entolateral sulcus (en; syn. paralimbic cleft). In correspondence, medially adjacent to these major fissures and proceeding into the interhemispheric cleft, follows the splenial sulcus or limbic cleft (Figs. 6.12 and 6.13). The sulci delimitate the major gyri of the telencephalic cortex: laterally the opercular lobule, and the inferior, intermediate and superior supralimbic lobules, which together constitute the very large supralimbic lobe (Fig. 6.13). Medially follow the paralimbic lobe (Fig. 6.12: PL) and the limbic lobe (L); the latter is bordered ventrally by the corpus callosum and probably corresponds to the gyrus cinguli (limbic lobe) of other mammals, representing a well-developed neocortical component of the limbic system.

The large size of the dolphin brain and thus the impressive area and volume of the neocortex in these animals is also reflected in the fact that the insular cortex is hidden in the lateral wall of the hemisphere at the bottom of the sylvian cleft

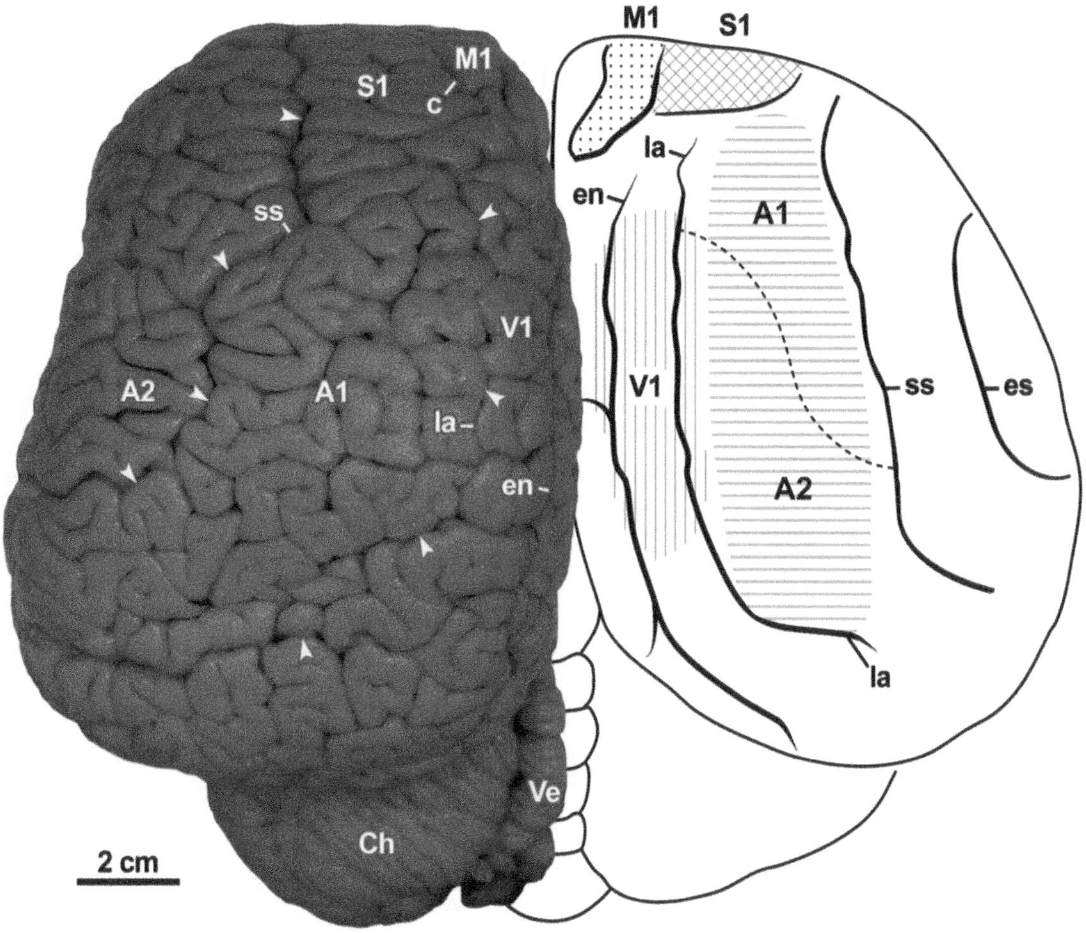

FIGURE 6.14   Dorsal view of the brain of the bottlenose dolphin with the gyrification pattern of the neocortex showing the neocortical motor and sensory projection fields (left) (Scale = 2 cm). *Arrowheads* mark the course of major neocortical sulci. Schematic drawing (right), after Morgane et al. (1986). Abbreviations: A1, A2, primary, secondary auditory neocortex; c, cruciate sulcus; Ch, cerebellar hemisphere; en, entolateral sulcus; es, ecto-sylvian sulcus; la, lateral sulcus; M1, motor neocortex; S1, somatosensory neocortex; ss, suprasylvian sulcus; V1, visual neocortex; Ve, vermis. *(Modified by Jutta Oelschläger from Morgane et al., 1986. Evolutionary morphology of the dolphin brain. In: Schusterman, R.J., Thomas, J.A., Wood, F.G. (Eds.), Dolphin Cognition and Behavior: A Comparative Approach. Lawrence Erlbaum Associates, Hillsdale, pp. 5–29; and Kern et al., 2011. Stereology of the neocortex in Odontocetes: qualitative, quantitative, and functional implications. Brain Behav. Evol. 77, 79–90.)*

(Fig. 6.9a: in). Highly proliferative neocortical areas around the cleft have "buried" the multisensory insula by the formation of superficial so-called "opercular" or perisylvian cortex (Fig. 6.7). In mammals, the insula is engaged in vegetative functions and regarded rather conservative by many authors.

Localization of Cortical Areas

As already mentioned, the accurate division of the telencephalic hemisphere into lobes as executed in other mammals is difficult. Obviously the neocortex of dolphins has undergone profound modifications in combination with a considerable size progression during evolution. However, electrophysiological data (evoked potential method[d]) show that, in principle, the primary projection fields of the neocortex are arranged in the same sequence as in many other mammals (Figs. 6.14–6.16; see the following).

Hence, the motor cortex is located in front (here rostroventrally) on the hemisphere, immediately followed caudally (dorsally) by the somatosensory cortex, and on the vertex of the hemisphere the auditory cortex situated more laterally and

---

d. An evoked potential or evoked response is a recording from the nervous system by electroencephalography (EEG) or other electrophysiological recording methods following the presentation of a stimulus.

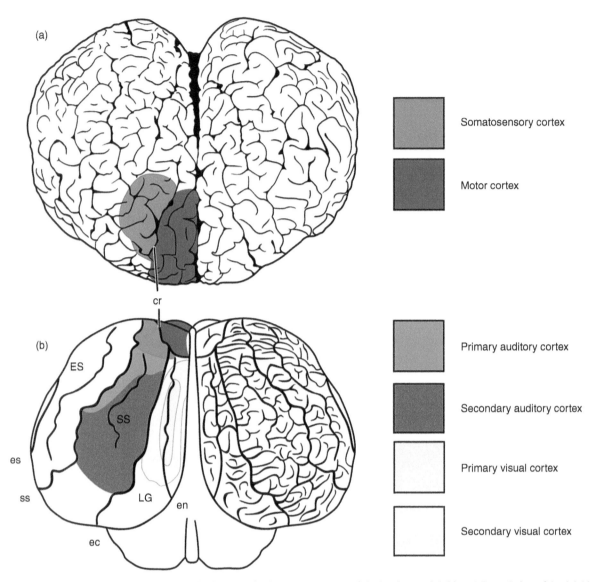

FIGURE 6.15 **Localization of the primary projection areas in the cerebral cortex of the bottlenose dolphin.** (a) Rostral view of the dolphin brain showing the natural sulcal pattern with the motor neocortex field (*red*) and the somatomotor cortex (*blue*). (b) Dorsal aspect of the dolphin brain, more schematic. On the right hemisphere, the pattern of sulci and gyri is replicated. On the left hemisphere, only first-order sulci and gyri are shown. Abbreviations: cr, cruciate sulcus; ec, lateral sulcus; en, entolateral sulcus; es, ectosylvian sulcus; ES, ectosylvian gyrus; LG, lateral gyrus; ss, suprasylvian sulcus; SS, suprasylvian gyrus. *(Modified and redrawn by Jutta Oelschläger; a, from Morgane et al., 1980. The anatomy of the brain of the bottlenose dolphin* (Tursiops truncatus). *Surface configurations of the telencephalon of the bottlenose dolphin with comparative anatomical observations in four other cetacean species. Brain Res. Bull. 5, 1–107; b, from Supin et al., 2001. The Sensory Physiology of Aquatic Mammals. Kluwer Academic Publishers, Boston.)*

the visual cortex medially. These experimental data in dolphins (Supin et al., 2001) have been confirmed by histological investigations (Kern et al., 2011; Kern, 2012; Morgane et al., 1986; van Kann et al., submitted) concerning the localization of the primary neocortical fields within the framework of gyri and sulci.

A landmark for such structural analyses in dolphins is the rather constant correlation of the cruciate sulcus (Figs. 6.13–6.15) with the primary motor and somatosensory cortices, which are separated by this sulcus. They are located above the orbital lobe of the hemisphere, the *motor cortex* (M1) more rostrally (ventrally) and medially and the *somatosensory cortex* (S1) more caudally (dorsally) and laterally. The cruciate sulcus seems to be a parallel "continuation" of the entolateral sulcus on the anterior aspect of the hemisphere (Figs. 6.14 and 6.15) and could have a counterpart in the ansate sulcus of hoofed animals and in the central sulcus of primates. The primary *auditory field* (A1) is large and seems to comprise most of the suprasylvian gyrus (SS) in dorsal aspect of the hemisphere. The *visual field* (V1) is located in the lateral gyrus (LG) and perhaps in the adjacent dorsocaudal part of the perilimbic lobe (PL; Fig. 6.12). Secondary areas of

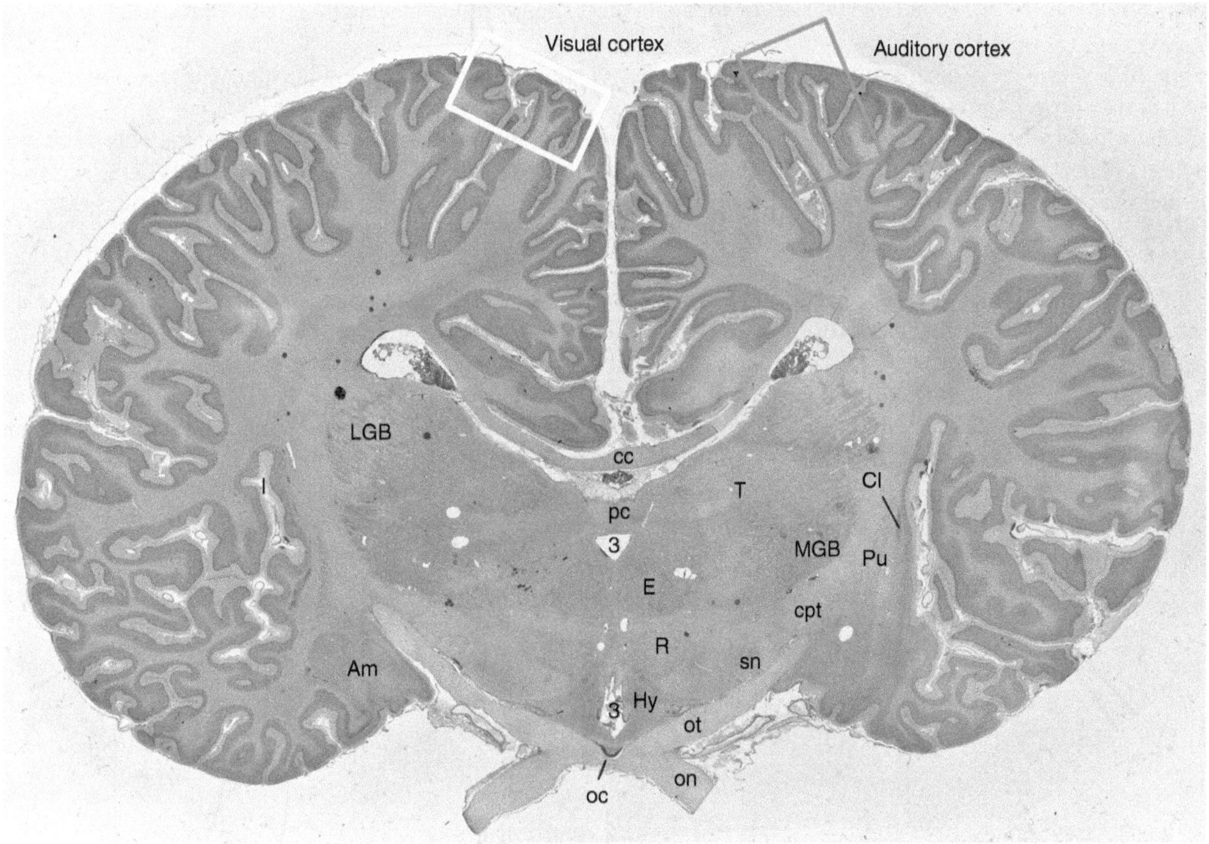

FIGURE 6.16   The visual and auditory cortices in the common dolphin. Survey. Coronal histological section of total common dolphin brain; cresyl violet stain for general orientation. Level of the diencephalon with the posterior commissural complex (pc), thalamus (T), lateral geniculate body (LGB), substantia nigra (sn), hypothalamus (Hy). Telencephalic structures and regions are the insula (I), amygdala (Am), claustrum (Cl). The elliptic nucleus (E) is already part of the mesencephalon. The *yellow* box shows a characteristic location of the visual cortex in the lateral gyrus (cf. Fig. 6.17). The *green box* stands for the auditory cortex in the suprasylvian gyrus (cf. Fig. 6.18). Abbreviations: 3, third ventricle; cc, corpus callosum; cpt, corticopontine tract (Tr. cortico-pontinus); MGB, medial geniculate body; oc, optic chiasm; on, optic nerve; ot, optic tract; Pu, putamen; R, red nucleus.

the auditory and visual cortices are given in Figs. 6.14 and 6.15. After Morgane et al. (1988), the auditory and visual fields are located in the parietal lobe of the dolphin, and this seems to correspond with the situation in cattle (Nickel et al., 2004). Morgane et al. (1988) identified visual neocortex along a "calcarine" fissure even deeper in the perilimbic lobe (Fig. 6.13). It is, however, uncertain whether this sulcus is a homolog of the calcarine sulcus of primates, which is characteristic for the visual cortex in the occipital lobe (area striata).

The actual arrangement and shape of the primary neocortical fields in dolphins differ much from those in other mammals. In dorsal aspect, these areas together comprise a broad belt across the surface of the telencephalic hemisphere. For this configuration, on formal grounds, several effects may be responsible: (1) the dolphin hemisphere seems to have experienced a strong rotation emphazising the temporal lobe; (2) the telescoping effect may not only have contributed to the towering of the hemispheres but, at the same time, to the arching of the primary fields in lateral aspect; (3) the size progression of the lateral hemispheric wall (perisylvian cortex, ectosylvian gyrus). Whether these large lateral cortices have to be regarded as higher integrative areas engaged in multisensory processing of information is not known but somehow likely.

Histology of the Neocortex

Cortical Thickness   Furutani (2008) investigated several specimens of Risso's dolphin, striped dolphins, and bottlenose dolphin with average brain masses of 2113, 970, and 2091 g, respectively. The author determined the thickness of the cortices M1, S1, V1, and A1 and found them to average for 1.85, 1.72, and 1.74 mm in the three species. Interestingly, in all animals investigated, the thicknesses of the various cortices increased in the series M1 – S1 – V1 – A1. Thus, the thinnest cortical area in all species was M1 (average: 1.64 mm), the thickest cortical area A1 (average: 1.92 mm). In comparison with other mammals, dolphins have cortical widths that are known from artiodactyls or even somewhat thinner (Hummel, 1975). Even large whales do not attain the average cortical thickness of higher primates (human: average 2.63 mm) (Kern et al., 2011).

Fig. 6.19 shows the comparison of the auditory cortices of the common dolphin and the human. Some important characteristics are obvious here, among them differences in the width of the cortical plate and the columnar arrangement of the neurons, which is distinct throughout most of the cortical layers in the human ("raindrop pattern") but only vague in the dolphin. There is no obvious sign of a layer IV in the dolphin's cortex. Another striking difference is the wide spectrum of size and shape of neurons in the human versus the more homogenous pattern in the dolphin with the obvious tendency to middle-sized and large pyramidal somata.

Layering and Cell Types The neocortex of dolphins is characteristic but difficult to understand. Its cytoarchitectural patterns are varied and complex (Hof et al., 2005). As we will see later, it may be interpreted as an important paradigm for the adaptation of terrestrial mammals to a new and dramatically demanding environment. Former papers on this topic were often convinced of its "primitiveness" (Glezer et al., 1988), but recent investigations show that this cortex may be regarded a modified ungulate cortex with important features of their terrestrial ancestors that could be adapted during the transition to fully aquatic life. Concluding from the size of the neocortex, this adaptation seems to have been a major factor in the evolution of the brain as a whole.

The cortical gray is five- to six-layered and similar to that in other mammals (Figs. 6.16–6.19). The overall thickness of the gray matter in dolphins is less than in other mammals with brains of the same dimension (Kern et al., 2011). Despite the

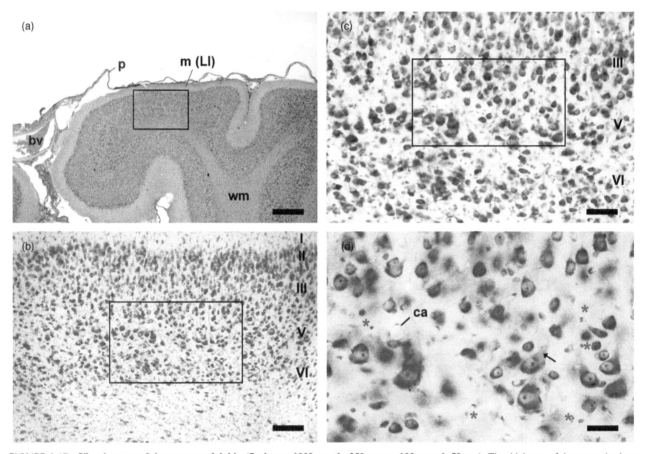

FIGURE 6.17 **Visual cortex of the common dolphin (Scales: a, 1000 μm; b, 250 μm; c, 100 μm; d, 50 μm).** The thickness of the cortex is about 1.45 mm (for rough orientation see yellow box in Fig. 6.16). (a) Overview with gyrus and pia mater; in the cortical plate (cortical grey; box), the large pyramidal neurons in layer V are already obvious as a fine dotted line. (b) Low magnification of box in a); cortical plate with the constituting layers I–VI; layer IV is indistinct or absent. (c) Higher magnification of box in b); most of the neurons are more or less of a pyramidal type. However, there are single very small round somata that could be so-called stellate cells (see text). (d) Box in (c) at maximal magnification (40×); here, different neuron populations are evident: (i) large to very large atypical pyramids that mostly occur in clusters. They have thick apical dendrites (*small arrow*); (ii) smaller pyramid-like neurons are more rounded; (iii) the smallest neuronal somata are multipolar to more or less spherical, with short radiating dendrites. For comparison, glia cells are evident by their extremely small nuclei (*asterisks*). Abbreviations: bv, blood vessel; ca, capillary; m, molecular layer; p, pia mater; wm, white matter.

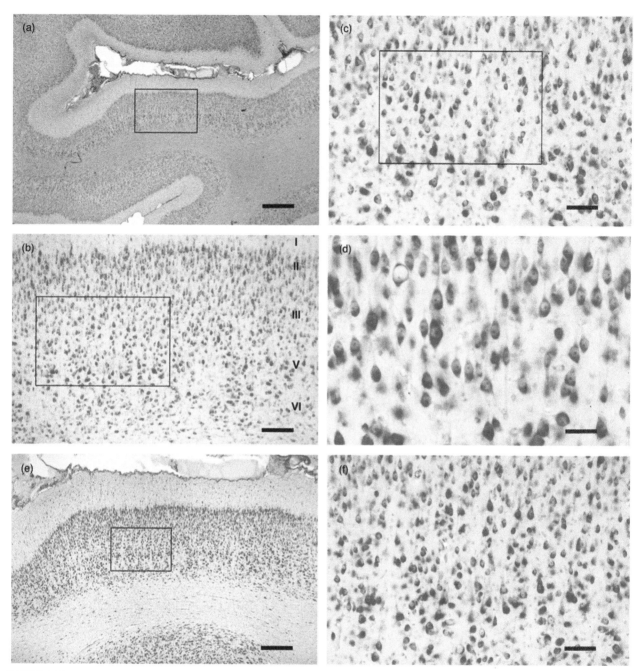

FIGURE 6.18   Auditory cortex of the common dolphin (Scales: a = 1000 μm; b = 250 μm; c = 100 μm; d = 50 μm; e = 500 μm; f = 100 μm). (a) Overview from the green box in Fig. 6.16; (b–d) cortical plate with a faint vertical parallel texture at different magnifications. (b) At lower magnification (box in a), layer V is obvious as a cluster of larger pyramidal neurons below the middle of the cortical plate. (c–d) the different layers are getting more and more distinct and many of the neuronal somata are bipolar with vertical processes. (d) This 20× magnification of auditory cortex differs markedly from the respective visual cortical detail in Fig. 6.17. In (d), the neurons are more slender, bipolar, and arranged in parallel. (e–f) in the vertical direction, the constituting cortical columns are obvious ('raindrop pattern' in humans). Cell density is higher and soma size tendentially smaller than in the other primary neocortices. (f) is magnified from the box in (e); here, many perikarya are pyramid-like, the smaller ones more like granular cells.

large size of the dolphin brain, the width of the neocortex is somewhat smaller than in terrestrial hoofed animals. Moreover, there is less regional variability across the telencephalic wall of dolphins and the layering pattern is less distinct, because, in most cases, a clear layer IV is lacking. In terrestrial mammals, this layer is often more or less granularized; that is, it may contain many small neurons with round perikarya and short processes. Whereas such neurons seem to be dispersed in smaller numbers across layers III to V, they can obviously be concentrated between layers III and V. Nickel et al. (2004) show a clear layer IV in the visual cortex of the pig and in the cingulate gyrus of cattle, and the visual cortices of the sheep

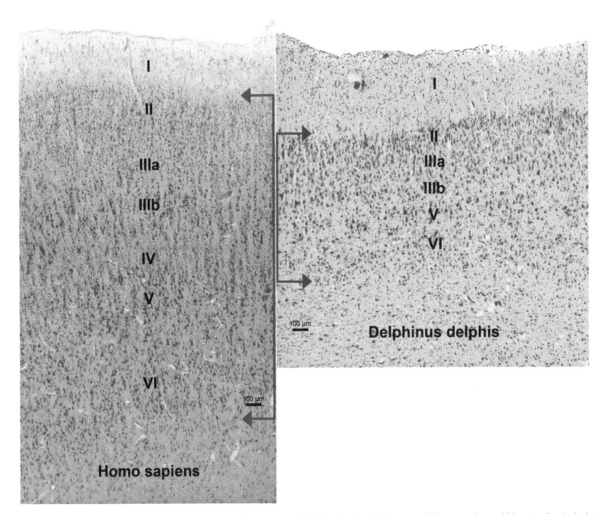

FIGURE 6.19 **Auditory cortices from the human and the common dolphin (Section thickness = 25 μm; scales = 100 μm).** Cortical plates are shown in preparations made after the same procedure. I–VI cortical layers. Layer IV is lacking in the dolphin cortex, which is much thinner than that in the human.

and some antelopes present the same picture (own observations). In primates, this fourth layer reaches its maximal development with innumerable very small multipolar neurons arranged in several sublayers (human). Earlier papers had often denied the existence of a layer IV in dolphins; however, there are indications that this layer may be present in the visual cortex in young postnatal dolphins but disappears later in the course of brain maturation (Garey and Leuba, 1986; Glezer and Morgane, 1990; Morgane et al., 1988).

Layer I as the *molecular layer* (consisting of a mass of unmyelinated fibers rich in synapses and poor in neurons) is relatively thick in dolphins; it comprises about one third of cortical gray thickness and shows about 10–18% of the neuron density in layer II. Layer I is regarded a major input recipient for thalamocortical fibers together with layer II as a counterpart to layer IV in many terrestrial mammals. In layers I and II the majority of all synapses (70%) were found. In dolphins, layer II is thus attributed to be the main relay element conveying information from the subcortical and intracortical afferent systems via layer I to the other cortical layers.

Layer II (*external granular layer* of terrestrial mammals) is thin but rich in densely packed small pyramid-like "everted neurons" and exhibits the highest neuron density among the cortical layers. In dolphins, layer II neurons were reported to account for at least 50% of all the neurons in a given cortical volume (Garey and Leuba, 1986). The "everted neurons" extend their dendrites obliquely into the molecular layer; that is, they show a clear predominance of apical (subpial) dendrites over basal dendrites.

Layer III (*external pyramidal layer*) is comparatively thick and characterized by a variety of pyramidal cells, which are smaller in the upper (outer) sublayer and large to very large in the inner sublayer.

Layer IV (*internal granular layer*) for the most part is not obvious in dolphins, in correlation with a general trend to moderate or even loss of granularization in these animals. It is not clear whether here neurons of layer IV have been displaced into the neighboring layers (III and V).

Layer V (*internal pyramidal layer*) is much narrower than layer III and contains large to very large pyramidal cells. The latter, however, may be challenged in size by those in layer III (inner sublayer) in some neocortical areas.

Layer VI is well developed in dolphins, sometimes shows various aggregations of neurons and fades out into the white matter.

In Figs. 6.16–6.18, cortical areas (boxed) from the rear parts of the visual and auditory primary projection fields are represented. They come from the middle of the common dolphin brain, a sectional plane through the posterior commissure (pc), elliptic nucleus (E), and the optic chiasm (oc) at the transition between the parietal and temporal lobes. Involving the electrophysiological results of Supin et al. (2001) they belong to the visual and auditory field, respectively. Histological analysis of the two boxed areas, however, show some deviation from the situation encountered in more anterior (rostral) areas of the brain (cf. the bottlenose dolphin in Fig. 6.20). In the common dolphin, the potential visual area (yellow box, Fig. 6.16) is characterized by stout and large pyramid-like somata whereas the potential auditory area (green box) shows more slender and largely bipolar neurons. These results indicate that, apart from some typical data on primary projection fields (Kern et al., 2011, van Kann et al., submitted, Knopf et al., in press), the dolphin neocortex is still largely unknown.

Neuron Density Although a number of papers have dealt with the quantitative morphology of the cetacean neocortex (eg, Butti et al., 2009; Kern et al., 2011; Walløe et al., 2010), our knowledge of it is still rather limited. Brains differ in size and cortical thickness; larger animals within a systematic group tend to have larger brains with a more convoluted cortex and a slightly thicker cortical gray. Allometric analyses of mammalian brains revealed that the percentage of cerebral cortex volume (gray and white matter) increases disproportionately with brain size (Hofman, 1989; Glezer et al., 1988; Manger, 2006; see also Oelschläger et al., 2010). In this respect, both odontocetes and simian primates exhibit corticalization indices (CI 1) higher than those in other mammals, and both on a similar level. In contrast, the percentage of cortical gray matter volume in total brain volume provides corticalization indices (CI 2) that are (a) distinctly lower and (b) separate the odontocetes from the simian primates. Obviously, there are two types of a maximal development of the neocortex, one seen in the odontocetes (maximal volume of white matter) and the other in simian primates including the human (maximal volume of gray matter). In other words, whereas the volume of the gray matter (per brain volume) is maximal in simian primates, it is distinctly lower in odontocetes, even lower than that in hoofed animals (Manger, 2006).

In several species of dolphins of different body sizes (common dolphin, bottlenose dolphin, pilot whale, false killer whale, killer whale), Poth et al. (2005) counted neurons below a standard (pial) surface in order to eliminate differences in cortical thickness. In the three primary cortical sensory areas (somatosensory, auditory, visual), they found that with increasing brain mass the subpial neuron number per cortical area unit decreased. Interestingly, in the auditory cortex, neuron numbers per subpial unit were not greater than in the somatosensory or visual cortex. Obviously the increase in functional significance in the auditory system of delphinids is associated with an enlargement of the sensory cortical area, and not necessarily with a higher neuron number per cortical unit (Huggenberger, 2008).

In the paper of Poth et al. (2005), the following data as to cortical (gray matter) thickness and cortical neuron density were presented: With increasing brain mass, total cortex thickness increased from 1.44 mm in the common dolphin (brain mass 835 g) to 2.05 mm in the largest delphinid, the killer whale (brain mass 6.052 g). The conversion of the subpial neuron counts to genuine neuron densities revealed values from 24,618 cells/mm^3 in the common dolphin via the bottlenose dolphin (brain mass 1302 g) with a cell count of 26,126 cells/mm^3 to 8,542 cells/mm^3 in the killer whale. In the human (brain mass 1350 g), cortex thickness was determined as 2.70 mm and neuron density as 44,000 cells/mm^3. Unfortunately, these values cannot strictly be compared to most data in the literature because of differences in methodology and there seem to exist variations between families of toothed whales. Thus, Kern et al. (2011) reported a lower average cortical neuron density (18,350/mm^3) for the smaller harbor porpoise with a brain mass of about 500 g than for the larger bottlenose dolphin (24,620/mm^3) with a brain mass of about 1800 g. Moreover, this study (Kern et al., 2011) revealed that, in contrast to the harbor porpoise, neuron density in the primary neocortical areas (M1, S1, A1, V1) of the bottlenose dolphin is always higher in layer III than in layer V and that neuron density in layer III is generally around 1.5 times higher in the bottlenose dolphin than in the harbor porpoise. The generally higher density in cortical layer III in the bottlenose dolphin suggests a higher tangential cortical connectivity of this layer in view of the more agile and complicated behavior of these gregarious animals in the open sea including versatile phonation via complex sound and ultrasound signals (whistles and click sounds).

In their neocortex, dolphins show a mixture of potential conservative and advanced features at the cortical architectonic level, as well as at neuronal and synaptic levels. When looking at the cortex as a whole, the aforementioned so-called conservative features occur mostly in the superficial layers I and II, whereas the deeper layers III–VI are characterized as more or less equivalent to those in advanced terrestrial mammals, with the exception of layer IV (see previous discussion). To date, adequate comparative data on ungulate groups or other aquatic mammals is not available, with the exception of the wild boar (*Sus scrofa*; van Kann et al., submitted). Recent analysis revealed that the total neuron densities of the primary neocortical projection fields (M1, S1, A1, V1) are always higher in dolphins than in the pig, with increasing values along this series, particularly in layer III. Thus, the highest density values are found in the visual cortex (V1). The density values for the human sensory neocortical fields are generally high, particularly in V1. Leaving aside the spiny stellate neurons of layer IV, however, the densities are generally similar in the human and dolphin, with the exception of the visual cortex (V1) where also layers III and V of the human show a maximal density among the three mammalian representatives and their sensory areas and layers (van Kann et al., submitted).

Synapses The synaptic parameters of the visual neocortex in dolphins show many of the qualitative and quantitative features of the generalized mammalian bauplan. For example, the quantitative relationships between the number of synapses contacting different components of cortical neurons in the dolphin such as perikaryon, dendritic shafts, and dendritic spines do not differ significantly from these parameters in most other mammals. The same holds true for the distribution of other synaptic parameters throughout the cortical layers such as the area and form factor of individual synaptic boutons, the length of active zones (densities) of synaptic membranes, and several parameters relating to synaptic vesicles (Glezer and Morgane, 1990; Morgane et al., 1988).

The majority of all synapses (70%) was found in cortical layers I and II. The latter seem to receive the bulk of cortical input as opposed to layer IV in many terrestrial mammals. Thus, in cetaceans, layer II seems to be the main relay element conveying information from the subcortical and intracortical afferents via layer I to the other cortical layers.

As to neuron density, the human shows higher values than the dolphin. However, with respect to the total number of synapses in their visual cortices (0.87×10^{14} vs 1.3×10^{14}), the dolphin and the human resemble each other more closely than other mammalian species (Glezer and Morgane, 1990). These data infer that information processing in dolphins may be as effective as in other progressive mammals (total number of synapses) but the number of neurophysiological units (neurons) is less and tend to be pyramidal cells. This kind of neuronal equipment seems to be adequate for quick and strong reactions but may be not so useful for delicate and complicated, more time-consuming integrative processes.

Glia There are only a few data available on gliocytes in the cetacean neocortex. In the bottlenose dolphin, glial density was found to vary in different cortical areas from 28,000 to 93,200 cells/mm^3, values rather similar to those in the human (average: 40,000–100,000 cells/mm^3) (Garey and Leuba, 1986). The number of gliocytes per number of neurons (glia/neuron ratio) is species-specific, and it varies among the mammalian groups and during ontogenesis in parallel to changes in neuron density. This basically implies that larger brains have higher glia/neuron ratios. Thus, the ratio rises from small rodents (mouse, rabbit: 0.35) via ungulate species (pig, cow, and horse: 1.1), the human (1.68–1.78), and the bottlenose dolphin (2–3.1), to large whales (fin whale: 4.54–5.85) (Oelschläger and Oelschläger, 2009). Also, the glia/neuron ratio within each species increases from birth to maturity, thereby signaling the importance of glia for growing neurons and, thus, for neocortical function. And this, again, implies the necessity of comparing only mature specimens. Accordingly, in species with similar neocortex volumes, the glia/neuron ratio may be interpreted both ways, that is, in favor of neurons and of glia, respectively, as both are of equal importance.

Comparison of Neocortex Areas

Fig. 6.20 exhibits a consequent comparison of the primary neocortical areas (M1, S1, A1, V1) with respect to low-magnification histology. Here the largely intermediate position of the pig between the bottlenose dolphin and the human is obvious.

In *Area M1* (Fig. 6.20a) no layer IV is obvious and a few very large neurons are present in layer V. In the human, they are called Betz cells. Because in the human the width of the cortical gray matter (cortical plate) is about the double that of the dolphin, layer VI is not exhibited in Fig. 6.20a.

Area S1 (Fig. 6.20b) shows the intermediate position of the pig as to the existence of a layer IV, which, however, is not very distinct here. The dolphin has no unequivocal layer IV and pyramid-like somata are dominating here. The human is again rather different from the other two mammals as to the thickness of the cortical plate, but also because of the high neuron density of the layers with a maximum in layer IV and a "raindrop pattern" (see the previous discussion) in layers III and IV.

Area A1 (Fig. 6.20c) exhibits rather different situations in the three species. Whereas in the dolphin this cortex is coarse, agranular (no typical layer IV), and characterized by large to very large pyramidal somata, the pig may exhibit a larger variety in neuron size and also no typical layer IV or "raindrop pattern." The latter is fully developed in the human together with a distinct general granularity and a characteristic layer IV.

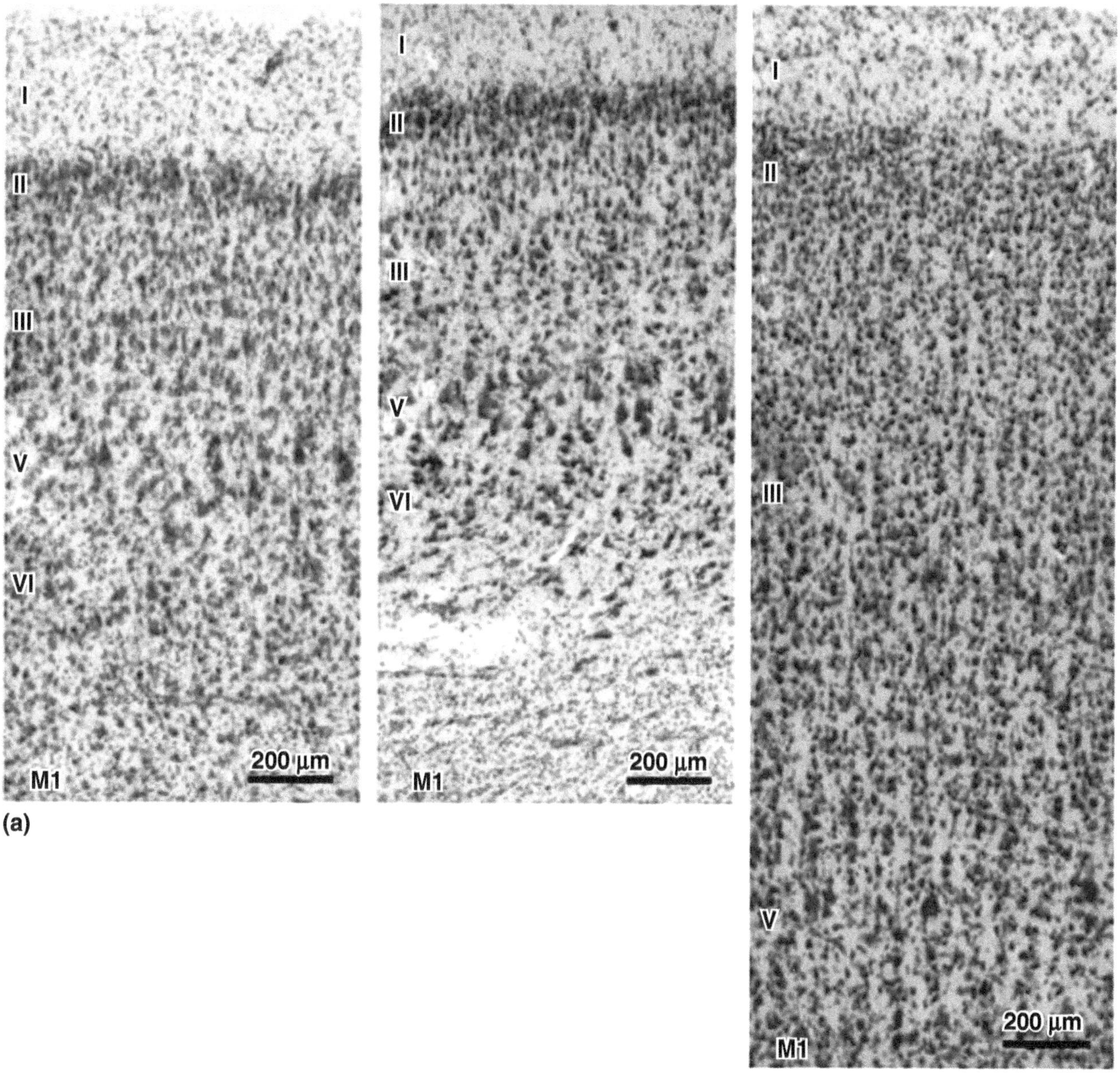

FIGURE 6.20   Comparison of the four primary neocortical areas in the bottlenose dolphin (left), the pig (middle), and the human (right). Layer VI in the human could not be shown in full extent because of the excessive thickness of the cortical plate. (a) Somatomotor cortex (M1). There is no layer IV and a few large to very large somata are to be seen. (b) Somatosensory cortex (S1). The pig and the human show the typical configuration of a koniocortex, with the layer IV much better developed in the human. In the dolphin, a pyramid-like neuron type is dominant; with a maximal soma size in layer V. (c) Auditory cortex (A1). The dolphin cortex is agranular (no layer IV) and characterized by small to very large pyramidal somata, particularly in layer V. The pig cortex is more uniform with only a few large perikarya in layer V. The human auditory cortex is an extreme koniocortex and mostly consists of small granular neurons: Layer IV is distinct and includes of even minute star-shaped (stellate) neurons. Throughout the layers II–VI, the columnar character of the cortex is evident, particularly in layers III and IV (raindrop pattern). (d) Visual cortex (V1). The visual cortex of the dolphin is rather similar to its auditory cortex; the somata are again coarse and agranular. The cortex of the pig is intermediate between the dolphin and the human: It is more homogeneous than in the dolphin and, at the same time, shows a faint layering pattern as in the human, including a layer IV. As its auditory cortex, the visual cortex of the human is koniocortical but there is no raindrop pattern and the layering is much more distinct. Layer IV shows a maximal development as to thickness and cell number; it is divided into four sublayers. I–VI, cortical layers I–VI. For more information, see text. *(From Kern, A., 2012. Der Neokortex der Säugetiere—Evolution und Funktion. Inaugural-Dissertation des Fachbereichs Medizin der Johann Wolfgang Goethe-Universität, Frankfurt am Main; and Kern et al., 2011. Stereology of the neocortex in Odontocetes: qualitative, quantitative, and functional implications. Brain Behav. Evol. 77, 79–90.)*

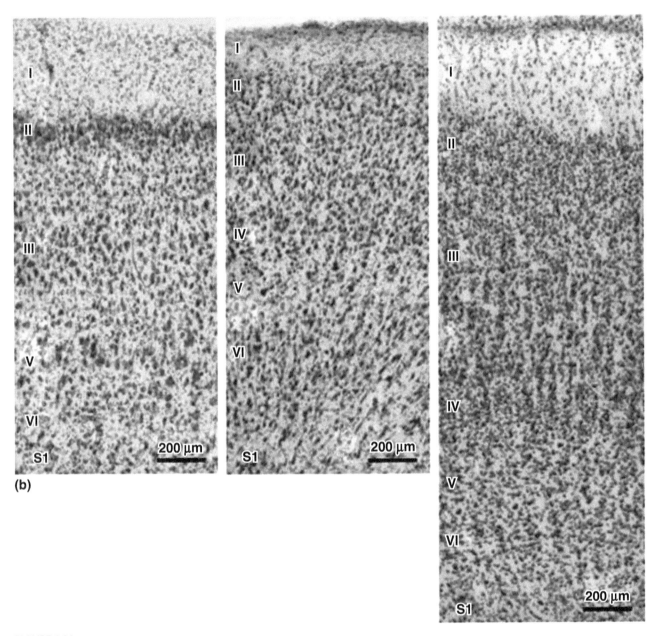

FIGURE 6.20   *(cont.)*

Area V1 (Fig. 6.20d) is characterized by graded changes from the dolphin to the pig and to the human. In the dolphin, the cortex is coarse and does not show a clear layering; the pyramidal neurons of layers III and V are dominant. The pig exhibits a more or less pronounced layering. However, a potential layer IV is obvious in some places (cf. van Kann et al., submitted). Interestingly, other artiodactyl hoofed animals (ie, cattle, *Bos taurus*; sheep, *Ovis aries*; blue duiker, *Philantomba* spec) have an area striata in the medial parietal cortex showing a clear layer IV. The paradigmatic situation in the visual cortex of Brodmann area 17[e] in the human is characterized by the extremely well-differentiated layer IV as a granular cortex or koniocortex (Greek *konios* = sand, grain): Here, the somata are so small and rounded that they evoke the picture of a track of fine sand or dust under the microscope.

---

e. Brodmann areas (Brodmann, 1909) are regions of the cerebral cortex, in the human or other primate brains, defined by their cytoarchitecture. Today it is known that these areas have their specific functional implications.

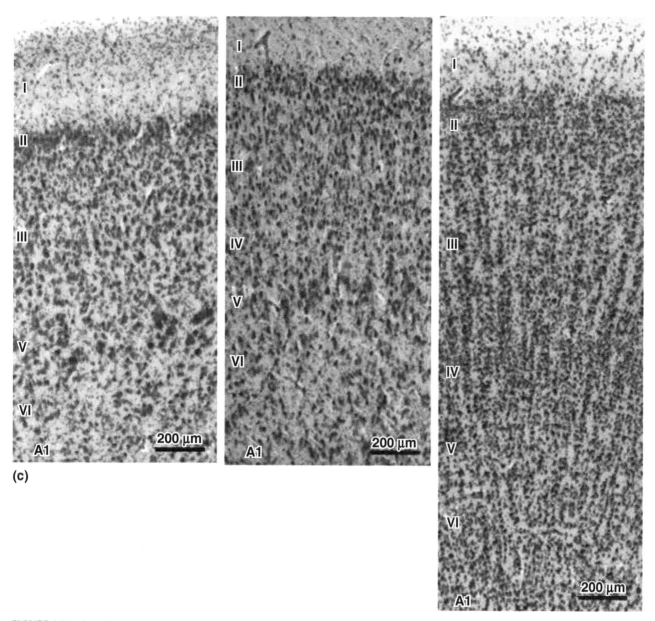

FIGURE 6.20   (*cont.*)

In principle, the neurons in the neocortex of mammals belong to several characteristic types, most obvious the pyramidal neurons, which are particularly large in layers III and V. Pyramidal neurons are mostly excitatory and exhibit spinous dendrites. In dolphins, many pyramidal neurons are atypical (Figs. 6.21 and 6.22a), for example, the small so-called "extraverted neurons" in layer II, which have weak basal dendrites and strong apical dendrites that widely spread their branches in an oblique direction into the molecular layer (Fig. 6.21, EX). Fundamental investigations on the neuron morphology of dolphins have been carried out by Glezer and Morgane (1990), who used Golgi silver stain that impregnates only single neurons but completely, as well as electron microscopy for the analysis of synaptic connectivity. Other methods for the identification of neuron classes (interneurons) in the dolphin neocortex used the immunocytochemistry of neurotransmitters like, for example, calcium-binding proteins (Parvalbumin, Calbindin, Calretinin), or cytochrome oxidase (Glezer et al., 1998; Hof et al., 2000; Revishchin and Garey, 1991).

Nonpyramidal cells represent the other main class of cortical neurons seen in the heterolaminar cortex (layer IV present) of the visual field which is located in the lateral gyrus of dolphins (Fig. 6.22b). However, those are relatively few in number and are mostly of the large stellate type. Small multipolar stellate cells are seen mainly in the incipient layer IV

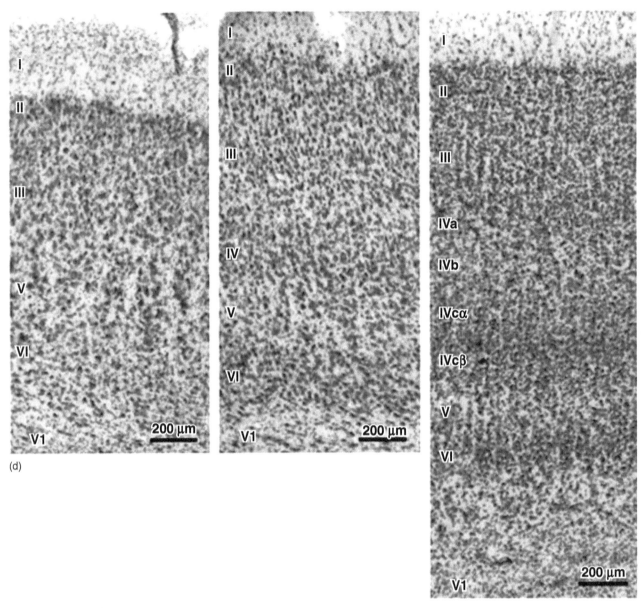

FIGURE 6.20 (*cont.*)

and are mostly aspinous or poorly spinous nonpyramidal cells. Obviously, both the pyramidal and nonpyramidal neurons are characterized by recurrent ascending branches to layer I, which presumably is the main input layer for intracortical and subcortical inputs in these animals.

An interesting functional aspect of interneurons is their inhibitory influence. They are responsible for intralaminar (horizontal) inhibition or for intracolumnar (vertical) inhibition of functionally related pyramidal neurons. Another functional property of interneurons is the inhibition of other inhibiting interneurons, thus leading to a disinhibition of directly related (excitatory) pyramidal neurons. By this short-term disinhibition, inhibited pyramidal cells can be activated indirectly and thus send slightly delayed excitatory, that is, outgoing (efferent) impulses in the form of temporal and spatial patterns.

Neocortex Function in Dolphins

The neurobiology of the dolphin neocortex has been investigated in many publications. Due to some similarities with "primitive" (plesiomorphic) mammals (insectivores), the cortex of dolphins was regarded to represent an inferior stage of development (Glezer et al., 1988). Recently, Kern et al. (2011) have presented data that characterize the dolphin cortex as highly evolved and extremely specialized within the mammalia.

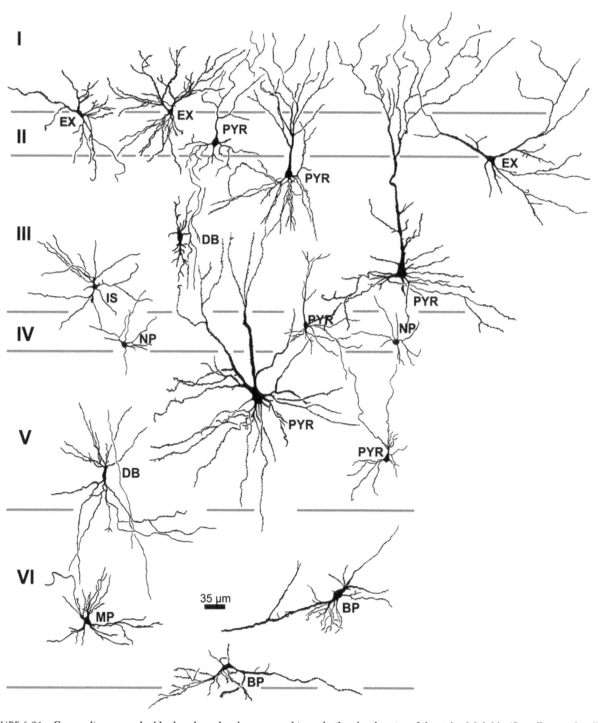

FIGURE 6.21   Composite camera lucida drawings showing neuronal types in the visual cortex of the striped dolphin (*Stenella coeruleoalba*).
This cortex is called heterolaminar because a weak layer IV is present, as an exception among the dolphin cortices. In layer II, extraverted neurons (EX) are shown with widespreading spinous apical dendrites. Note the extension of these dendrites into layer I. Layer III is characterized by the presence of atypical pyramidal neurons (PYR) of different shapes and sizes. Also, a nonpyramidal isodendritic neuron (IS) as well as a double bouquet (DB) cells is shown. Neurons in incipient layer IV are mostly aspinous or poorly spinous nonpyramidal cells (NP). In layer V, a large atypical pyramidal cell is shown (PYR); neurons of this type often have a branching apical dendrite, which usually extends to layer I. In this layer, also a smaller atypical pyramidal cell (PYR) is shown as well as a double bouquet (DB) neuron. Layer VI is characterized by many irregular neurons; they are often more or less bipolar (BP) and extend horizontally or sometimes vertically. MP, multipolar neuron Rapid Golgi impregnation. *(Redrawn by Jutta Oelschläger from Glezer and Morgane, 1990. Ultrastructure of synapses and Golgi analysis of neurons in neocortex of the lateral gyrus (visual cortex) of the dolphin and pilot whale. Brain Res. Bull. 24, 401–427.)*

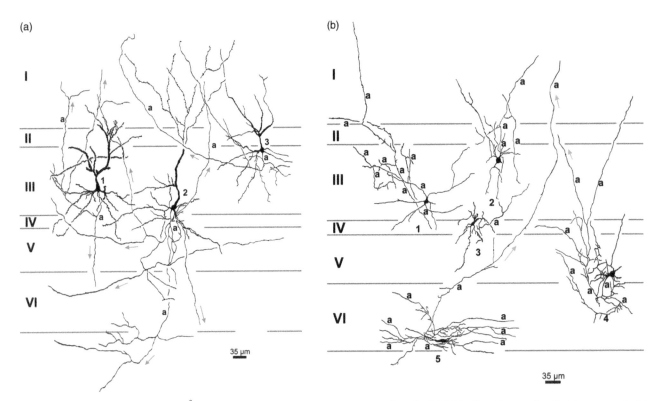

FIGURE 6.22 Camera lucida drawings[f] of characteristic neurons in the visual cortex of a striped dolphin (*Stenella coeruleoalba*), impregnated with the Rapid Golgi method, a silver stain. (a) Atypical pyramidal neurons (1–3) in layers II and III with richly branched axons. Many of these axons send ramifications (collaterals, *blue arrows*) to layer I above and/or in the direction of the white matter below; they also may extend horizontally into layers IV–VI. In dolphins, layer IV is narrow and mostly vague to nonexistent morphologically. (b) Nonpyramidal neurons in layers III–VI. They are large to medium-sized interneurons with a rich local branching of their axons. Each of the cells, however, also sends a long recurrent (ascending) collateral to layer I (molecular layer). The latter is the main recipient of incoming visual input from the diencephalon. All the interneurons 1–2 and 4–5 are isodendritic large stellate cells; interneuron 3 is a stellate cell. In visually oriented terrestrial mammals (eg, primates, hoofed animals), stellate cells may be very small (granular) and extremely numerous. Obviously, they are responsible for "fine-grain" analysis of visual stimuli. (a) and (b) demonstrate that in the dolphin visual neocortex both pyramidal and nonpyramidal neurons send their axonal collaterals or even entire axonal stem processes to layer I, which presumably is the main input layer for intra/subcortical inputs. *(Modified by Jutta Oelschläger from Glezer and Morgane, 1990. Ultrastructure of synapses and Golgi analysis of neurons in neocortex of the lateral gyrus (visual cortex) of the dolphin and pilot whale. Brain Res. Bull. 24, 401–427.)*

The major part of thalamic input (neural equivalents of signals from the body surface and sensory organs) into the dolphin neocortex enters the molecular layer via tangential fibers (Fig. 6.23, right). These thalamocortical fibers, generally, are excitatory. Within the neocortex, their information is mainly received by the characteristic "everted neurons" of layer II, which send their climbing dendrites into the molecular layer (as do the large pyramids in deeper layers; see Figs. 6.21 and 6.22a). The excitatory axons of the "everted neurons" descend vertically into the cortex and distribute their activating signals among deeper neurons and layers and additionally involve interneurons. Mechanisms of excitation and inhibition lead to short-term activation of large pyramidal neurons in layers III, V, and VI, which project to other brain regions. By this, adequate reactions of the dolphin to signals received from the environment are initiated.

The appreciation of the morphological and physiological information known on dolphin cortex neurobiology leads to the following conclusion: In contrast to the situation in primates (Fig. 6.23, left), where incoming information is analyzed intensively in a rather demanding and complicated operation by large numbers of single components (stellate cells, columns, local cortical loops), dolphins (toothed whales) seem to be characterized by a rather condensed way of information processing. In primates, the information is transmitted from layer IV upward to layers III and II for intrinsic analysis and then passed on to pyramidal cells in layers III and V for corticofugal projection. Dolphins, however, exhibit a rather immediate downward- cascading from tangential fiber systems in layer I via everted neurons in layer II to layers III and V, where pyramidal cells project corticofugally. Thus, for the most part, dolphins seem to be tuned to coarse and quick information processing in a thinner cortex.

---

f. The camera lucida is an optic tool and still used as the most common method among neurobiologists for drawing brain structures.

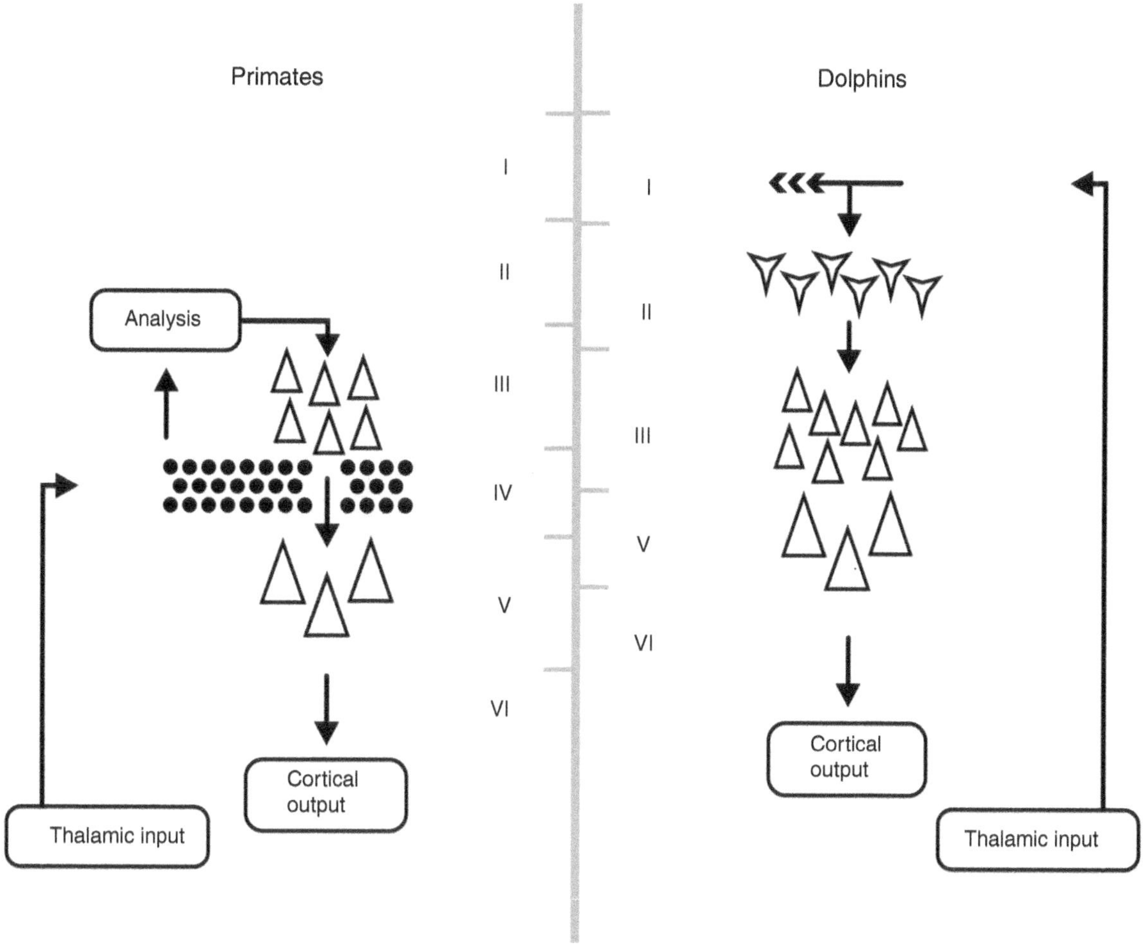

FIGURE 6.23   Simplified scheme of neocortical function in primates (left) and dolphins (right). (Left) Six-layered koniocortex of primates. Ascending (afferent) thalamic input from the sensory organs enters layer IV (granular layer; stellate cells, black dots). Myriads of stellate cells send the information to layers II and III for elaborate analysis. Descending (efferent) projections start predominantly from pyramidal neurons in layers III and V. (Right) Five-layered pyramidal cortex of dolphins. The incoming information from the thalamus enters the cortex at layer I (molecular layer) and proceeds first tangentially (series of *arrowheads*). Here the input is captured by so-called extraverted neurons in layer II (inverted triangles) and transferred to the layer III/layer V complex where descending (efferent) projections originate. Note that cortex function in primates seems to represent the left, fine-grain type of information processing, particularly in the visual cortex. The dolphin cortex instead is characterized by comparatively smaller numbers of larger and mostly pyramidal neurons. This cortex may represent a more coarse type of unidirectional information processing, with a mass activation of pyramidal neurons (no granular layer IV). It seems to be characterized by a less elaborate but faster information processing needed in water with a sound conduction velocity of about 1500 m/sec. I–VI: layers of cortex I bis VI. *(Modified by Jutta Oelschläger from Kern et al., 2012. Der Neokortex der Säugetiere— Evolution und Funktion. Inaugural—Dissertation des Fachbereichs Medizin der Johann Wolfgang Goethe-Universität, Frankfurt am Main.)*

Allocortex

The second major formation of cortex in the mammalian telencephalon is called allocortex (állos = strange, other). In contrast to the neocortex or isocortex, which is characterized by a six-layered architecture, the components of the allocortex (archicortex and paleocortex) are rather heterogeneous (Fig. 6.24). They comprise three- to five-layered areas in different regions of the telencephalic hemisphere. The paleocortex is situated at the rostral base of the hemisphere and the archicortex for the most part in the medial wall of the temporal lobe. They contact each other in the septal area rostroventrally and at the tip of the temporal lobe. Whereas the paleocortex is correlated to olfactory function in mammals, the archicortex is dedicated to the collection and integration of multisensory input and its conversion into top-down goal-directed motor behavior.

Paleocortex

Adult dolphins lack olfactory epithelia, nerves, bulbs, and peduncles. Only in embryos and early fetuses, these primary paleocortical structures are present but get reduced in later ontogenetic stages (see Chapter 5, Section "Olfaction and

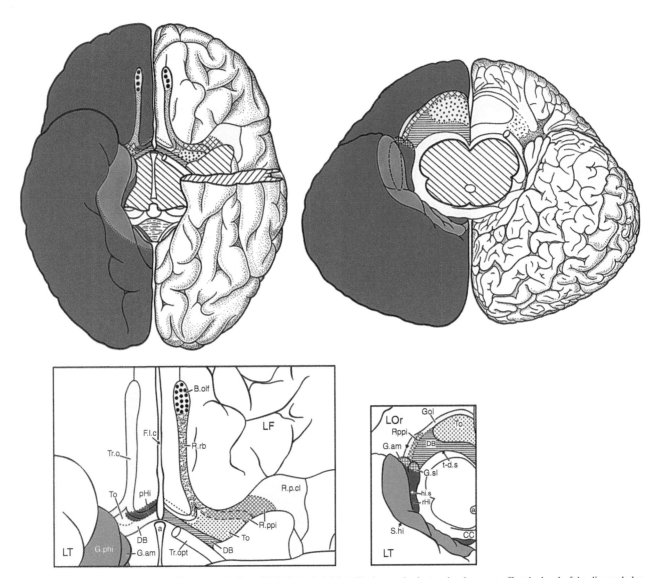

FIGURE 6.24 (Top) Basal aspects of the human (left) and dolphin brain (right). The human brainstem has been cut off at the level of the diencephalon, the dolphin brain at the level of the midbrain. (Cortices: *red*, neocortex; *yellow*, paleocortex; *dark blue*, archicortex; *middle blue*, periarchicortex). In the human, the anterior half of the left temporal lobe has been omitted to show the paleocortical formation (enlarged and labeled below). In the dolphin, the paleocortex is largely exposed and also enlarged and labeled below. In both brains, the archicortices are presented as to their basal aspects. Whereas the paleocortical formation is relatively small but complete in the human; its anterior part (olfactory bulb and tract) has been reduced in the dolphin. Here, the olfactory tubercle, diagonal band, and prepiriform cortex and so forth are present but their functional properties are unclear. Abbreviations: a, aqueduct; B.olf, olfactory bulb; CC, corpus callosum; DB, diagonal band of Broca; F.l.c, fissura longitudinalis cerebri; G.am, gyrus ambiens; G.phi, gyrus parahippocampalis; G.sl, gyrus semilunaris; LF, frontal lobe; LOr, orbital lobe; LT, temporal lobe; pHi, precommissural hippocampus; rHi, retrocommissural hippocampus; R.p.cl, regio peripalaeocorticalis claustralis; R.ppi, prepiriform region; R.rb, retrobulbary region; S.hi, sulcus hippocampi; t-d.s, telo-diencephalic sulcus; Tr.o, olfactory tract; Tr.opt, optic tract. *(Redrawn by Jutta Oelschläger from Morgane et al., 1980. The anatomy of the brain of the bottlenose dolphin* (Tursiops truncatus). *Surface configurations of the telencephalon of the bottlenose dolphin with comparative anatomical observations in four other cetacean species. Brain Res Bull 5, 1–107; and Stephan, 1975. Allocortex, Handbuch der Mikroskopischen Anatomie des Menschen. Springer-Verlag, Berlin.)*

Sense of Taste", and Chapter 6, Section "Cranial Nerve" 1—"Olfactory Nerve"). Therefore, adult animals do not possess a regular olfactory sense. Nevertheless, the secondary olfactory structures known from terrestrial mammals can still be found in dolphin brains: (1) olfactory tubercle, (2) diagonal band of Broca, (3) prepiriform cortex, and (4) periamygdalar area (Fig. 6.24) (Morgane et al., 1986, 1980; Morgane and Jacobs, 1972, 1986). The histological characteristics of these structures are rather different but are characterized by three to five cortical layers.

The olfactory tubercle seems to be well developed. In many cases, it cannot be separated clearly from the neighboring diagonal band and prepiriform cortex, together appearing as a configuration often called "olfactory lobe" (Figs. 6.7d and 6.24) (Morgane and Jacobs, 1972). In reality, it is the large corpus striatum (see Section "Basal Ganglia" later in this chapter) that bulges the basal surface of the brain as an hourglass-shaped "olfactory tubercle." In histological sections, the striatum is covered by an incomplete layer of thin paleocortex, a situation more or less similar to that in humans. However, in some places, the cortex of the olfactory tubercle shows characteristic "islands of Calleja." Within the paleocortical formation, the diagonal band is also rather extended. The prepiriform cortex of the dolphin appears as a narrow gyrus surrounding the olfactory tubercle and diagonal band rostrally and laterally. Jacobs et al. (1971) addressed this gyrus as a trajectory of the lateral olfactory tract (Voogd et al., 1998). It continues in the direction of the temporal lobe along the transverse insular gyrus and ends at the periamygdalar region (gyrus ambiens). The periamygdalar area, as the free paleocortical surface of the amygdala, is adequately developed in dolphins with respect to the situation in the human. Interestingly, the amygdala as a whole seems to have an even larger size in dolphins than in primates (Schwerdtfeger et al., 1984).

The secondary olfactory structures of dolphins are comparatively small in area and volume. In principle, they correspond well with their homologs in the human, both in topography and size dimensions. Therefore it was suspected that, in dolphins, the secondary olfactory structures are associated with nonolfactory functions (as perhaps in terrestrial mammals). Filimonoff (1966) had already argued that the secondary olfactory structures in dolphins are remnants of originally multifunctional character, but after the loss of the sense of olfaction. As in other mammals, such indirect nonolfactory connections may exist in dolphins between the olfactory tubercle and other paleocortical areas on the one hand and the parahippocampal gyrus (entorhinal area) and anterior insula on the other. Reciprocal connections between these areas on both sides may exist via the anterior commissure, which is rather thin. Additional structures with connections to paleocortical structures may comprise the cingulate gyrus, thalamus, hypothalamus, and the brainstem (central gray) (Benninghoff and Drenckhahn, 2004).

Archicortex

The archicortex is a central part of the limbic system. It extends along the medial margin (limbus) of the telencephalic hemisphere in the area of the corpus callosum and the medial temporal lobe, that is, from the septal area to the end of the hippocampus (Fig. 6.57: SA, H) and including the dentate gyrus (Schaller 1992). The hippocampus itself is often called "Ammon's horn" because of its three-dimensional configuration. In adult terrestrial mammals, this cortex areal is hidden for the most part in the depth of the medial temporal lobe. Here, it receives visual, auditory, somatosensory, and visceral inputs that are already highly preprocessed, but only indirectly olfactory information. Via its connections, for example, with the hypothalamus, the septal nuclei, and the cingulate gyrus, the hippocampus influences the endocrine, visceral, and emotional processing in the brain and is crucial for the aspects of learning and memory.

During ontogenesis, most of the archicortex is shifted from the dorsomedial wall of the embryonic telencephalic hemisphere in a half-circle around a transverse axis through the insular cortex as a rotation center. This shift occurs during the so-called "rotation of the hemisphere", which is caused by the extreme expansion of the neocortex and leads to the formation of the temporal lobe. In contrast to other mammals, the archicortex of dolphins is not really displaced from the surface of the telencephalon: Here, Ammon's horn is not curled in the inferior horn of the ventricular system but lies more or less flat at the medial surface of the temporal lobe. The fornix as the main efferent fiber tract of the hippocampus illustrates this embryonic and evolutionary shift of the archicortex: It runs from the hippocampus in a spiral along the ventral surface of the corpus callosum, exchanges fibers with its counterpart and enters the hypothalamus to terminate in the mamillary body (*inner archicortical circle*). Vestigial archicortical material (indusium griseum, syn. gray veil) is always found on the dorsal surface of the corpus callosum as a thin sheet, together with two thin fiber tracts (striae longitudinales). Caudally, the indusium becomes thicker and parallels the fornix in the direction of the temporal lobe. It ends adjacent to the hippocampus in a narrow archicortical formation called "dentate gyrus" because of its serial indentations (*outer archicortical circle*).

There are some characteristics of the dolphin limbic system that may help to understand their specific adaptations as well as their neurobiological capabilities. First of all, it is the particularly small size of the hippocampal formation that was realized by earlier investigators. Thus, Filimonoff (1966; cited after Morgane and Jacobs, 1972) reported that in the common dolphin only 0.8% of the total telencephalic cortical surface is comprised of archicortex, instead of 2.2% in the human. For more information, see Section "Limbic System" later in this chapter.

Periarchicortex

Immediately adjacent to the outer archicortical circle runs the third ring of the cortical limbic system, this time above the sulcus of corpus callosum, from the frontal to the temporal lobe. This long gyrus consists of the cingulate (supracallosal)

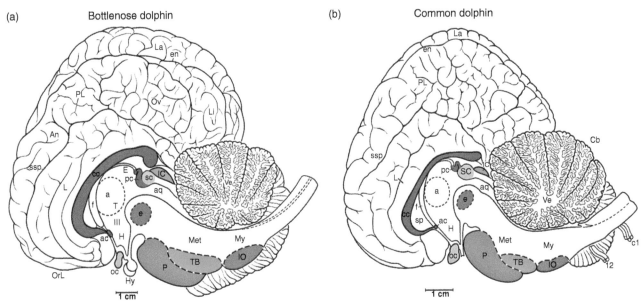

(a) Bottlenose dolphin

(b) Common dolphin

FIGURE 6.25 **Mediosagittal aspects of the brains of the bottlenose dolphin brain (left) and the common dolphin brain (right) (Scales = 1 cm).** Apart from their difference in size (1400 vs 800 g), the two brains are very similar. Brought to the same size, however, they differ in some characters. Thus, the bottlenose dolphin brain as a whole seems to be more stout, the telencephalic hemisphere as well as the metencephalon (cerebellum/pons) region more voluminous. The corpus callosum (cc, *red*) is slightly longer in the bottlenose dolphin but much thicker, which is enigmatic. Within the neocortical configuration, the entolateral sulcus is transferred more onto the medial side of the hemisphere; thus, the supralimbic lobe (La) seems to be more pronounced in the bottlenose dolphin. Some of the more important landmarks in the two brains are given in colors. Abbreviations: a, interthalamic adhesion; ac, anterior commissure; An, anterior lobule; aq, cerebral aqueduct; c1, first cervical nerve; Cb, cerebellum; cc, corpus callosum; e, elliptic nucleus; E, epithalamus; en, entolateral sulcus; f, fornix; H, hypothalamus; Hy, hypophysis; IC, caudal (inferior) colliculus; IO, inferior olive; L, limbic lobe; La, lateral gyrus (the dorsomedialmost part of the supralimbic lobe; cf. Figure 6.13); Li, lingual lobule; Met, metencephalon; My, myelencephalon; oc, optic chiasm; OrL, orbital lobe; Ov, oval lobule; P, pons; pc, posterior commissure; PL, paralimbic lobe; SC, rostral (superior) colliculus; sp, septum pellucidum; ssp, suprasplenial (limbic) sulcus; T, thalamus; TB, trapezoid body; Ve, vermis; 12, hypoglossal nerve; III, third ventricle. *(a, redrawn by Jutta Oelschläger from Langworthy, 1931. A description of the central nervous system of the porpoise* (Tursiops truncatus). *J Comp Neurol 54,437–499; modified after Morgane and Jacobs, 1972. Comparative anatomy of the cetacean nervous system. In: Harrison, R.J. (Ed.), Functional Anatomy of Marine Mammals. Academic Press, Londonl and Pilleri and Gihr, 1970. The central nervous system of the mysticete and odentocete whales. In: Pilleri, G. (Ed.), Investigations on Cetacea. Institute of Brain Anatomy, Bern; and Morgane and Jacobs, 1972. Comparative anatomy of the cetacean nervous system. In: Harrison, R.J. (Ed.), Functional Anatomy of Marine Mammals. Academic Press, London.) (b, modified by Jutta Oelschläger from Pilleri et al., 1980. Concise macroscopical atlas of the brain of the common dolphin* (Delphinus delphis; Linnaeus, 1758). *Brain Anatomy Institute, University of Berne, Waldau-Berne, Switzerland.)*

lobule anteriorly, the retrosplenial lobule posteriorly, and the parahippocampal lobule ventrally (cf. Fig. 6.13). In its entity, the long gyrus was called 'limbic lobe' by Peter Morgane and coworkers (Morgane et al., 1980) in dolphins and periarchicortex by Heinz Stephan (Starck, 1979; Stephan, 1975) in primates and other mammals. It represents an intermediate zone between the archicortex with three cortical layers and the neocortex with five to six layers. The periarchicortex, for the most part, comprises five layers (Morgane et al., 1982), with the exception of the entorhinal cortex; the latter exhibits at least 6-7 layers not being equivalent to those in the neocortex.

In dolphins, the cingulate gyrus (cingulum = belt; in Fig. 6.25 labeled as Limbic lobe (L) after Morgane and Jacobs), located in front and above the corpus callosum is much better developed than the parahippocampal gyrus below the corpus callosum, a situation also seen in artiodactyl species (Igarashi and Kamiya, 1972). Also, the dolphin cingulate gyrus (L) seems to be much more expanded than that in the human and may show up to two so-called 'intercalate' sulci.

Interestingly, in terrestrial mammals, the cingulate gyrus is rather consistent as to its topographical relations to both the hippocampal formation as well as the corpus callosum (see Section "Limbic System"). At the same time, there seems to be good correspondence concerning functional implications of this area throughout rodents and primates. The anterior (agranular) cingulate cortex (ACC) was reported to be engaged in cognitive and motor control of behavior. It receives input from the hippocampus, the amygdala, the septal area, and the thalamic limbic nuclei (Benninghoff and Drenckhahn, 2004). Efferent projections involve the motor and premotor cortex, the basal ganglia, the rostral (superior) colliculus, as well as the pontine nuclei and cerebellum, that is, the main elements of the pyramidal and extrapyramidal motor systems. The ACC is also closely connected to the posterior (granular) cingulate cortex (PCC), which is more sensory related, that is, it receives input from somatosensory, auditory, and visual association fields as well as from the visual thalamus and visual cortex. Both the ACC and PCC are closely associated with the so-called Papez circuit. The latter is regarded a representative circuitry (loop) within the limbic system and involved in all aspects of limbic information processing as, for example, orientation, learning, memory, emotion as well as the vegetative control of external cues from somatosensory, auditory, and visual association areas.

Within the parahippocampal gyrus, the entorhinal cortex is the most important entry for projections into the limbic system. In terrestrial mammals, they come from primary olfactory centers, the amygdala, and various neocortical areas. Efferent projections leave the entorhinal cortex mainly via the perforant tract and enter the hippocampal formation between the dentate gyrus and Ammon's horn. Practically all efferent fibers assemble in the alveus and fimbria hippocampi and leave the hippocampal formation via the fornix. The precommissural part of the fornix (rostral to the anterior commissure) runs to the septal area, the postcommissural part mainly to the medial nucleus of the mamillary body. Via the fibers of the mamillothalamic tract (bundle of Vicq d'Azyr), the latter projects to the anterior thalamic nucleus. From here the Papez circuit continues via the cingulate gyrus and its fiber tract (cingulum) back to the hippocampus. In the limbic system of mammals, this circuit is most prominent. It shows size correlations between the hippocampus, postcommissural fornix, and mamillary body (but not with the cingulate cortex or limbic lobe; see previous discussion). In artiodactyls, these structures are well developed, whereas in dolphins they seem to be more or less extensively reduced with the exception of the anterior cingulate gyrus.

Commissures

Commissures are fiber tracts that are defined to connect corresponding cortical areas of both telencephalic hemispheres with each other. They comprise the anterior commissure, the hippocampal commissure, and the corpus callosum. The size of the commissural systems is obviously specific for dolphins (Figs. 6.6c, 6.12, 6.13, 6.16, 6.25, and 6.57).

The *anterior commissure* (Fig. 6.7) connects neocortical and paleocortical areas of both temporal lobes. In dolphins it is rather thin, presumably due to the strong reduction of the olfactory system.

The *hippocampal commissure* (not shown) is relatively weak in dolphins due to the general reduction of the archicortex; it consists of fibers that are exchanged between the two fornices immediately below the corpus callosum. From here, on their way to the hippocampus in the temporal lobe, the fornices are called fimbriae.

The *corpus callosum* (cc) connects exclusively neocortical areas with each other. Near the mediosagittal plane, above the lateral ventricles, it forms a solid plate consisting of transverse nerve fiber bundles. To the sides, these fibers diverge in the direction of their target areas and form the three-dimensional radiation of corpus callosum. Rostrally on both sides, in the direction of the frontal lobes, the corpus callosum is extended into the smaller anterior forceps; caudally, in the direction of the temporal lobes, it extends into the larger posterior forceps. In terrestrial mammals, the size of the sagittal cross-sectional area of the corpus callosum is roughly correlated with the volume of the neocortex and thus the brain as a whole. More specifically, the midsagittal area of the corpus callosum depends on the number of neurons, which send transcallosal projections to the contralateral hemisphere. In dolphins and other toothed whales, the corpus callosum is relatively small and sometimes very thin with respect to brain size in comparison with other large mammals of similar body dimension (Fig. 6.7) (Tarpley and Ridgway, 1994). And there is an inverse relationship between the thickness of the cowrpus callosum plate and brain mass in dolphins; thus, the midsagittal sectional area decreases in larger-brained whales, indicating a reduced interhemispheral connectivity. In Fig. 6.25, the bottlenose dolphin seems to be an exception to this rule: its much larger brain seems to have a comparatively thicker corpus callosum than that of the smaller common dolphin brain.

Such differences in the interhemispheric connectivity via the corpus callosum are also obvious in the comparison of the brain of the largest delphinid, the killer whale, with that of the human. Whereas the whale and the human exhibit about the same midsagittal cross-sectional area of the corpus callosum, the brain of the killer whale is some five times heavier than that of the human.

As to regional differences of the corpus callosum in dolphins, there obviously is much variation (Fig. 6.25). In the bottlenose dolphin, the anterior and posterior parts (genu and splenium) are slightly thicker than the middle part (trunk). In the common dolphin, this tendency is much stronger, and the corpus callosum as a whole is extremely thin. Interestingly, the largest population of fibers with a thickness more than 5 μm was found in the posterior part of the corpus callosum and between the anterior third and anterior middle part of the corpus callosum (Keogh and Ridgway, 2008; Oelschläger et al., 2010). The latter region of the corpus callosum is anatomically close to the parietal lobes of the brain, which contain the primary auditory and visual functional cortices (Morgane et al., 1988). As to the auditory fibers crossing in that region, this observation, at first glance, is surprising because the surface area of the accessory primary auditory cortex (A1) is about as large as the other primary cortices (M1, S1, V1) together. However, in mammals, the auditory pathway from the cochlear nuclei to the thalamus runs both contralaterally and ipsilaterally (see Section "Auditory System" later in this chapter).

Thus, as to functional aspects, a relatively low number of auditory fibers crossing in the corpus callosum would make sense because much of the auditory computation is done subcortically (see Section "Inferior Colliculus").

From the standpoint of comparative and functional neurobiology, the small size of the corpus callosum and thus a weak linkage between both telencephalic hemispheres seems to indicate a relatively larger independence of each hemisphere from the other. This could also be favorable for the exotic kind of unihemispheric sleep, which seems to be typical for dolphins and other aquatic mammals. In electroencephalographic experiments, sleeping bottlenose dolphins have been reported to show signs of wakefulness in one hemisphere and sleep in the opposite hemisphere (unihemispheric short-wave sleep) (Lyamin et al., 2008; Ridgway, 2002, 1990; Manger et al., 2003; Ridgway et al., 2006b). The possibility of intermittent and alternate sleep for the one hemisphere and then the other could explain the ability of the bottlenose dolphin to maintain auditory vigilance for five continuous days (Ridgway et al., 2006a, 2009).

The *posterior commissure* (Commissura epithalamica; Figs. 6.16, 6.25, and 6.26) is not a commissure in the strict sense. Located in the diencephalon, it relates the two nuclei of the posterior commissure. They are situated below the pretectal area and at the sides of the anterior mesencephalic central gray. In the dolphin, both the nuclei and the commissure seem to be well developed. They have connections with the rostral (superior) colliculus, the interstitial nucleus of Cajal, and the elliptic nucleus (see Section "Mesencephalon" later in this chapter), the cerebellar nuclei, the mesencephalic and pontine reticular formation as well as the spinal cord. In dolphins, the posterior commissure is integrated into the commissural complex (cf. Oelschläger et al., 2008; Oelschläger and Kemp, 1998; see also Figs. 6.9, 6.16, 6.26, and 6.27), which is double hook-shaped in the sagittal section and also includes the habenular commissure. The latter is thin and situated slightly dorsorostral to the posterior commissure. The posterior commissure seems to be enlarged caudally by a rather thick transverse fiber bundle, which is triangular in cross section and joined by the distinct commissure of the superior colliculi.

DIENCEPHALON

The diencephalon of dolphins is highly voluminous and consists of four major parts, the small epithalamus, the very large thalamic complex, the subthalamus, and the hypothalamus. There are many correspondences with other mammals as to the structure of the dolphin diencephalon but also some features characteristic for toothed whales (Jansen and Jansen, 1969). In the mediosagittal aspect the shape of the diencephalon is often rather wedge-like in adult cetaceans, with the hypothalamus bending slightly caudally and tapering in the direction of the hypophysis. In the transverse section, the lateral extension of the thalami contributes much to the extraordinary width of the dolphin brain and there is no tapering but a gradual arching of its ventral border (optic tracts) in the direction of the optic chiasm (Fig. 6.16).

Epithalamus

This dorsal part of the diencephalon comprises the habenular complex, the pineal region, the pretectal area, and the paraventricular nucleus (Fig. 6.26). The habenular complex as a link between the basal ganglia and the limbic system is well developed, and both complexes are interconnected with each other by the habenular commissure. In terrestrial mammals, the habenulae receive input from the olfactory cortex, the preoptic area, and the amygdaloid nucleus via the stria medullaris thalami (Benninghoff and Drenckhahn, 2004). In dolphins, however, no functional olfactory structures are obvious and it is not known which kind of afferent information is conveyed by the stria in these animals.

The pineal organ is absent or at least reduced in adult dolphins. Morgane and Jacobs (1972) report that they never found any trace of it in any whale species that they have studied. Also microslide series of adult brains embedded with the meninges did not reveal clear traces of pineal tissue. In this context, repetitive changes of light intensity and spectral changes in animals diving in dim environments may be problematic for the establishment of a conventional diurnal cycle. Interestingly, a pineal anlage develops in the dolphin embryo but obviously disappears later. In a complete microslide series of a young fetal narwhal (*Monodon monoceros*) of 137 mm total length, no trace of a pineal body was found (Holzmann, 1992; Oelschläger and Kemp, 1998). In contrast, Lyamin et al. (2008) found a well-developed pineal organ in a pregnant bottlenose dolphin.

Thalamus

This major component of the diencephalon is rather large (Fig. 6.27); for the bottlenose dolphin, the maximal transverse diameter of the two thalami was reported to be 7 cm in the adult brain (Morgane and Jacobs, 1972), for the human 6 cm. In mammals, generally, the thalamus is often referred to as the gate to the neocortex because most of the ascending afferent systems innervate specific thalamic nuclei, which have reciprocal projections with neocortical areas in the hemispheric lobes. Figures 6.27A–E show the brain of the common dolphin (*Delphinus delphis*), coronal MRI scans and histological

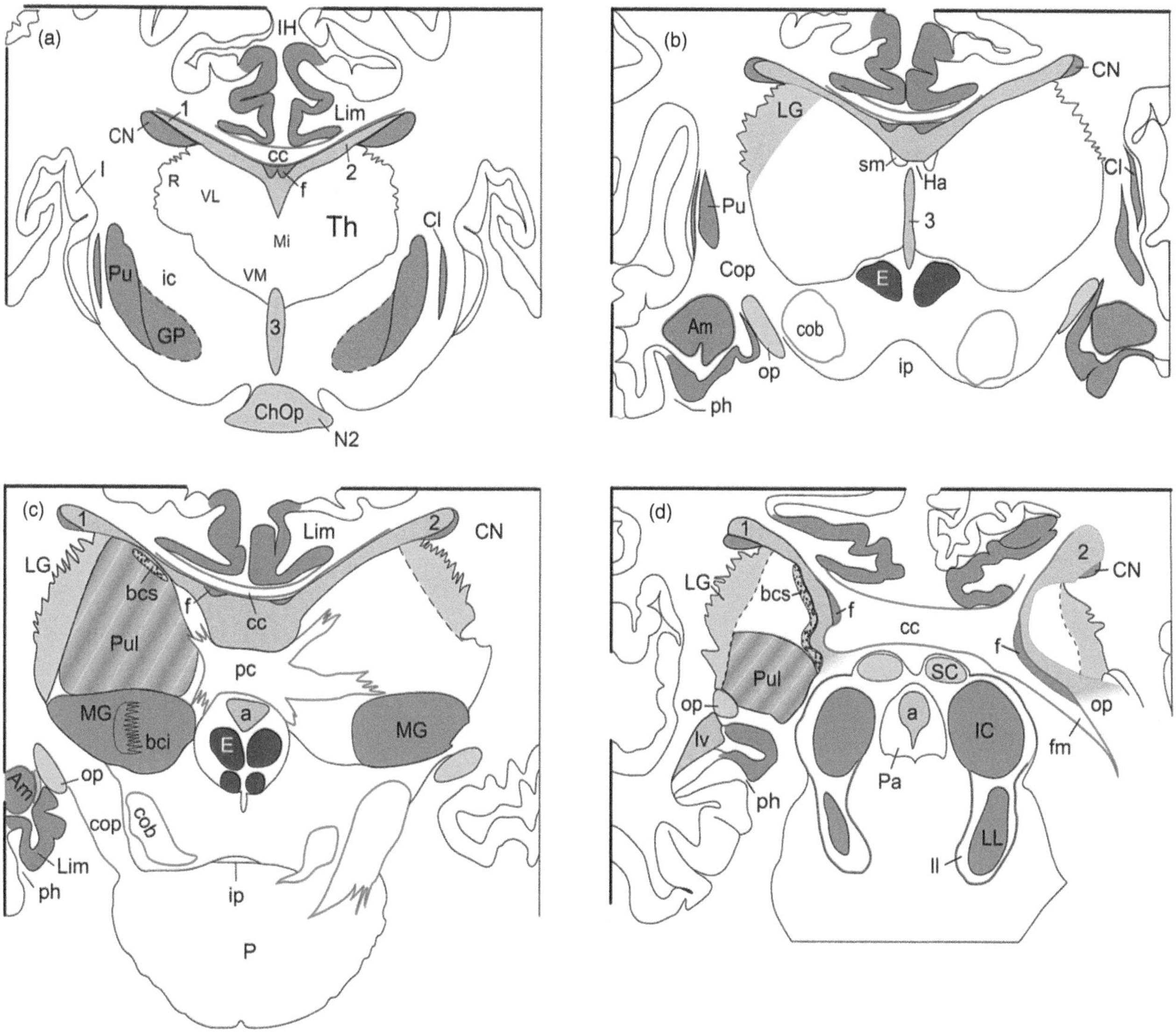

FIGURE 6.26   Rostrocaudal tracings of transverse sections through brain of the bottlenose dolphin. (a) Section passes through the basal ganglia; i.e. caudate nucleus (CN), putamen (Pu), claustrum (Cl), and the chiasmatic level (ChOp). (b) Section about 10 mm behind previous level cuts through midamygdalar (Am) region. (c) Section through caudal pole of amygdala and posterior commissure (pc) is about 5 mm behind previous level. The posterior commissural complex (pc) stands plate-like in the coronal plane; therefore, its sectional area is maximal. (d) Section about 3 mm caudal to previous level shows visual system in *yellow*, auditory system in *green*, limbic cortex (Lim) in *blue*, basal ganglia in *brown*, ventricular system (1–3, a) in gray, fiber tracts with *red* contour, elliptic nucleus (E) in magenta. The pulvinar (Pul) is an integration center for visual and auditory input. Other abbreviations: 1–3, ventricles 1–3; a, aqueduct; Am, amygdala; bci, brachium of colliculus inferior (caudalis); bcs, brachium of superior (rostral) collicle (part); cc, corpus callosum; ChOp, optic chiasm; Cl, claustrum; CN, caudate nucleus; cob, corticobulbar tract; cop, corticopontine tract; E, elliptic nucleus; f, fornix; fm, forceps major of corpus callosum; GP, globus pallicus; Ha, habenula; I, insula; ic, internal capsule; IC, inferior (caudal) colliculus; IH, interhemispheric sulcus; ip, interpeduncular fossa; LG, lateral geniculate body; Lim, limbic lobe; LL, nucleus of lateral lemniscus; ll, lateral lemniscus; MG, medial geniculate body; Mi, massa intermedia; op, optic tract; P, pons; Pa, periaqueductal gray; pc, posterior commissure; ph, parahippocampal sulcus; Pu, putamen; Pul, pulvinar; R, reticular thalamic nuclei; SC, superior (rostral) colliculus; sm, stria medullaris thalami; Th, thalamus; VL, ventrolateral thalamic nuclei; VM, ventromedial thalamic nuclei. *(Redrawn and modified by Jutta Oelschläger from Morgane and Jacobs, 1972. Comparative anatomy of the cetacean nervous system. In: Harrison, R.J. (Ed.),* Functional Anatomy of Marine Mammals. *Academic Press, London, pp. 117–244.)*

sections. The levels of the sectional planes in a rostro-caudal sequence are given in the pilot in Fig. 6.27A; the histological sections (Nissl and Heidenhain-Woelcke stain) correspond only roughly because they come from two different brains.

The nuclear composition of the dolphin thalamus is generally very similar to that in a variety of other mammals (Fig. 6.28). It comprises regions representing the anterior, medial, ventral, and lateral nuclear groups of terrestrial mammals, the medial geniculate and lateral geniculate nuclei as well as intralaminar nuclei and medial and midline divisions together with the reticular nucleus. Most interesting as to dolphin anatomy are the following areas.

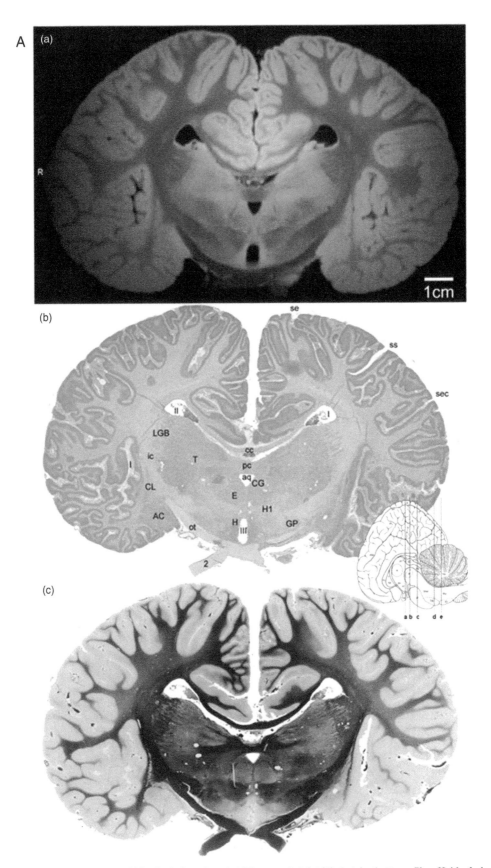

FIGURE 6.27A Coronal MRI scan (top) and histological sections (middle: cresyl violet Nissl stain; bottom: fiber Heidenhain- Woelcke stain) of the brain of common dolphin (*D. delphis*): level of the diencephalon (T, thalamus; 2, optic chiasm) (Scale = 1 cm). Other important landmarks (see text) are the corpus callosum (cc), posterior commissure (pc), the insula (I), claustrum (Cl), amygdaloid complex (AC), and the elliptic nucleus (E). Abbreviations in part b: I, left lateral ventricle; II, right lateral ventricle; III, third ventricle; 2, optic nerve/chiasm; AC, amygdaloid complex; aq, cerebral aqueduct; cc, corpus callosum; CG, central (periaqueductal) gray; CL, claustrum; E, elliptic nucleus; GP, globus pallidus; H, hypothalamus; H1, Forel's field; I, insula; ic, internal capsule; LGB, lateral geniculate body; ot, optic tract; pc, posterior commissure; se, sulcus entolateralis; sec, sulcus ectosylvius; ss, sulcus suprasylvius; T, thalamus. (*Modified from Oelschläger et al., 2008. Morphology and evolutionary biology of the dolphin (Delphinus sp.) brain— MR imaging and conventional histology. Brain Behav. Evol. 71, 68–86.*)

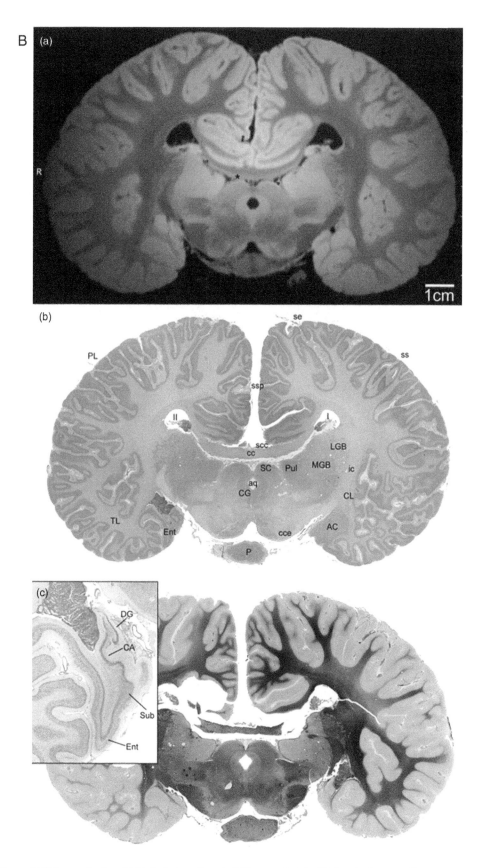

FIGURE 6.27B Coronal MRI scan (top) and histological sections (middle: cresyl violet Nissl stain; bottom: fiber Heidenhain- Woelcke stain) of the brain of common dolphin (*D. delphis*): level of the anterior mesencephalon with the superior collicles (SC) and geniculate bodies (LGB, MGB) (Scale = 1 cm). P, first cut of the pons. Inset: Enlarged detail of the hippocampus area (CA, DG, Ent, Sub) from adjacent section. *Asterisk* in a: fimbria hippocampi. Abbreviations in part b: I, left lateral ventricle; II, right lateral ventricle; AC, amygdaloid complex; aq, cerebral aqueduct; cc, corpus callosum; cce, crus cerebri; CG, central (periaqueductal) gray; CL, claustrum; Ent, regio entorhinalis; ic, internal capsule; LGB, lateral geniculate body; MGB, medial geniculate body; P, pons; PL, parietal lobe; Pul, pulvinar; SC, superior colliculus; scc, sulcus corporis callosi; se, sulcus entolateralis; ss, sulcus suprasylvius; ssp, sulcus suprasplenialis; TL, temporal lobe. (*Modified from Oelschläger et al., 2008. Morphology and evolutionary biology of the dolphin (Delphinus sp.) brain—MR imaging and conventional histology. Brain Behav. Evol. 71, 68–86.*)

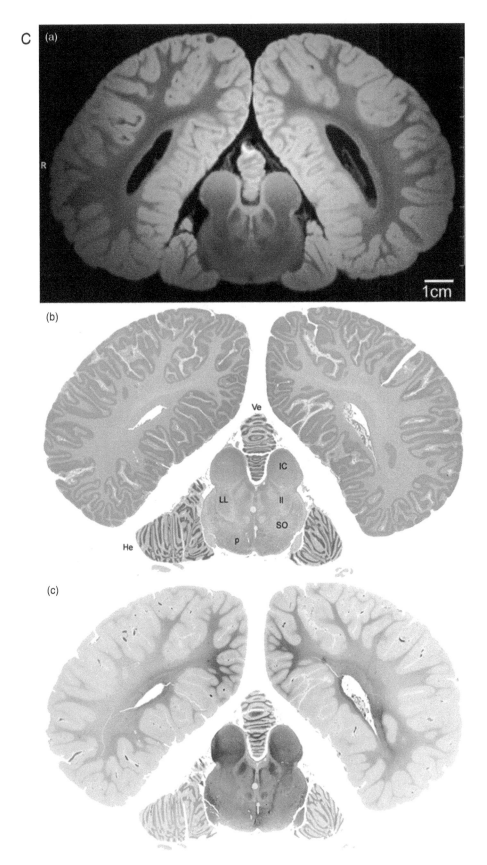

FIGURE 6.27C Coronal MRI scan (top) and histological sections (middle: cresyl violet Nissl stain; bottom: fiber Heidenhain- Woelcke stain) of the brain of common dolphin (*D. delphis*): Level of the posterior mesencephalon with the inferior collicles (IC), nucleus of lateral lemniscus (LL), pons (P), and the rostral extremities of the cerebellum (He, Ve) (Scale = 1 cm). Abbreviations in part b: He, hemisphere; IC, inferior colliculus; ll, lateral lemniscus; P, pons; SO, superior olive; Ve, vermis. *(Modified from Oelschläger et al., 2008. Morphology and evolutionary biology of the dolphin (Delphinus sp.) brain—MR imaging and conventional histology. Brain Behav. Evol. 71, 68–86.)*

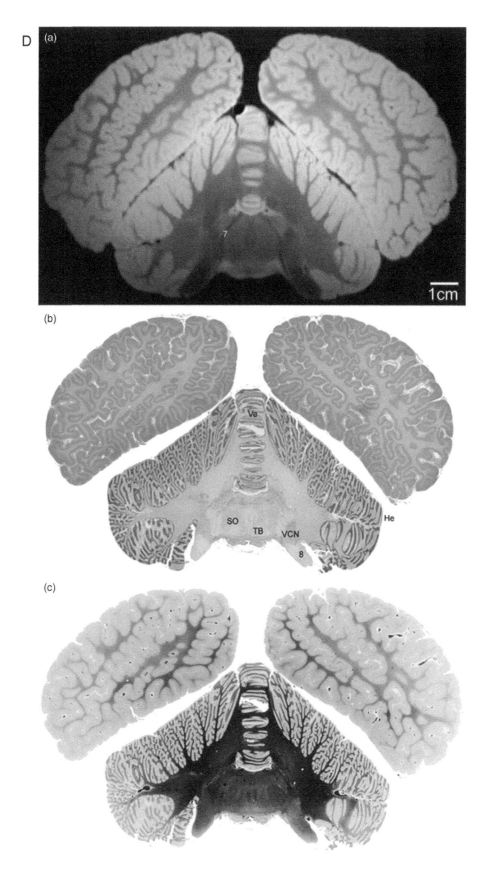

FIGURE 6.27D Coronal MRI scan (top) and histological sections (middle: cresyl violet Nissl stain; bottom: fiber Heidenhain- Woelcke stain) of the brain of common dolphin (*D. delphis*): level of the anterior medulla oblongata with seventh and eighth nerves (7, 8), ventral cochlear nucleus (VCN), nucleus of trapezoid body (TB), superior olive (SO), and through anterior part of cerebellum (He, Ve) (Scale = 1 cm). Abbreviations in part b: 8, vestibulocochlear nerve; He, hemisphere; SO, superior olive; TB, nucleus of trapezoid body; VCN, ventral cochlear nucleus; Ve, vermis. *(Modified from Oelschläger et al., 2008. Morphology and evolutionary biology of the dolphin (Delphinus sp.) brain—MR imaging and conventional histology. Brain Behav. Evol. 71, 68–86.)*

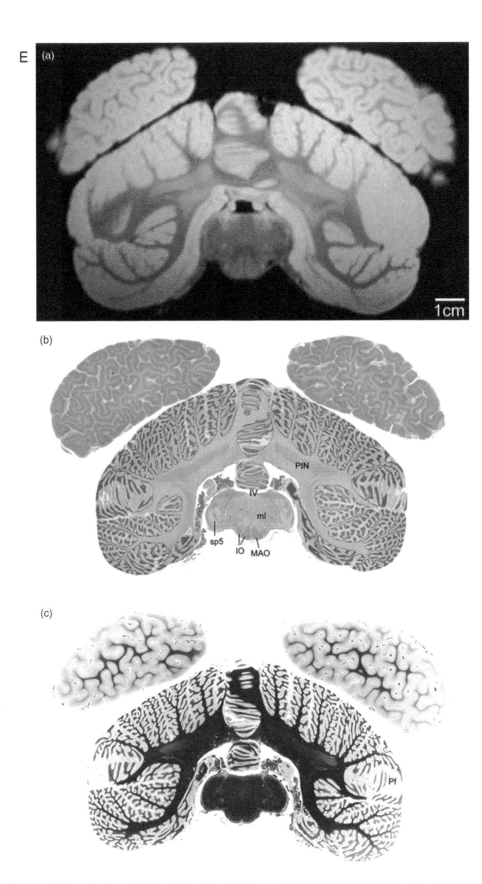

FIGURE 6.27E Coronal MRI scan (top) and histological sections (middle: cresyl violet Nissl stain; bottom: fiber Heidenhain- Woelcke stain) of the brain of common dolphin (*D. delphis*): level of the posterior medulla oblongata with the inferior olives (IO, MAO) and the center of the cerebellum with the posterior interpositus nucleus (PIN) (Scale = 1 cm). Abbreviations in part b: IV, fourth ventricle; IO, inferior olive; MAO, medial accessory nucleus of IO; ml, medial lemniscus; PIN, posterior interpositus nucleus; sp5, spinal tract of trigeminal nerve. *(Modified from Oelschläger et al., 2008. Morphology and evolutionary biology of the dolphin (Delphinus sp.) brain—MR imaging and conventional histology. Brain Behav. Evol. 71, 68–86.)*

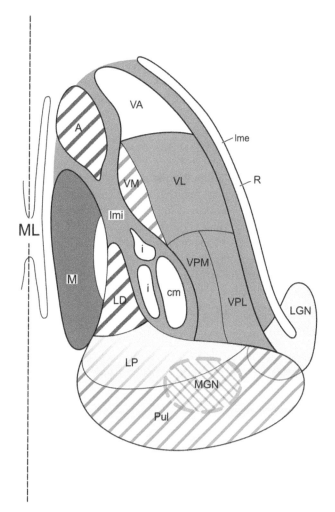

FIGURE 6.28   Schematic drawing of the right human thalamus and its main nuclear components, flattened out into a horizontal plane. Abbreviations: A, anterior nuclear group; cm, centromedian nucleus; i, intralaminar nuclei; LD, nucleus lateralis dorsalis; LGN, lateral geniculate nucleus; LP, nucleus lateralis posterior; lme, lamina medullaris externa; lmi, lamina medullaris interna; M, medial nucleus; ML, midline nuclei; MGN, medial geniculate nucleus; R, reticular nucleus; Pul, pulvinar nuclei; VA, nucleus ventralis anterior; VL, nucleus ventralis lateralis; VM, nucleus ventralis medialis; VPL, nucleus ventralis posterior lateralis; VPM, nucleus ventralis posterior medialis. *(Modified by Jutta Oelschläger after Van Dongen and Nieuwenhuys, 1998. Diencephalon. In: Nieuwenhuys, R., Ten Donkelaar, H.J., Nicholson, C. (Eds.).* The Central Nervous System of Vertebrates, *Vol. 3, Springer-Berlin-Heidelberg-New York.)*

The *anterior group* of nuclei, which is related to the cortex of the large cetacean limbic lobe, is well developed. The anteroventral nucleus, projecting to the well-developed anterior limbic cortex, dominates this group. In contrast, however, the mammillary body and the interconnecting mamillothalamic tract as parts of the Papez circuit are comparatively small and thin (see Section "Periarchicortex").

In the *medial group* of thalamic nuclei, the single mediodorsal or medial nucleus (Fig. 6.28: M) is large in most mammalian species investigated so far. In terrestrial mammals, the nucleus is connected with a variety of structures, among them the amygdaloid nucleus, the olfactory tubercle, the piriform cortex, and particularly the anterior frontal lobe of the telencephalic hemisphere (premotor, polar, and orbital cortices). In primates and whales, there seems to exist a strong size correlation between the mediodorsal nucleus and the (pre)frontal cortex; both the mediodorsal nucleus and (pre)frontal cortex are large. Interestingly, in the bottlenose dolphin, the nucleus was reported to comprise about 10% of the whole thalamus volume (Morgane and Jacobs, 1972). In dolphins, however, the strong development of the mediodorsal nucleus should be rather due to nonolfactory input (reduction of the olfactory system) and may be reciprocal connections to the large anterior cingulate cortex (ACC). (See "Periarchicortex" in this chapter.)

The *ventral and lateral groups* of nuclei account for the largest part of the thalamic mass in the dolphin brain (Morgane and Jacobs, 1972).

The *ventral group* comprises nuclei that are largely responsible for projections of proprioception, touch, and pain via the spinal cord and cerebellum. An interesting example for understanding correlations between peripheral and central sensory structures is the ventral posterior nucleus (VPN). In mammals, somatosensory fibers from the body and limbs ascend in the gracilis and cuneate nuclei and proceed in the medial lemniscus to the lateral subnucleus of the VPN (VPL). From here, they proceed to the respective fields in the somatosensory cortices SI and SII. Those fibers from the head and mouth region project via the lemniscus trigeminalis to the medial subnucleus of the VPN (VPM) and from there to the somatosensory cortical fields (Voogd et al., 1998). In dolphins, the VPN as a whole is relatively small, but as to the two subnuclei, the medial one, where the head is represented, is comparatively larger. In the smaller size of the VPL subnucleus, the limited somatosensory representation of the body due to the loss of hind limbs, the decrease in size and simplification of the forelimbs, and the concomitant acquisition of a spindle-shaped body are obvious. All of these changes result in a distinct reduction of the body surface per body volume, also reflected in the morphology and size characteristic of the spinal cord (see Section "Spinal Cord, Spinal Nerves, and Ganglia" in this chapter). In contrast, the trigeminal somatosensory innervation of the head (VPM) seems to be more intense in dolphins, probably because of the special requirements of the large nasal/forehead area in the emission of ultrasound signals for communication and orientation (see Section "Dolphin Head Characteristics").

As seen in higher primates, the *lateral group* of thalamic nuclei in the dolphin is dominated by what appears to be a massive pulvinar (Pul). In the human, this nuclear complex comprises the caudal third of the whole thalamus. In the bottlenose dolphin, the pulvinar was reported to represent the largest single nuclear complex within the thalamus. Here, it more or less merges in both the strongly protruding medial geniculate nucleus (MGN; auditory) and the large lateral geniculate nucleus (LGN; visual). In the human, the pulvinar is regarded a center of integration for projections from the lateral lemniscus (auditory system), as well as from the lateral geniculate nucleus (visual system) (Kahle and Frotscher, 2009). Moreover, the pulvinar was reported to have reciprocal projections with association areas in the parietal and occipital lobes of the telencephalic hemisphere. These areas are surrounded by somatosensory, visual, and auditory primary cortical fields and probably are engaged in complex processes like speech (Bähr and Frotscher, 2009).

Medial and Lateral Geniculate Nuclei

The *medial geniculate nucleus* (MGN) is impressively large in cetaceans (Figs. 6.26 and 6.28) and reflects the outstanding development of the auditory system in these animals (see Section "Medial Geniculate Body"). It can be supposed that areas in frontal and lateral parts of the cortical auditory projection field, which produce type-1 responses in evoked-potential studies, receive afferent signals from the MGN by a short direct way and through projections with high conduction velocities. In contrast, the area generating type-2 responses in the remaining part of the auditory projection field receives afferent signals from longer or slower pathways (Supin et al., 2001).

In the bottlenose dolphin, the *lateral geniculate nucleus* (LGN) is surprisingly well developed (Figs. 6.26 and 6.28), though less so than the MGN. The LGN projects to the visually excitable part of the lateral gyrus, but does not show the laminar (striate) organization usually associated with biretinal projections. This may be related to the fact that in cetaceans the fibers in the optic nerve show a complete or almost complete decussation, that is, at least most of the optic fibers cross the midline at the optic chiasm (Ridgway, 1990) and end in the opposite (contralateral) visual field. The LGN is easily identified by a system of coarse optic fascicles, which come from the optic nerve (brachium colliculi rostralis) and create a characteristic system of linear parallel streaks (Figs. 6.9a and 6.26).

Hypothalamus

The basal part of the diencephalon in dolphins exhibits an organization similar to that encountered in other mammals. The anterior, tuberal, and posterior hypothalamic nuclei are evident but not particularly prominent. The paraventricular and supraoptic nuclei are obvious because of their large hyperchromatic cells, the latter nucleus being especially well formed. As in other mammals, the supraoptic commissure of dolphins is well developed and well organized. The small size of the mammillary bodies, which in the postnatal animals do not protrude at the brain surface, correlates with the weak development of the hippocampus, postcommissural fornix, and mammillothalamic tract (Morgane and Jacobs, 1972).

Hypophysis

The adenohypophysis in dolphins is rather wide and seated not in a bony sella turcica but rather in a shallow cavity within the heavily vascularized dura mater. The adeno- and neurohypophyses are present as distinct structures, separated by a thin meningeal sheet. Several authors report that there is no pars intermedia nor an intraglandular cleft, a residual lumen of the hypophyseal pouch.[g] The neurohypophysis in dolphins is slender and consists of the usual components, and its topographical relationships to the optic chiasm are inconspicuous (Morgane and Jacobs, 1972).

MESENCEPHALON

As in other mammals, the mesencephalon of dolphins is a narrow central part of the brainstem (Figs. 6.6 and 6.9b). It is composed of three macroscopic levels: (1) the tectum (SC) dorsally, (2) the intermediate tegmentum (t), which includes the anterior (mesencephalic) part of the reticular formation, and (3) the *crus cerebri* (cc) ventrally (see also Fig. 6.27). Between the tectum and tegmentum is situated the mesencephalic part of the ventricular system (mesencephalic aqueduct; see Fig. 6.9b), which in dolphins is narrow and low in the rostral part and much higher and wider in the caudal part.

Tectum. The roof of the mesencephalon is characterized by the lamina tecti (quadrigeminal plate), which comprises two rostral and two caudal colliculi (superior and inferior colliculi of primates).[h] The superior colliculi of dolphins seem to be well developed, as other centers of the visual pathway are. The inferior colliculi, however, are much larger in comparison, in accordance with the overall impressive size of the auditory system. In histological sections as well as in MR scans, the dolphin superior colliculus exhibits a pattern of faint lines that is known from other mammals. In terrestrial mammals, the superior layers represent a sensory apparatus that gets its input from the retina and visual cortex, whereas the deeper layers comprise an integration apparatus for the processing of multimodal input (Benninghoff and Drenckhahn, 2004). In dolphins, the rostral and caudal colliculi so far have not been investigated systematically.

Tegmentum. In the mesencephalon of dolphins, the anterior most part of the tegmentum, also called pretectum, is well developed. In a transverse section at the level of the posterior commissure (Figs. 6.26c and 6.27c) lie two tegmental nuclei, the appearance of which is unique among mammals, perhaps with the exception of the elephant (Cozzi et al., 2001; Oelschläger et al., 2010). The two nuclei are called "elliptic nucleus" (E) and situated at the anterior end of and within the central gray, rostral and dorsal to the oculomotor nuclear complex and near the nucleus ruber (Figs. 6.9, 6.26, and 6.27). The elliptic nucleus seems to represent a strongly hypertrophied nucleus of Darkschewitsch (Fig. 6.48, D) that, during ontogeny, tightly associates with the adjacent nucleus interstitialis Cajal (I). Both nuclei are very large in comparison with those in the human, particularly the nucleus of Darkschewitsch. In terrestrial mammals, the interstitial nucleus is engaged in oculomotor function, that is, in vertical eye (and head) movements but also sends fiber bundles into the spinal cord (Barone and Bortolami, 2004). The nucleus of Darkschewitsch receives fiber systems from the cerebellum, vestibular nuclei, pretectal nuclei, and the motor cortex. Projections of the nucleus of Darkschewitsch run directly to the spinal cord and indirectly via the medial tegmental tract (mtt) to the inferior olivary complex (medial accessory nucleus) and from there to the cerebellum (Voogd et al., 1998). Strong cortical projections, including such from the auditory cortex, run along the pyramidal tract through the crus cerebri to the pons and are also redirected to the cerebellar cortex. Acoustic stimuli, however, were also reported to be conveyed via the inferior colliculus and the superior colliculus and further to the paramedian pontine reticular formation (Benninghoff and Drenckhahn, 2004).

The identity of the elliptic nucleus and its potential functional implications have been under discussion. Morgane and Jacobs (1972) indicate that the parallel hypertrophy of this nucleus with its two large subnuclei in distantly related mammals (dolphins and elephants) with diverse but extraordinarily modified nasal passages may stand for a particularly important role in controlling movement patterns within the facial musculature of both the blowhole in dolphins and the nasal trunk of elephants, respectively. (This topic will be discussed in "Cerebellum" in this chapter)

The red nucleus (nucleus ruber) of dolphins has been discussed conversely in the literature. In the bottlenose dolphin, it is located ventral to the elliptic nucleus and rather modest in size as compared with the situation in terrestrial mammals (primates; Morgane and Jacobs, 1972). Nevertheless, the nucleus exhibits a rostral parvocellular and a caudal magnocellular component. The substantia nigra has a rather extensive distribution dorsal to and intermingling with the fiber bundles of the cerebral peduncle. Both the red nucleus and substantia nigra (Fig. 6.29: Nr, Sn) are part of the basal ganglia (see Section "Extrapyramidal Motor System" in this chapter).

---

g. The hypophyseal pouch (= Rathke's diverticulum) is a tubular outgrowth of ectoderm from the stomodeum (primordium of the mouth) of the embryo; it grows dorsad to the infundibular process (pituitary stalk) of the diencephalon, around which it forms a cuplike mass the adenohypophysis.

h. Although the primate (human) nomenclature of the colliculi (superior and inferior colliculi) is adopted to the literature of laboratory mammals and to dolphins in the older literature, we try to conform to the veterinary nomenclature (rostral, caudal colliculi, etc.) as often as possible.

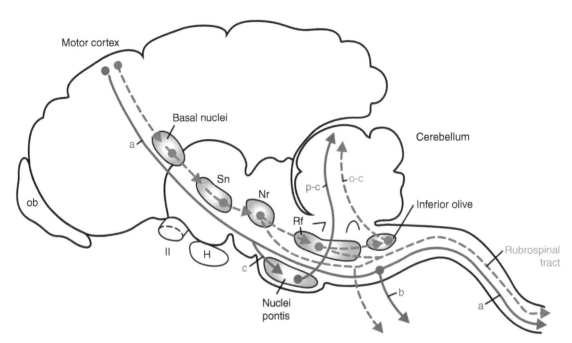

FIGURE 6.29 **Diagrammatic circuitry of the pyramidal system (*solid red lines*) and the extrapyramidal system (*dashed green lines*) in the mammalian brain.** The pyramidal feedback loop includes the pons, cerebellum, and the thalamus (not shown). In the multisynaptic extrapyramidal system (simplified), the numerous feedback loops provide for the necessary balance between stimulating and inhibitory influences. All these processes are governed by the cerebellum, which receives afferent fiber systems via the inferior olivary complex. Abbreviations: a, corticospinal fibers; b, corticobulbar fibers; c, corticopontine fibers; H, hypophysis; Nr, nucleus ruber; ob, olfactory bulb; o-c, olivocerebellar tract; p-c, pontocerebellar tract; Rf, reticular formation; Sn, Substantia nigra; II, optic nerve. *(Redrawn and modified by Jutta Oelschläger from Dyce et al., 1991. Anatomie der Haustiere. Ferdinand Enke Verlag Stuttgart.)*

Crus Cerebri

This thick strand of fibers comprises the long descending efferent projections (Figs. 6.26 and 6.27: cob, cop, cce). Beneath the basal surface of the mesencephalon, these fibers run over the red nucleus (nucleus ruber) and the substantia nigra, which are essential for motor function (see Section "Basal Ganglia" later in this chapter). Together, the fibers of the crus cerebri and the mesencephalic nuclei within the tegmentum form the so-called cerebral peduncle. The fibers of the crus cerebri originate in the neocortex of the telencephalic hemisphere. They form fiber tracts that project to the nuclei of the cranial nerves (corticonuclear fibers, or corticobulbar tract) (Figs. 6.26 and 6.29), to the pontine nuclei (corticopontine fibers) and to the spinal cord (corticospinal fibers). In the crus cerebri, the pyramidal tract as a whole is still well developed, but in the metencephalon (pons region), it gets much weaker because many fiber bundles leave the tract in order to synapse in the nuclei of the cranial nerves or in the pontine nuclei: The latter send strong projections to the cerebellum for further processing. In dolphins and other toothed whales, the pyramidal tract quickly vanishes along the pons and medulla; only a few axon bundles reach the upper cervical spinal cord (see Section "Important Fiber Systems").

METENCEPHALON (CEREBELLUM/PONS)

The metencephalon consists of three levels:

1. The cerebellum as the superior part.
2. The tegmentum with the massive reticular formation.
3. The strongly protruding pons.

Between the cerebellum and the tegmentum is situated the fourth ventricle (rhomboid fossa) that opens into the subarachnoid space via altogether three foramina: Paired ones are situated bilaterally at the brainstem near the exit of the facial group of cranial nerves (facial, vestibulocochlear nerve) and called Aperturae laterales ventriculi quarti (Luschka). The unpaired one is situated dorsally in the roof of the fourth ventricle and called Apertura mediana ventriculi quarti (Magendie). The pyramidal tract is already weak here.

In the literature, the tegmentum and pons of the metencephalon are sometimes called dorsal and ventral pons. This is, however, misleading because the two parts belong to different systems: The "dorsal pons" is part of the tegmentum that extends from the rostral border of the mesencephalon to the caudal border of the myelencephalon (medulla oblongata) and even further into the spinal cord. The tegmentum contains the nuclei of the cranial nerves and the reticular formation. The "ventral pons" instead consists of white matter representing long descending fiber systems, including the pyramidal tract (Fig. 6.29: a). They originate in the neocortex, run through the crus cerebri, and contact neurons in nuclei of cranial nerves (corticobulbar/corticonuclear tracts) and only sparsely reach the cervical part of the spinal cord (corticospinal tract). In dolphins, the majority of fibers (corticopontine tract) are redirected at specific nuclei, which are dispersed throughout the pons. Their axons cross contralaterally to the other side of the brainstem, form the middle cerebellar peduncles (Fig. 6.31), and terminate in the cortex of the cerebellar hemispheres. These crossing fiber masses and the dispersed nuclei together form the transverse bulge at the base of the brainstem, which is called pons (sensu stricto). Because of the tight connections between the pons and the cerebellum, both are very large in the dolphins.

Whereas the transverse fiber systems of the telencephalon (corpus callosum, anterior commissure) are moderately developed in marine dolphins compared to terrestrial mammals, their longitudinal, corticofugal projections to the pons and from there to the cerebellum are very well developed. In this respect, the dorsolateral group of pontine nuclei seems to be involved particularly in the transmission of auditory input to the cerebellum (Spangler and Warr, 1991).

Cerebellum

External morphology and size. In the past, Larsell (1970) presented a first qualitative description of the cerebellum in the bottlenose dolphin. Recently, Hanson et al. (2013) have published a thorough analysis of the cerebellum in this species both in qualitative and quantitative respect including magnetic resonance imaging (MRI) reconstructions. This paper can be regarded a solid basis for future morphological work on the dolphin cerebellum. In marine dolphins, the cerebellum and pons are very large in absolute and relative terms (Figs. 6.7, 6.9, 6.12, 6.25, 6.27, 6.30, 6.32, 6.35, 6.36, 6.39, and 6.60) and there is obviously a certain size correlation with the neocortex as seen in primates and the human (Schwerdtfeger et al., 1984). The dolphin cerebellum even challenges that of the human in size. Adult common dolphins of about 75 kg have slightly smaller cerebella than humans of the same body mass (Marino et al., 2000). However, because total brain mass is lower in this dolphin, the percentage of the cerebellum in the total brain is higher than in the human.

Lobular morphology. In general, the cerebellum of dolphins conforms to the overall mammalian pattern of organization (Fig. 6.33) (Voogd et al., 1998; cf. Hanson et al., 2013). And as far as the data in the literature are concerned, there seems to be a general plan of cerebellar morphology in whales and dolphins. Thus, it consists of two hemispheres and a vermis (Figs. 6.7, 6.9, 6.14, and 6.32). In comparison with terrestrial mammals, the hemispheres are very large, whereas the vermis appears to be narrow. These major parts in the cetacean cerebellum, in turn, consist of a set of lobes and lobules, which exhibit variations in development as compared to other mammals. These variations are unique to cetaceans and could be neural correlates of adaptations to the aquatic environment.

In general, two transverse fissures on each side separate three cerebellar lobes: the primary fissure separates the small anterior (very small in dolphins) from the very large posterior lobe, and the posterolateral fissure separates the posterior lobe from the flocculonodular lobe (Figs. 6.33 and 6.34). The latter is particularly small in dolphins (Breathnach, 1960; Oelschläger and Oelschläger, 2002). The flocculus is the target of direct and indirect projections from the mammalian vestibular apparatus and subserves the calibration of the vestibulo-ocular reflex (Glickstein et al., 2007). It is also involved in compensatory actions of the neck muscles in so-called "smooth pursuit" movements of the head and eyes in terrestrial mammals, together with the ventral paraflocculus (Voogd et al., 1998). Whether dolphins can perform "smooth pursuit" movements in hunting (smooth regulation of sound-induced and sound-controlled behavior; De Ribeaupierre, 1997) is not known. They have limited neck mobility, presumably can use their visual system only during daylight, and may thus have to rely mostly on the auditory system in order to follow their prey effectively.

The size relations between the cerebellar lobes and lobules are characteristic for cetaceans. In midsagittal section (Figs. 6.7, 6.32, and 6.34), the conventional subdivision of the vermis into nine lobules of the mammalian cerebellum is obvious. In Fig. 6.33, the organization of the primate and cetacean cerebellum is shown side by side. In the cetacean cerebellar hemisphere (right), the extremely small size of the anterior lobe (Morgane and Jacobs, 1972) may be explained by electrophysiological findings in other mammals. They indicate that the hemispheral parts of this lobe comprise the cortical representation of the fore and hind limbs, which are highly modified or even have vanished in these animals. Such physiological data also indicate a representation of the dolphin head in the caudally adjacent simple lobule (Lobus simplex) within the posterior lobe, which is rather large in the dolphin (Figs. 6.33 and 6.34b). The caudally following lobules (ansiform, paramedian lobule) are distinctly less developed than the human counterparts, particularly the ansiform complex, and not

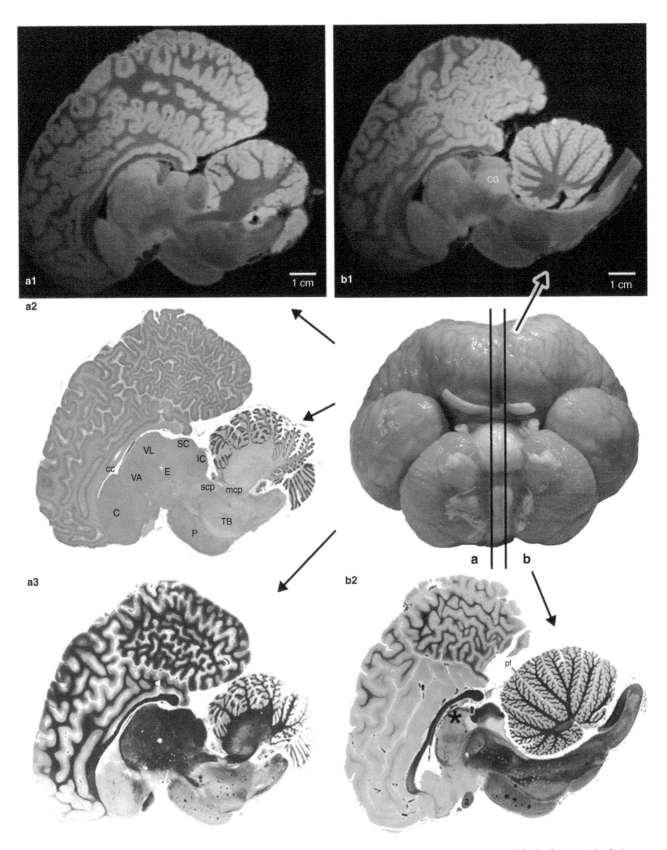

FIGURE 6.30 **Near-midsagittal scans and sections through the medial wall of the telencephalic hemisphere and the brainstem.** (a1–a3) A more lateral parasagittal plane against (b1–3) a more medial parasagittal plane. a1 and b1 are MR scans; a2, Nissl section, a3 and b2 are sections stained for fibers (Heidenhain-Woelcke). Important details are the caudate nucleus (C), subnuclei of the thalamus (VA, VL), elliptic nucleus (E), pons (P) as well as nuclei of the auditory system: trapezoid body and nucleus (TB), nucleus of inferior (posterior) colliculus (IC). White *asterisk* in b2: commissural complex. The cerebellum (Cer) is seen together with its superior (rostral) and middle peduncle (scp, mcp). cc, corpus callosum; SC, superior (rostral) colliculus. Black dots in histologic sections: staining artifacts. *(Modified from Oelschläger et al., 2008. Morphology and evolutionary biology of the dolphin (Delphinus sp.) brain—MR imaging and conventional histology. Brain Behav. Evol. 71, 68–86.)*

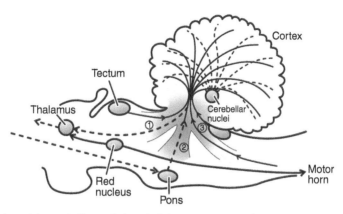

FIGURE 6.31   **The main connections of the cerebellum called cerebellar peduncles.** Rostral ①, middle ②, caudal ③ peduncle. The rostral peduncles, mainly composed of efferent fiber tracts, connect cerebellum and midbrain. A middle, or lateral, pair of peduncles rise straight upward from the pons along the sides of the metencephalon; these peduncles comprise fibers running from cerebral cortex to cerebellum, which are relayed ventrally in the pons and cross to the other side. Caudal peduncles rise up through the medulla and carry proprioceptive fiber bundles from the spinal cord. *(Redrawn and modified by Jutta Oelschläger from Romer and Parsons, 1986. The Vertebrate Body, sixth ed. Saunders College Publishing, Philadelphia.)*

so much the paramedian lobule. The latter has been related to body representation with respect to the enormous significance of the tail in cetaceans (cf. Oelschläger and Oelschläger, 2009). The paraflocculus is exceptionally large, particularly the ventral parafloccular lobule. The latter comprises almost the entire caudoventral surface of the cerebellum. In terrestrial mammals, the equivalent of the paraflocculus usually receives climbing fibers from the rostral part of the medial accessory inferior olive. In cetaceans, both structures are exceptionally large, indicating a functional relationship between the paraflocculus and the trunk and tail unit.

Histology of the Cerebellar Cortex

The histology of the cortex is uniform over the entire cerebellum and very similar to that in other mammals (Figs. 6.35–6.37), which was also demonstrated by the conservative pattern of immunoreactivity of calcium-binding proteins (calretinin, calbindin, and parvalbumin; Kalinichenko and Pushchin, 2008). There are, however, no detailed studies available so far as to the morphology and function of this cortex in dolphins. As usual, there are three layers of extremely different structure and appearance: (1) the molecular layer (stratum moleculare) immediately beneath the pia mater, (2) the Purkinje cell layer (stratum ganglionare), and (3) the granular layer (stratum granulosum).

1. The molecular layer is poor in cell somata and mainly consists of unmyelinated fibers. It also contains interneurons (stellate and basket cells) as well as the large two-dimensional dendritic trees of the Purkinje cells (piriform cells). The structure of the molecular layer indicates that here much of the incoming afferent information from the brain and spinal cord is processed with respect to pattern formation in space and time and with the help of cerebellar interneurons.
2. The Purkinje cell layer comprises very large neurons with pear-shaped somata (Fig. 6.37d) and a specific type of dendrites; the latter intensively branch in parallel regular rows into the molecular layer like antennae. Purkinje cells are the only efferent neurons in the cerebellar cortex and they inhibit the neurons of the cerebellar nuclei.
3. The granular layer contains huge numbers of minute neurons with round perikarya and several short dendritic processes. In mammals, their dendrites receive information from axons ascending from afferent excitatory neurons in the pontine nuclei, spinal cord, and vestibular nuclei (mossy fibers); their excitatory axons ascend into the molecular layer, divide, and send long collaterals (parallel fibers) through many rows of large Purkinje dendritic trees. A second type of ascending excitatory axons (climbing fibers) come from the inferior olivary nuclei; each of these fibers contacts only one piriform (Purkinje) cell. The ascending information is modulated by interneurons (Golgi cells, stellate, and basket cells). All of these interneurons can inhibit the Purkinje cells. If this happens, the inhibitory influence of the Purkinje cells on the neurons in the cerebellar nuclei is suspended (disinhibition). Then the information from the mossy and climbing fibers can ascend in the brain.

Longitudinal Projection Zones in the Cerebellar Cortex

In mammals, Purkinje cells projecting to specific target nuclei were found to be arranged in longitudinal zones that may extend over many, and in some cases, over all cerebellar lobules of hemisphere or vermis, crossing the interlobular

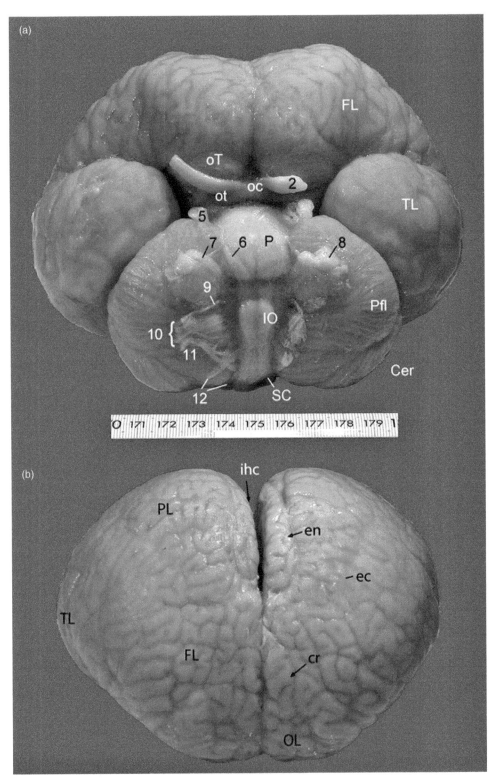

FIGURE 6.32 Five aspects of the intact common dolphin brain (scale: 2 cm). The brain is still in the leptomeninx (arachnoid and pia mater). In dolphin brains, the width exceeds the length, the telencephalic hemispheres are very well developed, the brainstem is thick and the cerebellum very large for brain size. (a) Basal aspect with cranial nerves (2–12), pons, inferior olivary complex, and cerebellum. (b) Rostral aspect shows that the dolphin brain is relatively narrow in front and widens constantly in the caudal direction, where the parietal lobe blends into the very large temporal lobe. (c) Lateral aspect, left. This view illustrates the globular shape of the dolphin brain. This was obviously achieved by the telescoping of the brain, which can easily be seen in the medio-sagittal section (Fig. 6.32e) and is repeated during fetal ontogenesis. (d) Occipital (dorsal) aspect showing the transition from the parietal lobe to the temporal lobe and the cerebellum (vermis, hemisphere) enclosed in the leptomeninx, which is similar to a spider's web. Spinal cord cut transversely at the level of the rostral cervical segments. (e) Mediosagittal aspect of the right half of the brain that facilitates the understanding of brainstem structure and the remarkable surface configurations of the cerebral cortex. The leptomeninx is still seen on the free surface of the hemisphere above the corpus callosum. Brain specimen was dedicated by Sam H. Ridgway (San Diego). Abbreviations: ⊗, leptomeninx (arachnoid and pia mater); cc, corpus callosum; Cer, cerebellum; ec, ecto-lateral sulcus; en, entolateral sulcus; f, fornix; FL, frontal lobe; g, gray substance; ihc, interhemispheric cleft; IO, inferior olive; ivf, interventricular foramen; l, lateral sulcus; oc, optic chiasm; ot, optic tract; oT, olfactory tubercle; P, pons; pc, posterior commissure; Pfl, paraflocculus; PL, parietal lobe; rsn, root of spinal nerve; SpC, spinal cord; spc, spinal canal; SS, suprasylvian gyrus; TB, nucleus of trapezoid body; TL, temporal lobe; V, vermis.

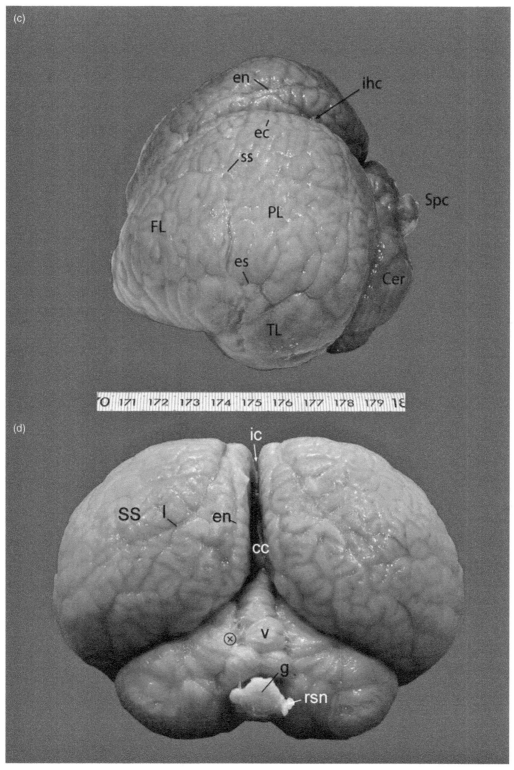

FIGURE 6.32   *(cont.)*

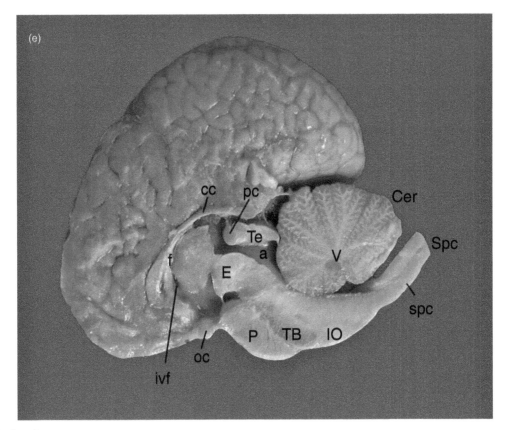

(e)

cc pc Cer

Te a Spc

f V

E spc

P TB IO

oc

ivf

FIGURE 6.32 (*cont.*)

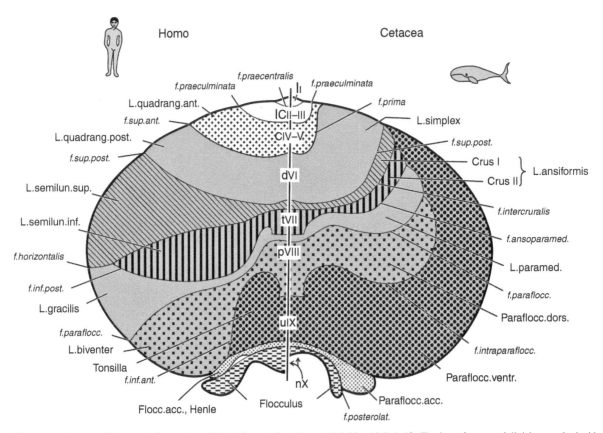

Homo   Cetacea

f.praecentralis
f.praeculminata   I i   *f.praeculminata*
L.quadrang.ant.   lCII–III   *f.prima*
f.sup.ant.   L.simplex
L.quadrang.post.   ClV–V
f.sup.post.   *f.sup.post.*
                                        Crus I
L.semilun.sup.   dVl   } L.ansiformis
                                        Crus II
L.semilun.inf.   tVII   *f.intercruralis*
f.horizontalis   pVIII   *f.ansoparamed.*
f.inf.post.   L.paramed.
L.gracilis   *f.paraflocc.*
f.paraflocc.   uIX   Paraflocc.dors.
L.biventer   *f.intraparaflocc.*
Tonsilla   Paraflocc.ventr.
f.inf.ant.   nX
Flocc.acc., Henle   Flocculus   Paraflocc.acc.
                        f.posterolat.

FIGURE 6.33 Diagram of the cerebellum of man (left half) and of whales and dolphins (right half). The homologous subdivisions on both sides are labeled with identical colors and symbols. The diagram indicates approximately the relative size of the cortical areas of the different cerebellar subdivisions. Folium vermis, separating declive and tuber vermis, fissura prepyramidalis, separating tuber vermis and pyramis, and fissura secunda, separating pyramis and uvula, are not labeled. Colors: anterior lobe–*yellow*; posterior lobe–*brown*, flocculonodular lobe–*green*. Gyri of the cerebellar cortex labeled in normal lettering, sulci in italics. The area presented here for each lobule stands for its total cortical surface in the two groups (human vs. whales and dolphins). Whereas in whales and dolphins, the anterior lobe of the cerebellar hemisphere is very small compared to that in the human, the paraflocculus dorsalis and ventralis in the posterior lobe are extremely large in the dolphin, the homologous lobulus biventer and the tonsilla in the human exhibit about the same cortical surface as the other lobuli. (*Redrawn by Jutta Oelschläger from Jansen, 1950. The morphogenesis of the cetacean cerebellum. J Comp Neurol 93: 341–400.*)

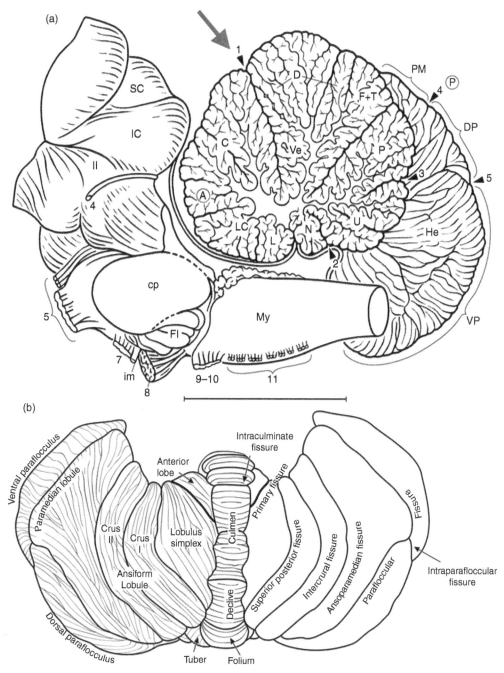

FIGURE 6.34 (a) Brainstem and cerebellum of the fin whale (*Balaenoptera physalus*) from the left side (Scale = 5 cm). Left half of cerebellum omitted to show the mediosagittal aspect of the vermis and the medial aspect of the right hemisphere. Abbreviations: numbers with arrowheads: 1, primary fissure; 2, posterolateral fissure; 3, secondary fissure; 4, paraflocccular fissure; 5, intraparaflocccular fissure; Ⓐ, anterior lobe of cerebellum; C, culmen; D, declive; DP, dorsal paraflocculus; Fl, flocculus; F + T, folium and tuber; He, cerebellar hemisphere; L, lingula; LC, lobus centralis; ll, lateral lemniscus; My, myelencephalon; N, nodulus; Ⓟ, posterior lobe of cerebellum; P, pyramis; PM, paramedian lobule; U, uvula; VP, ventral paraflocculus. 5–11, cranial nerves; *Arrow*, aspect on cerebellum seen in Fig. 6.34b. (b) Line drawing illustrating the lobular and fissuration pattern of the rostrodorsal cerebellar surface of the bottlenose dolphin. Note the smallness of the anterior lobe and the size of the paraflocculus. (*Redrawn and modified by Jutta Oelschläger (a) from Jansen and Jansen, 1969. The nervous system of cetacea. In: Anderson (Ed.). The Biology of Marine Mammals. Academic Press, London; and (b) from Morgane and Jacobs, 1972. Comparative anatomy of the cetacean nervous system. In: Harrison, R.J. (Ed.). Functional Anatomy of Marine Mammals. Academic Press, London.*)

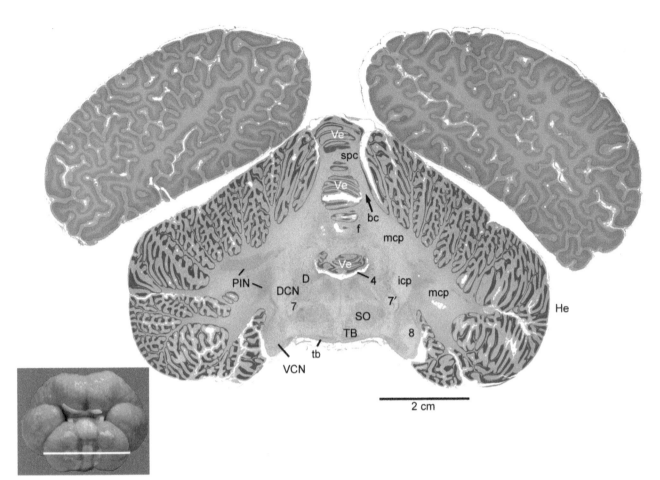

FIGURE 6.35 Common dolphin. Coronal section (inset) through the brainstem, cerebellum and the rear part of the temporal lobes [see pilot]. The cerebellum is cut at its maximal diameter, with the cortex of the vermis interrupted by bands of white substance. Note the inverse correlation between the thickness of the cortical plate and its folding density in the cerebral and cerebellar cortices. The flat part of the fourth ventricle (4) extends between the lower contour of the vermis and the upper contour of the myelencephalon. The cerebellar nuclei (PIN, f) are embedded in the white substance of the cerebellum. Abbreviations: 7, facial nerve; 7′, facial nucleus; 8, cochlear nerve; bc, brachium conjunctivum; D, Deiters' nucleus (lateral vestibular nucleus); DCN, dorsal cochlear nucleus; f, nucleus fastigii; icp, inferior (caudal) cerebellar peduncle (restiform body); He, cerebellar hemisphere; mcp, middle cerebellar peduncle; PIN, posterior interposed nucleus; scp, superior (rostral) cerebellar peduncle; SO, superior olive; tb, trapezoid bod; TB, nucleus of the trapezoid body; VCN, ventral cochlear nucleus; Ve, vermis.

fissures in their course. In terrestrial mammals, the evidence for the arrangement in the corticonuclear projections is based on the myeloarchitecture of the cerebellar white matter, on axonal tracer studies, and on the development and the chemoarchitecture of the cerebellar cortex. In cetaceans, it is obvious from ontogenetic and comparative studies that one longitudinal zone, the lateral intermediate cortical zone (C2 zone of other mammals) at the transition between the vermis and the hemisphere is particularly important for the understanding of cerebellar function. Following the interpretation of Voogd et al. (1998), this C2 zone is several times wider in these animals than in other (terrestrial) mammals (Fig. 6.38). This interpretation is supported by quantitative correlations between (1) the size of the C2 zone, (2) that of the posterior interposed nucleus into which it projects, as well as (3) that of the medial accessory inferior olive from which the C2 zone gets afferent projections. Immunohistochemical investigations that could further substantiate these correlations so far have not been done in dolphins.

Cerebellar nuclei. In the ontogenesis of the cetacean cerebellar cortex, the fundamental mammalian pattern of transverse and longitudinal zones is discernible. The longitudinal zones are obviously related topographically to the development of the cerebellar nuclei (anterior interposed, medial, posterior interposed, lateral cerebellar nucleus). These nuclei are homologous to the emboliform, fastigii, globose, and dentate nuclei in primates. In cetaceans, the lateral intermediate cortical zone of the cerebellar cortex (C2 zone; Fig. 6.38) is enormously developed, occupying about three-fourths of the

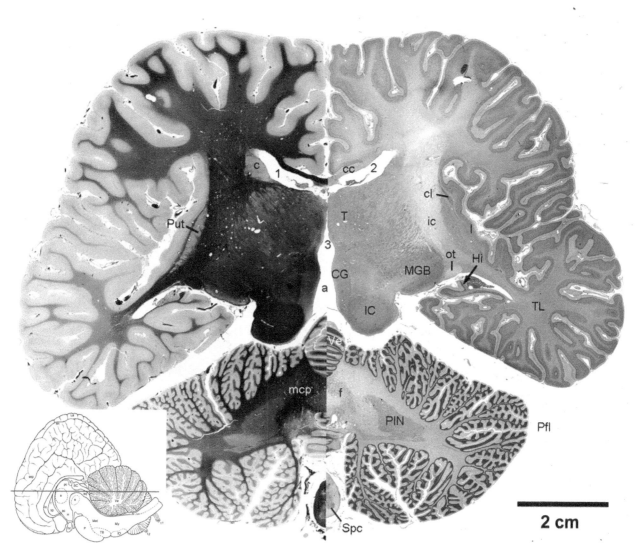

FIGURE 6.36   Horizontal section of common dolphin brain, half fiber stain (Heidenhain-Woelcke) and half Nissl stain for somata. Cerebellum with massive central fiber masses (middle cerebral peduncle, mcp) and intercalated cerebellar nuclei (PIN, f). Other important structures for orientation are the corpus striatum (c, Put), thalamus, inferior or caudal colliculus (IC), medial geniculate body (MGB), optic tract (ot), hippocampus (Hi), and spinal cord (SpC). Abbreviations: 1, 2, 3, ventricles 1–3; a, aqueduct; c, caudate nucleus; cc, corpus callosum; CG, central gray; cl, claustrum; f, fastigial nucleus; I, insula; ic, internal capsule; Pfl, paraflocculus; PIN, posterior interposed nucleus; Put, putamen; T, thalamus; TL, temporal lobe; Ve, vermis. Specimen of the Pilleri Collection in the Natural Museum and Research Institute Senckenberg, Frankfurt am Main.

cerebellar surface (paraflocculus) and correlating with the huge posterior interposed nucleus (PIN; Figs. 6.35 and 6.36) (Jansen and Jansen, 1969). This nucleus is a homolog to the small globose nucleus in primates (Voogd et al., 1998). Interestingly, the posterior interposed nucleus of dolphins completely dominates the cerebellar nuclear complex in much the same way that the dentate (syn. lateral cerebellar) nucleus dominates the nuclear complex in the human cerebellum.

In sum, the size of the dolphin cerebellum is considerable, in accordance with massive descending cortical fiber systems and the well-developed pontine nuclei, which serve as their main relay to the cerebellar cortex, particularly to the area of the very large paraflocculus. Thus, in dolphins, a quantitative correlation seems to exist between the impressive expansion of the neocortex, the size of the ventral pons, and the dimensions of the cerebellum, a phenomenon known from the so-called "ascending primate series" (Schwerdtfeger et al., 1984; Stephan et al., 1988). In dolphins, the very large elliptic nucleus and the inferior olivary complex also belong into this circuitry (Oelschläger, 2008) (see Section "Integrative Aspects—Audiomotor Navigation" in this chapter).

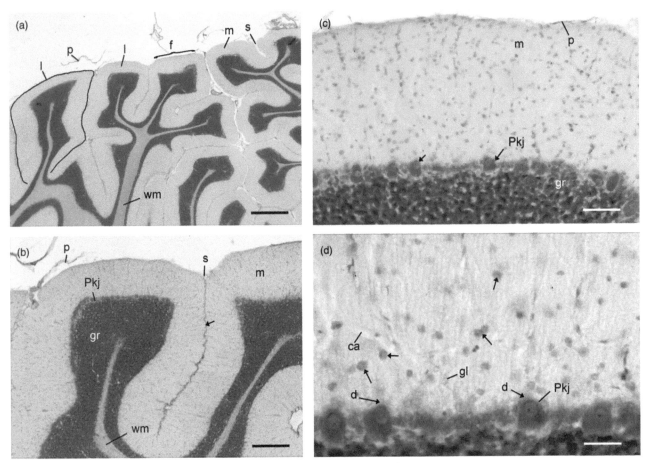

FIGURE 6.37 **Cerebellum of the common dolphin. (Scales: a = 1000 μm; b = 250 μm; c = 100 μm; d = 50 μm).** Nissl stain. (a) Survey on cortical folding at the surface of the hemisphere; the inner sheet of the leptomeninx (pia mater, p) is largely torn off during histological processing; the folding of the cerebellar cortex has produced gyri (folia or leaflets, f and l), which are separated by a labyrinth of sulci (s). The figure shows four gyri in transverse section. The left leaflet has been given a contour. The next leaflet (l) is enlarged in (b) to show its three layers, an outer pale layer (molecular layer, m) and a dark inner layer which is rich in small cells with round perikarya (granular cells, gr). The third (intermediate) layer is seen in (c); it consists of only few but large cells with pear-shaped bodies, the Purkinje cells (Purkinje cell layer, Pkj). Adjacent to the granular cell layer is the white matter (wm) of the cerebellar cortex, which consists of incoming (afferent) and outgoing (efferent) masses of nerve fibers. These sheets of white matter are confluent in the depth of the cerebellum (arbor vitae). (d) shows the scarcity of cells in the molecular layer; many of them are glial cells (gl). This highest magnification exhibits the faint but complicated neuropil consisting of dendritic branches of piriform (Purkinje) cells (d) and ascending afferent fibers. Small cells with a blue perikaryon (*arrows*) are presumably interneurons that are important for the function of the cerebellum, together with the Purkinje cells.

Pons and Cerebellar Connections

In dolphins, the pons represents the basal segment of the metencephalon (Figs. 6.31, 6.32, and 6.39). It is part of the long projection fiber systems coming from the cortex of the telencephalic hemisphere and running through the crus cerebri. All these structures are rather prominent, in correlation with the outstanding expansion of the neocortex and cerebellum. In mammals, one of the prototypical fiber strands is the pyramidal tract. The latter is divided into three components, the corticospinal, cortinuclear, and corticopontine subsystems. The corticopontine subsystem comprises projections from sensorimotor, visual, and auditory cortices to the cerebellum (Voogd et al., 1998). Relay stations are the pontine nuclei, which project to the cerebellum via the thick medial cerebellar peduncle. A main recipient of the visual and auditory cortico-ponto-cerebellar projections in dolphins is the paraflocculus.[i] This extremely large ventral part of the cerebellum (C2 zone of the intermediate cortex) sends Purkinje cell axons to the very large posterior interposed nucleus (PIN) in the white matter of the cerebellum (Figs. 6.35 and 6.36). Projections of this nucleus terminate in the deep layers of the superior colliculus,

---

i. The paraflocculus, which was shown to be the main target of auditory pontocerebellar projections in the rat, was estimated to receive three-fifths of the pontocerebellar fibers in the blue whale (*Balaenoptera musculus*) (Oelschläger and Oelschläger, 2009).

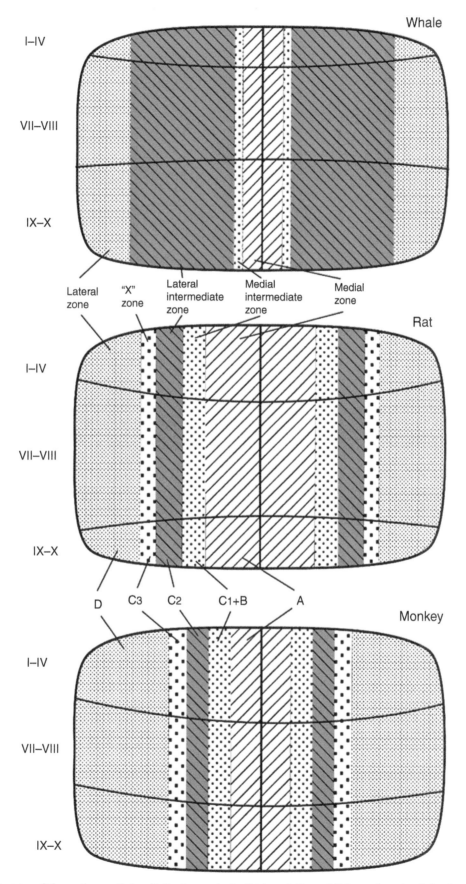

FIGURE 6.38 Diagrams of the corticogenetic longitudinal zones in the fetal cerebellum of Cetacea, the rat and the rhesus monkey. *Orange* intermediate C2 zones in the monkey most likely correspond with those of rat and whale (see text). While in cetaceans, the intermediate C2 zone is very strongly developed, both the rat and the monkey have only a weak development of this longitudinal zone. On the other hand, whereas the rat is somehow intermediate, the monkey exhibits a dominant lateral D zone (not labeled). *Redrawn and modified by Jutta Oelschläger from Korneliussen (1967, 1968), Kappel (1981) after Voogd et al. (1998).*

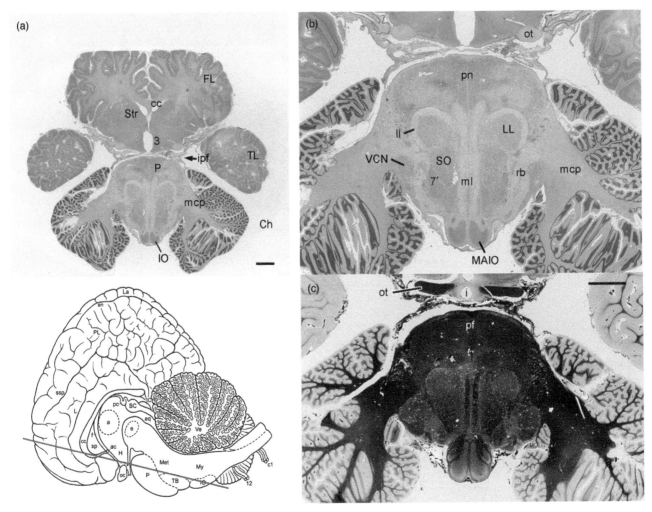

FIGURE 6.39 Common dolphin brain. (a) and (b), Nissl stain; (c), fiber stain. Deep horizontal sections (red line in diagram) at the level of the basal ganglia, pons, and myelencephalon: (a) survey, (b) and (c) details. Both staining methods correspond well with one another. (a) The two main parts of the brain are separated by the interpeduncular fossa (ipf): the prosencephalon on top and the rhombencephalon at the bottom; On both sides of the fossa, the basal tips of the two temporal lobes (TL) are shown. In the rhombencephalon (mesencephalon not sectioned), the cerebellar hemispheres are connected via the pons region (P); the two middle cerebellar peduncles (mcp) frame the nuclei of the myelencephalon. In (b) and (c), the pons region shows both of its components, the pontine nuclei (pn, light blue) on top and the arching fiber masses (pf, dark blue) at the bottom (c). Similarly, the medial lemniscus (ml), which contains ascending somatosensory fibers to the thalamus, is light in the Nissl stain (top) and dark blue in the fiber stain. More laterally, the components of the auditory pathway are seen; the lateral lemniscus and its nucleus (ll and LL), the superior olive (SO), and the ventral cochlear nucleus (VCN). More caudally, the nucleus of the facial nerve (7′), restiform body (rb, inferior cerebellar peduncle) and the inferior olivary complex (MAIO) are seen. Abbreviations: 3, third ventricle; cc, corpus callosum; Ch, cerebellar hemisphere; FL, frontal lobe; i, infundibulum; IO, inferior olivary complex; mcp, middle cerebellar peduncle; ot, optic tract, Str, striatum.

in the mesencephalic reticular formation, the periaqueductal grey, the nucleus of Darkschewitsch (elliptic nucleus of dolphins), and other nuclei at the mesencephalic-diencephalic border, which project to the inferior olive. The latter nucleus sends climbing fibers to the cortex of the paraflocculus.

MYELENCEPHALON (MEDULLA OBLONGATA)

Tegmentum

The myelencephalon is rather large in the common dolphin (see Chapter 10) because of the extended tegmentum with the well-developed nuclei of the cranial nerves (e.g. trigeminal nerve) and, particularly, the components of the hypertrophied auditory pathway (Figs. 6.27, 6.30, 6.35, and 6.39). Even in plesiomorphic dolphins with a low encephalization (La Plata dolphin, *Pontoporia blainvillei*), the medulla is distinctly larger than in terrestrial mammals (simian monkeys, including

TABLE 6.1 Fiber Counts in Single Cranial Nerves of Several Species (Glezer, 2002)

	I	II	III	IV	V	VI	VII	VIII	IX	X	XI	XII
T. truncatus	0	29.6	1.78	0.59	29.64	0.395	9.881	19.76	3.953	1.482	1.482	1.383
I. geoffrensis	0	4.0	0.39	0.0002	40.30	0.232	5.425	31.00	3.487	6.210	6.210	2.583
Balaenoptera spec.	?	37.8	1.89	0.18	33.24	0.988	3.998	13.84	1.662	3.145	2.695	0.539
Homo sapiens	0.83	75.5	1.81	0.26	10.57	0.498	0.981	3.77	0.755	2.589	1.887	0.604

They are given as percentages of the total fiber count of all cranial nerves in that species. As an example, the value 29.6 for the bottlenose dolphin (upper left) indicates that in this species the optic nerve contains 29.6 % of the total fiber count in all cranial nerves.

the human) with respect to brain and body size (Schwerdtfeger et al., 1984). In this caudal part of the brainstem, the reticular formation is again very well developed. In dolphins, the central vestibular complex is largely represented by the well-developed Deiters' lateral vestibular nucleus (Kern et al., 2009); the other vestibular nuclei are strongly regressive in dolphins (see Section "Vestibular System" in this chapter). Fig. 6.35 shows the very large ventral cochlear nucleus and thick eighth nerve (VCN, 8). In dolphins, the diameter of the vestibulocochlear nerve is maximal among all cranial nerves and the number of cochlear fibers is much larger than in the human. The vestibular nerve, however, is thin and comprises only about 5% of those in the cochlear nerve. In contrast to its ventral counterpart, the dorsal cochlear nucleus (DCN) is minute or lacking (see the following). The facial nerve, which is well developed in dolphins (Figs. 6.7, 6.9, 6.27, 6.32, 6.39, and 6.48; Table 6.1), seems to have a moderate diameter in comparison with the dominant vestibulocochlear nerve. The decussatio pyramidum at the caudal end of the brain is very weak; here, the corticospinal tracts fade out among the upper cervical segments of the dolphin spinal cord (Fig. 10.4).

Inferior Olive

In dolphins, the inferior olivary complex[j] (IO) bulges at the ventral surface of the caudal brainstem (Figs. 6.7, 6.12, 6.25, 6.27, 6.32, 6.39, 6.41, 6.49, and 6.60), a similar situation as in the human. In the median plane, both complexes contact each other; this is possible because the pyramids, thick axon bundles of the mammalian corticospinal tract, are nearly lacking. Here, the rudimentary corticospinal projection of the pyramidal tract runs diffusely and superficially and therefore is inconspicuous. Also, the hypoglossal nerve does not exit medial to the inferior olive as in terrestrial mammals but laterally. As the main components, a medial accessory inferior olive (MAO) and a lateral principal nucleus (PO) can be distinguished in dolphins (Fig. 6.40) (Jansen and Jansen, 1969; Morgane and Jacobs, 1972; Glickstein et al., 2007). In contrast to other (terrestrial) mammals, the medial accessory nucleus is much larger than the principal nucleus and rather voluminous in its rostral part (MAOr). The principal inferior olive of dolphins consists of two nuclear lamellae, the lateral one of which blends in the dorsal part of the dorsal accessory olive (DAO). Whereas in dolphins the medial accessory olive is the dominant subnucleus, primates (the human) exhibit a dominant principal olivary nucleus.

In dolphins, the rostral part of the medial accessory inferior olive (MAOr) receives massive input from the elliptic nucleus (nuclei of Darkschewitsch and Cajal) via the medial tegmental tract, which consists of thick axons (De Graaf, 1967; Jansen and Jansen, 1969). The elliptic nucleus itself receives input from the limbic system, extrapyramidal system as well as from the pyramidal tract and auditory system (see also Chapter 10, Fig. 10.3). In dolphins, both the elliptic nucleus and the inferior olive are exceptionally large. Obviously, their pronounced development in dolphins is related to the huge paraflocculus (see Section "Cerebellum" in this chapter) and the large size of the cerebellar posterior interposed nucleus (PIN). The medial tegmental tract (mtt) forms immediately ventral to the elliptic nucleus and is difficult to follow throughout the brainstem because of its neighborhood to other medial longitudinal fiber tracts (medial lemniscus, medial longitudinal fascicle). However it is conspicuous in the anterior medulla oblongata dorsal to the rostral part of the medial accessory

j. The Nucleus olivaris inferior was recently termed Nucleus olivaris. The "inferior" is no longer necessary because the term Nucleus olivaris superior has been changed to Nucleus dorsalis corporis trapezoidei (World Association of Veterinary Anatomists, 2012). However, as we expect the reader to be more familiar with the former nomenclature we will use the terms Nucleus olivaris inferior (inferior olive) and Nucleus olivaris superior (superior olive) (Federative International Programme on Anatomical Terminologies (FIPAT), 2011) here.

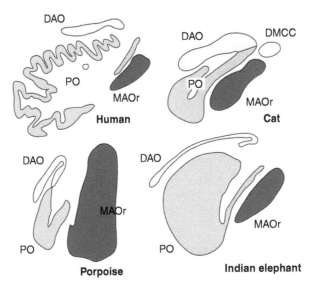

FIGURE 6.40 **Transverse sections through rostral levels of the inferior olivary complex in four mammals: human, cat, porpoise, and elephant.** In primates (human), the principal olivary nucleus (PO) is dominant and even strongly folded. The cat seems to show a comparatively unspecialized situation; here, the components are of about the same size. In the porpoise (a dolphin-like toothed whale), the rostral medial accessory olive (MAOr) is dominant; it is integrated into a premotor feedback loop together with the elliptic nucleus and the cerebellum with its C2 zone and posterior interposed nucleus (see text). The elephant, again, shows the strong dominance of the principal olivary component, which is solid here and not gyrated as in the human. Abbreviations: DAO, dorsal accessory olive; DMCC, dorsomedial cell column; MAOr, rostral medial accessory olive; PO, principal olive. *(Redrawn and modified by Jutta Oelschläger after Glickstein et al., 2007. Evolution of the cerebellum. In: Kaas, J. (Ed.), Evolution of Nervous Systems. A Comprehensive Reference, vol. 3, Mammals. Elsevier, Academic Press, Amsterdam.)*

inferior olive (MAIO), and lateral to the raphe nuclei. The MAIO projects via the olivocerebellar tract to the paraflocculus, from there to the posterior interposed nucleus and back to the elliptic nucleus and further to the thalamus and the neocortex. In addition to the projection from the elliptic nucleus via the mtt, the inferior olive receives motor input from the neocortex and, at the same time, sensory feedback from the spinal cord and the trigeminal system. As some kind of a parallel in primates, the red nucleus projects to the principal nucleus of the inferior olive via the central tegmental tract (Fig. 6.42: ctt). With respect to the coordination of motor behavior, the mtt seems to play a role as important in dolphins as is the case for the ctt in primates.

Summarizing this last paragraph, it seems likely that the cerebellum of dolphins (paraflocculus) plays an important role in the integration and processing of various cortical and subcortical input via (1) the pontocerebellar and spinocerebellar tracts, and (2) the input of the olivocerebellar tract.

The potential role of the elliptic nucleus, the pontine nuclei and the inferior olive in the dolphin brain is also shown in Fig. 6.41. Here, the input to the brainstem coming from the neocortex is relayed in the pontine nuclei (P), contralaterally innervating the cortex of the paraflocculus (Pfl), which, in turn, projects back to the neocortex via the cerebellar nuclei (posterior interposed nucleus, PIN) and the thalamus (T), with collaterals ending in the elliptic nucleus (E). Premotor projections descending from the limbic lobe (ACC) are relayed in the elliptic nucleus and proceed via the mtt further down to the inferior olive and back again to the neocortex via the cerebellar nuclei (PIN) and the thalamus/elliptic nucleus. It is therefore evident that a loop of connectivity may exist between the elliptic nucleus, the inferior olive, the paraflocculus/cerebellar nuclei and back to the elliptic nucleus. In the dolphin, they share a remarkable size progression, in parallel to that of the pontine nuclei. More schematically, this is shown in Figs. 6.42 and 6.43, with an indication on the reciprocity of the brain structures involved:

Primates: nucleus ruber → inferior olive (principal subnucleus) → cerebellum (lateral zone of hemisphere → dentate nucleus) → nucleus ruber.

Dolphins: nucleus ellipticus → inferior olive (medial accessory subnucleus) → cerebellum (intermediate C2 zone of paraflocculus → posterior interposed nucleus) → nucleus ellipticus.

From the extant literature, it appears that in dolphins the cerebellum may have some additional input from the red nucleus, which is comparatively smaller in dolphins than in primates. For further information see Section "Integrative Aspects – Audiomotor Navigation".

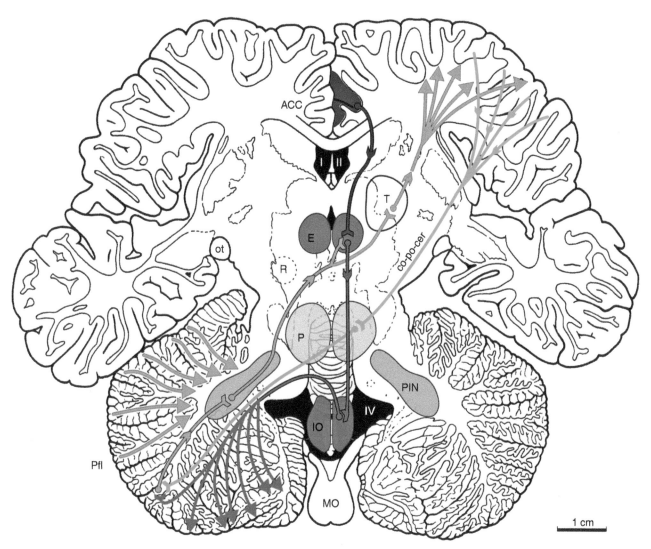

FIGURE 6.41   On a wider scale, the potential role of the inferior olive in the dolphin brain is visualized here. Input to the brainstem coming from the neocortex (corticopontine tract; *blue*) is relayed in the pontine nuclei (P), contralaterally innervating the cortex of the paraflocculus (Pfl). The latter projects back to the neocortex via the cerebellar nuclei (posterior interposed nucleus, PIN; *yellow*) and the thalamus (T). Collaterals terminate in the elliptic nucleus E. Premotor projections descend from the anterior limbic lobe (ACC) to the elliptic nucleus and further down to the inferior olive (IO) and back again to the ACC or neocortex via the cerebellar nuclei (PIN) and the elliptic nucleus/thalamus. It is therefore evident that there exists a loop of connectivity between the elliptic nucleus, the inferior olive, and the paraflocculus/cerebellar nuclei; they have in common a remarkable size progression, in parallel to that of the pontine nuclei (P). *(Modified by Jutta Oelschläger from Pilleri et al., 1980. Concise macroscopical atlas of the brain of the common dolphin (Delphinus delphis; Linnaeus, 1758). Brain Anatomy Institute, University of Berne, Waldau-Berne, Switzerland.)*

CRANIAL NERVES

In the head of vertebrates, there is a series of varied nerves that are, particularly at first sight, difficult to compare with those of the body (Romer and Parsons, 1977, 1991). They deserve close consideration for our further discussion on nervous function and sensory organs in dolphins. Table 6.1 gives fiber counts of single cranial nerves of four species as percentages of the total fiber count of all cranial nerves in the individual species.

Some of these cranial nerves are purely or largely somatomotor as the nerves of the eye muscles (oculomotor [III[k]], trochlear [IV], abducens [VI]), and the hypoglossal nerve [XII]. Others represent the mixed innervation (trigeminal [V], facial [VII], glossopharyngeal [IX], vagus [X], accessory [XI]) of the pharyngeal arches, which constitute the visceral part

---

k. The cranial nerves are the first (rostralmost) 12 pairs of nerves of the mammalian body. Classically, they are numbered I to XII. This system was introduced 1778 by S. Th. Soemmering (Starck, 1979) for the human and later adapted to vertebrates, in general. Therefore, an additional nerve (terminal nerve), discovered later in sharks and then in mammals, has been named Nervus 0.

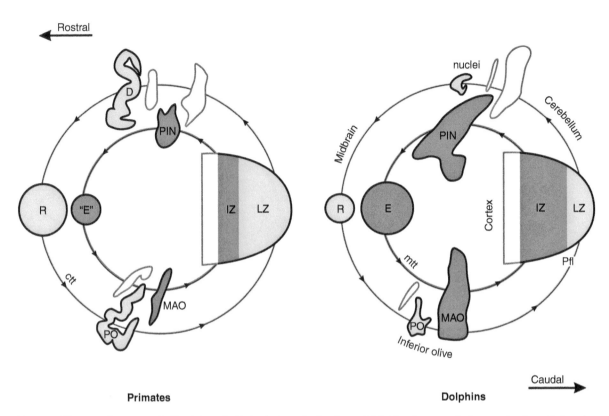

FIGURE 6.42 **The central premotor feedback loops in the brainstem of primates and dolphins.** In mammals, there are two loops that run in parallel in both groups of animals: the outer ring (in green) consists of the red nucleus (R), principal olive (PO), lateral cortical zone of the cerebellum (LZ), as well as its dentate nucleus (D). The inner ring here (in red) consists of the elliptic nucleus ("E" and E, respectively; nucleus of Darkschewitsch), median accessory olivary nucleus (MAO), intermediate C2 zone of the cerebellar cortex (IZ) and the posterior interposed nucleus (PIN). In both groups of animals, the components of one loop show an alternative quantitative correlation to those of the other loop; whereas primates have the red nucleus and the conjoined elements large, dolphins show hypertrophy in the elliptic nucleus and its associated elements. This principal deviation in the two groups of animals may be related to a different organization of the extrapyramidal system with respect to the motor innervation of the body stem (see also Fig. 6.52). *(Redrawn and modified by Jutta Oelschläger after Striedter, G.F., 2005. Principles of brain evolution. Sinauer Associates, Sunderland, MA.)*

of the head. Particularly interesting here are those nerves innervating the three main sensory organs of mammals (nose, eye, and ear): the olfactory [I], optic [II], and vestibulocochlear [VIII] nerves.

The "olfactory nerve" is not a typical cranial nerve. A normal sensory nerve (eg, trigeminal) usually has the cell bodies of its neurons located in a ganglion outside the brain and is part of the peripheral nervous system. In the olfactory nerve proper, the genuine nerve fibers are identical with axons of neurons which are located in the (olfactory) sensory epithelium of the nasal cavity (olfactory placode). Through the level of the later cribriform plate, the axons run inward to the brain (central nervous system) where they induce the olfactory bulb, which is a hollow protrusion of the rostrobasal telencephalic hemisphere. During embryonal growth of the head, the olfactory bulb is getting more and more slender and finally the "olfactory ventricle" is obliterated by the addition of fiber masses (olfactory tract) running to the olfactory tubercle.

The *optic nerve*, entering the brain at the optic chiasm, is not a genuine nerve either, but also a brain fiber tract, since the retina is formed as an outgrowth of the brain (diencephalon). Because, during embryogenesis, the optic cup originates as an outgrowth of the hypothalamus, its connection with the diencephalon remains a central part of the brain. In parallel to the situation in the olfactory tract, the elongation of the optic cup (optic stalk) loses its ventricle because of ingrowing fiber masses from the retina and is then a solid brain tract running from the retina to the lateral geniculate body.

The *vestibulocochlear nerve* is the third exclusively sensory nerve. It connects the inner ear with the brain but shows a somehow different situation than the preceding "cranial nerves I and II". On the one hand, the sensory part of the inner ear develops out of the otic placode which is invaginated. In its further development it is adjoined by material from the neural crest. The neurons in the vestibular and cochlear ganglia receive afferent fibers from the sensory epithelia and send axons via the vestibular and cochlear nerves to their central nuclei in the brainstem. Thus, on the other hand, the latter are typical peripheral nerves but their ganglia lie rather close to the sensory structures.

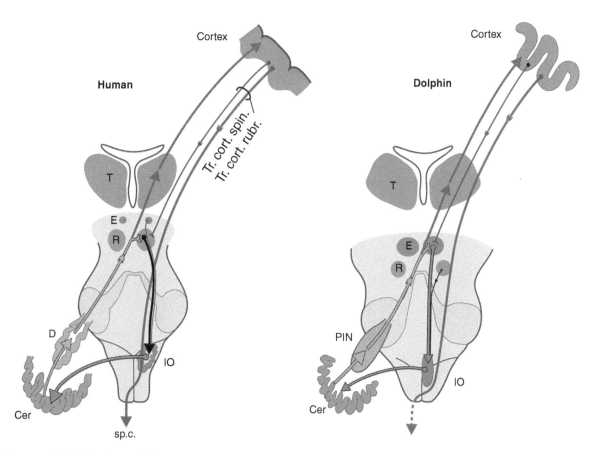

FIGURE 6.43 **Simplified scheme of the two central premotor feed-back loops in the brainstem of primates and dolphins (cf. Figs. 6.41 and 6.42).** In the human, the red nucleus (R) is the main gate for extrapyramidal (premotor) information processing. The large size of this nucleus is correlated with a similar dimension of the brain structures following in this loop, that is, the inferior olivary complex (IO, principal nucleus), the cerebellar cortex (Cer, lateral zone of hemisphere), dentate nucleus (D), and back to the red nucleus (R). In the dolphin, a sister loop running in parallel is engaged in the extrapyramidal innervation projecting to the motor neurons in the ventral horn of the spinal gray matter. Here, the elliptic nucleus (E) is the main gate for extrapyramidal (premotor) information processing; and the other members in this sister loop are the medial accessory subnucleus of the inferior olivary complex, the intermediate C2 zone of the cerebellar cortex (paraflocculus), which projects to the posterior intermediate nucleus (PIN, syn. globose nucleus of primates) and back to the elliptic nucleus. However, whereas in primates the size of the two "rivalling" gateways is extremely different (very large size of red nucleus, very small nucleus Darkschewitsch (ED), the situation is different in dolphins. Here the elliptic nucleus (E; syn. nucleus Darkschewitsch) is about as large as the red nucleus in primates, but their own red nucleus is comparatively well developed. It is possible that dolphins (like hoofed animals) use both main gates into extrapyramidal (premotor) information processing. They rely heavily on these systems with respect to mass movements of their body stem while their pyramidal system is underdeveloped to lacking in the spinal cord. A similar feedback loop in mammals is known using the nuclei pontis instead of the red nucleus or elliptic nucleus for the additional integration of sensory (particularly auditory) information. *(Modified by Jutta Oelschläger from Duus, 1976. Neurologisch-topische Diagnostik. Anatomie. Physiologie. Klinik. Georg Thieme Verlag Stuttgart).*

Cranial Nerve 0—Terminal Nerve

There are three functional systems in the nasal region of the mammalian head attached to the basal telencephalon: the olfactory system (see the following), the vomeronasal system, and the terminalis (terminal) system. In adult toothed whales including dolphins, only the terminal nerve is present (Figs. 6.44–6.46). The vomeronasal system is totally lacking and the olfactory system develops in the embryo, but then vanishes (Starck, 1979). While the olfactory and vomeronasal systems have been clearly defined as chemoreceptor organs, the functional significance of the terminal system in mammals is still largely unknown. During the last decades, however, there has been some activity as to the development of the terminal nerve in several mammalian groups. The terminal system arises medial to the olfactory bulb, which is induced by axons of the sedentary olfactory receptor cells in the olfactory placode that grow into the brain wall. In contrast, terminal neuroblasts as the second cell population in the olfactory placode, leave and migrate to attach to the telencephalic wall close to the olfactory nerve, which induces the ephemeral olfactory bulb (see the following). Whether the perikarya of the terminal neuroblasts enter the brain wall or simply send fibers to induce intrinsic neuroblasts of

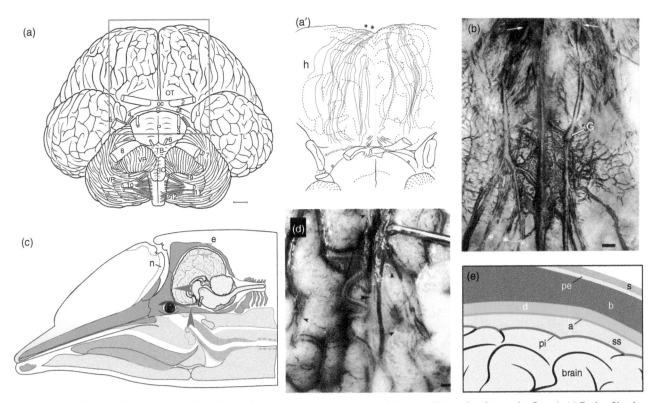

FIGURE 6.44   Terminal nerve topography and its distribution within the meninges (sales: a = 10 mm; b = 2 mm; d = 5 mm). (a) Brain of bottle-nose dolphin in basal aspect. Frame indicates the anterior central part of the brain (magnified in a′). For orientation and the course of the terminal nerve see (a′) and (c) (red lines and arrowheads; cf. Oelschläger and Buhl, 1985). (b) Anterior wall of skull in caudal aspect (*horizontal arrowhead* in c) with the covering dura mater in situ; the frontal lobes of telencephalic hemispheres have been removed. *Asterisks* in a′ and b indicate where the connecting central strands of the terminal nerve have been cut. In the dura, the terminal nerve contains large ganglia (G) on both sides of the falx cerebri. The ganglia receive multiple strands from the brain and the subarachnoid space, respectively (a′ and b). Dorsally, the ganglia send peripheral strands to the ethmoid region where they seem to penetrate the anterior wall of skull (former cribriform plate) in the direction of the nasal passage (n), which runs vertically in dolphins. (c) Survey on the course of the terminal nerve (*red*) shown in the mediosagittal section of the dolphin head. (d) Ventral aspect as in (a) with the basal surface of the brain. It is covered by arachnoid (a in e) and shows terminal nerve fiber strands (*small arrowheads*) in the subarachnoid space. The larger arrowhead indicates the continuation of strands in the subarachnoid space. Probe indicates one strand that has been dissected out of the arachnoid. Abbreviations: a, arachnoid; b, bone; d, dura mater; G, ganglion; h, telencephalic hemisphere; pe, periost; pi, pia mater; s, skin; ss, subarachnoid space. *(Redrawn by Jutta Oelschläger from Ridgway et al., 1987. The terminal nerve in odontocete cetaceans. Ann. N. Y. Acad. Sci. 519, 201–212.)*

the brain wall is not yet clear (Buhl and Oelschläger, 1986). As a result, the terminal nerve as a continuum of LHRH[l] (GnRH)-producing neurons from the placode to the brain wall forms the initial part of the so-called hypothalamo-hypophyseal-gonadal axis,[m] which is essential for the maturation of the gonads and thus for reproduction. In prenatal dolphins, this more or less diffuse pathway extends to the olfactory tubercle and septal region where fiber strands enter the brain wall. In a comprehensive study by Ridgway et al. (1987) on adult bottlenose dolphins and other delphinid species, macroscopic fiber strands with interspersed ganglia could be followed centrally and were seen to accompany branches of the cerebral arteries and to enter the brain wall about the anterior perforated substance (olfactory tubercle), the optic chiasm, and presumably the optic tract and prepiriform cortex (Fig. 6.44). In this impressive study, conservative estimates determined several thousands of large round neuronal perikarya encapsulated by a monolayer of satellite-like cells. In between these potential sensory or maybe autonomic neurons only a few fusiform (bipolar) perikarya showed LHRH-immunoreactivity (Fig. 6.45).

Interestingly, while in terrestrial mammals, including the human, the olfactory system persists and the terminal system is more or less strongly reduced up to the adult (usually to a few hundred neurons), toothed whales show the opposite

l. Gonadotropin releasing hormone (GnRH; synonym: luteinizing hormone releasing hormone, LHRH) is a trophic peptide hormone responsible for the release of follicle-stimulating hormone (FSH) and luteinizing hormone (LH) from the anterior pituitary.
m. The hypothalamo-hypophyseal-gonadal axis is a critical part in the development and regulation of a number of the body's systems, such as the reproductive and immune systems because these glands act in cooperation. Fluctuations in the hormones cause, in turn, changes in the hormones produced by each gland and have various effects on the body (Kallmann's syndrome).

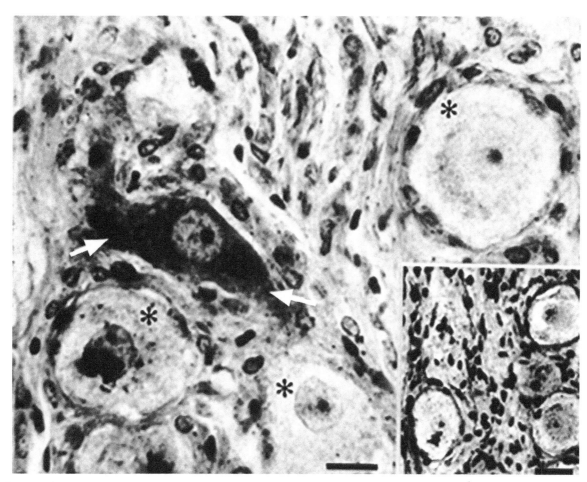

FIGURE 6.45   Section through an intracranial terminal nerve ganglion in the common dolphin (scale bar = 15 μm; inset = 25 μm). The tissue was osmicated, embedded in paraffin, labeled for luteinizing hormone–releasing hormone (LHRH, GnRH) and counterstained with cresyl violet. The numerous large round perikarya (*asterisks*) seem to have a layer of satellite glia (not labeled) and may represent pseudounipolar (somatosensory) neurons with highly myelinated axons. The few fusiform perikarya (*white arrows*) obviously belong to bipolar neurons. Some of them are immunopositive for LHRH; immunolabeled nerve fibers are associated with nonreactive perikarya and were seen to course with large fascicles of the terminal nerve. (*Modified from Ridgway et al., 1987. The terminal nerve in odontocete cetaceans. Ann. N. Y. Acad. Sci. 519, 201–212.*)

situation: They totally reduce the olfactory nerve and bulb from early fetal stages onward and keep a maximum of terminal neurons within the mammalia. The situation that in the adult bottlenose dolphin the terminal ganglia exhibit only very few LHRH immunoreactive neurons, among a large majority of pseudounipolar neurons, is somehow enigmatic. In this respect, it may be noteworthy that this nerve was called "general somatic sensory" by Romer and Parsons (1986).

On the basis of the information available, it has been speculated that the terminal nerve in adult dolphins could still have a function serving the nasal organ. In these animals, the latter has been transformed into a pneumatically driven system to produce sounds and ultrasound with high air pressure. Sensory nasal epithelia, therefore could not be found by systematic histologic investigation of several individuals of the harbor porpoise (Prahl et al., 2009). An autonomic component in the terminal nerve could be important for the maintenance of optimal conditions in the nasal mucosa as to the generation and emission of echolocation and communication sounds. Here, the regulation of the subepithelial connective tissue via the innervation of local blood vessels could change conditions as to the physical properties of the sound-generating apparatus, last not least during diving maneuvers. Another idea as to potential functions of the terminal nerve in the adult dolphin again refers to the upper respiratory tract. Thus, a dense sensory innervation of the mucosa and underlying connective tissue by means of the terminal nerve (Huber, 1934) could serve as the afferent pathway of a sensorimotor loop, monitoring pressure within the nasal air sacs and communicating this information via the brainstem and the reticular formation to the facial motor nucleus and its nerve. The latter, as the efferent pathway, should be able to prompt quick adjustments of the facial musculature, and thus the acoustic properties of associated tissues in the blowhole region.

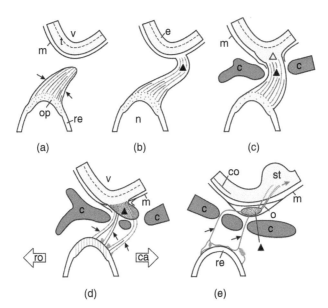

FIGURE 6.46   **Schematic representation of the ontogenetic development of olfactory bulb and terminalis ganglion in toothed whales (*Phocoena* type).** Sagittal reconstructions. (a) 10 mm crown-rump length (CRL) stage. Olfactory fila (*black arrows*) growing out of the olfactory placode (op). (b) and (c) 14 and 22.4 mm CRL stage. Olfactory fila form a filamentous trunk (solid triangle), dissolve the primitive meninx (m), and induce a primordial olfactory bulb (light triangle). (d) 28.6 mm CRL stage. Olfactory bulb anlage in slight regression. e) 45 mm CRL stage. The ganglion of the placodal component (solid triangle) is uncoupled from the telencephalon. Abbreviations: c, cribriform plate; ca, caudal; co, cortical layer; e, ependymal layer; m, primitive meninx; o, nervus terminalis; re, respiratory mucosa; ro, rostral; st, primitve striatum; t, telencephalic wall; v, ventricle; *asterisk*, nasal cavity. (*Redrawn and modified by Jutta Oelschläger after Buhl and Oelschläger, 1986. Ontogenetic development of the nervus terminalis in toothed whales. Evidence for its non-olfactory nature. Anat. Embryol. 173, 285–294.*)

"Cranial Nerve" 1—Olfactory Nerve

This nerve is still little known but very important for the development and adult life of mammals and somehow even spectacular as to its ephemeral existence in toothed whales.

Out of the three functional systems in the nasal region of the mammalian head, only the terminal nerve is present in the postnatal dolphin; the vomeronasal system is totally lacking and the olfactory system develops in the embryo but then vanishes (Fig. 6.46). In the adult dolphin, the anterior olfactory system (olfactory mucosa, olfactory nerve, and the olfactory bulb including the olfactory tract) are totally lacking. As to the remaining olfactory system, the olfactory tubercle is large but highly modified and perhaps no more involved in olfaction and the functional implications of the prepiriform and other paleocortical areas is unknown (see Section "Olfactory System/Paleocortex" in this chapter).

"Cranial Nerve" 2—Optic Nerve

In most cetaceans, the visual system is reported to be fairly well developed (Figs. 6.16, 6.27, 6.32, 6.49, and 6.54). The optic nerve in dolphins may be thicker than in terrestrial mammals of the same body dimension (eg, bottlenose dolphin: 7 mm; common dolphin: 5 mm; horse: 5.5 mm; human: 4.5 mm) (Blinkov and Glezer, 1968; Nickel et al., 2004; Pilleri and Gihr, 1970). In the adult harbor porpoise, the optic nerve contains 81,700 axons and in the bottlenose dolphin 147,000–390,000 which is low compared to axon counts in the human (see Section "Visual System", and Chapter 5, Section "Vision: Anatomy and Optics of the Eye"). In comparison with the bottlenose dolphin, a highly active marine species, the Amazon River dolphin (*I. geoffrensis*) shows a rather low axon count (15,500), and the optic nerve of the Indian river dolphin (Susu; *Platanista gangetica*), whose eye lacks a lens and may be capable of serving as a light receptor only, was reported to contain between a few hundred and 16,000 axons. In contrast, the mean fiber diameter in a few cetacean species was assessed as 1.6 to 2.5 times greater than in the cat and 3 to 5 times greater than in humans (Gao and Zhou, 1992; Morgane and Jacobs, 1972; Supin et al., 2001).

The low density of ganglion cells in the cetacean retina corresponds to a low density of axons in the optic nerve. In dolphins, cross sections of the optic nerve revealed a fiber density of 48,000 fibers/mm^2 (Supin et al., 2001). For comparison, in humans and monkeys, the optic fiber density approaches 220,000 fibers/mm^2; thus, while the cross-sectional area of the optic nerve in dolphins is larger than in terrestrial mammals, the total number of optic fibers is markedly lower. As an explanation, more than 50% of the cross-sectional area of the dolphin optic nerve is occupied by extraneural space (compared with 12–20 % in terrestrial mammals).

Bottlenose dolphins have a thick retina comprised of rods and cones. However, there is no distinct fovea centralis but a rostral and a temporal fovea. The retina is thick and shows a layer of giant optic ganglion cells (up to 150 µm in diameter), with thick dendrites and myelinated axons of up to 9 µm diameter. Cetaceans exhibit laterally placed eyes and, if at all, one small binocular visual field rostrally and ventrally as well as another one dorsally and slightly caudally (see Chapter 5, Section "Vision: Anatomy and Optics of the Eye"). The optic fibers show a complete or almost complete decussation, and the lateral geniculate body is not laminated (see Section "Visual System" in this chapter). The superior colliculus is large in most cetaceans but dominated in size by the inferior colliculus in toothed whales, which is an equally important center in the ascending auditory pathway (see Section "Auditory System" in this chapter). In toothed whales, the size dimensions of the two colliculi (caudal/rostral; measured in collicular length x width) may attain even 7:1 in the Susu (*Platanista gangetica*). In this respect, the Indian river dolphin *Platanista* is an extreme because of the strong reduction of its visual system (cf. Knopf et al., in press).

Marine dolphins with their well-developed visual system (eg, common dolphin) show a faint stratification of the superior colliculi in stained histological sections as well as in MR scans, where a layering typical of that in terrestrial mammals (cat) is recognizable (Fig. 6.9; Oelschläger et al., 2008). In terrestrial mammals, the superior layers are dedicated to the processing of visual input, but layer IV (intermediate gray layer) shows a whole set of modules that belong to sensory or motor-associated systems. It is here, the site for integration of many incoming and outgoing fiber systems, where seems to be located to allow eye and head movements as well as navigation after sensory cues (visual, auditory, etc.) (Platt et al., 2004; Harting, 2004). Recently, some indications have been found in a subterranean rodent (Ansell's mole-rat, *Fukomys anselli*) that the superior colliculus may be involved in the processing of magnetic information (Nemec et al., 2001; see also Chapter 5, Section "Magnetosensation"). Keeping this in mind, it seems probable that in dolphins, as in terrestrial mammals, the superior colliculus not only serves as an important relay station in the visual pathway but, at the same time, represents a capable superior association center of all kind of sensorimotor integration. Particularly important for dolphins is the navigation by auditory cues coming "passively" from the environment or actively as reflections of their own sonar (sound) emissions in the sense of acousticomotor behavior (see Chapter 5, Section "Ear: Sense of Balance and Hearing", and Chapter 10, Section "Audition").

All in all, dolphins obviously have some limitations as to vision in their environment with respect to water quality, time of day, and weather conditions. On the other hand, their orientation by visual cues may be much better than was formerly thought. Jansen and Jansen (1969) wrote "When one considers the stunts of which the captive porpoises [particularly bottlenose dolphins] are capable and the accuracy with which they will jump out of the water to pick a fish from the caretaker's hand, it seems obvious that vision at least in these species must be very good."

GENUINE PERIPHERAL CRANIAL NERVES

Concerning the characteristic mammalian organization of the brainstem (mesencephalon to myelencephalon) as to cranial nerves proper and their nuclei, we can follow the columnar pattern of grey matter which derives from the basal and alar plates (Romer and Parsons, 1986) in the neural tube of the embryo (Figs. 6.5 and 6.47). Each of the columns represents a specific neurobiological system, which, in principle, is homologous with the columnar organization of the spinal cord (Fig. 6.6). In the brainstem, however, the individual columns may divide into secondary columns, and successively be interrupted by ingrowing fiber tracts, thus forming more or less longitudinal chains of single nuclei of the respective physiological quality (Figs. 6.47, and 6.48).

These secondary columns and chains, respectively, are from ventral to dorsal and from medial to lateral (Fig. 6.47): somatic motor (red), special visceral motor (violet), general visceral motor (yellow), general visceral sensory (light green), special visceral sensory (dark green), general somatic sensory (light blue), and special somatic sensory (dark blue). More information as to the identity, size, function, and topographic localization of the cranial nerves and their nuclei is given in the following.

Unfortunately, there are two different but similar nomenclatures commonly used for the various physiological qualities of the columns. They use the terms *sensory* for *afferent* and *motor* for *efferent* and vice versa. In Figs. 6.5 and 6.47 we closely follow the nomenclature used by Romer and Parsons (1986). Figure 6.48 has a more complicated color coding with respect to the conditions in cattle and dolphin and additional structures are included in the dolphin for general orientation and functional interpretation in the text (see figure legend there).

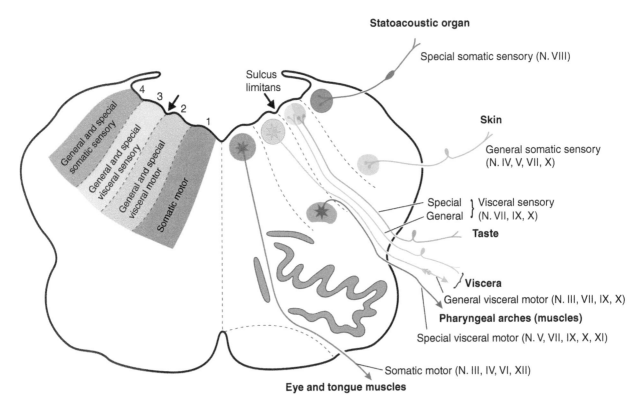

FIGURE 6.47 Schematic cross section through the brainstem showing the functional nuclear columns on the left side. The neurophysiological composition of the genuine cranial nerves and their respective longitudinal chains of nuclei on the right side, together with their peripheral structures. (*Redrawn and modified by Jutta Oelschläger after Benninghoff and Drenckhahn 2004. Anatomie—makroskopische Anatomie, Histologie, Embryologie, Zellbiologie, 16., völlig neu bearb. Aufl. ed. Elsevier Urban & Fischer, München.*)

Cranial Nerves 3, 4, and 6—Oculomotor, Trochlear, Abducens Nerves

Within the brainstem of dolphins, the locations of the somatic motor nuclei of the third, fourth, and sixth cranial nerves are generally typical for these nuclei in other mammals (Fig. 6.48). In bottlenose dolphins, all of these nuclei contain large multipolar neurons with rather coarse but evenly distributed Nissl material. The oculomotor nucleus is prominent and divided into cell clusters, which may exhibit a somatotopic pattern similar to that in terrestrial mammals. The general visceral efferent (parasympathetic) nucleus Edinger-Westphal, responsible for the accommodation reflex and the narrowing of the pupillary aperture (innervation of the ciliary muscle and the sphincter pupillary muscle, respectively) so far was not identified in dolphins but described to be small in the black finless porpoise (*Neomeris phocaenoides*) (Morgane and Jacobs, 1972).

Hosokawa (1951) showed a typical mammalian innervation of all extrinsic ocular muscles by somatic efferent fibers of the oculomotor nerve, except for the superior oblique and lateral rectus muscles, which are innervated by the trochlear and abducens nerves, respectively. In cetaceans, generally, the majority of rectus muscle fibers insert in the eyelids and are more correctly termed palpebral muscles (Pütter, 1902). Moreover, cetaceans have well-developed retractor muscles that insert into the sclera and are innervated by abducens fibers. Although no systematic study of the oculomotor mechanisms of cetaceans has been undertaken, it appears likely that the retractor bulbi and the oblique muscles are primarily involved in retraction and the movement of the eyeball because of their scleral attachments.

In the bottlenose dolphin, the oculomotor nerve comprises the highest number of axons among the three nerves mentioned (Morgane and Jacobs, 1972). Nevertheless, the oculomotor nerve comprises only about one third of the fibers found in the human, who is grossly of about the same body dimension, followed by the much smaller trochlear and abducens nerves. This comparatively poor neural equipment of the outer eye muscles of dolphins is potentially indicative of a moderate moveability of the eye. This argument and the fact that the optic nerve is obviously well developed leads to the presumption that the dolphin's eye may serve as a light collector for fast reactions to visual cues rather than as a device for the detection and smooth pursuit of prey seen in carnivorous mammals.

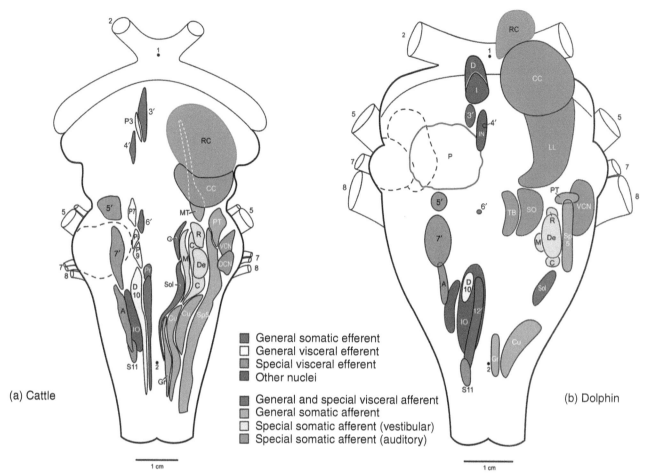

FIGURE 6.48 Schematic view of the brainstem showing the functional nuclear columns in the cattle (*Bos taurus*, a) and in the dolphin (*Tursiops truncatus*, b), brought to the same length (Scales: 1 cm). (1, 2, reference points in optic chiasm and pyramidal decussation). The brainstem of the cattle is a little longer but narrower than that of the dolphin which is rather compact in the mesencephalon, pons and anterior myelencephalon due to the extreme size of the auditory system. The width of the dolphin brainstem also seems to be due to the extraordinary development of fiber systems including the cranial nerves (trigeminal, cochlear), the voluminous reticular formation and the inferior olivary complex. In contrast, the vestibular nuclei in the dolphin are much reduced with the exception of the lateral (Deiters') vestibular nucleus (see text). In dolphins, there are no traces of parasympathetic nuclei for the salivary glands which are absent. In both species, some of the nuclei could not be included because of lacking information. Brainstem nuclear columns in a) left side, from medial to lateral: general somatic efferent (red); general visceral efferent (yellow); special visceral efferent (orange); other nuclei (brown). Right side, from medial to lateral and from caudal to rostral: general and special visceral afferent (violet); general somatic afferent (dark blue); special somatic afferent (vestibular, light blue); special somatic afferent (auditory; green). In contrast to that in the cattle, the situation in the bottlenose dolphin is rather different. In most cases, the relative size of the nuclei can be correlated with the functional significance of the relevant systems. Some nuclei reconstructed in the fetal narwhal by Holzmann (1991) were extrapolated to the situation in the adult bottlenose dolphin whenever possible. 2, optic nerve; 5, trigeminal nerve; 7, facial nerve; 8, vestibulocochlear nerve; 3', motor nucleus of oculomotor nerve; 4', nucleus of trochlear nerve; 5', motor nucleus of trigeminal nerve; 6', nucleus of abducens nerve; 7', motor nucleus of facial nerve; 12', nucleus of hypoglossal nerve; A, nucleus ambiguus; C, caudal vestibular nucleus; CC, caudal (inferior) collicle; Cu, cuneate nucleus; D10, dorsal nucleus of vagus nerve; DCN, dorsal cochlear nucleus; De, Deiters' nucleus (Nucleus vestibularis lateralis); G, gustatory nucleus; Gr, gracile nucleus; I, interstitial nucleus of Cajal; IN, interpeduncular nucleus; IO, inferior olive; LL, nucleus of lateral lemniscus; M, medial vestibular nucleus; MT, mesencephalic nucleus of trigeminal nerve; P, pontine nuclei; P3, parasympathetic nucleus of oculomotor nerve (Edinger-Westphal); P7, parasympathetic nucleus of facial nerve; P9, parasympathetic nucleus of glossopharyngeal nerve; Pi, parasympathetic nucleus of nervus intermedius; Pr, prepositus nucleus; PT, pontine nucleus of trigeminal nerve; R, rostral (superior) vestibular nucleus; RC, rostral (superior) collicle; S11, spinal nucleus of accessory nerve; SO, superior olivary complex; Sol, Nucleus solitarius; Sp5, spinal nucleus of trigeminal; TB, nucleus of the trapezoid body; VCN, ventral cochlear nucleus. Broken lines, sectioned cerebellar peduncles (horizontal plane). *((a) redrawn by Jutta Oelschläger from Barone and Bortolami (2004). Anatomie Compareé des Mammifères Domestiques. Vol. 6. Neurologie I, Système nerveux central. Vigot Frères, Editeurs, Paris.)*

Cranial Nerve 5—Trigeminal Nerve

The trigeminal nerve is the thickest cranial nerve in baleen whales and sometimes in the sperm whale. In all of the other toothed whales, the vestibulocochlear nerve has the maximal diameter (Figs. 6.32, 6.35, 6.48, and 6.49). Whereas the maxillary and mandibular branches are well developed, the ophthalmic branch is rather thin (Fig. 6.50) (Buhl and Oelschläger, 1986; Langworthy, 1931). The diameters of the axons tend to be thin in the trigeminal nerve. In the bottlenose dolphin, the trigeminal nerve contains 151,000 axons (human: 140,000) (Morgane and Jacobs, 1972). At first sight, this large axon number for the bottlenose dolphin may be unexpected. An answer to this might be that the trigeminal nerve is involved in the sensory innervation of the highly derived and extremely important blowhole region in the dolphin's forehead with its secondary accessory nasal air sacs, secondary "vocal cords," and acoustic fat bodies (see Chapter 5, Section "Dolphin Head Characteristics"). This area (apart from skin sensitivity) is essential for the dolphin because here sound and ultrasound signals for communication and echolocation are produced, focused, and emitted into the surrounding water. These functional implications may also qualify the trigeminal nerve (sensory part) as a counterpart of the facial nerve (motor part), which operates the blowhole muscles in sonar production. In the dolphin, all the associated facial musculature is concentrated in the blowhole area and the number of axons in the facial nerve is about sevenfold higher than in the human (see the following).

The trigeminal nerve is equipped with a series of brainstem nuclei (Fig. 6.48) located in different gray matter columns. Apart from one special visceral efferent (motor) nucleus (5'), there are three general somatic afferent (sensory) nuclei (principal nucleus, PT; mesencephalic nucleus; spinal nucleus, Sp5), and one special visceral afferent nucleus (solitary nucleus, Sol). All these nuclei have been identified in the respective areas in toothed whales, including dolphins, by a series of investigators. However, the authors sometimes disagreed as to the specifics of the single nuclei. Thus, for example, the motor nucleus of the trigeminal was described as large or small by different authors (De Graaf, 1967; Jansen and Jansen, 1969; Langworthy, 1931; Morgane and Jacobs, 1972). Unfortunately, in the literature, there are only very few reliable data as to the size of structures both in absolute and relative terms. In a fetal narwhal of 137 mm total length, which was carefully investigated by three-dimensional reconstruction from a complete microslide series, the motor nucleus of the trigeminal nerve was distinctly smaller than the motor nucleus of the facial nerve (Holzmann, 1991). An explanation for this seems to be the feeding mode of dolphins. They grasp fish with their teeth and swallow them whole; thus, there is no mastication as in related animals like artiodactyls, which chew thoroughly with their highly specialized masticatory apparatus. For this, artiodactyls need a strong masticatory musculature, comparatively large motor nucleus, and a strong motor root of the trigeminal nerve.

As to the sensory trigeminal structures in the brainstem of the bottlenose dolphin, the spinal trigeminal nucleus and its tract are very well developed and extend to spinal levels caudally (Morgane and Jacobs, 1972; Oelschläger et al., 2008, 2010). In the adult animal, the main sensory nucleus and the motor nucleus are prominent and have a similar rostrocaudal extension of 5 mm in reconstructions from transverse microslide series (Morgane and Jacobs, 1972). The remarkable size of the main sensory nucleus is also reflected in the development of the thalamic ventral posteromedial nucleus, which was reported to be larger in the dolphin than in the human. In mammals, this system is involved in the tactile innervation of the face. And the size of the trigeminal system correlates well with the fact that the dolphin forehead houses the nasal sound/ultrasound generator and transmitter used for communication and for orientation and echolocation in hunting (see Chapter 5, Section "Dolphin Head Characteristics").

Cranial Nerve 7—Facial Nerve

The facial nerve of cetaceans is very well developed. Using data given in Morgane and Jacobs (1972), the harbor porpoise and bottlenose dolphin have axon counts in the facial nerve five- to seven-fold higher than the human. As to the composition of the nerve in the bottlenose dolphin, the motor root innervating the facial muscles comprises two-thirds of the 49,100 fibers. The remainder fibers belong to the intermediate nerve, which consists of sensory and secretory fibers. At the skull base, after the passage through the facial canal in the middle ear region, the motor component (nerve) runs without further branching from the stylomastoid foramen at the side of the skull along the zygomatic arch (Fig. 6.50). At the antorbital notch, the nerve turns to the surface of the skull roof and divides to innervate the blowhole musculature. These muscles represent the main portion of the maxillonasolabial muscle and are responsible for the operation of the blowhole as well as nasal vocalization and ultrasound emission of dolphins and other toothed whales, a feature not known in baleen whales.

In correlation with the high number of fibers in the facial nerve, its motor nucleus is rather large; in the common dolphin it even bulges the ventral surface of the brain (Tuberculum faciale) (Breathnach, 1960; Hatschek and Schlesinger, 1902).

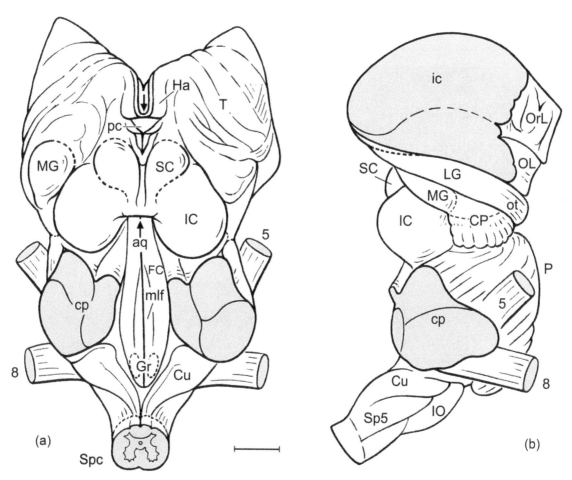

FIGURE 6.49 **Brainstem of the bottlenose dolphin (*T. truncatus*); dorsal (a) and lateral (b) aspects (scale = 1cm).** (a) Dorsal, (b) lateral aspect. Abbreviations: cp, cerebellar peduncles; CP, cerebral peduncle; Cu, cuneate nucleus; FC, facial colliculus; Gr, gracile nuclei; Ha, habenula; ic, internal capsule; IC, inferior (caudal) colliculus; IO, inferior olive; LG, lateral geniculate body; MG, medial geniculate body; mlf, medial longitudinal fascicle; OL, olfactory lobe (olfactory tubercle); OrL, orbital lobe; ot, optic tract; pc, posterior commissure; P, pons; SC, superior (rostral) colliculus; Spc, spinal cord; Sp5, spinal nucleus of trigeminal nerve; T, thalamus; 5, trigeminal nerve; 8, vestibulocochlear nerve. *Arrows* in (a) pointing into cerebral aqueduct. Scale: 1 cm. *(Redrawn by Jutta Oelschläger from Langworthy, 1931. A description of the central nervous system of the porpoise* (Tursiops truncatus). *J Comp Neurol 54,437–499; modified after Pilleri and Gihr, 1970. The central nervous system of the mysticete and odontocete whales. In: Pilleri, G. (Ed.), Investigations on Cetacea. Institute of Brain Anatomy, Bern; and Morgane and Jacobs, 1972. Comparative anatomy of the cetacean nervous system. In: Harrison, R.J. (Ed.), Functional Anatomy of Marine Mammals. Academic Press, London.)*

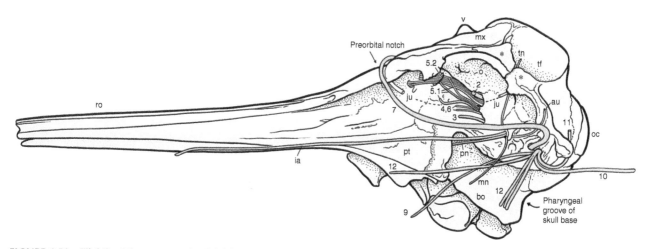

FIGURE 6.50 **Slightly oblique aspect of a dolphin skull (Stenella spec.) from the left, lateral, and ventral.** Tip of rostrum slightly rotated to the right in order to get better view on the exits of the cranial nerves out of the skull base. Tympanoperiotic complex (ear bones) is lacking. Abbreviations: 3, oculomotor nerve; 4, trochlear nerve; 5.1, ophthalmic branch of trigeminal nerve; 5.2, maxillary branch of trigeminal nerve; 6, abducens nerve; 9, glossopharyngeal nerve; 10, vagus nerve; 11, accessory nerve; 12, hypoglossal nerve; au, auriculotemporal nerve; bo, basioccipital bone; ia, inferior alveolar nerve; ju, jugal bone; mn, monogastric nerve; mx, maxillary bone; my, mylohyoid nerve; o, orbit; oc, occipital condyle; pn, pterygoid nerve; pt, pterygoid bone; tf, temporal fossa; tn, temporal nerve; v, vertex of skull; *, secondary zygomatic arch. *(Image courtesy of G. Klauer, Frankfurt; from Rauschmann, 1992. Morphologie des Kopfes beim Schlanken Delphin Stenella attenuata mit besonderer Berücksichtigung der Hirnnerven. Makroskopische Präparation und moderne bildgebende Verfahren. Inaugural-Dissertation des Fachbereichs Humanmedizin der Johann Wolfgang Goethe-Universität, Frankfurt am Main. Identification of skeletal elements with the help of Mead and Fordyce, 2009.)*

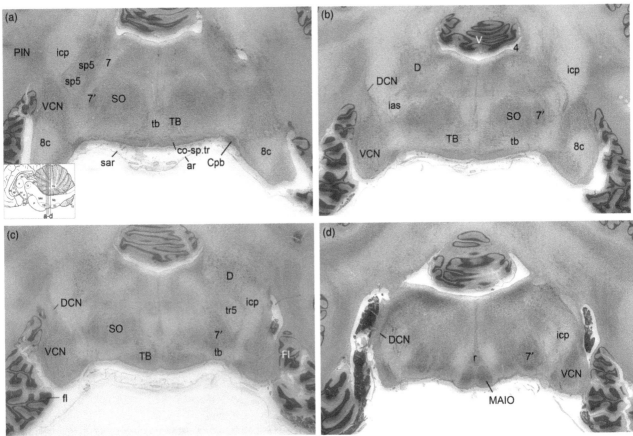

FIGURE 6.51 **Topographical representation of the vestibular and auditory systems of the common dolphin in a rostrocaudal series of transverse (coronal) sections through the anterior myelencephalon.** (Nissl cell stain) The four sections span the distance between the middle of the ventral cochlear nucleus (VCN) with the cochlear nerve (8c) and the superior olive (SO) rostrally (a) and the anterior part of the inferior olivary complex (MAIO) caudally (d). As to the vestibular nuclei, only Deiters' nucleus (D) is seen here. It is very large and most prominent in (b) and particularly in (c). The other three vestibular nuclei are very much reduced and therefore difficult to detect (see Fig. 6.48). The auditory system is well represented in this area of the brainstem. The cochlear nerve enters the ventral cochlear nucleus, which is very prominent (acoustic tubercle). The following components of the ascending auditory pathway are also very large with one exception, the dorsal cochlear nucleus (DCN). The trapezoid body and nucleus (tb, TB) are well developed and seen dorsal to the thin corticospinal tract (co-sp.tr), which covers the basal surface of the myelencephalon. It disappears in the upper cervical segments of the spinal cord. Dorsal to the trapezoid structures the very large superior olive (SO) is seen, it is maximally extended in (a) and (b). The facial motor nucleus is also very large and adjoins the SO laterally. Its maximal diameter is attained in (c) and (d). Like Deiters' nucleus, the facial motor nucleus is easily to identify by means of its large somata; the same is true for the spinal nucleus of the trigeminal nerve adjacent to its fiber tract (Sp5, sp5). Abbreviations: 7, facial nerve; 7', facial nucleus; ar, arachnoid mater; Cpb, corpus pontobulbare; Fl, flocculus; ias, intermediate acoustic stria; icp, inferior (caudal) peduncle; r, raphe nucleus; sar, subarachnoid space with blood vessels; tr5, spinal tract of trigeminal nerve. *(From Malkemper et al. (2012). The dolphin cochlear nucleus: topography, histology and functional implications. J. Morphol. 273, 173–185.)*

Among the nuclei within the dolphin brainstem from the rostral border of the mesencephalon to the caudal border of the myelencephalon, the facial nucleus belongs to the middle-sized nuclei (Figs. 6.39 and 6.51) and seems to be several times larger than the motor nucleus of the trigeminal nerve. In domestic animals (cattle), both nuclei seem to be similar in size (Fig. 6.48). The motor nucleus of the facial nerve is composed of subgroups of neurons, which, as in other mammals, may be dedicated to the motor supply of individual areas of the facial musculature (Barone and Bortolami, 2004; Oelschläger et al., 2010).

As to functional implications, the large size of the facial motor nucleus and the high number of motor axons are indicative of a low muscle fiber/nerve fiber ratio in the motor units. This may be favorable for precision movements within the sheets of the blowhole musculature concerning nasal echolocation and communication.

The existence of an intermediate nerve and chorda tympani in dolphins was published several times by different authors; however, so far no superior salivatory nucleus was found in their brainstem. In mammals, this parasympathetic nucleus innervates the submandibular and sublingual glands as well as the lacrimal gland; in cetaceans, salivary glands are lacking.

Tarpley and Ridgway (1991) have thoroughly investigated the so-called "Harderian gland" in the socket of the bottlenose dolphin eye (see Chapter 5, Section "Vision: Anatomy and Optics of the Eye"). The innervation of this gland is probably taken over by the intermediate nerve (greater petrosal nerve). Thus, the large number of fibers in the intermediate nerve of the bottlenose dolphin could support the assumption that this nerve conveys general somatic afferent as well as general visceral efferent (parasympathetic) fibers as it does in other mammals. Centrally, the special visceral afferent (gustatory) fibers of the facial nerve represent the most rostral contribution to the solitary tract in mammals (Benninghoff and Drenckhahn, 2004). This tract is prominent in the bottlenose dolphin, as it appears to be in other cetacean species (De Graaf, 1967), although taste reception in the regularly sense may not be present in these animals.

Cranial Nerve 8—Vestibulocochlear Nerve

The superior relevance of the auditory system for the life of whales and dolphins was recognized by Spitzka (1886) and by Kükenthal and Ziehen (1893). Since that time, the extraordinary development of the cochlear nerve (Figs. 6.48 and 6.49) and the structures of the ascending auditory pathway (nuclei and fiber tracts) in the brain could be confirmed repeatedly by later publications (eg, Malkemper et al., 2012; Oelschläger et al., 2010; Oelschläger and Oelschläger, 2009; Ridgway, 1990; Supin et al., 2001 and citations therein) (see Section "Auditory System" in this chapter).

The central role of the auditory system in the life of dolphins is also obvious in quantitative relations between auditory structures and the brain as a whole. However, apart from their (mostly) excessive size, the components of the auditory pathway exhibit structural features known from other mammals including the human. Moreover, they show many specific adaptations reflecting the structural and functional requirements necessary for survival. Thus, for example, the ancestors of whales and dolphins faced the problem of impedance matching again during their evolutionary "reentry" into the aquatic environment (Vater and Kössl, 2004). Their terrestrial ears were inappropriate for function underwater and needed to be (re)adapted to the acoustic properties of the latter (see Chapter 5, Section "Head and Senses", and Chapter 10, Section "Neurobiology and the Evolution of Dolphins"). These evolutionary changes are obvious in the quantitative morphology of the auditory pathway from the cochlea, where sound is perceived, up to the primary auditory neocortex. One of the driving factors is the parameter time in a medium (water) that propagates sound about five times faster than air; this fact being reflected in the fiber spectrum of the cochlear nerve and maximal diameters of the axons and their myelin sheaths (see the following).

From morphological, electrophysiological, as well as behavioral and psychoacoustic investigations it is clear that the structure and function of the dolphin's ear and brain, in principle, correspond to those in terrestrial mammals and strictly follow "universal" demands: In dolphins, they are the result of an accelerated evolution of ancient (fossil) hoofed animals in a new and challenging environment. Unfortunately, there is only little relevant information on extant artiodactyls (eg, pig, cattle, and sheep). This information could be useful as a basis for better understanding of how dolphins became as they are today.

Cranial Nerves 9–11—Glossopharyngeus Nerve, Vagus Nerve, and Accessory Nerve

These cranial nerves form the vagus group, a collection of nerves that belong into the set of pharyngeal (or gill) arches in the mammalian embryo and that are associated with specific brainstem nuclei (Kardong, 2015). In the bottlenose dolphin, the nerves leave the brainstem via rootlets along a longitudinal line on the lateral surface of the medulla oblongata, lateral to the inferior olivary complex (Figs. 6.7a, 6.32a, and 6.47). Morgane and Jacobs (1972) wrote that the fundamental organization of these nerves is similar to that of most terrestrial mammals; they also suggested that life in the aquatic environment has had but little effect on the basic neural pattern, most of which is concerned with the maintenance of the visceral core of the body in connection with vital circulatory, gular, and digestive as well as respiratory activities. Thus, the dorsal motor nucleus of the vagus, the solitary tract and its nucleus, and the nucleus ambiguus are well developed in the bottlenose dolphin and have essentially the same relation to medullary landmarks such as the medial longitudinal fasciculus, hypoglossal nucleus, and inferior olive as in other mammals.

The special visceral efferent (motor) column, which has a middle position among the nuclear columns on both sides of the brainstem, exhibits four nuclei arranged in a rostrocaudal sequence. Trigeminal motor, facial motor, ambiguus, and accessory nucleus (Fig. 6.48). In this sequence, it is interesting to see how much larger the facial motor nucleus is in comparison with the other nuclei in this column (see Section "Facial Nerve"). Via the three nerves (9-11), the ambiguus nucleus controls the striated mucles of the soft palate, the pharynx, larynx, and esophagus. In the general visceral efferent (motor) column of mammals (parasympathetic), which in mammals runs along the brainstem rather closely to the median plane, only two nuclei may be present. The existence of the accessory oculomotor nucleus

(Edinger- Westphal) is doubtful and the rostral (superior) and caudal (inferior) salivatory nuclei are obviously lacking, as are the lacrimal gland and the salivary glands (sublingual, submandibular, and parotid). The dorsal motor nucleus of vagus is present and normally developed; it contains preganglionic neurons that innervate thoracic and abdominal organs, including the esophagus, heart, and respiratory system.

Cranial Nerve 12—Hypoglossus Nerve

The hypoglossal nuclei and nerves have a special position along the nervous system. They mediate between, rostrally, the somatic (motor) nuclei and the nerves of the external eye muscles, and caudally, the ventral somatic (motor) horns of the spinal grey matter and the ventral roots of the spinal nerves. In the literature on cetaceans, all the rostral structures are reported to be well developed. In contrast to terrestrial mammals (primates), however, the rootlets of the nerve do not exit medial to the inferior olive, but lateral to this nuclear complex (Igarashi and Kamiya, 1972) (Fig. 6.32) (see Section "Important Fiber Systems" in this chapter). In the embryo, the first two to four (occipital) segments disintegrate and form the occiput of skull, the muscles of the tongue, and the hypoglossal nerve innervating them (Starck, 1979). In the dolphin, the cell masses of the hypoglossal nucleus gradually descend within the brainstem and sometimes even become continuous with the anterior horn of the cervical spinal cord (Morgane and Jacobs, 1972).

Quantitative Aspects of Cranial Nerves

The fiber counts of individual cranial nerves in the total fiber count of a species (Table 6.1) confirm that the optic nerve (II) is well developed in genuine dolphins (delphinids; representative: the bottlenose dolphin) but not in river dolphins (Amazon river dolphin, *I. geoffrensis*). In contrast, the fiber proportion of the optic nerve is very high in the human. As to the trigeminal nerve (V), the dolphins, in general, seem to be superior to the human, and even more so in the fiber proportion of the facial nerve (VII). The vestibulocochlear nerve of dolphins is again rich in fibers, particularly in *Inia*. This is certainly due to the dominance of the cochlear fibers in these animals. The vestibular fibers of dolphins are only marginal in number, in contrast to the situation in the human, where the percentage of cochlear and vestibular fibers is more or less equal.

A few more data from the literature may even better illustrate the morphological and functional trends in the vestibular and auditory systems of dolphins. They are separate units and represent two different sensory organs for equilibrium and hearing (see Chapter 5, Section "Senses"). As a consequence, their evolutionary implications concerning qualitative and quantitative aspects are also different (see Chapter 10, Section "Neurobiology and the Evolution of Dolphins").

As in other mammals, the cochlear and vestibular nerves enter the brainstem separately but adjacent to each other in the cerebello-pontine angle. In five dolphin-like and delphinid species, the total number of axons in the vestibular nerve ranged from 3213 to 4091. In the Yangtze river dolphin (*Lipotes vexillifer*) and the humpback dolphin (*Sousa chinensis*), these figures were found to be very close to the number of neurons in the vestibular ganglion (Gao and Zhou, 1992). In the same five species, however, the total number of axons in the cochlear nerve ranged much higher (77,171-109,400), a factor of about 20 times higher than that in the vestibular nerve (Gao and Zhou, 1995). In the human, the number of cochlear fibers is given as 30,000 (Wever, 1949; Supin et al., 2002).

The *vestibular fiber diameters* in these odontocetes range from 0.6–18.3 µm and their modal (weighted average) values from 8.5–10.5 µm, respectively. The largest vestibular fibers detected in terrestrial mammals (guinea pig, cat, monkey) were 8–10 µm in diameter, with a modal value of 2.5–3.5 µm (Gacek and Rasmussen, 1961). Comparison of the vestibular fiber data found in dolphin-like odontocetes (*Lipotes, Neophocaena, Sousa*) and delphinids (*Delphinus, Tursiops*), respectively, with those in the land mammals just mentioned indicates that (1) the diameter range of the vestibular fibers in dolphins is about two times, (2) that of the modal diameter three times, and (3) the estimated conduction velocity[n] about two to three times higher than in the guinea pig, cat, and monkey (Gao and Zhou, 1995). So, the vestibular nerves of small odontocetes are characterized by low fiber numbers and large fiber diameters.

The *cochlear fiber diameters* of three dolphins mentioned above (*Neophocaena, Delphinus, Tursiops*) ranged from 0.5–34 µm and the modal diameters 10.5–12.5 µm, respectively. The thickest cochlear fibers detected in these species ranged from 34.7–50.2 µm. The largest nerve fiber among the vertebrates with a diameter of 54.9 µm was recorded in the cochlear nerve of the finless porpoise (*Neophocaena phocoenoides*). In contrast, the largest cochlear fibers in the guinea pig, cat, and monkey are 7–8 µm in diameter, with a modal value of 3–4 µm (Gacek and Rasmussen, 1961). This range of maximal diameters for the cochlear nerve fibers in dolphins comes close to the dimension of the somata of large neurons, that is, the width of pyramidal cells of Betz in the human and large carnivores (Blinkov and Glezer, 1968). Such giant cells, however, have fiber diameters of 10–20 µm only (Benninghoff and Drenckhahn, 2004). So, the cochlear nerves of small odontocetes are characterized by high fiber numbers, and large to very large fiber diameters. The data from Gao and Zhou

n. Conduction velocity of impulses in nerve fibers can be estimated using a rule of thumb: velocity in meters = about 6 times the fiber diameter in micrometers.

(1992) reveal that the range of fiber diameter, modal diameter, and maximum diameter of cochlear nerves in these small odontocetes are 2–5 times those in guinea pig, cat, and monkey.

Almost all of these cochlear fibers in dolphins are heavily myelinated; thus, the ratio of the inner diameter (axon cylinder) to the total diameter (axon cylinder plus myelin sheath) of the fibers averages between 0.6 and 0.7. As a whole, the high percentage of large fibers and the wide range of fiber diameters in many smaller odontocetes indicate that their cochlear nerve is specialized for high-speed communication with the brain (Gao and Zhou, 1991). The maximum diameters of the myelinated cochlear fibers reported in their paper so far are the largest recorded for the vestibulocochlear nerve, or for any other nerve among vertebrates. The potential conduction velocity in fibers with diameters larger than 30 μm in *Neophocoena* and *Sousa* was estimated for more than 180 m/s, that is, about four to five times higher than in terrestrial mammals. This capacity for ultra-high conduction velocity in the cochlear fibers of these dolphins is presumably essential for transmitting sound impulses at a high rate during echolocation.

Among the mammalia, the eighth nerve of toothed whales is particularly thick. Also, within the ensemble of cranial nerves in dolphins, the vestibulocochlear nerve has the largest diameter. At the same time, however, its fiber density is rather low. Thus, whereas the mean cross-sectional area of the eighth nerve (in descending order) is about 25 mm^2 in delphinid species, 1.4 mm^2 in cat, and 0.9 mm^2 in monkey, fiber density (fibers/mm$^2 \times 10^3$) in these animals is in a reverse order: cat 38, monkey 34, and 3–4 in delphinids. Obviously, in dolphins, there is a similar inverse ratio of cross-sectional area (high) to fiber density (low) in the cochlear nerve as reported for the optic nerve (see the previous discussion). In part, this is attributed to the high percentage of large fibers with a strong myelination in both nerves not known of other mammals. Whereas the maximum diameters found in optic nerve fibers of dolphin and dolphin-sized cetaceans ranged from about 26–45 μm, cochlear fibers ranged from 35–50 μm. The largest (cochlear) fiber found so far had a diameter of about 55 μm. These figures are on a distinctly higher level here than in other mammals.

In their electron microscopic study on the optic nerve of the bottlenose dolphin, Dawson et al. (1982, 1983) explained the outstanding thickness of the optic nerve of dolphins (1) with a "giant" nerve cell system in the retina ("giant" with respect to the remarkable size of the ganglion cell perikarya and the thickness of their axons), (2) a considerable glial-myelin elaboration, and (3) slightly more than 50% extraneuronal space in the optic nerve. For comparison, in terrestrial mammals, the normal extraneuronal space content in the optic nerve was estimated to comprise only 12–20% (van Harrefeld, 1972). Thus, in marine dolphins like the bottlenose dolphin, the unusual thickness of the retina could also relate to such a large extraneuronal space. They exhibit larger optic nerves and thicker optic fibers than the human, but fiber counts of about one eighth only at about the same dimension of the brain. Finally, concluding from the structure of the retina to the remaining central nervous system, Dawson et al. (1983) speculated that the dolphin brain as a whole could owe much of its volume to a comparatively large extraneuronal space that could serve as a metabolically relevant compartment for the adequate vegetative support during long periods of apnea in diving.

Important Fiber Systems

Medial lemniscus. This fiber system comprises the most important afferent fiber tracts of the exteroceptive sensitivity from the spinal cord and brainstem. It is subdivided into the lemniscus spinalis and the lemniscus trigeminalis and combines (1) the epicritic sensitivity[o] (bulbothalamic tract) as the extension of the dorsal (posterior) funiculi of the spinal cord, (2) the protopathic sensitivity[p] (spinothalamic tracts) including nociception, and (3) tracts to tectum and face including gustatory fibers.

In cetaceans, the afferent spinal system is moderately developed in accordance with the reduction of the hind limbs and pelvic girdle. In these animals, the dorsal funiculi (gracile and cuneate fascicles) are strikingly small; they are thought to convey input predominantly from the flippers and the tail (sense of position: proprioception[q]). Nevertheless, cutaneous sensitivity in the trunk was reported to be high (Slijper, 1973). The gracile and cuneate nuclei are small and the medial lemniscus is still weak in the caudal medulla oblongata, but becomes considerably stronger at more rostral medullary levels, due to the addition of well to extremely well-developed afferent systems of the head (trigeminal, auditory systems).

Trigeminothalamic tract. The dorsal part of the principal sensory trigeminal nucleus gives rise to the ipsilateral dorsal trigeminothalamic tract (Voogd et al., 1998). The latter joins the trigeminal lemniscus and terminates in the medial part of the ventral posterior thalamic nucleus as the main somatosensory thalamic nucleus (see Section "Thalamus" in this chapter). The projection from the main sensory trigeminal nucleus to the medial part of the ventral posterior thalamic nucleus (Wallenberg's tract) is remarkably large in cetaceans. In dolphins, it may be involved in the operation of the blowhole (epicranial) complex with respect to sound and ultrasound generation and emission (Oelschläger and Oelschläger, 2002).

Medial tegmental tract (mtt). The elliptic nucleus of dolphins (Figs. 6.9, 6.25, 6.26, 6.27, 6.41, and 6.48; see Section "Mesencephalon" in this chapter), located in the rostral mesencephalon, gives rise to the strong mtt that proceeds to the very

---

o. Epicritic sensitivity is for somatic sensations of fine discriminative touch, vibration, two-point discrimination, stereognosis, and conscious and unconscious proprioception.

p. Protopathic sensitivity is for the general sensation of pain and temperature that is poorly localized.

q. Proprioception is the sense of the relative position of neighboring parts of the body and of the strength of effort being employed in movement.

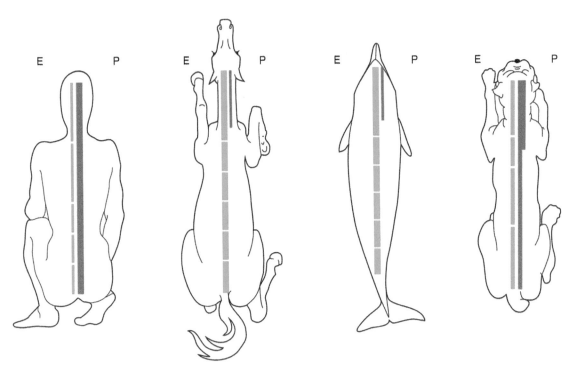

FIGURE 6.52 **Comparison of the pyramidal system (P, *red*) and extrapyramidal (E, *green*) systems in different mammals.** From left to right: human, horse, dolphin, dog. Spinal apparatus brought to about the same length. Note that these spinal systems are very similar in the horse (hoofed animal) and the dolphin (cetacean), in contrast to the situation in the human and the dog. The latter two mammals are rather similar to each other, although the pyramidal tract is weaker in the dog and the extrapyramidal system somewhat more pronounced in the human. *(Redrawn and modified by Jutta Oelschläger from Dyce et al., 1991. Anatomie der Haustiere. Ferdinand Enke Verlag Stuttgart.)*

large medial accessory inferior olive. The inferior olive also receives afferents from the spinal cord (spino-olivary tract) and projects to the C2 zone in the extremely large paraflocculus (see Section "Cerebellum" in this chapter). The paraflocculus, as the main target of pontocerebellar projections, has a massive projection to the posterior interposed nucleus of the cerebellum (PIN). From here, ascending fibers run back to the elliptic nucleus and other nuclei at the diencephalic/mesencephalic border, which, in turn, project to the inferior olive via the medial tegmental tract (Figs. 6.41, 6.42, and 6.50). In conjunction with the medial accessory olive, the paraflocculus has been associated with mass movements of the posterior trunk and tail, the only region where the axial skeleton possesses a reasonable range of motion. The medial tegmental tract thus seems to be part of a recurrent circuit (elliptic nucleus, inferior olive, paraflocculus, posterior interposed nucleus, elliptic nucleus), which combines sensory input (mainly acoustic in toothed whales) with locomotor activity (acousticomotor navigation) (Oelschläger, 2008).

Pyramidal tract. In the dolphin, this important tract originates in the neocortical motor area rostral to the "cruciate" sulcus (Figs. 6.14 and 6.15) and runs through the internal capsule in the caudal direction (Fig. 6.29; see also Chapter 10, Fig. 10.4). At mesencephalic levels, the pyramidal tract is seen in the corticobulbar and corticopontine tracts (Fig. 6.26, cob, cop) but is already very small at anterior medullary levels (Fig. 6.43b). Typical compact fiber bundles (pyramids), as seen in terrestrial mammals behind the pons, are not present in cetaceans. Here, the pyramidal tract is weak and situated lateral to the inferior olivary complex so that both inferior olives join midsagittally (Figs. 6.27, 6.32a, and 6.51d). Obviously, most of the pyramidal motor fibers are relayed in the pontine nuclei, which have strong projections to the cerebellum (see Section "Pons and Cerebellar Connections" in this chapter). Perhaps the extremely well-developed inferior olives, more specifically, the rostral medial accessory olive, which occupies the ventromedial area in the medulla, have pushed the residual pyramidal tracts aside during development. In dolphins, the crossing of the pyramidal tracts (decussation) is indistinct and the proceeding of the tracts hardly visible. Also, it is unlikely that in these animals, corticospinal fibers descend more than a few cervical segments in the spinal cord (Fig. 6.52). This pattern resembles much that seen in hoofed animals and the elephant.[r] In many terrestrial mammals (ungulates), the corticospinal tract is small and the rubrospinal tract

---

r. In terrestrial mammals relevant for our scope, there is an inverse relationship between the size of the pyramidal tract and that of the rubrospinal tract. In primates the pyramidal tract is very well developed and the rubrospinal tract weak (not the red nucleus which is very large). Perissodactyls and artiodactyls have small pyramidal tracts and their rubrospinal tracts are large (Verhaart, 1970).

is strongly developed. In cetaceans, both the corticospinal and rubrospinal tracts are small (Verhaart, 1970), which seems to be a specialty of these aquatic animals. As kind of a compensation, their mass movements of the body stem may rather be generated and maintained by the projections of Deiters' nucleus (lateral vestibulospinal tract), the olivospinal tract as well as the reticulospinal tract.

SPINAL CORD, SPINAL NERVES, AND GANGLIA

Within the nervous system, the spinal cord of dolphins is responsible for the innervation of the trunk and tail, that is, from the anterior part of the neck to the flukes including the pectoral and pelvic girdles together with their appendages. The pelvic girdle is strongly reduced but still present in the adult and the hind limb only to be seen as a bud in the embryonal period. As to innervation, a few spinal segments send branches of spinal nerves to the rudimentary pelvis and the muscles that suspend and operate the genital system.

As a whole, the spinal cord is a two-way mediator between the locomotory apparatus of the trunk and the brain, as it transmits sensory and motor information. The body musculature is very strong, contributes to the spindle shape of the animal, and serves powerful locomotion (propulsion) via the trunk-and-fluke complex. This is particularly obvious in marine delphinids, which hunt actively by performing quick and versatile maneuvers.

Remarkable macroscopic features of the spinal cord are its size dimensions. The relation of body length to spinal cord length in dolphins should be around 3.75:1 as found in the harbor porpoise; large whales, however, have values of up to 6.7:1 (right whale, *Eubalaena sp.*; Jansen and Jansen, 1969). The ratio of brain mass to spinal cord mass in the bottlenose dolphin (40:1) is higher than in any other species measured so far, except for the human (50:1). The bottlenose dolphin was reported to possess a total of 40–44 spinal segments, 8 cervical, 13 thoracic, and 23 lumbocaudal (Morgane and Jacobs, 1972); the cord ends about the third lumbar vertebra (Ridgway, 1968). In the bottlenose dolphin, the spinal cord is nearly cylindrical (Fig. 6.53). A distinct cervical enlargement (intumescence) for the innervation of the pectoral girdle and

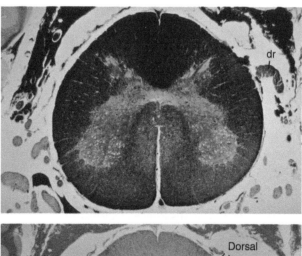

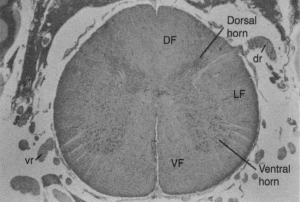

FIGURE 6.53   **Histological section of the cervical spinal cord of the common dolphin.** Top: Heidenhain-Woelcke stain for myelinated nerve fibers; Bottom: Cresyl violet stain (Nissl) for somata. White matter: DF, dorsal funiculus; LF, lateral funiculus; VF, ventral funiculus; dr, dorsal root of cervical spinal nerve; vr, ventral root of spinal nerve. Width of spinal cord about 1 cm. *Specimens from the Pilleri Collection in the Natural History Museum and Research Institute Senckenberg, Frankfurt am Main.*

flipper has been described in different odontocetes, whereas the lumbar intumescence is much less prominent (Jansen and Jansen, 1969), obviously in correlation with the reduction of the pelvic girdle and hind limb. A generally consistent feature remarked on by several authors is that the dorsal (somatosensory) roots appear to be relatively thin in comparison to the ventral (somatomotor) roots, an unusual condition in mammals.

As to the internal structure of the spinal cord, the dorsal white columns (gracilis and cuneate fascicles), and the dorsal horns of the gray matter are relatively smaller in whales and dolphins (Morgane and Jacobs, 1972). Interestingly, the substantia gelatinosa (layer II of the dorsal horn in the gray matter) was reported to be poorly developed or even lacking. In mammals, this layer is involved in internal and external nociception. All this seems in accordance with studies that the posterior medial lemniscus and the spinothalamic tracts are also small (see the previous discussion). These data may be correlated with the reduction in body surface and the general loss of sensory input, particularly from the reduced or simplified extremities. In contrast, the ventral horns of the gray substance, responsible for motor action of the very strong body musculature, are much more developed (Fig. 6.53).

FUNCTIONAL SYSTEMS

Olfactory System/Paleocortex

In comprehensive reviews of the last decades (eg, Jansen and Jansen, 1969; Morgane and Jacobs, 1972), the olfactory system has sometimes been defined in a broader sense by including the archicortex (eg, hippocampus) and accessory structures in a superior "rhinencephalon" (Figs. 6.13, 6.24, 6.32, 6.57, and 10.2). The components of the rhinencephalon are those telencephalic cortical areas, which are demarcated from the neocortex by the rhinal sulcus. Although in mammals, the two systems (olfactory system and hippocampal formation) have very close relations with each other both topographically and functionally, it seems more adequate to avoid the term "rhinencephalon" here, because the olfactory function in the conventional sense has been lost in dolphins. An accessory olfactory system comprising the vomernonasal organ, vomeronasal nerve, and accessory olfactory bulb, known from many terrestrial mammals, including hoofed animals, is completely lacking in cetaceans.

In contrast to other mammals and to some degree also to baleen whales, the olfactory system of toothed whales including dolphins is rather special. Obviously in parallel with profound changes in the structure and function of the nasal region in toothed whales, generally, the peripheral and the anterior central olfactory system are lacking in the adult animal. In baleen whales, the olfactory system persists complete but is miniaturized. In the ontogenesis of the toothed whales species investigated so far, all the components of the system are present but in later embryonal stages the olfactory epithelium, olfactory nerve, and olfactory bulb dwindles away and disappears, while the anterior olfactory nucleus and nucleus of the olfactory tract are vestigial (Breathnach, 1960; Oelschläger and Buhl, 1985; Oelschläger and Kemp, 1998). Only very rarely, an olfactory tract persists in adult large toothed whales like the sperm whale (*Physeter macrocephalus*) and the bottlenose whale (*Hyperoodon ampullatus*). The associated secondary paleocortex on the anterior basal surface of the telencephalic hemisphere is obviously also reduced and it is not clear to what degree it can serve "olfactory" function.

In adult terrestrial mammals, the olfactory system (Fig. 6.24) comprises the olfactory epihelium with the chemoreceptor cells, the bundles of the fila olfactoria that together represent the olfactory nerve, the main olfactory bulb, accessory olfactory bulb, the olfactory tract, stria olfactoria lateralis et medialis, anterior olfactory nucleus, the olfactory tubercle (1), the diagonal band (2), prepiriform cortex (3), and the cortical areas of the amygdaloid complex (4). These paleocortical parts (1–4) have cortical plates of different thickness and layering; none of them shows the typical six-layered configuration of the neocortex. In dolphins, the mammalian bauplan has been profoundly changed: thus, the olfactory epithelium is only present in the embryo as are the fila olfactoria and a neuroepithelial bud at the mediobasal tip of the telencephalic hemisphere (Figs. 5.12 and 6.46). This bud seems to represent a mixture of (i) the olfactory bulb induced by the ingrowing olfactory fibers and (ii) the LHRH (GnRH)-immunoreactive material of the terminal nerve that accumulates in this area and invades the brain wall in order to establish the hypothalamo-hypophyseal-gonadal axis (see Section "Cranial Nerve 0" in this chapter). The latter is essential for the expression of the sexual character. In the adult harbor porpoise, neither an olfactory epithelium nor fila olfactoria, olfactory bulb, and olfactory tract have been found by systematic investigation. However, a large terminal ganglion remains in the meninges for life, the largest terminal ganglion found so far in mammals. The ganglion contains a few LHRH-immunoreactive neurons and a high number of large potentially somatosensory perikarya (Buhl and Oelschläger, 1986, 1988; Oelschläger et al., 1987; Ridgway et al., 1987). The latter may serve as the afferent pathway of a loop that includes the motor nucleus of the facial nerve as an efferent pathway for the quick monitoring and regulation of pressure in the nasal cavity during sound and ultrasound emission (see Dolphin Head Characteristics).

Particularly interesting among the remaining paleocortical areas is the situation of the olfactory tubercle in dolphins. In the older literature it was reported that the tubercle is rather large and ovoid. Recent histological and MRI investigations have shown, however, that the basal surface of the tubercle in question is more or less devoid of a cortical plate (see Section "The Cortices" in this chapter). Thus, there seems to exist a paradox between the size of the tubercle and the quality of the paleocortex. Obviously the underlying thick round connection between the caudate nucleus and the putamen (corpus striatum) strongly bulges ventrally like a watchglass (fundus striati). By doing so, the paleocortical plate is expanded so much that it thins out and partially exposes the striatum. With regards to the general trend of a reduction of the olfactory system in toothed whales, this scenario seems plausible. The situation in dolphins, to some degree, parallels that in the microsmatic human (Stephan, 1975) concerning the reduction of the olfactory system and the large size of the corpus striatum.

Visual System

In dolphins, the visual system is well represented in the midbrain (superior colliculi), diencephalon (optic chiasm and optic tract, lateral geniculate bodies), and in the visual cerebral cortex (see Figs. 6.6–6.9, 6.15, 6.16, 6.26, 6.27, 6.32, 6.48, 6.49, and 6.54). However, these visual centres are much less in volume than corresponding parts of the auditory system (inferior colliculi, medial geniculate bodies, auditory cerebral cortex). In the cerebral cortex of dolphins, visual representation was found with the evoked potential method (see Section "The Cortices" in this chapter). This area occupies the cortical area named lateral gyrus; it is located near the midplane of the brain at the transition of the vertex into the medial wall of the cerebral hemisphere (Figs. 6.7, 6.14, 6.15, and 6.25). Within this area, there are two subfields: one zone generates short-latency evoked potentials (primary projection field, V1) and the other zone generating evoked potentials of longer latency, a secondary (nonprimary) belt zone (V2) (Mass and Supin, 2009). The first subfield is located in the depth of the entolateral sulcus (Fig. 6.15), a second-order sulcus in the lateral gyrus; the second subfield occupies the remainder of the lateral gyrus, on both sides of the entolateral sulcus. In dolphins, these two zones differ in cytoarchitectonic features: V1 contains a residual layer IV, where visual thalamocortical fibers terminate; in V2, this layer IV is absent. For topographical reasons it could be argued that the entolateral sulcus may represent a parallel to the calcarine sulcus in primates.

In general, the visual system of dolphins exhibits a rather high degree of efficiency in structural development and performance: good visual acuity, well-developed visual brain centers, precisely aimed and visually driven behavior as well as multisensory processing of cortical input. Special features are adaptations to both aquatic and aerial environments: These are particularly the two local concentrations of retinal ganglionic layer neurons in areas of best vision (Supin et al., 2001). The latter correspond with features in pupil and cornea structure and allow for good sight in both air and water as well as the focussation in the directions of highest functional significance for the dolphin, with respect to visually guided capture of prey (see Chapter 5, Section "Vision: Anatomy and Optics of the Eye").

As to the route of the visual axons in the direction of the visual cortex, most mammals have the majority of them project to the opposite (contralateral) cerebral hemisphere, whereas a considerable number of axons run to the cerebral hemisphere on the same (ipsilateral) side. In cattle, for example, around 83% of optic axons cross in the optic chiasm (pig 88%, sheep 89%). Here, dolphins may be an exception to the rule: Neuroanatomical investigations of dolphins that accidentally had lost one eye or were enucleated experimentally revealed a total decussation of the optic nerve fibers (Fig. 6.54), a condition atypical for mammals but common in other vertebrates such as birds (Nickel et al., 2004). Noninvasive electrophysiological investigations reported that large evoked potential responses from visual cortex were obtained when light flashes were presented to the contralateral eye but not if presented to the ipsilateral eye (Ridgway, 1990).

This anatomical and physiological evidence suggests that dolphins do not have binocular vision. However, Ridgway (1990) stated that bottlenose dolphins have some degree of overlap in the visual field ventrally and rostrally as well as dorsally and slightly caudally. Despite these small areas of overlap in the visual field, the eyes appear to move independently, that is, one eye might look forward and dorsally, while the other might look rearward and ventrally.

Vestibular System

The life of aquatic mammals in the three-dimensional space requires sophisticated body control and a distinct sense of position. In terrestrial mammals, the equilibrium is maintained by the interaction of sensory epithelia in the vestibular apparatus with the brainstem.

The vestibular system of dolphins is composed of semicircular canals and the utriculus-sacculus complex. The semicircular canals are particularly interesting in terms of comparative anatomy because they are exceptionally small in toothed whales (see Chapter 5, Figs. 5.21 and 5.22). This situation is not found either in other mammals or other vertebrates investigated so far (Spoor et al., 2002). And it is in strong contrast to the situation in the auditory system, which is extremely well developed in dolphins.

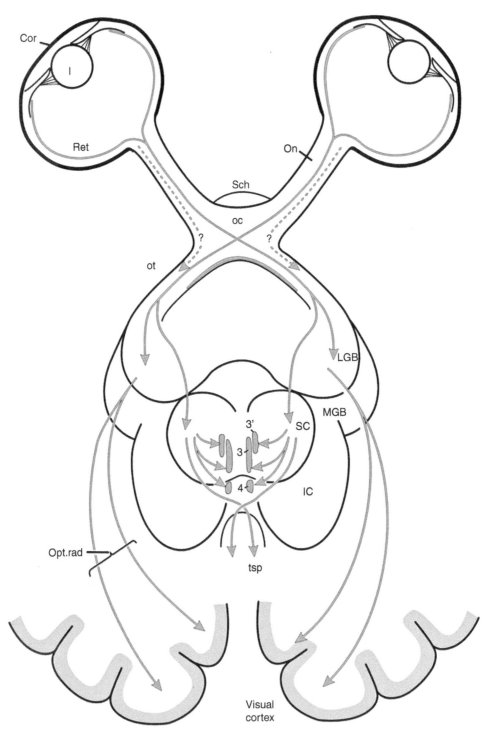

FIGURE 6.54 Visual system of mammals, adapted to the situation in dolphins. Only a small part of the optic input does not cross to the opposite side (contralaterally; stippled). In comparison with those in the auditory system, components of the visual system as, for example, the lateral geniculate body and the superior (rostral) colliculus are of a moderate size. The pineal organ is mostly lacking. Abbreviations: 3, motor nucleus of oculomotor nerve; 3', parasympathetic nucleus of oculomotor nerve; 4, nucleus of trochlear nerve; Opt.rad, optic radiation; Cor, cornea; IC, inferior (caudal) colliculus; l, lens; LGB, lateral geniculate body; MGB, medial geniculate body; oc, optic chiasm; On, optic nerve; ot, optic tract; Ret, retina; SC, superior (rostral) colliculus; Sch, suprachiasmatic nucleus; tsp, tectospinal tract. *(Modified and redrawn by Jutta Oelschläger from Barone and Bortolami, 2004. Anatomie comparée des mammifères domestiques, vol. 6, Neurologie I, Systéme nerveux central. Vigot Frères, Paris.)*

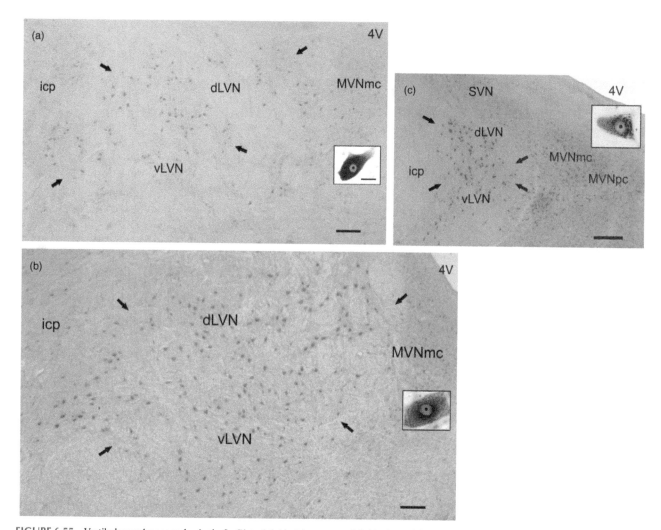

FIGURE 6.55 Vestibular nuclear complex in the La Plata dolphin (a), common dolphin (b), and the human (c) (scale = 400 μm). *Arrows* delimit the border of the Deiters' nucleus (lateral vestibular nucleus, LVN). Insets show representative cell bodies of the relevant species (scale = 25 μm; for diameters, see text). Note that the sectional thickness is several times higher in the human section (100 μm) than in those of the two dolphins (a, 20 μm; b, 30 μm); therefore, the cell density misleadingly seems to be higher. In the dolphins, Deiters' nucleus is 3–5 times larger than in humans and was reported to be even 9–16 times larger (Zvorykin, 1975). In the literature, Deiters' nucleus is considered a "precerebellar" nucleus because of its close functional correlations with cerebellar nuclei. Apart from multiple diffuse projections into the reticular formation, this nucleus sends efferents into the anterior motor horn of the spinal cord via the lateral vestibulospinal tract, thus providing a major source for spinal activation and control. Abbreviations: 4 V, 4th ventricle; dLVN, dorsal part of the lateral vestibular nucleus; icp, inferior cerebellar peduncle; MVNmc, magnocellular part of the medial vestibular nucleus; MVNpc, parvocellular part of the medial vestibular nucleus; SVN, superior (rostral) vestibular nucleus; vLVN, ventral part of lateral vestibular nucleus. *(From Kern et al., 2009. The central vestibular complex in dolphins and humans: Functional implications of Deiters' nucleus. Brain Behav. Evol. 73, 102–110; inset in part c, modified from Olszewski and Baxter, 1954. Cytoarchitecture of the human brain stem. Karger, Basel.)*

In toothed whales, the small size of the semicircular canals seems to be correlated to a likewise reduction of the vestibular nerve. In dolphins, less than 5% of the fibers in the VIIIth cranial nerve are devoted to vestibular function as compared to approximately 40% in other mammals (see Chapter 5, Section "Ear: Sense of Balance and Hearing", and Section "Cranial Nerve 8—Vestibulocochlear Nerve" in this chapter). Furthermore, the adaptation to an exclusively aquatic lifestyle was obviously correlated with profound modifications of the vestibular brainstem nuclei in the cetacean ancestors. In extant dolphins, all of the "genuine" vestibular nuclei (rostral, medial, caudal [descending] nucleus) are miniaturized (Fig. 6.55); they essentially depend on input from the vestibular sensory neurons into the brainstem. The only exception to this is seen in the lateral (Deiters') vestibular nucleus (Fig. 6.51). In pelagic (ie, off-shore) dolphins, the dimensions of this nucleus are remarkable in both absolute and relative terms (Kern et al., 2009). This may have to do with a considerable functional independence of Deiters' nucleus from the remaining vestibular system (see the following).

Concerning the "genuine" vestibular nuclei, their small size in dolphins (with respect to the situation in humans) may be explained by some observations:

1. In primates, the vestibular nuclei are integrated into the neural control of the vestibuloocular reflex[s] and optokinetic nystagmus.[t] They are responsible for horizontal gaze holding and smooth pursuit (Benninghoff and Drenckhahn, 2004). For toothed whales, however, daylight vision in their natural habitats is often limited by murky water or during diving in dim light or below the photic zone. In marine dolphins, the eyes are well developed and the optic nerves are rather thick, although axon numbers are only a fraction of those found in the human (see Section "Cranial Nerve 2—Optic Nerve" in this chapter). In general, the small size of the "genuine" vestibular nuclei in dolphins seems to be correlated with a decrease in oculomotor capacity, which is obvious in low axon numbers and small nuclei, respectively, of the extraocular muscle nerves (Morgane and Jacobs, 1972).

2. Another criterion for the small size of the "genuine" vestibular nuclei refers to the mode of dolphin locomotion: these animals move by vertical undulation of the powerful body stem and caudal fin (fluke). The dolphin cranium is integrated in the undulation process of the spinal apparatus by means of well-developed atlantooccipital joints (Oelschläger, 1990). The atlantoaxial joints, however, were reduced in the course of an evolutionary fusion of the first cervical vertebrae (see Chapter 3, Section "Vertebral Column"). In this particular propulsion, only repetitive vertical angular movements of the head within the atlantooccipital joints occur. These forced nodding movements may have led to a decrease in vestibular (semicircular) canal size and sensitivity in order to prevent a continuous stimulation of the system.

   A similar problem might have arisen with the detachment of the periotic bones from the skull base in order to avoid bone conduction and thus acoustic overstimulation by "white noise" of the animal's own sounds (see Chapter 5, Section "Ear: Sense of Balance and Hearing"). In adult animals, the maximally dense and heavy and thus mechanically inert ear bones are largely suspended via soft tissues, a condition presumably counterproductive for extremely sensitive static organs such as the semicircular canals of mammals (Oelschläger, 1990).

3. In consequence, the size reduction of the semicircular canals and of the associated brainstem nuclei seem to be correlated with the lack of adequate sensory input due to restricted movability of the head in these animals (Spoor et al., 2002) as a major concession to powerful propulsion by vertical undulations of the locomotory system (head and body stem).

Among the vestibular nuclei, Deiters' nucleus is the most remarkable in the dolphins investigated so far. Whereas the genuine vestibular nuclei are more or less reduced in both absolute and relative terms, this nucleus is distinctly larger in dolphins than in the human (Kern et al., 2009). In the literature (Voogd et al., 1998), Deiters' nucleus is considered a "precerebellar" nucleus because of its close functional correlations with cerebellar nuclei. The nucleus sends efferent projections into the anterior motor horn of the spinal cord, thus providing muscular activation and control. Concomitantly, the pyramidal tract of toothed whales fades out in the upper cervical segments of the spinal cord. Thus, the simultaneous size increase of both Deiters' nucleus and the cerebellum indicates their close functional relationship in odontocetes with respect to three-dimensional high-speed maneuvers (hunting, playing) as well as for efficient target detection in the open sea.

Auditory System

Biosonar systems are remarkable in their capacity to precisely extract spectral and temporal information from echoes of emitted pulses. These animals use acoustic information to image the external world with an accuracy that challenges or exceeds the capacity of the mammalian visual system (Vater and Kössl, 2004). One example is the precision with which such systems can resolve target distance through the encoding of time delays between the pulse and echo with a resolution of better than 1 µs. Moreover, biosonar can obtain information that is invisible to the eye, as evidenced by the dolphin's ability to discriminate the internal structure and composition of a metal cylinder. Advantages for the animal are (1) that the biosonar is an active process that generates much of its own acoustic environment and (2) that it can tune the modality used for orientation and hunting. However, this kind of information gathering may be rather demanding as to computation capacity of the brain, particularly if there is a second acoustic system serving communication on a much lower frequency level.

As to studies of hearing in echolocating dolphins, experimental approaches are much restricted due to ethical reasons. Hearing abilities and mechanisms therefore have to be studied with psychophysical (behavioral), noninvasive

---

s. The vestibuloocular reflex elicits eye movements during stimulation of the vestibular system; head movement results in eye movement in the direction opposite to head movement, thus preserving the image on the center of the visual field(s).

t. The optokinetic nystagmus is a combination of a saccade (quick and simultaneous movement of both eyes between two phases of fixation of picture details) and a smooth pursuit (slow movements allow the eyes to closely follow a moving object during fixation). This nystagmus is seen when the eyes of an individual follow an object that moves out of the field of vision. The eyes move quickly back to the position they were in when they first focused the object.

evoked-potential (EP) techniques, all of them avoiding harm for the animals, or with postmortem morphological methods. Because of the high conduction velocity of sound in water,[u] a considerable amount of information can be received by a dolphin per time unit from its natural surroundings and/or by its own sonar information system. Although the function of the central auditory system of dolphins is not fully understood, many intriguing relevant features have been reported in recent years. In these animals, temporal processing and resolution of acoustic information largely depends on the so-called "integration time," which is needed for the discrimination of pairs of successive sound pulses in echolocation. This means that the temporal resolution of sound signals is increasing with decreasing integration time. In this respect, the integration times of dolphins seem to be a magnitude shorter than that of humans (Fuzessery et al., 2004).

Interestingly, the factor time in the comparison of dolphins with other mammals is also seen if the acoustic/auditory capacity of a dolphin is transformed to that of the human. Thus, Ridgway (1990) wrote: "If given enough time, the human auditory system seems to perform as well on some echolocation tasks. Indeed, when pulses similar to dolphin echolocation pulses were projected at targets by instrumented divers and the received echoes stretched 128 times (tantamount to a slowed-down tape recording and therefore reduced equivalently in frequency), human divers performed with as few errors as bottlenose dolphins in distinguishing metal targets of copper, brass, or aluminum and geometrical aluminum shapes covered with neoprene rubber."

Major Components of the Auditory System

In mammals, the *ascending (afferent) auditory pathway* conducts acoustic signals from the cochlea as far as to the auditory cortex (Fig. 6.56). For the most part, it represents a network of auditory nuclei and associated fiber tracts within the CNS. The nuclei are arranged in a double chain along the brainstem, connected with each other on the same side of the brain (ipsilaterally) and across the midline (contralaterally). The auditory nuclei may also be addressed as part of (i) the direct auditory pathway or (ii) the indirect auditory pathway (Fig. 6.56: legend). In terrestrial mammals, both limbs have the cochlear spiral ganglion (first neuron of the auditory pathway) and the cochlear part of the vestibulocochlear nerve in common as well as the DCN (second neuron) and the ventral cochlear nucleus (VCN; second neuron).

Along the ascending auditory pathway, three tasks have to be completed:

1. The preservation of tonotopy, that is, the frequency pattern in the organ of Corti, along the pathway up to the auditory neocortex.
2. The extraction of acoustic characteristics from complex sound patterns at all levels of the acoustic pathway. The sensitivity to such features is particularly high in the auditory cortex.
3. The localization of a sound source by directional hearing is essential for an adequate reaction of the animal. In this respect, differences in sound runtime and in the amplitudes of both sides are analyzed. For directional hearing, the superior olivary complex (SOC) is particularly important.

Along the ascending auditory pathway and within the auditory cortex, the sound pattern of an acoustic signal is fragmented into bits of information: There are a variety of specialized neuron populations that are activated by specific frequencies and an increase or decrease in frequency, respectively; others react to changes in sound intensity, duration, and so forth. The neuronal "extraction" of sound characteristics begins at the level of the cochlea (mainly frequency information) and then proceeds in the cochlear nuclei (second neuron) in the medulla oblongata as well as in a series of nuclei on both sides of the brainstem (Fig. 6.56). More rostrally, most of the auditory brainstem projections run to the inferior colliculi (convergence of interconnections). From here, strong fiber tracts (brachium of the inferior colliculus, bic) run to the thalamus (and medial geniculate body, MGB) and from there via the acoustic radiation to the auditory cortex. In the thalamus and telencephalon, the acoustic radiation gives off several branches that project to the striatum, hippocampus, preoptic area, and amygdala (not shown). Thus, auditory information gets direct access to motor coordination (striatum; particularly for vocalization), to the associative processing of sensory stimuli (cortex) as well as to the coordination of instinct behavior (basal telencephalon and limbic system).

The *descending (efferent) auditory pathway* of mammals permits the hierarchical control of information processing in inferior auditory centers by such in superior auditory levels. As a whole, the auditory system is able to adapt to changing conditions in sound perception by the modification of neuronal sensitivity, frequency filters, temporal patterns of response, directional hearing, and contrast. One example of such an efferent feedback loop is the olivocochlear bundle that projects from the surroundings of the superior olive to the cochlea (Benninghoff and Drenckhahn, 2004) and controls the gateway between the inner and outer hair cells and thus the origin of the ascending auditory pathway.

---

u. Sound travels about 4.4 times as fast in water ($\sim$1,500m/s) than in air ($\sim$340m/s).

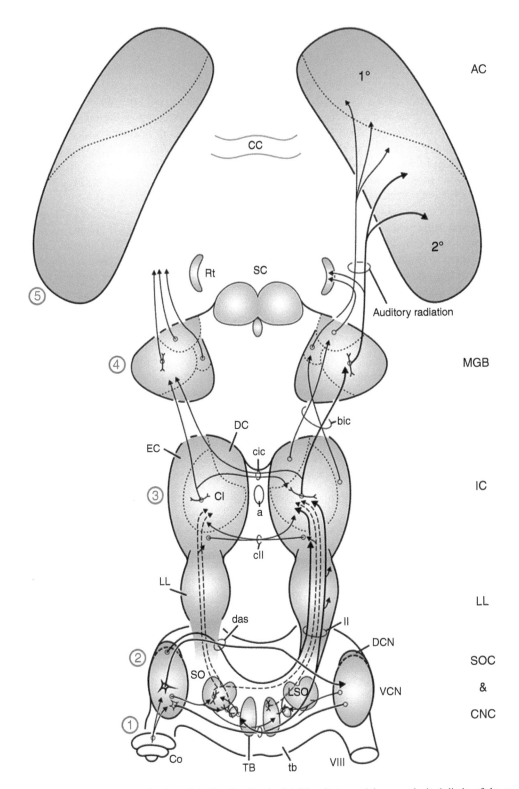

FIGURE 6.56 Auditory system of mammals adapted to the situation in dolphins. In terrestrial mammals, both limbs of the ascending auditory pathway have the cochlear spiral ganglion (first neuron) and the cochlear nerve in common as well as the dorsal cochlear nucleus (DCN; second neuron) and the ventral cochlear nucleus (VCN; second neuron). In dolphins, the DCN is nearly completely reduced (see text). Numbers of the neurons in the ascending auditory pathway in *red*. The direct auditory pathway proper originates in the DCN and runs directly to the inferior (caudal) colliculus (IC; third neuron), thus bypassing both the superior olivary nucleus (SO) and the nucleus of the lateral lemniscus (LL). The indirect pathway crosses contralaterally via the trapezoid body (TB), the superior olivary nucleus (SO, LSO) and the LL to the IC. Nearly all the fibers converge to the IC and from there they continue to the medial geniculate body (MGB; fourth neuron) and the auditory cortex (AC; fifth neuron). Abbreviations: 1°, primary auditory cortex; 2°, secondary auditory cortex; a, mesencephalic aqueduct; AC, auditory cortex; bic, brachium of inferior colliculus; cc, corpus callosum; CI, central nucleus of IC; cic, commissure of the inferior colliculi; cll, commissure of the lateral lemniscus; CNC, cochlear nuclear complex; Co, cochlea; das, dorsal acoustic stria; DC, dorsal cortex of IC; DCN, dorsal cochlear nucleus; EC, external cortex of IC; MGB, medial geniculate body; ll, lateral lemniscus; LL, nucleus lemnisci lateralis; tb, trapezoid body; TB, nucleus of trapezoid body; Rt, auditory sector of reticular thalamic nucleus; SC, superior (rostral) colliculi; SO, superior olive; SOC, superior olivary complex; VCN, ventral cochlear nucleus. VIII, vestibulocochlear nerve; 1–5, neurons of the ascending auditory pathway. *(Redrawn and modified by Jutta Oelschläger from Malmierca and Merchán, 2004. Auditory System. In: Paxinos, G., (Ed.). The Rat Nervous System, third ed. Elsevier Academic Press, Amsterdam.)*

As an entity, the auditory system of dolphins with its components is homologous to that in terrestrial mammals. At the same time, it is extremely well developed. This is not only seen in the absolute and relative size of many structures (with respect to the brain as a whole) but also in their architecture. All this is easily explained by the considerable and well-known auditory capacities of the odontocetes. Concomitantly, however, single components deviate from the general trend to hypertrophy and show marked reduction. All these trends, though, are part of a neurobiological "syndrome" and have to be interpreted as specializations signalling adaptations of dolphins to their aquatic habitat. Thus, regressive structures may be paradigms of nonresponsiveness between the auditory neuron populations and the multisensory input coming from the peripheral nervous system. In the following, examples of such regressive auditory structures (DCN, medial superior olive) shall be referred to in the course of our further presentation of the ascending auditory pathway.

According to the impressive diameter of the vestibulocochlear nerve and the high total number of axons in the cochlear nerve (Morgane and Jacobs, 1972; see also Sections "Cranial Nerve 8—Vestibulocochlear Nerve" and "Quantitative Aspects of Cranial Nerves" in this Chapter), the dolphin ventral cochlear nucleus (VCN) and other auditory centers are very large. The secondary auditory fiber tracts (trapezoid body, acoustic striae) are also well developed. In the La Plata dolphin (franciscana syn. *Pontoporia blainvillei*), the harbor porpoise and the common dolphin, the absolute volume of the cochlear nuclei is six to ten times larger than both in the cat and the human (Schulmeyer, 1992; Oelschläger and Oelschläger, 2009; Malkemper et al., 2011). Numbers of neurons in the cochlear nuclei of harbor porpoises range between 583,000 and 716,900; that is, 6–7 times that in the human (Jansen and Jansen, 1969). The ratio of primary cochlear nerve fibers to secondary cochlear neurons is 1:6.5 (in the human 1:4). In absolute terms, the MGB is about seven times larger, the inferior (caudal) colliculus twelve times, and the lateral lemniscus 250 times larger in the dolphin than the equivalent structures in the human brain (Bullock and Gurevich, 1979). Concomitant with this hypertrophy of the auditory system, physiological measurements of auditory brainstem responses in dolphins show high amplitudes—up to 30 times higher than in the human (Popov and Supin, 2007). In the La Plata dolphin, multiples of the cat's nuclei are found in the nucleus of the trapezoid body (17x) and in the intermediate nucleus of the lateral lemniscus (39x) (Schulmeyer, 1992).

Morphological and Functional Aspects

Ventral Cochlear Nucleus

The central auditory pathway starts caudally with the thick cochlear nerve (Figs. 6.7, 6.9, 6.27, 6.32, 6.35, 6.48, 6.49, 6.51, and 6.56), the comparatively very large ventral cochlear nucleus (Fig. 6.51: VCN) as well as the trapezoid body and nucleus (tb, TB) that bulge at the basal surface of the brainstem between the pons and the inferior olivary complex (P, IO). The auditory input ascends via the large superior olivary complex (Fig. 6.56: SO) and the thick lateral lemniscus and nucleus, which blends in the voluminous inferior colliculus (ll, LL, IC). From here, the auditory pathway proceeds via the brachium colliculi caudalis (bic) to the well-developed MGB and through the internal capsule (Fig. 6.27) to the extended neocortical auditory projection field in the suprasylvian gyrus (Figs. 6.15, and 6.16) between the suprasylvian and lateral sulci.

The dolphin VCN comprises four subunits consisting of specific neuron populations that are also found in terrestrial mammals. This has been proved by cytological investigations in the common dolphin and the La Plata dolphin involving the morphology of axon terminations (Malkemper et al., 2012; Schulmeyer, 1992). The same set of main neuron types were found in the dolphin CN as in other mammals, including spherical cells, globular cells, giant cells, octopus cells, and multipolar cells. However, so-called large spherical cells were present in the La Plata dolphin but, enigmatically, absent in the common dolphin.

In terrestrial mammals, VCN neurons may project bilaterally or ipsilaterally to the medial and lateral subnuclei of the superior olivary complex and contralaterally to the medial nucleus of the trapezoid body. Other targets of VCN neurons are the contralateral lateral lemniscus, the periolivary region of the superior olivary complex, the central nucleus of the inferior colliculus on both sides, the medial geniculate nucleus and the reticular pontine nucleus (Fig. 6.56) (Cant and Benson, 2003).

The axons of giant cells in the VCN constitute the dorsal acoustic stria (das) and seem to inhibit the contralateral VCN and the inferior colliculus. In contrast, axons of the multipolar cells have an excitatory influence on the contralateral inferior colliculus and an inhibitory influence on the contralateral VCN. The octopus cells give rise to the intermediate acoustic stria (not shown), which is rather thick in toothed whales (in contrast to the human). Octopus cells project mainly into the contralateral ventral nucleus of the lateral lemniscus (LL), which sends the major inhibitory input into the IC. Such hypertrophied systems of mutual excitatory and inhibitory auditory influence among neurons and nuclei in one neural framework (including the inhibition of inhibitory neurons, the so-called "disinhibition") are the substrate for information processing with respect to the creation of topographical and temporary windows of activity. This may be an important point for mammals that heavily depend on thorough analysis of complicated auditory patterns. Particularly interesting are the (small) spherical bushy cells and the globular bushy cells, both of which can be interpreted functionally (Paxinos, 2004). The spherical bushy cells project to both subnuclei of the superior olive, whereas axons of the globular bushy cells terminate in the medial nucleus of the trapezoid body. Both neuron types seem to be specialized for transmitting precise temporal information for sound localization.

Dorsal Cochlear Nucleus

The dorsal cochlear nucleus (DCN) is associated dorsally with the caudal part of the VCN (Fig. 6.47). In the area of the lateral recess of the fourth ventricle (see Section "Ventricular System" in this chapter), it is part of a slightly prominent acoustic tubercle together with the dorsal acoustic stria (Fig. 6.51: das). Medially the DCN borders on the lateral vestibular nucleus (D). The DCN does not show a layering as in some other mammals. In comparison to the VCN, the DCN is very small in toothed whales. Some authors have even doubted the existence of a DCN in toothed whales (Morgane and Jacobs, 1972). Histological investigations (La Plata dolphin, harbor porpoise, common dolphin), however, showed rudiments of this nucleus more or less "in place" (Malkemper et al., 2012; Osen and Jansen, 1965; Schulmeyer et al., 2000). In terrestrial mammals, the volume ratio of the DCN in proportion to the whole CN complex is at least four times higher in comparison with the common dolphin and the La Plata dolphin. The DCN in the human is approximately six times larger, and 12 times larger in the sheep.

The DCN consists of small multiform neurons with interspersed single pyramid-like perikarya. These pyramidal cells are also interpreted as fusiform cells. Giant cells, which are known from the DCN of other mammals (eg, cat; Osen, 1969), are not found in the DCN of dolphins (La Plata dolphin and common dolphin) and the dorsal acoustic stria originates from "giant cells" in the ventral and not in the DCN (Schulmeyer et al., 2000).

In terrestrial mammals, the DCN is supposed to be engaged in the assessment and/or elimination of "auditory artifacts" caused by positional changes of the head and pinnae of an animal toward a sound source. Proper use of the information provided by the external ear requires coordination of the auditory processing of signals from the ear with the position of the pinna. The convergence of auditory and somatosensory information in the DCN thus could represent a form of sensorimotor coordination for optimizing auditory processing (Young and Davis, 2001).

In contrast to dolphins, cats have big moveable pinnae and, in relation to brain size and the volumes of the other auditory nuclei, a large, distinctly laminated DCN. In humans and other apes, however, the pinnae are only slightly moveable. Their DCN is moderately developed and differentiated, without lamination, as a result of the loss of interneurons. If we compare the toothed whales with bats, a terrestrial group of mammals with similar acoustic/auditory capabilities due to echolocation, a drastic difference in the morphology of the DCN is obvious. Bats have a well-developed, slightly laminated DCN, and, in contrast to toothed whales, they have large moveable pinnae (Malkemper et al., 2012; Schulmeyer et al., 2000).

The ascending projections of the mammalian cochlear nuclear complex are generally tonotopically organized so that the isofrequency laminae of this complex (ie, areas where the neurons are receptive to the same frequencies) are connected with those of higher-order centers (inferior colliculus, primary auditory neocortex). Apart from this, the cochlear nucleus is involved in the so-called "acoustic startle reflex"[v] (Paxinos, 2004), which could not be demonstrated for dolphins so far.

In mammals, the cochlear nuclear complex also receives descending projections from the auditory cortex, inferior colliculus, ventral complex of lateral lemniscus, and the superior olivary complex. A large proportion of these fibers may be inhibitory, but there are also excitatory descending fibers that can influence cochlear function but also the cochlear nuclear complex and the inferior colliculus.

Superior Olivary Complex

The mammalian superior olivary complex (Fig. 6.51) is located in the caudal pons area and comprises mainly the lateral superior olive, the medial superior olive, and the nucleus of the trapezoid body. The three nuclei are surrounded by diffuse neuron populations, collectively referred to as the periolivary region. Interestingly, there is a general size correlation between the lateral superior olive and trapezoid body, whereas the medial superior olive seems to be independent. Thus, the lateral superior olive and trapezoid body are well developed in the rat and cat, whereas they are quite small in the human (Paxinos, 2004). In contrast, the medial superior olive is diminutive in rat, while it is well developed in both cat and human. These differences are obviously related to body size and the range of frequencies in hearing.

The superior olive of dolphins is very large. In the common dolphin (brain mass about 800 g) it is as voluminous as the ventral cochlear nucleus, that is, 150 times larger than in the human, and it contains 15 times more neurons

---

v. The acoustic startle reflex or response is a brainstem reflectory reaction of large muscle groups of the whole body (fright reaction) or the eyes (eyeblink).

(Zvorykin, 1963). In the smaller La Plata dolphin (brain mass about 220 g), the superior olive is about 50 times larger than in the human and 25 times larger than in the cat (Schulmeyer, 1992)

Apart from a volume increase during evolution, whales and dolphins seem to have modified the typical structure of the mammalian superior olive. Cytological analysis in the La Plata dolphin and harbor porpoise using silver stain preparations and including synapse morphology indicate that in these two species only one nuclear unit exists that should be homologous to the lateral superior olive of the terrestrial mammals. In the largest delphinid species (killer whale), there are two distinct superior olivary subnuclei, a medial and a lateral one. This is paralleled by a similar observation in the North Atlantic bottlenose whale (*Hyperoodon ampullatus*) (De Graaf, 1967), where the medial nucleus is larger than the lateral one.

Within the ascending auditory pathway, the superior olivary complex is the first nuclear configuration that receives afferent input from both ears for a comparison. Thus, the complex is the most important structure for the spatial analysis of sound sources (directional hearing) (Benninghoff and Drenckhahn, 2004). In cats and dogs, the medial superior olive was found to have many neurons sensitive to low-frequency cues, whereas the lateral superior olive has many neurons sensitive to high-frequency input. It seems highly probable that, at least in part, size differences in the subnuclei of the mammalian superior olivary complex are correlated with differences in body dimension, ecological niche, feeding mode, and thus in the auditory spectrum of these animals. In most mammals, the lateral superior olive is organized in an S-shape, with the long axis also representing the tonotopic axis. Low frequencies are located laterally, and high frequencies medially. A great mass of fibers ascending in the auditory pathway passes from the VCN to the trapezoid body and proceeds to the ipsilateral and contralateral nuclei of the superior olive and further to the lateral lemniscus and the inferior colliculus. In general, the lateral superior olivary neurons are excited by ipsilateral sounds and inhibited by contralateral sounds via inhibitory neurons of the contralateral trapezoid body. This allows the faithful encoding of interaural intensity differences and enables mammals to localize high-frequency sounds with considerable spatial precision (Grothe and Koch, 2011).

In cetaceans, the situation seems to be different from that in terrestrial mammals. Thus, in smaller toothed whales (La Plata dolphin, harbor porpoise, bottlenose dolphin), there is no neuron population that could be addressed as a medial superior olive. The narrow band of neurons that was regarded the medial superior olive by some authors was identified histologically and cytologically as part of the dorsomedial periolivary cell group (Schulmeyer, 1992). In larger toothed whales such as the killer whale and the North Atlantic bottlenose whale, there is a medial superior olive (see the previous discussion). Probably these differences in the size and structure of the superior olive are correlated to hearing function and head size (frequency range and sound travel time between ears; see the following) in the various whales and dolphins. However, up to date there is no clear conception about the functional implications of the very large superior olivary complex of dolphins with its relatively low neuron density in an excessively well-developed neuropil. It is certainly essential in the reliable analysis of sound and ultrasound signals during echolocation and orientation. This large nuclear complex may represent a neural basis for diverse adaptational processes in different species with respect to body size, feeding, and communication.

Nucleus of the Lateral Lemniscus

There is agreement in the literature that among mammals, such a high relative size of the respective nuclear and fiber masses in toothed whales (with respect to brain size) is only seen in bats. In dolphins, very strong fiber bundles run from the superior olivary complex dorsorostrolaterally to the inferior colliculus (Figs. 6.9, 6.27, 6.39, 6.48, 6.56, and 6.60). They are called lateral lemniscus and enclose three nuclei of different size: the smaller ventral nucleus of the lateral lemniscus (VNLL), which borders the superior olivary complex; the dorsally adjacent and extremely large intermediate nucleus (INLL); and the magnocellular dorsomedial nucleus (DNLL). In dolphins, the nuclei of the lateral lemniscus, in sum, are the second largest nuclear complex in the auditory system after the inferior colliculus (see the following). These nuclei have been homologized with counterparts in terrestrial mammals using cytological criteria (Schulmeyer, 1992; Zvorykin, 1976). The neurons of the ventral nucleus are uniform with round perikarya and calyces of Held[w] and the intermediate nucleus comprises mainly smaller multipolar neurons with different shapes of somata. The dorsomedial nucleus contains large multipolar perikarya, which belong to the largest within the auditory system and is traversed by the medial limb of the lateral lemniscus. The DNLL, which receives input from both ears and projects to the ipsilateral and contralateral colliculus through the commissure of the lateral lemniscus (Fig. 6.56: cll), is moderately developed in toothed whales when compared to the situation in bats. In the La Plata dolphin, it is as large as that in the cat and rather different in size among the toothed whale species. There are some indications that the dorsomedial nucleus and its wiring are not only involved in

w. The calyx of Held is a particularly large synapse found only in the mammalian auditory central nervous system (named by Hans Held (Held, 1893) which resembles the calyx of a flower around the soma of the postsynaptic neuron. The latter neurons are found in nuclei of the trapezoid body and the lateral lemniscus.

the function of the sonar system in dolphins, more or less in parallel to the situation of the DCN. As the latter nucleus, the DNLL was suggested to connect the auditory system with those motor centers which, in terrestrial mammals, are engaged in the coordination of head rotation and the orientation of the head and pinna toward a sound source. Here, the loss of the atlantoaxial joints (see Chapter 5, Section "Ear: Sense of Balance and Hearing") and the strong modification of the cervical vertebral column may explain the comparatively small size of the dorsomedial nucleus in toothed whales by means of restrictions in head mobility.

Inferior Colliculus

In mammals, the caudal (inferior) colliculus has a key position in the auditory system as an obligatory relay center for nearly all ascending auditory tracts. It is subdivided into the central nucleus (Fig. 6.56: CI), a laterally and rostrally placed external cortex (EC), and a dorsal cortex (DC). These three subnuclei have been further subdivided on the basis of their neuropil and neuron populations in the cat or bat.

The inferior colliculus receives fibers from both lower and higher auditory centers as well as from nonauditory structures. Ascending input originates in the lower auditory centers and tends to terminate more densely in the ventral portions of the inferior colliculus, whereas the descending input from the auditory cortex and the commissural input concentrate in the dorsal portions of the inferior colliculus. In other words, neurons in the ventral part of the inferior colliculus are mainly under the influence of the lower centers, and neurons in the dorsal part are mainly under the influence of the descending pathways; between these two main inputs, an area of overlap seems to exist, including commissural projections. The inferior colliculus has ascending projections to the MGB and descending projections to the superior olivary complex and the cochlear nuclear complex. In addition, the inferior colliculus possesses well-developed commissural and intrinsic fiber systems.

Many neurons in the central nucleus of the inferior colliculus carry information about the location of sound sources. The majority of these cells are sensitive to differences in interaural time and intensity, which are known to be essential cues for localizing sound in the horizontal plane (azimuth). Other neurons are sensitive to spectral cues that may localize sound in the vertical plane (zenith).

Neurons in the central nucleus project to the thalamus, the external cortex of the inferior collicle and the superior colliculus. Auditory, visual, and somatosensory neurons in the deep layers of the superior colliculus all converge on output pathways in the same structure that controls orienting movements of the eyes, head, and external ears. The circuits of the superior colliculus are probably mapped with respect to targets in space and are aligned with the sensory maps. Such a sensorimotor interface facilitates the sensory guiding of directional movements.

In dolphins, the auditory input ascends via the very large lateral lemniscus in the pons region to the voluminous inferior colliculus in the midbrain. This superior-most center in the brainstem auditory system has a considerable size in dolphins. Thus, the inferior colliculus of the common dolphin is several times larger than the superior colliculus. It stands upright in parasagittal sections (Figs. 6.9, 6.26, 6.27, 6.30, 6.36, 6.49, and 6.56). In the adult common dolphin, the inferior colliculus is ovoid and the whole lemnisco-collicular complex is reminiscent of a thick club or barbell (Fig. 6.48). The height of the inferior colliculus is about 13 mm and the collicular width of the midbrain (between the lateral surfaces of both colliculi) is about 36 mm (Oelschläger et al., 2010). In comparison with the human, which is of about the same body size dimension, the inferior colliculus of the common dolphin is 12 times larger. From here, the auditory pathway proceeds via the brachium of the inferior colliculus to the well-developed MGB and from there through the internal capsule via the acoustic radiation to the extended neocortical projection fields in the suprasylvian gyrus of the telencephalic hemisphere.

The intrinsic organization of the inferior colliculus in toothed whales is largely unknown. Interestingly, however, there are some correspondences with the nuclear equivalents of other mammals. MR scanning of the inferior colliculus in the pigmy sperm whale gives first hints on the structural composition (Oelschläger et al., 2010). At the transition to the rostral colliculi, the organization of the caudal colliculus seems to correspond well with the situation in the cat (Oliver and Huerta, 1992).

Medial Geniculate Body

The medial geniculate body (MGB) (Figs. 6.9, 6.16, 6.26, 6.27, 6.28, 6.49, 6.56, and 6.60) is the diencephalic (thalamic) interface of the ascending auditory pathway. It lies somewhat apart at the ventrocaudal surface of the thalamus, medial to the lateral geniculate body. The nucleus receives afferent fibers from the ipsilateral caudal colliculus via the brachium colliculi caudalis, but also from the nucleus of the trapezoid body and the ipsilateral and contralateral cochlear nuclei.

It has been suggested that the dorsal medial geniculate complex (with a major projection to the primary auditory cortex; see Section "The Cortices" in this chapter) is especially large in whales and bats, animals with a specialized sonar system

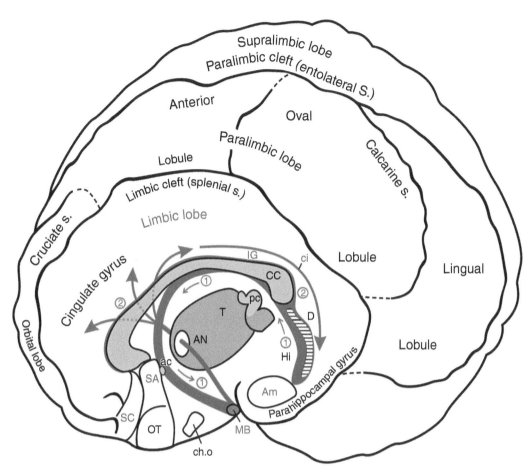

FIGURE 6.57   Archicortex and Papez circuit drawn into the mediosagittal aspect of the bottlenose dolphin brain (right hemisphere). In all mammals, the archicortex is arranged in an inner archicortical circle (1) and an outer archicortical circle (2); (*solid red* structures and *arrows*). During development they are formed by the division of the originally homogeneous archicortical area into two cortical bands. This process is driven by the semicircular outgrowth of the corpus callosum (cc) from the area of the anterior commissure (ac). The more or less semicircular structures (1) and (2) owe their spatial configuration to an unusual size increase of the neocortical areas. The inner archicortical ring (1) consists of the hippocampus and fornix, which is relayed in the mamillary body (MB), and the anterior nucleus of the thalamus (AN). From here, the stimuli from the hippocampus enter the outer archicortical circle (2) via the corpus callosum and proceed in the cingulate gyrus and cingulum back to the hippocampus. The Papez circuit consists of two half-circles which overlap each other; they are involved in phenomena like orientation, memory, emotion, and reproduction. See also Fig. 6.58 for a schematic representation of the Papez circuit within the limbic system. Commissures are in *yellow*. Abbreviations: Am, amygdala; AN, anterior nucleus of thalamus; ci, cingulum; cc, corpus callosum; ch.o, optic chiasm; D, dentate gyrus; Hi, hippocampus; IG, indusium griseum; MB, mamillary body; OT, olfactory tubercle; pc, posterior commissure; SA, septal area; SC, subcallosal area; T, thalamus. *Artwork by Jutta Oelschläger.*

for echolocation. In whales, and especially in toothed whales, the medial geniculate complex is the largest sensory nuclear group of the thalamus (Kruger, 1959), which reflects the importance of hearing for whales.

Limbic System

This system is comprehensive and rather complicated (Figs. 6.24, 6.27, 6.57, and 6.58). Therefore, apart from a general approach to the system as a whole, we shall focus on those limbic components which show distinct deviations from the general mammalian bauplan. Such deviations can help elucidate the potential functional properties and special significance of brain structures in dolphins in the light of those in terrestrial mammals.

The term "limbic system" goes back to Paul Broca (1824–1880) and his description of the "grand lobe limbique" (limbus means margin). The limbic system comprises cortical and subcortical structures, which are intensively interconnected and important for complex associative functions. These are emotional experience and behavior, the integration of emotional processes with somatic, endocrine and autonomic functions, learning and memory, attention, activity, impulse, and reproduction.

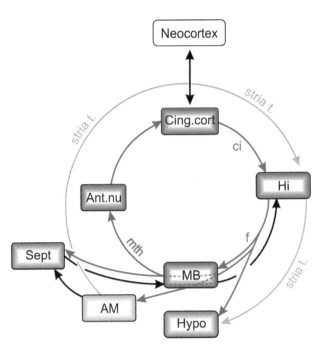

FIGURE 6.58   **The limbic system in schematic presentation.** The Papez circuit (*red*) can be regarded a core loop with the hippocampus (Hi) as its central structure. Note that here the two overlapping half-circles in Fig. 6.57 are folded apart into one circle or loop. The hippocampus receives input from all sensory systems, the septum and the neocortex via the terminal part of the parahippocampal gyrus and cingulum (ci), called entorhinal cortex. Nearly all efferent fiber systems of the hippocampus run within the fornix (f). The latter gives off fiber bundles to the septum (Sept), amygdaloid complex (AM), and hypothalamus (Hypo). The majority of fornix fibers ends in the mamillary body (MB). From here, the mamillothalamic tract (mth) runs to the cingulate cortex (Cing.cort) via the anterior nucleus of thalamus (Ant.nu). The cingulum (ci) proceeds to the hippocampus and thus closes the Papez circuit. In humans, the destruction of one member of the Papez circuit leads to handicaps in memory formation. An interesting addition to the Papez circuit is the amygdaloid complex (AM), which, in terrestrial mammals, is engaged in the emotional modulation of vegetative parameters, in the control of fear and rage and in the learning of emotional memory. Via the stria terminalis (stria t.; blue) it projects to the hypothalamus (Hypo) but also directly to the septum (Sept). The septum itself projects back to the hippocampus (*black arrows*), as another contribution to the Papez circuit. *Artwork by Jutta Oelschläger.*

Although the terms "limbic system," "limbic structures," and "limbic functions" are well established in neurobiological research, it is difficult to give a generally accepted definition of the system. This is due to the fact that the limbic system in a more general sense shares subcortical structures with the basal ganglia (BG) and the extrapyramidal system (EPS), including a wide spectrum of activities and a certain overlapping of functional implications. However, as in the case of the EPS, the limbic concept still has didactic and heuristic relevance essential for understanding brain function.

The following structures are widely accepted as components of the limbic system:

1. the Archicortex (see Section "The Cortices" in this chapter), which is situated at the margin of the telencephalic hemisphere, together with...
2. the Periarchicortex, a transition zone to the neighboring neocortical areas; and...
3. the Telencephalic and Diencephalic nuclei: septal area, amygdaloid complex, thalamic nuclei (Ncll. anteriores), the hypothalamic mamillary body (medial subnucleus), and connections between these areas. These structures are considered part of "reverberating circuits" or "loops" in which excitatory signals can circle during processing.

Other nuclei with intensive connections to and functional interaction with limbic structures mentioned above are: additional thalamic nuclei, the habenular nuclei in the epithalamus, mesencephalic nuclei (interpeduncular nucleus), as well as ventral components of the basal ganglia, particularly the nucleus accumbens. They all are attributed to the limbic system.

The advantage of such a complicated system of diverse functional components with their peculiar relationships distributed throughout the brain may be the following: they represent a considerable number of variables for modulatory effects as to information processing with respect to challenges and chances within the environment of a species. Thus, each component with its specifics in location, structure, capacity, and properties can be regarded member in a "steering committee" that organizes and coordinates responses of an individual to requirements from outside the body.

The main component and didactic center of the limbic system is the hippocampus; in the adult terrestrial mammal, it is located in the depth of the medial temporal lobe. Due to strong proliferation of the neocortex during ontogenesis, this part of the archicortex is transferred from the dorsomedial wall of the telencephalic hemisphere in three-quarters of a circle around the insular cortex (rotation of the hemisphere). Consequently, the fornix as the main efferent fiber tract of the hippocampus is much elongated and illustrates this embryonic shift of the archicortex (Fig. 6.57, seen in a mediosagittal aspect on the right hemisphere). Thus, the fornix runs from the hippocampal formation (Hi) in a spiral course along the ventral surface of the corpus callosum and enters the hypothalamus to terminate in the mamillary body ("inner archicortical circle"; Fig. 6.57: 1). Vestigial archicortical material (indusium griseum; means grey veil) is always found on the dorsal surface of the corpus callosum as a thin sheet, together with two thin fiber tracts (stria longitudinalis medialis et lateralis). Caudally, at the splenium of the corpus callosum (cc), the indusium becomes thicker and runs now as a narrow gyrus in parallel to the fornix to the hippocampus ("outer archicortical circle"; Fig. 6.57: 2). Because of its serial indentations, it is called dentate gyrus (D).

Immediately adjacent to the outer archicortical circle on each side of the median plane runs a "third circle of the limbic system" along (but above) the corpus callosum from the subcallosal area (SA) to the medial surface of the temporal lobe. This cortex was referred to as "limbic lobe" by Morgane and Jacobs (1972) (syn. gyrus limbicus; Figs. 6.7, 6.13, 6.25, and 6.57). This limbic lobe is regarded an intermediate zone (periarchicortex; Stephan, 1975) between the archicortex and the neocortex. It consists of a rostral or dorsal part (cingulate gyrus) and a caudal or ventral part (parahippocampal gyrus). In dolphins, the cingulate gyrus (CG) is much more expanded than the parahippocampal gyrus (bottlenose dolphin; Figs. 6.7 and 6.12: L), in contrast to the situation in humans (Fig. 6.13). Thus, the dolphin cingulate gyrus may even show two so-called "intercalate" sulci that separate three smaller gyri. The terminal (ventral) part of the parahippocampal gyrus (Figs. 6.27 and 6.57) is denominated "entorhinal area" and closely connected to the adjacent hippocampus; in primates, the latter is curled and invaginated into the lateral ventricle. This ensemble of archicortex and entorhinal area represents the "hippocampal formation" as a functional unit.

In terrestrial mammals (human), the anterior part of the cingulate cortex (Figs. 6.7 and 6.60, ACC) is closely related functionally to the pyramidal and extrapyramidal motor systems, whereas the posterior part (posterior cingulate cortex, PCC) receives input from the somatosensory, auditory, and visual association areas. The entorhinal cortex is the most important entry into the hippocampal formation: it receives fiber bundles coming from primary olfactory centers, the amygdala, and various neocortical areas. Practically all efferent fibers leave the hippocampal formation via the fornix. The precommissural part of the fornix (rostral to the anterior commissure) runs to the septum region, the postcommissural part mainly to the medial nucleus of the mamillary body. The latter projects to the anterior thalamic nucleus. The hippocampus and the postcommissural fornix as well as the mamillary body are part of the *Papez circuit*, which proceeds further to the cingulate gyrus and back to the hippocampus (Figs. 6.57 and 6.58: red). In the limbic system of mammals, this circuit is most prominent. It shows a certain size correlation between the hippocampus, the postcommissural fornix, and the mamillary body. In artiodactyls, these structures are well developed, whereas in dolphins they seem to be distinctly reduced.

As far as is known today, the terrestrial ancestors of whales and dolphins moved to a new (aquatic) environment characterized by dramatically different life conditions. Seen from the terrestrial situation, they had to tolerate the loss or simplification of sensory systems and thus a reduced spectrum of incoming information. However, there was one modality that could guarantee stability in orientation, feeding, and communication needed for survival—the auditory system. It seems likely that the decrease in size of the hippocampus in dolphins is correlated to such a narrowed sensory basis. However, this may not necessarily indicate restrictions in motor learning because this is largely exerted by the basal ganglia, brainstem and cerebellum (extrapyramidal system). Bats, which share important commonalities with dolphins as to sonar orientation and hunting as well as acoustic communication, were reported to have archicortices developed and sized (for the total brain) as other mammals (Baron et al., 1996).

The Role of the Hippocampus and Related Areas

In the literature, the hippocampal formation of dolphins is characterized as being organized like in other mammals. In spite of some specifics as a consequence of their adaptation to aquatic habitats, they show the same fundamental traits as large primates and ungulates (see the following). At the same time, also the topography of the hippocampus is quite similar. It is located in the medial wall of the large temporal lobe of the telencephalic hemisphere.

Careful allometric studies have shown that, while the hippocampus in primates is a progressive structure, this brain part has obviously lost much of its volume during cetacean (dolphin) evolution (Filimonoff, 1966; Morgane and Jacobs, 1986; Schwerdtfeger et al., 1984). Because the hippocampal formation receives input from the sensory organs, Morgane and Jacobs (1986) wrote that "a large hippocampus in a particular species may well relate to a high level of the processing of polysensory information..., in contrast to dolphins which have lost substantial parts of their presumed original sensory diversity." From this, the authors concluded that "of all species investigated so far, the dolphin appears to have the smallest hippocampus relative to other brain components." The idea that, in mammals, the hippocampus as a whole is closely related functionally to the olfactory system (constituting a rhinencephalon in the wider sense) seems to be misleading; in primates, the olfactory structures on the one hand and the hippocampus on the other show opposing trends in size progression (Schwerdtfeger et al., 1984). And the situation is even more complicated; in dolphins, the hippocampus is small but obviously functional, whereas the olfactory system has lost all of its peripheral structures. Secondary "olfactory" structures in anosmatic[x] dolphins are suspected to have also nonolfactory functional implications. In higher (microsmatic) primates including the human, the olfactory structures, in principle, are preserved but rather small (Stephan and Manolescu, 1980) whereas the hippocampus ranks on a high developmental level among mammals.

Investigations using routine histological preparations as well as silver impregnation and sectioning of the hippocampus in the bottlenose dolphin have shown that, in contrast to the situation in terrestrial mammals, the hippocampus of dolphins is not only rather small in size but, at the same time, lacks the typical convolutions of the cornu ammonis[y] (CA) (Jacobs et al., 1979). Thus, the fields CA1–5 as well as the dentate gyrus are not invaginated into the ventricular system but lie open at the medial surface of the temporal lobe. In their comprehensive monograph, Morgane and Jacobs (1986) concluded from older papers that the absolute volume of the fully differentiated hippocampus is about five times larger in sheep (in spite of its smaller brain) and about ten times larger in the human than in the harbor porpoise. In contrast, however, the adjacent ring of periarchicortex (limbic lobe) is comparatively large in dolphins. In the human, this cortex formation is equivalent (1) to the cingulate gyrus rostrally and dorsally, which is located above the corpus callosum, and (2) to the parahippocampal gyrus located at the medial surface of the temporal lobe. The temporal part of the periarchicortex adjacent to the hippocampal formation (entorhinal region) (Figs. 6.27 and 6.57) is slender but obviously well developed in the dolphin, although more pronounced in the human. During the last decade, this entorhinal part of the mammalian parahippocampal gyrus has gained much attention as an important multisensory association area.

Neuroethological investigations in terrestrial mammals have discovered several neuron populations engaged in spatial orientation and navigation. They were found in the hippocampus of rodents, bats, monkeys, and humans but also in the cingulate and entorhinal cortices (Burger et al., 2010). These neurons form populations that monitor the locomotion of an animal with respect to its place, head position, as well as the parameters speed, distance and time; and all of these neuron populations have been detected in entorhinal cortices. Seen from the situation in terrestrial mammals, dolphins in the open sea, with their restrictions as to afferent information (olfactory, vestibular, visual, somatosensory), could make use of additional cues for navigation beyond their sensory equipment which is dominated by the auditory system. Experimental data as to the processing of magnetic information in the brain of mammals have only been presented for the superior (rostral) colliculus in subterranean rodents (Nemec et al., 2001). No data are available on the situation in hoofed animals. And finally it is not clear whether migrating dolphins, in large-scale navigation, could benefit from a multidimensional frame of reference as terrestrial mammals do.

Basal Ganglia

The term "basal ganglia" refers to a group of closely connected cell masses forming a loose continuum extending from the basis of the telencephalon via the central part of the diencephalon and into the tegmentum of the mesencephalon (Nieuwenhuys, 1998; see also Fig. 6.59, asterisks). Primarily, this complex encompasses the striatum (caudate nucleus and putamen), and the globus pallidus. These basal nuclei are part of a highly complicated control system for motor activity, which comprises additional structures like the nucleus accumbens, olfactory tubercle, amygdaloid complex, claustrum, substantia innominata (basal nucleus of Meynert), subthalamic nucleus, red nucleus and substantia nigra. As a whole, these structures and their interconnecting fiber tracts have been called "extrapyramidal system," because it is organized in a way complementary to the pyramidal system. The latter is a more direct and fast counterpart of the extrapyramidal system, but both converge on the bulbar and spinal motor apparatuses. Both systems can be seen as two sides of a coin, with their different cortices being an interface or common substrate for many brain functions. The complexity of the two systems as an ensemble obviously reflects the considerable need for adaptive strategies under different circumstances and in situations "new" to the individual.

Among the basal ganglia, the corpus striatum seems to have some superiority functionally (Figs. 6.26, 6.36, and 6.59). Together with the nucleus accumbens, olfactory tubercle, amygdaloid complex, claustrum, and substantia innominata, the striatum arises from the ganglionic eminence in the embryonal telencephalon.

All of these nuclei have been found in dolphins and as far as they have been investigated they show topographical relationships as well as histological characteristics rather similar to those in other mammals. The basal ganglia mentioned

---

x. Anosmia is the loss of the sense of smell.
y. The human hippocampus as a whole has the shape of a curved tube, which resembles a seahorse in shape or a ram's horn (cornu ammonis).

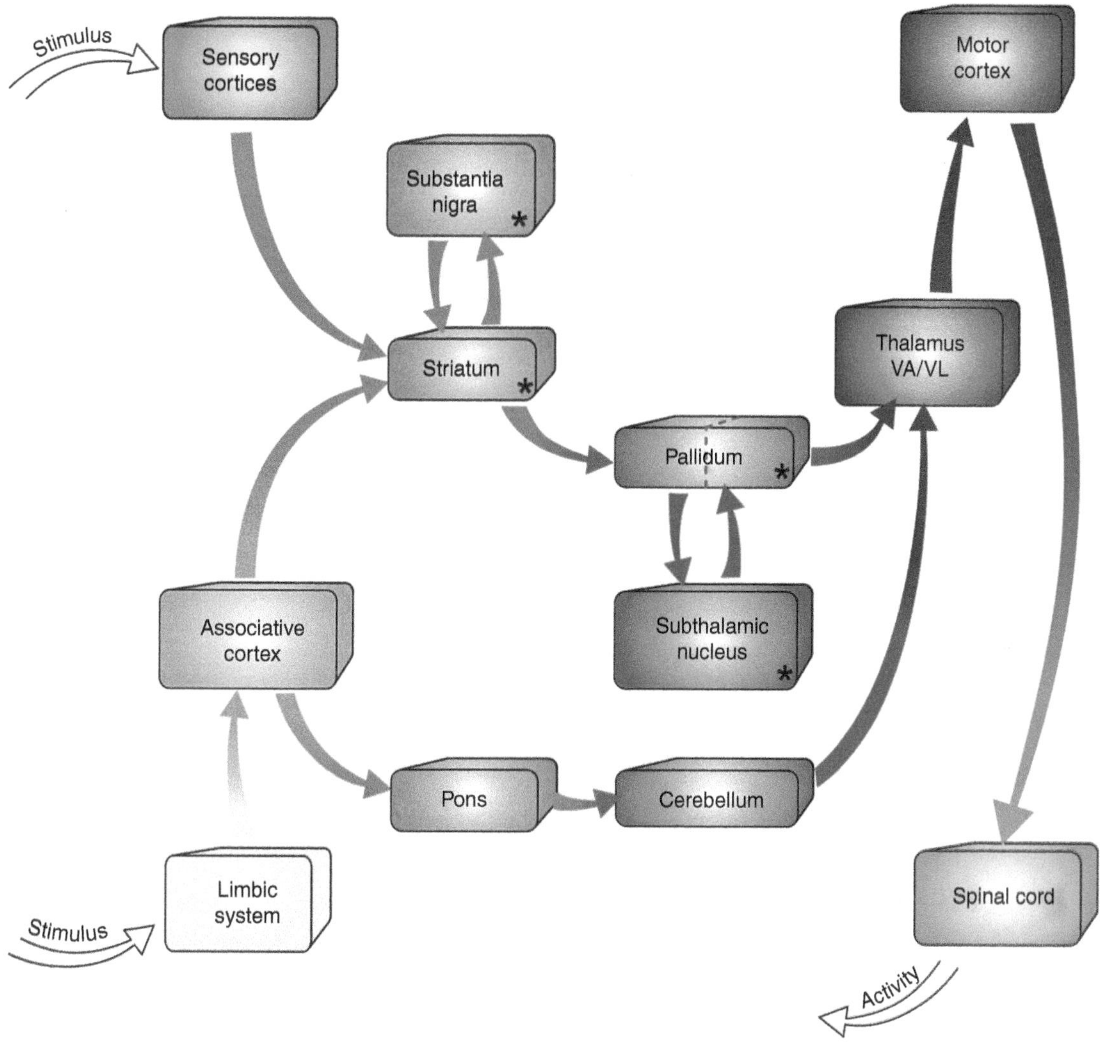

FIGURE 6.59   Simplified circuitry of some basal ganglia and the cerebellum within the central regulation of the motor system. The process from the initiation of motor activity via the modulation of the outline and finally the execution of motor behavior is symbolized by the transition from *yellow* via *red* to *blue*. There are two ways from the sensory input to motor execution. Sensory, limbic, and associative cortices project either to the striatum (above) and the pons (below) or to both (associative cortex). The upper pathway continues via the pallidum (*Globus pallidus*), the lower via the cerebellum; both ways converge in the direction of the thalamus, which stimulates the motor cortex for motor accomplishment in the spinal cord. Abbreviations: *, nuclei belonging to the basal ganglia; VA/VL, nucleus ventralis anterior and ventralis lateralis of thalamus. *Redrawn and modified by Jutta Oelschläger after Trepel, 2015. Neuroanatomie. Struktur und Funktion, sixth ed. Elsevier/Urban und Fischer, München.*

earlier are connected with additional premotor structures such as nuclei of the thalamus, the reticular formation, cerebellum, vestibular nuclei, and some cortical areas.

The *corpus striatum* or caudoputamen complex is well developed in dolphins (Fig. 6.39). Its components (caudate nucleus, putamen; Figs. 6.26, 6.30, and 6.36) are separated from each other during ontogenesis by the so-called "internal capsule," a thick aggregation of long ascending and descending fiber tracts. This feature is generally found in mammalian brains (including those of dolphins). From rostral to caudal sectional planes (Fig. 6.27), the fiber bundles increasingly perforate and divide the striatum on their way to and from the neocortex during ontogenesis. As a consequence, in the anterior striatum, both subnuclei are still in contact via bridges of ganglionic cell masses (striae). Rostrobasally and medially, the head of the caudate nucleus and the putamen are connected via the nucleus accumbens (see the following). Cytoarchitectonically, the two components of the striatum (caudate nucleus, putamen) are identical. Obviously, in dolphins as in other mammals, there is a correlation between the size of the striatum and that of the neocortex (Schwerdtfeger et al., 1984). The head of the caudate nucleus and the nucleus accumbens distinctly bulge ventrally in the area of the "olfactory tubercle." In mammals, generally, the basal ganglia are covered here by paleocortex (true olfactory tubercle). In dolphins, however, with their extensive reduction of the olfactory system, this tubercle has little to do with olfaction (see the following). At the same time, the "olfactory tubercle" is rather pronounced in dolphins and more or less watchglass-shaped (olfactory lobe) because of the size of the striatum (fundus striati). From here, the caudate nucleus runs along the lateral ventricle from the anterior horn in the direction of the inferior horn, which is situated in the temporal lobe of the hemisphere. Here, the tip of the tapering caudate nucleus is sometimes continuous with the amygdaloid complex. The semicircular shape of the caudate nucleus, lateral ventricle, choroid plexus, and other structures in this area, as well as the considerable tapering of this nucleus from head to tail, can be explained by the so-called "rotation" of the hemisphere. The transverse "rotation axis" runs through the center of the insula. This phenomenon takes place during the ontogenesis of large-brained mammals including dolphins, is caused by strong growth of the neocortex, and leads to the formation of a temporal lobe.

The *nucleus accumbens* is a medial expansion of the head of the caudate nucleus, and extends around the ventral wall of the lateral ventricle into the medial wall of the hemisphere. There are indications that the accumbens nucleus and part of the olfactory tubercle together represent the ventral sector of the striatum (ventral striatum). As to functional implications of the accumbens nucleus, Pennartz et al. (1994) reported that this cell mass, comprising a collection of specific subregions and projecting to the mesencephalic tegmentum (mesencephalic locomotor region), is preferentially innervated by fibers of specific limbic territories. Together with the striatum proper, the accumbens nucleus receives dopaminergic projections from each of the major cortical areas and the ventral midbrain. Via the mesencephalic tegmentum, that is, the ventral tegmental area, it influences locomotor behavior.

In mammals, generally, the *olfactory tubercle* is situated directly behind the anterior olfactory nucleus and ventrally covers the head of the caudate nucleus and the accumbens nucleus; here, aggregations of mostly small neurons (islands of Calleja) are characteristic. In large-brained mammals (dolphin; Figs. 6.7, 6.13, and 6.32, human) with a strongly reduced or small olfactory system, the three-layered paleocortex may be rudimentary and totally lacking in some places of the "olfactory tubercle" (Oelschläger et al., 2008).

The structure of the *amygdaloid complex* in dolphins (Figs. 6.9, 6.16, and 6.26–6.28) closely resembles that of other mammals. It comprises a basal ganglia component and an olfactory (paleocortical) component. With respect to the reduced state of the dolphin olfactory system as a whole, it is surprising how large the amygdala is in these animals. Its remarkable size in delphinids has been related to the considerable development of the temporal lobe and to audition (Morgane and Jacobs, 1972; Stephan and Andy, 1977). Interestingly, the corticomedial part of the amygdala in the anosmatic harbor porpoise has the same proportion in the whole amygdaloid complex as in the macrosmatic sheep. This phenomenon may perhaps be explained with the multisensory function of the amygdaloid nuclear complex: In terrestrial mammals, its subnuclei are targets for olfactory, gustatory, somatosensory, visual, and auditory input. In this respect, the auditory system may have a particularly high representation in the amygdala of dolphins. In terrestrial mammals, the amygdaloid complex has extensive connections with frontal, temporal, insular, and occipital cortices. As a main output channel, it gives rise to a highly organized system of pathways to many hypothalamic and brainstem areas. The concept of the so-called "extended amygdala" with some associated nuclei (bed nucleus of stria terminalis, substantia innominata, nucleus accumbens) is ideally suited to explain endocrine, autonomic, and somatomotor aspects of emotional and motivational states (flight and defense, social interaction and communication). Many authors have therefore included the amygdala into the limbic system. Other important aspects are the visual (human: face) and other sensory recognition and long-term memory processing, for example, learning associations of stimuli in different modalities. The latter is apparently in concert with the hippocampal formation, but not obligatory. This is seen in dolphins and other toothed whales, where the hippocampal formation is very much reduced and the amygdaloid complex very large.

The *claustrum* (Figs. 6.16, 6.26, 6.27, and 6.36) is a sheet of gray matter underlying the cortex in the lateral wall of the hemisphere. It is separated from the insular cortex laterally by the capsula extrema (a leaflet of white matter), and from the putamen medially by the capsula externa. Its connections are mainly with the neocortex, the hippocampal formation, the entorhinal cortex, and the amygdala. In dolphins, the claustrum is always present. Immunohistochemical studies (Cozzi et al., 2014) confirmed the general topography of the claustrum in the bottlenose dolphin. However, it seems to have been transferred here more rostrally during evolution, perhaps due to some rearrangement of the primary cortical areas caused by differential growth of the sensory systems.

Substantia innominata (basal nucleus of Meynert). This flattened cell mass is rather ill- defined and located between the following structures: putamen and globus pallidus, olfactory tubercle, anterior commissure, hypothalamus and amygdala. The Substantia innominata is mainly composed of loosely arranged medium-sized cells but also contains large fusiform and multipolar cells. Taken together, the large cells are addressed as the nucleus basalis of Meynert. Obviously there is a size correlation between Substantia innominata mass and brain mass of the respective individual; thus, the nucleus shows a maximal development in large-brained primates and cetaceans (dolphins). The large cells in the "nucleus" are widely projecting throughout the neocortex (Voogd et al., 1998).

Globus pallidus (Figs. 6.26 and 6.27). The globus pallidus is of diencephalic origin and situated immediately medial to the putamen from which it is separated by a thin leaflet of white matter only. In large brains, the globus pallidus was formerly summarized with the adjacent putamen as the lentiform nucleus because of its shape. This denomination, however, is unsatisfactory because both nuclei originate from different brain parts (putamen: telencephalon; globus pallidus: diencephalon) and were only secondarily conjoined by the establishment of the internal capsule. Both nuclei also differ much in structure from one another as well as in color. Because many bundles of myelinated fibers traverse the globus pallidus, the nucleus has a paler color in fresh brain preparations than the telencephalic putamen or caudate nucleus. The cells of the medial pallidal segment, also called nucleus entopeduncularis in nonprimates, are intercalated in the course of the ansa lenticularis, the principle stream of efferent fibers from the basal ganglia.

Subthalamic nucleus. This nucleus is situated in the caudal part of the diencephalon, dorsomedial to the internal capsule at its transition into the crus cerebri. It is moderately developed in carnivores and dolphins but well developed in primates. In terrestrial mammals, the subthalamic nucleus is closely attached functionally to the globus pallidus. More precisely, it is an interface between the lateral and medial part of the pallidum and important for the activation of the motor cortical areas (Fig. 6.59).

Substantia nigra. In the bottlenose dolphin, the nucleus has a rather extensive distribution dorsal to and intermingling with the fibers of the crus cerebri (Fig. 6.16). It is composed of scattered medium- to large-sized multipolar cells, some of which contain a very small amount of melanin as compared to characteristic nigral cells in terrestrial mammals. The dopaminergic projections of the substantia nigra are assumed to be involved in the stabilization of the circuitry in the basal ganglia. They are also influential for learning processes.

The *red nucleus* (nucleus ruber; Figs. 6.16, 6.29, and 6.41–6.43) as the main origin of the central tegmental tract (Fig. 6.29; ctt) is discussed later together with the *elliptic nucleus* (see Section "Integrative Aspects—Audiomotor Navigation" in this chapter).

Extrapyramidal System

According to many authors, a number of the nuclei mentioned in Section "Basal Ganglia" are incorporated in the so-called "extrapyramidal (premotor) system" (Nieuwenhuys 1998). Other authors have abandoned the term and definition but obviously the terms are difficult to replace as to their didactic value for understanding brain function (Figs. 6.29 and 6.59). In mammals, this configuration is an interesting counterpart to the pyramidal system in view of motor action, including, for example, fight-or-flight behavior.

The pyramidal system of mammals (PS) is defined as a comparatively clear-cut ensemble of parallel fiber tracts that originate in the motor cortex as well as in other neocortical areas, and that travel together through the internal capsule and the brainstem. They divide and follow separate courses to motor and/or sensory nuclei of cranial nerves at various levels of the brain (tractus corticonuclearis mesencephali, pontis, bulbi; see Terminologia Anatomica, FCAT 1998) and along the ventral (motor) horns of the spinal cord (tractus corticospinalis). An important part of the pyramidal tract (tractus corticonuclearis pontis) is relayed in the pontine nuclei to the cortex of the cerebellum. Consequently, this part of the pyramidal tract is often referred to as "cortico-ponto-cerebellar tract" (see Section "Pons and Cerebellar Connections" in this chapter). Characteristic of the pyramidal system is the largely compact and direct course of intentional and high-precision motor projections, particularly in primates. In the spinal cord of terrestrial mammals, the tract activates predominantly the neurons of the flexor muscles and, at the same time, inhibits those of extensor muscles. Complementary, for example, within the extrapyramidal system, the vestibulospinal tract and fibers of the pontine reticular formation inhibit neurons of the flexor muscles and activate extensor muscles that work against gravity and are particularly important for body posture and equilibrium (see Section "Vestibular System" in this chapter).

Interestingly, the pyramidal tract of dolphins ends in the rostralmost cervical segments of the spinal cord (Fig. 6.52), a situation also seen in artiodactyls (cow, sheep, deer). As in these close relatives, the extrapyramidal efferent projections of dolphins (Figs. 6.60, and Chapter 10, Figs. 10.3, 10.4, 10.10, and 10.11) seem to substitute the pyramidal system in the organization of spinal locomotor behavior (in dolphins: mass movements by the body stem and fluke).

In contrast to the pyramidal system, the extrapyramidal system of mammals (EPS) is largely involuntary, highly complicated, and integrative as to the structures involved and to their functional implications. It contains a wide spectrum of different nuclei that are interconnected in diverse recurrent circuits (loops). At least part of the basal ganglia are involved in such loops with the cerebral cortex (see Section "Basal Ganglia" this Chapter). Also, the EPS is regarded "older" than the pyramidal system in an evolutionary context and consists of multisynaptic chains of neurons. The considerable variety of nuclei suggests that different loops operate in relation to different types of cortical function. Thus, the scope of the extrapyramidal system as a (pre-) motor system has to be supplemented by multisensory and integrative relationships to

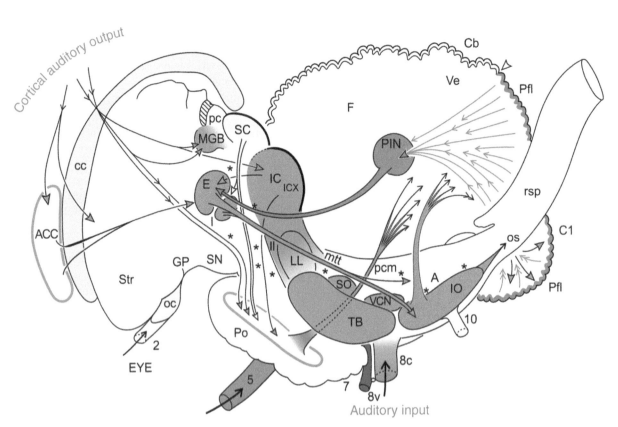

FIGURE 6.60 Simplified scheme of the feedback loops in the dolphin brainstem most important for premotor information processing. The auditory system (*green*) as the dominant sensory system is the didactic center here. Other sources of sensory information are the visual system (*yellow*) and the trigeminal (*blue*), which, for the most part, are left out here. The auditory brainstem (cochlear nuclei, trapezoid body and nucleus, superior olive, lateral lemniscus, and nucleus) gets information from the vestibulocochlear nerve (8) as well as from the auditory cortex in the telencephalic hemisphere. This extended cortex feeds into the pontine nuclei, the anterior cingulate cortex (ACC), an integrative part of the limbic system as well as in the elliptic nucleus, which is associated with Cajal's interstitial nucleus, a center involved in rotational axial body movements (turns) in terrestrial mammals. The auditory input into the pontine nuclei is projected into the cerebellar cortex (paraflocculus) and to cerebellar nuclei (PIN, posterior interposed nucleus) and via the superior cerebellar peduncle to the elliptic nucleus. From here, multisensory input from various sources is projected to the inferior olivary complex and further to the cerebellar cortex, and returned to the elliptic nucleus via the cerebellar nuclei. A, nucleus ambiguus; ACC, anterior cingulate cortex; C1, first cervical nerve; Cb, cerebellum; cc, corpus callosum; E, elliptic nucleus (Darkschewitsch); F, nucleus fastigii; GP, globus pallidus; I, interstitial nucleus of Cajal; IC, inferior (caudal) colliculus; ICX, external cortex of IC; IO, inferior olive; ll, lateral lemniscus; LL, nucleus of the lateral lemniscus; MGB, medial geniculate body; mtt, medial tegmental tract; oc, optic chiasm; os, olivospinal tract; pc, posterior commissure; pcm, pedunculus cerebellaris medius; Pfl, paraflocculus; PIN, posterior interposed nucleus; Po, pons; rsp, reticulospinal tract; SC, superior (rostral) colliculus; SN, substantia nigra; So, superior olive; Str, striatum; TB, trapezoid body; VCN, ventral cochlear nucleus; Ve, vermis; 2, optic nerve; 5, trigeminal nerve; 7, facial nerve; 8c, cochlear nerve; 8v, ventricular nerve; 10, vagus nerve; ***, periaqueductal gray and reticular formation. *Modified by Jutta Oelschläger.*

many brain structures. Such structures important for motor function, without belonging to the motor system proper, are the limbic system (Fig. 6.59), thalamus, cerebellum, reticular formation, vestibular nuclei, and some cortical areas. In other words, this puzzling extrapyramidal system of many diverse nuclei and fiber tracts is involved in a whole series of functional aspects, all of which, however, are related to motor function. For instance, in terrestrial mammals, paleocortical and subcortical structures of the basal telencephalon are important relay stations within the limbic system and influence higher cognitive functions. The amygdaloid complex and associated nuclei are essential for emotional and motivational aspects of perception and behavior and influence the accompanying visceral and endocrine reactions. The striatum, accumbens nucleus, and ventral pallidum take part in somatomotor aspects of this behavior (emotional motor system) but also influence activities such as learning and memory. This mechanism is contrasted by hippocampus-based learning, which builds memories that depend on particular cues in the environment. Also, in terrestrial mammals, loops interconnecting the cortex, the striatum and the dopamin-containing cell groups in the midbrain, particularly the ventral tegmental area (VTA), play a role in reward-based learning and memory.

Input systems of the extrapyramidal system (EPS). The caudate-putamen complex (striatum; Figs. 6.36, 6.39, and 6.59) receives a massive and direct projection from the entire neocortex (including prefrontal, prelimbic, and anterior cingulate cortices), and this complex projects back to the motor and premotor cortices via the globus pallidus and the thalamus. Additional important input to the corpus striatum and its circuitry comes from the reticular formation of the brainstem. Particularly the mesencephalic portion of the reticular formation is one of the principal sources of afferent fibers to the thalamus and to both the caudate nucleus and putamen.

Output systems of the extrapyramidal system (EPS). The main outflow from the striatum and the nuclei related to it converges via the globus pallidus and the thalamus upon the motor, premotor, and supplementary motor cortical areas. Although much of the output is directed to the frontal cortex, there are several connections by which this set of subcortical cell masses have access to the somatic motor systems of the brainstem and spinal cord. The largest of these projections run from the substantia nigra to the superior colliculus and the reticular formation, from the globus pallidus to the lateral habenula and periaqueductal gray as well as from the "mesencephalic extrapyramidal area" to the medullary reticular formation. An interesting basal circuit in mammals exists in the brainstem and serves premotor processing of all kinds of afferent information from neocortical areas: In dolphins, particularly auditory input seems to be fed into a circuit from the elliptic nucleus via the medial tegmental tract to the inferior olive and cerebellum (see Section "Integrative Aspects—Audiomotor Navigation" in this chapter).

Reticular Formation

The reticular formation is one of the most underestimated structures in the mammalian brain. Although extremely important in many ways, it is rather diffuse and therefore often ignored by students of anatomy and neurobiology. The reticular (syn. netlike) formation is phylogenetically very old and accounts for a substantial part of the tegmentum.

In cross sections through the brainstem and thus the tegmentum of dolphins, nuclei belonging to the cranial nerves and derived from the rhombencephalic nuclear columns can be distinguished (Figs. 6.27, 6.47, and 6.48). In the myelencephalon (syn. medulla oblongata), which is rather flat posteriorly (Fig. 6.51d), these nuclei are located in the periphery (the large ones laterally and ventrally). The dorsal central part is occupied by the reticular formation. In the direction of the dolphin pons and mesencephalon, strongly hypertrophied nuclei and fiber tracts of the auditory and extrapyramidal systems successively displace the reticular tissue (McFarland et al., 1969; Fig. 6.39). Phylogenetically older tracts run through the reticular formation, whereas "new" tracts (rudimentary corticospinal tract in dolphins) appear to be attached to the brainstem ventrally/superficially. As the size progression of the auditory system in the brainstem of terrestrial mammals is much weaker than in dolphins (see Section "Auditory System" this chapter), their reticular formation is generally more prominent in the myelencephalic, pontine and mesencephalic parts of the tegmentum. This seems to be in contrast to the situation in the human, where the reticular formation is comparatively small in the medulla and pons (Barone and Bortolami, 2004). In histological appearance, the reticular formation is characterized as a complex pattern of gray substance that forms a homogeneous three-dimensional network. For the most part, it comprises diffusely arranged intrinsic neuronal somata together with neuropil consisting of their dendrites and axons together with afferent and efferent fibers from ascending and descending fiber tracts. The network extends within the tegmentum from the rostral border of the mesencephalon in the caudal direction to the end of the myelencephalon and further into the spinal cord. Locally, the somata of the neurons are compacted to various nuclei of the reticular formation, which, however, are often difficult to delimit. Only some nuclei are well discernible, particularly the raphe nuclei, which form the median zone of the reticular formation. Throughout the brainstem, specific intrinsic neurons of the reticular formation are associated in populations that can generate patterns of autonomic and motor response subserving simple, stereotyped, coordinated functions ranging from facial expression to feeding and breathing (Kandel et al., 2013).

The reticular formation receives afferent projections from nearly all parts of the nervous system, among them, the spinal cord, myelencephalon, pons area, cerebellum, mesencephalic tegmentum, auditory and visual pathways, as well as cortical areas. In turn, the reticular formation sends efferent projections to a comparable variety of brain areas, among them the centers of somatomotor control, the autonomic system, spinal cord, brainstem, cerebellum, and so forth.

Within this framework of connectivity, functional implications of the reticular formation are to coordinate the wiring of single or more brainstem nuclei with respect to reflexes and simple behaviors necessary for homeostasis and survival (pupillary light reflex, corneal reflex, vestibulo-ocular reflexes, stapedial reflex, mastication, deglutition, emesis, respiration, circulation, etc.) but also the sleep/wake cycle, the regulation of cortex activity, as well as extrapyramidal motor activity and arousal (Benninghoff and Drenckhahn 2004).

Integrative Aspects—Audiomotor Navigation

As a whole, the dolphin brain is characterized by a series of adaptations ranging from strong hypertrophy to massive reduction and even loss of structures. Such a size increase or decrease of nuclei and/or fiber tracts may affect rather different areas of the brain and often cannot be understood as isolated phenomena. Also, within a neurobiological system like the ascending auditory pathway, size progression may be very different for single components, and this cannot be explained in view of the overall developmental state of the system. Such a deviation can be understood by implicating the functional role of the individual structure in its specific circuitry with other areas outside the auditory system (typical examples are centers with an integrative character). Thus, while most parts of the subcortical auditory system are exceptionally well developed in dolphins, others are very small and sometimes even not found by investigators (DCN, medial superior olive; see Section "Auditory System" in this chapter). In this respect, the strong reduction or even disappearance of the DCN cannot be understood without considering its functional implications as a relay nucleus used for the integration of auditory input with such from sensory systems innervating the head/neck and external ear regions. This correlation is an important aspect for many terrestrial mammals, which can use the sensory and motor innervation of these areas in view of directional hearing and "smooth pursuit" (ie. when it is necessary to dampen or omit auditory artifacts caused by changes in head and pinna position during orientation reactions and for keeping track of prey). Dolphins, however, have drastically restricted their head mobility in favor of powerful and efficient locomotion and reduced the external ears (and other appendages) in order to become hydrodynamic. Concerning the external ears, this reduction process must have led to the loss of essential sensory input (auditory, somatosensory), which no longer could be relayed to the facial and other motor nerves. All this meant a functional trend into insignificance for the DCN because there was no obvious other input avoiding this. Concomitantly, sound reception was shifted to the lower jaw and associated acoustic fat bodies (see Chapter 5, Section "Head and Senses").

Additional examples for quantitative correlations of negative growth are the limbic Papez circuit in dolphins (Figs. 6.57 and 6.58) and the loss of the anterior olfactory system as well as the reduction of the vestibular system and the somatosensory innervation in the spinal cord (see related sections in this chapter).

A quantitative correlation, this time in a positive sense, is the central loop in the brainstem constituted by nuclei and a single cortical area: the elliptic nucleus, medial accessory inferior olive, paraflocculus and the posterior interposed nucleus (Figs. 6.41–6.43 and 6.60). In dolphins, these structures are all hypertrophic with respect to those in generalized terrestrial mammals. They are, however, not immediately evident as belonging to the same major neurophysiological ensemble. A structure like the cerebellar cortex (paraflocculus) of dolphins, with extended cortical representations of their sensory modalities, is more impressive than relay structures as, for example, the elliptic nucleus, which collects and redirects sensory projections. In this example of feedback loop, the elliptic nucleus (1) sends a strong projection to the medial accessory inferior olive (2) which integrates afferent input from the spinal cord (spino-olivary tract) and projects to the lateral intermediate zone of the paraflocculus (3), and the latter to the posterior interposed nucleus (4) and then back to (1) the elliptic nucleus (Figs. 10.3, 10.4, 10.10, and 10.11). In dolphins, all the stations along this loop are hypertrophic compared to those in generalized terrestrial mammals, and some of them show a similar size dimension.

Concerning the circuitry in the mammalian brainstem with respect to the cerebellum as its largest superior center of integration, it seems that there are two nuclei of equal functional importance both situated at the gate to the tegmentum and thus the brainstem as a whole: the red nucleus in primates and the elliptic nucleus in dolphins. More or less adjoining each other in the rostral part of the mesencephalon, the two nuclei seem to have similar connections.

Red Nucleus (Nucleus ruber)

This structure is the largest nucleus in the mesencephalic tegmentum of primates. It is penetrated by numerous bundles of myelinated fibers from the superior cerebellar peduncle, but also by rootlets of the oculomotor nerve and fibers of the habenulointerpeduncular tract.

Afferents: (1) This nucleus and the nuclei of the mesencephalic-diencephalic junction are the main link in the projection from neocortex onto the inferior olive; (2) input from the cerebellum (dentate nucleus; Fig. 6.42); both of the afferent fiber systems are organized somatotopically.

Efferents: (1) Directly to spinal cord (rubrospinal tract); (2) via central tegmental tract (ctt) to inferior olive (principal nucleus; Fig. 6.42); from there to cerebellum (posterior lobe of hemisphere, paraflocculus, lateral zone in Fig. 6.38). This loop continues via the dentate nucleus back to red nucleus and from there to the thalamus and the motor and premotor neocortex.

Elliptic Nucleus (Nucleus of Darkschewitsch)

Afferents: (1) In terrestrial mammals (primates), the nucleus and other nuclei at the mesencephalic-diencephalic junction receive fiber systems from motor and premotor cortices (Brodmann areas 4, 6, 8) including area 7 in the parietal cortex, auditory system, vestibular nuclei, and pretectal nuclei; (2) input from cerebellum (posterior interposed nucleus; Fig. 6.42).

Efferents: Via medial tegmental tract (mtt) to inferior olivary complex (medial accessory nucleus; Figs. 6.42 and 6.60), from there to cerebellum (posterior lobe of hemisphere, paraflocculus, intermediate C2 zone); this loop proceeds to the posterior interposed nucleus and the elliptic nucleus.

The red nucleus and the elliptic nucleus both receive a major projection from the cerebellar nuclei. They are also under the influence of the premotor, motor, and parietal cortex. Obviously, they are integrated in two parallel feedback loops that project to the cerebellum (Voogd et al., 1998). The third important descending projection to the cerebellum is that from the neocortex via the pontine nuclei (Figs. 6.29, 6.41, and 6.60; and Chapter 10, Figs. 10.3, 10.4, 10.10, and 10.11). Interestingly, the two nuclei (ruber, elliptic) more or less seem to replace each other in dolphins and primates (Fig. 6.43). They also show reciprocal quantitative relations: Whereas in primates (human), the parvocellular red nucleus is very large and obviously the most important relay structure for projections from the cortex to the inferior olive (very large principal olive), dolphins have a very large elliptic nucleus (Darkschewitsch), which also projects to the inferior olive but to another subnucleus (very large medial accessory olive). In consequence, the projections of the two nuclei and their targets are different in size reciprocally: In primates, the central tegmental tract is very strong (human), but it is weak in carnivores and absent in ungulates. In contrast, the medial tegmental tract is weak in primates and carnivores, present in ruminants, and strong in dolphins. For a considerable part of their course, the two fiber tracts may run together. Likewise, in the two loops (Fig. 6.42), the projections from the inferior olive to the cerebellum terminate in different parts of the cerebellar cortex. In primates, the fibers of the principal subnucleus of the inferior olive end in a moderately developed area of the cerebellar hemisphere (with respect to the whole very large cerebellum: lobulus biventer and tonsil), which, in turn, projects to the folded nuclear masses of the very large dentate nucleus (homologous to the small lateral nucleus in dolphins). In dolphins instead, the fibers of the medial accessory olive end in the very large dorsal and ventral paraflocculus, which, in turn, project to the very large posterior interposed nucleus (homologous to the small globose nucleus in primates). In other words, in the cerebella of the two groups, which show two extremes as to cerebellar size, structure and circuitry within mammals, the posterior interposed nucleus of dolphins completely dominates the cerebellar nuclear complex in much the same way that the dentate (syn. lateral) nucleus does in the human cerebellum. Finally, the two very large cerebellar nuclei project to the nucleus ruber and the nucleus ellipticus, respectively: Fibers from the dentate nucleus (primates) end in the nucleus ruber, and those from the posterior interposed nucleus terminate in the nucleus ellipticus (Figs. 6.40–6.43).

This double-loop construction in the brainstem of mammals may have to do with locomotion. The nuclei involved are attributed to premotor function. As is known in the literature on primates, the pyramidal tract is particularly important for the motor activity of the distal limb, whereas the proximal limb and trunk are activated by the so-called "extrapyramidal system." In dolphins, the latter system very much reminds of the closely related hoofed animals (Fig. 6.52). All of these animals lack the pyramidal system from the middle cervical segments caudally. They have to rely on their extrapyramidal system, which in dolphins is highly specialized. Here, the very large elliptic nucleus may serve as the main entrance into the locomotory apparatus of the body and flukes. Via this nucleus, the input from the neocortex, and thus from the extremely well-developed ascending auditory pathway is transferred into a central premotor loop that is characterized by similar relative size progression of the succeeding components. From here, the assimilated information is probably transferred to the reticular formation and its projections into the spinal cord (reticulospinal, vestibulospinal, and olivospinal tracts). It is therefore likely that all these premotor structures (elliptic nucleus, inferior olivary nucleus, paraflocculus, posterior interposed nucleus) are largely engaged in the processing of auditory input delivered by the powerful sonar system in these animals (Fig. 6.60; and Chapter 10, Figs. 10.3, 10.10, and 10.11). Dolphins and other toothed whales may thus move by audiomotor navigation, using the sensory system that is by far the most reliable source of information in their environment and integrating the input of other sensory systems as available.

Concerning the aspect of sensorimotor integration within the cerebellum, it is interesting that in rats, the paraflocculus receives the major part of the pontocerebellar projection with the output from the auditory neocortex. In cetaceans, the paraflocculus was thought to be associated functionally with the locomotor apparatus of the trunk and tail. The components of the basic premotor circuits described here use the cerebellum as an association center. Two fiber tracts are connected to the paraflocculus, one via the pontine nuclei (cortico-ponto-cerebellar tract) and the other via the elliptic/red nucleus. They are not only involved in the integration of auditory and other sensory inputs (visual, somatosensory) via the pons but also in their transformation into motor activity in the direction of the spinal cord, that is, the locomotory apparatus of the body stem and fluke. Thus, efferent projections from the very large inferior olive and Deiters' nucleus, together with more or less diffuse caudal projections along the well-developed reticular formation descend to innervate the motor horns of the spinal gray matter. In a wider sense, this reticular system is a counterpart of the pyramidal innervation of the blowhole region (epicranial complex) via the facial nucleus and nerve (see Chapter 5, Head of Dolphins as an Entity). In conclusion, while the blowhole is engaged in sonar (sound) emission and the auditory system receives the relevant information important for navigation, the body stem executes adequate maneuvers to follow, for example, fast-moving prey.

REFERENCES

Bähr, M., Frotscher, M., 2009. Neurologisch-topische Diagnostik: Anatomie, Funktion, Klinik, 9, überarb. Aufl. ed. Thieme, Stuttgart.

Baron, G., Stephan, H., Frahm, H.D., 1996. Comparative Neurobiology in Chiroptera: Brain Characteristics in Functional Systems, Ecoethological Adaptation, Adaptive Radiation, and Evolution. Birkhäuser Verlag, Basel.

Barone, R., Bortolami, R., 2004. Neurologie I. Système Nerveux Central. In: Anatomie Comparée des Mammifères Domestiques, vol. 6, Vigot Frères, Paris.

Benninghoff, A., Drenckhahn, D. (Eds.), 2004. Anatomie—makroskopische Anatomie, Histologie, Embryologie, Zellbiologie, 16., völlig neu bearb. Aufl. ed. Elsevier Urban & Fischer, München.

Blinkov, S.M., Glezer, I.I., 1968. The Human Brain in Figures and Tables: A Quantitative Handbook. Basic Books, New York.

Blix, A.S., Walløe, L., Messelt, E.B., 2013. On how whales avoid decompression sickness and why they sometimes strand. J. Exp. Biol. 216, 3385–3387.

Boddy, A.M., McGowen, M.R., Sherwood, C.C., Grossman, L.I., Goodman, M., Wildman, D.E., 2012. Comparative analysis of encephalization in mammals reveals relaxed constraints on anthropoid primate and cetacean brain scaling. J. Evol. Biol. 25, 981–994.

Boenninghaus, G., 1903. Das Ohr des Zahnwales, zugleich ein Beitrag zur Theorie der Schalleitung: Eine biologische Studie. Fischer.

Brauer, K., Schober, W., 1976. Catalogue of Mammalian Brains. VEB Gustav Fischer, Verlag, Jena.

Breathnach, A.S., 1960. The cetacean central nervous system. Biol. Rev. 35, 187–230.

Brodmann, K., 1909. Vergleichende Lokalisationslehre der Grosshirnrinde in ihren Prinzipien dargestellt aufgrund des Zellenbaues. Barth, Leipzig.

Buhl, E.H., Oelschläger, H.A., 1986. Ontogenetic development of the nervus terminalis in toothed whales. Evidence for its non-olfactory nature. Anat. Embryol. 173, 285–294.

Buhl, E.H., Oelschläger, H.A., 1988. Morphogenesis of the brain in the harbour porpoise. J. Comp. Neurol. 277, 109–125.

Bullock, T.H., Gurevich, V.S., 1979. Soviet literature on the nervous system and psychobiology of Cetacea. Int. Rev. Neurobiol. 21, 47–127.

Butti, C., Sherwood, C.C., Hakeem, A.Y., Allman, J.M., Hof, P.R., 2009. Total number and volume of Von Economo neurons in the cerebral cortex of cetaceans. J. Comp. Neurol. 515, 243–259.

Cant, N.B., Benson, C.G., 2003. Parallel auditory pathways: projection patterns of the different neuronal populations in the dorsal and ventral cochlear nuclei. Brain Res. Bull. 60, 457–474.

Cozzi, B., Spagnoli, S., Bruno, L., 2001. An overview of the central nervous system of the elephant through a critical appraisal of the literature published in the XIX and XX centuries. Brain Res. Bull. 54, 219–227.

Cozzi, B., Roncon, G., Granato, A., Giurisato, M., Castagna, M., Peruffo, A., Panin, M., Ballarin, C., Montelli, S., Pirone, A., 2014. The claustrum of the bottlenose dolphin *Tursiops truncatus* (Montagu 1821). Front. Syst. Neurosci. 8, 42.

Dawson, W.W., Hawthorne, M.N., Jenkins, R.L., Goldston, R.T., 1982. Giant neural systems in the inner retina and optic nerve of small whales. J. Comp. Neurol. 205, 1–7.

Dawson, W., Dawson, W.W., Hope, G.M., Ulshafer, R.J., Hawthorne, M.N., Jenkins, R.L., 1983. Contents of the optic nerve of a small cetacean. Aquat. Mamm. 10, 45–56.

De Felipe, J., Alonso-Nanclares, L., Arellano, J., Ballesteros-Yanez, I., Benavides-Piccione, R., Munoz, A., 2007. Specializations of the cortical microstructure of humans. In: Kaas, J.H., Preuss, T.M. (Eds.), Evolution of the Nervous Systems. A Comprehensive Reference. Elsevier Academic Press, Amsterdam.

De Graaf, A.S., 1967. Anatomical Aspects of the Cetacean Brain Stem. Van Gorcum, Amsterdam.

De Ribeaupierre, F., 1997. Acoustical information processing in the auditory thalamus and cerebral cortex. In: Ehret, G., Romand, R. (Eds.), The Central Auditory System. Oxford University Press, New York, pp. 317–397.

Duus, P., 1976. Neurologisch-topische Diagnostik. Anatomie. Physiologie. Klinik. Georg Thieme Verlag, Stuttgart, 431 pp.

Dyce, K.M., Sack, W.O., Wensing, C.J.G., 1991. Anatomie der Haustiere. Ferdinand Enke Verlag, Stuttgart.

Federative International Programme on Anatomical Terminologies (FIPAT), 2011. Terminologia Anatomica, second ed. Thieme, Stuttgart.

Filimonoff, I.N., 1966. On the so-called rhinencephalon in the dolphin. J. Für Hirnforsch. 8, 1–23.

Furutani, R., 2008. Laminar and cytoarchitectonic features of the cerebral cortex in the Risso's dolphin (*Grampus griseus*), striped dolphin (*Stenella coeruleoalba*), and bottlenose dolphin (*Tursiops truncatus*). J. Anat. 213, 241–248.

Fuzessery, Z.M., Feng, A.S., Supin, A.Y., 2004. Central auditory processing of temporal information in bats and dolphins. In: Thomas, J.A., Moss, C.F., Vater, M. (Eds.), Echolocation in Bats and Dolphins. University of Chicago Press, Chicago, IL, USA, pp. 115–122.

Gacek, R.R., Rasmussen, G.L., 1961. Fiber analysis of the statoacoustic nerve of guinea pig, cat, and monkey. Anat. Rec. 139, 455–463.

Gao, G., Zhou, K., 1991. Fiber analysis of the optic and cochlear nerves of small cetaceans. Can. J. Zool. 69, 2360–2364.

Gao, G., Zhou, K., 1992. Fiber analysis of the optic and cochlear nerves of small cetaceans. In: Thomas, J.A., Kastelein, R.A., Supin, A.Y. (Eds.), Marine Mammal Sensory Systems. Plenum Press, New York, pp. 39–52.

Gao, G., Zhou, K., 1995. Fiber analysis of the vestibular nerve of small cetaceans. In: Kastelein, R.A., Thomas, J.A., Nachtigall, P.E. (Eds.), Sensory Systems of Aquatic Mammals. De Spil Publishers, Woerden, The Netherlands, pp. 447–453.

Garey, L.J., Leuba, G., 1986. A quantitative study of neuronal and glial numerical density in the visual cortex of the bottlenose dolph: evidence for a specialized subarea and changes with age. J. Comp. Neurol. 247, 491–496.

Geisler, J.H., Luo, Z., 1998. Relationship of Cetacea to terrestrial ungulates and the evolution of cranial vasculature in Cete. In: Thewissen, J.G.M. (Ed.), The Emergence of Whales. Plenum Press, New York, pp. 163–212.

Glezer, I.I., 2002. Neural morphology. In: Hoelzel, A.R. (Ed.), Marine Mammal Biology—An Evolutionary Approach. Blackwell Science, Oxford, pp. 98–115.

Glezer, I.I., Morgane, P.J., 1990. Ultrastructure of synapses and golgi analysis of neurons in neocortex of the lateral gyrus (visual cortex) of the dolphin and pilot whale. Brain Res. Bull 24, 401–427.

Glezer, I.I., Jacobs, M.S., Morgane, P.J., 1988. Implications of the "initial brain" concept for brain evolution in Cetacea. Behav. Brain Sci. 11, 75–89.

Glezer, I.I., Hof, P.R., Morgane, P.J., 1998. Comparative analysis of calcium-binding protein-immunoreactive neuronal populations in the auditory and visual systems of the bottlenose dolphin (*Tursiops truncatus*) and the macaque monkey (*Macaca fascicularis*). J. Chem. Neuroanat. 15, 203–237.

Glickstein, M., Oberdick, J., Voogd, J., 2007. Evolution of the cerebellum. In: Kaas, J.H. (Ed.), Evolution of Nervous Systems. Academic Press, Oxford, pp. 413–442.

Grothe, B., Koch, U., 2011. Dynamics of binaural processing in the mammalian sound localization pathway—the role of GABAB receptors. Hear. Res. 279, 43–50.

Harting, J.K., 2004. Puffs and patches: a brief chronological review. In: Hall, W.C., Moschovakis, A.K. (Eds.), The Superior Colliculus: New Approaches for Studying Sensory Motor Integration. CRC Press, pp. 83–105.

Hatschek, R., Schlesinger, H., 1902. Der Hirnstamm des Delphins (*Delphinus delphis*). Arb. Aus Dem Neurol. Inst. Wien. Univ. 9, 1–117.

Held, H., 1893. Die centrale Gehörleitung. Arch. Anat. Physiol. Anat. Abt. 17, 201–248.

Herculano-Houzel, S., 2011. Not all brains are made the same: new views on brain scaling in evolution. Brain Behav. Evol. 78, 22–36.

Hof, P.R., Glezer, I.I., Nimchinsky, E.A., Erwin, J.M., 2000. Neurochemical and cellular specializations in the mammalian neocortex reflect phylogenetic relationships: evidence from primates, cetaceans, and artiodactyls. Brain Behav. Evol. 55, 300–310.

Hof, P.R., Chanis, R., Marino, L., 2005. Cortical complexity in cetacean brains. Anat. Rec. A 287, 1142–1152.

Holzmann, T., 1991. Morphologie und mikroskopische Anatomie des Gehirns beim fetalen Narwal, *Monodon monoceros* (Inaugural-Dissertation, Fachbereich Medizin). Johann Wolfgang Goethe-Universität, Frankfurt am Main.

Hosokawa, H., 1951. On the extrinsic eye muscles of the whale with special remarks upon the innervation and function of the musculus retractor bulbi. Sci. Rep. Whales Res. Inst. Tokyo 6, 1–31.

Huber, E., 1934. Contribution to palaeontology IV: anatomical notes on pinnipedia and cetacea. Publ. Carnegie Inst. Wash. 447, 105–136.

Huggenberger, S., 2008. The size and complexity of dolphin brains—a paradox? J. Mar. Biol. Assoc. UK 88, 1103–1108.

Hummel, G., 1975. Lichtmikroskopische, elektronenmikroskopische und enzymhistologische Untersuchungen an der Großhirnrinde von Rind. Schaf und Ziege. J. Für Hirnforsch. 16, 245–285.

Igarashi, S., Kamiya, T., 1972. Atlas of the Vertebrate Brain: Morphological Evolution from Cyclostomes to Mammals. University Park Press, Baltimore.

Jacobs, M.S., Morgane, P.J., McFarland, W.L., 1971. The anatomy of the brain of the bottlenose dolphin (*Tursiops truncatus*). Rhinic lobe (rhinencephalon). I. The paleocortex. J. Comp. Neurol. 141, 205–271.

Jacobs, M.S., McFarland, W.L., Morgane, P.J., 1979. The anatomy of the brain of the bottlenose dolphin (*Tursiops truncatus*). Rhinic lobe (rhinencephalon): the archicortex. Brain Res. Bull. 4 (Suppl. 1), 1–108.

Jansen, J., 1950. The morphogenesis of the cetacean cerebellum. J. Comp. Neurol. 93 (3), 341–400.

Jansen, J., Jansen, J.K.S., 1969. The nervous system of Cetacea. In: Andersen, H.T. (Ed.), The Biology of Marine Mammals. Academic Press, New York, pp. 175–252.

Kahle, W., Frotscher, M., 2009. Nervensystem und Sinnesorgane, 10., überarb. Aufl. ed, Taschenatlas der Anatomie. Thieme, Stuttgart.

Kalinichenko, S.G., Pushchin, I.I., 2008. Calcium-binding proteins in the cerebellar cortex of the bottlenose dolphin and harbour porpoise. J. Chem. Neuroanat. 35, 364–370.

Kandel, E.R., Schwartz, J.H., Jessel, T.M., Siegelbaum, S.A., Hudspeth, A.J. (Eds.), 2013. Principles of Neural Science, fifth ed. McGraw-Hill, New York.

Kappel, R.M., 1981. The development of the cerebellum in macaca mulatta. A study of regional differences during corticogenesis. Thesis, Leiden.

Kardong, K.V., 2015. Vertebrates: Comparative Anatomy, Function, Evolution, seventh ed. McGraw-Hill Education, New York, NY.

Keogh, M.J., Ridgway, S.H., 2008. Neuronal fiber composition of the corpus callosum within some odontocetes. Anat. Rec. Adv. Integr. Anat. Evol. Biol. 291, 781–789.

Kern, A., 2012. Der Neokortex der Säugetiere—Evolution und Funktion. Inaugural-Dissertation des Fachbereichs Medizin der Johann Wolfgang Goethe-Universität, Frankfurt am Main, 249 pp.

Kern, A., Seidel, K., Oelschläger, H.H.A., 2009. The central vestibular complex in dolphins and humans: functional implications of Deiters' nucleus. Brain Behav. Evol. 73, 102–110.

Kern, A., Siebert, U., Cozzi, B., Hof, P.R., Oelschläger, H.H.A., 2011. Stereology of the neocortex in Odontocetes: qualitative, quantitative, and functional implications. Brain Behav. Evol. 77, 79–90.

Knopf, J.P., Hof, P. R., Oelschläger, H.H.A., In press. The neocortex of Indian river dolphins (Genus *Platanista*): comparative qualitative and quantitative analysis. Brain Behav. Evol.

König, H.E., Liebich, H.-G., 2009. Veterinary Anatomy of Domestic Mammals. Schattauer, Stuttgart.

Köpf-Maier, P. (Ed.), 2000. Wolf-Heidegger's Atlas of human Anatomy. fifth ed. Karger, Basel, New York.

Korneliussen, H.K., 1967. Cerebellar corticogenesis in Cetacea, with special references to regional variations. J. Hirnforsch. 9, 151–185.

Korneliussen, H.K., 1968. On the morphology and subdivisions of the cerebellar nuclei of the rat. J. Hirnforsch. 10, 109–122.

Kruger, L., 1959. The thalamus of the dolphin (*Tursiops truncatus*) and comparison with other mammals. J. Comp. Neurol. 111, 133–194.

Kükenthal, W., Ziehen, T., 1893. Über das Centralnervensystem der Cetaceen nebst Untersuchungen über die vergleichende Anatomie des Gehirns bei Placentaliern. Denkschr. Med. -Naturwissenschftlichen Ges. Zu Jena 3, 77–198.

Langworthy, O.R., 1931. A description of the central nervous system of the porpoise (*Tursiops truncatus*). J. Comp. Neurol. 54, 437–499.

Lyamin, O., Manger, P., Ridgway, S., Mukhametov, L., Siegel, J., 2008. Cetacean sleep: an unusual form of mammalian sleep. Neurosci. Biobehav. Rev. 32, 1451–1484.

Malkemper, E.P., Oelschläger, H.H.A., Huggenberger, S., 2012. The dolphin cochlear nucleus: topography, histology and functional implications. J. Morphol. 273, 173–185.

Malmierca, M.S., Merchán, M.A., 2004. Auditory system. In: Paxinos, G. (Ed.), The Rat Nervous System. third ed. Elsevier Academic Press, Amsterdam, pp. 997–1082.

Manger, P.R., 2006. An examination of cetacean brain structure with a novel hypothesis correlating thermogenesis to the evolution of a big brain. Biol. Rev. Camb. Philos. Soc. 81, 293–338.

Manger, P.R., Ridgway, S.H., Siegel, J.M., 2003. The locus coeruleus complex of the bottlenose dolphin (*Tursiops truncatus*) as revealed by tyrosine hydroxylase immunohistochemistry. J. Sleep Res. 12, 149–155.

Marino, L., Rilling, J.K., Lin, S.K., Ridgway, S.H., 2000. Relative volume of the cerebellum in dolphins and comparison with anthropoid primates. Brain Behav. Evol. 56, 204–211.

Marino, L., McShea, D.W., Uhen, M.D., 2004. Origin and evolution of large brains in toothed whales. Anat. Rec. A 281, 1247–1255.

Mass, A.M., Supin, A.Y., 2009. Vision. In: Perrin, W.F., Würsig, B., Thewissen, J.G.M. (Eds.), Encyclopedia of Marine Mammals. Academic Press, San Diego, CA, pp. 1200–1211.

McFarland, W.L., Morgane, P.J., Jacobs, M.S., 1969. Ventricular system of the brain of the dolphin, *Tursiops truncatus*, with comparative anatomical observations and relations to brain specializations. J. Comp. Neurol. 135, 275–368.

McFarland, W.L., Jacobs, M.S., Morgane, P.J., 1979. Blood supply to the brain of the dolphin, *Tursiops truncatus*, with comparative observations on special aspects of the cerebrovascular supply of other vertebrates. Neurosci. Biobehav. Rev. 3, 1–93.

Mead, J.G, Fordyce, R.E., 2009. The Therian Skull: A Lexikon with Emphasis on the Odontocetes. Smithsonian Contributions Zoology, No. 627, 248 pp.

Miller, G.S., 1923. The telescoping of the cetacean skull. Smithson. Misc. Collect. 76, 1–71.

Montgomery, S.H., Geisler, J.H., McGowen, M.R., Fox, C., Marino, L., Gatesy, J., 2013. The evolutionary history of cetacean brain and body size. Evolution 67, 3339–3353.

Morgane, P.J., Jacobs, M.S., 1972. Comparative anatomy of the cetacean nervous system. In: Harrison, R.J. (Ed.), Functional Anatomy of Marine Mammals. Academic Press, London, pp. 117–244.

Morgane, P.J., Jacobs, M.S., 1986. A morphometric Golgi and cytoarchitectonic study of the hippocampal formation of the bottlenose dolphin, *Tursiops truncatus*. In: Isaacson, R.L., Pribram, K.H. (Eds.), The Hippocampus. Plenum Press, New York and London, pp. 369–432.

Morgane, P.J., Jacobs, M.S., McFarland, W.L., 1980. The anatomy of the brain of the bottlenose dolphin (*Tursiops truncatus*). Surface configurations of the telencephalon of the bottlenose dolphin with comparative anatomical observations in four other cetacean species. Brain Res. Bull. 5, 1–107.

Morgane, P.J., McFarland, W.L., Jacobs, M.S., 1982. The limbic lobe of the dolphin brain: a quantitative cytoarchitectonic study. J. Für Hirnforsch. 23, 465–552.

Morgane, P.J., Jacobs, M.S., Galaburda, A.M., 1986. Evolutionary morphology of the dolphin brain. In: Schusterman, R.J., Thomas, J.A., Wood, F.G. (Eds.), Dolphin Cognition and Behavior: A Comparative Approach. Lawrence Erlbaum Associates, Hillsdale, pp. 5–29.

Morgane, P.J., Glezer, I.I., Jacobs, M.S., 1988. Visual cortex of the dolph: an image analysis study. J. Neurol. Comp. 273, 3–25.

Nagel, E.L., Morgane, P.J., McFarlane, W.L., Galliano, R.E., 1968. Rete mirabile of dolphin: its pressure-damping effect on cerebral circulation. Science 161, 898–900.

Nemec, P., Altmann, J., Marhold, S., Burda, H., Oelschläger, H.H.A., 2001. Neuroanatomy of magnetoreception: the superior colliculus involved in magnetic orientation in a mammal. Science 294, 366–368.

Nickel, R., Schummer, A., Seiferle, E., 2004. Nervensystem, Sinnesorgane, endokrine Drüsen, 8., unveränd. Aufl. ed, Lehrbuch der Anatomie der Haustiere. Parey, Stuttgart.

Nieuwenhuys, R., 1998. Telencephalon. In: Nieuwenhuys, R., ten Donkelaar, H.J., Nicholson, C. (Eds.), The Central Nervous System of Vertebrates, vol. 3, Springer, Berlin-Heidelberg, New York, pp. 1871–2023.

Oelschläger, H.A., 1990. Evolutionary morphology and acoustics in the dolphin skull. In: Thomas, J.A., Kastelein, R. (Eds.), Sensory Abilities of Cetaceans. NATO ASI Series, vol. 196, Plenum Press, New York, pp. 137–162.

Oelschläger, H.H.A., 2008. The dolphin brain—a challenge for synthetic neurobiology. Brain Res. Bull. 75, 450–459.

Oelschläger, H.A., Buhl, E.H., 1985. Development and rudimentation of the peripheral olfactory system in the harbor porpoise *Phocoena phocoena* (Mammalia: Cetacea). J. Morphol. 184, 351–360.

Oelschläger, H.H.A., Kemp, B., 1998. Ontogenesis of the sperm whale brain. J. Comp. Neurol. 399, 210–228.

Oelschläger, H.H.A., Oelschläger, J.S., 2002. Brain. In: Perrin, W.F., Würsig, B., Thewissen, J.G.M. (Eds.), Encyclopedia of Marine Mammals. Academic Press, San Diego, CA, pp. 133–158.

Oelschläger, H.H.A., Oelschläger, J.S., 2009. Brain. In: Perrin, W.F., Würsig, B., Thewissen, J.G.M. (Eds.), Encyclopedia of Marine Mammals. Academic Press, San Diego, CA, pp. 134–149.

Oelschläger, H.A., Buhl, E.H., Dann, J.F., 1987. Development of the nervus terminalis in mammals including toothed whales and humans. Ann. N. Y. Acad. Sci. 519, 447–464.

Oelschläger, H.H.A., Haas-Rioth, M., Fung, C., Ridgway, S.H., Knauth, M., 2008. Morphology and evolutionary biology of the dolphin (*Delphinus sp.*) brain—MR imaging and conventional histology. Brain Behav. Evol. 71, 68–86.

Oelschläger, H.H.A., Ridgway, S.H., Knauth, M., 2010. Cetacean brain evolution: dwarf sperm whale (*Kogia sima*) and common dolphin (*Delphinus delphis*)—an investigation with high-resolution 3D MRI. Brain Behav. Evol. 75, 33–62.

Oliver, D.L., Huerta, M.F., 1992. Inferior and superior colliculi. In: Webster, D.B., Popper, A.N., Fay, R.R. (Eds.), The Mammalian Auditory Pathway: Neuroanatomy, The Springer Handbook of Auditory Research. Springer Science & Business Media, New York, pp. 168–221.

Olszewski, J., Baxter, D., 1954. Cytoarchitecture of the human brain stem. Karger, Basel.

Osen, K.K., 1969. Cytoarchitecture of the cochlear nuclei in the cat. J. Comp. Neurol. 136, 453–484.

Osen, K.K., Jansen, J., 1965. The cochlear nuclei in the common porpoise, *Phocaena phocaena*. J. Comp. Neurol. 125, 223–257.

Paxinos, G. (Ed.), 2004. The Rat Nervous System. third ed. Elsevier Academic Press, Amsterdam; Boston.

Pennartz, C., Groenewegen, H., Lopesdasilva, F., 1994. The nucleus accumbens as a complex of functionally distinct neuronal ensembles: an integration of behavioural, electrophysiological and anatomical data. Prog. Neurobiol. 42, 719–761.

Pilleri, G., Chen, P., Shao, Z., 1980. Concise macroscopical atlas of the brain of the common dolphin (*Delphinus delphis*; Linnaeus, 1758), 25 pls. Brain Anatomy Institute, University of Berne, Waldau-Berne, Switzerland. 16 pp.

Pilleri, G., Gihr, M., 1970. The central nervous system of the mysticete and odontocete whales. In: Pilleri, G. (Ed.), Investigations on Cetacea. Institute of Brain Anatomy, Bern, pp. 89–128.

Platt, M.L., Lau, B., Glimcher, P.W., 2003. Situating the superior colliculus within gaze controle network. In: Hall, W.C., Moschovakis, A.K. (Eds.), The Superior Colliculus: New Approaches for Studying Sensorimotor Integration. CRC Press, Boca Raton, pp. 1–34.

Popov, V.V., Supin, A.Y., 2007. Analysis of auditory information in the brains of cetaceans. Neurosci. Behav. Physiol. 37, 285–291.

Poth, C., Fung, C., Güntürkün, O., Ridgway, S.H., Oelschläger, H.H.A., 2005. Neuron numbers in sensory cortices of five delphinids compared to a physeterid, the pygmy sperm whale. Brain Res. Bull. 66, 357–360.

Prahl, S., Huggenberger, S., Schliemann, H., 2009. Histological and ultrastructural aspects of the nasal complex in the harbour porpoise, *Phocoena phocoena*. J. Morphol. 270, 1320–1337.

Pütter, A., 1902. Die Augen der Wassersäugetiere. Zool. Jahrb 17, 99–402.

Rauschmann, M., 1992. Morphologie des Kopfes beim Schlanken Delphin *Stenella attenuata* mit besonderer Berücksichtigung der Hirnnerven. Makroskopische Präparation und moderne bildgebende Verfahren. Inaugural-Dissertation des Fachbereichs Humanmedizin der Johann Wolfgang Goethe-Universität, Frankfurt am Main.

Revishchin, A.V., Garey, L.J., 1991. Laminar distribution of cytochrome oxidase staining in cetacean isocortex. Brain Behav. Evol. 37, 355–367.

Ridgway, S.H., 1968. The bottlenose dolphin in biomedical research. In: Gay, W.I. (Ed.), Methods of Animal Experimentation. Academic Press, New York.

Ridgway, S.H., 1990. The central nervous system of the bottlenose dolphin. In: Leatherwood, S., Reeves, R.R. (Eds.), The Bottlenose Dolphin. Academic Press, San Diego, CA, pp. 69–97.

Ridgway, S.H., 2002. Asymmetry and symmetry in brain waves from dolphin left and right hemispheres: some observations after anesthesia, during quiescent hanging behavior, and during visual obstruction. Brain Behav. Evol. 60, 265–274.

Ridgway, S.H., Hanson, A.C., 2014. Sperm whales and killer whales with the largest brains of all toothed whales show extreme differences in cerebellum. Brain Behav. Evol.

Ridgway, S.H., Harrison, R.J., 1986. Diving dolphins. In: Bryden, M.M., Harrison, R.J. (Eds.), Research on Dolphins. Clarendon Press, Oxford, pp. 33–58.

Ridgway, S.H., Howard, R., 1979. Dolphin lung collapse and intramuscular circulation during free diving: evidence from nitrogen washout. Science 206, 1182–1183.

Ridgway, S.H., Demski, L.S., Bullock, T.H., Schwanzel-Fukuda, M., 1987. The terminal nerve in odontocete cetaceans. Ann. N. Y. Acad. Sci. 519, 201–212.

Ridgway, S., Carder, D., Finneran, J., Keogh, M.J., Kamolnick, T., Todd, M., Goldblatt, A., 2006a. Dolphin continuous auditory vigilance for five days. J. Exp. Biol. 209, 3621–3628.

Ridgway, S., Houser, D., Finneran, J., Carder, D., Keogh, M.J., Bonn, W.G.V., Smith, C.R., Scadeng, M., Dubowitz, D., Mattrey, R., Hoh, C., 2006b. Functional imaging of dolphin brain metabolism and blood flow. J. Exp. Biol. 209, 2902–2910.

Ridgway, S., Keogh, M., Carder, D., Finneran, J., Kamolnick, T., Todd, M., Goldblatt, A., 2009. Dolphins maintain cognitive performance during 72 to 120 hours of continuous auditory vigilance. J. Exp. Biol. 212, 1519–1527.

Romer, A.S., Parsons, T.S., 1977. The Vertebrate Body, fifth ed. Saunders, Philadelphia.

Romer, A.S., Parsons, T.S., 1986. The Vertebrate Body, sixth ed. Saunders College Publishing, Philadelphia.

Romer, A.S., Parsons, T.S., 1991. Vergleichende Anatomie der Wirbeltiere, fifth ed. Paul Parey, Berlin.

Schaller, O. (Ed.), 1992. Illustrated Veterinary Anatomical Nomenclature. Ferdiand Enke Verlag, Stuttgart, 614 pp.

Schulmeyer, F.J., 1992. Morphologische Untersuchungen am Hirnstamm der Delphine unter besonderer Berücksichtigung des La Plata-Delphins, *Pontoporia blainvillei* (Inaugural-Dissertation, Fachbereich Medizin). Johann Wolfgang Goethe-Universität, Frankfurt am Main.

Schulmeyer, F.J., Adams, J.C., Oelschläger, H.H.A., 2000. Specialized sound reception in dolphins—a hint for the function of the dorsal cochlear nucleus in mammals. Hist. Biol. 14, 53–56.

Schwerdtfeger, W.K., Oelschläger, H.H.A., Stephan, H., 1984. Quantitative neuroanatomy of the brain of the La Plata dolphin, *Pontoporia blainvillei*. Anat. Embryol. 170, 11–19.

Slijper, E.J., 1973. Die Cetaceen, Reprinted from Capita Zoologica Bd. VI and VII, 1936. ed. Asher, Amsterdam.

Spangler, K.M., Warr, W.B., 1991. The descending auditory system. In: Altschuler, R.A., Bobbin, R.P., Clopton, B.M., Hoffman, D.W. (Eds.), Neurobiology of Hearing: The Central Auditory System. Raven Press, New York, pp. 27–45.

Spitzki, D.E.C., 1886. Notes on the anatomy of the dolphin's brain. J. Comp. Med. Surg., 224–229.

Spoor, F., Bajpai, S., Hussain, S.T., Kumar, K., Thewissen, J.G.M., 2002. Vestibular evidence for the evolution of aquatic behaviour in early cetaceans. Nature 417, 163–166.

Starck, D., 1979. Vergleichende Anatomie der Wirbeltiere auf evolutionsbiologischer Grundlage, first ed. Springer, Berlin.

Stephan, H., 1975. Allocortex, Handbuch der Mikroskopischen Anatomie des Menschen. Springer-Verlag, Berlin.

Stephan, H., Andy, O.J., 1977. Quantitative comparison of the amygdala in insectivores and primates. Acta Anat. 98, 130–153.

Stephan, H., Manolescu, J., 1980. Comparative investigations on hippocampus in insectivores and primates. Z. Für Mikrosk. -Anat. Forsch. 94, 1025–1050.

Stephan, H., Baron, G., Frahm, H.D., 1988. Comparative size of brains and brain components. Comp. Primate Biol. 4, 1–38.

Striedter, G.F., 2005. Principles of Brain Evolution. Sinauer Associates, Sunderland, MA.

Supin, A.Y., Popov, V.V., Mass, A.M., 2001. The Sensory Physiology of Aquatic Mammals. Kluwer Academic Publishers, Boston.

Tarpley, R.J., Ridgway, S.H., 1991. Orbital gland structure and secretions in the Atlantic bottlenose dolphin (*Tursiops truncatus*). J. Morphol. 207, 173–184.

Tarpley, R.J., Ridgway, S.H., 1994. Corpus callosum size in delphinid cetaceans. Brain Behav. Evol. 44, 156–165.

Terminologia Anatomica, 1998. International Anatomical Terminology (FCAT, Federative Committee on Anatomical Terminology). Thieme, Stuttgart, New York.

Trepel, M., 2015. Neuroanatomie. Struktur und Funktion, sixth ed. Elsevier/Urban und Fischer, München.

Tyson, E., 1680. Phocaena or the Anatomy of a Porpess. Benj, Tooke.

van Dongen, P.A.M., Nieuwenhuys, R., 1998. Diencephalon. Nieuwenhuys, R., ten Donkelaar, H.J., Nicholson, C. (Eds.), The Central Nervous System of Vertebrates, vol. 3, Springer-Berlin-Heidelberg, New York, pp. 1845–1871.

van Harrefeld, A., 1972. The extracellular space in the vertebrate central nervous system. In: Bourne, G. (Ed.), The Structure and Function of Nervous Tissue. Academic Press, New York, p. 461.

van Kann, E., Cozzi, B., Hof, P.R., Oelschläger, H.H.A., Submitted. Qualitative and quantitative analysis of primary neocortical areas in mammals—ecology and evolution. Brain Behav. Evol.

Vater, M., Kössl, M., 2004. Introduction: the ears of whales and bats. In: Thomas, J.A., Moss, C.F., Vater, M. (Eds.), Echolocation in Bats and Dolphins. University of Chicago Press, Chicago, IL, USA, pp. 89–98.

Verhaart, W.J., 1970. Comparative Anatomical Aspects of the Mammalian Brain Stem and the Cord. Studies in Neuroanatomy 9, Vol. I (Text) 338pp., Vol. II (Illustrations and Tables) 312pp. van Gorcum & Comp. N.V. Assen.

Viamonte, M., Morgane, P.J., Galliano, R.E., Nagel, E.L., McFarland, W.L., 1968. Angiography in living dolphin and observations on blood supply to the brain. Am. J. Physiol. 214, 1225–1249.

Vogl, A.W., Fisher, H.D., 1982. Arterial retia related to supply of the central nervous system in two small toothed whales– narwhal (*Monodon monoceros*) an beluga (*Delphinapterus leucas*). J. Morphol. 174, 41–56.

Vogl, A.W., Todd, M.E., Fisher, H.D., 1981. An ultrastructural and fluorescence histochemical investigation of the innervation of retial arteries in *Monodon monoceros*. J. Morphol. 168, 109–119.

Voogd, J., Nieuwenhuys, R., Van Dongen, P.A.M., Ten Donkelaar, H.J., 1998. Mammals. In: Nieuwenhuys, R., Donkelaar, H.J.ten, Nicholson, C. (Eds.), The Central Nervous System of Vertebrates. Springer, Berlin, pp. 1637–2097.

Walløe, S., Eriksen, N., Dabelsteen, T., Pakkenberg, B., 2010. A neurological comparative study of the harp seal (*Pagophilus groenlandicus*) and harbor porpoise (*Phocoena phocoena*) brain. Anat. Rec. Adv. Integr. Anat. Evol. Biol. 293, 2129–2135.

Wever, E.G., 1949. Theory of Hearing. John Wiley and Sons, New York, pp. 290–336.

World Association of Veterinary Anatomists, 2012. Nomina Anatomica Veterinaria (NAV) [WWW Document]. Nomina Anat. Vet. URL http://www.wava-amav.org.

Young, E.D., Davis, K.A., 2001. Circuitry and function of the dorsal cochlear nucleus. In: Oertel, D., Popper, A.N., Fay, R.R. (Eds.), Integrative Functions in the Mammalian Auditory Pathway. Springer, New York, pp. 160–206.

Zvorykin, V.P., 1963. Morphological substrate of ultrasonic and locational capacities in the dolphin. Arkh. Anat. Gistol. Embriol. 45, 3–17.

Zvorykin, V.P., 1976. Specific features of the structural organization of the external geniculate body in the dolphin *Phocaena phocaena* as compared with other dolphins. Arkh Anat. Gistol. Embriol. 70, 58–66.

Chapter 7

Body Control: The Endocrine System and the Peripheral Nervous System

The control of the mammalian body is based on a number of mechanisms that rely on continual monitoring of the internal milieu and constant feedback. Part of these mechanisms are described in Chapter 5 (Head and Sense), Chapter 8 (Feeding and Digestive System), and Chapter 9 (Urinary System and Genital Systems). Here we deal with the inner secretions and the organs that produce them, and with what we know of the peripheral nervous system of dolphins.

The endocrine production of the gonads (and placenta) are described in Chapter 9, in the general framework of the reproductive functions.

The endocrine system is getting increasingly difficult to define. In the anatomical tradition, endocrine glands pour their hormonal secretions directly into the bloodstream. However, in the last decades, the concept of endocrine organ has extended to single cells, such as those of the diffuse neuroendocrine system, often producing more than one hormone. So it turns out that several parts of the body—even the gut—contain endocrine components.

The endocrine system of cetaceans has received unequal attention: some organs have been thoroughly studied, while others have raised but little interest in the scientific community. The lack of detailed knowledge about some glands leaves several questions unanswered about cetacean physiology and may limit therapeutic strategies in the sick animal. The information available on the peripheral nervous system (PNS) of dolphins (and cetaceans in general) is also scarce, and, for the most part, concerns the cranial nerves. Since cranial nerves have been described in Chapter 6, here we'll present a short summary of the data available on peripheral nerves outside the head.

THE ENDOCRINE SYSTEM

The Hypophysis

Anatomy

The earliest account in the literature about the cetacean pituitary gland (or hypophysis) (Figs. 7.1 and 7.2) dates back to the observations by Guldberg (1885). Then in the 1930s a few authors carried out some seminal studies on morphological and histological aspects of this organ in cetaceans, mostly of large species of commercial interest. Very few works dealt with small odontocetes, the first being that by Wislocki (1929), who described the pituitary of the bottlenose dolphin (*Tursiops truncatus*). After his publication, the fishing of porpoises was soon banned in Cape Hatteras (North Carolina, USA), where Wislocki purchased his samples, so collecting pituitaries from small cetaceans became challenging.

One peculiar feature of cetacean pituitaries, observed since the beginning (Wislocki, 1929; Valsø, 1934; Geiling, 1935), was the complete separation of the two main parts of the gland, the *pars nervosa* (neurohypophysis) and the *pars distalis* (adenohypophysis) (Fig. 7.3), which in most mammals are juxtaposed.[a] As a common rule, the adenohypophysis is much bigger than the neurohypophysis, generally twice as wide, and embraces it with the upper *pars tuberalis*, which actually is in contact with the *pars nervosa* antero-medially, forming the hypophyseal stalk (Cowan et al., 2008). There is no hypophyseal cleft and no distinct *pars intermedia* in cetaceans, unlike in most terrestrial mammals.[b] Cowan et al. (2008) described an equivalent of the *pars intermedia*, called "dorsal shoulder" and characterized by a "variant configuration of

---

a. This situation is closely similar to the avian pituitary (Dellman, 1998). The river dolphins of the genus *Platanista* seem to be an exception, having just a thin layer of connective tissue separating the two lobes (Arvy and Pilleri, 1974).

b. During ontogeny, the Rathke's pouch evaginating from the oral cavity (*stomodeum*) grows toward the infundibular process of the third ventricle. The anterior wall of the pouch will form the adenohypophysis, while the posterior wall will differentiate into an intermediate lobe in the fetus, upon contact with the developing neurohypophysis. However, in mammals in which the *pars intermedia* fails to develop, the *pars nervosa* and *pars distalis* remain separated by dural layers. The intermediate lobe is lacking also in Indian elephants (Oboussier, 1948), dugongs (Fernand, 1951), and some Xenarthra (Oldham, 1941; Herlant, 1958).

Anatomy of Dolphins. http://dx.doi.org/10.1016/B978-0-12-407229-9.00007-5

Pituitary stalk   Adenohypophysis

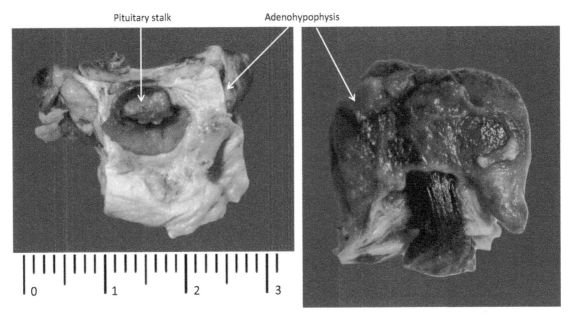

FIGURE 7.1   Whole pituitary gland of *T. truncatus* (posterodorsal view, on the left) and *G. griseus* (dorsal view, on the right).

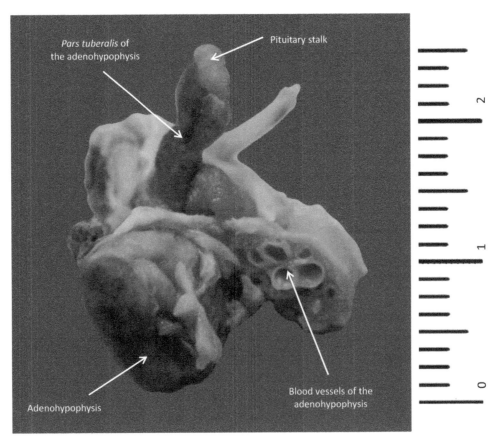

FIGURE 7.2   **Side view of the pituitary gland of *T. truncatus*.**

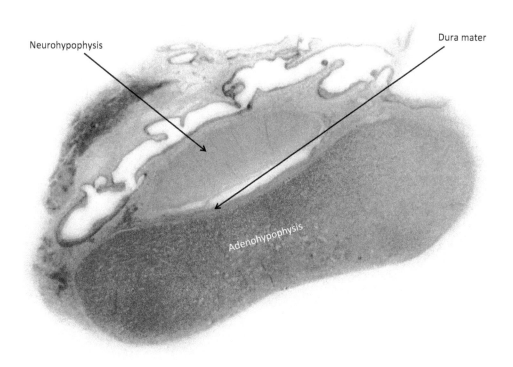

Neurohypophysis

Dura mater

Adenohypophysis

FIGURE 7.3   **Section of the whole pituitary gland of *T. truncatus*.**

cords and cell distribution." However, no clear cytological and histological evidence was provided. In our laboratory we studied several individuals belonging to four species of dolphins, but did not identify any structure with a morphological aspect different from the rest of the adenohypophysis (Panin et al., 2013), in agreement with previous literature.

Another unusual characteristic of cetaceans is the almost complete absence of a *sella turcica*, a cup-shaped depression in the basisphenoid, hosting the gland, which is missing also in elephants, Sirenians and hyraxes (Cave, 1993). In most mammals, the *sella turcica* is delimited by two bony ridges, called olivary eminence (rostral) and *dorsum sellae* (caudal). The cetacean pituitary lies on the cranial floor in no delimited *sella*, but Cowan et al. (2008) observed a deep *fossa* in few individuals of some dolphin species (eg, *Lagenodelphis hosei*, *Steno bredanensis*, and *Stenella frontalis*). Absence of a *sella* is seemingly due to the agenesis of an independent ossification center, called pulvinosphenoid (Cave, 1993), which in the early embryo of most mammals is continuous with the otic capsules.[c]

The vascularization is generally similar to that of other mammals, with pial arteries penetrating the pituitary stalk, first forming a capillary plexus, then merging into portal vessels and finally spreading again into a capillary bed (Harris, 1947). Drager (1944, 1953) first described the innervation in the hypophysis of the porpoise (*Phocoena phocoena*), noting nerve fibers accompanying the portal vessels and entering into the *pars tuberalis*, whereas the adenohypophysis had no innervation. Harris (1950) confirmed Drager's findings but observed also a silver-stained fiber framework around sinusoids of the *pars distalis*, possibly corresponding to what we now know to be the projections from the supraoptic and paraventriculr hypothalamic nuclei.

The absolute sizes and weights of the pituitary glands of selected dolphin species are reported in Table 7.1. The gland weight/body weight index (BWI) [(0.002–0.042) (Cowan et al., 2008)] is generally comparable to terrestrial mammals, for example, cats (0.03–0.04), dogs (0.006), and humans (0.01–0.14) (Harrison, 1969). According to earlier studies, there is no apparent correlation between gland weight and body weight, at least in the common (*Delphinus delphis*) and the striped dolphin (*Stenella coeruleoalba*) (Gihr and Pilleri, 1969).[d]

---

c. The same process might be responsible also for the laxity of attachment of the ear bones to the cranium in cetaceans.

d. A substantial increase in the pituitary gland mass has been documented in relation with pregnancy and lactation in Mysticetes (Jacobsen, 1941; Hennings, 1950).

TABLE 7.1 Weight and Size of the Pituitary Gland of Cetacean Species Available in the Literature, with Given Reference Name and Sample Size (Unless Specified, Data Refers to One Individual)

Species	References	Weight	Size	Notes
G. griseus	Pilleri and Gihr (1969)	1.4 g		Pars nervosa
	Cowan et al. (2008)	1.84 g	28 mm width 18 mm length 16 mm height	
T. truncatus	Wislocki (1929)		2 cm width 0.5 cm height	Pars distalis
	Wislocki (1929)		1 cm width 3 mm height	Pars nervosa
	Cowan et al. (2008)	0.61–3.44 g		n = 72
S. coeruleoalba	Gihr and Pilleri (1969)	400–800 mg		n = 7
	Cowan et al. (2008)	0.53 g	24 mm width 9 mm height	
D. delphis	Gihr and Pilleri (1969)	400–800 mg		n = 12
S. attenuata	Cowan et al. (2008)	0.96–1.21 g		n = 3
S. longirostris	Cowan et al. (2008)	0.8–1.08 g		n = 2
P. crassidens	Cowan et al. (2008)	1.2 g		
L. hosei	Cowan et al. (2008)	1.35 g		
S. bredanensis	Cowan et al. (2008)	2.57 g	27 mm width 11 mm length 12 mm height	

Early authors detected three main types of cells in the adenohypophysis with histochemistry (eg, Slidder's or Mallory's trichrome staining, Figs. 7.4–7.6), namely, acidophils (that generally stain yellow), basophil (staining blue to violet), and chromophobic ones (which do not stain) (Wislocki, 1929; Valsø, 1936; Jacobsen, 1941; Hanström, 1944; Vuković et al., 2011). By electron microscopy instead, Harrison and Young (1969a, 1970a) recognized six cell types according to several ultrastructural features. A histological feature common to most mammals is the presence of colloid-filled follicles,

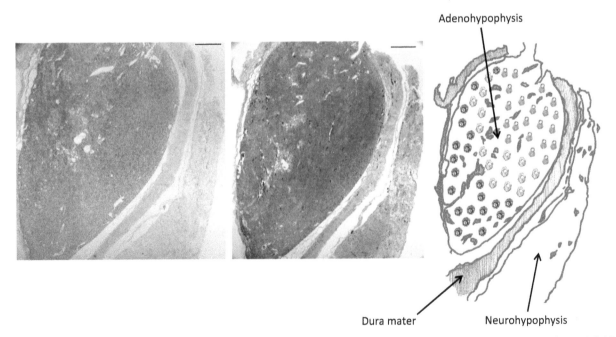

FIGURE 7.4 **Section of the pituitary gland of *G. griseus* (scale bar = 1 mm).** Left, Mallory trichrome stain; center, Slidder stain; right, subdivision of the gland.

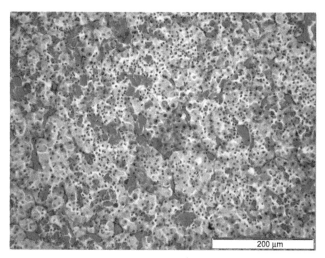

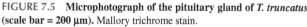

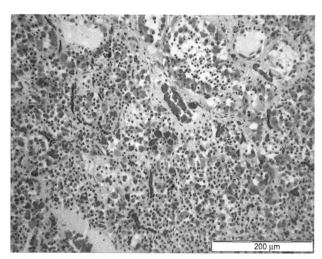

FIGURE 7.5   Microphotograph of the pituitary gland of *T. truncatus* (scale bar = 200 μm). Mallory trichrome stain.

FIGURE 7.6   Microphotograph of the pituitary gland of *G. griseus* (scale bar = 200 μm). Mallory trichrome stain.

whose abundance varies according to the species and to the region of the gland, but seems to be positively correlated with age. Presence of follicles, however, does not occur in all the individuals of a given species, as reported for bottlenose dolphins and *Kogia* spp. (Cowan et al., 2008)[e]; the same authors observed also the absence of immunoreactivity of the colloid to several antisera against pituitary hormones, in eight species.

The hormonal content and activity of the cetacean hypophysis have been investigated by several authors since the first reports. Early immunohistochemical studies identified neuro-secretory cells but no specific hormones in *T. truncatus* (Drager, 1953) and *Delphinapterus leucas* (Pilleri, 1963). The main products of the pituitary gland are nine hormones, two of which produced by the hypothalamus and released by the neurohypophysis (vasopressin and oxytocin), one secreted by the *pars intermedia* (melanocyte-stimulating hormone), and six by the adenohypophysis (growth hormone, prolactin, thyroid-stimulating hormone, luteinizing hormone, follicle-stimulating hormone, adrenocorticotropic hormone). The knowledge about each of these hormones in cetaceans is quite heterogeneous and will be summarized shortly for each of them.

Hormones of the Hypophysis

Most of the early studies on the hormonal activity of the cetacean pituitary were performed on Mysticetes or sperm whales during the whaling campaigns. The following is a description of the findings relative to the endocrine production of the pituitary in the dolphin species.

Neurohypophyseal Hormones

Circulating ADH levels were firstly determined in five *T. truncatus* and one *Orcinus orca* by Malvin et al. (1971) who reported undetectable amounts in all but three samples, where values were less than 0.3 μU/mL. On the contrary, in healthy dogs (Bonjour and Malvin, 1970) or humans (Segar and Moore, 1968) involved in similar studies, values were 0.9–1.7 μU/mL and 0.4–3.1 μU/mL, respectively.[f] Recently Ortiz and Worthy (2000) found a mean ADH level of 3.3 ng/mL in the blood of 31 bottlenose dolphins and curiously no correlation between vasopressin levels and plasma osmolality (similarly to Malvin et al., 1971). Ballarin et al. (2011) in the same species detected much lower mean values for plasma ADH (1.1 pg/mL), with a positive quadratic correlation between ingested freshwater, assumed either by hydration or with food. The results showed that water uptake up to 5 liters led to a rise in vasopressin, whereas a higher water load was followed by a decreased ADH concentration. A possible explanation is that when hydration exceeds a definite level, vasopressin is no longer required to retain liquids in the bottlenose dolphin. These data on circulating vasopressin, however, should be compared with values from wild animals, to exclude possible biases due to the maintenance in "brackish" water (with salinity often much lower than 35–38% of seawater) in the pools of the dolphinariums hosting them.

---

e. The content of the follicles is still poorly understood, although Ogawa et al. (1996) in pigs found sialic acid, N-acetyl galactosamine, galactose, and glycoproteins.

f. The old unit of measure "U" (and its submultiples) is not directly comparable to other quantitative measures, since it was derived from bioassays that, although standardized, can be subjected to a number of variables and might not provide constant reliable outcomes.

α-Melanocyte-Stimulating Hormone (α-MSH)

Among the different melanotropins produced in the hypophysis, the predominant and most studied form is α-MSH (others include β-seryl-, β-glutamyl-, and γ-MSH), which regulates pigmentation (Holder and Haskell–Luevano, 2004), but also has a role in regulating energy homeostasis and food intake at the level of the hypothalamus as a neurotransmitter (Benoit et al., 2000). It is produced mainly by melanotrophs of the intermediate lobe in the terrestrial mammals that have it. As mentioned previously, however, there is no *pars intermedia* in the cetacean pituitary gland, so α-MSH might be synthesized elsewhere. A melanin-dispersing activity higher than that of human or bovine glands, and a higher content of the melanotropic agent in the *pars distalis* than in the *pars tuberalis* were described in the bottlenose (*T. truncatus*) and the Atlantic spotted dolphin (*S. frontalis*) by Geiling et al. (1940). Oldham et al. (1940) further compared the content of a "melanophore-dispersing hormone" in different regions of the *pars distalis* of the armadillo and of three cetaceans, including the bottlenose dolphin.[g] Interestingly, in all four species, the highest activity of the hormone was found in the antero-ventral region, whereas the lowest in the posterodorsal (juxtaneural) one. This was a remarkable observation since the latter would be the position of the intermediate lobe in most mammals.

Only recently, Cowan et al. (2008) identified α-MSH immunohistochemically in *T. truncatus*, finding abundant and faintly labeled cells often associated with follicles. Actually, the sequence of α-MSH is identical to the first 13 amino acids of ACTH, and in the hypophysis of birds (that lack a *pars intermedia*), both hormones are produced by the same cells called corticomelanotrophs, which occur only in the rostral half of the adenohypophysis (Iturriza et al., 1980). To clarify if the α-MSH-immunoreactive (-ir) cells in cetaceans would overlap with ACTH-ir ones, our group performed a double-label immunofluorescent investigation in four odontocetes (*T. truncatus, D. delphis, Grampus griseus, S. coeruleoalba*) (Panin et al., 2013). The results suggested that in these species, and probably other odontocetes, there seems to be a population of true melanotrophs, distinct from corticotrophs, not organized in an intermediate lobe but dispersed throughout the adenohypophysis, with no particular distribution pattern. This seems to contradict the previous findings by Oldham et al. (1940) of a higher content of α-MSH in the antero-ventral region of the gland. The actual role of this hormone has yet to be elucidated since seasonal changes in pigmentation do not occur in cetaceans, and probably its main function may be linked to the regulation of appetite stimulus during periods of food deprivation, for example, migrations (Ortiz et al., 2010). Quantification of blood levels remains a minimal requirement to shed light on the melanocortin system of cetaceans.

Growth Hormone (GH)

The first immunohistochemical detection of cetacean growth hormone was performed by Schneyer and Odell (1984), who also detected prolactin in the bottlenose dolphin. Later, Cowan et al. (2008) identified GH-cells in several other odontocete species, some showing a widespread diffusion over the gland (*G. griseus, Globicephala macrorhynhcus, L. hosei, S. bredanensis, Physeter macrocephalus*), while others showing cords of GH cells alternating with other cell types (*Pseudorca crassidens, Mesoplodon europaeus*). In dolphins, the complete growth hormone sequence has been determined only in the common dolphin *D. delphis* (Maniou et al., 2002). It is worth noting that GH, prolactin, human chorionic gonadotropin (hCG), and other placental hormones seem to derive all from a common ancestral gene by successive duplication events (Maniou et al., 2002). This common origin might be also reflected by the fact that in histochemistry, only growth hormone and prolactin cells are acidophilic, whereas cells producing other hormones are basophilic.

Prolactin

Relatively few information are available on the cetacean prolactin, but its primary physiological effects have been extensively reported for several cetacean species, that is, regulation of lactation periods and rates of milk production. Lactation period varies depending on the species (see Table 9.1 in Chapter 9; for a detailed review, see Oftedal, 1997).

Immunohistochemical detection of PRL cells was first described in the bottlenose dolphin by Schneyer and Odell (1984), as previously mentioned. Cowan et al. (2008) detected diffuse and sparse immunoreactive cells in bottlenose, Risso's and rough-toothed dolphins, and in pilot, sperm, and Gervais's beaked whales, whereas in the striped dolphin PRL-ir cells were often associated with follicles.

Adrenocorticotropic Hormone

A heterogeneous distribution of ACTH-ir cells was described by Cowan et al. (2008) in *T. truncatus, G. griseus, L. hosei*, and *S. bredanensis*. Positive cells were generally large, diffuse over the whole parenchyma, but concentrated in the ventral adenohypophysis, and often aggregated in clusters or cords. In other species, corticotrophs were more abundant in the

---

g. The other cetaceans were the beluga and the fin whale.

central region (*Stenella* spp.) or occurred in alternating stripes (*P. crassidens*). Our group reported relatively few ACTH-ir cells in bottlenose, Risso's striped, and common dolphins, without a particular spatial distribution, but occurring often in small groups of 1 to 8 cells (Panin et al., 2013).

Plasma ACTH levels of 44–216 pg/mL were measured by Reidarson and McBain (1999) in six healthy captive bottlenose dolphins. St. Aubin (2002) assessed the stressing effects of chase and encirclement in wild pantropical spotted dolphins (*Stenella attenuata*), and found an altered hypothalamic-pituitary axis (HPA) response in this species. They measured the levels of ACTH, cortisol, and aldosterone in blood samples drawn during the different stages of encirclement operations. Usually, in repeated samplings over a time series after the administration of a stressful stimulus, ACTH is higher in the early samples and lower in the later, while cortisol follows the reverse pattern. This is due to a well-known negative feedback of circulating cortisol on the release of ACTH by the hypophysis. However, in *S. attenuata* the levels of ACTH were persistently high (range 90.1–1532 pg/mL), with neither correlation with time, nor with cortisol or aldosterone concentration. Mean ACTH values for spotted dolphins (457.7 pg/mL) were much higher than for bottlenose dolphins (246.2 pg/mL) subjected to similar "chase and encirclement" situations. This would suggest a lack of, or a reduced accommodation to stress in the former species. Anyway, the difference between the two delphinids might partly be explained by different sample design conditions, since samples from spotted dolphins were not collected any sooner than 100 min of chasing, while those from bottlenose dolphins always within 60 min, after little or no chasing. Schmitt et al. (2010) indicated the presence of a circadian rhythm of plasma ACTH in three long-term captive belugas, with significantly higher baseline levels in the morning than in the evening, with a mean concentration of 8.41 ± 5.8 pg/mL. ACTH was found to be a more reliable indicator of stress than both cortisol or aldosterone, as evaluated after placement on a stretcher ($P < 0.1$) or during endoscopy procedure ($P < 0.5$). It should be kept in mind that the type of collection heavily affects the interpretation of results, particularly important if ACTH is used as a marker of stress. Indeed, recently Lewis et al. (2013) tested different anticlotting collection tubes and found that EDTA ones provided significantly higher hormone levels than those with heparin or serum separator.

Thyroid-Stimulating Hormone (TSH)

TSH is maybe the least known hypophyseal hormone of cetaceans. Cowan et al. (2008) could not find an antiserum reacting against TSH in none of the twenty-two specimens (belonging to eight species) they examined. This suggests that the sequence of the β-subunit, which is the molecule-specific moiety of the glycosylated pituitary hormones (ie, FSH, LH, hCG, and TSH), might be quite different compared to other mammals. Previous attempts of quantifying TSH using human-designed commercial kits proved unsuccessful in bottlenose dolphins and belugas (St. Aubin, 2001). Nonetheless, Villanger et al. (2011) were able to quantify TSH in belugas using a commercial kit designed for dogs, thus the sequence of the β-subunit might have a substantial similarity at least with the canine hormone. They found a mean level of 57 and 110 pg/mL in a group of male adults and of subadults, respectively, which were comparable to those of healthy sledge dogs (Kirkegaard et al., 2011). However, those belugas were heavily contaminated by organochlorine compounds and there were no control specimens, so plasma levels of healthy cetaceans remain to be determined, as well as the sequence of the β-subunit.

Follicle-Stimulating Hormone (FSH) and Luteinizing Hormone (LH)

Gonadotropins are of fundamental importance for the study of reproductive physiology, since their circulating levels and fluctuations allow to infer the reproductive stage of an individual. In the case of cetaceans, this aspect is particularly difficult to assess, because many species show a considerable flexibility in the seasonality of reproduction (Thayer et al., 2003).

Both FSH and LH were quantified by Schneyer et al. (1985) for the first time in the bottlenose dolphin by radioimmunoassay (RIA). Mean levels of FSH and LH for all individuals were 0.22 and 0.37 ng/mL, respectively, although they had no purified standards for comparing relative potency of human and cetacean hormones. A significantly higher LH concentration was found in adults than in juveniles, while both hormones were higher in females than in males, and in summer than in autumn (pooling both genders). However, sampling was not performed on a regular basis and was scattered over 3 years. Both FSH and LH were not correlated with age classes; however, they were significantly correlated with each other. In the Yangtze finless porpoise (*Neophocaena phocaenoides asiaeorientalis*) FSH and LH were negatively correlated in wild adult males (Hao et al., 2007). The former was higher in December and the latter in March/April, suggesting a lag between the maturation of sperms during the cold seasons and the production of testosterone during the reproductive one. In the females instead, no seasonality was investigated, but LH levels were significant higher in adult nonpregnant females than in juveniles, although FSH were not. The immunohistochemical detection of gonadotropic cells performed by Cowan et al. (2008) failed to reveal FSH-ir cells except in two bottlenose dolphin females (among 22 individuals of eight species), whereas LH-ir cells were found to be sparse to abundant in several species and often associated with follicles in the *pars tuberalis*.

TABLE 7.2 List of Dolphin Species with References Reporting Absence/Presence of a Pineal Gland

	Species	References
Absence	*D. delphis*	Oelschläger et al. (2008); our observations
	L. obliquidens	Arvy (1971)
	S. longirostris	Arvy (1971)
	S. coeruleoalba	our observations
	T. truncatus	Ridgway (1990); McFarland et al. (1969); our observations
	G. griseus	our observations
Presence	*T. truncatus*	Morgane and Jacobs (1972); Lyamin et al. (2008)

The Pineal Gland

Anatomy

The pineal gland (*epiphysis cerebri*) of mammals is a pine-like structure of the dorsal thalamus (epithalamus) placed over a recess of the third ventricle.

The pineal gland in mammals is a neuroendocrine structure and the main source of circulating melatonin, the hormone that regulates the circadian and seasonal rhythms. Melatonin is generally secreted with a nadir during daytime and a peak during night time, in response to an internal rhythm generated by the suprachiasmatic nucleus of the hypothalamus. The presence of a pineal gland in cetaceans is somewhat controversial and there is still no consensus about it (see Table 7.2). Some species have been reported to lack one, like the Amazon river dolphin (*Inia geoffrensis*) (Gruenberger, 1970), the Pacific white-sided dolphin (*Lagenorhynchus obliquidens*) (Arvy, 1971), the spinner dolphin (*Stenella longirostris*) (Arvy, 1971), the narwhal (*Monodon monoceros*) (fetal) (Holzmann, 1991), the short-beaked common dolphin (*D. delphis*) (Oelschläger et al., 2008), and the dwarf sperm whale (*Kogia sima*) (Oelschläger et al., 2010).[h] On the other hand, while Fuse (1936) described the pineal gland as "rudimentary" in the finless porpoise (*N. phocaenoides*), some authors reported the presence of an actual epiphysis in fetuses and adults of larger species (see Panin et al., 2012, for details).

The presence of a pineal gland apparently is not constant among all the individuals of a determinate cetacean species.[i] The common bottlenose dolphin, *T. truncatus*, represents an emblematic case. Neither McFarland et al. (1969) nor Ridgway (1990) were able to detect a pineal body in several brains of this species they dissected. On the contrary, Morgane and Jacobs (1972) observed a pineal gland "present up to the adult stage," and recently Lyamin et al. (2008) reported an unmistakable image of a large pineal gland in a pregnant female bottlenose dolphin. The authors attributed the large size of the gland to pregnancy.[j] However, the gland does not appear or disappear completely in a given species. Our group examined a series of bottlenose dolphin brains by gross dissection to evaluate the presence of a pineal gland, but failed to detect one in 29 brains (Panin et al., 2012). We also did not find the gland in a series of Risso's dolphins, striped dolphins, and in a single common dolphin. In one instance, we saw what macroscopically looked like a gland in the correct epithalamic position (Fig. 7.7). However, the structure was later found under the microscope to be composed of pial vessels without pinealocytes.

Melatonin

Curiously the circulating levels of melatonin have never been assessed until very recently, in captive Indo-Pacific (Funasaka et al., 2011) and common bottlenose dolphins (Panin et al., 2012). In *Tursiops aduncus*, the concentrations were found to be very low, generally below 10 pg/mL even during the nighttime, whereas the highest levels were reported in the morning, and no significant daily or seasonal variation was found. On the contrary, in *T. truncatus*, we detected higher circulating levels of melatonin, with a daytime range of 15–21 pg/mL. Although limited by the conditions of the sampling design, a significant difference was found both on a daily and a seasonal basis. Clearly a much higher sample size is needed, both as to the number

---

h. Some terrestrial mammals also lack a pineal gland, like the armadillo (Phillips et al., 1986), the elephant (Haug, 1972) and the dugong (Ralph et al., 1985). Another puzzling fact is that pinnipeds, which are marine mammals as well, have among the largest pineal glands and the highest melatonin levels of all mammals (Turner, 1888; Cuello and Tramezzani, 1969; Stokkan et al., 1995; Montie et al., 2009).

i. For example, in six humpback whales (*Megaptera novaeangliae*), Gersch (1938) found one pineal gland, whereas both Breathnach (1955) and Pilleri (1966a) did not find any, although they stated that the gland might have been removed with the meninges during the dissection. In the sei whale (*Balaenoptera borealis*), Pilleri (1966b) described a pineal gland "intimately bound with chorioid plexus under the splenium of the corpus callosum," but Arvy (1971) reported it to be absent. Duffield et al., 1992 were able to find a pineal body only in one out of eleven specimens of bowhead whale (*Balaena mysticetus*) purchased by Alaskan Eskimos. In the blue whale (*Balaenoptera musculus*), Pilleri (1965) could not find any pineal gland in adult individuals.

j. In bats (Haldar et al., 2006), squirrels (Bishnupuri and Haldar, 2000), and humans (Yadav et al., 2008), an enlarged gland has been associated with pregnancy.

Corpus callosum

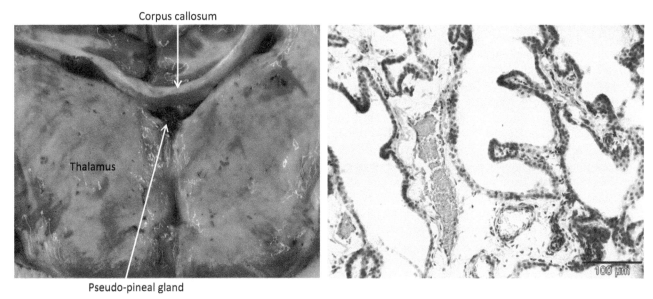

Thalamus

Pseudo-pineal gland

FIGURE 7.7   **Psueudopineal gland of** *T. truncatus* **(scale bar = 100 μm).** Its structure reveals it to be instead composed by the choroid plexus. Nissl stain.

of individuals and of blood samples, to shed light on melatonin production and variation in cetaceans, along with a strict sampling design, which is difficult to perform, due to the particular type of conditions of dolphins in captivity. The source of circulating melatonin remains obscure, and probably there might be alternative sources other than the pineal gland, which are still to be ascertained. Actually several tissues in mammals are able to produce it, none of which are endocrine tissues, particularly the retina (Tosini et al., 2008), the gastrointestinal tract (Konturek et al., 2007), and the Harderian gland (Payne 1994). We performed an immunohistochemical study in these tissues of the bottlenose dolphin (Panin et al., 2012), taking as a marker the enzyme hydroxy-indol-O-methyl-transferase (HIOMT), the ultimate enzyme of the biosynthetic pathway of the hormone. All of these tissue proved immunoreactive for the human HIOMT, indicating a possible melatonin production.

The Thyroid

Anatomy

The thyroid gland is made by two lobes placed lateral to the larynx connected by a bridge of thyroid tissue (Fig. 7.8). This gland possesses the classical follicular organization (Fig. 7.9) and is generally very well developed in cetaceans, including the bottlenose dolphin. In this latter species, the thyroid may show other forms. In some animals, the bridge between the two lobes becomes prevalent and larger than the lateral parts, and the organ appears as a single gland ventral to the respiratory

Thyroid gland

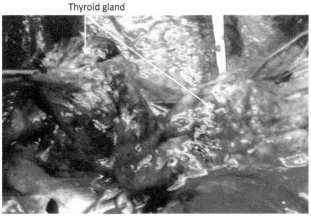

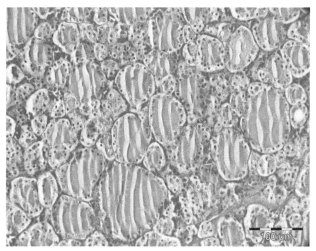

FIGURE 7.9   **Microphotograph of the thyroid of** *Ziphius cavirostris* **(scale bar = 100 μm).** HE stain.

FIGURE 7.8   **Thyroid gland of** *T. truncatus***.**

tract. Intermediate forms may also be observed. The follicles are small in young animals with minimal quantity of colloid and a general adenomatous aspect. Follicles become larger with age, and the colloid color varies from light to dark pink. Follicles of the thyroid are smaller in cetaceans than in the human. The presence of parafollicular cells has been considered highly likely, even though confusion with tangentially cut follicles is possible (Simpson and Gardner, 1972).

There are no lymphocytes in the normal gland.

The thyroid gland in mammals plays a central role in the regulation of development and basal metabolism. There is comparably more literature available on this gland than on other endocrine components in cetaceans, probably owing to the old common belief that the thyroid, which is involved in thermoregulation, would be particularly active in these hairless mammals.[k] Since cetaceans live in an aquatic medium with a high thermal dispersion, it was long thought that they would have higher metabolic rates to preserve body temperature, and so larger thyroid glands, than terrestrial counterparts (Snyder, 1983).[l] Crile and Quiring (1940), for example, found a relative weight of thyroid tissue in the beluga three times higher than that of a thoroughbred horse. While some cetaceans actually have larger glands than terrestrial mammals of similar size (Tables 7.3 and 7.4), apparently neither exists a direct relationship between thyroid weight and body weight, body length or age (Gihr and Pilleri, 1969; Kot et al., 2008), nor marine mammals seem to have higher metabolic rates than terrestrial ones (Lavigne et al., 1986). Early descriptions focused mainly on gross morphological aspects of the thyroid gland. Its shape is generally the same as in most mammals, that is, a median body lying ventrally along the trachea, cranial to the bifurcation of the primary bronchi. A certain variability has been observed, with four different configurations: (1) two large lobes connected by a thin isthmus, (2) no definite isthmus with substantial fusion of the two lobes, (3) complete separation of the two lobes, and (4) irregular cluster of nodules. Rarely the thyroid can be found associated with the thymus and encapsulated with it (Shimokawa et al., 2002; Cowan and Tajima, 2006). The lobulation of the gland by collagen septa seems to increase in older individuals, at least in *T. truncatus*, but size and general morphology do not change significantly with age, gender, sexual maturity, body weight, or body length (Cowan and Tajima, 2006; Kot et al., 2008), also in other species (Rosa et al., 2007; Schnitzler et al., 2008). However, Turner et al. (2006) and Kot et al. (2012a) did find a positive correlation between body and thyroid weight in bottlenose dolphins, possibly due to a higher sample size than in the previous studies. Kot et al. (2012b) obtained a mixed model equation comparing the effect of four reproductive events (estrus, anestrus, lactation, pregnancy) on thyroid volume and found a higher positive effect of lactation compared to both estrus and anestrus. Despite this, they failed to confirm significant differences with sonographic measurements between any of the reproductive events, maybe due to the reduced sample size.

Histological features show a higher degree of variability in relation to different factors (Cowan and Tajima, 2006). A normal thyroid is made up by follicles filled with colloid and lined by simple cuboidal epithelial cells (thyrocytes or principal cells). Thyrocytes secrete the colloid, whose main component is thyroglobulin (Tg), a protein from which thyroid hormones are synthesized. Among follicles, there are so-called "light cells" or "C cells," which instead secrete the hormone calcitonin; sometimes C cells have been found in cetaceans also along the follicular epithelium, but within the basal lamina and so not communicating with the colloid (Harrison and Young, 1969b,c; Harrison and Young, 1970b; Shimokawa et al., 2002). As a general rule, active follicles are smaller with hypertrophied (columnar) epithelium, whereas inactive ones are larger with a flattened epithelium. The conformational change is more evident in species with seasonal peak of activity of the gland. Shimokawa et al. (2002) reported also follicles with invaginated epithelium in *G. griseus*, possibly indicative of colloid depletion. It has also been noted that colloid is less chromophile and more hydrated in younger individuals than in older ones (Cowan, 1966). Recently, in the bottlenose dolphin, both the mean colloid volume of follicles and the nuclear to cytoplasmic (N/C) ratio of follicular cells were found significantly lower in the bottlenose dolphin than in man (Kot et al., 2013). The authors claimed that these two parameters indicated a higher metabolic rate in dolphins, although the N/C ratio (a measure of the metabolic state) was not found correlated with colloid volume by themselves. Harrison and Young (1970b) described some peculiar ultrastructural features of thyrocytes in the common dolphin, *D. delphis*. The apical surface showed some irregularly arranged microvilli and occasional cilia; between adjacent thyrocytes desmosomes delimited intercellular channels, in which sometimes the endoplasmic reticulum opened and nerve endings could be found. Occasionally also multivescicular bodies and membrane whorls were observed, whose significance is still unknown (Young and Harrison, 1969). C cells are recognizable by the absence of endocytosed dark colloid droplets, the presence of many little electron-dense granules, and well-developed rough endoplasmic reticulum (Young and Harrison, 1969; Shimokawa et al., 2002). Recently they have been labeled with an antibody against synaptophysin in the bottlenose dolphin (Cowan and Tajima, 2006).

---

k. The thyroid is also important to assess welfare in captive dolphins.

l. In general, the thyroid gland is more developed in marine than in terrestrial mammals, although this is unrelated to iodine availability in the food. This is important to remember to avoid wrong diagnosis of goiter or hyperplasia of the organ in dolphins.

TABLE 7.3 Relative Weights of the Thyroid Gland in Selected Dolphin Species

Species	Relative Weight (mg/kg Body Weight)	References
D. delphis	129.7–500	Gihr and Pilleri (1969)
Globicephala macrorhynchus	200	Harrison (1969)
L. obliquidens	600	Ridgway and Patton (1971)
S. coeruleoalba	93.8–171.8	Gihr and Pilleri (1969)
T. truncatus	140.7–517.9	Kot et al. (2013)

TABLE 7.4 Absolute Size and Weights of Thyroid Glands in Selected Dolphin Species Reported in the Literature

Species	Reference	Weight	Size	Notes
D. delphis	Gihr and Pilleri (1969)	4.8–25 g		n = 5 males n = 2 females
	Harrison and Young (1970a)	37–58 g		n = 10 immature females
		4.8 g		newborn
		84 g		adult male
G. macrorhynchus	Harrison (1969)	32.6 g		juvenile female
Globicephala melaena	Cowan (1966)	300–600 g		n = 55
	Harrison (1969)	22 g		n = 2 immature females
G. griseus	Pilleri and Gihr (1969)	90 g		
	Hayakawa et al. (1998)		11.5 cm length 4.6 cm width 2.3 cm thickness	n = 3
	Shimokawa et al. (2002)		5–15 cm length 1.5–2 cm width	n = 7
L. obliquidens	Harrison (1969)	26.45 g		n = 2
Orcella brevirostris	Anderson (1878)			
S. attenuata	Harrison (1969)	3.1–11.6 g		n = 4
S. coeruleoalba	Gihr and Pilleri (1969)	6–12.2 g		n = 1 male n = 3 females
	Harrison (1969)	9.75g		
S. longirostris	Harrison (1969)	4.5–8.9 g		n = 4
S. bredanensis	Neuville (1928)		9 cm length 3 cm width 2 cm thickness	left lobe of thyroid
T. truncatus	Hayakawa et al. (1998)		11.9 cm length 4.3 cm width 2.2 cm thickness	n = 3
	Neuville (1928)			
	Cowan and Tajima (2006)	11–58 g		n = 49
	Harrison (1969)	3.3–31.5 g		

Length Is Meant to Indicate Maximum Size Along the Longitudinal Axis, Width on the Perpendicular Axis of the Gland.

TABLE 7.5 Thyroid Hormone Concentrations of Cetaceans Reported in the Literature

Species	tT4 (nmol/L)	fT4 (pmol/L)	tT3 (nmol/L)	fT3 (pmol/L)	rT3 (nmol/L)	References	Notes
G. macrorhynchus	54.8	51.4				Ridgway et al. (1970)	$n = 2$ ♂
L. obliquidens	33.2	22.3				Ridgway et al. (1970)	$n = 2$ ♂
	47.9	29.6				Ridgway et al. (1970)	$n = 5$ ♀
O. orca	77.9	35.8				Ridgway et al. (1970)	$n = 2$ ♂
S. bredanensis	108.24	13.1	1.646	1.62		West (2002)	$n = 7$
S. attenuata	85.1	3.96	1.61	2.22		St. Aubin (2002)	$n = 50$
T. truncatus	105.3	44.8				Ridgway et al. (1970)	$n = 15$ ♂
	94.9	46.1				Ridgway et al. (1970)	$n = 16$ ♀
	158–261		0.93–4.14			Greenwood and Barlow (1979)	$n = 29$
	138.3		1.27			Orlov et al. (1988)	
	175.1	20.9	2.01	1.98	2.78	St. Aubin et al. (1996)	$n = 72$
	125		0.95			West et al. (2001)	$n = 11$
	147.5–165.4	25.9–26.8	0.84–0.99			Fair et al. (2011)	$n = 36$–47 ♂
	160.2–177.5	28.1	0.9–1.03			Fair et al. (2011)	$n = 7$–18 ♀

Abbreviations: *tT4*, total thyroxine; *fT4*, free thyroxine; *tT3*, total triiodothyronine; *fT3*, free triiodothyronine; *rT3*, reverse triiodothyronine; ♂, males; ♀, females. Values Extrapolated by Each Study are Meant to Represent Only Adult Healthy Nonpregnant Individuals, as Far as Possible.

Thyroid Hormones

Currently there are no reports in cetaceans about the production of calcitonin, which is involved in calcium and phosphate balance. On the contrary, a relatively high number of studies are available on thyroid hormones (THs) in marine mammals (for dolphins, see Table 7.5). The main effects of thyroid hormones[m] are to increase the metabolic rate (also for keeping body temperature) and to regulate body growth, but they are also involved in early development, especially of several brain areas in the fetus (Anderson, 2001). Mean and maximum concentrations of thyroid hormones in bottlenose dolphins are apparently higher than in some mammals, including men, bovines, and elephants (Fair et al., 2011). Mean levels do not seem to change significantly year-round in bottlenose dolphins (St. Aubin, 2001).[n]

A third parameter is often considered when evaluating thyroid function, that is, the level of reverse-triiodothyronine (rT_3). After acute stressful conditions, the adrenal glands secrete excess amounts of cortisol, which inhibits both TSH secretion from the pituitary, and the normal conversion of T_4 to T_3, shunting conversion of T_4 to the inactive rT_3 by the enzyme iodothyronine deiodinase (ID) Type III (Köhrle, 2000). If stress is prolonged, the excess reverse-T_3 prevents the action of T_3, leading to a so-called "rT_3 dominance" even after the stress has ceased (Felice et al., 2006). Therefore, rT_3 levels are an indirect but useful stress marker comparable to ACTH in cetaceans.[o]

---

m. THs exist in two active forms, thyroxine (T_4) and triiodothyronine (T_3), the former with much higher circulating levels than the latter (99.7% vs. 0.71% in bottlenose dolphins (West et al., 2014), but with much less affinity for thyroid hormone receptors. T_4 is, in fact, considered a prohormone of the active T_3, and it is converted into T_3 in peripheral tissues (mainly liver and kidney) by the enzymes iodothyronine deiodinase (ID) Type I and II. The two hormones are evaluated both as the free forms (fT_4 or fT_3) or as the total forms (tT_4 or tT_3), that is, the sum of free and protein-bound hormones, since in the blood three main types of proteins (thyroxin-binding globulin, transthyretin and albumin) carry the bulk of THs.

n. On the contrary, in bowhead whales (Rosa et al., 2007) and belugas, there is a typical peak in summer, associated in the latter with a true molt, a unique phenomenon among cetaceans (St. Aubin and Geraci, 1989; St. Aubin et al., 2001).

o. St. Aubin and Geraci (1992) found exceptionally high levels of reverse T_3 at capture in free-ranging belugas compared to most mammals (6.3 nmol/L), probably owing to the summer peak of T_4 levels that might have saturated ID Type I and II, increasing conversion by ID Type III.

The first quantification of THs in cetaceans was performed in heterogeneous conditions in five species (killer and pilot whale; Pacific white-sided and bottlenose dolphin; Gange's river dolphin) (Ridgway et al., 1970). Differences of 6–9°C in the environmental temperature did not affect THs levels in *T. truncatus* and *O. orca*, but notably, in the bottlenose dolphin a significant increase in T_4 was observed after a 72 h fast (Ridgway and Patton, 1971). Since some species undergo periods of food deprivation (eg, due to long migrations), thyroid hormones may be useful to evaluate how metabolism changes during this stressful condition. The only other study about the effect of fast on THs was carried out by Ortiz et al. (2010) in the bottlenose dolphin, confirming increased T_4 levels up to 38 h of fast, whereas re-feeding rapidly restored baseline values. Levels of rT_3 instead decreased 30% within 24 h of fast, returned to initial values within 38 h, and decreased further 32% after re-feeding. T_3 levels never changed significantly over time, indicating that a preferential deiodination occurred in favor of rT_3 during fasting to protect the dolphins from the energetic burden on cellular metabolism (otherwise stimulated by T_3) during a period of caloric restriction.

Besides fasting, a major source of stress for cetaceans is handling and/or captivity, whose effects may bias basal THs concentrations and somewhat limit health status assessment if their levels are considered alone.[p] West (2002) evaluated thyroid function in four rough-toothed dolphins (*S. bredanensis*) found stranded and held in an indoor facility for rehabilitation. Only two of them survived, but weekly samples over 3–4 months showed a significant increase of T_3 and T_4, indicating a successful recover after a heavily stressful event (the stranding). Another subgroup of the same study was made up of three free-ranging dolphins, held in semicaptivity outdoor for 6 years and sampled only occasionally. In these groups, a generalized decrease of THs over the years was attributed to an age-dependent effect, since they were all considered immature at the beginning. Given the reduced sample size, it is more likely that the decreased levels over such a long period could be ascribable to the acclimation of the individuals to the new condition, after a stressful event (the confinement). St. Aubin (1987) similarly reported the lack of responsiveness to bovine TSH in semicaptive belugas with prolonged time of confinement.

Sex-related differences have not been demonstrated clearly, since in the bottlenose dolphin some authors found them (St. Aubin et al., 1996; West et al., 2014), while others did not (Greenwood and Barlow 1979; Fair et al., 2011). In the striped dolphin, a sex-related effect was reported only for rT_3, with higher levels in males than in females (St. Aubin, 2002). A significant decrease of THs levels with age was reported in bottlenose dolphins (St. Aubin et al., 1996; Fair et al., 2011; West et al., 2014), comparing juveniles with adults, though a decrease after sexual maturity was not investigated. Unlike THs, rT_3 seems to increase with age in bottlenose dolphins (St. Aubin et al., 1996). Recently, a significant interaction between sex and age was reported in *T. truncatus*, with higher tT_3 levels in female than in male juveniles, whereas fT_3 was higher in male than female adults (West et al., 2014).

Other factors influencing the levels of thyroid hormones are pregnancy and lactation. Lactating female bottlenose dolphins had higher concentrations of THs than juveniles (Fair et al., 2011), whereas pregnant ones did not show significant differences compared with nonpregnant ones (West et al. 2001; Fair et al., 2011). A recent detailed analysis of pregnancy in the bottlenose dolphin revealed significant higher values of all THs in early-pregnant (first quadrimester) than in nonpregnant females; moreover, differences were significant also for tT_4 and fT_3 compared to late-pregnant (third quadrimester) dolphins (West et al., 2014), similarly to that observed by Fair et al. (2011) in the same species. Interestingly, a mild and transient hyperthyroidism during the first trimester (roughly one third of pregnancy) is a common occurrence in humans, due to two principal mechanisms: (1) a direct stimulation of the thyroid follicular cells by chorionic gonadotropin (hCG) that shares a high structural homology with TSH, and (2) a reduced clearance of TBG mediated by estrogens, leading to a higher production of free T_4 to compensate for the higher amount of TBG-bound hormone (positive feedback). Furthermore, the human fetal thyroid gland does not actively secrete THs until the 18–20 gestational week, then this mechanism might also act to counterbalance the inactive fetal gland (de Escobar et al., 2004). A similar mechanism might exist in dolphins, since both Fair et al. (2011) and West et al. (2014) found a surge in THs in the first third of pregnancy like in humans. The lack of significant differences between pregnant and nonpregnant females in previous studies might be due to the absence of early-pregnant females in the sampled population (gestational stage was not evaluated) or simply to the smaller sample size of pregnant individuals.

Human TSH proved effective to induce TH secretion in bottlenose dolphins, with maximum increase of 29% in tT_3 and 51% in fT_3 by 6 h and 12 h, respectively (West et al., 2014). A positive response to TSH between far-related species (man and dolphin) suggests a high degree of conservation of thyrotropin across mammals.

---

p. St. Aubin and Geraci (1988) quantified THs in 24 restricted free-ranging belugas within 5 h from capture and found mean plasma T_3 and T_4 levels of 2.7 nmol/L and 247 nmol/L, respectively. In six of those specimens, held in captivity for 10 weeks, T_3 declined 80% and T_4 35–65% by days 2–4. After 5 weeks, THs levels were further depressed by ACTH injection (due to the action of cortisol previously described) or saline injection (due to handling stress), and remained low for the rest of the confinement. Long-term captive belugas compared with free-ranging ones showed significantly lower values of THs (St. Aubin et al., 1996), as reported also in manatees (Ortiz et al., 2000) and sea lions (Myers et al., 2006).

The Parathyroid Glands

The parathyroid glands are usually embedded within the thyroid gland, and in most mammals they are composed of two type of cells, the principal ones (secreting parathormone) and the oxyphil ones (whose function is still poorly understood), arranged in clusters. Their main function is to counteract the effects of thyroid calcitonin, so to raise calcium plasma concentration. The number and position of these small glands in cetaceans vary both between species and between individuals, and often they are very difficult to detect. In a Risso's dolphin (Pilleri and Gihr, 1969), two parathyroids were found.[q] Important histological observations were made by Kamiya et al. (1978) on the striped dolphin, in which two to four parathyroids were found in six individuals, also around the thymus. No oxyphil cells were found.[r] In Risso's and bottlenose dolphins, two to four parathyroids were found encapsulated on the dorsal surface of the thyroid (Cowan and Tajima, 2006; Hayakawa et al., 1998). In all the individuals examined, no oxyphil cells were found, and the glands were reported to be proportionally smaller than in human beings. The authors noted that the thyroids were often surrounded by tiny lymph nodes, which could be easily mistaken for parathyroids. Electron microscopy of the principal cells (with pale cytoplasm, electron-dense vesicles, and occasional lipid droplets) revealed ill-developed endoplasmic reticulum and Golgi apparatus, symptoms of a substantially inactive gland. The authors suggested that this might be due to the aquatic habits of cetaceans, analogously to what experienced by astronauts.[s] In cetaceans, the adaptation to a low gravity aquatic environment might have led to a similar decreased significance of parathyroid glands, and might explain also the observed lack of oxyphil cells. Nonetheless, mean plasma calcium levels of both Risso's (8.84 mg/dL) and bottlenose dolphins (8.76 mg/7dL) were within the normal range of most mammals. The authors suggested that cetaceans might regulate calcium and phosphate homeostasis and bone metabolism in alternative ways than land mammals. The absolute size and weight of the parathyroid glands in dolphins are reported in Table 7.6.

The Adrenal Glands

Anatomy

The adrenal glands of small toothed whales were first described in the harbor porpoise (Kolmer, 1918; Slijper, 1936). The first account was attributed to Hunter and Banks (1787), although the author never cited that gland in his description of seven different species. The adrenals are located in the typical mammalian position rostral to the anterior pole of the kidney (Fig. 7.10) and their shape varies slightly according to the species (Fig. 7.11), from oval (Gihr and Pilleri, 1969) to bean-shaped or pyramidal (Arvy, 1971). In adult odontocetes, the glands lie close to the kidneys. Ligaments connect the adrenals caudally to the rostral pole of the kidneys, and medially to the descending colon. The right gland is connected also to the posterior lobe of the liver and is in a slightly more advanced position than the left one.

The weight of the glands can vary considerably between individuals (Table 7.7) (Cowan, 1966). In the bottlenose dolphin, both the absolute and the relative weight of the gland was not found to be correlated with sex or sexual maturity, nor the right gland had significantly different weight than the left (Clark et al., 2005, 2006).

TABLE 7.6 Absolute Size and Weights of Parathyroid Glands in Selected Dolphin Species and of Man as a Comparison

Species	Size (mm)	No. of Individuals	No. of Glands	References
G. griseus	2 cm	n = 1	n = 2	Pilleri and Gihr (1969)
	0.8–3 × 2.5–3.2 mm	n = 3	n = 2	Hayakawa et al. (1998)
S. coeruleoalba		n = 6	n = 2–4	Kamiya et al. (1978)
T. truncatus	0.8–2 × 1.5–2.5 mm	n = 3	n = 2–4	Hayakawa et al. (1998)
Homo sapiens	4–9 × 2.5–6 mm			Isono et al. (1993)

The First Measure of Size is the Maximum Diameter, the Second is the Minimum.

q. A total of three glands were reported in the now extinct baiji (*Lipotes vexillifer*) (Zhongjie, 1985).

r. The same study (Kamiya et al., 1978) described also the parathyroid glands of the Gange's river dolphin. Individuals of this species had two or three glands located around the thyroid, with abundant interlobular connective tissue and a little heterotopic thymic tissue.

s. Absence of gravity leads to a reduction of parathyroid activity and of parathormone plasma levels, with a subsequent increase of plasma calcium concentration.

FIGURE 7.10   **Position of the adrenal gland of *S. coeruleoalba* rostral to the kidney.**

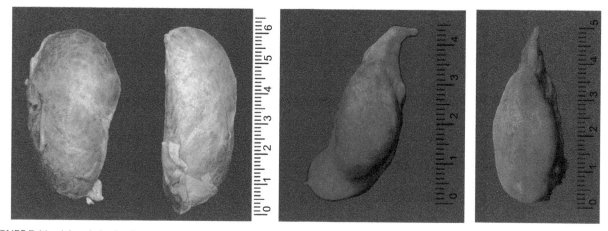

FIGURE 7.11   Adrenal glands of *T. truncatus* (left), *S. coeruleoalba* (center), and *G. griseus* (right).

TABLE 7.7 Absolute Size and Weights of Adrenal Glands in Selected Dolphin Species Reported in the Literature

Species	Weight	Relative Weight	References	Notes
D. delphis		0.09–0.15‰	Gihr and Pilleri (1969)	Extrapolated*
G. griseus		0.008‰	Pilleri and Gihr (1969)	Extrapolated*
S. coeruleoalba		0.05–0.21‰	Gihr and Pilleri (1969)	Extrapolated*
T. truncatus	1.4–15.5 g	0.08–0.14‰	Clark et al. (2005)	$n = 31$

Extrapolated data are calculated by authors based on reported body and gland weights.

The adrenal glands can be often mistaken for lymph nodes having a retroperitoneal position and a similar external appearance, but in half section the distinction between cortex and medulla is generally patent (Fig. 7.12), while lymph nodes are more uniform.

The cortex (Figs. 7.13 and 7.14) has the typical mammalian subdivision into *zona glomerulosa* (secreting mineralocorticoids), *z. fasciculata* (for glucocorticoids) (Fig. 7.15), and *z. reticularis* (for androgens), and is penetrated by numerous incomplete septa of connective tissue projecting from the thick capsule, resulting in a pseudolobulated appearance. The zona fasciculata is generally the thickest part of the cortex, with considerable variation among specimens.

The medulla (Figs. 7.14 and 7.15) is characterized by an outer layer (or medullary band) with adrenalin-producing cells (E-cells) that stain darkly also with hematoxylin-eosin, and a central region with lighter stained cells (N-cells) producing noradrenaline, a pattern shared with ungulates (Clark et al., 2008). A difference with the latter seems to be the cooccurrence of E- and N-cells in the medullary band, where E-cells have been detected also by IHC (immunocytochemistry) against PNMT (penylethanolamine N-methyltransferase), the last enzyme of the biosynthetic pathway leading to adrenalin (Clark et al., 2005). The adrenal medulla is somewhat richer in blood vessels than the cortex. The cells are cuboidal and form cords or nests separated by thin connective septa in which blood vessels run, as in most mammals.

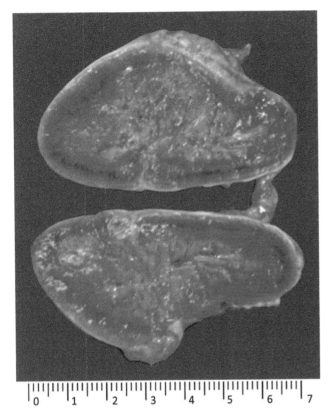

FIGURE 7.12   Section of the adrenal gland of *G. griseus.*

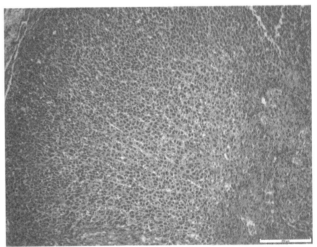

FIGURE 7.13   **Adrenal cortex of *G. melas* (scale bar = 200 µm).** HE stain.

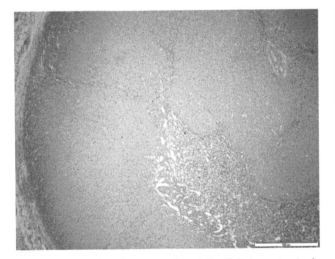

FIGURE 7.14   **Adrenal cortex and medulla of *T. truncatus* (scale bar = 500 µm).** HE stain.

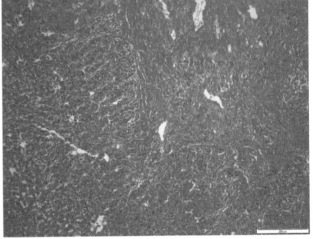

FIGURE 7.15   **Adrenal pars fasciculate and medulla of *D. delphis* (scale bar = 200 µm).** HE stain.

Age lipofusion pigments have not been observed in cetaceans.

A peculiar feature is the presence of projections of the inner medulla into the cortical tissue, 1–3 in *T. truncatus* (Clark et al., 2005), at most 1 in *S. attenuata* and *S. longirostris* (Clark et al., 2008), whereas in *S. coeruleoalba* nor cortical nor medullary projections were noticed (Vuković et al., 2010).[t] Similarly, invaginations of the cortex into the medulla were

---

t. These medullary protrusions were also described in terrestrial Cetartiodactyla, including the camel (Abdalla and Ali, 1989). The adrenal glands of the camel show also the "cortex-inversus" disposition described in the next lines.

TABLE 7.8 Composition of the Adrenal Gland of Selected Dolphin Species Expressed as Cortex to Medulla Ratio and Percent Cross-Sectional Area

Species	C/M Ratio	PCA	Reference
Stenella attenuate	1.46–3.71	64% cortex + 29.4% medulla + 6.2% other	Clark et al. (2008)
S. longirostris	1.19–3.28 (males) 2.85–3.89 (females)	65% cortex + 27.3% medulla + 7.7% other (males) 71.7% cortex + 22.7% medulla + 5.6% other (females)	Clark et al. (2008)
T. truncatus	0.71–4.42 0.85–1.73 (acute stress) 0.73–2.82 (chronic stress)	48% cortex + 41% medulla + 11% other 48% cortex + 41% medulla + 11% other (acute stress) 53% cortex + 36% medulla + 11% other (chronic stress)	Clark et al. (2005) Clark et al. (2006)

C/M–cortex to medulla ratio; PCA–percent cross-sectional area ("other" indicates the sum of blood vessels, connective tissue, and capsule of the adrenals).

found in the bottlenose dolphin, and in striped dolphin, small islands of cortical tissue were observed entrapped in the connective tissue of the capsule (Vuković et al., 2010). Another atypical conformation of the adrenals is the "cortex inversus," that is, the presence of cortical tissue unsheathing only arterioles penetrating the medulla and comprising all the three *zonae*. The functional significance of this arrangement is still poorly understood.

The cortex/medulla (C/M) ratio can vary greatly according to the species (Table 7.8). It is about 1:1 in the bottlenose dolphin, and is constant across the whole gland, with no difference according to sex or sexual maturity (Clark et al., 2005). In the genus *Stenella*, the C/M ratio was at least twofold higher than in bottlenose dolphins (2.42–2.90), with no correlation with sex or sexual maturity in *S. attenuata*, whereas in *S. longirostris* males had significantly higher C/M ratio and cortical mass than females (Clark et al., 2008). Similarly, the percent cortical area (a proxy for the cortical mass) ranges between an average 48% in bottlenose dolphins (Clark et al., 2005) and more than 64% in *Stenella* spp. (Clark et al., 2008).

The morphological parameters of the adrenals have been used to assess the stress status (acute vs. chronic) of stranded or bycaught specimens. In the bottlenose dolphin, the total weight of the glands, the C/M ratio and the percent area of both cortex and medulla in cross-section, and even the number of A-cells were found significantly higher in animals died from causes related to chronic stress than to acute stress (Clark et al., 2006). This is due to the fact that the adrenal cortex is the source of cortisol and other stress-related hormones; therefore under the effects of continuous and prolonged stressful stimuli, the cells of the *zona fasciculata* undergo hypertrophy to secrete excess amounts of these hormones.

Even the color of the cortical tissue has been used to assess the degree and extent of the stress effects in a couple of species, the pantropical spotted (*S. attenuata*) and the spinner dolphin (*S. longirostris*) (Myrick and Perkins, 1995). These species were subjected to a high mortality rate due to entrapment and/or entanglement in fishing nets during the so-called "dolphin fishery" that has tuna as target species, but exploits the strict cooccurrence of dolphins and tunas in the eastern tropical Pacific (ETP). Many of the reported dolphin deaths were not attributable to entanglement,[u] so deceases were ascribed to stress-related mechanisms, such as capture myopathy, well documented in terrestrial mammals (Spraker, 1993). Since the carcasses of ETP dolphins could not be accessed for up to 2 h after death, the blood might have been compromised and stress hormone quantification not reliable. The color of the adrenal cortex, combined with its histological features, was used to search for evidence of peri-mortem continuous acute stress (CAS), which would occur in a stereotyped manner as studied in rats (Symington, 1969). In dolphins and porpoise, the healthy cortical color is light yellow to beige as in most mammals (Slijper, 1962), due to the presence of abundant intracellular lipid droplets rich in cholesterol, which is the precursor of all steroid hormones. During CAS, the cholesterol stores are used up in *zona fasciculata* cells to synthesize cortisol under the stimulation of ACTH, and with the concurrent hyperemia, there is a progressive darkening of the cortex that turns reddish-brown. In rats, this process begins within 1 h from the beginning of the CAS, and the darkening starts at the interface between *z. fasciculata* and *z. reticularis* and progresses radially outward, as observed in both *Stenella* spp. The darkening degree was found significantly correlated with chase time (from speedboat launch to the moment the net entered the water) and encirclement time (from net entry into water to full closure of the net) only for adult spotted dolphins. In both neonates and young of *S. attenuata* and all age classes of *S. longirostris*, the darkness had no correlation with the phases of net deployment. This lack of correlation was attributed to a higher susceptibility to CAS of both younger spotted dolphins and all spinner dolphins that are smaller and more hyperactive than spotted dolphins (Pryor and Norris, 1978). In these cases, the cortical response would

---

u. In recent years, new protective measures introduced in the fishery procedures have reduced (but not fully abolished) ETP dolphin mortality.

be so rapid (or already ongoing) that even at the shortest observed preconfinement times, cortices were nearly saturated at maximum darkness.

Hormones of the Adrenal Gland

The adrenal cortex of cetaceans has the typical mammalian functional subdivision into three *zonae*, and the zonation was first demonstrated in the sperm whale by paper and column chromatography (Race and Wu, 1961). The subcellular compartmentalization of the steroidogenic enzymes is the same of all mammals. While the general physiology of the adrenal glands is the same as in most mammals, two main features apparently set cetaceans apart from them. They have an unusually high content of noradrenaline and dopamine in the medulla, and a lesser responsiveness of cortisol in the stress response in favor of corticosterone and probably aldosterone (ALDO). Studying the stress response in cetaceans is a controversial matter, since a "baseline" range of values may be hard to assess. Free-ranging specimens need to be captured in some way, with chasing and handling operations that might cause discomfort to wild individuals, and in the case of long-term captive animals, the captivity itself could be a stressful condition that could alter a normal response of the hypothalamic-pituitary-adrenal axis. Other possible confusing factors may reside in the sampling design, because different durations of the stressful stimuli and different timing of blood draws may lead to inconsistent outcomes, for example, if a blood sample is drawn out of phase with respect to the time of physiological response of an individual. Furthermore, within a social group, it has been suggested that the rank position of an individual could be an additional factor influencing the observed "baseline" values. It has been widely reported in fact in other mammals with a hierarchical social organization often experience chronic stress, exhibiting increased cortisol levels (Sapolsky 1992).

Catecholamines

Among catecholamines, adrenaline (or epinephrine) is considered the only one acting as an endocrine hormone, while noradrenaline (or norepinephrine) and dopamine act as neurotransmitters. Both epinephrine (EPI) and norepinephrine (NOR) are involved in the immediate fight-or-flight response after an acute exposure to stressful stimuli. Their circulating levels increase rapidly, inducing peripheral vasoconstriction, increasing heart rate, and mobilizing energy stores in the form of glucose from hepatic glycogen and free fatty acids from adipocytes.

In most mammals, the main catecholamine produced by the chromaffin cells of the medulla is epinephrine (Hadley and Levine, 2006).[v]

Relatively few studies addressed the catecholamines in the blood of dolphin species (Table 7.9). Stress may increase blood concentration of the hormones, as in spotted dolphins (*S. attenuata*) subjected to long chases and encirclement (St. Aubin and Dierauf, 2001), in which the blood catecholamines showed persistent elevated levels and no correlation with time. Samples were not drawn any sooner than 100 min after the beginning of the chase, which in some cases lasted over 4 h, so this indicated a continuous sympathetic stimulation of the chromaffin cells during the sampling operations. Dopamine was found very high as well (46–461 pg/mL), a fact the authors could not explain (St. Aubin, 2002). Neither dopamine nor norepinephrine were correlated to epinephrine levels, and this is curious as EPI is the sympathetic effector for EPI release from the medulla. Baseline values for such highly active specimens in the wild are quite difficult, if not impossible, to estimate, given their pelagic habits and nervous behavior, which hamper the possibility of confining them into semi-captivity for acclimation.

In the bottlenose dolphin, Suzuki et al. (2012) found a higher daily mean of both EPI and NOR in the winter than in the summer solstice, and also a significantly higher variance of their levels. The body temperature as well was significantly higher in winter, and this prompted to suggest a role of catecholamines (especially norepinephrine) in the thermoregulation of cetaceans. Further, all the three catecholamines showed a significant positive correlation with each other. Mean levels of EPI and NOR were exceptionally high compared to those found in other studies (see Table 3), but the authors did not comment about it; probably the high levels could be due to the tight sample design, with blood drawn every 3 h throughout an entire day.

In the same study, no circadian rhythm of catecholamines was found both in winter and in summer, contrarily to what was observed in other mammals including man, where catecholamines show a nocturnal nadir (Suzuki et al., 2012). No significant effect of the time of the day was noticed also in two populations of bottlenose dolphins of the East Coast of the United States by Fair et al. (2014). In the same individuals, a significantly higher level of norepinephrine was found in males (2.8 nmol/L) than in females (4 nmol/L), whereas epinephrine did not vary between sexes.

---

v. An unusual feature of the adrenal glands of whalebone whales seems to be the high content of noradrenaline, with its prevalence over adrenaline.

TABLE 7.9 Hormonal Production from the Adrenal Glands of Selected Dolphin Species

Species	EPI	NOR	Cortisol	ALDO	References	Notes
G. melas			441 nmol/L		Geraci and St. Aubin (1987)	n = ? stranded
O. orca			6.9 nmol/L		Suzuki et al. (1998)	n = 3 captive
			6.1 nmol/L		Suzuki et al. (2002b)	n = 2 captive
S. attenuata	27.3–8692.3 pmol/L	1.6–25.5 nmol/L	70.9–219.4 nmol/L	13.85–846 pmol/L	St. Aubin (2002)	n = 61 wild
S. coeruleoalba			210–490 nmol/L		Gales (1992)	n = 8 stranded
S. bredanensis			27.6 nmol/L		Suzuki et al. (2002a)	n = 1 wild
			0–358.8 nmol/L		West (2002)	n= 4 stranded, n = 3 wild
T. truncatus					Seal and Doe (1965)	n = 1
			46.9–82.8 nmol/L		Medway et al. (1970)	n = 8 captive
			29.8 nmol/L	280 pmol/L	Thomson and Geraci (1986)	n = 3 captive
			90 nmol/L		Orlov et al. (1988)	n = 40 wild
			30-40 nmol/L		Bossart and Dierauf (1990)	n = ?
			71.8 nmol/L		St. Aubin et al. (1996)	n = 36 wild
			10.5 nmol/L		Suzuki et al. (1998)	n = 2 captive
			35.9–91.1 nmol/L		Reidarson and McBain (1999)	n = 2 captive
			27.6–154.6 nmol/L	69.3–1263.1 pmol/L	Ortiz and Worthy (2000)	n = 31 wild
			552 nmol/L		Orlov et al. (1991)	n = 1 captive
			13.8 nmol/L		Suzuki et al. (2002b)	n = 2 captive
	170.6 pmol/L	5.5 nmol/L			Romano et al. (2004)	n = 1 captive
			6.6–64.5 nmol/L		Pedernera-Romano et al. (2006)	n = 7 captive
			10.2 nmol/L		Naka et al. (2007)	n = 5 captive
			33.1 nmol/L		Noda et al. (2007)	n = 5 captive
			15.5–18.2 nmol/L		Suzuki et al. (2008)	n = 3 captive
	240.2–431.3 pmol/L	3.3–4.8 nmol/L	12.2–18.2 nmol/L		Ridgway et al. (2006)	n = 1 captive
	382.2 pmol/L	3.3 nmol/L	3.3 nmol/L		Ridgway et al. (2009)	n = 1 captive
			8.6–16.8 nmol/L		Funasaka et al. (2011)	n = 4 captive
			90 nmol/L		Copland and Needham (1992)	n = 6 captive
			15.5 nmol/L		Houser et al. (2011)	n = 2 captive
			2.8–11.2 nmol/L		Tizzi et al. (2010)	n = 1 captive
			11.9–20.4 nmol/L		Blasio et al. (2012)	n = 10 captive
	1–1.4 nmol/L	7.7–16.3 nmol/L			Suzuki et al. (2012)	n = 4 captive
	0.2–4.5 nmol/L	0.9–5.4 nmol/L	2.8–198.7 nmol/L	15.2–907.4 pmol/L	Fair et al. (2014)	n = 168 wild

Note: "Wild" animals are defined as those captured and maintained both in the wild or in transient captivity for the experiments; "captive" ones are permanently held in facilities. Concentrations reported are those concerning the least disturbed status of the specimens, if discernible from the sampling design. In case hormone values were obtained from different sample sizes (n is the number of individuals), a subscript identifies the corresponding column. Single values indicate means of multiple measures. Values expressed as µg/g or ng/g refer to quantification from tissue extracts instead from blood.

Glucocorticoids

Glucocorticoids are responsible for the long-term response to stress, with a slower action than catecholamines but longer lasting effects. They generally have permissive actions on catecholamine synthesis, sustaining their action and enhancing their catabolic effects on hepatocytes, adipocytes, and muscle fibers, upregulating the expression of gluconeogenic genes to increase the availability of energetic substrates in case of emergency. The prolonged and higher than physiological concentrations in case of protracted stress may have immunosuppressive effects, as well as also antiinflammatory actions. While in most mammals the main glucocorticoid is cortisol, considered the best indicator of chronic or prolonged stress, in cetaceans there are some indications that it may not play a central role in the fight-or-flight response. Circulating levels generally fall within a narrow range (30–40 nmol/L) (Bossart and Dierauf, 1990) and their elevations are considered to be modest, limiting the utility of cortisol as a stress indicator (St. Aubin and Dierauf, 2001).

Corticosterone is another major glucocorticoid, which is usually considered as a precursor of aldosterone, and apparently it does not have a specific role. Seal and Doe (1965) measured its levels in marine mammals, including the bottlenose dolphin (187.9 nmol/L), finding much higher levels than in humans (58.4 nmol/L) (Raubenheimer et al., 2006), while in another group of wild bottlenose dolphins the range was lower (6.1–51.1 nmol/L) (Ortiz and Worthy, 2000). Cortisol to corticosterone ratio in the blood is about 5:1 in the latter species (Thomson and Geraci, 1986; Ortiz and Worthy, 2000), instead of 10:1 as reported in humans (Raubenheimer et al., 2006). Such a different ratio could be explained by a different affinity of both corticosterone and cortisol for the corticotropin-binding globulin (CBG), which carries the bulk of adrenal steroid hormones in the blood (Pedernera-Romano et al., 2006). Its carrying capacity is about 50% in the bottlenose dolphin and 94% in man (Hadley and Levine, 2006). The apparent limited response to stress of cortisol could be linked to a higher responsiveness of corticosterone, which may play a more important role than expected in the stress response of cetaceans.

Suzuki et al. (1998) raised the possibility that another corticosteroid, 21-deoxycortisol (21DOC), might be of physiological importance and partly explain the relatively low amplitude variations of cortisol, because of a neglected cross-reactivity of the antibodies employed in radioimmunoassays (RIA). In fact, they incidentally discovered that in killer whales and bottlenose dolphins, 21DOC was as highly concentrated as cortisol, while in most mammals, including man, it occurs only in traces.

A wealth of studies focused on cortisol in relation to stress induced by handling operations and different sampling conditions in the wild and in controlled environments. In captive bottlenose dolphins, Thomson and Geraci (1986) reported that the same animals subjected to 3 h of pursuit exhibited cortisol levels twice as high as those found after sampling with no chase (25 ng/mL). In both cases, the dolphins were then held on a stretcher (to simulate transport) and sampled out of water, but after the calm capture a significant three-fold rise was observed over the next hour (29–43.5 ng/mL), whereas after the chase the cortisol did not rise significantly (21.7–36.2 ng/mL). This indicated that both chase and transport induced a stress response of the same amplitude but they did not cumulate their effects. After some days, administration of synthetic ACTH after 20 min of chase on the same dolphins did not induce higher cortisol levels than those observed in the previous treatments. In the same study, the authors measured cortisol in a group of wild dolphins subjected to 5 h of pursuit or administration of an ACTH analog, and in both treatments, the cortisol levels were never significantly different than the maximum found in captive individuals. However, the observed results might have been biased, since the same three captive dolphins were subjected to the different treatments on alternated days, with a problem of pseudo-replicates, because they were not independent groups. Furthermore, such a tight sampling design on the same animals might have masked their actual stress response. The authors expected a cumulative effect of chase and ACTH administration, which did not occur. However, this lack of responsiveness might be due to a compromised health status of the individuals, two of which died after the analog administration and were found with congested adrenals. The authors noticed a rise in cortisol levels within 1–2 hr after ACTH injection, while Orlov et al. (1988) reported a peak of cortisol 3–6 h after ACTH administration in the same species, although individuals were not allowed to swim freely but kept for 44 h on stretchers. In belugas, similarly to the bottlenose dolphin, exogenous ACTH induced an elevation of cortisol within 1–2 h that returned to resting values within 4–5 h (Orlov et al., 1991).

The duration of the acclimation period after a perturbation can affect the observed values, with longer times leading more likely to resting levels ("baseline"), and likewise longer capture processes (with chasing, entrapment, manipulation) may lead to more significant differences of the values with respect to the baseline.

In semidomesticated bottlenose dolphins, St. Aubin et al. (1996) found lower mean cortisol levels (49.7 nmol/L) than in wild ones (71.8 nmol/L), but still higher than what found in long-term captive specimens by Suzuki et al. (1998) (6.9 nmol/L). In the rough-toothed dolphin (*S. bredanensis*), a significant decrease in blood cortisol was reported in two wild individuals (both stranded and rehabilitated) over 3 months of confinement, while other conspecifics in the same conditions did not show significant variations. In all the animals, the levels were below 138 nmol/L after fifty days of confinement, and in three of them held for 6 years cortisol never exceeded 16 nmol/L (West 2002).

Ortiz and Worthy (2000) assessed the effect of capture and restraint on several blood parameters in wild bottlenose dolphins. They found no correlation between capture or restraint time within 40 min and the level of both cortisol and corticosterone, suggesting a lack of a neuroendocrine response after short periods of manipulation, similarly to what reported by Thomson and Geraci (1986). Fair et al. (2014) observed a significant rise in cortisol in wild individuals after a 45–100 min of manipulation (127.5 nmol/L), compared to levels in blood drawn immediately after capture (73.14 nmol/L). Ortiz and Worthy (2000) also reported a positive and significant correlation between cosrtcosterone and cortisol; the former, but not the latter, was correlated with aldosterone levels. This could indicate a more important secretion of corticosterone than cortisol, as mentioned previously, since ACTH, although secondarily, can induce the synthesis of both corticosteroids and aldosterone (Hattangady et al., 2012). Curiously, no correlation between ACTH and cortisol levels was found in spotted dolphins after chasing and encirclement for up to 4 h (St. Aubin, 2002). These observations support a diffuse idea that in cetaceans hypothalamic-pituitary-adrenal axis may function in different ways compared to most mammals. Bottlenose dolphins however responded positively to oral dexamethasone administration by suppressing circulating cortisol within 8 h (Medway et al., 1970) and also ACTH and cortisol up to 36; both cortisol and ACTH reversed to baseline values within 48 h (Reidarson and McBain, 1999).

Cortisol in bottlenose dolphins was not correlated with serum glucose, a fact Ortiz and Worthy (2000) attributed to the absence of an acute response for short (<40 min) restraint periods. However, glucose levels in cetaceans are always very low, as they rely mostly on free fatty acid oxidation for their metabolism. Actually they can be considered diabetic (Venn-Watson and Ridgway, 2007), so probably gluocse is not a proper parameter to consider. In five captive animals sampled within 10 min after draining the pool, Naka et al. (2007) observed higher levels of cortisol compared to resting values, but not significantly; however, blood was drawn way too soon to obtain a significant physiological response. Noda et al. (2007) studied the effect of handling and transportation of long-term captive bottlenose dolphins moved to a new facility; both the operations led to significantly higher values (154.6 and 171.1 nmol/L, respectively) of cortisol compared to the resting stage (33.1 nmol/L), with not much difference between them. Interestingly, the authors also noted a significant negative correlation between cortisol and three immunologic parameters (total white blood cell, number of eosinophils and rate of production of reactive oxygen species), which also showed significant decreases between resting values and those after both handling and transportation, highlighting the stressful and immunodepressive effects of these operations even in long-term captive specimens. Suzuki et al. (2008) tested two different high-performance mattresses to evaluate their potential for reduction of stress during transport, but after 30 min of being carried on both mattresses, cortisol increased three-fold (64–92.5 nmol/L) compared to resting levels (19.6–27.6 nmol/L), so even short periods of transport can induce a dolphin's stress steadily. Confinement into open or closed facilities apparently did not influence mean cortisol levels in captive bottlenose dolphins according to Blasio et al. (2012), but Ugaz et al. (2013) found the opposite. The latter tested four groups kept in four different pools, two open and two closed, and found significantly lower cortisol levels in the group held in the larger open pool (0.09 nmol/L), compared to those of both the small (1.4 nmol/L) and the large closed pool (0.65 nmol/L). Interestingly, the reported values of all the groups were the lowest ever recorded for resting captive bottlenose dolphins (Table 3). Bottlenose dolphins have been reported to endure 5 days of continuous auditory vigilance without effects on circulating levels of adrenal steroids or catecholamines (Ridgway et al., 2006, 2009).

Cortisol is known to have a circadian rhythm in mammals (including man) with a peak just before awakening and a nadir few hours after the onset of sleep (Hadley and Levine, 2006). The diurnal and seasonal variations in cetaceans were investigated in several studies. Judd and Ridgway (1977) found a peak of cortisol at midnight in captive *T. truncatus*, and similarly Suzuki et al. (2002b) observed a significant peak at 3:00 am in two *T. aduncus* with a nadir at 6:00 pm. The same authors assessed how different sampling frequencies over 24 h influenced the detection of an actual circadian rhythm and its characteristics (Suzuki et al., 2003). They obtained inconsistent outcomes, with a peak at 01:00 and 03:00 am and a nadir at 06:00 and 09:00 pm if samples were drawn every 3 or 4 h, respectively, while the peak was at 09:00 am if blood was drawn every 6 or 8 h. The increase in the sampling frequency apparently induced a higher stress in the animals, with a higher peak at 03:00 (3 h frequency) than at 1:00 (4 h frequency). The shift, however, could be ascribed also to seasonal factors, as the daily samplings were performed in different months. Curiously, a week later during the same month, the peak was found at 03:00 am with a 4 h frequency and at 09:00 am with a 6 h frequency. In the Indo-Pacific dolphin, a daily rhythm was evident at the spring equinox and at the summer and winter solstices, with a peak at 09:00 am and low levels for the rest of the day (Funasaka et al., 2011). Wild bottlenose dolphins living in the coastal waters of Florida and South Carolina instead did not show a diurnal variation in cortisol values, or a correlation between cortisol and the time of blood draw, although whole-day profiles were not assessed (Fair et al., 2014).

Suzuki and colleagues sampled three captive killer whales between September 1996 and December 1997 (a mature female, a mature male, and an immature male). They failed to observe a significant daily rhythm over a single day in all the individuals (Suzuki et al., 2002b, 2003). However, considering the average values spanning some months, significantly higher values were found in the morning than in the afternoon only in the female over 3 months in autumn/winter (Suzuki et al., 1998). The difference was significant also in the mature male taking into account the averages over the whole year

(Suzuki et al., 2003). In the mature individuals, they found also a significant negative correlation between cortisol and testosterone or progesterone in the morning samples, but not in those drawn in the afternoon (Suzuki et al., 2003).

As to the seasonal variations in the bottlenose dolphin, significantly higher levels of cortisol were found in winter/spring than in summer/autumn by Orlov et al. (1988), but St. Aubin et al. (1996) did not find significant seasonal variations in both wild and semidomesticated specimens. In the Indo-Pacific species (*T. aduncus*) instead, a significantly higher daily mean of cortisol was found at the spring equinox than at the winter or summer solstices, with no seasonal difference in the amplitude of cortisol fluctuations (Funasaka et al., 2011).

In the killer whale, slightly higher cortisol levels in winter than in summer were reported in two male individuals, but not significantly, whereas in one pregnant female, cortisol varied significantly with a decline rate of once pre 4.7 months (Suzuki et al., 2002b). In one rough-toothed dolphin (*S. bredanensis*), the concentrations did not change all the year round (Suzuki et al. 2002a). Seasonal fluctuations of cortisol may not be dependent only upon photoperiod and social cues (eg, male–male interactions during breeding season, hierarchical dominance), but directly on physical environmental inputs like water temperature. Interestingly, Funasaka et al. (2011) found a significant negative correlation between circulating cortisol and water temperature in *T. aduncus*. Likewise, Houser et al. (2011) exposed two bottlenose dolphins to progressively lower water temperature over 10 days, from 17–20°C to 4.6°C, and reported a three-fold rise in morning cortisol levels corresponding to the coldest water exposure compared to normal temperatures.

Despite the fact that cortisol is considered to have low amplitude variations in response to stress, very high values and/or wide ranges have been reported in several cases. Exceptional values of cortisol were reported in the bottlenose dolphin (552 nmol/L) (Orlov et al., 1991) from individuals kept out of water for 44 h. Gales (1992) found very high serum cortisol levels (210–490 nmol/L) in striped dolphins that got mass stranded and moved to a recovery facility. Samples were taken 18 h after they were found on the beach, and some of the individuals were even freeze-branded, so those values were probably very distant from baseline values. Similar blood levels were found also in long-finned pilot whales (*Globicephala melas*) stranded for more than 6 h (Geraci and St. Aubin, 1987) (Table 7.9).

Apart from stranding cases, a compromised health status can be a considerable source of stress and alter normal circulating levels of cortisol. High cortisol values (82.25 nmol/L) have been reported in a chronically ill bottlenose dolphins in captivity (Waples and Gales, 2002), with levels comparable to those observed in wild dolphins just after capture. In the same species, comparing two resident populations inhabiting two different sites of the US Atlantic coastal waters, a significantly higher blood content of catecholamines was found in the group living in the more polluted waters (Fair et al., 2014). A recent study suggests an attempt at establishing diagnostic reference intervals for cortisol and aldosterone in the bottlenose dolphin considering different potential perturbation factors (Hart et al., 2015).

Neither age (St. Aubin et al., 1996), nor sex (Ortiz et al., 2000), seem to affect cortisol values in *T. truncatus*. However, Fair et al. (2014) found significantly higher cortisol levels in young individuals (31.7 nmol/L) than in adults (22.4 nmol/L) in two populations of wild bottlenose dolphins.

Since blood sampling with venipuncture is in itself a potential source of stress that can alter the physiological levels of cortisol and other stress hormones, other biological matrices have been investigated, to allow less invasive sampling techniques (Amaral, 2010). In captive bottlenose dolphins, a significant correlation was found between plasma and salivary cortisol, measured for the first time by Pedernera-Romano et al. (2006). On average, its values in the saliva (5.43 nmol/L) were about 27% those in the blood (19.7 nmol/L), similarly to what reported in man (6.7–34%).

In the killer whale, fecal glucocorticoids (fGC) were employed to model the trend of stress status in different pods over a 3-year period, and a significant negative correlation was found between fGC and the fishing effort on Chinook salmon in British Columbia. The average values of all the wild and presumably healthy individuals (1000 ng/g) were about 27 times lower than those found in a severely emaciated male killer whale that stranded and was euthanized in Hawai'i and taken for comparison (Ayres et al., 2012).

Cortisol levels were evaluated also in the blow (or sputum), where a highly significant positive correlation was found with values in the blood in one captive bottlenose dolphin (Tizzi et al., 2010). Interestingly, in a pregnant female, blow cortisol progressively rose over the course of pregnancy, with a tenfold rise 2 days before delivery (446.2 pg/mL) compared to initial values a week postconception (8.1 pg/mL).

Mineralocorticoids

Generally, aldosterone is not considered a stress hormone in mammals, but several studies reported significant variations after stressful conditions in some species. Its role as part of the stress response in marine mammals is thought to be an adaptation to maintain fluid and electrolyte balance under duress. Aldosterone secretion in mammals is mainly under control of the renin-angiotensin system; however, there are some indications that in cetaceans ACTH may have a stronger importance than in other mammals in the control of its release, as previously described. In spotted dolphins subjected to chase and encirclement, aldosterone was significantly and positively correlated with cortisol and ACTH levels, suggesting a direct stimulation by the hypophyseal hormone. Both corticosteroids were elevated with a lack of a temporal pattern, suggesting that aldosterone was an important actor during the stress response (St. Aubin, 2002). Angiotensin II, which in mammals is the direct effector of aldosterone secretion, was measured for the first time in the bottlenose dolphin by Naka et al. (2007). They observed a rise in captive individuals sampled within 10 min after completely draining the pool (64.3 pg/mL) compared to resting values (48.1 pg/mL), but elevations were not significant, probably due to the too short period between the end of the draining and the blood sampling.

Thomson and Geraci (1986) found a sixfold rise in plasma aldosterone within 3 h in captive bottlenose dolphins held on stretchers out of water, similarly to what reported by St. Aubin et al. (1996) in wild animals after 4 h of chase. On the contrary, for periods shorter than 40 min, aldosterone was not correlated with total capture or restraint time in the same species, and curiously it was not correlated neither with Na^+ nor with $Na^+: K^+$ ratio (Ortiz and Worthy, 2000). Curiously, 1 h of exposure to high sound pressures (>200 dB) induced a significant elevation of aldosterone but not cortisol in a captive bottlenose dolphin (Romano et al., 2004).

In two populations of wild bottlenose dolphins living in the Atlantic coast of the United States, a significantly higher aldosterone level was found in young than in adult individuals (Fair et al., 2014). This suggests that immature cetaceans might be more susceptible to alterations in the electrolyte homeostasis during stressful conditions.

Interestingly, a recent phylogenetic analysis of several osmoregulatory genes has shown that in cetaceans there might have been a positive selection toward cetacean-specific sequences. In particular, the positive selection concerned a urea transporter (SLC14), aquaporin-2 (AQP), angiotensinogen (AGT), and angiotensin-converting enzyme (ACE) (Xu et al., 2013). AGT and ACE have a direct role in the secretion of aldosterone: The first is the precursor of angiotensin I, which is converted by ACE into angiotensin II, which directly induces the secretion of aldosterone. All the mutations are localized close to or within functional domains of these proteins, so this might imply a possible difference in the physiological regulation of their function, as a result of selection pressure to maintain the water and salt balance in a hyperosmotic environment. This also suggests that cetaceans might have evolved different mechanisms underlying aldosterone secretion and osmoregulation, and could partly explain the unusual responsiveness of aldosterone to stress, as observed in several studies cited earlier.

The Endocrine Pancreas

A macroscopic description of the organ (Fig. 8.52), and the microscopic anatomy of the exocrine pancreas are reported in Chapter 8.

The main endocrine products of the pancreas are insulin, glucagon, somatostatin, and pancreatic polypeptide. These hormones are produced by cells of the endocrine pancreas grouped into the cords of polygonal cells called pancreatic islets (of Langerhans). Within the islets, α cells secrete glucagon, β cells produce insulin,[w] δ cells secrete somatostatin, and γ cells secrete the pancreatic polypeptide. Glucagon increases the levels of blood sugars, and insulin reduces them, while somatostatin (among other neuroendocrine functions) interacts with the production of insulin and glucagon.

Pancreatic islets of endocrine significance are dispersed among the parenchyma of the organ (Fig. 7.16). Their disposition changes with age in the bottlenose dolphin: Young animals show small islets and endocrine cells scattered among the pancreatic lobules (Fig. 7.17), whereas older animals (Fig. 7.18) have larger islets, a condition that would be associated to pathology in humans. (Colegrove and Venn-Watson, 2015). In the bottlenose dolphin, the architecture of the islet appears to be most similar to the pig, where α and β cells are localized to the center or periphery of the islet, respectively, or are well dispersed throughout the islet (Colegrove and Venn-Watson, 2015). However, the same study found that large islets (greater than 10,000 μm^2) are common in dolphins, unlike in the pig but similarly to what is found in humans. The conclusion is that the architecture of the islets in dolphins may be similar to that of terrestrial Cetartiodactyla, but with the potential of larger islet sizes found in humans (Colegrove and Venn-Watson, 2015).

Pancreatic Hormones

The role of the endocrine pancreas in dolphins has raised some interesting questions, mainly because the levels of blood sugars in these animals mimic human prediabetic conditions, with hyperinsulinemia, hyperlipidemia,[x] and elevated concentration

---

w. Before the industrial synthesis of insulin became available, large whales were hunted also for extraction of insulin for therapeutic use in humans.

x. Bottlenose dolphins with high insulin levels also show high cholesterol levels, specifically high-density lipoprotein (HDL) and very low-density lipoprotein (VLDL) cholesterol (Venn-Watson et al., 2013).

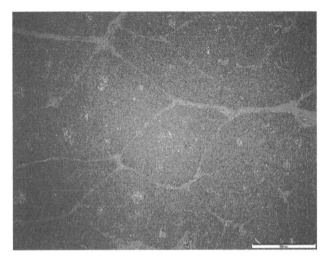

FIGURE 7.16 Microphotograph of the pancreas of *S. coeruleoalba* with the pancreatic islets within the parenchyma (scale bar = 500 μm). HE stain.

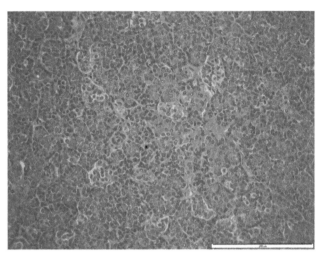

FIGURE 7.17 Microphotograph of the pancreas of a newborn *T. truncatus* (scale bar = 200 μm). HE stain.

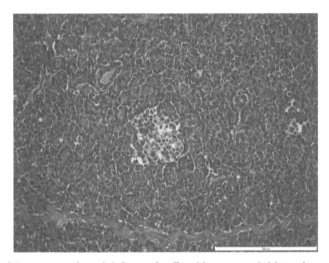

FIGURE 7.18 Microphotograph of the pancreas of an adult *S. coeruleoalba* with a pancreatic islet at the center (scale bar = 200 μm). HE stain.

of blood sugar (Venn-Watson and Ridgway, 2007; Venn-Watson et al., 2011, 2013). Specifically, normal healthy bottlenose dolphins have a sustained postprandial hyperglycemia, producing a transient diabetes mellitus-like state during 6–72 h of fasting (Venn-Watson and Ridgway, 2007). When dolphins were fasted for 72 h, plasma glucose levels remained at prefast levels, and even after dolphins ingested glucose, plasma levels remained high for many hours (Ridgway, 2013). During prolonged fasting states, dolphins do not develop ketosis.[y] Interestingly, plasma glucose must exceed 300 mg/dL (about twice as high as the human threshold) before glucose appears in urine (Ridgway, 2013). The high levels of glucose in a high-protein, low-carbohydrate diet may be related to the demand for glucose from the central nervous system.

The Diffuse Neuroendocrine System

The components of the diffuse neuroendocrine system of dolphins that are active on the intestine are described in Chapter 8. A single article describes the presence and concentration of natriuretic peptides in *L. obliquidens*, *T. truncatus*, and other cetacean species (Naka et al., 2007). The structure of atrial natriuretic peptide (ANP) precursor was conserved, but β-type natriuretic peptides (BNP) precursor was more variable and consisted of 26 instead of 32 amino acids (Naka et al., 2007).

---

y. Meaning that the metabolism does not switch to the use of ketone bodies acetoacetate and β-hydroxybutyrate as would happen in human fasting or following low-carbohydrate diets.

THE PERIPHERAL NERVOUS SYSTEM

Most of the articles about the PNS of dolphins are focused on cranial nerves (see Chapter 6). The origin and course of spinal nerves follow the general outline of terrestrial mammals, with a ventral and a dorsal division. Spinal nerves (Fig. 7.19) are easily recognized close to their origin from the spinal cord when they exit the vertebral column and travel through the thick locomotor muscles of the back, or turn ventrally into the muscles of the thorax and abdomen.

Cervical and Thoracic Nerves

Brachial Plexus

The ventral divisions of the first pair of cervical nerves innervates the deep muscles of the rostral spine and may anastomose with branches of the last cranial nerves. At least in *Kogia breviceps* (von Schulte and de Forest Smith, 1918), it is possible to distinguish a lesser and a major brachial trunk, constituted by C_{3-6}, and $C_{7-8} + T_{1-2}$, respectively. Nerves of the two trunks travel toward the scapula, leaving branches for the *mm. scaleni*, and then merge to innervate the shoulder region and constitute the brachial plexus as it is known in other mammals. Older references describe a cervical (C_{1-3}), and brachial or axillary plexus ($C_{4-8} + T_1$) in the striped dolphin (Cunningham, 1877) (Fig. 7.20). The same author states that in this latter species the phrenic nerve originates from C_3, and describes three nerves directed to the muscles of the forelimb (Fig. 7.21).

The contribute of some rostral cervical nerves to the plexus would represent a topographical difference with terrestrial mammals, perhaps due to the shortness of the cetacean cervical spine (and to the virtual absence of an external neck). Nerves bound to the muscles of the forearm are extremely reduced due to the diminished importance of the whole muscular system distal to the arm (Fig. 7.22).

FIGURE 7.19 Lumbar spinal nerves of *S. coeruleoalba* (top) and schematic representation of the spinal nerves (bottom). *(From Cunningham, 1877. The spinal nervous system of the porpoise and dolphin. J. Anat. Phys. 11, 209–228.)*

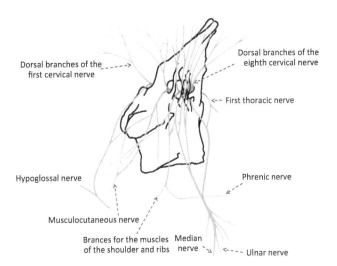

FIGURE 7.20 **Cervical vertebrae and brachial plexus.** *(Redrawn by Massimo Demma, from Cunningham, 1877. The spinal nervous system of the porpoise and dolphin. J. Anat. Phys. 11, 209–228.)*

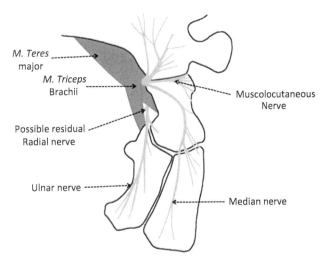

FIGURE 7.21 **Distribution of the nerves to the forelimb.** *(Redrawn by Massimo Demma, from Cunningham, 1877. The spinal nervous system of the porpoise and dolphin. J. Anat. Phys. 11, 209–228.)*

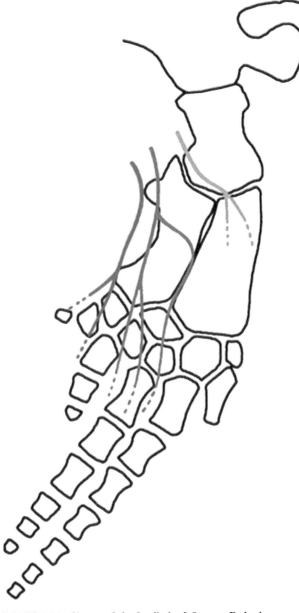

FIGURE 7.22 **Nerves of the forelimb of *O. orca*. Red, ulnar nerve; blue, median nerve; green, musculocutaneous nerve.** *(Redrawn by Massimo Demma, from Cooper et al., 2007. Neuromuscular anatomy and evolution of the cetacean forelimb. Anat. Rec. 290, 1121–1137.)*

Ventral divisions of the thoracic and lumbar nerves reach the muscular walls of the thorax and abdominal sector. The innervation of the *m. iliocostalis* has been described in dolphin embryos and found to have certain similitude with that of the goat, but less with the human (Nomizo et al., 2005).

The absence of the pelvic limb restricts the scope of the lumbo-sacral plexus. In *Kogia breviceps* (von Schulte and de Forest Smith, 1918), harbor porpoise, and striped dolphin (Cunningham, 1877), the remains of the plexus are characterized by the formation of two longitudinal trunks, placed respectively, above and below the transverse vertebral processes. The ventral division enters the hypaxial muscular complex, that is, the motor equivalent of the hip and leg muscles of terrestrial mammals. According to Cunningham (1877), in the harbor porpoise and in the striped dolphin it is also possible to recognize an internal pudic (= pudendal) nerve, derived from L_7, and a genital nerve, but the obturator, femoral, and sciatic nerves are absent.

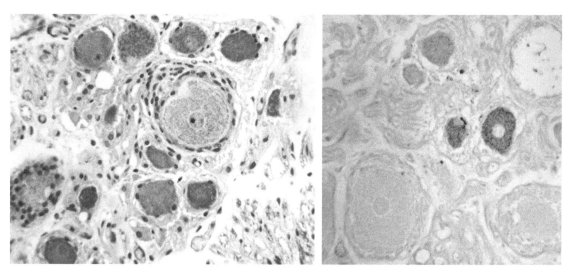

FIGURE 7.23   **Pseudo-unipolar neurons in the spinal ganglia of *S. coeruleoalba*.** (left: Klüver-Barrera; right: Substance P immunocytochemistry).

Spinal Ganglia

A recent series of paper have analyzed the chemical neuroanatomy of the spinal ganglia of dolphins (Bombardi et al., 2010, 2011) (Fig. 7.23). Substance P, cholecystokin, nitric oxide and serotonin were described. The nature of the substances present in the dorsal root ganglia may suggest pain modulation in the dorsal horn of the spinal cord, and vasomotor control of the adjacent spinal *retia mirabilia*.

The Visceral Nervous System

Parasympathetic cranial nerves are described in Chapter 6. The paravertebral chain of viscera ganglia (Fig. 7.24) follows the spinal cord outside the dura mater and close to spinal *retia mirabilia*. The ganglia are easily identified.

Visceral nerves are probably travelling with ventral divisions of C_{3-8} directed to the brachial plexus (Cunningham, 1877; von Schulte and de Forest Smith, 1918). The visceral innervation of the heart, and the relative ganglia, were described by Ogawa (1952) in the striped dolphin *Prodelphinus coeruleoalbus* (= *S. coeruleoalba*) (Figs. 7.25 and 7.26) and other cetacean species. An impressively detailed description of the visceral nervous system of *T. truncatus*, *D. delphis*,

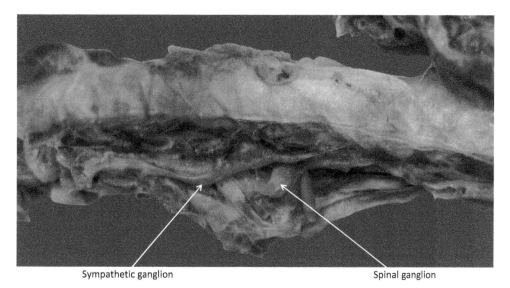

Sympathetic ganglion   Spinal ganglion

FIGURE 7.24   **Sympathetic and spinal ganglia of *T. truncatus*.**

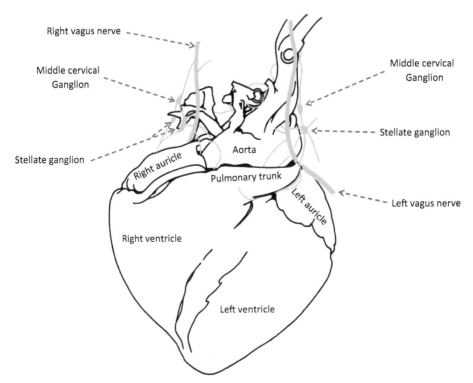

FIGURE 7.25 **Visceral innervation of the heart of *S. coeruleoalba* (ventral view).** *(Redrawn by Massimo Demma from Ogawa, 1952. On the cardiac nerves of some cetacea, with special reference to those of Berardius bairdii Stejneger. Sci. Rep. Whale Res. Inst. Tokyo 4, 1–22.)*

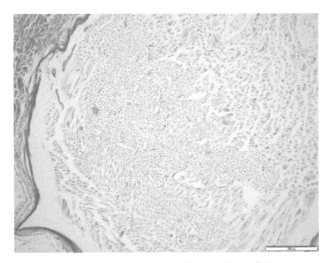

FIGURE 7.26 **Branch of the vagal nerve in the pericardium of *S. coeruleoalba* (scale bar = 200 μm).**

and other cetacean species is reported in Agarkov and Veselovsky (1987). According to these authors, in the bottlenose dolphin the sympathetic trunk[z] is placed ventrally to the intercostal muscles and ribs, and in the thorax below the subvertebral *retia mirabilia*. This latter position is valid also in the common dolphin. In both bottlenose and common dolphins, in the abdominal cavity, visceral nerves ramify into the greater and lesser splanchnic nerve (Agarkov and Veselovsky, 1987).

---

z. Interestingly, Agarkov and Veselovsky (1987) clearly state that were not able to dissect and identify cervical ganglia in the majority of bottlenose and common dolphins that they examined. In some individuals, the ganglia were found only on one side. In no individuals cervical ganglia were present in both sides. However, the beautifully illustrated article may suffer from concepts lost in translation.

REFERENCES

Abdalla, M.A., Ali, A.M., 1989. Morphometric and histological studies on the adrenal glands of the camel, *Camelus dromedarius*. Acta Morphol. Neerl.-Scand. 26, 269–281.

Agarkov, G.B., Veselovsky, M.V., 1987. Investigations on the peripheral nervous system in cetaceans. Invest. Cetacea 20, 192–227.

Amaral, R.S., 2010. Use of alternative matrices to monitor steroid hormones in aquatic mammals: a review. Aquatic Mammals 36, 162–171.

Anderson, G.W., 2001. Thyroid hormones and the brain. Front. Neuroendocrinol. 22, 1–17.

Anderson, J., 1878. Anatomical and Zoological Researches: Comprising an Account of the Zoological Results of the Two Expeditions to Western Yunnan in 1868 and 1875, and a Monograph of the Two Cetacean Genera, *Platanista* and *Orcella*. Bernard Quaritch, London.

Arvy, L., 1971. Endocrine glands and hormonal secretion in cetaceans. Invest. Cetacea 3, 229–300.

Arvy, L., Pilleri, G., 1974. On the hypophysis of *Platanista indi* Blyth, 1859. Invest. Cetacea 5, 83–92.

Ayres, K.L., Booth, R.K., Hempelmann, J.A., Koski, K.L., Emmons, C.K., Baird, R.W., Balcomb-Bartok, K., Bradley Hanson, M., Ford, M.J., Wasser, S.K., 2012. Distinguishing the impacts of inadequate prey and vessel traffic on an endangered killer whale (*Orcinus orca*) population. PLoS One 7, e36842.

Ballarin, C., Corain, L., Peruffo, A., Cozzi, B., 2011. Correlation between urinary vasopressin and water content of food in the bottlenose dolphin (*Tursiops truncatus*). Open Neuroendocrinol. J. 4, 9–14.

Benoit, S., Schwartz, M., Baskin, D., Woods, S.C., Seeley, R.J., 2000. CNS melanocortin system involvement in the regulation of food intake. Horm. Behav. 37, 299–305.

Bishnupuri, K.S., Haldar, C., 2000. Impact of photoperiodic exposures during late gestation and lactation periods on the pineal and reproductive physiology of the Indian palm squirrel, *Funambulus pennanti*. J. Reprod. Fertil. 118, 295–301.

Blasio, A.L., Valdez Perez, R., Romano Pardo, M., Galindo Maldonado, F., 2012. Maintenance behaviour and cortisol levels in bottlenose dolphins (*Tursiops truncatus*) in closed and open facilities. Veterinaria México 43, 103–112.

Bombardi, C., Grandis, A., Nenzi, A., Giurisato, M., Cozzi, B., 2010. Immunohistochemical localization of substance P and cholecystokinin in the dorsal root ganglia and spinal cord of the bottlenose dolphin (*Tursiops truncatus*). Anat. Rec. 293, 477–484.

Bombardi, C., Cozzi, B., Nenzi, A., Mazzariol, S., Grandis, A., 2011. Distribution of nitrergic neurons in the dorsal root ganglia of the bottlenose dolphin (*Tursiops truncatus*). Anat. Rec. 294, 1066–1073.

Bonjour, J.-P., Malvin, R.L., 1970. Plasma concentrations of ADH in conscious and anesthetized dogs. Amer. J. Physiol. 2184, 1128–1132.

Bossart, B.D., Dierauf, L.A., 1990. Marine mammal clinical laboratory medicine. In: Dierauf, L.A. (Ed.), Handbook of marine mammal medicine: health, disease and rehabilitation. CRC Press, Boca Raton, pp. 1–53.

Breathnach, A.S., 1955. The surface features of the brain of the humpback whale (*Megaptera novaeangliae*). J. Anat. 89, 343–354.

Cave, A.J.E., 1993. On the morphological constitution of the cetacean pituitary region. Invest. Cetacea 24, 253–258.

Clark, L.S., Pfeiffer, D.C., Cowan, D.F., 2005. Morphology and histology of the Atlantic bottlenose dolphin (*Tursiops truncatus*) adrenal gland with emphasis on the medulla. Anat. Histol. Embryol. 34, 132–140.

Clark, L.S., Cowan, D.F., Pfeiffer, D.C., 2006. Morphological changes in the Atlantic bottlenose dolphin (*Tursiops truncatus*) adrenal gland associated with chronic stress. J. Comp. Pathol. 135, 208–216.

Clark, L.S., Cowan, D.F., Pfeiffer, D.C., 2008. A morphological and histological examination of the pan-tropical spotted dolphin (*Stenella attenuata*) and the spinner dolphin (*Stenella longirostris*) adrenal gland. Anat. Histol. Embryol. 37, 153–159.

Colegrove, K.M., Venn-Watson, S., 2015. Histomorphology of the bottlenose dolphin (*Tursiops truncatus*) pancreas and association of increasing islet β-cell size with chronic hypercholesterolemia. Gen. Comp. Endocrinol. 214, 17–23.

Cooper, L.N., Dawson, S.D., Reidenberg, J.S., Berta, A., 2007. Neuromuscular anatomy and evolution of the cetacean forelimb. Anat. Rec. 290, 1121–1137.

Copland M.D., Needham D.J., 1992. Hematological and Biochemical changes associated with transport of dolphins (*Tursiops truncatus*). AAAM Conferences, May18–22. Hong Kong, p. 23.

Cowan, D.F., 1966. Observations on the pilot whale *Globicephala melaena*: organ weight and growth. Anat. Rec. 155, 623–628.

Cowan, D.F., Tajima, Y., 2006. The thyroid gland in bottlenose dolphins *Tursiops truncatus* from the Texas coast of the Gulf of Mexico: normal structure and pathological changes. J. Comp. Pathol. 135, 217–225.

Cowan, D.F., Haubold, E.M., Tajima, Y., 2008. Histological, immunohistochemical and pathological features of the pituitary gland of odontocete cetaceans from the western Gulf of Mexico. J. Comp. Pathol. 139, 67–80.

Crile, G.C., Quiring, D.P., 1940. A comparison of the energy releasing organs of the white whale and the thoroughbred horse. Growth 4, 291–298.

Cuello, A.C., Tramezzani, J.H., 1969. The epiphysis cerebri of the Weddell seal: its remarkable size and glandular pattern. Gen. Comp. Endocrinol. 12, 154–164.

Cunningham, D.J., 1877. The spinal nervous system of the porpoise and dolphin. J. Anat. Phys. 11, 209–228.

de Escobar, G.M., Obregón, M.J., del Rey, F.E., 2004. Maternal thyroid hormones early in pregnancy and fetal brain development. Best Pract. Res. Clin. Endocrinol. Metab. 18, 225–248.

Dellman, H.D., 1998. Endocrine system. In: Dellman, H.D., Eurell, J. (Eds.), Textbook of Veterinary Histology, fifth ed. Lippincott, Williams and Wilkins, Philadelphia, pp. 287–302.

Drager, G.A., 1944. A comparative study of the innervation of the *pars distalis* of the *hypophysis cerebri*. Anat. Rec. 8, 428.

Drager, G.A., 1953. The innervation of porpoise pituitary gland with special emphasis on the adenohypophysis. J. Comp. Pathol. 99, 75–89.

Duffield, D.W., Haldiman, J.T., Henk, W.G., 1992. Surface morphology of the forebrain of the bowhead whale, *Balaena mysticetus*. Mar. Mammal Sci. 8, 354–378.

Fair, P.A., Montie, E., Balthis, L., Reif, J.S., Bossart GD, 2011. Influences of biological variables and geographic location on circulating concentrations of thyroid hormones in wild bottlenose dolphins (*Tursiops truncatus*). Gen. Comp. Endocrinol. 174, 184–194.

Fair, P.A., Schaefer, A.M., Romano, T.A., Bossart, G.D., Lamb, S.V., Reif, J.S., 2014. Stress response of wild bottlenose dolphins (*Tursiops truncatus*) during capture–release health assessment studies. Gen. Comp. Endocrinol. 206, 203–212.

Felice, F., Gombos, F., Esposito, V., Nunziata, M., Scully, C., 2006. Burning mouth syndrome (BMS): evaluation of thyroid and taste. Med. Oral Pathol. Oral Cir. Bucal 11, E22–E25.

Fernand, V.S.V., 1951. The histology of the pituitary and of adrenal glands in the dugong (*Dugong dugong*). Ceylon J. Med. Sci. 8, 57–62.

Funasaka, N., Yoshioka, M., Suzuki, M., Ueda, K., Miyahara, Uchida, S., 2011. Seasonal difference of diurnal variations in serum melatonin, cortisol, testosterone, and rectal temperature in Indo–Pacific bottlenose dolphins (*Tursiops aduncus*). Aquat. Mamm. 37, 433–442.

Fuse, G., 1936. Sober die Epiphyse bei einigen wasserbewohnenden Saugetieren. Arb. Anat. Inst. Sendai. 18, 241–341.

Gales, N.J., 1992. Mass stranding of striped dolphin, *Stenella coeruleoalba*, at Augusta, Western Australia: notes on clinical pathology and general observations. J. Wild Dis. 28, 651–655.

Geiling, E.M.K., 1935. The hypophysis cerebri of the finback (*Balaenoptera physalus*) and the sperm whale (*Physeter megalocephalus*). Bull. Johns Hopkins Hosp. 57, 123–142.

Geiling, E.M.K., Vos, B.J., Oldham, F.K., 1940. The pharmacology and anatomy of the hypophysis of the porpoise. Endocrinology 27, 309–316.

Geraci, J.R., St. Aubin, D.J., 1987. Cetacean mass strandings: a study into stress and shock. Miami, FL. Proceedings of the 7th Biennial Conference on the Biology of Marine Mammals. p. 25.

Gersch, I., 1938. Note on the pineal gland of the humpback whale. J. Mammal. 19, 477–480.

Gihr, M., Pilleri, G., 1969. On the anatomy and biometry of *Stenella styx* Gray and *Delphinus delphis* L. (Cetacea, Delphinidae) of the Western Mediterranean. Invest. Cetacea 1, 15–65.

Greenwood, A.G., Barlow, C.E., 1979. Thyroid function in dolphins: radioimmunoassay measurement of thyroid hormones. Brit. Vet. J. 135, 96.

Gruenberger, H.B., 1970. On the cerebral anatomy of the Amazon dolphin *Inia geoffrensis*. Invest. Cetacea 2, 129–144.

Guldberg, G., 1885. Über die Bildung der Hypophyse bei Säugetieren. Christiana Videnskabs Salskaps Forhande 9, 1–56.

Hadley, M.E., Levine, J.E., 2006. Endocrinology, sixth ed. Prentice Hall, Inc., Upper Saddle River, NJ.

Haldar, C., Yadav, R., Alipreeta, A., 2006. Annual reproductive synchronization in ovary and pineal gland function of female short-nosed fruit bat, *Cynopterus sphinx*. Comp. Biochem. Physiol. A Mol. Integr. Physiol. 144, 395–400.

Hanström, B., 1944. Zur Histologie und vergleichenden Anatomie der Hypophyse der Cetaceen. Acta Zool. 25, 1–25.

Hao, Y.J., Chen, D.Q., Zhao, Q.Z., Wang, D., 2007. Serum concentrations of gonadotropins and steroid hormones of *Neophocaena phocaenoides asiaeorientalis* in middle and lower regions of the Yangtze River. Theriogenology 67, 673–680.

Harris, G.W., 1947. The hypophysial portal vessels of the porpoise (*Phocaena phocaena*). Nature 159, 874–875.

Harris, G.W., 1950. Hypothalamo-hypophysial connexions in the cetacea. J. Physiol. 111, 361–367.

Harrison, R.J., 1969. Endocrine organs: hypophysis, thyroid and adrenal. In: Andersen, H.T. (Ed.), The Biology of Marine Mammals. Academic Press, New York, pp. 350–359.

Harrison, R.J., Young, B.A., 1969a. Stellate cells in the delphinid adenohypophysis. J. Endocrinol. 43, 323–324.

Harrison, R.J., Young, B.A., 1969b. Ultrastructural characteristics of the dolphin thyroid. J. Anat. 104, 173–174.

Harrison, R.J., Young, B.A., 1969c. Ultrastructure of light cells in the dolphin thyroid. Z Zellforsch 96, 222–228.

Harrison, R.J., Young, B.A., 1970a. Ultrastructure of the dolphin adenohypophysis. Z Zellforsch 96, 475–482.

Harrison, R.J., Young, B.A., 1970b. The thyroid gland of the common (Pacific) dolphin, *Delphinus delphis bairdi*. J. Anat. 106, 243–254.

Hart, L.B., Wells, R.S., Kellar, N., Balmer, B.C., Hohn, A.A., Lamb, S.V., Rowles, T., Zolman, E.S., Schwacke, L.H., 2015. Adrenal hormones in common bottlenose dolphins (*Tursiops truncatus*): influential factors and reference intervals. PLoS One 10, e0127432.

Hattangady, N.G., Olala, L.O., Bollag, W.B., Rainey, W.E., 2012. Acute and chronic regulation of aldosterone production. Mol. Cell Endocrinol. 350, 151–162.

Haug, H., 1972. Die Epiphyse und die circumventrikulären strukturen des epithalamus im gehirn des elefanten (*Loxodonta africana*). Z Zellforsch 129, 533–547.

Hayakawa, D., Chen, H., Emura, S., Tamada, A., Yamahira, T., Terasawa, K., Isono, H., Shoumura, S., 1998. The parathyroid glands of two species of dolphin—Risso's dolphin, *Grampus griseus*, and bottlenose dolphin, *Tursiops truncatus*. Gen. Comp. Endocrinol. 110, 58–66.

Hennings, H., 1950. The whale hypophysis with special reference to its ACTH content. Acta Endocrinol. 5, 376–386.

Herlant, M., 1958. L'hypophyse et le système hypothalamo–hypophysaire du pangolin (*Manis tricuspis* Raf. et *Manis tetradactyla* L.). Archs. Anat. Microsc. Morph. Exp. 47, 1–23.

Holder, J.R., Haskell–Luevano, C., 2004. Melanocortin ligands: 30 years of structure–activity relationship SAR studies. Med. Res. Rev. 24, 325–356.

Holzmann, T., 1991. Morphologie und mikroskopische Anatomie des Gehirns beim fetalen Narwal, *Monodon monoceros*. PhD Thesis. University of Frankfurt am Main, Frankfurt am Main.

Houser, D.S., Yeates, L.C., Crocker, D.E., 2011. Cold stress induces an adrenocortical response in bottlenose dolphins (*Tursiops truncatus*). J. Zoo. Wild Med. 42, 565–571.

Hunter, J., Banks, J., 1787. Observations on the structure and oeconomy of whales. By John Hunter, Esq. FRS; Communicated by Sir Joseph Banks, Bart. PRS. Phil. Trans. R. Soc Lond. 77, 371–450.

Isono, H., Shoumura, S., Emura, S., 1993. The parathyroid gland under normal and experimental conditions. Kaibogaku Zasshi Acta Anat. Nippon 68, 5–29.

Iturriza, F.C., Estivariz, F.E., Levitin, H.P., 1980. Coexistence of α–melanocyte–stimulating hormone and adrenocorticotrophin in all cells containing either of the two hormones in the duck pituitary. Gen. Comp. Endocrinol. 42, 110–116.

Jacobsen, A.P., 1941. Endocrinological studies in the blue whale. Hvalrådets Skr. 24, 1–84.

Judd, H.L., Ridgway, S.H., 1977. Twenty-four hour patterns of circulating androgens and cortisol in male dolphins. In: Ridgway, SH., Benirscheke, K. (Eds.) Breeding Dolphin: Present Status, Suggestion for the Future, Natl. Tech. Info. Serv., pp. 269–277.

Kamiya, T., Yamasaki, F., Komatsu, S., 1978. A note on the parathyroid glands of Ganges susu. Sci. Rep. Whales Res. Inst. 30, 281–284.

Kirkegaard, M., Sonne, C., Dietz, R., Letcher, R.J., Jensen, A.L., Hansen, S., Jenssenf, B.M., Grandjean, P., 2011. Alterations in thyroid hormone status in Greenland sledge dogs exposed to whale blubber contaminated with organohalogen compounds. Ecotoxicol. Environ. Saf. 74, 157–163.

Köhrle, J., 2000. The deiodinase family: selenoenzymes regulating thyroid hormone availability and action. Cell Mol. Life Sci. 57, 1853–1863.

Kolmer, W., 1918. Zur vergleichenden histologie, zytologie und entwicklungsgeschichte der säugernebenniere. Arch. Mikrosk Anat. 91, 1–139.

Konturek, S.J., Konturek, P.C., Brzozowska, I., Pawlik, M., Sliwowski, Z., Cześnikiewicz-Guzik, M., Kwiecień, S., Brzozowski, T., Bubenik, G.A., Pawlik, W.W., 2007. Localization and biological activities of melatonin in intact and diseased gastrointestinal tract (GIT). J. Physiol. Pharmacol. 58, 381–405.

Kot, B.C.W., Ying, M.T.C., Brook, F.M., Fernando, N., Kinoshita, R., Martelli, P., 2008. Difference in age, gender, body weight, body length and sexual maturity in association with variations in thyroid size and morphology of Indo-Pacific bottlenose dolphin, *Tursiops aduncus*. Proc. Int. Ass. Aquatic Anim. Medicine. 39, 62–64.

Kot, B.C., Ying, M.T., Brook, F.M., Kinoshita, R.E., Cheng, S.C., 2012a. Ultrasonographic assessment of the thyroid gland and adjacent anatomic structures in Indo-Pacific bottlenose dolphins (*Tursiops aduncus*). Am. J. Vet. Res. 73, 1696–1706.

Kot, B.C., Ying, M.T., Brook, F.M., Kinoshita, R.E., Kane, D., Chan, W.K., 2012b. Sonographic evaluation of thyroid morphology during different reproductive events in female Indo-Pacific bottlenose dolphins, *Tursiops aduncus*. Mar. Mamm. Sci. 28, 733–750.

Kot, B.C.W., Lau, T.Y.H., Cheng, S.C.H., 2013. Stereology of the thyroid gland in indo-pacific bottlenose dolphin (*Tursiops aduncus*) in comparison with human (*Homo sapiens*): quantitative and functional implications. PLoS One 8, e62060.

Lavigne, D.M., Innes, S., Worthy, G.A.J., Kovacs, K.M., Schmitz, O.J., Hickie, J.P., 1986. Metabolic rates of seals and whales. Can. J. Zool. 64, 279–284.

Lewis, L., Lamb, S.V., Schaefer, A.M., Reif, J.S., Bossart, G.D., Fair, P.A., 2013. Influence of collection and storage conditions on adrenocorticotropic hormone (ACTH) measurements in bottlenose dolphins (*Tursiops truncatus*). Aquat. Mamm. 39, 324–329.

Lyamin, O.I., Manger, P.R., Ridgway, S.H., Mukhametov, L.M., Siegel, J.M., 2008. Cetacean sleep: an unusual form of mammalian sleep. Neurosci. Biobehav. Rev. 32, 1451–1484.

Malvin, R.L., Bonjour, J.P., Ridgway, S.H., 1971. Antidiuretic hormone levels in some cetaceans. Proc. Soc. Exp. Biol. Med. 136, 1203–1205.

Maniou, Z., Caryl Wallis, O., Wallis, M., 2002. Cloning and characterisation of the GH gene from the common dolphin (*Delphinus delphis*). Gen. Comp. Endocrinol. 127, 300–306.

McFarland, W.L., Morgane, P.J., Jacobs, M.S., 1969. Ventricular system of the brain of the dolphin, *Tursiops truncatus*, with comparative anatomical observations and relations to brain specializations. J. Comp. Neurol. 135, 275–368.

Medway, W., Geraci, J.R., Klein, L.V., 1970. Hematologic response to administration of corticosteroid in the bottle-nosed dolphin (*Tursiops truncatus*). Am. J. Physiol. 157, 563–565.

Montie, E.W., Pussini, N., Schneider, G.E., Battey, T.W., Dennison, S., Barakos, J., Gulland, F., 2009. Neuroanatomy and volumes of brain structures of a live California sea lion (*Zalophus californianus*) from magnetic resonance images. Anat. Rec. 292, 1523–1547.

Morgane, P.J., Jacobs, M.S., 1972. Comparative anatomy of the cetacean nervous system. Harrison, R.J. (Ed.), Functional Anatomy of Marine Mammals, vol. 1, Academic Press, London, New York, pp. 117–244.

Myers, M.J., Rea, L.D., Atkinson, S., 2006. The effects of age, season and geographic region on thyroid hormones in Steller sea lions (*Eumetopias jubatus*). Comp. Biochem. Physiol. Part A 145, 90–98.

Myrick, A.C., Perkins, P.C., 1995. Adrenocortical color darkness and correlates as indicators of continuous acute premortem stress in chased and purse-seined captured male dolphins. Pathophysiology 2, 191–204.

Naka, T., Katsumata, E., Sasaki, K., Minamino, N., Yoshioka, M., Takei, Y., 2007. Natriuretic peptides in cetacean: identification, molecular characterization and changes in plasma concentration after landing. Zool. Sci. 24, 577–587.

Neuville, H., 1928. Recherches sur le genre "*Steno*" et remarques sur quelques autres cétacés. Arch. Mus. Natl. Hist. Nat. 3, 69–240.

Noda, K., Akiyoshi, H., Aoki, M., Shimada, T., Ohashi, F., 2007. Relationship between transportation stress and polymorphonuclear cell functions of bottlenose dolphins, *Tursiops truncatus*. J. Vet. Med. Sci. 69, 379–383.

Nomizo, A., Kudoh, H., Sakai, T., 2005. Iliocostalis muscles in three mammals (dolphin, goat and human): their identification, structure and innervation. Anat. Sci. Int. 80, 212–222.

Oboussier, H., 1948. Über die grössenbeziehungen der hypophyse und ihrer teile bei säugetieren und vögeln. Arch. Entw. Mech. Org. 143, 181–274.

Oelschläger, H.H.A., Haas-Rioth, M., Fung, C., Ridgway, S.H., Knauth, M., 2008. Morphology and evolutionary biology of the dolphin (*Delphinus* sp.) brain—MR imaging and conventional histology. Brain. Behav. Evol. 71, 68–86.

Oelschläger, H.H.A., Ridgway, S.H., Knauth, M., 2010. Cetacean brain evolution: dwarf sperm whale (*Kogia sima*) and common dolphin (*Delphinus delphis*)—an investigation with high-resolution 3D MRI. Brain Behav. Evol. 75, 33–62.

Oftedal, O.T., 1997. Lactation in whales and dolphins: evidence of divergence between baleen-and toothed-species. J. Mammary Gland Biol. 2, 205–230.

Ogawa, T., 1952. On the cardiac nerves of some cetacea, with special reference to those of *Berardius bairdii* Stejneger. Sci. Rep. Whale Res. Inst. Tokyo 4, 1–22.

Ogawa, S., Couch, E.F., Kubo, M., Sakai, T., Inoue, K., 1996. Histochemical study of follicles in the senescent porcine pituitary gland. Arch. Histol. Cytol. 59, 467–478.

Oldham, F.K., 1941. The development of the hypophysis of the armadillo. Am. J. Anat. 68, 293–315.

Oldham, F.K., Last, J.H., Geiling, E.M.K., 1940. Distribution of melanophore-dispersing hormone in anterior lobe of cetaceans and armadillo. Proc. Soc. Exp. Biol. Med. 43, 407–410.

Orlov, M.M., Mukhlia, A.M., Kulikov, N.A., 1988. Hormonal indices of the normal dolphin *Turpsiops truncatus* and in the dynamics of experimental stress. Zh. Evol. Biokhim. Fiziol. 24, 557–563.

Orlov, M.M., Mukhlya, A.M., Kuzmin, A., 1991. Hormonal and electrolyte changes in cetacean blood after capture and during experimental stress. J. Evol. Biochem. Physiol. 27, 151–156.

Ortiz, R.M., Worthy, G.A., 2000. Effects of capture on adrenal steroid and vasopressin concentrations in free-ranging bottlenose dolphins (*Tursiops truncatus*). Comp. Biochem. Physiol. A. Mol. Integr. Physiol. 125, 317–324.

Ortiz, R.M., MacKenzie, D.S., Worthy, G.A., 2000. Thyroid hormone concentrations in captive and free-ranging West Indian manatees (*Trichechus manatus*). J. Expl. Biol. 203, 3631–3637.

Ortiz, R.M., Long, B., Casper, D., Ortiz, C.L., Williams, T.M., 2010. Biochemical and hormonal changes during acute fasting and re-feeding in bottlenose dolphins *Tursiops truncatus*. Mar. Mamm. Sci. 26, 409–419.

Panin, M., Gabai, G., Ballarin, C., Peruffo, A., Cozzi, B., 2012. Evidence of melatonin secretion in cetaceans: plasma concentration and extrapineal HIOMT-like presence in the bottlenose dolphin *Tursiops truncatus*. Gen. Comp. Endocrinol. 177, 238–245.

Panin, M., Giurisato, M., Peruffo, A., Ballarin, C., Cozzi, B., 2013. Immunofluorescence evidence of melanotrophs in the pituitary of four odontocete species. An immunohistochemical study and a critical review of the literature. Ann. Anat. 195, 512–521.

Payne, A.P., 1994. The harderian gland: a tercentennial review. J. Anat. 185, 1–49.

Pedernera-Romano, C., Valdez, R.A., Singh, S., Chiappa, X., Romano, M.C., Galindo, F., 2006. Salivary cortisol in captive dolphins (*Tursiops truncatus*): a non-invasive. Anim. Welf. 15, 359–362.

Phillips, J.A., Harlow, H.J., McArthur, N.H., Ralph, C.L., 1986. Epithalamus of the nine-banded armadillo, *Dasypus novemcinctus*. Comp. Biochem. Physiol. A. Comp. Physiol. 85, 477–481.

Pilleri, G., 1963. Zur vergleichenden morphologie und randordnung des gehirns von *Delphinapterus* (Beluga) *leucas* Pallas (*Cetacea Delphinapteridae*). Rev. Suisse Zool. 70, 569–586.

Pilleri, G., 1965. Morphologie des Gehirnes des Blauwals *Balaenoptera musculus* Linnaeus (Cetacea, Mysticeti, Balaenopteridae). Jahrb Naturhist. Mus. Bern 1963–1965, 187–203.

Pilleri, G., 1966a. Morphologie des gehirnes des seiwals *Megaptera novaeangliae* Borowski (Cetacea, Mysticeti, Balaenopteridae). J. Hirnforsch. 8, 447–491.

Pilleri, G., 1966b. Morphologie des gehirnes des seiwals *Balaenoptera borealis* Lesson (Cetacea, Mysticeti, Balaenopteridae). J. Hirnforsch. 8, 221–267.

Pilleri, G., Gihr, M., 1969. On the anatomy and behaviour of Risso's dolphin (*Grampus griseus* G. Cuvier). Invest. Cet. 1, 74–93.

Pryor, K., Norris, K.S., 1978. The tuna/porpoise problem: behavioral aspects. Oceanus 21, 31–37.

Race, G.J., Wu, H.M., 1961. Adrenal cortex functional zonation in the whale (*Physeter catodon*). Endocrinology 68, 156–158.

Ralph, C.L., Young, S., Gettinger, R., O'Shea, T.J., 1985. Does the manatee have a pineal body? Acta Zool. 66, 55–60.

Raubenheimer, P.J., Young, E.A., Andrew, R., Seckl, J.R., 2006. The role of corticosterone in human hypothalamic-pituitary-adrenal axis feedback. Clin. Endocrinol. 65, 22–26.

Reidarson, T.H., McBain, J.F., 1999. Hematologic, biochemical, and endocrine effects of dexamethasone on bottlenose dolphins (*Tursiops truncatus*). J. Zoo. Wildlife Med. 30, 310–312.

Ridgway, S.H., 1990. The central nervous system of the bottlenose dolphin. In: Leatherwood, S., Reeves, R. (Eds.), The Bottlenose Dolphin. Academic Press, New York, pp. 69–97.

Ridgway, S.H., 2013. A mini review of dolphin carbohydrate metabolism and suggestions for future research using exhaled air. Front. Neuroendocrinol. 4,152.

Ridgway, S.H., Patton, G.S., 1971. Dolphin thyroid: some anatomical and physiological findings. Zeit Vergleich Physiol. 71, 129–141.

Ridgway, S.H., Simpson, J.G., Patton, G.S., Gilmartin, W.G., 1970. Hematologic findings in certain small cetaceans. J. Am. Vet. Med. Ass. 157, 566–575.

Ridgway, S.H., Carder, D., Finneran, J., Keogh, M., Kamolnick, T., Todd, M., Goldblatt, A., 2006. Dolphin continuous auditory vigilance for five days. J. Expl. Biol. 209, 3621–3628.

Ridgway, S.H., Keogh, M., Carder, D., Finneran, J., Kamolnick, T., Todd, M., Goldblatt, A., 2009. Dolphins maintain cognitive performance during 72 to 120 hours of continuous auditory vigilance. J. Expl. Biol. 212, 1519–1527.

Romano, T.A., Keogh, M.J., Kelly, C., Feng, P., Berk, L., Schlundt, C.E., Carder, D.A., Finneran, J.J., 2004. Anthropogenic sound and marine mammal health: measures of the nervous and immune systems before and after intense sound exposure. Can. J. Fish. Aquatic Sci. 61, 1124–1134.

Rosa, C., O'Hara, T.M., Hoekstra, P.F., Refsal, K.R., Blake, J.E., 2007. Serum thyroid hormone concentrations and thyroid histomorphology as biomarkers in bowhead whales (*Balaena mysticetus*). Can. J. Zool. 85, 609–618.

Sapolsky, R., 1992. Cortisol concentrations and the social significance of rank instability among wild baboons. Psychoneuroendocrinol. 17, 701–706.

Schmitt, T.L., St. Aubin, D.J., Schaefer, A.M., Dunn, J.L., 2010. Baseline, diurnal variations, and stress-induced changes of stress hormones in three captive beluga whales, *Delphinapterus leucas*. Mar. Mamm. Sci. 26, 635–647.

Schneyer, A.L., Odell, D.K., 1984. Immunocytochemical identification of growth hormone and prolactin cells in the cetacean pituitary. In: Perrin, W.F., Brownell, R.L., DeMaster, D.P. (Eds.), Reproduction in Whales, Dolphins and Porpoises. International Whaling Commission, Cambridge, p. 495, Special Issue 6.

Schneyer, A., Castro, A., Odell, D., 1985. Radioimmunoassay of serum follicle-stimulating hormone and luteinizing hormone in the bottlenosed dolphin. Biol. Reprod. 33, 844–853.

Schnitzler, J.G., Siebert, U., Jepson, P.D., Beineke, A., Jauniaux, T., Bouquegneau, J.M., Das, K., 2008. Harbor porpoise thyroids: histologic investigations and potential interactions with environmental factors. J. Wild. Dis. 44, 888–901.

Seal, U.S., Doe, R.P., 1965. Vertebrate distribution of corticosteroid-binding globulin and some endocrine effects on concentration. Steroids 5, 827–841.

Segar, W.E., Moore, W.W., 1968. The regulation of antidiuretic hormone release in man: I. Effects of change in position and ambient temperature on blood ADH levels. J. Clin. Invest. 47, 2143–2151.

Shimokawa, T., Nakanishi, I., Hondo, E., Iwasaki, T., Kiso, Y., Makita, T., 2002. A morphological study of the thyroid gland in Risso's dolphin, *Grampus griseus*. J. Vet. Med. Sci. 64, 509.

Simpson, J.G., Gardner, M.B., 1972. Comparative microscopic anatomy of selected marine mammals. In: Ridgway, S.H. (Ed.), Mammals of the Sea. Biology and Medicine. Charles C. Thomas, Springfiels, IL, pp. 298–418.

Slijper, E., 1936. Die Cetacea vergleichend-anatomish und systematisch. Capita Zool. 7, 1–590.

Slijper, E.J., 1962. Whales. Hutchinson, London.

Snyder, G.K., 1983. Respiratory adaptations in diving mammals. Resp. Physiol. 54, 269–294.

Spraker, T.R., 1993. Stress and capture myopathy in artiodactylids. Fowler, M.E. (Ed.), Zoo and Wild Animal Medicine: Current Therapy, vol. 3, W.B. Saunders, Philadelphia, pp. 481–488.

St. Aubin, D.J., 1987. Stimulation of thyroid hormone secretion by thyrotropin in beluga whales, *Delphinapterus leucas*. Can. J. Vet. Res. 51, 409–412.

St. Aubin, D.J., 2001. Endocrinology. In: Dierauf, L.A., Gulland, F. (Eds.), CRC Handbook of Marine Mammal Medicine: Health, Disease, and Rehabilitation,. second ed. CRC Press, Boca Raton, FL, pp. 165–192.

St. Aubin, D.J., 2002. Hematological and serum chemical constituents in pantropical spotted dolphins (*Stenella attenuata*) following chase and encirclement. Southwest Fisheries Science Center, La Jolla, CA. Administrative Report LJ-02-37C.

St. Aubin, D.J., Deguise, S., Richard, P.R., Smith, T.G., Geraci, J.R., 2001. Hematology and plasma chemistry as indicators of health and ecological status in beluga whales, *Delphinapterus leucas*. Arctic 54, 317–331.

St. Aubin, D.J., Dierauf, L.A., 2001. Stress and marine mammals. CRC Handbook of Marine Mammal MedicineCRC Press, Boca Raton, FL, pp. 253–271.

St. Aubin, D.J., Geraci, J.R., 1988. Capture and handling stress suppresses circulating levels of thyroxine (T4) and triiodothyronine (T3) in beluga whales *Delphinapterus leucas*. Physiol. Zool. 61, 170–175.

St. Aubin, D.J, Geraci, J.R., 1989. Adaptive changes in hematologic and plasma chemical constituents in captive beluga whales, *Delphinapterus leucas*. Can. J. Fish. Aquatic Sci. 46, 796–803.

St. Aubin, D.J., Geraci, J.R., 1992. Thyroid hormone balance in beluga whales, *Delphinapterus leucas*: dynamics after capture and influence of thyrotropin. Can. J. Vet. Res. 56, 1–5.

St. Aubin, D.J., Ridgway, S.H., Wells, R.S., Rhinehart, H., 1996. Dolphin thyroid and adrenal hormones: circulating levels in wild and semidomesticated *Tursiops truncatus*, and influence of sex, age, and season. Mar. Mamm. Sci. 12, 1–13.

Stokkan, K.A., Vaughan, M.K., Reiter, R.J., Folkow, L.P., Mårtensson, P.E., Sager, G., Lydersen, C., Blix, A.S., 1995. Pineal and thyroid functions in newborn seals. Gen. Comp. Endocrinol. 98, 321–331.

Suzuki, M., Tobayama, T., Katsuma, E., Fujise, Y., Yoshioka, M., Aida, K., 1998. Serum cortisol levels in captive killer whale and bottlenose dolphin. Fish Sci. 64, 643–647.

Suzuki, M., Ishikawa, H., Otani, S., Tobayama, T., Katsumata, E., Ueda, K., Uchida, S., Yoshioka, M., Aida, K., 2002a. The characteristics of adrenal glands and its hormones in cetacean. Fish. Sci. 68 (suppl), 272–275.

Suzuki, M., Tobayama, T., Katsumata, E., Uchida, S., Ueda, K., Yoshioka, M., Aida, K., 2002b. Secretory patterns of cortisol in Indo-Pacific bottlenose dolphins and killer whales. Fish. Sci. 68 (suppl), 451–452.

Suzuki, M., Uchida, S., Ueda, K., Tobayama, T., Katsumata, E., Yoshioka, M., Aida, K., 2003. Diurnal and annual changes in serum cortisol concentrations in Indo-Pacific bottlenose dolphins *Tursiops aduncus* and killer whales *Orcinus orca*. Gen. Comp. Endocrinol. 132, 427–433.

Suzuki, M., Hirako, K., Saito, S., Suzuki, C., Kashiwabara, T., Koie, H., 2008. Usage of high-performance mattresses for transport of Indo-Pacific bottlenose dolphin. Zoo. Biol. 27 (4), 331–340.

Suzuki, M., Nozawa, A., Ueda, K., Bungo, T., Terao, H., Asahina, K., 2012. Secretory patterns of catecholamines in Indo-Pacific bottlenose dolphins. Gen. Comp. Endocrinol. 177, 76–81.

Symington, T., 1969. Functional Pathology of the Human Adrenal Gland. Livingston, Edinburgh, UK.

Thayer, V., Read, A., Friedlaender, A.S., Colby, D.R., Hohn, A.A., McLellan, W.A., Pabst, D.A., Dearolf, J.L., Bowles, N.I., Russell, J.R., Rittmaster, K.A., 2003. Reproductive seasonality of western Atlantic bottlenose dolphins off North Carolina, USA. Mar. Mamm. Sci. 19, 617–629.

Thomson, C.A., Geraci, J.R., 1986. Cortisol, aldosterone, and leucocytes in the stress response of bottlenose dolphins, *Tursiops truncatus*. Can. J. Fish. Aqu. Sci. 43, 1010–1016.

Tizzi, R., Accorsi, P.A., Azzali, M., 2010. Non-invasive multidisciplinary approach to the study of reproduction and calf development in bottlenose dolphin (*Tursiops truncatus*): the Rimini Delfinario experience. Int. J. Comp. Psychol. 23.

Tosini, G., Pozdeyev, N., Sakamoto, K., Iuvone, P.M., 2008. The circadian clock system in the mammalian retina. Bioessays 30, 624–633.

Turner, J.P., Clark, L.S., Haubold, E.M., Worthy, G.A., Cowan, D.F., 2006. Organ weights and growth profiles in bottlenose dolphins (*Tursiops truncatus*) from the northwestern Gulf of Mexico. Aquatic Mammals 32, 46–57.

Turner, W., 1888. The pineal body (*Epiphysis cerebri*) in the brains of the walrus and seals. J. Anat. Physiol. 22, 300–303.

Ugaz, C., Valdez, R.A., Romano, M.C., Galindo, F., 2013. Behavior and salivary cortisol of captive dolphins (*Tursiops truncatus*) kept in open and closed facilities. J. Vet. Behav. Clin. Appl. Res. 8, 285–290.

Valsø, J., 1934. Der Hormongehalt der Hypophyse des blauwals (*Balaenoptera sibbaldii*). J. Mol. Med. 13, 1819–1820.

Valsø, J., 1936. Die Hypophyse des blauwals (*Balaenoptera sibbaldii*). Anat. Embryol. 105, 715–719.

Venn-Watson, S.K., Ridgway, S.H., 2007. Big brains and blood glucose: common ground for diabetes mellitus in humans and healthy dolphins. Comp. Med. 57, 390–395.

Venn-Watson, S., Carlin, K., Ridgway, S., 2011. Dolphins as animal models for type 2 diabetes: sustained, postprandial hyperglycemia and hyperinsulinemia. Gen. Comp. Endocrinol. 170, 193–199.

Venn-Watson, S., Smith, C., Stevenson, S., Parry, C., Daniels, R., Jensen, E., Cendejas, V., Balmer, B., Janech, M., Neely, B.A., Wells, R., 2013. Blood-based indicators of insulin resistance and metabolic syndrome in bottlenose dolphins (*Tursiops truncatus*). Endo. Front. 4, 136.

Villanger, G.D., Lydersen, C., Kovacs, K.M., Lie, E., Skaare, J.U., Jenssen, B.M., 2011. Disruptive effects of persistent organohalogen contaminants on thyroid function in white whales (*Delphinapterus leucas*) from Svalbard. Sci. Total Environ. 409, 2511–2524.

von Schulte, H.W., de Forest Smith, M., 1918. The external characters, skeletal muscles, and peripheral nerves of Kogia breviceps (Blainville). Bull. Am. Mus. Nat. Hist. 38, 7–72.

Vuković, S., Lucić, H., Živković, A., Đuras Gomerčić, M., Gomerčić, T., Galov, A., 2010. Histological structure of the adrenal gland of the bottlenose dolphin (*Tursiops truncatus*) and the striped dolphin (*Stenella coeruleoalba*) from the Adriatic Sea. Anat. Histol. Embryol. 39, 59–66.

Vuković, S., Lucić, H., Đuras Gomerčić, M., Galov, A., Gomerčić, T., Ćurković, S., Škrtić, D., Domitran, G., Gomerčić, H., 2011. Anatomical and histological characteristics of the pituitary gland in the bottlenose dolphin (*Tursiops truncatus*) from the Adriatic Sea. Vet. Arhiv. 81, 143–151.

Waples, K.A., Gales, N.J., 2002. Evaluating and minimising social stress in the care of captive bottlenose dolphins (*Tursiops aduncus*). Zoo. Biol. 21, 5–26.

West, K.L., 2002. Ecology and biology of the rough-toothed dolphin (*Steno bredanensis*). Doctoral Dissertation, University of Hawaii at Manoa.

West, K.L., Atkinson, S., Shwetz, C., Sweeney, J., Stone, R., 2001. Thyroid hormone concentrations during different reproductive stages in adult female bottlenose dolphins (*Tursiops truncatus*). Proc. Int. Ass. Aquatic An. Med. 32, 122–123.

West, K.L., Ramer, J., Brown, J.L., Sweeney, J., Hanahoe, E.M., Reidarson, T., Proudfoot, J., Bergfelt, D.R., 2014. Thyroid hormone concentrations in relation to age, sex, pregnancy, and perinatal loss in bottlenose dolphins (*Tursiops truncatus*). Gen. Comp. Endocrinol. 197, 73–81.

Wislocki, G.B., 1929. The hypophysis of the porpoise (*Tursiops truncatus*). Arch. Surg. 18, 1403–1412.

Xu, S., Yang, Y., Zhou, X., Xu, J., Zhou, K., Yang, G., 2013. Adaptive evolution of the osmoregulation-related genes in cetaceans during secondary aquatic adaptation. BMC Evol. Biol. 13, 189.

Yadav, R., Haldar, C., Pandey, L.K., 2008. Peripheral melatonin concentration and its correlation with estradiol and progesterone levels during different months of pregnancy and after delivery in women. J. Endocrinol. Reprod. 12, 80–86.

Young, B.A., Harrison, R.J., 1969. Ultrastructure of light cells in the dolphin thyroid. Zeit. Zellforsch. Mikrosk. Anat. 96, 222–228.

Zhongjie, L., 1985. A preliminary study on the thyroid and parathyroid gland of chinese river dolphin (*Lipotes vexillifer*). Acta Hydrobiol. Sinica 9, 171–175.

Chapter 8

Feeding and the Digestive System

Feeding in water requires grabbing or sucking, or filtering,[a] but virtually no chewing. Accidental ingestion of water—and especially salt water—during food intake must be avoided (to make a long story short, consider the inconvenience of eating a snack while sitting on the bottom of a swimming pool!). Furthermore, as we have seen in Chapter 4, the digestive apparatus of cetaceans virtually bears no communication with the respiratory tract.[b] Why so?

Chewing is the act of dividing and grinding the food in the mouth, following the general principle that smaller chunks of food are easier to digest because reduced mass allows a more rapid enzymatic processing in the postdiaphragmatic part of the digestive apparatus. Chewing contributes to trigger the production of the enzyme-rich saliva by major (parotid, mandibular, and submandibular) and minor salivary glands located in the walls of the jaws. The presence of food into the mouth is also functional to taste and discrimination of flavors, a key factor in the choice of aliments. All of these structures and their functions are reduced, absent, or at least heavily modified in dolphins.

Top terrestrial predators (large cats, wolves, bears, and the like) rely heavily on protein (and fat)-rich meat and their masticatory apparatus (teeth, jaw muscles, temporomandibular joint) is devised to grab, break, tear the body of the prey, but chewing is limited. Consumption of carbohydrates (grass, roots) is restricted to specific needs. In this sense, dolphins can be considered as super-predators whose nourishment consists almost exclusively of fish and cephalopods (squids, cuttlefish, octopuses), depending on the dolphin's species, geographical peculiarities of the sea, and obviously availability of the prey.[c] Since rhythmic mastication is not required, the temporal muscle (that shuts the jaws and keeps them serrated) is extremely powerful, but the masseter, the chief agent of repeated food grinding, is relatively weak. So the shape of the dolphin's mouth is remarkably different from that of land mammals, and also its constituent organs and their structure, including that of the hyolingual apparatus,[d] change and adapt to underwater foraging. In fact, the whole face and even the nonmasticatory muscles are considerably modified.

THE MOUTH AND UPPER DIGESTIVE TRACT

Shape of the Mouth and Face

The mouth of several dolphin species (common, striped, bottlenose, etc.) is elongated and characterized by an extended rostrum, principally made up by paired maxillary and incisive bones and unpaired vomer in the upper jaw, and by the mandible in the lower jaw. The maxillary bones and the mandible each support a relatively long row of small pointed teeth (Fig. 8.1). Other species, including the Rissos's dolphin and the killer, false killer, and pilot whale, have a more blunt head profile (Fig. 8.2). Their bony rostrum is still elongated, but relatively less pronounced than in smaller dolphins. The rather convex dorsal outline of the face is partially due to the melon and connective tissue. Species with a long rostrum generally possess a higher number of teeth. Toothed whales who prey preferably on squids have generally less teeth or even lack teeth in the upper jaw[e] (see Table 8.1).

---

a. This is the case of whalebone whales, whose baleens act precisely as one-directional filters. For the complex mechanic of suction feeding in baleen- and toothed-whales see Heyning and Mead (1996), and Werth (2000, 2006), respectively. For recent insights into the functions of beleens see Werth and Potvin, 2016; Werth et al., 2016.

b. Contrary to most terrestrial mammals, in which the respiratory and digestive apparatuses cross paths in the pharynx. Dolphins can apparently voluntarily dislocate their larynx to ingest large preys (Dold and Ridgway, 2007). In a peculiar case, a pilot whale, maintained in a marine theme park, "could be fed a squid, then blow the entire squid out its blowhole" (Sam Ridgway, personal communication); in some instances, however, ingestion of fish may cause incidental asphyxiation and drowning (IJsseldijk et al., 2015). Water must never enter the respiratory apparatus also in land mammals. Furthermore, some mammals (including the horse) cannot use the mouth for breathing due to the position and shape of their epiglottis.

c. Dolphins may also prey on crustaceans and even shellfish. However, digestion of the chitinous exoskeleton rich of calcium carbonate may be difficult especially in large quantities. Baleen whales (Mysticeti) feed on small crustaceans (krill) and the anatomy of their stomach complex is different from that of dolphins, and bears a certain superficial similitude to the gastric apparatus of terrestrial ruminants.

d. The hyoid complex has been described in Chapters 3 and 5. For a complex discussion of its role in dolphin (and other cetaceans) feeding see Bloodworth and Marshall (2007), and Werth (2007).

e. Dental buds may appear during fetal life but then disappear. Sperm whales and beaked whales also lack teeth in the upper jaw in the extra-uterine life.

Anatomy of Dolphins. http://dx.doi.org/10.1016/B978-0-12-407229-9.00008-7

FIGURE 8.1   Mouth of *T. truncatus*.

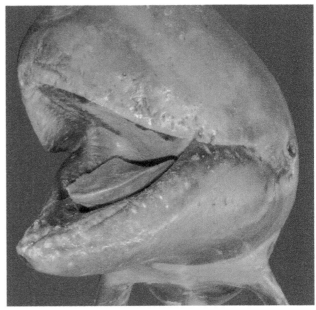

FIGURE 8.2   Mouth of *G. griseus*.

TABLE 8.1 Dental Formulas of Selected Dolphins Species

Species	T. truncatus	S. coeruleoalba	D. delphis	G. griseus	G. melas	G. macrorhynchus	O. orca	P. crassidens	L. obliquidens
½ Upper jaw	18–26	39–53	45–60	0	8–13	7–8	10–12	7–12	23–26
½ Lower jaw	18–26	39–55	45–60	2–7	15–18	7–8	10–12	7–12	23–26
Total	72–104	156–216	180–240	4–14	46–62	28–32	40–48	28–48	92–104

Data from Miyazaki 2002. Teeth. In: Perrin, W.F., Würsig, B., Thewissen, J.G.M. (Eds.), Encyclopedia of Marine Mammals. first ed. Academic Press, San Diego, pp. 1227–1232.

The lips are thin and not independently mobile and the oral rim is quite wide to allow opening of the mouth to an extensive vertical angle.[f] The jaws open only along a single plane with no lateral movement possible. This is also due to the reduction or absence of the coronoid process in the mandible.

In most terrestrial mammals, the external integumentary epithelium of the lips is very different from the epithelium that lines the oral cavity (Menon and Pfeiffer, 2002). In dolphins (and cetaceans in general), the oral cavity is exposed to water, just like the skin covering all the external surface of the body, and the two epithelia share many morphological features.

Teeth

The teeth of mammals are in most species (including man) constituted by (1) an erupted part visible over the gums (crown); (2) a neck, that separates the crown from the root; (3) one or more roots that are placed in the tooth socket below the gingival surface; and (4) a pulp cavity containing connective, vascular, and nervous tissue that fills also the root canal: Sensitivity of the tooth is carried by trigeminal fibers that enter the pulp from ramification of the superior and inferior alveolar nerves.

Each tooth is connected to its socket by connective tissue full of collagen fibers that forms the periodontal ligament.

The teeth of dolphins have a conical, pointed, and sometimes slightly curved shape, with a single root. Distinction between crown, neck, and root is not as clear as in most terrestrial mammals, also because the pulp cavity is open toward the mandibular or maxillary bony socket in young individuals, but reduces its volume gradually over time.

---

f. This latter is also a striking difference to terrestrial mammals. The horse, for instance, possesses a very long mouth but the oral rim is reduced. In this way the animal may open the mouth only a few degrees. The long oral cavity is functional to chewing (and to allow a long space to the nasal cavities above). On the contrary dolphins may open their jaws quite widely.

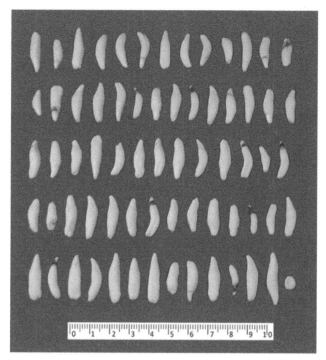

FIGURE 8.3   **Teeth of an adult** *T. truncatus*.

The teeth of an individual are more or less all alike (so odontocetes are called *homodonts*, meaning of almost identical teeth) (Fig. 8.3), without the differentiation in incisive, canine, premolar, or molar of terrestrial *heterodont* mammals. The teeth of species with a comparable head profile and body mass are very similar. Discrimination of two species with a similar shape of the head based on analysis of a few teeth can be tricky (ie, to tell a striped from a common dolphin by its teeth requires a well-versed specialist).

The teeth erupt only once (odontocetes have no distinction between deciduous and permanent teeth) and therefore these mammals are also called *monophyodonts*, meaning of a single eruption.[g] For dental formulas of selected dolphin species, see Table 8.1.

The teeth are constituted by three hard tissue: enamel, dentine, and cementum (Fig. 8.4). In dolphins, enamel (Ishiyama, 1987) is present only as a reduced layer casing the tip of the external crown,[h] and wears out with age. Consequently, the highly mineralized, acellular,[i] prismatic enamel is best seen in young individuals. Enamel is followed by dentine,[j] the principal constituent of the tooth. Dentine is cellular and the odontoblasts synthesize the mineralized matrix from the inside of the radiating tubules. A superficial dentine layer (prenatal dentine) is separated by the neonatal line from the postnatal dentine whose deposition shows layered patterns (GLGs, or growth layer groups) and may be used to determine the age of the specimen (see the following). The cementum is also a calcified tissue whose structure is produced by cementoblasts. The cementum forms layers around the root of the tooth, but in several dolphin species it protrudes out of the gingival surface and with time covers great part of the tooth.

The dental cavity is cone-shaped, opens toward the gum and diminishes its volume with age, its space becoming filled by dentine.

Dentine layers increase with the age of the individual, even if the situation in older individuals is more complex and distinction between adjacent dentine layers is sometimes unclear. However, the count of dentine layers remains one of the more reliable methods to tell the age of stranded dolphins.[k]

---

g. Most mammals (including humans) are *diphyodonts*, while a few species (including elephants and manatees) are *polyphyodonts*, meaning that have multiple dental eruptions. This latter condition is often associated with extensive chewing of hard varieties of grass, roots, bark or the like.

h. On the contrary, the crown of human teeth is covered by enamel.

i. The cells responsible for the formation of enamel, called adamantoblasts, soon disappear.

j. Dentine is commonly called ivory in the narwhal, walrus and elephant tusks.

k. For details on age determination using dentine counts see Hohn (2009) and also the relatively old but still authoritative volume edited by Perrin and Myrick (1980). For alternate methods of age counting not based on dentine layers see also the description of the thoracic limb in Chapter 3.

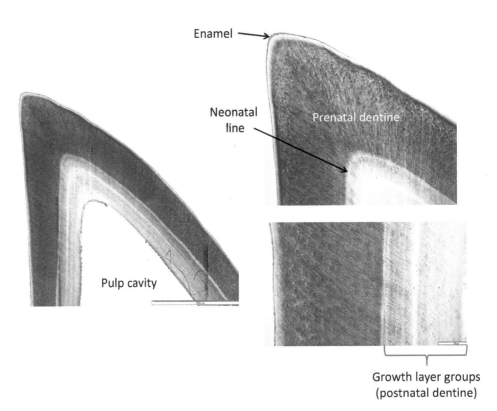

FIGURE 8.4 **Section of a tooth of** *T. truncatus.*

The Lips, Tongue, and Mouth Cavity

The lips have a thick, rubber-like aspect and are unable to move (stir or stretch) independently. There is no *m. orbicularis oris* (as in humans) and mimic muscles that move the lips and the rest of the face are absent or vestigial[l] (see Chapter 3).

The lips are hard and heavily keratinized, and, on the inner side, turn into the mucosa of the oral vestibule. The color of the mucosa varies from pink to dark gray, sometimes retaining the same color of the external surface. The oral vestibules are quite shallow, possibly in relation to the absence of mastication that nullifies the need for a functional space outside the vestibular (outer) surface of the teeth from which food moves to and from of the center of the oral cavity during rhythmical chewing. The mucosa of the vestibules continues inward with the thick gum surrounding the base of each tooth.

The tongue (Fig. 8.5) is short and wide and possesses only limited mobility. The tip of the tongue is capable of some movement, and, in young dolphins, including *Tursiops truncatus* (Sokolov and Volkova, 1973; Kastelein and Dubbeldam, 1990), *Lagenorhynchus acutus* (Werth, 2007), and *Stenella coeruleoalba* (Yamasaki et al., 1976), shows anterolateral papillae (anterolateral fimbria) consisting of compact outgrows placed along a single row. According to Ferrando et al. (2010), these latter papillae may help create a tight seal between the tongue and the roof of the oral cavity, and so help suction feeding. They tend to be reduced or disappear with age, although they may persist in some individuals. Small grooves (or *fossae*) are present on the surface of the body of the tongue (Sokolov and Volkova, 1973), and at the linguopharyngeal junction.

In the tongue of the bottlenose dolphin, there are no filiform,[m] fungiform, foliate, and circumvallate papillae (Arvy and Pilleri, 1972; Pfeiffer et al., 2001; for a thorough review of the older literature on the subject, see Donaldson, 1977).

The mucosa is covered by flat squamosal epithelium highly keratinized, less so at the apex (Werth, 2007) (Fig. 8.6). The corium of the tongue is rich in collagen but lacks elastin typical of terrestrial mammals, making the tongue poorly flexible and deformable (Werth, 2007). The submucosa forms papillae that interdigitate with the outer layers (Simpson and Gardner, 1972). Mucous glands have been described in the tongue of the common dolphin (Sokolov and Volkova, 1973). We noted the presence of seromucous glands in the tongue of the common dolphin (Figs. 8.7 and 8.8). Tubuloalveolar, probably holocrine, glands have been described in the tongue of the bottlenose dolphin (Donaldson, 1977). The extrinsic musculature of the tongue (Fig. 8.9) has been described in Chapters 3 and 5. According to Werth (2007), the intrinsic muscle fibers are oriented along the three planes and are more prominent in newborn individuals than in adults.

---

l. The human smile is due to retraction of the orbicular muscle of the mouth together with the action of several other small muscles that drag, depress or lift the lips, with the general effect of showing the incisors and possibly the canine teeth. When we see a dolphin "smiling", the animal is in fact opening widely the jaws and thus showing the rows of teeth, giving the superficial impression of retracting the lips.

m. Filiform papillae have been described in the tongue of the Franciscana dolphin *Pontoporia blainvillei* (Guimarães et al., 2012).

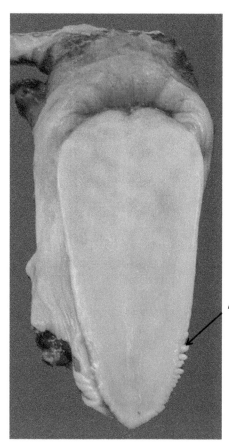

FIGURE 8.5   **Tongue of a young *T. truncatus* with anterolateral papillae.**

Anterolateral
papillae

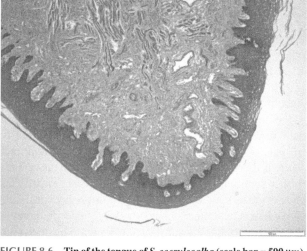

FIGURE 8.6   **Tip of the tongue of *S. coeruleoalba* (scale bar = 500 μm).** HE stain.

Seromucous glands

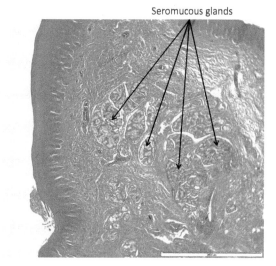

FIGURE 8.7   **Section of the tongue of *D. delphis* showing the presence of glands (scale bar = 2 mm).** HE stain.

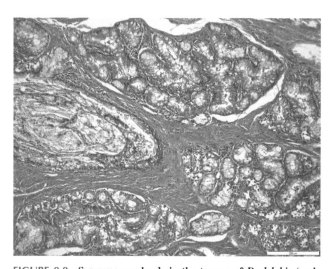

FIGURE 8.8   **Seromucous glands in the tongue of *D. delphis* (scale bar = 200 μm).** HE stain.

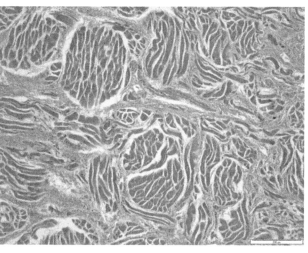

FIGURE 8.9   **Muscle fibers in the tongue of *S. coeruleoalba* (scale bar = 200 μm).** HE stain.

The absence of taste buds (Sokolov and Volkova, 1973; Donaldson, 1977; Pfeiffer et al., 2001) is quite puzzling. Menon and Pfeiffer (2002) provide a detailed list of evidence supporting the presence of taste receptors on the tongue or elsewhere in the oral cavity. However, they too failed to identify these supposed structures in the tongue of *Globicephala melas* and other species, leaving the question still open. The grooves present on the surface of the tongue may in fact be the location of the hitherto unveiled taste receptors (Sokolov and Volkova, 1973). Sparse reports describe the presence of modified taste buds in the tongue of *Delphinus delphis* and *T. truncatus* (Suchowskaja, 1972) and *S. coeruleoalba* (Yamasaki et al., 1978).[n] According to these latter studies, the receptors are placed at the bottom of small cavities located between the body and the root of the tongue. However, notwithstanding previous experimental evidence on taste reception (Nachtigall and Hall, 1984), data from a recent study (Feng et al., 2014) suggest extensive losses of sweet, umami, bitter, and sour taste receptor genes in selected cetacean species (including bottlenose dolphin, pilot whale, white-beaked dolphin, Atlantic white-sided dolphin), whereas the salty taste receptor gene (at least in the bottlenose dolphin) is maintained. The loss of taste receptor genes is a further confirmation of the anatomical failure to identify taste buds in the tongue, but leaves more questions open (the presence of genes for salty receptors is not surprising considering the external environment in which dolphins live).

Dolphins (and other odontocetes) show specific predilections for certain varieties of food,[o] and do not grab or swallow their prey casually. But since there is no sense of smell (see Chapters 5 and 6), and no visible taste organ, how do they perceive flavors and develop preference for specific food? (See Chapter 5 for behavioral studies on taste discrimination.) Food is ingested as a whole, thus further limiting the possibility of developing preferences by tasting the progressive break-up of the prey induced by chewing. As of today, no clear answers can be given to these questions. Dolphins can be very gentle when they grab delicate objects (including human hands during training in captivity), thus showing high trigeminal sensibility and perfect motor control in the jaw-closing process.[p] We know that the lips, tongue, and mouth of dolphins are extremely sensitive to touch and possibly pressure, due to the dense innervation. Dolphins virtually interact with the world through the mouth.

The Salivary Glands

There are no major or minor organized salivary glands in dolphins, since the production of saliva has no function in animals that swallow their prey whole. A few serous glands were detected in the tongue of the common dolphin (Arvy and Pilleri, 1972; Sokolov and Volkova, 1973). Other studies (Guimarães et al., 2011), performed in the estuarine dolphin *Sotalia guianensis*, described salivary glands in the tongue. Mucous and serous lingual glands are present (Slijper, 1979; Sokolov and Volkova, 1973; Donaldson, 1977; Werth, 2007; Ferrando et al., 2010).

Since dolphins open their mouth also underwater, any large amount of saliva would be washed out. It also possible that, giving the modalities of feeding, a certain quantity of mucus may help in gulping the food as a whole.

THE PHARYNX AND ESOPHAGUS

At the back end of the mouth, the faucal isthmus (Fig. 8.10) allows food to pass through the pharynx but limits the quantity of water that reaches the highly muscular esophagus and stomachs.

The pharynx of cetaceans, in fact, is made up by two channels, connected dorsally, that envelop the larynx (Lawrence and Schevill, 1965) and then merge into the esophagus and travel dorsal to the trachea. The pharynx allows food to enter the esophagus through the combined movements of the genioglossus, hyoglossus, and styloglossus muscles. In the bottlenose dolphin, the lateral submucosa of the oropharynx contains tonsil-like lymphoid nodules that open close to pits (crypts) of the mucosa. The beginning of the trachea after the larynx corresponds to the passage of the pharynx into the esophagus (Harrison et al., 1970).

The esophagus has the structure typical of the organ in all mammals. Basically it is a long tube of spiral smooth and skeletal musculature that conveys food from the pharynx to the forestomach. The esophagus of the bottlenose dolphin has a relatively thin mucosal layer lined by layers of nonkeratinized squamous epithelium (Fig. 8.11) with no submucosal glands (Simpson and Gardner, 1972). The *lamina propria* shows collagen fibers, and papillae of connective tissue that reach the interdigitate squamous epithelium (Harrison et al., 1970). The thick longitudinal *muscularis mucosae* contains many blood vessels. The outer wall of the esophagus is made up by muscle fibers, divided into an inner circular and an outer tangential layers. The first half of the esophagus contains striated musculature, the second half smooth musculature.

---

n. Taste buds were described in the tongue of a newborn Stejneger's beaked whale, *Mesoplodon stejnegeri* (Shindo et al., 2008).

o. An easy example is the preference shown by captive bottlenose dolphins for the capelin *Mallotus villosus*.

p. The consistence of food (or of any object in the mouth) is not perceived by taste organs, but instead by the interaction between the tooth and its gingival alveolus and by perception of muscular force applied during chewing in terrestrial mammals. Specific trigeminal nerve endings convey information to the brain relative to the stretch conditions of muscle spindles in the masticatory muscles and the consequent pressure enforced by the jaws and perceived by the alveoli. Thus dolphins can carry very fragile objects in their mouth while swimming without breaking them, but on the other hand they are perfectly capable of applying tremendous force while clenching.

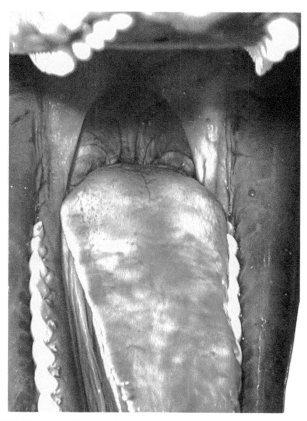

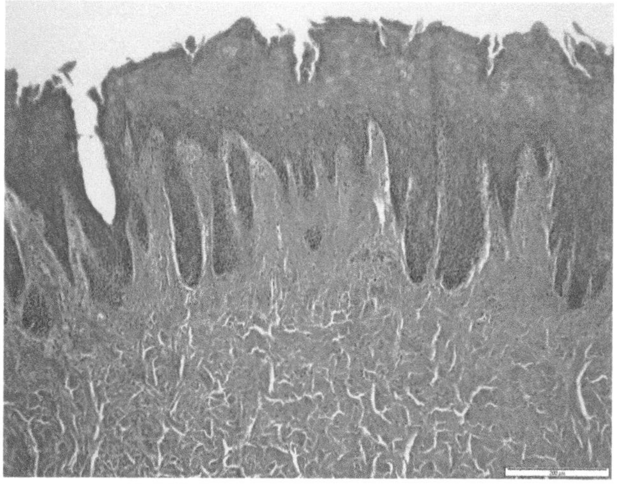

FIGURE 8.11 **Epithelial surface of the esophagus of *T. truncatus* (scale bar = 200 µm).** HE stain.

FIGURE 8.10 **Faucal isthmus with tongue and pharyngeal folds of an adult *T. truncatus*.**

One has also to consider that the "food" is practically alive, since there is no chewing in the mouth. Small dolphins prey on herrings and other small fish, but killer whales prey also on seals and small dolphins (and even attack large baleen and sperm whales for their tongue and other parts). So the esophagus of large dolphin species must accommodate huge volumes of food that is pushed forcibly inward. According to Harrison et al. (1970), the esophagus of an adult bottlenose dolphin is over 50 cm in length and 5 cm in width. However, the esophagus can easily stretch to accommodate larger prey, but the real limit is given by the diameter of the faucal isthmus and of the pharyngeal passage (Harrison et al., 1970).

THE STOMACHS

The stomach of mammals is a hollow organ placed in the peritoneal cavity, generally composed by three major layers (*tunicae*). Dolphins are polygastric animals, meaning that they possess more than one stomach (three in *T. truncatus*, *S. coeruleoalba*, and *D. delphis*). As all the stomachs derive from a single enlargement of the primitive gastric tube,[q] it could also be said that the single stomach of dolphins is divided into several chambers, so that is also called multichambered stomach.[r] The purpose of this anatomical structure is to compensate for the absence of chewing by grinding and compressing in the first chamber, before absorption takes place in the subsequent compartments.[s]

The three stomachs of dolphins (Figs. 8.12–8.14) are composed by a first chamber or forestomach (aglandular, for rough squeezing of food; probably no absorption occurs in this part), a second chamber or true stomach (lined with glandular

q. At least this has been demonstrated in terrestrial mammals with multichambered stomachs.

r. Herbivore mammals need specialized digestive compartments where bacterial fermentation may break down cellulose. Ruminants (bovine and the like) and *Tylopoda* (camels and the like) are closely related terrestrial Cetartiodactyla that also possess a multichambered stomach. However, the resemblance with odontocetes is not functional, as the large stomachs of big-sized terrestrial herbivores digest grass and plants with the aid of bacteria suspended in solution, with an ambient pH of approx. 6.7. The stomachs of dolphins have evolved to grind food and then process proteins of animal origin in a highly acid environment. On the other hand baleen whales may use microbial fermentation to digest the chitin and wax-esters of krill (Herwig et al., 1984; Olsen et al., 1994; Nordøy, 1995).

s. Other toothed cetaceans (*Physeteridae*, *Monodontidae*, etc.) may possess even more stomachs (or have stomachs with a different structure, as in *Ziphius cavirostris*).

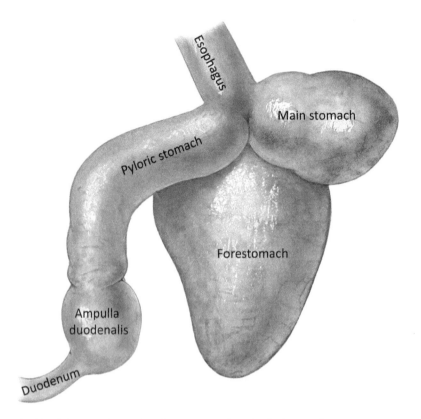

FIGURE 8.12 **Diaphragmatic side of the stomach complex of _T. truncatus_.** *(Drawing by Massimo Demma).*

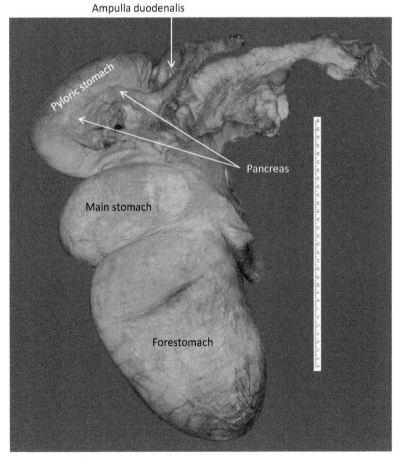

FIGURE 8.13 **Isolated stomach complex of _T. truncatus_.**

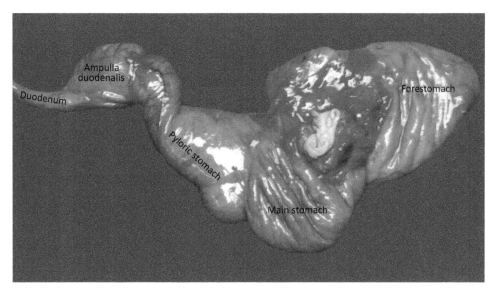

FIGURE 8.14   **Isolated stomach complex of** *G. griseus*.

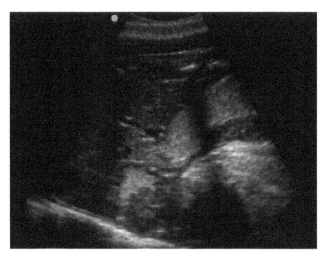

FIGURE 8.15   **Ultrasound image of the stomach complex of the bottlenose dolphin.** *Courtesy of Pietro Saviano.*

epithelium, with production of gastric juice), and a third chamber, a sort of pyloric stomach. The stomach complex can be observed in live animals with ultrasound techniques[t] (Fig. 8.15). These three chambers are followed by an enlarged duodenal ampulla. The connections between the chambers and the ampulla are rather complex (Fig. 8.16), and the mucosas of the different chambers are different (Fig. 8.17).

Table 8.2 reports some physical data on the three main stomachs of selected dolphin species.

Forestomach

Apparently there is no sphincter (*cardia*) between the esophagus and the forestomach (Harrison et al., 1970). The first chamber is the largest of the stomachs and has a conical shape. Its walls are thick with muscular layers. Its inner surface is whitish and presents longitudinal folds that can be observed also in vivo with ultrasound techniques (Fig. 8.18).

The mucosa (Figs. 8.19–8.20) is lined by stratified keratinized squamous epithelium and contains no glands. The *lamina propria* is relatively thin (Harrison et al., 1970), contrarily to the *muscularis mucosae*. The tunica muscularis (Fig. 8.21) is quite developed, as expected given the role of the forestomach in mechanic digestion. A detailed description of its

---

t. For reference images and values of ultrasound imaging of the intestinal apparatus of the bottlenose dolphin see Saviano (2013) and Fiorucci et al. (2015).

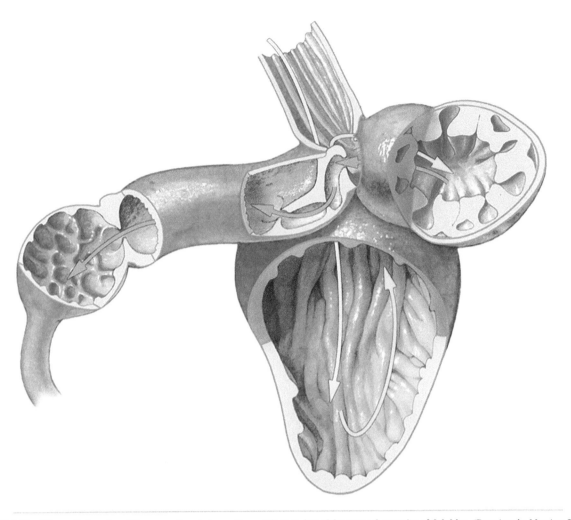

FIGURE 8.16   Schematic drawing of the connections between the gastric chamber of the stomach complex of dolphins. *(Drawings by Massimo Demma).*

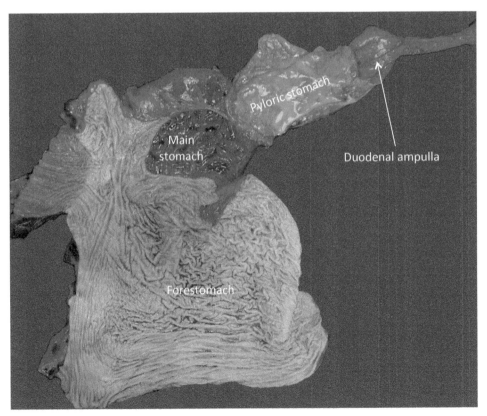

FIGURE 8.17   Mucosa of the stomach complex of *G. griseus*.

TABLE 8.2 Dimensions of the Stomachs (Empty Status) in Adult Specimens

Species/Stomach	Forestomach			Main Stomach			Pyloric Stomach		
	Length (cm)[a]	Width (cm)[a]	Capacity (l)[b]	Length (cm)[a]	Width (cm)[a]	Capacity (l)[b]	Length (cm)[a]	Width (cm)[a]	Capacity (l)[b]
T. truncatus	24	15	3	14	11	2	22	3–8	1.5
D. delphis	16	12	1.5	10–12	9–10	1–1.5	20–25	3–4	1
Stenella coeruleolba	18[c]	12[c]	2	10–12[c]	8–10[c]	1–1.5	20–22[c]	3–4[c]	1

[a]After Harrison et al., 1970 except for Stenella.
[b]Capacity, all species: personal unpublished observations.
[c]Length and width for Stenella: personal unpublished observations.

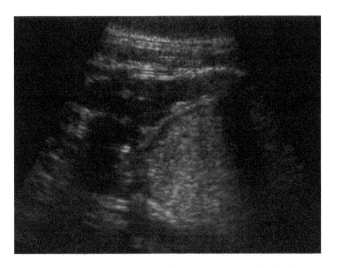

FIGURE 8.18   Ultrasound image of the forestomach of the bottlenose dolphin. *Courtesy of Pietro Saviano.*

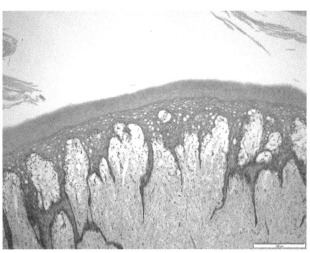

FIGURE 8.19   Mucosa of the first gastric chamber (forestomach) of *T. truncatus* (scale bar = 200 μm). HE stain.

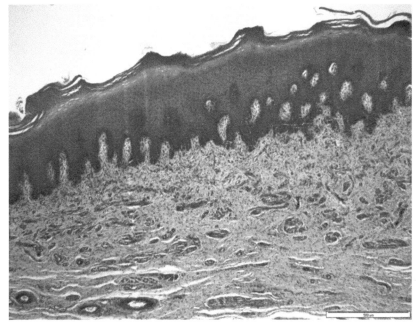

FIGURE 8.20   Mucosa of the first gastric chamber (forestomach) of an adult *T. truncatus* (scale bar = 500 μm). HE stain.

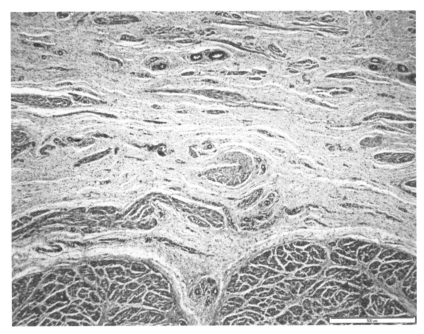

FIGURE 8.21   Longitudinal and circular layers of the tunica muscularis of the forestomach of *T. truncatus* (scale bar = 500 μm). HE stain.

ultrastructure is contained in Johnson and Harrison (1969) and especially in Harrison et al. (1970), who had access to fresh material adapt to electron microscopy.

The connection between the forestomach and the main stomach (Fig. 8.22) shows no specific sphincter and is placed on the rostral part of the stomach complex, adjacent to the opening of the esophagus into the forestomach. Gastric juice apparently can flow from the main stomach into the first chamber, to help digestion of food. This explains the possible

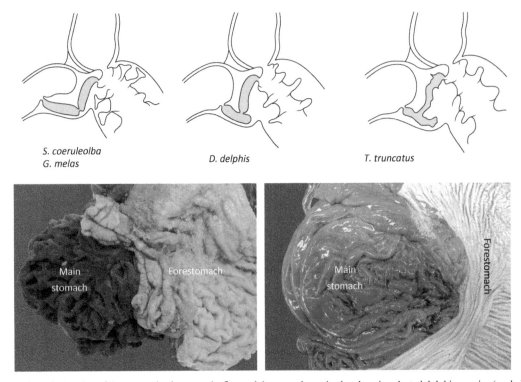

FIGURE 8.22   Schematic drawing of the connection between the first and the second gastric chambers in selected dolphin species (avobe), and connection between the first and the second gastric chambers in *T. truncatus* (left) and *G. griseus* (right). *(Redrawn after Harrison et al., 1970. The oesophagus and stomach of dolphins* (Tursiops, Delphinus, Stenella). *J. Zool. 160, 377–390).*

presence of ulcers also in the forestomach, considered a puzzling fact in the past (Harrison et al., 1967). Undigested remains (otoliths, cephalopod beaks) are common findings in the forestomach of dolphins.

Main Stomach

The main stomach is often darker than the forestomach at necropsy, possibly depending also on time after death and autolysis. This chamber is roundish and smaller than the first one, to which the forestomach is always closely connected on the right side at the top. The main stomach is visible in vivo with ultrasound imaging (Fig. 8.23).

The forestomach and the main stomach appear strikingly different when opened. This latter chamber is lined by a classic gastric mucosa of pink to violet color, that abruptly substitutes the squamous epithelium at the passage from the forestomach. There are folds, with irregular directions, often covered by mucus.

The structure of the gastric mucosa (Figs. 8.24–8.26) is similar to that of monogastric mammals (including man), and contains gastric glands with neck cells (secreting mucous), parietal cells (secreting HCl), and chief cells (secreting pepsinogen). Harrison et al. (1970), who described the fine structure of the mucosa, reported the absence of mucous neck cells. There are no argentaffin cells.

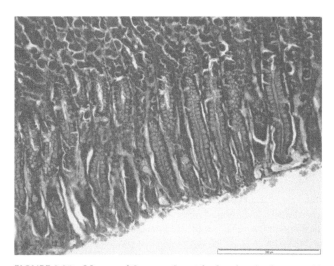

FIGURE 8.23   **Ultrasound image of the second gastric chamber (main stomach) of the bottlenose dolphin.** *Courtesy of Pietro Saviano.*

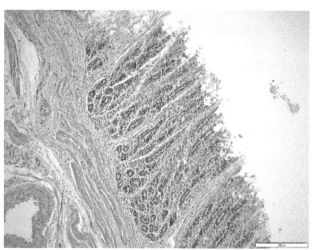

FIGURE 8.24   **Mucosa of the second gastric chamber (main stomach) of *T. truncatus* (scale bar = 200 μm).** HE stain.

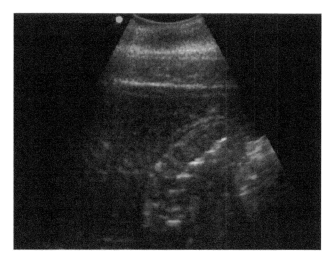

FIGURE 8.25   **Mucosa of the second gastric chamber (main stomach) of *D. delphis* (scale bar = 200 μm).** HE stain.

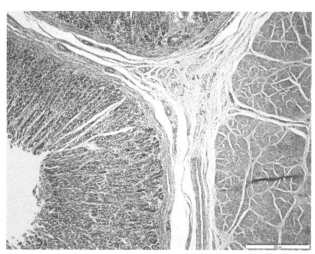

FIGURE 8.26   **Section of the second gastric chamber (main stomach) of *T. truncatus* (scale bar = 500 μm).** HE stain.

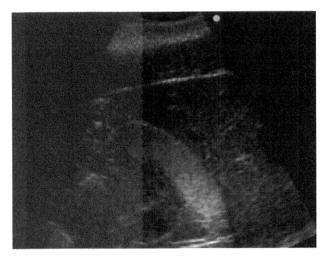

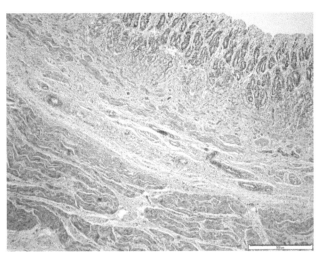

FIGURE 8.27   **Ultrasound image of the pyloric stomach of the bottle-nose dolphin.** *Courtesy of Pietro Saviano.*

FIGURE 8.28   **Structure of the third gastric chamber (pyloric stomach) of *T. truncatus* (scale bar = 500 μm).** HE stain.

Connecting Chambers

As represented in Figs. 8.16 and 8.22, the passage between the main and the pyloric stomach takes place through a series of connecting narrow chambers. The small chambers open one into the other through "nozzle-like" valvular arrangements (Harrison et al., 1970), but there is no complete sphincter.

The morphology of the connecting chambers varies slightly in the different species. The only information that we have on larger dolphin species states that in *Globicephala* the connecting chambers between the main stomach and the pyloric stomach are similar to those of *Stenella* (Harrison et al., 1967).

The structure of the connecting chambers is that of the pyloric stomach. The change from the structure of the main stomach is abrupt.

Pyloric Stomach

The pyloric stomach is a long tube whose long curved axis is transverse to that of the forestomach. The connecting chambers open into the pyloric stomach with an incomplete sphincter (Harrison et al., 1970). The third pyloric chamber is less muscular than the preceding two. Once open, its mucosa is purple or dark pink. In our experience it is lighter than that of the main stomach, but hard to distinguish from that of the connecting chambers. The pyloric stomach is also visible in vivo with ultrasound techniques (Fig. 8.27).

The mucosa (Fig. 8.28) is lined by columnar epithelium with mucous cells and gastric pits with tubular nonbranching glands. Argentaffin cells are frequent. The fine structure of this chamber was described by Johnson and Harrison (1969) and by Harrison et al. (1970).

This stomach terminates with a pyloric sphincter that separates it from the duodenal ampulla.

Duodenal Ampulla

The duodenal *ampulla* (Fig. 8.29) is just an initial enlargement of the duodenum, of which it has the same structure (see the following). The *ampulla* may superficially be mistaken at first sight for an additional stomach chamber.[u] The duodenal ampulla is separated from the remaining part of the duodenum by a *septum duodeni* that includes a sphincteric valve (Cave, 1982). The duodenal ampulla receives the joint termination of the pancreatic and hepatic ducts (*ampulla*

---

u. The distinction between a true stomach and a dilation of the initial part of the duodenum is based on the presence of a sphincter between the pyloric part of the stomach and the duodenum, and on the fact that the pancreatic and hepatic ducts open in the duodenum (either independently as in some mammals or together as in dolphins). The definition of the duodenal ampulla of dolphins as a part of the intestine and not of the stomach complex is therefore at least debatable, since the *ampulla duodenalis* possesses a sphincter-like structure of its own that separates it from the rest of the duodenum. Furthermore, the joint termination of the common hepatic duct and pancreatic duct (see below) opens into the duodenal ampulla in some dolphin species, but in the postampulla duodenum in others.

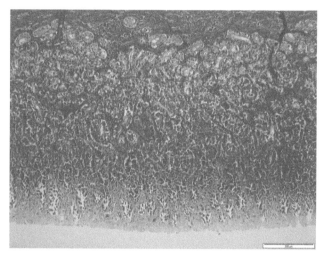

FIGURE 8.29   Mucosa of the duodenal ampulla of *T. truncatus* (scale bar = 200 μm). HE stain.

hepatopancreatica, see the following) in the common dolphin, while in the Risso's dolphin and long-finned pilot whale the duct opens into the postsphincteric part of the duodenum (Cave, 1982).

Peritoneal Cavity, Position of the Stomachs and Relationships to the *Greater* and *Lesser Omentum*

The peritoneal cavity is extremely small relative to the size of dolphins (for clinical implications, see also the following description of the intestine), and is shaped like an elongated oval. In fact, the dimensions of the cavity cannot be immediately perceived observing dolphins from the outside, because the huge musculature of the tail makes up most of second half of the body.

The peritoneum itself is relatively thick and can be easily seen during careful removal of the abdominal walls at necropsy. The serous membrane is even more evident (and sometimes whitish) in younger specimens.

The stomachs are placed immediately after the diaphragm (Figs. 8.30–8.32), in the part of the abdominal cavity that is covered by the ribs.[v] Since the diaphragm itself is dome-shaped, the three stomachs are placed one behind and above the other, caudal and dorsal to the liver. The pillars of the diaphragm run along the dorsal surface of the stomach (mainly dorsal to the first stomach), and may contribute to gastric motility.

The thin and translucent *greater omentum*[w] (superficial layer) originates from the greater curvature of the stomachs (generally placed ventrally and facing caudally). The short *lesser omentum* connects the dorsal edge of the liver with the rostral edge (lesser curvature) of the stomachs.

The greater omentum of the bottlenose dolphin does not cover the whole intestinal mass but is instead rather short and does not extend caudally much further than the caudal extremity of the first two chambers.

Topographical Relationships With Other Organs

The kidneys (Fig. 8.32) are immediately dorsal and caudal to the stomach complex. A large mesenteric lymph gland is located caudal to the stomachs (and particularly caudal to the pyloric stomach and duodenal ampulla), and just below the rostral pole of the right kidney. The small thick spleen is adjacent to the first segment of the stomach (see the description in Chapter 4).

---

v. Due to the convexity of the diaphragm, part of the abdominal viscera positioned immediately after it are placed within the rib cage in what is called the intrathoracic part of the abdominal cavity. This latter intrathoracic part is further subdivided into a central region (or *epigastrium*, meaning that is above the center of the abdomen) and two lateral regions (left and right *hypochondrium*, so named because of their position below the ribs in man).

w. The *greater* and the *lesser omentum* are peritoneal folds enveloping and connecting the stomachs and most of the other viscera. They derive respectively from the rotation of the primitive dorsal and ventral mesentery induced by growth of the developing stomach chambers. In many terrestrial mammals (including man) the *greater omentum* covers the majority of the content of the abdominal cavity when approached ventrally and laterally.

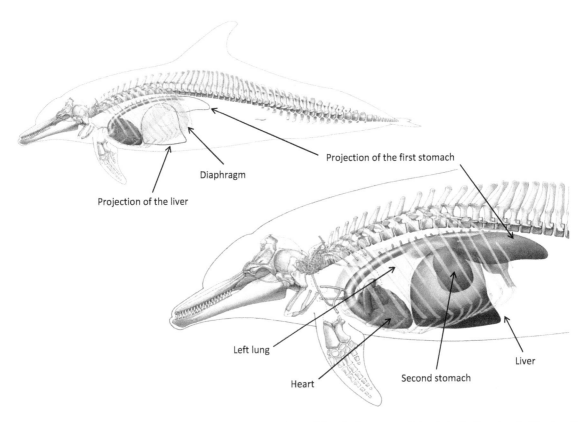

FIGURE 8.30 Schematic drawing of the position of the stomach complex and liver in *T. truncatus* (lateral view). *(Drawings by Massimo Demma).*

THE INTESTINE

A key macroscopic characteristic of the intestine of *Delphinidae* is the general absence of external subdivisions among the various tracts. Following the stomachs, the intestine appears as a continuous convoluted tube that terminates in the anus with minimal changes of diameter or external specialization (Fig. 8.32), observable also in vivo with ultrasound techniques (Fig. 8.33). According to Harrison et al. (1977), the length of the intestine in adult specimens of *Tursiops*, *Stenella*, *Sousa*, and *Delphinus*, varies from 885 to 1680 cm. There are no gross distinctions between small and large intestine, there is no caecum[x] or vermiform appendix. This apparent simplified disposition results from the diet, which includes mostly proteins, all of animal origin (generally fish and cephalopods[y]). Since there is no discernible gross intestine, some authors (Harrison et al., 1977) called small intestine the whole intestine of dolphins. In fact, Simpson and Gardner (1972) noted that the mucosa of the whole gut of dolphins is similar to that of the human small intestine.

The duodenum is recognizable because of the connection of the initial ampulla (Fig. 8.34) to the pyloric stomach, and because (at least in *D. delphis*) it receives the united pancreatic and hepatic ducts (see also footnote u). A sphincteric valve separates the *ampulla duodenalis* from the rest of the duodenum (Cave, 1982). The mesentery that suspends the intestine is thin and long.

Macroscopic distinction among the different tracts of the intestine is impossible once topographical clues are no longer present (ie, when the intestine is removed at necropsy, and no other organ, ligament, or connection are left for orientation).

The structure of the intestine (Fig. 8.35) varies only slightly from the duodenum to the rectum. The mucosa shows relevant folds[z] (Fig. 8.36) that increase its inner surface. According to Harrison et al. (1977), there are no well-developed

---

x. River dolphins (*Platanistidae*) present a distint caecum (Takahashi and Yamasaki, 1972).

y. The killer whale *Orcinus orca* (and possibly the false killer whale *Pseudorca crassidens*) attacks and eats also other marine mammals, from seals to the large baleen and sperm whales.

z. called *plica circulares* by Simpson and Gardner (1972).

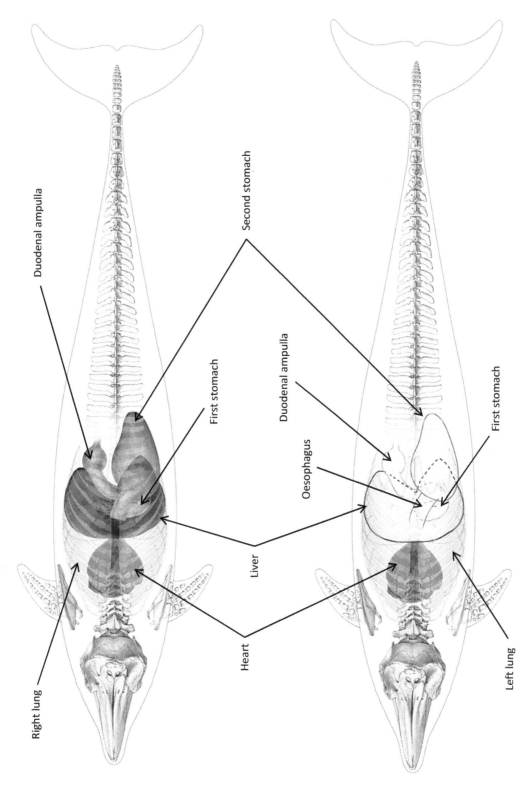

FIGURE 8.31 **Schematic drawings of the position of the stomach complex and liver in *T. truncatus* (dorsal view).** (*Drawings by Massimo Demma*).

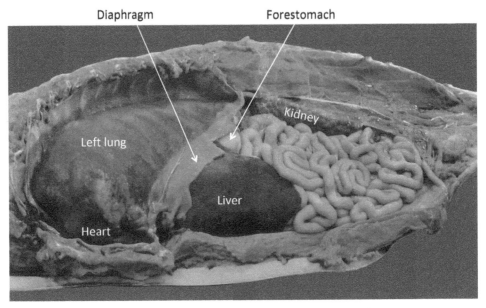

FIGURE 8.32 **Diaphragm and position of left lung, liver, apex of the forestomach, and left kidney of** *T. truncatus*.

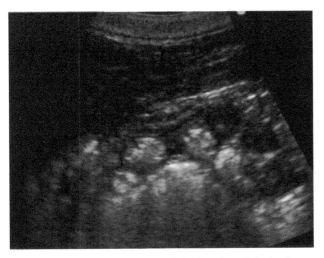

FIGURE 8.33 **Ultrasound image of the intestine of the bottlenose dolphin.** *Courtesy of Pietro Saviano.*

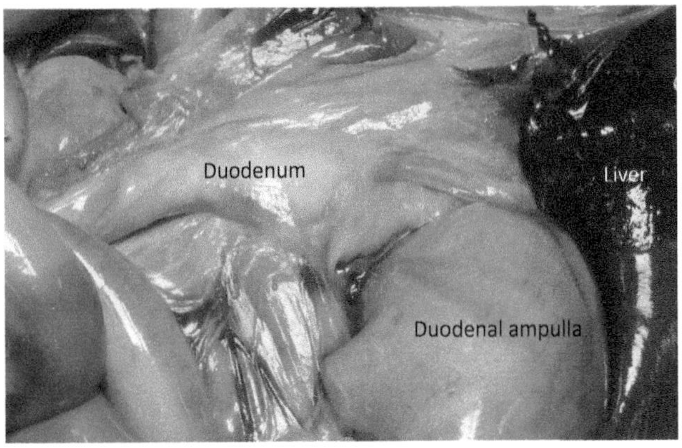

FIGURE 8.34 **Duodenal ampulla and duodenum of** *T. truncatus*.

villi. However, the same authors describe them a few pages later in the same article. This apparent contradiction is not so easily solved, since the intestinal mucosa degenerates rapidly after death. In our experience, villi are relatively short or thin, and may be lost during sampling and processing of the specimens. Simpson and Gardner (1972) stated that villi appear approximately 30 cm below the pylorus and disappear close to the end of the gut, where the structure becomes somewhat similar to that of the human gross intestine. Following this observation, the same authors (Simpson and Gardner, 1972) called colon this terminal tract of the intestine (approximately 30 cm long in the bottlenose dolphin), characterized by absence of villi and increased presence of goblet cells. Other reports (Russo et al., 2012) also cited the absence of villi in the distal part of the gut. Paneth cells[aa] are well evident in the intestinal glands (crypts of Lieberkün) (Harrison et al., 1977). Undifferentiated (stem) cells are present only at the bottom of the intestinal glands (Harrison et al., 1977).

---

aa. Other Authors (Simpson and Gardner, 1972) affirm that there are no Paneth cells in the gut of porpoises. Our own observations instead confirm their presence as described by Harrison et al. (1977). Interestingly, the presence of Paneth cells is uncertain in the closely related ruminant Cetartiodactyla.

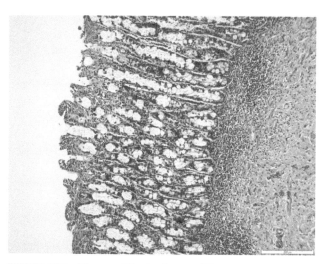

FIGURE 8.35   Mucosa of the duodenum of *T. truncatus* (scale bar = 200 μm). HE stain.

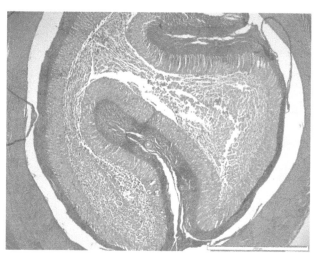

FIGURE 8.36   Intestinal mucosa with folds of *T. truncatus* (scale bar = 2 mm). HE stain.

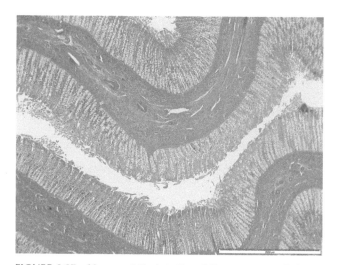

FIGURE 8.37   Mucosa of the intestine of *T. truncatus* (scale bar = 2 mm). HE stain.

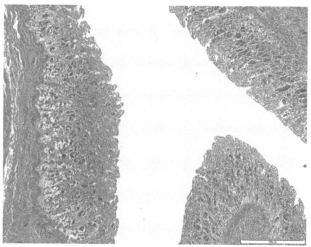

FIGURE 8.38   Mucosa of the intestine of *Stenella coeruloalba* (scale bar = 500 μm). HE stain.

The mucosa shows the typical lining found in terrestrial mammals, with absorptive cells endowed with microvilli (Figs. 8.37 and 8.38). There are also goblet and enterochromaffin cells. A detailed description of the fine structure of intestinal cells is present in Harrison et al. (1977).

The lamina propria contains small lymphocytes, plasma cells, and capillaries. Lymphatic nodules are evident (Fig. 8.39).

Harrison et al. (1977) report the presence of a myenteric plexus (Auerbach's plexus) between the layers of the *muscularis externa*; a submucosal plexus (Meissner's plexus) on the internal surface of the *muscularis externa*; and a mucosal plexus at the level of the *muscularis mucosae* (Fig. 8.40). Fibers (mostly unmyelinated) or cells containing neuroendocrine peptides were described in the myenteric plexus (β-endorphin, leu-enkephalin, NPY, substance P, CGRP, met-enkephalin, GRP, bombesin, somatostatin) and submucosal plexus (VIP, NPY, substance P, CGRP, met-enkephalin, GRP, bombesin, somatostatin) (Domeneghini et al., 1997). In the same study, no serotonin-containing element was found. The presence of a rich network of neuroendocrine elements in the gut was considered potentially indicative of rapid response of the target structures[bb] (Domeneghini et al., 1997). The presence of orexin-A and B (Gatta et al., 2014), and of leptin-like peptides (Russo et al., 2012) in the gut of the bottlenose dolphin also suggest a complex regulation of intestinal motility and feeding

---

bb. Other findings in *Ziphiidae* and *Delphinapteridae* (Pfeiffer, 1993) described modification of the innervation of the myenteric plexus and specializations of the muscularis externa with the presence of intercalation-like striations in the musculature of the gut, thus potentially supporting voluntary (and involuntary) gut movements associated with suction-feeding.

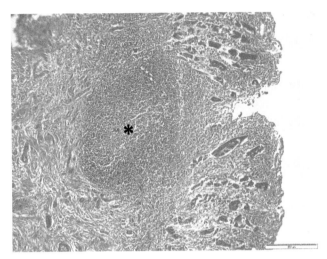

FIGURE 8.39   Intestinal plaques (*asterisk*) of *Stenella coeruloalba* (scale bar = 200 μm). HE stain.

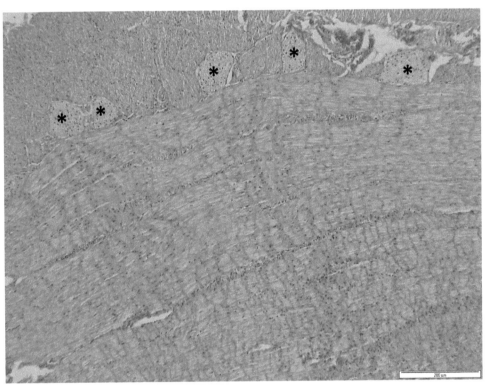

FIGURE 8.40   Nervous plexuses (*asterisks*) in the smooth muscle layers of the gut in *T. truncatus* (scale bar = 200 μm). HE stain.

strategy. Studies on captive bottlenose dolphins showed that orexin-A plasma levels are higher in the morning before feeding (Gatta et al., 2014).

Water absorption takes place in the dolphin intestine through aquaporin-1, a protein that acts as a water selective channel identified in the apical membrane of the mammalian enterocytes, and especially in the deep intestinal glands (crypts of Lieberkün) (Suzuki, 2010).

The terminal tract of the intestine that precedes the anal sphincter is characterized by an increased thickness of the walls (Fig. 8.41) and associated lymphatic structures (Fig. 8.42).

The peculiar morphological characteristics of the digestive tract of dolphins, and specifically the shortage of any bowel-like structure in the gut, suggest minimal or no storage of undigested residue (Simpson and Gardner, 1972). In fact, evacuated feces are generally rather liquid and may contain unprocessed materials, especially when shrimps are part of the diet.

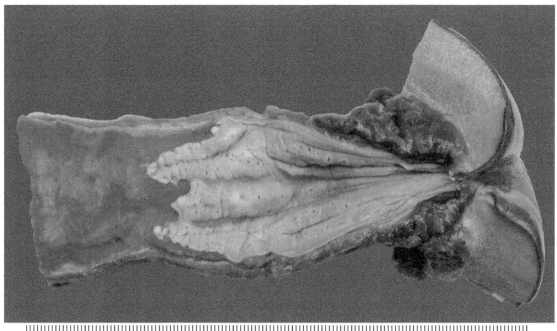

FIGURE 8.41   **Anal canal of an adult *T. truncatus*.**

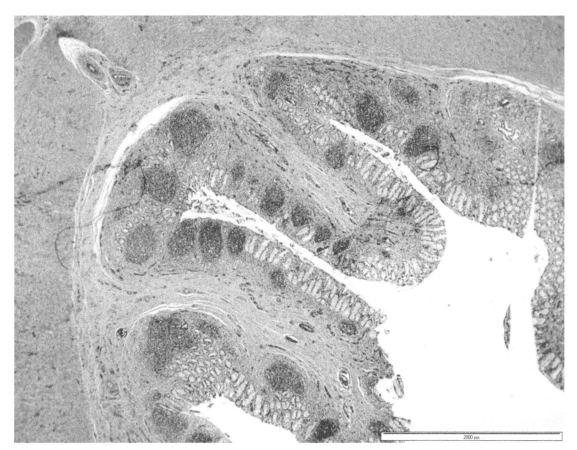

FIGURE 8.42   **Microscopic image of the rectum of *T. truncatus* (scale bar = 2mm).** HE stain.

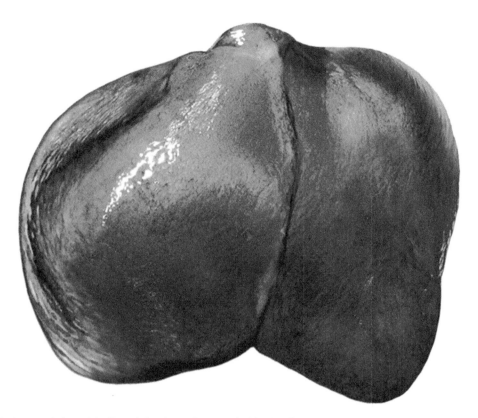

FIGURE 8.43 **Diaphragmatic face of the liver of** *G. griseus*. *(Drawings by Massimo Demma).*

THE LIVER

The liver is the largest gland of the body[cc] and is located in the peritoneal cavity, adjacent to the visceral surface of the diaphragm. The weight of the liver averages 2–3% of total body weight in dolphins. So the liver of a male adult specimen of *T. truncatus* of 300 kg may weigh from 6 to 9 kg. According to Slijper (1979), the liver of Cetacea (and of dolphins, in particular) is definitely heavier and larger than can be expected based on comparative values in terrestrial mammals. On the contrary, the liver of larger dolphins and whales is comparatively smaller than that of dolphins and porpoises as referred to body mass.

The liver of the bottlenose dolphin can be reached at the level of the lower third of the 6th to 9th intercostal space, in the right hypochondriac region, the epigastric region, and the left hypochondriac region of the abdomen, in the so-called intrathoracic part of the abdominal cavity (see also footnote v).

The liver of dolphins has a dark blue-gray to violet color and shows a cleft in the ventral margin indicative of a macroscopic subdivision into lobes.[dd]

The liver has two sides, one adhering to the diaphragm (diaphragmatic side, Fig. 8.43) and one facing the viscera in the peritoneal cavity (visceral side, Fig. 8.44). The diaphragmatic surface adheres completely to the whole surface of the diaphragm to which it is connected by a peritoneal folding. The diaphragmatic side of the liver follows exactly the shape of the diaphragm. The dorsorostral part of the liver is quite thick and almost protruding. In fact, one of the characteristic features of the liver in dolphins is the pronounced convexity of the diaphragmatic surface and dorsal margin, due to the specific shape and position of the diaphragm and the relative low volume of the peritoneal cavity. The liver can be explored in the live animal using ultrasound technology (Fig. 8.45).

---

cc. The liver plays a key role in the animal metabolism, and damage to the hepatic structure deeply affects the health of the individual and its chances of survival. Dolphins are on the top of the food chain and therefore their liver accumulates eventual toxic substances. This is one of the reasons why the study of pollutants (including heavy metals) in the cetacean liver is so important to understand the condition of the marine environment.

dd. The liver of many terrestrial Cetartiodactyla (including the pig) is macroscopically divided into lobes, generally with and independent ventral margin. However, ruminants (and man) show no external macroscopical subdivision into lobes, although functional compartments (also called lobes) exist based on the internal ramification of the main vessels.

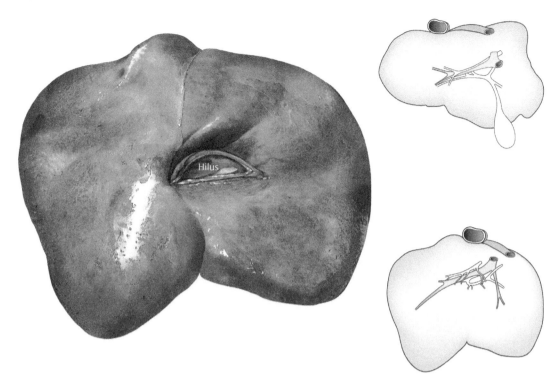

FIGURE 8.44   Visceral face of the liver of Grampus griseus (left) and schematic representations of the visceral surface of the livers of Bos taurus (top right) and *G. griseus* (bottom right) livers, showing the organization of the common bile duct with and without gallbladder. *(Drawings by Massimo Demma; top right modified after Nickel et al., 1980. Trattato di anatomia degli animali domestici, Volume II Splancnologia. Ambrosiana Editrice, Milan).*

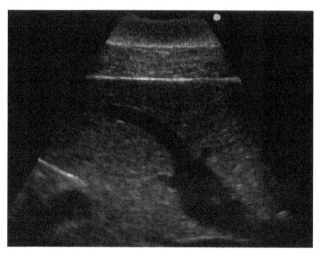

FIGURE 8.45   Ultrasound image of the liver of the bottlenose dolphin. *Courtesy of Pietro Saviano.*

The visceral surface shows the hilum of the organ with the entrance of the hepatic artery and portal vein, the exit of the common bile duct (*ductus choledocus*), and the passage of the lymphatic vessels and nerves.

In an often-cited description of the liver of an adult female *Grampus griseus*, Richard and Neuville (1896) reported that the caudal vena cava, while adhering to the dorsal surface of the organ for about 20 cm, received two large intrahepatic sinuses, besides several additional hepatic veins. The right sinus was further divided into two parts, one of which, the right one, was markedly larger. The sinuses had an ovoid shape. Although similar structures are well known in Phocids, the presence of large hepatic sinuses in the normal liver of dolphins remained doubtful awaiting further investigation. Harrison and Tomlinson (1956) reported no hepatic sinus in their description of the vena cava in the fetuses of several cetacean species, including *D. delphis* and *Globiocephala melaena* (= *G. melas*). However, the presence of very large, ramified but possibly

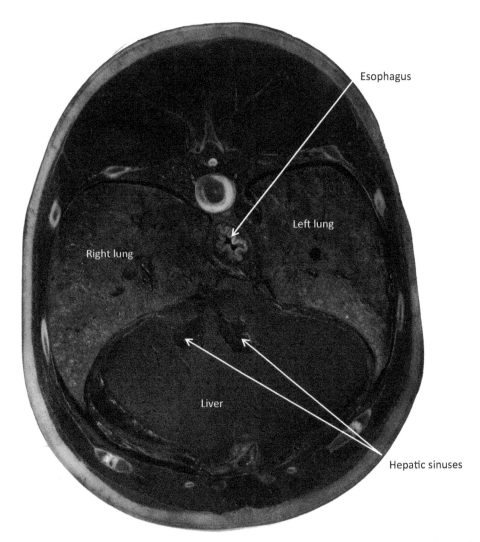

FIGURE 8.46 **Transversal section of the body of an adult *T. truncatus* at the level intrathoracic part of the abdominal cavity.** Two large hepatic sinuses are well evident close to the dorsal margin of the liver.

continuous sinus-like macroscopic vascular cavities in the liver (Fig. 8.46) is evident in the photographs of the serial sections of the body of an adult male bottlenose dolphin (for a discussion of the potential functional role of the hepatic sinuses, see Chapter 4).

The liver of cetaceans, like that of many terrestrial species,[ee] lacks a gallbladder and the bile produced in the hepatic stroma is carried into the duodenum by common hepatic duct.[ff] The hepatic duct and the pancreatic ducts have a common termination (see the following). The absence of an extra-hepatic storage organ for the bile is possibly related to the continuous ingestion of food and consequent frequent presence of food in the proximal intestine in the living dolphin.

Structure of the Liver

The general architecture of the liver of dolphins (Figs. 8.47–8.49) is similar to that of the liver of terrestrial mammals. However, connective septa between adjacent lobules are not always marked, and sometimes absent, and the central veins of the hepatic lobules not always well evident and easy to identify. In their description of the liver of bottlenose, common,

ee. Including the horse and the rat (but not the mouse).
ff. In man and the other mammals, the cystic duct that carries the bile from the gallbladder joins the common hepatic duct to form the common bile duct or *ductus choledochus*. Since in dolphins the gall bladder is absent, the *ductus choledochus* does not exists, and the bile is conveyed to the duodenum by the common hepatic duct.

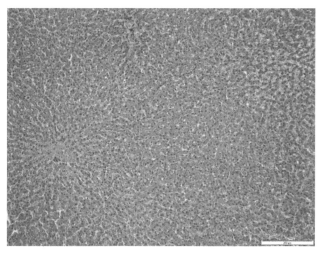

FIGURE 8.47   Microscopic image of the liver of *T. truncatus* (scale bar = 200 μm). HE stain.

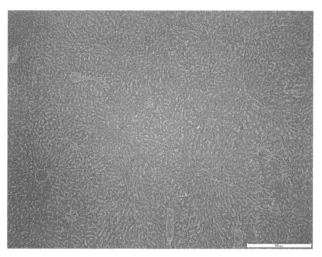

FIGURE 8.48   Microscopic image of the liver of *D. delphis* (scale bar = 500 μm). HE stain.

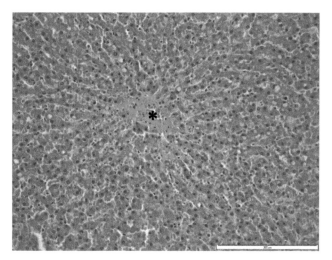

FIGURE 8.49   Centrolobular vein (asterisk) in the liver of *T. truncatus* (scale bar = 200 μm). HE stain.

and spinner dolphins, Simpson and Gardner (1972) reported the presence of muscular sphincters in the branches of the portal veins.[gg] We observed the liver of several specimens of *T. truncatus*, *S. coeruleoalba*, and *G. griseus*, and the liver of a newborn (perhaps stillborn) *G. melas*. The central veins are not placed at regular interval and the connective septa are sometimes difficult to find. The sinusoids are sometimes separated by large spaces (Fig. 8.50). We also observed that several veins and perilobular arteries (Fig. 8.51) are apparently surrounded by sphincters, a feature possibly related to regulation of blood flow during diving. A striking feature is the presence of large vascular sinuses in the parenchyma, sometimes so large that their presence can be seen in the section without the microscope. The existence of a complex regulatory mechanism of blood flow in the liver of dolphins is highly probable, and perhaps deserves further investigations.

THE PANCREAS

The pancreas (Fig. 8.52) is a large exocrine and endocrine gland placed in the intrathoracic part of the abdominal cavity, in the ventral part of the epigastric region. The enzymatic secretions of the gland are carried to the duodenum by the pancreatic duct.

---

gg. Simpson and Gardner (1972) briefly explained the sphincters as functional to pooling of blood during prolonged diving. Unfortunately this interesting observation received no further attention in subsequent studies and remains unproved.

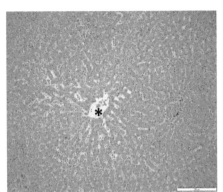

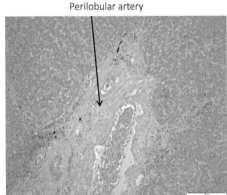

FIGURE 8.50 **Microscopic images of the liver of *D. delphis*: centrolobular vein (*asterisk*, left) and perilobular artery with thick muscular wall (right) (scale bars = 200 μm). HE stain.**

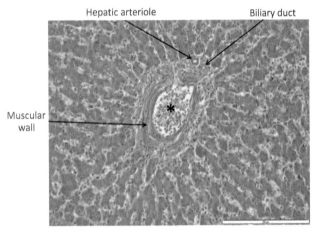

FIGURE 8.51 **Thick muscular wall surrounding a perilobular vein (*asterisk*), part of the portal triad (including also a biliary duct and a hepatic arteriole) in the liver of *T. truncatus* (scale bar = 200 μm). HE stain.**

FIGURE 8.52 **Pancreas of *T. truncatus*.**

The pancreas of dolphins is similar to the human pancreas (Colegrove and Venn-Watson, 2015), flat and irregularly shaped. The two sides of the organ correspond to the *laminae* of the initial part of the *mesoduodenum*[hh] that involves the cranial flexure of the duodenum. The pancreas, therefore is related to the *lesser omentum* and closely associated to the visceral surface of the liver. The hilum of the liver is always close to the body of the pancreas, and represents a topographic landmark at necropsy. The color of the pancreas is pink and the surface translucent. However, the color may rapidly darken after death.

The weight of the pancreas averages 0.1–0.2% of total body weight in dolphins (Slijper, 1979). So the pancreas of a male adult specimen of *T. truncatus* of 300 kg may weigh 0.3–0.6 kg.

The majority of dolphins possesses a single pancreatic duct that conveys the exocrine secretion of the gland into the lumen of the duodenum. According to Slijper (1979), there may be exceptions to the single-duct rule. The pancreatic duct joins the common terminal hepatic duct before entering the duodenum. The union of the hepatic and pancreatic duct(s) is called *ampulla hepatopancreatica*. The ampulla of *D. delphis* presents a conspicuous external bulbar part (*pars bulbaris ampullae* according to Cave, 1982), and a *pars papillaris ampullae* that enters the wall of the duodenum (Cave, 1982) (Fig. 8.53). According to the same author (Cave, 1982), in *G. griseus* and *G. melas* the *ampulla hepatopancreatica* enters

---

hh. The mesentery that links the first part of the intestine to the wall of the peritoneal cavity.

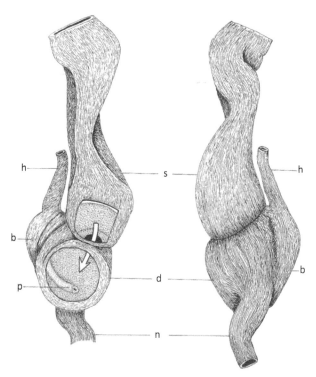

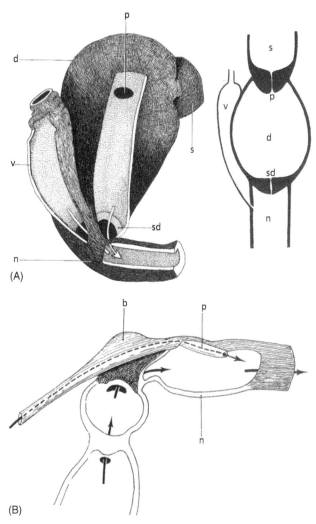

FIGURE 8.53 Schematic representation of the *ampulla hepatopancreatica* (common hepatic duct and pancreatic duct) in *D. delphis*. Common dolphin (*D. delphis*). Showing (left) ventral and (right) dorsal aspects of gastroduodenal region with mucosal surface partially exposed. An arrow traverses the pyloric canal, a bristle the hepatic duct. In relation to the duodenal wall the pars bulbaris ampullae is epimural, the pars papillaris intramural. b, pars bulbaris ampullae; d, dilated portion of duodenum; h, hepatic duct; n, nondilated portion of duodenum; p, pars papillaris ampullae; s, stomach. *(From Cave, 1982. The Vaterian ampulla of certain cetaceans. Invest. Cetacea 14, 131–147).*

FIGURE 8.54 Schematic representation of the *ampulla hepatopancreatica* (common hepatic duct and pancreatic duct) in *G. griseus* (left) and *G. melas* (right). (A) Risso's dolphin (*G. griseus*) duodenal region in (left) norma ventralis; (right) longitudinal section. d, dilated portion of duodenum; n, nondilated portion of duodenum; p, pylorus; s, stomach; sd, septum duodeni; V, vaterian ampulla. (B) *Globicephala*, Vaterian ampulla; the bulbar (b) and papillary (p) components are equally developed and the latter opens into the nonspecialised portion (n) of the duodenum. *(From Cave, 1982. The Vaterian ampulla of certain cetaceans. Invest. Cetacea 14, 131–147); and Turner, 1868. A Contribution to the Anatomy of the Pilot Whale (Globiocephalus Svineval, Lacépede). J Anat Physiol. 2, 66–79).*

the duodenum after the septum that separates the ampulla with the remaining part of the duodenum with a dedicated valve (Fig. 8.54).

Structure of the Pancreas

The pancreas is divided into lobules separated by bands of fibrous connective tissue (Colegrove and Venn-Watson, 2015). The structure of the acini that constitute the exocrine pancreas of cetaceans (Figs. 8.55 and 8.56) is very similar to that of terrestrial mammals, with dark-stained polygonal cells and ramified intercalated and interlobular ducts. These latter ducts will finally converge into the pancreatic duct described earlier.

The exocrine pancreas produces enzymes necessary for digestion. The characteristics of the endocrine pancreas, also in relation to the production of insulin, glucagon, and other hormones involved in carbohydrate metabolism, are described in Chapter 7.

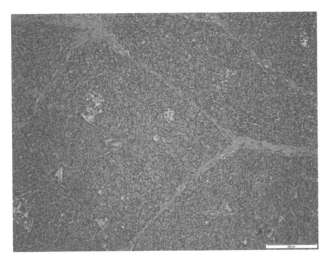

FIGURE 8.55 Microscopic image of the pancreas of *S. coeruleolba* (scale bar = 200 μm). HE stain.

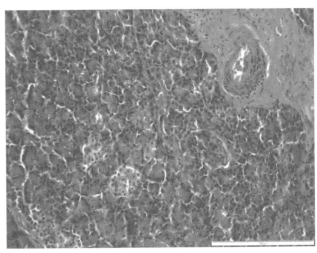

FIGURE 8.56 Microscopic image of the pancreas of *S. coeruleolba* (scale bar = 200 μm). HE stain.

DEVELOPMENT OF THE DIGESTIVE SYSTEM

Developmental phases (stages) of selected dolphin species have been reported in Table 9.6 (see Chapter 9 for details). According to Štěrba et al. (2000), the first anlage of the esophagus, stomach, gut, and rudimentary liver can be distinguished at Stage 3. During Stage 4, rudiments of the dental *laminae* appear and the liver anlage fill most of the primitive abdominal cavity (together with the mesonephros). The embryonic pancreas also appears.

The development of all the digestive organs, and their relative distinct features, become more evident during Stage 5. Since the liver still occupies most of the abdominal cavity, the lengthening gut is displaced into the wide umbilical stalk (physiological umbilical hernia). Dental buds develop during Stage 7, and the intestine now moves back from the umbilical stalk into the abdominal cavity in which space is available because of the increased size of the fetus. At Stage 8, the structure of the organs starts to differentiate into the adult form. In Stage 9, dental buds start to change into dental cups, and tonsillae develop in the walls of the pharynx. In later stages, the gastrointestinal apparatus gradually acquires the adult shape (Fig. 8.57).

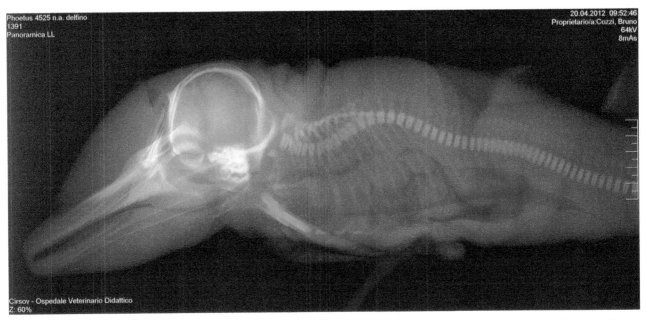

FIGURE 8.57 Fetus of *S. coeruleoalba* of 7 months of age showing the progressive organization of the gastrointestinal tract. *(Museum of Natural History G. Doria of Genova; specimen # 4525; X-ray courtesy of Prof Alessandro Zotti, Department of Animal Medicine, Production and Health of the University of Padova).*

REFERENCES

Arvy, L., Pilleri, G., 1972. Comparisons of the tongues of some odontocetes: *Pontoporia, Neomeris* and *Delphinus*. Invest. Cetacea 4, 191–200.

Bloodworth, B.E., Marshall, C.D., 2007. A functional comparison of the hyolingual complex in pygmy and dwarf sperm whales (*Kogia breviceps* and *K. sima*), and bottlenose dolphin (*Tursiops truncatus*). J. Anat. 211, 78–91.

Cave, A.J.E., 1982. The Vaterian ampulla of certain cetaceans. Invest. Cetacea 14, 131–147.

Colegrove, K.M., Venn-Watson, S., 2015. Histomorphology of the bottlenose dolphin (*Tursiops truncatus*) pancreas and association of increasing islet b-cell size with chronic hypercholesterolemia. Gen. Comp. Endocrinol. 214, 17–23.

Dold, C., Ridgway, S., 2007. Cetaceans. In: West, G., Heard, D., Caulkett, N. (Eds.), Zoo animal and wildlife immobilization and anesthesia. Wiley-Blackwell, Ames, Iowa, pp. 485–496.

Domeneghini, C., Massoletti, P., Arrighi, S., 1997. Localization of regulatory peptides in the gastrointestinal tract of the striped dolphin, *Stenella coeruleoalba* (Mammalia: cetacea). An immunohistochemical study. Eur. J. Histochem. 41, 285–300.

Donaldson, J., 1977. The tongue of the bottle-nosed dolphin (*Tursiops truncatus*). Harrison, R.J. (Ed.), Functional Anatomy of Marine Mammals, vol. 3, Academic Press, New York, pp. 175–197.

Feng, P., Zheng, J., Rossiter, S.J., Wang, D., Zhao, H., 2014. Massive losses of taste receptor genes in toothed and baleen whales. Genome Biol. Evol. 6, 1254–1265.

Ferrando, T., Caresano, F., Ferrando, S., Gallus, L., Wurtz, M., Tagliaferro, G., 2010. The tongue morphology and lingual gland histochemistry of Ligurian Sea odontocetes. Mar. Mamm. Sci. 26, 588–601.

Fiorucci, L., Garcia-Parraga, D., Macrelli, R., Grande, F., Flanagan, C., Rueca, F., Busechian, S., Bianchi, B., Arbelo, M., Saviano, P., 2015. Determination of the main reference values in ultrasound examination of the gastrointestinal tract in clinically healthy bottlenose dolphins (*Tursiops truncatus*). Aquat. Mammal. 41, 284–294.

Gatta, C., Russo, F., Russolillo, M.G., Varricchio, E., Paolucci, M., Castaldo, L., Lucini, C., de Girolamo, P., Cozzi, B., Maruccio, L., 2014. The orexin system in the enteric nervous system of the bottlenose dolphin (*Tursiops truncatus*). PLOS ONE 9, e105009.

Guimarães, J.P., de Britto Mari, R., Marigo, J., Rosas, F.C., Watanabe, I.S., 2011. Light and scanning electron microscopic study of the tongue in the estuarine dolphin (*Sotalia guianensis* van Bénéden, 1864). Zoolog. Sci. 28, 617–622.

Guimarães, J.P., Mari, R.B., Marigo, J., Rosas, F.C.W., Watanabe, I., 2012. Gross and microscopic observations on the lingual structure of the Franciscana (*Pontoporia blainvillei*—Gervais and d'Orbigny, 1844). Microsc. Res. Technique 75, 737–742.

Harrison, R.J., Tomlinson, J.D.W., 1956. Observations on the venous system in certain *Pinnipedia* and *Cetacea*. Proc. Zool. Soc. Lond. 126, 205–233.

Harrison, R.J., Johnson, F.R., Tedder, R.S., 1967. Underwater feeding, the stomach and intestine of some delphinids. J. Anat. 101, 186–187.

Harrison, R.J., Johnson, F.R., Young, B.A., 1970. The oesophagus and stomach of dolphins (*Tursiops, Delphinus, Stenella*). J. Zool. 160, 377–390.

Harrison, R.J., Johnson, F.R., Young, B.A., 1977. The small intestine of dolphins (*Tursiops, Delphinus, Stenella*). Harrison, R.J. (Ed.), Functional Anatomy of Marine Mammals, vol. 3, Academic Press, New York, pp. 297–331.

Herwig, R.P., Staley, J.T., Nerini, M.K., Braham, H.W., 1984. Baleen whales: preliminary evidence for forestomach microbial fermentation. Appl. Environ. Microbiol. 47, 421–423.

Heyning, J.E., Mead, J.G., 1996. Suction feeding in beaked whales: morphological and observational evidence. Nat. Hist. Mus. LA County Contrib. Sci. 464, 1–12.

Hohn, A.A., 2009. Age estimation. In: Perrin, W.F., Würsig, B., Thewissen, J.G.M. (Eds.), Encyclopedia of Marine Mammals, second ed. Academic Press, San Diego, pp. 11–17.

IJsseldijk, L.L., Leopold, M.F., Bravo Rebolledo, E.L., Deaville, R., Haelters, J., IJzer, J., Jepson, P.D., Gröne, A., 2015. Fatal asphyxiation in two long-finned pilot whales (*Globicephala melas*) caused by common soles (*Solea solea*). PLoS ONE 10 (11), e0141951.

Ishiyama, M., 1987. Enamel structure in odontocete whales. Scanning Microsc. 1, 1071–1079.

Johnson, F.R., Harrison, R.J., 1969. Ultrastructural characteristics of the dolphin stomach. J. Anat. 104, 173–174.

Kastelein, R.A., Dubbeldam, J.L., 1990. Marginal papillae on the tongue of the harbour porpoise (*Phocoena phocoena*), bottlenose dolphin (*Tursiops truncatus*), and Commerson's dolphin (*Cephalorhynchus commersonii*). Aquat. Mammal. 15, 158–170.

Lawrence, B., Schevill, W.E., 1965. Gular musculature in delphinids. Bull. Mus. Comp. Zool. Harv. 133, 1–65.

Menon, G.K., Pfeiffer, C.J., 2002. Cetacean oral and lingual epithelium: cell and subcellular structure. In: Pfeiffer, C.J. (Ed.), Molecular and Cell Biology of Marine Mammals. Krieger Publishing Co., Malabar, FL, pp. 412–423.

Miyazaki, N., 2002. Teeth. In: Perrin, W.F., Würsig, B., Thewissen, J.G.M. (Eds.), Encyclopedia of Marine Mammals. first ed. Academic Press, San Diego, pp. 1227–1232.

Nachtigall, P.E., Hall, R.W., 1984. Taste reception in the bottlenosed dolphin. Acta Zool. Fennica. 172, 147–148.

Nickel, R., Schummer, A., Seiferle, E., 1980. Trattato di anatomia degli animali domestici, vol. II, Splancnologia. Ambrosiana Editrice, Milan.

Nordøy, E.S., 1995. Do minke whales (*Balaenoptera acutorostrata*) digest wax esters? Br. J. Nutr. 74, 717–722.

Olsen, M.A., Aagnes, T.H., Mathiesen, S.D., 1994. Digestion of herring by indigenous bacteria in the minke whale forestomach. Appl. Environ. Microbiol. 60, 4445–4455.

Perrin, W.F., Myrick, Jr, AC, 1980. Growth of Odontocetes and Sirenians: Problems in Age Determination. International Whaling Commission, Cambridge, UK, pp. 1–230, special issue 3.

Pfeiffer, C.J., 1993. Neural and muscular control functions of the gut in odontocetes: morphologic evidence in beaked whales and beluga whales. J. Physiol. 87, 349–354.

Pfeiffer, D.C., Wang, A., Nicolas, J., Pfeiffer, C.J., 2001. Lingual ultrastructure of the long-finned pilot whale (*Globicephala melas*). Anat. Histol. Embryol. 30, 359–365.

Richard, J., Neuville, H., 1896. Foie et sinus veineux intra-hepatiques du *Grampus griseus*. Bull. Mus. Hist. Nat. Paris 2, 335–337.

Russo, F., Gatta, C., De Girolamo, P., Cozzi, B., Giurisato, M., Lucini, C., Varricchio, E., 2012. Expression and immunohistochemical detection of leptin-like peptide in the gastrointestinal tract of the South American sea lion (*Otaria flavescens*) and the bottlenose dolphin (*Tursiops truncatus*). Anat. Rec. 295, 1482–1493.

Saviano, P., 2013. Handbook of ultrasonography in dolphins: abdomen, thorax & eye. Pietro Saviano, Parma.

Shindo, J., Yamada, T.K., Yoshimura, K., Kageyama, I., 2008. Morphology of the tongue in a newborn Stejneger's beaked whale (*Mesoplodon stejnegeri*). Okajimas Folia Anat. Jpn. 84, 121–124.

Simpson, J.G., Gardner, M.B., 1972. Comparative microscopic anatomy of selected marine mammals. In: Ridgway, S.H. (Ed.), Mammals of the Sea. Biology and Medicine. Charles C. Thomas, Springfield, IL, pp. 298–418.

Slijper, E.J., 1979. Whales, II edition Hutchinson, London, pp. 1–512.

Sokolov, V.E., Volkova, O.V., 1973. Structure of the dolphin's tongue. In: Chapskii, K.K., Sokolov, V.E. (Eds.), Morphology and Ecology of Marine Mammals. Halsted Press (John Wiley & Sons), New York, pp. 119–127, Israel Program for Scientific Translation (Jerusalem).

Štěrba, O., Klima, M., Schildger, B., 2000. Embryology of dolphins—staging and ageing of embryos and fetuses of some cetaceans. Adv. Anat. Embryol. Cell Biol. 157, 1–133.

Suchowskaja, L.I., 1972. The morphology of taste organs in dolphins. Invest. Cetacea 4, 201–204.

Suzuki, M., 2010. Expression and localization of aquaporin-1 on the apical membrane of enterocytes in the small intestine of bottlenose dolphins. J. Comp. Physiol. B 180, 229–238.

Takahashi, K., Yamasaki, F., 1972. Digestive tract of the Ganges dolphin, *Platanista gangetica*. II. Small and large intestines. Okajimas Folia Anat. Jpn. 48, 427–452.

Turner, W.M., 1868. A contribution to the anatomy of the pilot whale (*Globiocephalus svineval*, Lacépede). J. Anat. Physiol. 2, 66–79.

Werth, A.J., 2000. A kinematic study of suction feeding and associated behaviors in the long-finned pilot whale, *Globicephala melas* (Traill). Mar. Mammal. Sci. 16, 299–314.

Werth, A.J., 2006. Odontocete suction feeding: experimental analysis of water flow and head shape. J. Morphol. 267, 1415–1428.

Werth, A.J., 2007. Adaptations of the cetacean hyolingual apparatus for aquatic feeding and thermoregulation. Anat. Rec. 290, 546–568.

Werth, A.J., Potvin, J., 2016. Baleen hydrodynamics and morphology of cross-flow filtration in Balaenid whale suspension feeding. PLoS ONE 11 (2), e0150106.

Werth, A.J., Straley, J.M., Shadwick, R.E., 2016. Baleen wear reveals intraoral water flow patterns of mysticete filter feeding. J. Morphol. 277, 453–471.

Yamasaki, F., Komatsu, S., Kamiya, T., 1976. An observation on the papillary projections at the lingual margin in the striped dolphin. Sci. Rep. Whales Res. Inst. 28, 137–140.

Yamasaki, F., Komatsu, S., Kamiya, T., 1978. Taste buds in the pits at the posterior dorsum of the tongue of *Stenella coeruleoalba*. Sci. Rep. Whales Res. Inst. 30, 285–290.

Chapter 9

Urinary System, Genital Systems, and Reproduction

The urinary (excretory) system and the genital (male and female) systems are often considered together for many reasons, including embryological origin and persistence of common anatomical traits (ie, the urethra). Therefore, we decided to present them together in a single chapter. External differences between the two sexes are illustrated at the beginning of the description of the genital system.

URINARY SYSTEM AND WATER BALANCE

"The whale has … 14,000 kidneys" (Oliver, 1968)

The main function of the urinary system is excretion and water balance. Cetaceans have developed a specialized kidney to handle the large volume of electrolytes and water they process due to their marine environment. Dolphins may gain water out of sea water because they can concentrate their urine in a way that they can excrete more than twice as many chloride in their urine as humans (for review see Ortiz, 2001; Venn-Watson et al., 2008; Ridgway and Venn-Watson, 2009). As in other mammals, the osmolality of the urine of dolphins is mainly controlled by the hormones vasopressin (Ballarin et al., 2011) and rennin. The urine is transported from the kidneys via the ureters into the small bladder of dolphins. As in other mammals, the ovoid kidneys in dolphins are in retroperitoneal position. In dolphins, the kidneys are specialized as reniculate (multilobed) kidneys, where each of the 300–1200 lobes (*lobus renalis*) has all the components of a complete metanephric kidney. Unique to cetaceans is a fenestrated muscular basket (*sporta perimedullaris musculosa*) between the cortex and the medulla, of unknown function. The nephrons of dolphins resemble the mammalian bauplan. The renal arteries of cetaceans enter the kidneys medially at the cranial poles but not in the renal hilus because the kidneys are drained by separate ureters leaving the kidneys at the caudal end.

The urinary system is topographically and developmentally related to the reproductive system. Accordingly, both systems share common terminal structures located on the level of the pelvic rudiments. The urinary system includes the kidneys (*renes*), the ureters, the bladder, and the urethra (Fig. 9.1).

The function of the urinary system is excretion and water balance. The kidneys remove azotic excretes and metabolism end products from the blood by filtering the plasma. This ultrafiltrate is the primary urine and is thus isotonic to the blood. In a second step, the kidneys reabsorb useful substances from the primary urine (eg, water, electrolytes, glucose) generating the secondary urine which is transported via the ureter into the bladder. In large dogs approximately 1500 L of blood are filtered daily in the kidneys from which 150 L primary urine are produced which result, in turn, into approximately 1.5 L secondary urine. Unfortunately there are no comparable numbers for dolphins available but these numbers should approximate the same range depending on water (and food) intake. Water balance and thus the balance of internal osmolarity in dolphins is of particular interest because most of these species live in a hypertonic medium.[a] Accordingly, to maintain the normal (physiological) composition of body fluids the dolphin kidney needs to reabsorb most of the water from the urine. Additionally to their function of excretion and water balance, the kidneys have endocrine functions producing the hormones renin, erythropoietin, and bradykinin (see later).

---

a. The sea water has greater amounts of solutes than the body fluids.

Anatomy of Dolphins. http://dx.doi.org/10.1016/B978-0-12-407229-9.00009-9

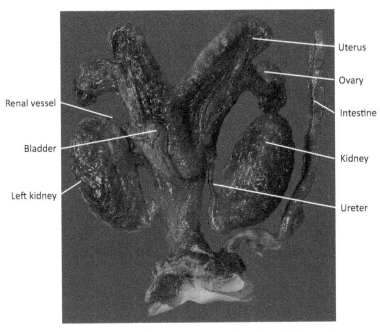

FIGURE 9.1 Ventral view of the female genital apparatus, urinary apparatus and rectum of *T. truncatus*.

The Kidney

Position, Shape, and Size

As in other mammals, the kidneys in dolphins lie in a retroperitoneal position (Figs. 9.2–9.4), typically against the *m. psoas major* at or near the dorsal midline attachment of the diaphragm (*crura*), with its cranial pole corresponding to the first lumbar vertebrae, and therefore more caudal than in terrestrial Cetartiodactyla. Due to the oblique position of the diaphragm, the caudal pole of the kidney reaches nearly the plane of the anus (Arvy, 1973–1974a). However, the right kidney is usually slightly more rostral than the left. Both kidneys are separated dorsally by the *v. cava caudalis* and ventrally by the colon descendens and the intestine position (Figs. 9.2–9.4). The kidneys can be explored in the live dolphin using ultrasound technology (Fig. 9.5).

The dolphin kidney is ovoid in shape (Fig. 9.6; not bean-shaped as the human or sheep kidneys), dark red to brown, and some authors reported impressions of neighboring structures. It is, however, not clear whether these impressions are

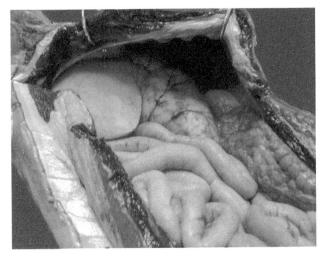

FIGURE 9.2 In situ preparation of retro-diaphragmatic organ systems of *Grampus griseus*.

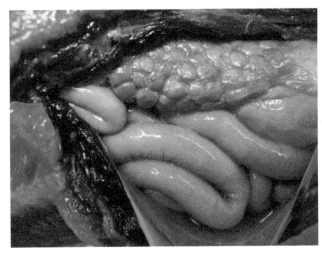

FIGURE 9.3 In situ image of the left kidney of *G. griseus*.

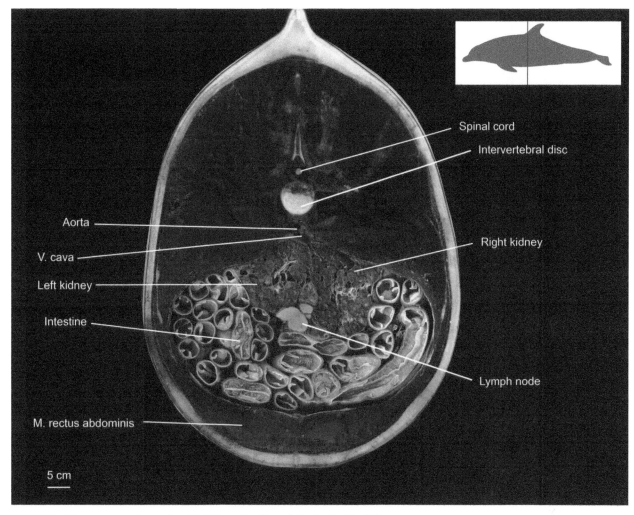

FIGURE 9.4   Transverse sawing section of a *T. truncatus* through the level of the kidneys.

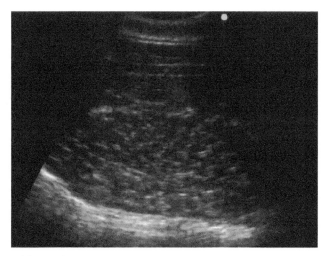

FIGURE 9.5   Ultrasound image of the kidney of the bottlenose dolphin. *Courtesy of Pietro Saviano.*

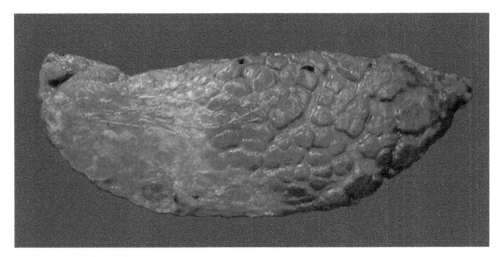

FIGURE 9.6   **Dissected complete kidney of** *T. truncatus*.

physiological or the result of post mortem artifacts. Topographically, the left kidney is related to the stomach, intestinal loops, and liver and the right one, to the liver, duodenal ampulla, and intestine. A single kidney of common dolphins is approximately 13 cm long and 6 cm wide, weights approximately 186 g (Arvy, 1973–1974a) and does not show preferential asymmetry because the left kidney is in mean only 15 g lighter than the right one. For comparison, the weights of kidneys of some dolphin species are summarized in Table 9.1. The density of a fresh specimen of common dolphin kidney was close to water (1.03) (Gihr and Pilleri, 1969).

Morphological and Functional Divisions

The kidney as a whole is covered by the renal fascia which is thin and fibrous in dolphins and it does not adhere very close to the *reniculi* (Fig. 9.3) (Arvy, 1973–1974a). Interestingly, the renal bed was reported to be very rarely adipose. Accordingly, the adipose capsule should be thin in dolphins (Fig. 9.2).

All mammals have metanephric kidneys (ie, containing cortex, medulla, calyces, and blood supply). In many marine mammals such as dolphins, the kidneys are specialized as reniculate (multilobed) kidneys, where each lobe (*lobus renalis*) has all the components of a complete metanephric kidney (Fig. 9.7). This characteristic is thus similar to the multilobed kidney of bovines but the dolphin kidney has more and smaller lobes (see later). Accordingly, each lobe was termed *reniculus* and each *reniculus* is equivalent to a unipapillary kidney of terrestrial mammals. These reniculi lay in several layers, so that

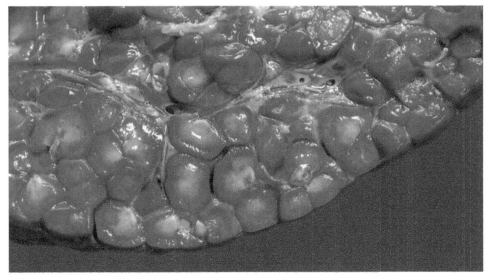

FIGURE 9.7   **Kidney of** *T. truncatus* **cut longitudinally.**

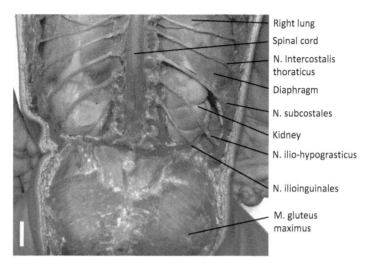

Right lung
Spinal cord
N. Intercostalis thoraticus
Diaphragm
N. subcostales
Kidney
N. ilio-hypograsticus
N. ilioinguinales
M. gluteus maximus

FIGURE 9.8 **Dorsal view of a historical preparation of a late-term human fetus showing the reniculated kidney on the right (scale bar = 1 cm).** The left kidney is still covered by its adipose capsule. *(Collection Center of Anatomy, University of Cologne).*

TABLE 9.1 Weight of Kidney and Approximate Size and Numbers of Dolphin Reniculi (Arvy, 1973–1974a)

Species	Weight of Kidneys (g)	Number of Reniculi per Kidney	Size of Reniculi (Diameter at Kidney's Surface in mm)	Weight of Reniculus (g)
T. truncatus	310	375	Up to 11	0.65
D. delphis	185	~410	9.3	0.34
G. melas	1150	641		1.08
O. orca	3040	1217		1.47
G. griseus	790	~515	14.5	1.91
Stenella sp.	~210	~310	9.6	0.43

there is one layer of reniculi at the kidney's surface and further reniculi interior to them (Fig. 9.7). In the field of comparative anatomy, the reniculated kidney had attracted some attention because only large or aquatic mammals [Cetacea, Pinnipedia, Ursidae (*Ursus arcticos*), Lutrinae (*Lutra lutra*), Proboscidae, Bovinae, Rhinicerotidae, Sirenia (*Trichechus manatus*)] have such kidney type although many mammalian fetuses such as humans develop a reniculated kidney first and the *reniculi* fuse secondarily (Fig. 9.8).

Starting with only few reniculi, the number and size of individual *reniculi* increase progressively during fetal development of cetaceans. The kidney of adult dolphins has approximately 300–1200 reniculi, depending on the species' size and individual age (Table 9.1).

Each *reniculus* appears conical with its tip directed inward (Fig. 9.9). The *reniculi* are covered and are loosely connected by a thin fibrous capsule. In the Risso's dolphin the *reniculi* were reported to be linked to each other by trabeculae of fatty connective tissue (Gihr and Pilleri, 1969). The renal cortex, which includes the majority of the Bowman's capsules and convoluted tubules (*partes convolutae distales et proximales*), is situated superficially in each reniculus. The cortex is approximately 2 mm thick in the bottlenose dolphin and covers the pyramid-shaped renal medulla (*pyramides renales*) containing the majority of collecting ducts and erected tubuli (*partes rectae distales et proximales*). As in all mammals, the *tubuli* are accompanied by blood vessels (Fig. 9.10).

The renal pyramids are approximately 5 mm wide in an 11 mm diameter *reniculus* of a bottlenose dolphin. The height of a *reniculus* from the renal papilla to the superficial capsule, which covers each *reniculus*, is approximately 7 mm in the bottlenose dolphin. The cetacean *reniculus* is surrounded by a thin fibrous capsule composed of loosely arranged collagenous and reticular fibers (Pfeiffer, 1997). Accordingly, they are easily separated from each other by stump preparation and there is no tendency for a fusion of *reniculi*. The base of each pyramid is covered by a funnel-shaped calyx which opens in an excretory canal (Fig. 9.9). Next to these single *reniculi* there are also *reniculi* in which up to three renal pyramids open into one calyx (Fig. 9.10) (Arvy, 1973–1974a).

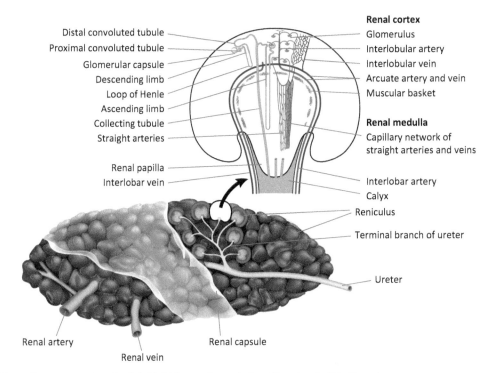

Distal convoluted tubule
Proximal convoluted tubule
Glomerular capsule
Descending limb
Loop of Henle
Ascending limb
Collecting tubule
Straight arteries

Renal papilla
Interlobar vein

Renal cortex
Glomerulus
Interlobular artery
Interlobular vein
Arcuate artery and vein
Muscular basket

Renal medulla
Capillary network of
straight arteries and veins

Interlobar artery
Calyx
Reniculus
Terminal branch of ureter

Ureter

Renal artery
Renal vein
Renal capsule

FIGURE 9.9 Schematic representation of a dolphin kidney and a *reniculus*. *(Drawings by Uko Gorter).*

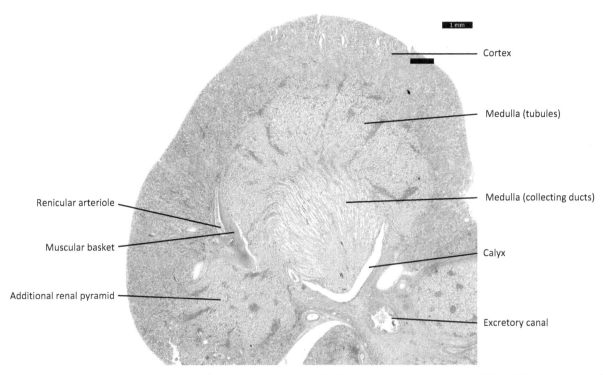

1 mm

Cortex
Medulla (tubules)
Medulla (collecting ducts)
Calyx
Excretory canal

Renicular arteriole
Muscular basket
Additional renal pyramid

FIGURE 9.10 Histological overview of a *reniculus* of *T. truncatus*. HE stain. The *reniculus* contains two pyramids (additional renal pyramid to the left).

FIGURE 9.11   **Detail of the muscular basket,** *sporta perimedullaris musculosa* **(detail of Fig. 9.10) (scale bar = 200 µm).** HE stain.

In all toothed whales, the *reniculus* is characterized by a small renal cortex and a relatively voluminous medulla (Fig. 9.10). Thus, in most delphinid species the cortex is thinner relative to the medulla in comparison to terrestrial species (Pfeiffer, 1997). Between the cortex and the medulla there is a fenestrated muscular basket (*sporta perimedullaris musculosa*) which is unique to cetaceans (Cave and Aumonier, 1962, 1967). However, this muscular basket consists of only a few muscle fibers in dolphins but resembles a fibrous sheath of loose connective tissue with collagenous and elastic fibers aligned unidirectionally (Fig. 9.11). The fenestrations of this basket vary from 66 to 333 µm in the bottlenose dolphin, but are wider in *Stenella longirostris* (735–1335 µm). According to its position the muscular basket is between the medulla and the reniculate vessels (see later) at the medullary pyramids. Consequently, this basket is an intrarenicular extension of the muscular and collagen tissue of the calyx' wall. The function of the muscular basket is not known. However, due to its position within the *reniculi*, it should be able to empty rapidly the medulla of urine (Pfeiffer, 1997).

As in all mammals, the functional units of the kidney are the nephrons which produce the urine. Each nephron is composed of a filtering component in the Bowman's capsule, situated in the cortex, and a system of tubules specialized for reabsorption and secretion. Within the Bowman's capsule there is a glomerulus composed of a capillary tuft that receives blood from an afferent arteriole of the renal circulation (Figs. 9.12 and 9.13). The glomerular blood pressure provides the driving force so that water and solutes are filtered into the lumen of the Bowman's capsule. The remaining blood (approximately 1/5 of plasma is filtered through the glomerular wall into the Bowman's capsule) including big molecularly substances passes into the efferent arteriole.

The following renal tubule reabsorbs and secrets various solutes such as ions, carbohydrates (eg, glucose), and amino acids (eg, glutamate). The components of the renal tubule are the proximal convoluted tubule in the cortex, the loop of Henle (U-shaped and lies in the medulla), and the distal convoluted tubule (Fig. 9.14). The subsequent collecting tubules convey the urine to the renal pelvis (Fig. 9.15).

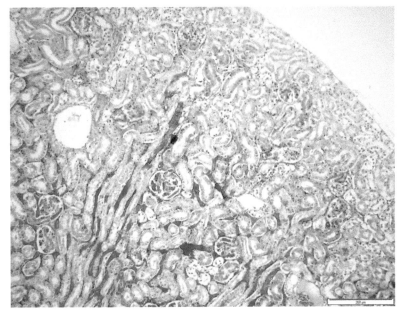

FIGURE 9.12 Cortex of the kidney of *D. delphis* with glomeruli (scale bar = 200 μm). Modified Azan stain (note that the superficial part of the cortex, on upper right side, was not stained accidentally).

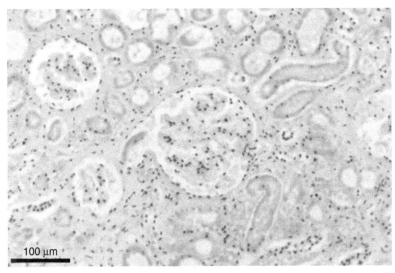

FIGURE 9.13 Detail of the renal cortex with Bowman's capsules (center) filled by a spherical plexus of blood capillaries, the glomerulus, in the renal medulla of the *reniculus* (detail of Fig. 9.10). HE stain. The capsule's urinary pole points left. The Bowman's capsules are surrounded by proximal tubules.

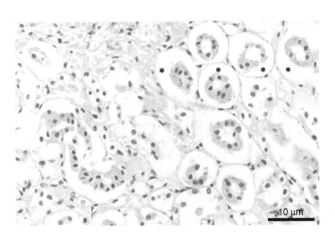

FIGURE 9.14 Detail of the renal medulla showing cuts of tubules of the *reniculus* (detail of Fig. 9.9). HE stain. The epithelium of the tubules shows significant shrinkage artifacts (*) due to fixation.

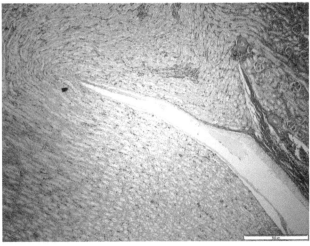

FIGURE 9.15 Renal pelvis of *D. delphis* (scale bar = 500 μm). Modified Azan stain.

The nephrons of dolphins resemble this mammalian bauplan and do not show remarkable differences in the delphinid species analyzed so far (*Delphinus delphis*, *Stenella coeruleoalba*). However, in general it is known that the cetacean nephron is short (Sperber, 1944). The mean diameter of the Bowman's capsule was 153 μm in a bottlenose dolphin and 125 μm in *S. longirostris*. In the harbor porpoise this capsule has a mean diameter of 163 μm. In comparison, the Bowman's capsule of dolphins are smaller than those in the adult horse and cow (~300 μm) as well as sheep (~260 μm). The nephrons of the harbor porpoise was studied in some detail by Inouye and by Sperber (Inouye, 1927; Sperber, 1944): The proximal tubules are approximately 5240 μm long, starting with a short neck and 47 μm in width. The thin segments of the loop of Henle are 1600–8500 μm long and only 13 μm thick. The thick segment is 3400–3700 μm long and 28 μm thick. The distal tubules are around 1000 μm long and narrower than the proximal tubules (32 μm). Accordingly, these structures are just large enough to be visible with the naked eye in sections of the kidney and approximate the measurements of the nephrons of young calves. Typical glomeruli and thick proximal convoluted and thin distal convoluted tubules are observed in the cortex (Pfeiffer, 1997).

As know from terrestrial mammals, an array of straight, collecting ducts and loops of Henle can be seen in the deepest portion of the cortex at the corticomedullary junction, and the collecting ducts and tubules can be identified in the medulla. As a delphinid peculiarity, prominent and tight bundles of vasa recta can be seen in the outer medulla resembling a *rete mirabile* (Pfeiffer, 1997). As a further peculiarity it was reported that the juxtaglomerular macula densa is exceptionally large in comparison to terrestrial mammals (Arvy, 1973–1974a).

The cause why marine mammals have reniculate kidneys is uncertain. The reniculate kidney of cetaceans, in general, has shorter nephrons than kidneys of terrestrial mammals. However, kidneys of terrestrial mammals divided into *reniculi* have relatively larger areas of a thin cortex, more glomeruli, and larger papillary volumes. Whether these facts are correlated with the marine environment is largely unknown because for example, there are indications on the number of reniculi, but no numbers of glomeruli in the cetacean kidney in literature. However, the relative large papillary volume can be considered as an adaptation for great efficiency in concentrating urine due to enhanced area for tubular reabsorption (Pfeiffer, 1997). Moreover, dolphins have a relatively high medulla thickness indicating a high efficiency of water utilization (Suzuki et al., 2008).

Vessels and Innervation

The renal arteries of cetaceans enter medially at the cranial poles of the kidneys but not in the renal hilus which is typical of most mammals. The renal vein leaves the kidney paralel to the artery, usually some centimeters further cranial (Fig. 9.9). Each kidney usually has one artery and vein but there are reports that two arteries may supply a single kidney in dolphins (Arvy, 1973–1974a).

The renal arteries originate at the aorta as in other mammals. After entering the kidney, the renal artery divides repeatedly into dichotomous bifurcations so that each reniculus is supplied by arterioles. The renicular arterioles, situated externally of the muscular basket (Fig. 9.9), are arched and run usually in pairs in direction of the cortex (*vasae interlobulares*). Here, they form the arcuate arteries (*arteriae arcuatae*). The arcuate arteries feed, on the one hand, the glomeruli of the Bowman's capsules and form an arterial plexus in the cortex which merges into a venous plexus. On the other hand, the arcuate arteries divide at the corticomedullary junction to form the straight arteries of the medulla. These arteries feed the straight veins. The straight veins as well as the venous plexus of the cortex both merge into the renicular veins that accompany the renicular arteries. Moreover, in some species, such as the common dolphin or some *Stenella* species, a series of radial veins form a subcapsular plexus that open independently into the renal vein. As a peculiarity in the bottlenose dolphin, the blood vessels enter and leave, respectively, the *reniculi* at the level of the corticomedullary junction. In this species, both the arcuate arteries and veins form an arcuate plexus. A subcapsular plexus is absent in the bottlenose dolphin but a few fine veins (15 μm in diameter) pass to the neighboring *reniculi*.

Nerves were reported to loop around each renal vein near the inferior vena cava.

Urinary Tracts and Bladder

The kidneys are drained by separate ureters (Fig. 9.9), which carry urine to a medially and relatively ventrally positioned urinary bladder (Fig. 9.1). The ureter is formed by the junction of the excretory canals of the individual *reniculi*. These excretory canals are in close contact to renicular vessels. Intrarenaly, these branches of the excretory canals meet in paramedial collector canals which open into the ureter. Because the ureter leaves the kidney near or at the caudal pole, the collector canals run nearly the whole length through the kidney (Fig. 9.9). Accordingly, there is no extended renal pelvis in dolphins but a dendritic renal pelvis type.

The dolphin ureter (Fig. 9.16) is only 15 mm thick and the urinary bladder is small (Gaskin, 1986). It lies on the floor of the caudal abdominal cavity and, when distended, may extend as far forward as the umbilicus in some dolphin species. The bladder of female common dolphins and *Stenella styx* is about 4.5 cm long and, when contracted, 1.3 cm wide. A transverse

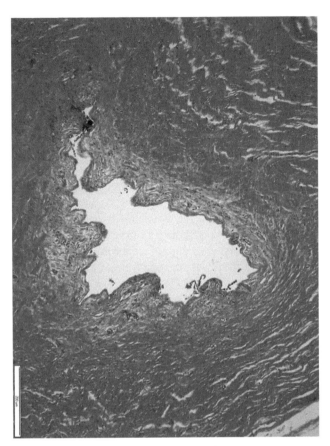

FIGURE 9.16   **Lumen of the ureter of** *S. coeruleoalba* **(scale bar = 200 μm).** HE stain.

vesicle fold stretches from the bladder along both sides of the *ligamentum teres*. Below this ligamentum, the ureters run on both sides to the bladder (Gihr and Pilleri, 1969).

The structure of the bladder (Figs. 9.17 and 9.18) is not different from that of terrestrial mammals, and shows the typical pseudostratified epithelium and thick smooth muscle with striated circular muscle close to the urethral opening.

Development

In embryos of terrestrial mammals, and thus presumably in embryos of dolphins, the earliest kidney is the mesonephros, comprised of ducts and tubules. As the mesonephros regresses, a second kidney structure, the metanephros, develops around the metanephric duct and is retained as the final kidney. The embryonic metanephric duct, which buds off the mesonephric duct (*ductus mesonephricus*, Wolff's duct), becomes the ureter. The urinary bladder develops from the proximal portion of the allantois and the urethra from the urogenital sinus (deriving from the embryonic cloaca).

Osmoregulation in the Marine Environment

Dolphins do not sweat and thus they do not loose salt across the skin. Unlike sea birds, dolphins lack specialized glands to excrete salts. Accordingly, salt excretion must be through the kidney, and cetaceans have developed a specialized kidney to handle the large volume of electrolytes and water they process. Dolphins may gain water out of seawater because they can concentrate their urine in a way that they can excrete more than twice as many chloride in their urine as humans (Costa, 2009).

Dolphins lose water with their feces and, because they live in a hyperosmotic environment, through their skin. Common dolphins lose as much as 4 L water per day which is approximately 70% of their total water intake. Thus extracting water from their prey and drinking limited quantities of seawater is necessary to prevent dehydration. This requires the processing of large urine volumes at high urine concentrations, and cetaceans have a specialized reniculate kidney that enables them to do this (Costa, 2009; Ortiz, 2001). In contrast, dolphins and cetaceans in general reduce water loss through respiration because they breathe less frequently than terrestrial mammals of equivalent size. The amount of oxygen and carbon dioxide

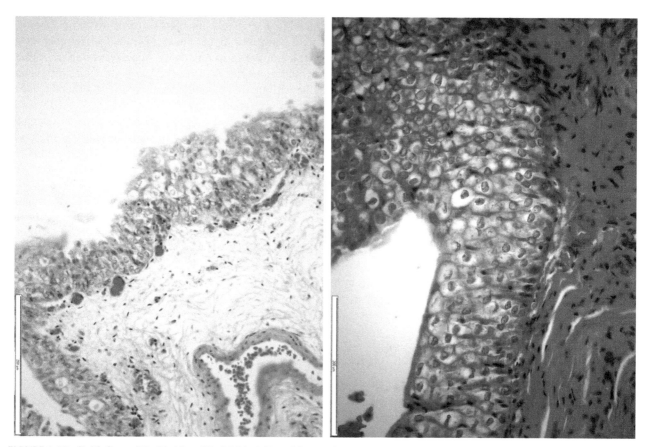

FIGURE 9.17   **Epithelium of the bladder of *D. delphis* (left) and *S. coeruleoalba* (right) (scale bars, left = 200 μm; right = 140 μm).** HE stain.

exchange per respiration cycle is as twice as high as in terrestrial mammals of equivalent size. Accordingly, dolphins breathe only half often in mean than terrestrial mammals.

Hormonal Osmoregulation

In mammals, the osmolality of the urine is mainly controlled by the hormones vasopressin and renin. Vasopressin, in dolphins as in other mammals mainly arginine vasopressin (AVP), is produced in the paraventricular nuclei of the hypothalamus to be stored in the neurosecretory terminals of the neurohypophysis (posterior pituitary). If osmoreceptors in the

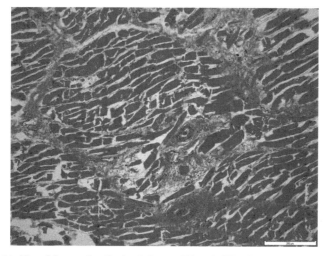

FIGURE 9.18   **Musculature of the bladder of *S. coeruleoalba* (scale bars = 200 μm).** HE stain.

hypothalamus record a high blood plasma osmolality, which should be the default situation in marine dolphins, this AVP would be released into the blood. Then, AVP causes vasoconstriction and thus higher blood pressure. On the other hand, AVP increases the number of aquaporins within the apical membrane of the collecting ducts. Aquaporins transport water from the urine into the plasma so that the expulsion of water is reduced (water is reabsorbed, urine becomes more concentrated). In parallel, the atrial natriuretic peptide (ANP) may be activated by pressure receptors in the atrium (volume increase) which causes a decrease of Na^+ reabsorption (increase of Na^+ excretion).

If the osmolality in the blood is to low (in relation to the urine) granular cells of the juxtaglomer apparatus excrete renin. In short, renin converts angiotensinogen to angiotensin I, which is quickly converted to angiotensin II (the most potent angiotensin) and finally to angiotensin III by angiotensin converting enzyme. Subsequently, angiotensin II stimulates the release of aldosterone from the adrenal gland, which in turn induces the reabsorption of Na^+ in the distal tubule of the nephron, resulting in a decrease in excreted Na^+.

Circulating concentrations of the two primary hormone systems (AVP and renin) involved in osmoregulation have been reported in usual concentration in dolphins in comparison to other mammals (Aubin, 2001). The concentration in the bottlenose dolphin was found to be about 3.3 pg/mL and aldosterone 234 pg/mL blood plasma (Ortiz and Worthy, 2000). In the bottlenose dolphin, the concentration of uinary AVP is related to the content of water of the ingested food (Ballarin et al., 2011).

In addition, urea seems to play an important role in the AVP-dependent mechanism of urine concentration in dolphins because urea develops the gradient for NaCl absorption against an osmotic gradient (Birukawa et al., 2005; Janech et al., 2002; Xu et al., 2013).

GENITAL SYSTEMS AND REPRODUCTION

In dolphins the need for a streamlined hydrodynamic body, and the virtual absence of mimic muscles of the face, greatly reduce the somatic differences between the two sexes. Males are consistently larger than females only in few species (notably in larger dolphins like killer whales), but since the genitalia are maintained within the body, immediate attribution of sex to a given individual observed from above the water is almost impossible.[b]

A way to discriminate is to observe the ventral region of the body including the genital slit (that hides the tip of the penis in males and closes the vulva in females) and the anus (Fig. 9.19). Females have an almost direct continue line between the genital slit (or groove) and the anal slit, while the distance between the two is more pronounced in males. In fact in males the distance between the genital slit and the anal opening is longer than the genital slit itself. Furthermore, adult females have two parallel additional grooves on the side of the genital slit, which correspond to the external opening of the nipples. These latter grooves are absent in males (Arvy, 1973–1974b).

The anatomy of the reproductive system of dolphins[c] is strikingly similar to that of their closest terrestrial counterparts, that is, the ruminants. As an example the penis of these even-toed mammals is kept by the *retractor penis* muscles inside a long preputial fold along the ventral edge of the abdomen and pelvis, except during mating. Many aspects of the female anatomy of ruminants also resemble those of dolphins, including the shape and structure of the uterus and ovaries.On the other hand the estrus cycle and reproductive physiology of dolphins in general have several points in common with that of odd-toed herbivores, such as the horse.[d]

ANATOMY OF THE FEMALE REPRODUCTIVE SYSTEM

The female reproductive apparatus of dolphins consists of paired oval-shaped ovaries and related oviducts, connected to a bicornuate uterus with a relatively long cervix opening into the vagina (Fig. 9.1 and 9.20). The external opening of the vulva appears as an elongated slit in which the clitoris is hidden. The modified labia, and the tight folding of the genital slit, prevents water from entering the vulva and the rest of the genital system.

---

b. Unless of course the animal is observed while performing a sex-specific behavior, such as copulation, lactation or swimming in close association to a young/newborn of the same species, or having an erection.

c. For a comprehensive multiauthor volume dedicated to several aspects of the reproductive anatomy and physiology of dolphins and other cetaceans see Miller (2007). For a recent description of the principal variations in the morphology of the female genital system of the bottlenose dolphin see Orbach et al. (2016).

d. The order Perissodactyla includes equine mammals (horse, donkey, zebra), the rhino and tapirs.

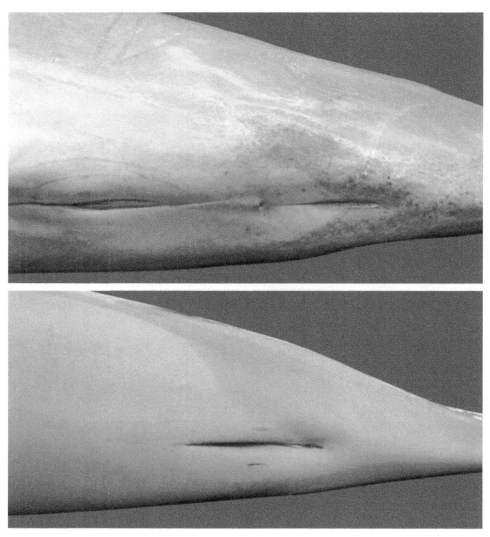

FIGURE 9.19 **Genital slit in a male (top) and female (bottom)** *T. truncatus*. Note the two grooves parallel to the genital slit in the female, that represent the opening for the nipples of the mammary glands.

The Ovaries and Uterine Tubes (Oviducts)

The ovaries of dolphins (Figs. 9.20 and 9.21) generally have an ovoid shape, but may also appear folded and U-shaped. Variations occur with age and subsequent pregnancies. In most species, the left ovary is the active gonad (or at least the first to become active) although the right ovary is also fully functional and often ovulates later in life or may take over in case of damage or failure of the left one (for detailed classification and review see Ohsumi, 1964; Plön and Bernard, 2007a).

Ultrasonographic imaging shows that the length of the left ovary varies with the length of the body (Brook, 2001), and may change during the estrous cycle[e].

The ovaries are located dorsally in the abdominal cavity, enveloped by an expansion of the *mesosalpinx* that constitutes the *bursa ovarica* (Rankin, 1961) close to the expanded opening of the uterine tubes, *ampulla tubae uterinae* (Slijper, 1966), and held in place by a specific ligament (*mesovarium*) connected to the broad ligament of the uterus. The ovary can be examined in vivo using ultrasound techniques (Fig. 9.22).

The architecture of the ovaries of dolphins (Figs. 9.23–9.25) (Harrison and Mc Brearty, 1977; Brook et al., 2002; Fukui, 2007) is very similar to that of terrestrial mammals. Macroscopically it is possible to recognize an ovarian fossa (*fossa ovarii*) in the active ovary (Fig. 9.23), like in the mare. The ovarian cortex is approximately 2 mm thick (Simpson and Gardner, 1972). Its basophilic stroma contains intertwining connective cells. The fetal ovary includes several primary

---

e. For measures of the ovary in the long-finned pilot whale, see Harrison, (1949).

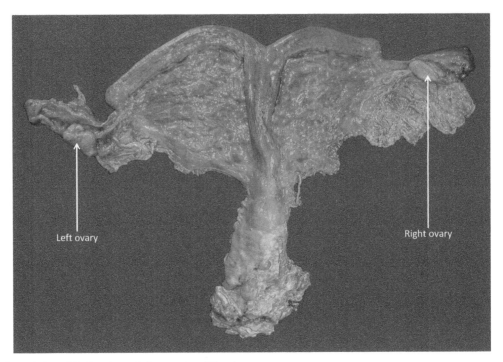

FIGURE 9.20 **Dorsal view of the female genital apparatus of *G. griseus*.**

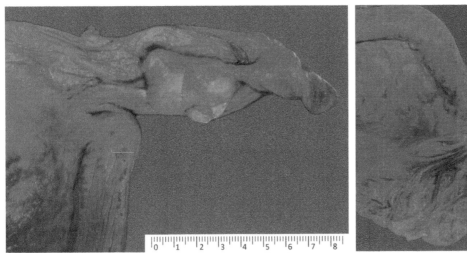
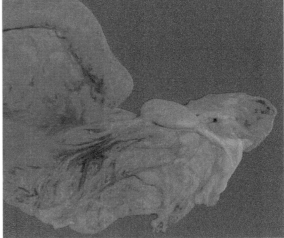

FIGURE 9.21 **Left (left) and right (right) ovary of *G. griseus*.**

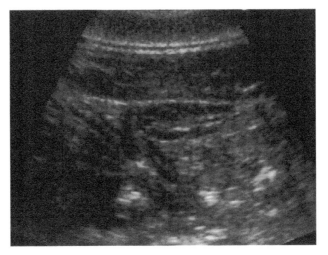

FIGURE 9.22 **Ultrasound image of the ovary of the bottlenose dolphin.** *Courtesy of Pietro Saviano.*

Follicles

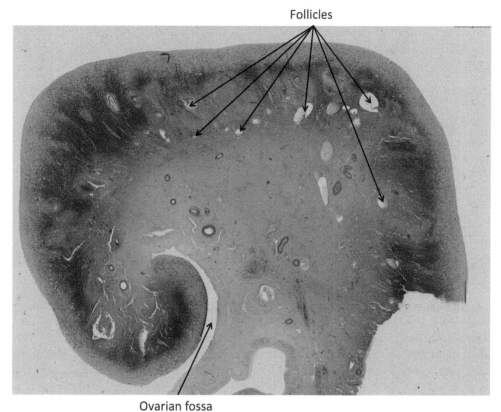

Ovarian fossa

FIGURE 9.23   **Microphotograph of the whole ovary of *T. truncatus.*** HE stain.

Follicles at different stages

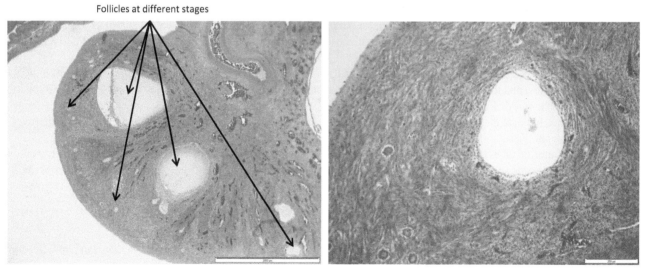

FIGURE 9.24   **Microphotographs of the cortex of the ovary of *T. truncatus* (scale bars, left = 2 mm; right = 200 μm).** HE stain.

oocytes and shows also remnants of the *rete ovarii* (Harrison, 1949). Follicles at various stages (and sizes, depending on the species) are easily seen in the cortex.

The medulla of the ovary contains less cells and more blood vessels and collagenous fibers (Simpson and Gardner, 1972). No interstitial cells have been observed in the medulla or the cortex of the ovary, at least in *Globicephala melas* (Harrison, 1949). Interestingly, a rare occurrence of true hermaphroditism has been recently signaled in the common dolphin (Murphy et al., 2011).

Dolphins have a single ovulation for each cycle. There are several spontaneous ovulations for each breeding season (possibly induced by day length), at least in the common dolphin (Dabin et al., 2008). More ovulations may occur before

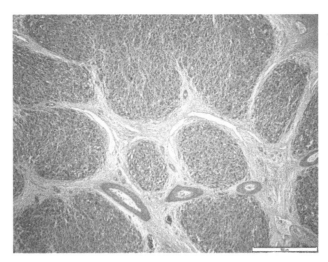

FIGURE 9.25   **Microphotograph of the stroma of the ovary of *T. truncatus* (scale bar = 500 μm).** HE stain.

pregnancy, and the conception rate is difficult to assess in wild animals. Apparently common dolphins show more estrous cycles for each breeding season than bottlenose dolphins (Dabin et al., 2008), although comparisons are hard to assess. When the ovarian follicle matures, the *corpus luteum* develops at the site of ovulation (Pomeroy, 2011). The markedly yellow *corpus luteum* then proceeds to secrete progesterone to maintain pregnancy if fertilization occurs, or recedes slowly in absence of pregnancy and leaves a sort of white-grayish scar, the *corpora albicantia* (Hirose et al., 1970; Collet and Harrison, 1981). Since the inactive remnants of the *corpora lutea* and the subsequent *corpora albicantia* seem to be maintained through life and remain well evident when present, the count of permanent scars was considered useful to determine the age of the animal at necropsy (although the presence of accessory *corpora lutea* may confound the count). The persistence of *corpora albicantia* was thought to be a distinctive trait of cetaceans. However, contrasting evidence suggests that, in the common dolphin, *corpora albicantia* heal quickly and their number does not increase with age after sexual maturity, and therefore cannot be considered a trustworthy element for age determination (for a thorough discussion see Dabin et al., 2008; for alternate or supporting mechanisms of persistence of *corpora albicantia* see Takahashi et al., 2006). According to other descriptions, *corpora albicantia* in bottlenose and striped dolphins are maintained only when derived from *corpora lutea* in pregnancy (for discussion see Robeck et al., 2001).

A distinction between the variety of different structures found in the ovary of cetaceans (*corpora atretica*; *corpora aberrantia*; cystic follicles) and the physiological signs of past pregnancies is not so easy. For a concise but clear description of the different ovarian *corpora*, see Pomeroy (2011).

Some evidence indicate that (at least in *Globicephala macrorhynchus*) ovaries may cease to function with old age (over 50 years) (Marsh and Kasuya, 1984).

The flexuous uterine tubes (*tuba uterina*, once called oviducts), seldom described in cetaceans, are much smaller in caliber that the uterine horns and terminate with a wide *ampulla* partially enveloping the ovary. *Fimbriae* have been described in the pilot whale (Slijper, 1966).

The Uterus and Cervix

The uterus of the bottlenose (Fig. 9.1), striped and Risso's (Fig. 9.20) dolphins is bicornuate (*uterus bicornis*) and closely resembles that of terrestrial hoofed mammals. The semicylindrical part is the body of the uterus (*corpus uteri*). It splits rostrally into the two uterine horns (*cornua uteri*) that continue with the uterine tubes. In the long-finned pilot whale the uterine horns are more independent and the fused part of the uterus is only 10 cm long (Harrison, 1949). The normal uterus presents a rather amorphous, soft tissue mass on ultrasonography (Brook, 2001).

The conformation of the dolphin uterus indicates a phylogenetic derivation from a double uterus. A remnant of this primitive divided structure may be found as a septum that runs along the axis of the body of the organ and partially divides the uterine cavity in two (*velum uteri*).

The implantation of the fertilized egg and the development of the placenta take place in the endometrium of the uterine horns (Rommel et al., 2007). In most dolphin (and other odontocete) pregnancies, the fetus is located in the left horn (Figs. 9.26 and 9.27), and the right one contains only the tail and part of the allantois (see later). The inner surface of the

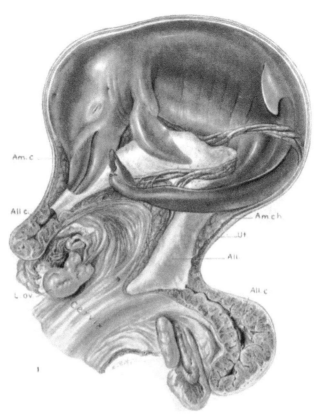

FIGURE 9.26   **Detailed drawing of the pregnant uterus of *T. truncatus*.** *(From Wislocki and Enders, 1941. The placentation of the bottle-nosed porpoise* (Tursiops truncatus). *Am. J. Anat. 68, 97–114).*

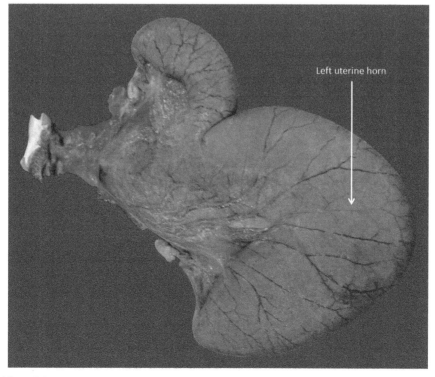

FIGURE 9.27   **Ventral face of the pregnant uterus of *T. truncatus*.** The left uterine horn is on the right.

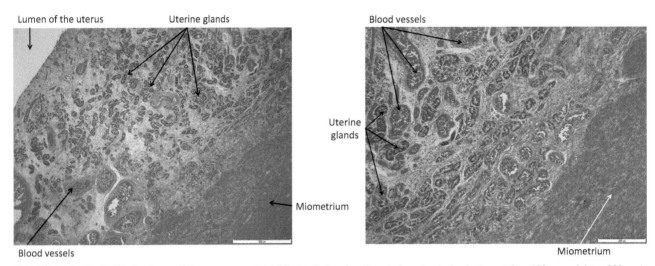

FIGURE 9.28 (Left) Uterine horn of *T. truncatus* and (right) detail showing the uterine glands (scale bars, left = 500 μm; right = 200 μm). HE stain.

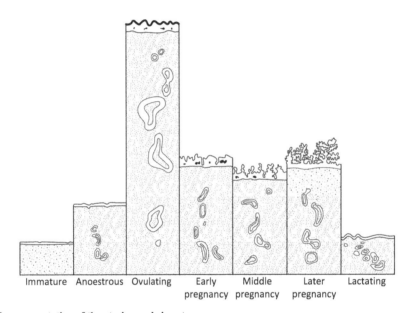

FIGURE 9.29 Schematic representation of the uterine cycle in cetaceans.

cornua and *corpus uteri* is characterized by several longitudinal folds, possibly functional to allow greater expansion of the uterine wall during pregnancy.

Structure

The uterus is composed by the external *serosa*,[f] the muscular layer (*myometrium*), and internal *endometrium* (Figs. 9.28 and 9.29). The *endometrium* (inner layer or mucosa) of the uterus is lined with pseudo-stratified cuboidal (in the uterine horns, Harrison, 1949) or columnar epithelium (Simpson and Gardner, 1972). The outer subepithelial layer (called *stratum compactum* by Harrison, 1949) contains the ducts of the uterine tubular glands and a superficial capillary plexus. The deeper layer (= *stratum spongiosum*) includes the convoluted parts of the uterine glands and other deeper vascular networks. As in most mammals, during the estrus cycle the superficial layer varies its structure and thickness and proliferates (Fig. 9.29).

---

f. The serosa that envelops the uterus is continue with the peritoneum that lines the walls of the abdominopelvic cavity. In fact the absence of a coxal (hip) bone and sacral vertebrae in dolphins makes it somewhat difficult to distinguish the passage between the abdominal and the pelvic cavity. The end of the peritoneal cavity is marked by the mesentery that folds on the genital organs, respectively called *mesovarium* (referred to the ovary), *mesosalpinx* (referred to the uterine tube) and *mesometrium* (referred to the uterus). The three folds together are also called broad ligament of the uterus.

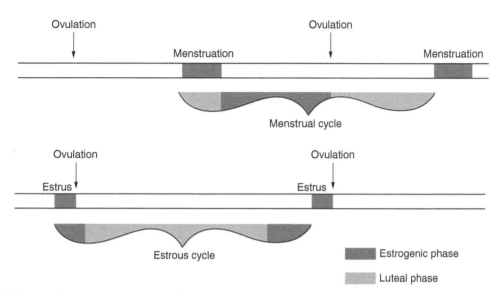

FIGURE 9.30   Differences between a menstrual cycle (top) and an estrous cycle (bottom).

When stimulated by ovarian steroids, the deeper layer of the uterine mucosa also increases its glandular content and the thickness of the vascular network.

The *myometrium* consists of two muscle layers (inner circular and outer longitudinal) separated by blood vessels (Slijper, 1966). The organization of the musculature thus suggests that uterine contractions may be rhythmic.

The growing phase of the uterine cycle is reversed if no pregnancy occurs and the uterine mucosa gradually returns to its quiescent state. However, there is no menstruation with blood loss and shedding of epithelial cells, as in primates. In fact the whole cycle is called estrus cycle in cetaceans, and not menstrual cycle (Fig. 9.30).

The *cervix uteri* (Fig. 9.31) is made up by the mucous membrane (Fig. 9.32), with the same epithelial characteristics of the uterus, and by a connective lamina propria (Simpson and Gardner, 1972), surrounded by smooth muscle fibers and a diffuse network of vessels. According to Simpson and Gardner (1972), this latter area shows the structure of erectile tissue.

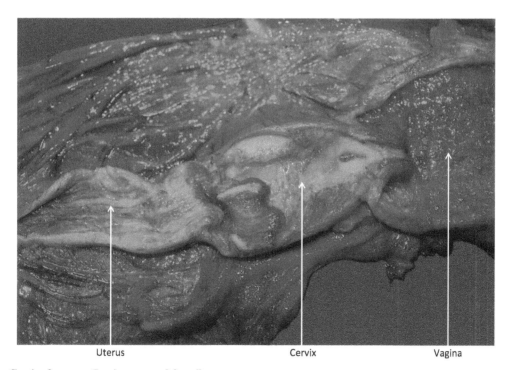

FIGURE 9.31   Cervix of a young *G. griseus* opened dorsally.

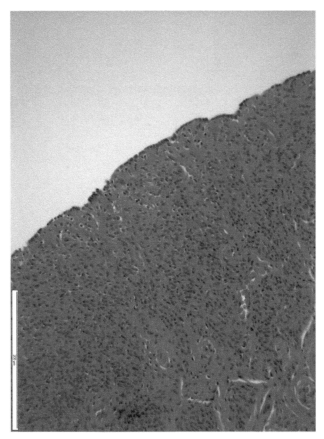

FIGURE 9.32   **Lumen of the cervix of** *S. coeruleoalba* **(scale bars = 200 μm).** HE stain.

The *cervix* opens into the vagina with a shape grossly similar to that of the human or equine cervix. The walls of the cervix are very thick with muscle bundles, and although the external diameter appears almost the same in the uterus, cervix, and vagina, the inner lumen of the cervix is greatly reduced. It has also been considered a *pseudocervix*, since its structure is indeed vaginal and not uterine. The cervix has several inner folds (*plicae circulares*) and is often filled with mucus[g] whose fluidity varies during the estrous cycle. The cervix of the long-finned pilot whale (*G. melas*) shows two complete circular folds that create a narrow connection with the uterine cavity (Harrison, 1949).

The possible functions of the folds and muscular bundles of the dolphin cervix include active accommodation of the penis, conveying of the seminal fluid, and immediate postcoital sealing of the uterine entrance (Harrison, 1949; Schroeder, 1990).[h]

Lactation reduces the thickness and glandular activity of endometrium and myometrium. Cyclicity is prevented by lactation and suckling, an evolutionary mammalian feature protective for the newborn. However, when suckling decreases below a certain time, the ovaries start to act again (West et al., 2000).

Temperature

The uterus and ovaries of dolphins are placed below the strong musculature of the back, and thus potentially exposed to the increased temperature produced during continuous exercise, considering also the insulating effect of the blubber.[i]

---

g. The mucus may form vaginal or cervical plugs depending also on the phase of the estrous cycle.

h. These functions further support the hypothesis that the muscular and vascular area of the dolphin *cervix uteri* have erectile properties (Simpson and Gardner, 1972). Variations in thickness of the inner lumen induced by blood would contribute to accommodation of the penis or sealing of the intrauterine cavity from the outside, an essential feature since salt water must not enter the uterine cavity (Claudia Gili, personal communication).

i. The muscles of the limbs are principally responsible for locomotion in terrestrial mammals, while swim in dolphins is based on repeated vertical movements of the flukes induced by the strong muscles of the spine (Chapter 4).

High temperature may alter the correct development of the fetus, since its metabolism may be twice that of the mother (Power et al., 1984). Maintenance of the correct fetal temperature requires that approximately 85% of the heat produced by the fetus is transferred by convection to the placenta (Power et al., 1984; Rommel et al., 2007).

To this effect, a countercurrent heat exchanger (or CCHE, Rommel et al., 1993, 1998, 2007) made up by lumbocaudal venous plexus (cool part) and uterine vessels (warm part) ensures that the uterus of cetaceans does not suffer from excessive heat, generated by the thermogenic lumbosacral locomotor muscles and maintained by the insulating blubber (see also the vascular cooling mechanism related to the testes under section Temperature).

The Vagina and Vestibulum

The vagina (Fig. 9.31) follows the cervix caudally and allows space for the penis during mating. Muscular and relatively short, its structure is also similar to that of the cow and other ruminants.

The tunica mucosa displays several circular folds (peculiar of cetaceans) apparently expanding from the cervix into the lumen of the vaginal canal. In dolphins there are 2–4 large folds (for further references see Yablokov et al., 1972), possibly related to muscular movements of the vagina during mating. Except for the part close to the *cervix uteri*, the vaginal walls have no glands.

Longitudinal folds are evident close to the external opening. The shape and disposition of these longitudinal folds may vary with the species and even with the single individuals.

The *vestibulum* is short and characterized by the opening of the urethra.[j] The presence of vaginal bands (analogous to the human hymen), that separate the *vestibulum* of the vagina from the vulva have never been described in dolphins (Plön and Bernard, 2007a).[k]

The Vulva (External Opening) and Clitoris

The external opening of the female genital system is transformed into a genital slit closed by thick rims of skin with powerful underlying musculature. This disposition is obviously an adaptation to life in the water.

The clitoris (Fig. 9.33) is characterized by a thick *tunica albuginea*, a *corpus cavernosum* and a specific *m. ischiocavernosus* (erector of the clitoris) (Slijper, 1966; Yablokov et al., 1972). It is comparatively large and covered by specific folds. Correspondence of these folds with the *labia minora* of terrestrial mammals has been proposed (Slijper, 1966). The vulva is characterized by an *m. dilator vulvae* and surrounded by robust longitudinal folds (corresponding to the *labia majora*). The *m. dilator vulvae* consist of two parts (Yablokov et al., 1972): one part is closely linked to the pelvic bones and the tail musculature (possibly the true *m. dilator vulvae*); the other part of the muscles consists of circular fibers (*m. constrictor vulvae*). The genital slit becomes progressively engorged during the last phase of the pregnancy (Fig. 9.34).

Development of the Female Genital System

Based on the observation of Štěrba et al. (2000), the first evidence of the genital ridge (still not differentiated) can be identified in Stage 4 (see Table 9.6 for staging). The genital tubercle appears in Stage 5, and the gonads differentiate into ovaries and testes in the following Stage (Stage 6). The phallus is evident in both sexes and points caudally. Paramesonephric (müllerian) ducts then appear in Stage 7, but distinction between the two sexes is still difficult. Histological differentiation of the ovary begins in the following stages, and the tip of the phallus becomes gradually enfolded by the external labia.

The Placenta

Dolphins have a diffuse epitheliochorial placenta[l] (Wislocki and Enders, 1941; Zhemkova, 1967; Simpson and Gardner, 1972; Benirschke and Cornell, 1987; da Silva et al., 2007; Miller et al., 2007b), in which the villi are evenly distributed (Table 9.2). In an epitheliochorial placenta the uterine epithelium, the connective tissue of the mucosa, and the walls of the maternal vessels, separate maternal blood from the fetal chorionic *villi*, that is, the fetal structures that exchange respiratory gasses

---

j. Calcareous concretions (calcium phosphate compounds without uric acid or fluoroxalate) of the size of a pebble may be found in the vagina of different species of dolphins (Yablokov et al., 1972; Sawyer and Walker, 1977) and may obviously hinder their reproductive capabilities.

k. The only toothed whale in which vaginal bands (= hymen) were described is Dall's porpoise, *Phocoenoides dalli*, belonging to the family Phocoenidae (Morejohn and Baltz, 1972).

l. Although the close evolutionary and genetic relationship between terrestrial Artiodactyla and Cetacea is a well-documented fact, the structure of the cetacean placenta (and the physiology of the female reproductive system) is closer to that of Perissodactyla (horse, zebra, and rhinos).

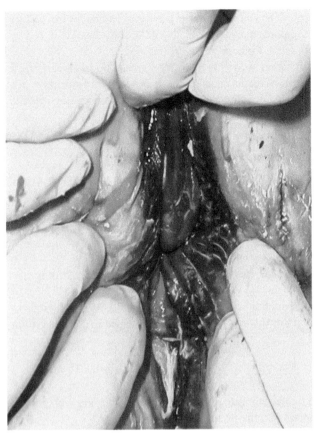

FIGURE 9.33 **Clitoris of an adult female *T. truncatus*.**

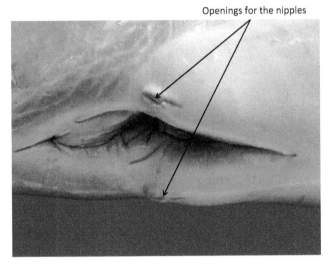

Openings for the nipples

FIGURE 9.34 **External genitalia of a pregnant *T. truncatus*.**

TABLE 9.2 Characteristics of the Placenta of Dolphins

Species	Weight of the Placenta (g)	Distribution of Villi	Cotyledons	Implantation	Barrier	Position of the Fetus	Hippomanes[a]
T. truncatus	2200–2800	Diffuse	No	Mesometrial	Epitheliochorial complete	Tail first	Common
O. orca	1700–2650	Diffuse	No	Unknown	Epitheliochorial complete	Tail first	Never reported

[a]*The so called* hippomanes *are small dark green or brown smooth bodies derived from the wall of the allantois that may be found in the allantoic fluids of mares and cows.*

and substances dissolved in the blood with the circulatory system of the mother.[m] Terrestrial Cetartiodactyla and Perissodactyla also possess an epitheliochorial placenta, but with different structural characteristics (large cotyledons in Ruminants and microcotyledons in Equids). In species that possess an epitheliochorial placenta, immunity is transferred to the fetus mainly with the colostrum.

This kind of placenta is generally associated with a long gestation period and the birth of a well-developed, almost independent offspring (for discussion see Enders and Carter, 2004). A different relationship between the fetal and maternal vessels has been suggested in a short description referred to the pilot whale *Globicephala melaena* [= *melas*] (Morton and Mulholland, 1961), perhaps suggesting an endochorial structure for this latter species.[n]

---

m. The human placenta is a hemochorial structure in which the fetal chorion is in direct contact with maternal blood.

n. In a short passage of their description of their paper on the placenta of the river dolphin *Sotalia fluvialis* da Silva et al. (2007) confirmed the observation for *Globicephala*.

Umbilical cord

FIGURE 9.35   **Fetus of *T. truncatus* seen through the allantois sac.**

The time of implantation (the moment in which the embryo and its annexes become attached to the uterus) is presently unknown. The fetus generally occupies the left horn, but the tail enters the contralateral one. A large nonvascularized allantoid sac envelopes the amniotic cavity (Figs. 9.26 and 9.35). The epithelium of the allantoid is flat to squamosal, and the epithelium of the amnion may also contain keratinized pigmented cells. Both allantois and amnion are contained within the chorionic sac (Benirschke, 2007).

The dolphin placenta is also an endocrine organ (like in primates and equids), and the production of chorionic gonadothrophin is higher than in man (Hobson and Wide, 1986). LH-like substances have been identified in the villosities of the placenta of the bottlenose dolphin (Watanabe et al., 2007), thus confirming that hormonal control of ovarian functions during pregnancy is important for this species.

Twin pregnancies are extremely rare in cetaceans. This may be explained also with the diffuse nature of the epithelio-chorial placenta. In twin pregnancies, the two placentas are fused at least in some parts and therefore the chorial *villi* cannot develop all over the placenta surface. This may led to an incomplete support to the developing fetuses and impossibility to further progress with the pregnancy.

The Estrus Cycle

Dolphins are highly specialized mammals with an especially convoluted brain. As with terrestrial mammals of similar characteristics, the physiological (and subsequently behavioral) relationship between the mother and the offspring is prolonged, with a long pregnancy and extensive lactation period.

Dolphins show a varying degree of seasonal reproductive activity,[o] depending on the geographical location: seasonality is necessary especially in migrating species to ensure that newborns face the best environmental conditions (and less risks) at birth. We now know that dolphins are polyestrous, spontaneous ovulators, with a single ovulation for each estrous cycle, leading to singleton pregnancies.

---

o. Early behavioral observations at sea, or microscopic studies based on stranded specimens, led to uncertain conclusions and contradictions, later (at least partially) solved with research on captive animals (for an old review see Harrison et al., 1969). The present techniques, including RIA (Atkinson and Yoshioka, 2007; Bergfelt et al., 2011) allow steroid hormone dosage on a number of biological fluids, feces or skin samples, thus contributing to much more precise analyses of the reproductive behavior and physiology of cetaceans (Biancani et al., 2009). Pregnancy may also be diagnosed on the basis of progesterone levels in the blubber samples (*T. truncatus* and *G. melas*: Perez et al., 2011; *Stenella attenuata*: Kellar et al., 2013; *D. delphis*: Kellar et al., 2014).

TABLE 9.3 Characteristics of the Estrous Cycle in the Bottlenose Dolphin (From Robeck et al., 2005)

Phase	Follicular (Days)	Luteal (Days)	Preovulatory EC Rise (Days)	EC Peak to LH Peak (H)	LH Surge to LH Peak (H)	Duration of LH Surge (H)	LH Peak to Postovul. UP Increase (Days)
Duration (Days)	8.1 ± 3.0	19.3 ± 2.8	2.4 ± 1.8	7.5 ± 10.6	9.4 ± 3.1	20.3 ± 5.1	2.1 ± 0.6

LH, luteinizing hormone; *EC*, urinary estrogen conjugate; *UP*, urinary progesterone metabolites.

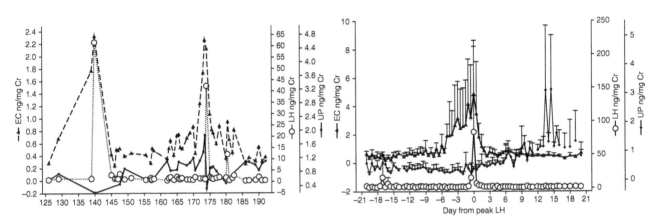

FIGURE 9.36 **Naturally occurring estrous cycle of *T. truncatus*.** EC, urinary estrogen conjugate; UP, urinary progesterone metabolites; LH, luteinizing hormone . *(From Robeck et al., 2001. Reproduction. In: Dierauf and Gulland (Eds.), CRC Handbook of marine mammal medicine, second ed. CRC Press, Boca Raton, Florida, pp. 193–236).*

TABLE 9.4 Timing of Ovulation in the Bottlenose Dolphin (After Robeck et al., 2005)

	Hours After EC Peak	Hours After LH Onset	Hours After LH Peak
Ovulation	26.8 ± 7.1	32.1 ± 8.9	24.3 ± 7.0

LH, luteinizing hormone; *EC*, urinary estrogen conjugate.

In the bottlenose dolphin, the length of the estrous cycle (Table 9.3; Fig. 9.36) varies between 21 and 42 days (Boness, 2009), and specific studies, performed also with transabdominal ultrasound on captive dolphins to characterize the cycle for artificial insemination, identified an average of 36 days (Robeck et al., 2005). The follicular phase lasts 8.1 ± 3.0 days, ending with ovulation (Table 9.4), followed by a luteal phase (diestrous) of 19.3 ± 2.8 days.

Estrous cycles may follow one another (polyestrous phase), with seasonal peaks[p] that vary according to the geographic location and other unclear factors. Common dolphins may cycle as much as seven times a year, and bottlenose dolphins three times (Boness, 2009). Female pilot whales are also seasonally polyestrous, with three cycles for breeding season (Harrison, 1969). Indications from captive false killer whales suggest that *Pseudorca crassidens* is a seasonally polyestrous species, with spontaneous ovulations and increased concentration of plasma progesterone in spring and summer (Atkinson et al., 1999). Captive Pacific white-sided dolphins (*Lagenorhynchus obliquidens*) ovulated from August to October, the estrous cycle was 31 days long with 10 days of follicular phase and 21 days of luteal phase (Robeck et al., 2009).

The mean length of the estrous cycle in killer whales is 41.2 days, with a follicular phase of approximately 18 days and a luteal one of approximately 20 days (Robeck et al., 2001). For the latter species other sources (Kusuda et al., 2011) indicate the length of the cycle at 44 days (44.9 ± 4 days based on plasma progesterone and 44.6 ± 5.9 days based on rectal temperature).

---

p. Some of these characteristics are shared by the horse and related mammals whose reproductive cycle is influenced by light and other circannual factors. Some even-toed ruminants are also seasonal breeders (sheep, goat), whereas others (ie, *Bos taurus*) are not.

TABLE 9.5 Reproductive Data of Selected Dolphin Species

Species	♀ Sexual Maturity (Years)	Body Length at Sexual Maturity (cm)	Pregnancy (Months)	Lactation (Months)	Newborn Length (cm)	Polyestral	Seasonal Prevalence	Reproductive Senescence (Years)
T. truncatus	7–10	210–238	12	>18 up to 24–36	84–140	y	y (?)	>45?
D. delphis	>7	155–190	11–12	10–19	75–100	y	y	>30?
S. coeruleoalba	5–13	210–220	12–13	8–20	92–100	y	y	>30?
G. griseus	11	277–332	14	?	110–150	y	y	38
G. melas	>8	>375	12–16	24 (or more)	177	y	?	>50?
O. orca	9.8–12[a]	>300 (460–540)	17–18	18–24	183–236	y	(y)	*
P. crassidens	8–14 (5?)	366–457	14–15?	12–18?	160–190	y	y (?)	45
Lagenorhynchus albirostris	?	>170	11–12?	>18	110–120	y	y	?

The data reported from different authors are often contradictory. *, see text. Completed with data derived from Harrison et al. (1972); Perrin and Reilly (1984); Evans (1994); Perrin et al. (1994); Atkinson et al. (1999); Bernard and Reilly (1999); Boyd et al. (1999); Dahlheim and Heyning (1999); Kruse et al. (1999); Odell and McClune (1999); Reeves et al. (1999); Wells and Scott (1999); Robeck et al. (2001); Raduán et al. (2007); Westgate and Read (2007); Jefferson and Webber (2007); Katsumata (2010); Pomeroy (2011)[b]; O'Brien and Robeck (2012); Robeck et al. (2015).
[a]There are considerable differences between free-ranging and captive killer whales. For the first group mean age at first conception is 12.1 years, whereas it is 9.8 years for captive killer whales, who had their first estrus at 7.5 years (Robeck et al., 2015).
[b]The length of lactation reported by Pomeroy (2011) is extremely shorter than what described by other authors and therefore has not been taken into account for the present table.

Seasonal anestrous has a variable length while postpartum lactational anestrus may be very long in most dolphin species, also because lactation is extremely extended in time (Table 9.5). Consequently, the interestrous interval may well surpass a year[q] (Boness, 2009).

Table 9.5 contains a summary of the key facts on pregnancy and reproduction in selected dolphin species. A thorough review of the older literature is reported in Harrison (1969) and Harrison et al. (1972). For a recent analysis of the profile of steroid hormones during the pregnancy in the bottlenose dolphin see Steinman et al., 2016.

Studies on captive bottlenose or killer whales yielded important information (O'Brien and Robeck, 2010), but whether these conditions apply also to free-ranging individuals remains to be confirmed. According to O'Brien and Robeck (2012) the duration of pregnancy in captive T. truncatus varies between 355 and 395 days (mean 376.1 ± 11.0). The range is wide enough to generate doubts or discordant data when referred to wild dolphins. Several issues (including length and age at puberty, and duration of pregnancy) are in fact difficult to assess in the wild.

Reproductive Senescence

A peculiar fact concerns the so-called reproductive senescence in dolphin species. The estrous cycles of the majority of mammalian females never really stop with age, but diminish their frequency (there are fewer ovulations) and become progressively far apart, contrarily to human menstrual cycles that stop at menopause. The number of pregnancies and successful birth also decrease. This applies to dolphin species too, except in *Orcinus orca* (Olesiuk et al., 1990), and in *Globicephala macrorhyncus* (Kasuya and Marsh, 1984; Marsh and Kasuya, 1984, 1986).[r] Killer whales over 40 years of age experience a full cessation of estrus cycles and have a *very long* post-reproductive lifespan[s] (Olesiuk et al., 1990; Foster et al., 2012). Besides humans and some primates, only the elephant, the short-finned pilot whale and the killer whale experience this phenomenon (Croft et al., 2015).

---

q. Bottlenose dolphin: 3–4 years; Risso's dolphin 2.4 years; killer whales: 2–12 years; false killer whales: 2.5–3.5 years (Robeck et al., 2001).

r. However, short-finned pilot whales and killer whales differ in one respect because in the former the estimated calving intervals increased progressively with age whereas in the latter there is no evidence of a further decline in the fecundity of reproductive females after age 30.5 till cyclicity stops (Olesiuk et al., 1990).

s. This fact has a lot of interesting implications for behavioral studies and the understanding of population dynamics and pod structure. If female killer whales (who may reach the age of 90) are no more fit to reproduce after the age of 40, their extended life span may have evolutionary advantages, including an important role in social structure and rearing of the cubs of other mothers (Foster et al., 2012).

FETAL DEVELOPMENT

A detailed description of the different stages of development of *S. attenuata, S. longirostris,* and *D. delphis* is reported in the seminal study of Štěrba et al. (2000), that examined a considerable number of museum specimens.[t] Table 9.6 summarizes some relevant features of the progressive fetal development of selected dolphin species. The fetus increases its maturation and development (and of course its length) tremendously in the last period of the pregnancy, so that almost all the morphological features become evident in the last phases.

A description of the development of the various organs has been given (when available) in the relative specific chapters.

The length of the newborn obviously varies with the different species, and is somewhat related to the size of the adults, although not directly proportionate to it. Individuals of the same species may differ greatly in size, depending on genetics and environmental factors. Therefore, geographically separate populations of the same species may show considerable somatic variations in size. Smaller species (including the striped and the common dolphin) tend to grow rapidly during the early postnatal phase, that is, during the first months of the first year.

As shown in Table 9.6, the early phases of the embryonic differentiation of the striped, common, and bottlenose dolphin are difficult to study for a number of reasons, including scarcity of specimens, difficulty in their precise aging, and classification. However, the substantial growth that takes place during the second half of the pregnancy makes it easier to document the progressive transformation of fetal structures into their adult form and organization (Figs. 9.37–9.39).

THE MALE REPRODUCTIVE SYSTEM

The external male genitalia of mammals that live in the water appear different from those of terrestrial ones. To distinguish males from females is not straightforward to the unexperienced eye, since the testis and the penis are hidden within the body. The organ of copulation and the gonads are located inside a cutaneus pocket and the peritoneal cavity, respectively, to avoid drop in the temperature of the sperm, water friction, and general hindrance in swimming.

The male reproductive organ of cetaceans is a retractile fibroelastic penis with a sigmoid flexure. The testicles of dolphins and whales are placed within the peritoneal cavity instead of being located externally in the scrotum as in most terrestrial mammals. The change in position not only responds to obvious hydrodynamic purposes but also requires major changes in the vascular bed supplying the gonads, as their temperature must remain lower than that of the surrounding organs to allow sperm maturation. The male genital apparatus of dolphins (Fig. 9.40) closely resembles that of terrestrial Cetartiodactyla.

The Testes and Epididymis

The testes of dolphins are located within the abdominal cavity[u] (Fig. 9.41). The testes of cetaceans are comparatively larger in respect to body mass than in most terrestrial mammals (Kenagy and Trombulak, 1986; Aguilar and Monzon, 1992).

The testes do not have the common ovoid aspects typical of several terrestrial mammals, including man. The gonads of dolphins are rather long cylinders (Fig. 9.42), with slender extremities, and can be examined in vivo using ultrasound technology (Fig. 9.43). The epididymis surrounds the testis for all its length and is more conspicuous at the two poles. The cranial pole of the testes is connected to the caput of the epididymis, and the caudal pole to its tail. The epididymis is in fact made up by long network of continuous tubes, whose total length may easily reach tens of meters.[v] There are no references on the length of the epididymis in dolphins. Testicular veins are well evident through the translucent *tunica albuginea* of the gonads (Fig. 9.44).

Size of the testes in the bottlenose dolphin varies from 4.4 to a maximum of 22.8 cm, and volume from 4 to 531 cm^3 (Brook et al., 2000), depending on age and body dimensions (Table 9.7). In the Pacific white-sided dolphin (*L. obliquidens*) the size of the testes varies according to the season, being larger during July and August, approximately corresponding to testosterone peaks (see later) and the onset of reproductive activities (Robeck et al., 2009).

---

t. Although fetal material is relatively easy to find in Museum and University collections all other the world, precise identification of the species might be rather difficult if data of the specimens were not recorded precisely from the start. In fact even the difference between toothed- and whalebone-whale fetuses is not immediately evident in the very early embryonic stages.

u. Other mammals possess internal testes, including terrestrial species like the elephant.

v. The human epididymis seems to be only a few cm long, but the folded duct that constitutes it is in fact several meters long in a normal male. The equine epididymis is approximately 25–30 m long.

TABLE 9.6 Stages of Fetal Development

Stage	Definition[a]	Stenella spp.[b] TL[c] (mm)	CRL[d] (mm)	Age (Days)	Bone Density of Tympanoperiotic Complex (g/cm^2)[e]	D. delphis TL	CRL	Age	Bone Density	T. truncatus TL	CRL	Age	Bone Density
#1	Primitive streak			13				13				—	
#2	First somites (1–7)			17				17				—	
#3	Four branchial bars, anterior and posterior limb buds, tail bud	21–25	8–13	22–28		24–31	8–10	22–28					
#4	Eye pigmented, handplate present	28–39	11–13	27–36	Below detection limits	32–39	13	27–36	Below detection limits				Below detection limits
#5	Handplate partially indented, nostrils move caudally	33–58	14–23	32–42		42–46	19–21	32–42					
#6	Palate fused, ossification begins	42–79	22–33	41–52		52–69	22–30	41–52					
End of the embryonic phase, beginning of the fetal phase													
#7	Eyelids fused, umbilical hernia repositioned	76–110	35–52	51–66		82–101	34–40	51–66					
#8	Skin thickens	112–134	53–73	62–78	Below detection limits	89–144	50–120	62–78	Below detection limits				Below detection limits
#9	Tactile hairs erupted	151–225	80–112	78–90		163–194	72–77	78–85					
#12[f]	Eyelids open	215–245	107–132	95–108		237	116	85–100					
#10	Brown pigment in the head and dorsum	244–477	108–230	100–175	n.d–0.491	298–480	128–210	110–175	n.d–0.596	540		≈150	0.312–0.371
#11	Black to bluish skin pigmentation	590–945	230–510	190–290	0.530–1.480	580–830	240–420	>175	> 0.440	920		≈300	0.591–0.660
#13	Birth	750–1000	—	≈365	1.270–1.841	750–1000	—	330–365	—	840–1400	—	≈365	1.173–1.537

[a]The definition of the subsequent fetal stages in mammals is standardized following the Carnegie system of staged human embryos (O'Rahilly, 1972) adapted to selected dolphin species by Štěrba et al. (2000). Here we substituted the general definition of the human stages with the one specific for dolphins. The definition of the Carnegie system is based on staged human embryos (O'Rahilly, 1972), and according to Štěrba et al. (2000), this system cannot always be applied to dolphins for a number of evolutionary and morphological differences (ie, the question of the hair). Hence the modification of the Carnegie staging system adopted in their study. In a subsequent publication, Thewissen and Heyning (2007) have reevaluated the original Carnegie system of staging, and used it for dolphin embryos and fetuses. However, since the paper of Štěrba et al. (2000) remains the most widely recognized reference text (as acknowledged also by Thewissen and Heyning, 2007), we based our description on the classification proposed by Štěrba et al. (2000), but encourage the interested reader to consult both articles.

[b]The original paper of Štěrba et al. (2000) described measurements relative to S. longirostris and S. attenuata, while the paper of Cozzi et al. (2015) described specimens of S. coeruleoalba. Since the three species are similar in dimensions, their data have been grouped together.

[c]TL = Total length of the specimen

[d]CRL = Crown-rump length of the specimen, a universal measure of fetal length.

[e]Bone density is expressed as g/cm^3 of calcium salts, following densitometric measures reported by Cozzi et al. (2015). For principles and significance see Zotti et al. (2009).

[f]Stage #12 in dolphins (eyelids open) precedes stages 10 and 11 of the Carnegie system (Štěrba et al., 2000).

Modified from Štěrba et al. (2000); bone density of the tympanoperiotic complex from Cozzi et al. (2015).

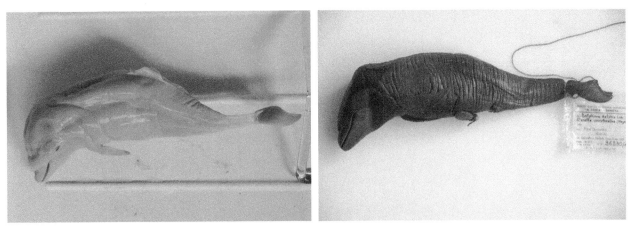

FIGURE 9.37 (Left) Fetus of *S. coeruleoalba* or *D. delphis* of approximately 4 months of age. (Right) 5 month old fetus of *T. truncatus. (Left, Museum of Natural History G. Doria of Genova; specimen # 36930a. Right, Museum of Natural History G. Doria of Genova; specimen # 36930a, initially classified as* S. coeruleoalba *or* D. delphis *and subsequently reclassified).*

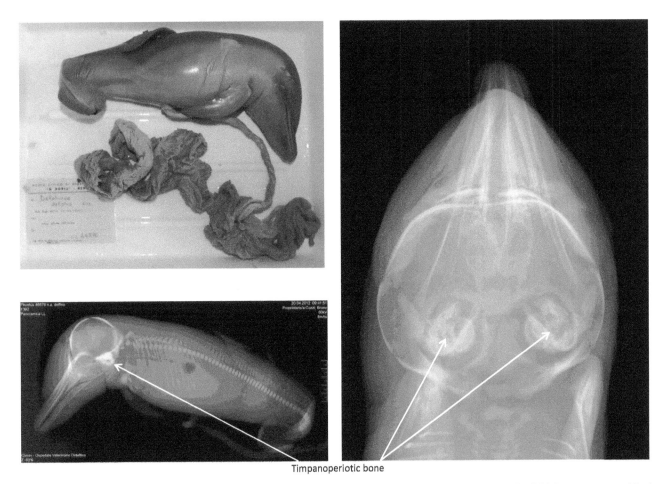

Timpanoperiotic bone

FIGURE 9.38 **Fetus of *D. delphis* of 6–7 months of age.** *Museum of Natural History G. Doria of Genova; specimen # 46876; X-rays courtesy of Prof Alessandro Zotti, Department of Animal Medicine, Production and Health of the University of Padova.*

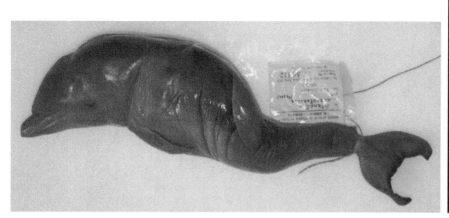

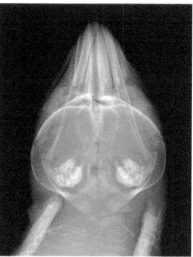

FIGURE 9.39   **Fetus of *S. coeruleoalba* of 8 months of age.** *Museum of Natural History G. Doria of Genova; specimen # 36932; X-rays courtesy of Prof Alessandro Zotti, Department of Animal Medicine, Production and Health of the University of Padova.*

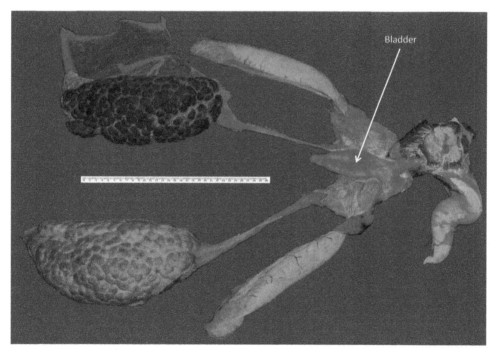

FIGURE 9.40   Ventral view of the male genital apparatus and urinary apparatus of *T. truncatus*.

Structure

The testes of dolphins have the typical structure of the testes of all mammals (Figs. 9.45 and 9.46). The testicular parenchyma is made up by seminiferous tubules and by islets of tissue among the tubules. Seminiferous tubules contain Sertoli cells and germinal cells at various stages of maturation[w]: spermatogonia, spermatocytes, spermatids and spermatozooa[x] (see also Section "Spermatogenesis, Sexual Maturity, and Seasonality"). During fetal development, Sertoli cells are responsible for the regression of the paramesonefric ducts (the anlage of female genital organs), by secreting the Anti-Müllerian Hormone (AMH, a glycoprotein). However, there are no specific descriptions of AMH in dolphins or other cetacean species.

---

w. The maturation of spermatozoa is completed in the epididymis.
x. For a thorough description of the spermatozoa of dolphins and whales, see Miller et al. (2007a).

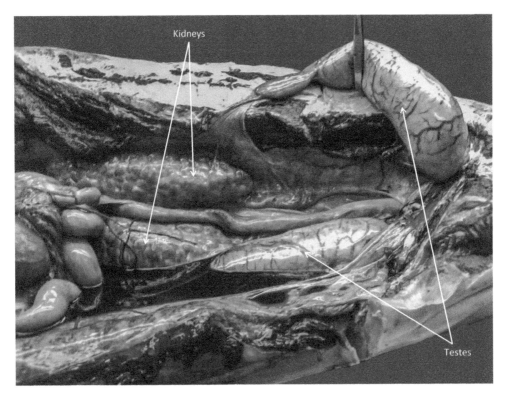

FIGURE 9.41 **Testes of *T. truncatus* in place.**

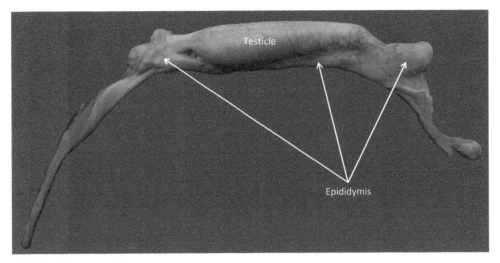

FIGURE 9.42 **Testicle and epididymis of *T. truncatus*.**

The intertubular islets contain the interstitial cells (also called Leydig cells) that are responsible for the productions of testosterone (see later for notes on male reproductive physiology). The interstitial cells of dolphins are generally more slender if compared to those of terrestrial mammals, and contain no pigment. They are less conspicuous than in other mammals, and their presence is somewhat difficult to detect (Simpson and Gardner, 1972). They may be grouped together in few (but not all) of the spaces between adjacent tubules.

In *T. truncatus aduncus*, the diameter of the tubules in the epididymis was determined by ultrasonography to be between 2–3 and even 3–5 mm at the caudal pole (Brook et al., 2000). In the same study, the diameter of the tubules decreased considerably (up to 2 mm) following ejaculation.

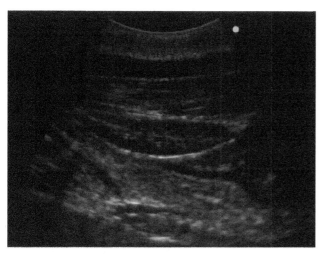

FIGURE 9.43 **Ultrasound image of the testicle of a young adult bottlenose dolphin.** *Courtesy of Pietro Saviano.*

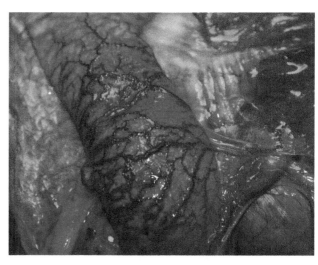

FIGURE 9.44 **Testicular veins of *T. truncatus* seen throughout the tunica albuginea.**

TABLE 9.7 Dimensions of the Testes in the Bottlenose Dolphin Based on Ultrasonography (Brook et al., 2000)

Maturity Stage	Dimensions (cm)	Volume (cm³)	Ultrasonography Notes
Immature males	4.4–6.4	4–4.6	Poorly differentiated Markedly hypoechoic in relation to the *m. hypaxialis lumborum*
Subadult males	8.8–15.9	10.3–45.7	Homogeneous echopattern Less echogenic than the *m. hypaxialis lumborum*
Mature males	14–22.8	147–531	Mid- to high-level intensity; isoechoic or slightly hyperechoic compared to the *m. hypaxialis lumborum*

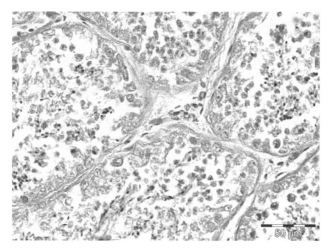

FIGURE 9.45 **Microphotograph of the testes of *T. truncatus* (scale bar = 50 μm).** HE stain.

Temperature

Temperature is a key factor for the maturation of spermatozoa, and one of main reasons why in most mammals the testes are outside the body cavity (at least during the reproductive season). A temperature of 37°C may be fatal for proper maturation and lead to infertility or even cause pathologies. As already outlined, the testes of dolphins are placed inside the peritoneal cavity and reduction of their temperature is essential. As for the ovary, a specific vascular plexus provides counter-exchange cooling of the blood that reaches the gonads (Rommel et al., 1992, 1994, 1998). Contrarily to terrestrial mammals, there is no single testicular artery, but there are 20–40 arteries vessels leaving the aorta towards the testicles (TAP, or testicular

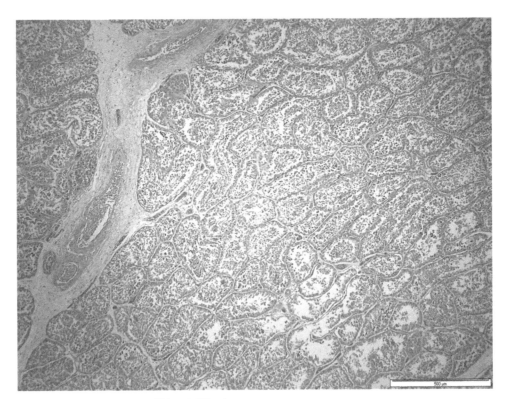

FIGURE 9.46 **Testicle of *G. griseus* (scale bar = 500 μm).** HE stain.

arterial plexus). They form a comb of parallel arteries that surround the testicle and then coalesce into a single artery that enters the organ from the caudal pole (Rommel et al., 2007). Venous drainage occurs through the testicular venous plexus (TVP) that drains blood into the caudal vena cava. TAP and TVP are encompassed by two layers of the peritoneum corresponding to the *mesorchium* of terrestrial mammals (Rommel et al., 2007). Heat exchange takes place by the lumbocaudal venous plexus (LCVP), formed by thin-walled veins and located ventrally to the hypaxial muscles. LCVP is supplied with blood that derives from the flukes and the dorsal fin, and is therefore cooler than that bound to (or returning from) the gonads (Pabst et al., 1995). LCVP is drained by the caudal vena cava.

The Ductus Deferens

The ductus deferens departs from the testis and epididymis in a caudal direction, travels close to the ureter, crosses over it, and enters the penile urethra with the seminal collicle at the base of the bladder (Fig. 9.47). The ductus is somewhat serpentine in its course. In contrast to many terrestrial mammals, there are no well-evident ampulla along its course.

Apart from its direction—rostrocaudal along the main axis of the peritoneal cavity instead of ventrodorsal from the scrotum—there are no differences between the *ductus deferens* of cetaceans and that of terrestrial mammals.

The Accessory Male Glands

The only accessory male gland present in Cetacea is the prostate (Fig. 9.48). According to Matthews (Matthews, 1950, cited by Yablokov et al., 1972), the prostate of dolphins consists of a diffuse part and a lobular part. The dispersed tissue is located around the seminal collicle, close to the seminal ducts and the beginning of the urethra. The (multi)lobular part of the gland is located close to the roots of the penis and related to a well-developed *m. bulbocavernosus.*[y]

The absence of seminal vesicles in dolphins may be substituted by a dilated tubular structure abutting the distal two-thirds of the testis (Brook et al., 2000). Since the same authors report that tubule diameters at the distal end of the testis decreased by as much as 2 mm after ejaculation, it is possible that the large vas deferens allows storage of seminal fluid in the absence of seminal vesicles in dolphins (Matthews, 1950).

---

y. Matthews (Matthews, 1950, cited by Yablokov et al., 1972), refers to the *m. bulbocavernosus* also as *m. compressor prostatae.*

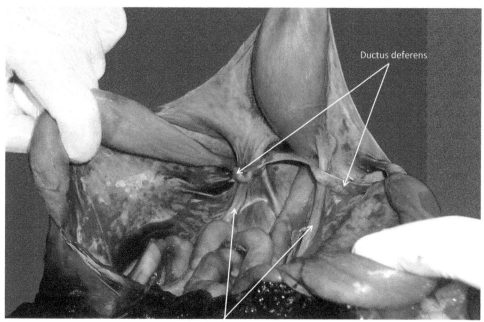

FIGURE 9.47   **Ductus deferens of *S. coeruleoalba*.** The green color in the lateral ligaments of the bladder is due to an injected dye.

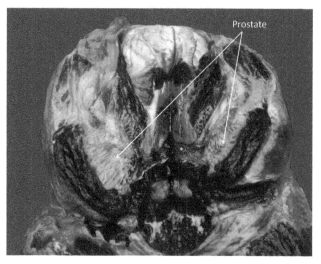

FIGURE 9.48   **Prostate gland of *T. truncatus*.**

The Penis

Dolphins and whales possess a fibroelastic penis, similar to that of ruminants. The copulatory organ displays an S-shape at rest due to the paired retractor muscles (Figs. 9.49–9.52), and achieves erection by blood influx into the *corpora cavernosa*, relaxation of the retractor muscles, and consequent distension of the sigmoid flexure. There is no definite glans penis in cetaceans. The foremost part of the penis (apex) is cone-shaped (Fig. 9.53) and contains the external urethral opening.

One major advantage of the S-shaped fibroelastic penis is the total concealment of the organ at rest, a feature extremely useful for hydrodynamics. The penis becomes extroverted during mating but also during play between young males.[z] There is no os penis (*baculum*).

---

z. The penis is often extroverted in large stranded whales, such as sperm whales—as a consequence of gas production in the abdominal cavity during decomposition.

FIGURE 9.49   Position of the penis in *S. coeruleoalba*.

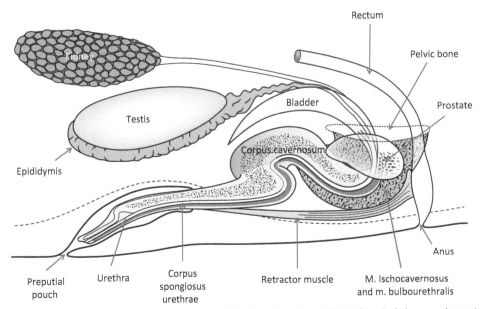

FIGURE 9.50   **Schematic drawing of the male genital apparatus of dolphins.** *(From Arvy, 1973–1974). The kidney, renal parasites and renal secretion in cetaceans. In: Pilleri, G. (Ed.). Investigations on Cetacea 5. Institute of Brain Anatomy, University of Berne, Berne, pp. 231–310).*

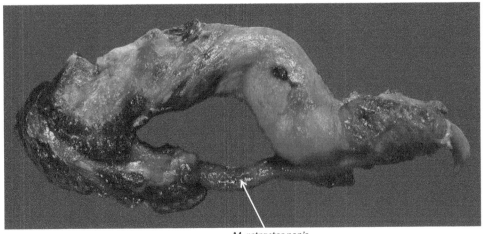

FIGURE 9.51   **Penis and retractor muscles of** *S. coeruleoalba.*

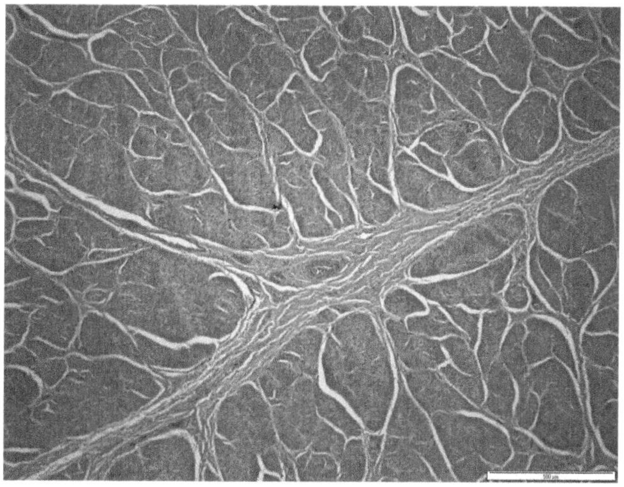

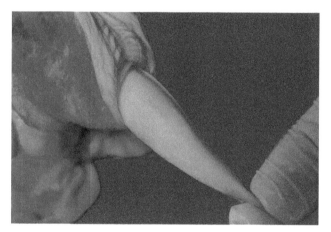

FIGURE 9.52   M. *retractor penis* of *S. coeruleoalba* (scale bar = 500 μm). HE stain.

FIGURE 9.53   **Apex of the penis of *T. truncatus*.**

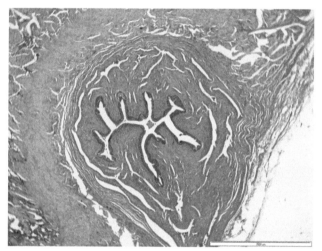

FIGURE 9.54   **Lumen of the urethra and *corpora spongiosa* of the urethra of *S. coeruleoalba* (scale bar = 2mm).** HE stain.

Structure

As all fibroelastic penises, the copulatory organ of dolphins has a thick *tunica albuginea*, and relatively small *corpus spongiosum* that surround the penile urethra (Figs. 9.54 and 9.55) and continues for all the length of the organ, although reduced at the rostral extremity. There is no *glans penis*,[aa] and the rostral extremity of the organ, or *apex*, contains a *corpus spongiosum*, and numerous nerve endings (Fig. 9.56).

The thick external *albuginea* hinders swelling of the diameter of the penis. The erection is attained by combined relaxation of the retractor muscles and blood engorgement of the erectile tissue of the *corpus spongiosum*. When the retractor muscles relax, the S-shaped penis straightens and protrudes from the genital pouch.

Development of the Male Genital System

The development of the male genital system follows the same steps already described for the female system (see aforementioned). Shortly, distinction between the two sexes is difficult till Stage 6 and 7, during which the gonads differentiate into ovaries and testes (Štěrba et al., 2000). During Stage 8 the structure of the testes shows the formation of an external tunica

---

aa. A *glans penis* (as in man and horse) possesses an independent *corpus cavernosum* and no surrounding limiting *tunica albuginea*, and is therefore able to achieve a larger diameter then the rest of the shaft during the erection. On the contrary, the structure of the *apex* of the penis is just a continuation of the *corpus cavernosum* and *corpus spongiosum* surrounded by the tunica albuginea.

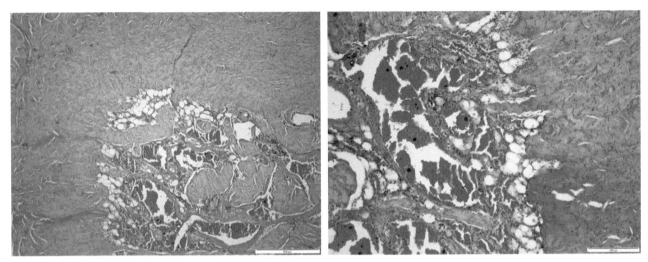

FIGURE 9.55   *Corpora spongiosa* **in the penile apex of** S. *coeruleoalba* **(scale bars, left = 500 μm; right = 200 μm).** HE stain.

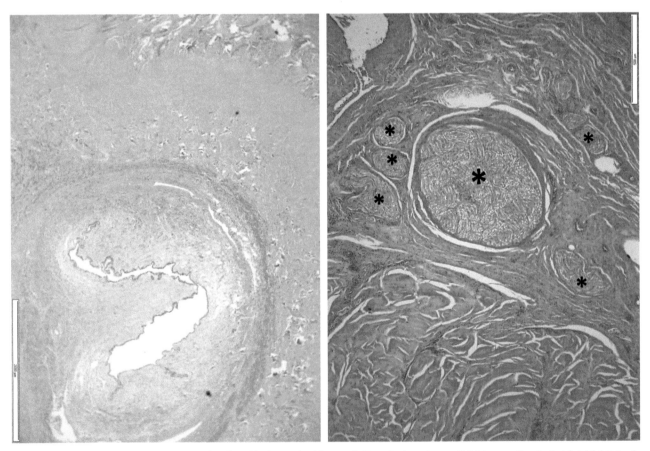

FIGURE 9.56   **Apex of the penis of** S. *coeruleoalba* **with the urethral lumen (left) and relevant superficial innervation (*asterisks*) (right) (scale bars, left = 2 mm; right = 500 μm).** HE stain.

albuginea, and the distinction of internal septa that separate lobules that contain the seminifeous tubules and the rete testis appears. The phallus initially points caudally, and becomes subsequently covered by the prepuce.

Some data indicate that during development the testes of Cetacea pass through stages similar to those of terrestrial mammals with testes in the scrotum, as a fetal *gubernaculums testis* has been described in several species of Cetacea, including *D. delphis* and *Phocæna phocæna* (Van der Schoot, 1995). This evidence suggests that a vestigial mechanism for testicular descent is still present during the fetal life of dolphins and whales. The hypaxial musculature of the caudal

TABLE 9.8 Sexual Maturation of Male Killer Whales (Values in Years of Age) According to Robeck and Monfort (2006)

Juvenile	Pubertal	Mature	Notes
1–7	8–12	>13	Height of the dorsal fin is related to sexual maturity
serum T	0.7 ± 0.7 ng/mL	6.0 ± 3.3 ng/mL	

T, testosterone ng/mL.

TABLE 9.9 Season Cyclicity in Male Dolphins

Species	Location	Period
T. truncatus	Captive	September–October/April–May
D. delphis	Eastern North Atlantic	September–October/December–July
S. longirostris	Eastern tropical Pacific	February–August
S. attenuata	Eastern tropical Pacific	[April] July–August
Lagenorhynchus obscurus	Western South Atlantic	August–November
G. melas	North Atlantic	March–September
G. macrorhynchus	North-Western Pacific	Year-round

The period reported includes the largest time interval.
Modified from Plön and Bernard (2007b).

region contributes to prevent the development of a complete vaginal process necessary for testicular migration (Van der Schoot, 1995).

Spermatogenesis, Sexual Maturity, and Seasonality

The physiological age of a male dolphin may be truly defined by the maturation of its gonads, that is, the ability to generate fully functional spermatozoa. Categories are hard to define (Plön and Bernard, 2007b), and are generally based on the differentiation of seminiferous tubules and on the presence of spermatocytes in the lumen.

Maturation of young individuals of the different species apparently varies according to geographical location, environmental characteristics, water temperature and other factors, possibly including light cycles.

Observations on noncaptive species may present divergent data. Table 9.8 summarize some available information on *O. orca*.

In captive killer whales, serum testosterone levels were higher from March to June in pubertal individuals and from September to December in mature whales (Robeck and Monfort, 2006). In the same study, sperm concentration in the ejaculate did not vary with the season. Whether the situation applies also to free-ranging individuals of the same species remains to be assessed. Captive male bottlenose dolphins showed peak sperm densities in September and October (Schroeder and Keller, 1989), and serum testosterone peaks in June (54.4 ng/mL), with declining values in the following months.[bb] Breeding activities peaked in September and October (during highest sperm density but with lowest serum concentration of testosterone). Captive male *L. obliquidens* showed spermatozoa in the ejaculate only from July to October, and their highest testosterone serum concentration was reached in the warm months (February: 0.09 ± 0.02 ng/mL; June 5.6 ± 3.8 ng/mL; July: 24.3 ± 1.3 ng/mL) (Robeck et al., 2009). Studies on melatonin concentration (Panin et al., 2012) also suggested seasonality for captive bottlenose dolphins.

Observations of wild populations of different dolphin species, based mostly on weight and histology of the testes sampled from stranded individuals, further suggest some degree of seasonality in male dolphins, as summarized in Table 9.9. Histological and morphometric analyses of the testes of wild bycaught *Lagenorhynchus acutus* suggest the presence of seasonality in males of this species as well (Neuenhagen et al., 2006).

---

bb. In the bottlenose dolphin, reported endogenous concentrations of testosterone not in the serum were 9.73–23 ng/mL in the saliva and 14.71–86.20 ng/mL in the blow, respectively (Hogg et al., 2005).

REFERENCES

Aguilar, A., Monzon, F., 1992. Interspecific variation of testis size in cetaceans: a clue to reproductive behavior? Eur. Res. Cetaceans 6, 162–164.

Arvy, L., 1973–1974a. The kidney, renal parasites and renal secretion in cetaceans. In: Pilleri, G. (Ed.), Investigations on Cetacea 5. Institute of Brain Anatomy, University of Berne, Berne, pp. 231–310.

Arvy, L., 1973–1974b. Notes on the mammary glands of male cetaceans. Invest. Cetacea 5, 219–224.

Atkinson, S., Yoshioka, M., 2007. Endocrinology of reproduction. In: Miller, D.L. (Ed.), Reproductive Biology and Phylogeny of Cetacea. Whales, Dolphins and Porpoises. Science Publishers, Enfield, NH, USA, pp. 171–192.

Atkinson, S., Combelles, C., Vincent, D., Nachtigall, P., Pawloski, J., Breese, M., 1999. Monitoring of progesterone in captive female false killer whales, *Pseudorca crassidens*. Gen. Comp. Endocrinol. 115, 323–332.

Aubin, D.J.S., 2001. Endocrinology. In: Dierauf, L.A., Gulland, F.M.D. (Eds.), CRC Handbook of Marine Mammal Medicine. CRC Press, Boca Raton, pp. 165–192.

Ballarin, C., Corain, L., Peruffo, A., Cozzi, B., 2011. Correlation between urinary vasopressin and water content of food in the bottlenose dolphin (*Tursiops truncatus*). Open Neuroendocrinol. J. 4, 9–14.

Benirschke, K., 2007. Atlantic bottlenose dolphin *Tursiops truncatus*. In: Benirschke, K. (Ed.), Comparative Placentation. International Veterinary Information Service, Ithaca NY, (www.ivis.org), A4101.0607.

Benirschke, K., Cornell, L.H., 1987. The placenta of the killer whale, *Orcinus orca*. Mar. Mammal Sci. 3, 82–86.

Bergfelt, D.R., Steinetz, B.G., Lasano, S., West, K.L., Campbell, M., Adams, G.P., 2011. Relaxin and progesterone during pregnancy and the post-partum period in association with live and stillborn calves in bottlenose dolphins (*Tursiops truncatus*). Gen. Comp. Endocrinol. 170, 650–656.

Bernard, H.J., Reilly, S.B., 1999. Pilot whales - *Globicephala*. Ridgway, S.H., Harrison, R. (Eds.), Handbook of marine mammals, vol 6. Academic Press, London, pp. 245–279.

Biancani, B., Da Dalt, L., Lacave, G., Romagnoli, S., Gabai, G., 2009. Measuring fecal progestogens as a tool to monitor reproductive activity in captive female bottlenose dolphins (*Tursiops truncatus*). Theriogenology 72, 1282–1292.

Birukawa, N., Ando, H., Goto, M., Kanda, N., Pastene, L.A., Nakatsuji, H., Hata, H., Urano, A., 2005. Plasma and urine levels of electrolytes, urea and steroid hormones involved in osmoregulation of cetaceans. Zool. Sci. 22, 1245–1257.

Boness, D.J., 2009. Estrus and estrous behavior. In: Perrin, W.F., Würsig, B., Thewissen, J.G.M. (Eds.), Encyclopedia of Marine mammals. second ed. Academic Press, Amsterdam, pp. 392–396.

Boyd, I.L., Lockyer, C., Marsh, H.D., 1999. Reproduction in marine mammals. In: Reynolds, III, J.E., Rommel, S.A. (Eds.), Biology of marine mammals. Smithsonian Institution Press, Washington; London, pp. 218–286.

Brook, F.M., 2001. Ultrasonographic imaging of the reproductive organs of the female bottlenose dolphin, *Tursiops truncatus aduncas*. Reproduction 121, 419–428.

Brook, F.M., Kinoshita, R., Brown, B., Metreweli, C., 2000. Ultrasonographic imaging of the testis and epididymis of the bottlenose dolphin, *Tursiops truncatus aduncas*. J. Reprod. Fertl. 119, 233–240.

Brook, F.M., Kinoshita, R., Benirschke, K., 2002. Histology of the ovaries of a bottlenose dolphin, *Tursiops aduncus*, of known reproductive history. Mar. Mammal Sci. 18, 540–544.

Cave, A.J.E., Aumonier, F.J., 1962. Morphology of the cetacean reniculus. Nature 193, 799–800.

Cave, A.J.E., Aumonier, F.J., 1967. The reniculus of *Tursiops truncatus, Stenella longirostris* and other cetaceans. J. R. Mic. Soc. 86, 323–342.

Collet, A., Harrison, R.J., 1981. Ovarian characteristics, corpora lutea and corpora albicantia in *Delphinus delphis* stranded on the Atlantic coast of France. Aquat. Mammal. 8, 69–76.

Costa, D.P., 2009. Osmoregulation. In: Perrin, W.F., Würsig, B., Thewissen, J.G.M. (Eds.), Encyclopedia of Marine Mammals. Academic Press, San Diego, CA, pp. 801–806.

Cozzi, B., Podestà, M., Vaccaro, C., Poggi, R., Mazzariol, S., Huggenberger, S., Zotti, A., 2015. Precocious ossification of the tympanoperiotic bone in fetal and newborn dolphins: an evolutionary adaptation to the aquatic environment? Anat. Rec. 298, 1294–1300.

Croft, D.P., Brent, L.J.N., Franks, D.W., Cant, M.A., 2015. The evolution of prolonged life after reproduction. Trends Ecol. Evol. 30, 407–416.

da Silva, V.M., Carter, A.M., Ambrosio, C.E., Carvalho, A.F., Bonatelli, M., Lima, M.C., Miglino, M.A., 2007. Placentation in dolphins from the Amazon River Basin: the Boto, *Inia geoffrensis*, and the Tucuxi, *Sotalia fluviatilis*. Reprod. Biol. Endocrinol. 5, 26.

Dabin, W., Cossais, F., Pierce, G.J., Ridoux, V., 2008. Do ovarian scars persist with age in all Cetaceans: new insight from the short-beaked common dolphin (*Delphinus delphis* Linnaeus, 1758). Mar. Biol. 156, 127–139.

Dahlheim, M.E., Heyning, J.E., 1999. Killer whale - *Orcinus orca*. Ridgway, S.H., Harrison, R. (Eds.), Handbook of marine mammals, vol 6. Academic Press, London, pp. 281–322.

Enders, A.C., Carter, A.M., 2004. What can comparative studies of placental structure tell us? A review. Placenta 25 (suppl) A, Trophoblast. Res. 18, S3–S9.

Evans, W.E., 1994. Common dolphin—*Delphinus delphis*. Ridgway, S.H., Harrison, R. (Eds.), Handbook of marine mammals, vol 5. Academic Press, London, pp. 191–224.

Foster, E.A., Franks, D.W., Mazzi, S., Darden, S.K., Balcomb, K.C., Ford, J.K.B., Croft, D.P., 2012. Adaptive prolonged postreproductive life span in killer whales. Science 337, 1313.

Fukui, Y., 2007. Ovary, oogenesis, and ovarian cycle. In: Miller, D.L. (Ed.), Reproductive Biology and Phylogeny of Cetacea. Whales, Dolphins and Porpoises. Science Publishers, Enfield, NH, USA, pp. 193–214.

Gaskin, D.E., 1986. Kidney and water metabolism. In: Bryden, M.M., Harrison, R. (Eds.), Research on Dolphins. Oxford University Press, New York, pp. 129–148.

Gihr, M., Pilleri, G., 1969. On the anatomy and biometry of *Stenella styx* Gray and *Delphinus delphis* L. (Cetacea, Delphinidae) of the Western Mediterranean. In: Pilleri, G. (Ed.), Investigations on Cetacea. Institute of Brain Anatomy, University of Berne, Berne, pp. 15–65.

Harrison, R.J., 1949. Observations on the female reproductive organs of the ca'aing whale *Globicephala melaena* Traill. J. Anat. 83, 238–253.

Harrison, R.J., 1969. Reproduction and reproductive organs. In: Andersen, H.T. (Ed.), Biology of Marine Mammals. Academic Press, New York, pp. 253–348.

Harrison, R.J., Mc Brearty, D.A., 1977. Ovarian appearances in captive delphinids (*Tursiops* and *Lagenorhynchus*). Aquat. Mammal. 5, 57–66.

Harrison, R.J., Boice, R.C., Brownell, Jr, R.L., 1969. Reproduction in wild and captive dolphins. Nature 222, 1143–1147.

Harrison, R.J., Brownell RIJr, Boice, R.C., 1972. Reproduction and gonadal appearances in some odontocetes. Harrison, R.J. (Ed.), Functional anatomy of marine mammals, Vol 1, Academic Press, London, pp. 361–429.

Hirose, K., Kasuya, T., Kazihara, T., Nishiwaki, M., 1970. Biological study of the corpus luteum and the corpus albicans of blue white dolphin (*Stenella coerulo-alba*). J. Mammal Soc. Jpn. 5, 33–40.

Hobson, B.M., Wide, L., 1986. Gonadotrophin in the term placenta of the dolphin (*Tursiops truncatus*), the Californian sea lion (*Zalophus californianus*), the grey seal (*Halichoerus grypus*) and man. J. Reprod. Fertil. 76, 637–644.

Hogg, C.J., Vickers, E.R., Rogers, T.L., 2005. Determination of testosterone in saliva and blow of bottlenose dolphins (*Tursiops truncatus*) using liquid chromatography-mass spectrometry. J. Chromatogr. B 814, 339–346.

Inouye, M., 1927. Über die Harnkanälchen des Rindes und des Tümmlers. In: Peter, K. (Ed.), Untersuchungen Über Bau Und Entwickelung Der Niere. Fischer, Jena, pp. 359–446.

Janech, M.G., Chen, R., Klein, J., Nowak, M.W., McFee, W., Paul, R.V., Fitzgibbon, W.R., Ploth, D.W., 2002. Molecular and functional characterization of a urea transporter from the kidney of a short-finned pilot whale. Am. J. Physiol. 282, R1490–R1500.

Jefferson, T.A., Webber, M.A., 2007. Marine Mammals of the World: A Comprehensive Guide to Their Identification. Academic Press, San Diego, pp. 1–592.

Kasuya, T., Marsh, H., 1984. Life history and reproductive biology of the short-finned pilot whale, *Globicephala macrorhynchus,* off the Pacific coast of Japan. Rep Int Whal Commn (Special Issue 6), 259–310.

Katsumata, E., 2010. Study on reproduction of captive marine mammals. J. Reprod. Dev. 56, 1–8.

Kellar, N.M., Trego, M.L., Chivers, S.J., Archer, F.I., 2013. Pregnancy patterns of pantropical spotted dolphins (*Stenella attenuata*) in the eastern tropical Pacific determined from hormonal analysis of blubber biopsies and correlations with the purse-seine tuna fishery. Mar. Biol. 160, 3113–3124.

Kellar, N.M., Trego, M.L., Chivers, S.J., Archer, F.I., Perryman, W.L., 2014. From progesterone in biopsies to estimates of pregnancy rates: large scale reproductive patterns of two sympatric species of common dolphin, *Delphinus* spp. off California, USA and Baja, Mexico. Bull. Southern California Acad. Sci. 113, 58–80.

Kenagy, G.J., Trombulak, S.C., 1986. Size and function of mammalian testes in relation to body size. J. Mammal. 67, 1–22.

Kruse, S., Caldwell, D.K., Caldwell, M.C., 1999. Risso's dolphin—*Grampus griseus*. Ridgway, S.H., Harrison, R. (Eds.), Handbook of marine mammals, vol 6. Academic Press, London, pp. 183–212.

Kusuda, S., Kakizoe, Y., Kanda, K., Sengoku, T., Fukumoto, Y., Adachi, I., Watanabe, Y., Doi, O., 2011. Ovarian cycle approach by rectal temperature and fecal progesterone in a female killer whale, *Orcinus orca*. Zoo Biol. 30, 285–295.

Marsh, H., Kasuya, T., 1984. Changes in the ovaries of the short-finned pilot whale, *Globicephala macrorhynchus*, with age and reproductive activity. Rep. Int. Whaling Commission (Special Issue 6), 331–335.

Marsh, H., Kasuya, T., 1986. Evidence for reproductive senescence in female cetaceans. Rep. Int. Whaling Commission, 57–74, (Special Issue 8).

Matthews, L.H., 1950. The male urogenital tract in *Stenella frontalis* (G. Cuvier). Atlantide Report, Danish Sciences Press, LTD., Copenhagen 1, 221–247.

Miller, D.L., 2007. Reproductive Biology and Phylogeny of Cetacea. Whales, Dolphins and Porpoises, Science Publishers, Enfield, NH, USA, pp. 1–428.

Miller, D.L., Styer, E.L., Kyta, S., Menchaca, M., 2007a. The mature cetacean spermatozoon. In: Miller, D.L. (Ed.), Reproductive Biology and Phylogeny of Cetacea. Whales, Dolphins and Porpoises. Science Publishers, Enfield, NH, USA, pp. 245–280.

Miller, D.L., Styer, E.L., Menchaca, M., 2007b. Placental structure and comments on gestational ultrasonographic examination. In: Miller, D.L. (Ed.), Reproductive Biology and Phylogeny of Cetacea. Whales, Dolphins and Porpoises. Science Publishers, Enfield, NH, USA, pp. 331–348.

Morejohn, V.G., Baltz, D.M., 1972. On the reproductive tract of the female Dall porpoise. J. Mammal. 53, 606–608.

Morton, W.R.M., Mulholland, H.C., 1961. The placenta of the Ca'ing whale, *Globicephala melaena* (Traill). J. Anat. 95, 605.

Murphy, S., Deaville, R., Monies, R.J., Davison, N., Jepson, P.D., 2011. True hermaphroditism: first evidence of an ovotestis in a cetacean species. J. Comp. Path. 144, 195–199.

Neuenhagen, C., Hartmann, M.G., Greven, H., 2006. Histology and morphometrics of testes of the white-sided dolphin (*Lagenorhynchus acutus*) in by-catch samples from the northeastern Atlantic. Mammalian Biol. 72, 283–298.

O'Brien, J.K., Robeck, T.R., 2010. The value of *ex situ* cetacean populations in understanding reproductive physiology and developing assisted reproductive technology for *ex situ* and *in situ* species management and conservation efforts. Int. J. Comp. Psychol. 23, 227–248.

O'Brien, J.K., Robeck, T.R., 2012. The relationship of maternal characteristics and circulating progesterone concentrations with reproductive outcome in the bottlenose dolphin (*Tursiops truncatus*) after artificial insemination, with and without ovulation induction, and natural breeding. Theriogenology 78, 469–482.

O'Rahilly, R., 1972. Guide to staging of human embryos. Anat. Anz. 130, 556–559.

Odell, K.K., McClune, K.M., 1999. False killer whale—*Pseudorca crassidens*. Ridgway, S.H., Harrison, R. (Eds.), Handbook of marine mammals, vol. 6. Academic Press, London, pp. 213–243.

Ohsumi, S., 1964. Comparison of maturity and accumulation rate of *corpora albicantia* between the left and right ovaries in cetacea. Sci. Rep. Whales Res. Inst. Tokyo 18, 123–148.

Olesiuk, P.F., Bigg, M.A., Ellis, G.M., 1990. Life history and population dynamics of resident killer whales (*Orcinus orca*) in the coastal waters of British Columbia and Washington State. Rep. Int. Whal. Commn. (Special Issue 12), 209–243.

Oliver, J., 1968. Nephrons and Kidneys: A Quantitative Study of Development and Evolutionary Mammalian Renal Architectonics. Harper & Row, New York.

Orbach, D.N., Marshall, C.D., Würsig, B., Mesnick, S.L., 2016. Variation in female reproductive tract morphology of the common bottlenose dolphin (*Tursiops truncatus*). Anat. Rec. 299, 520–537.

Ortiz, R.M., 2001. Osmoregulation in marine mammals. J. Exp. Biol. 204, 1831–1844.

Ortiz, R.M., Worthy, G.A., 2000. Effects of capture on adrenal steroid and vasopressin concentrations in free-ranging bottlenose dolphins (*Tursiops truncatus*). Comp. Biochem. Physiol. A 125, 317–324.

Pabst, D.A., Rommel, S.A., McLellan, W.A., Williams, T.M., Rowles, T.K., 1995. Thermoregulation of the intra-abdominal testes of the bottlenose dolphin (*Tursiops truncatus*) during exercise. J. Expl. Biol. 198, 221–226.

Panin, M., Gabai, G., Ballarin, C., Peruffo, A., Cozzi, B., 2012. Evidence of melatonin secretion in cetaceans: plasma concentration and extrapineal HIOMT-like presence in the bottlenose dolphin *Tursiops truncatus*. Gen. Comp. Endocrinol. 177, 238–245.

Perez, S., García-López, A., De Stephanis, R., Giménez, J., García-Tiscar, S., Verborgh, P., Mancera, J.M., Martínez-Rodriguez, G., 2011. Use of blubber levels of progesterone to determine pregnancy in free-ranging live cetaceans. Mar. Biol. 158, 1677–1680.

Perrin, W.F., Reilly, S.B., 1984. Reproductive parameters of dolphins and small whales of the family Delphinidae. Rep. Int. Whal. Commn. (Special issue 6), 97–133.

Perrin, W.F., Wilson, C.E., Archer, II, F.I., 1994. Striped dolphin—*Stenella coeruleoalba*. Ridgway, S.H., Harrison, R. (Eds.), Handbook of marine mammals, vol 5. Academic Press, London, pp. 129–159.

Pfeiffer, C.J., 1997. Renal cellular and tissue specializations in the bottlenose dolphin (*Tursiops truncatus*) and beluga whale (*Delphinapterus leucas*). Aquat. Mammal. 23, 75–84.

Plön, S., Bernard, R., 2007a. Anatomy with particular reference to the female. In: Miller, D.L. (Ed.), Reproductive Biology and Phylogeny of Cetacea. Whales, Dolphins and Porpoises. Science Publishers, Enfield, NH, USA, pp. 147–170.

Plön, S., Bernard, R., 2007b. Testis, spermatogenesis, and testicular cycles. In: Miller, D.L. (Ed.), Reproductive Biology and Phylogeny of Cetacea. Whales, Dolphins and Porpoises. Science Publishers, Enfield, NH, USA, pp. 215–244.

Pomeroy, P., 2011. Reproductive cycles of marine mammals. Animal Reprod. Sci. 124, 184–193.

Power, G.G., Schroder, H., Gilbert, R.D., 1984. Measurement of fetal heat production using differential calorimetry. J. Appl. Physiol. 57, 17–22.

Raduán, A., Blanco, C., Fernández, M., Raga, J.A., 2007. Some aspects of the life history of the Risso's dolphins *Grampus griseus* (Cuvier 1812) in the western Mediterranean Sea. Eur. Res. Cetaceans 21, 71.

Rankin, J.J., 1961. The bursa ovarica of the beaked whale, *Mesoplodon gervaisi Deslongcharnps*. Anat. Rec. 139, 379–385.

Reeves, R.R., Smeenk, C., Kinze, C., Brownell R.L., Jr, Lien, J., 1999. White-beaked dolphin—*Lagenorhynchus albirostris*. Ridgway, S.H., Harrison, R. (Eds.), Handbook of marine mammals, vol 6. Academic Press, London, pp. 1–30.

Ridgway, S.H., Venn-Watson, 2009. Effects of fresh and seawater ingestion on osmoregulation in Atlantic bottlenose dolphins (*Tursiops truncatus*). J. Comp. Physiol. B 180 (4), 563–576.

Robeck, T.R., Monfort, S.L., 2006. Characterization of male killer whale (*Orcinus orca*) sexual maturation and reproductive seasonality. Theriogenology 66, 242–250.

Robeck, T.R., Atkinson, S.K.C., Brook, F., 2001. Reproduction. In: Dierauf, L.A., Gulland, F.M.D. (Eds.), CRC Handbook of marine mammal medicine, second ed. CRC Press, Boca Raton, Florida, pp. 193–236.

Robeck, T.R., Steinman, K.J., Yoshioka, M., Jensen, E., O'Brien, J.K., Katsumata, E., Gili, C., McBain, J.F., Sweeney, J., Monfort, S.L., 2005. Estrous cycle characterisation and artificial insemination using frozen–thawed spermatozoa in the bottlenose dolphin (*Tursiops truncatus*). Reproduction 129, 659–674.

Robeck, T.R., Steinman, K.J., Greenwell, M., Ramirez, K., Van Bonn, W., Yoshioka, M., Katsumata, E., Dalton, L., Osborn, S., O'Brien, J.K., 2009. Seasonality, estrous cycle characterization, estrus synchronization, semen cryopreservation, and artificial insemination in the Pacific white-sided dolphin (*Lagenorhynchus obliquidens*). Reproduction 138, 391–405.

Robeck, T.R., Willis, K., Scarpuzzi, M.R., O'Brien, J.K., 2015. Comparisons of life-history parameters between free-ranging and captive killer whale (*Orcinus orca*) populations for application toward species management. J. Mammal. 96 (5), 1055–1070.

Rommel, S.A., Pabst, D.A., McLellan, W.A., Mead, J.G., Potter, C.W., 1992. Anatomical evidence for a countercurrent heat exchanger associated with dolphin testes. Anat. Rec. 232, 150–156.

Rommel, S.A., Pabst, D.A., McLellan, W.A., 1993. Functional morphology of the vascular plexuses associated with the cetacean uterus. Anat. Rec. 237, 538–546.

Rommel, S.A., Pabst, D.A., McLellan, W.A., Williams, T.M., Friedl, W.A., 1994. Temperature regulation of the testes of the bottlenose dolphin (*Tursiops truncatus*): evidence from colonic temperatures. J. Comp. Physiol. B 164, 130–134.

Rommel, S.A., Pabst, D.A., McLellan, W.A., 1998. Reproductive thermoregulation in marine mammals. Am. Scient. 86, 440–450.

Rommel, S.A., Pabst, D.A., McLellan, W.A., 2007. Functional anatomy of the cetacean reproductive system, with comparisons to the domestic pig. In: Miller, D.L. (Ed.), Reproductive Biology and Phylogeny of Cetacea. Whales, Dolphins and Porpoises. Science Publishers, Enfield, NH, USA, pp. 127–146.

Sawyer, J.E., Walker, W.A., 1977. Vaginal calculi in the dolphin. J. Wildlife Dis. 13, 346–348.

Schroeder, 1990. Breeding bottlenose dolphins in captivity. In: Leatherwood, S., Reeves, R.R. (Eds.), The Bottlenose Dolphin. Academic Press, San Diego, pp. 435–446.

Schroeder, J.P.1, Keller, K.V., 1989. Seasonality of serum testosterone levels and sperm density in *Tursiops truncatus*. J. Exp. Zool. 249, 316–321.

Simpson, J.G., Gardner, M.B., 1972. Comparative microscopic anatomy of selected marine mammals. In: Ridgway, S.H. (Ed.), Mammals of the Sea. Biology and Medicine. Charles C. Thomas, Springfiels, IL, pp. 298–418.

Slijper, E.J., 1966. Functional morphology of the reproductive system in Cetacea. In: Norris, K.S. (Ed.), Whales, Dolphins and Porpoises. University of California Press, Berkeley, pp. 277–318.

Sperber, I., 1944. Studies on the mammalian kidney. Zoologiska Bidrag fran Uppsala 22, 249–432.

Steinman, K.J., Robeck, T.R., O'Brien, K., 2016. Characterization of estrogens, testosterone, and cortisol in normal bottlenose dolphin (*Tursiops truncatus*) pregnancy. Gen. Comp. Endocrinol. 226, 102–112.

Štěrba, O., Klima, M., Schildger, B., 2000. Embryology of dolphins—staging and ageing of embryos and fetuses of some cetaceans. Adv. Anat. Embryol. Cell Biol. 157, 1–133.

Suzuki, M., Endo, N., Nakano, Y., Kato, H., Kishiro, T., Asahina, K., 2008. Localization of aquaporin-2, renal morphology and urine composition in the bottlenose dolphin and the Baird's beaked whale. J. Comp. Physiol. B 178, 149–156.

Takahashi, Y., Ohwada, S., Watanabe, K., 2006. Does elastin contribute to the persistence of corpora albicantia in the ovary of the common dolphin (*Delphinus delphis*). Mar. Mammal. Sci. 22 (4), 819–830.

Thewissen, J.G.M., Heyning, J., 2007. Embryogenesis and development in *Stenella attenuata* and other cetaceans. In: Miller, D.L. (Ed.), Reproductive Biology and Phylogeny of Cetacea. Whales, Dolphins and Porpoises. Science Publishers, Enfield, NH, USA, pp. 307–330.

Van der Schoot, P., 1995. Studies on the fetal development of the *gubernaculum* in cetacea. Anat. Rec. 243, 449–460.

Venn-Watson, S., Smith, C.R., Dold, C., Ridgway, S.H., 2008. Use of a serum-based glomerular filtration rate prediction equation to assess renal function by age, sex, fasting, and health status in bottlenose dolphins (*Tursiops truncatus*). Mar. Mamm. Sci. 24, 71–80.

Watanabe, N., Hatano, J., Asahina, K., Iwasaki, T., Hayakawa, S., 2007. Molecular cloning and histological localization of LH-like substances in a bottlenose dolphin (*Tursiops truncatus*) placenta. Comp. Biochem. Physiol. A 146, 105–118.

Wells, R.S., Scott, M.D., 1999. Bottlenose dolphin—*Tursiops truncatus*. Ridgway, S.H., Harrison, R. (Eds.), Handbook of Marine Mammals, vol 6. Academic Press, London, pp. 137–182.

West, K.L., Atkinson, S., Carmichael, M.J., Sweeney, J.C., Krames, B., Krames, J., 2000. Concentration of progesterone in milk from bottlenose dolphin during different reproductive states. Gen. Comp. Endocrinol. 117, 218–224.

Westgate, A.J., Read, A.J., 2007. Reproduction in short-beaked common dolphins (*Delphinus delphis*) from the western North Atlantic. Mar. Biol. 150, 1011–1024.

Wislocki, G.B., Enders, R.K., 1941. The placentation of the bottle-nosed porpoise (*Tursiops truncatus*). Am. J. Anat. 68, 97–114.

Xu, S., Yang, Y., Zhou, X., Xu, J., Zhou, K., Yang, G., 2013. Adaptive evolution of the osmoregulation-related genes in cetaceans during secondary aquatic adaptation. BMC Evol. Biol. 13, 189.

Yablokov, A.V., Bel'kovich, V.M., Borisov, V.I., 1972. Whales and dolphins (Kity I del'fini), vols. 1 and 2. Moscov (English translation 1974, US Joint Publications Research Service, Arlington, Virginia, USA).

Zhemkova, Z.P., 1967. Cetacean placentae. Folia Morphologica 15, 104–107.

Zotti, A., Poggi, R., Cozzi, B., 2009. Exceptional bone density DXA values of the rostrum of a deep-diving marine mammal: a new technical insight in the adaptation of bone to aquatic life. Skeletal Radiol. 38, 1123–1125.

Chapter 10

Neurobiology and the Evolution of Dolphins

GENERAL ASPECTS

Dolphins and other toothed whales seem to be perfectly adapted to their aquatic environment, and sometimes it is hard to believe that they originated from terrestrial mammals. They are most remarkable because of the way they accomplished such an extreme change of habitat. This transgression was correlated to a variety of modifications in the general mammalian bauplan (the construction of an organism or group of organisms concerning homolog characters) as to morphological and physiological aspects of every organ system. Their ancestors were generalized (basic) four-legged terrestrial mammals at the origin of even-toed hoofed animals (artiodactyls) in the Paleogene [about 65 million years ago; Fig. 10.1 (Gatesy et al., 2013)]. Extant (living) artiodactyls that are close to this origin comprise the mouse deer (tragulids), the pig (*Sus scrofa*), cattle (*Bos taurus*), and hippos (*Hippopotamus amphibius*).

So far, we can only speculate about a detailed scenario for this remarkable transition of the cetacean ancestors into aquatic habitats, which led to some of the largest, biggest brained, fastest swimming, loudest, and deepest diving mammals. As to the smaller odontocetes, dolphins (delphinids) are the most successful and most diverse and speciose family of marine mammals; they comprise at least 33 species (LeDuc, 2009). Many species are found far offshore in deep water, and the specifics of their ecological requirements are poorly known. For most species of delphinids, even basic aspects of their morphology, physiology, behavior, and ecology are virtually unknown.[a]

Delphinids likely arose in the mid- to late Miocene (11–12 million years ago). At present, there is much uncertainty about the evolutionary relationships among the delphinids. Until today, only very few dolphin species have been investigated appreciably; the amount of information available on single species is more or less reflected in the degree of their representation in this book. The standard dolphin for most aspects of biological information is the common bottlenose dolphin (*Tursiops truncatus*); it is undeniably the best known of all dolphins and cetaceans even. Bottlenose dolphins are found worldwide in temperate and tropical marine waters, both inshore and offshore (see Chapter 1).

As a whole, dolphins are highly interesting from an evolutionary perspective because they were subjected to a clear-cut "experiment in nature." They had to face the physical properties of the medium water and the fact that not all of their sensory systems could be maintained; some of them were either reduced or eliminated. One sensory input channel, however, the hearing organ became the utmost important window to the world of dolphins. The sense of hearing hypertrophied as kind of a "compensation" for the complete loss of the olfactory organ which, in most terrestrial mammals, is essential for navigation, communication, reproduction as well as the detection and analysis of food or prey (Fig. 10.2). Interestingly, signs of these transformations in the sensory organs as well as the locomotor system are found reflected in the peripheral and central nervous system. On the reverse, investigations of the nervous system can deliver arguments for the reconstruction of evolution. Therefore in the following, we shall try to combine relevant characters of the head and body with the morphology of the brain in order to get an idea of what dolphins are really like and how they became what they are. This is done before the ensemble of terrestrial mammals; the biological information known of these animals can serve as a standard for the interpretation of the dolphins as to functional and evolutionary aspects.

There are some obstacles, however, in understanding brain evolution in aquatic mammals. First, we are only marginally familiar with brain morphology of a few species, and here mainly the bottlenose dolphin. Second, the brain itself does not fossilize; only the outer shape can be studied in natural endocasts. Here, no information on the internal morphology as, for example, the ventricular system, the size and shape of brain nuclei and cortex structure as well as the thickness and course of fiber tracts is available. Thus the tracing of brain evolution in fossils is difficult or even impossible and has to be complemented or even replaced by phylogenetic reconstruction on the basis of extant relatives and closely related groups of mammals. Fortunately, today this is possible and effective; recent synthetic publications (Boisserie et al., 2011;

---

a. A good overview of the numerous adaptational features is given in Gatesy et al. (2013).

Anatomy of Dolphins. http://dx.doi.org/10.1016/B978-0-12-407229-9.00010-5

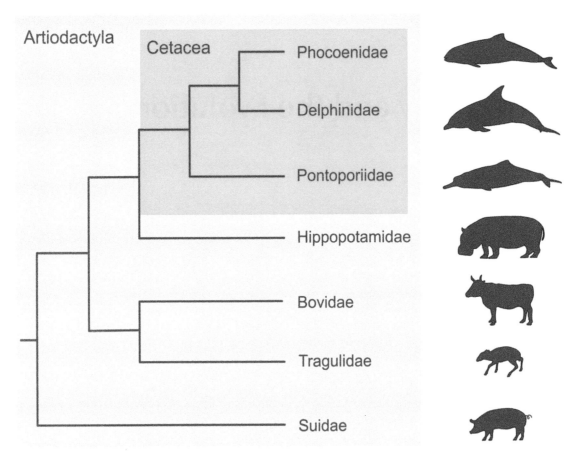

FIGURE 10.1   Simplified phylogenetic tree that was originally created on the basis of both morphological and genetic criteria. Tragulids are small even-toed hoofed animals that are considered the most plesiomorphic (basic) living artiodactyls. Dolphin-like and delphinid families in *blue*; actual phylogenetic trees regard cetaceans as an aquatic group of even-toed hoofed animals. Cetaceans are shown at the same body length. *(Modified by Jutta Oelschläger from Gatesy et al., 2013. A phylogenetic blueprint for a modern whale. Mol. Phylogenet. Evol. 66, 479–506.)*

Gatesy et al., 2013) have confirmed that the closest relatives to cetaceans are hoofed animals and that cetaceans can be regarded strongly derived and most highly encephalized artiodactyls (Cetartiodactyla, see Chapter 1). In smaller and middle-sized dolphins this means that, with the enlargement of the brain, the cranial vault was widened and the skull as a whole changed its shape considerably, a moderate parallel to the situation in the human where the neurocranium dominates the skull (see Chapter 3).

MAIN SENSORY SYSTEMS: LIFE UNDER CONSTRAINTS

To some degree, dolphins in their aquatic environment can be understood by reviewing the conditions of their input channels. They inform the animals about the parameters of their habitat to be met within their physiologial limits. Compared to air as the surrounding medium of terrestrial mammals, water is about 58 times more viscous and its specific gravity is about 770 times higher (at 20°C). Sound propagation velocity is about 5 times higher in water than in air. In the ancestors of dolphins, all these striking physical differences of the two media must have had profound impact on locomotion (Chapter 3) and behavior such as diving (Chapter 4), hunting, and feeding (Chapter 8) as well as orientation and communication (Chapter 5). In order to become fast swimmers, they diminished their body surface to a relative minimum and thus low water resistance (Chapter 2). The hind limbs were downsized until no longer visible externally. Also, the pelvic girdle was lessened and detached from the vertebral column. Whereas in dolphin embryos external limb buds are present at the base of the tail, they disappear in early fetuses (Reidenberg and Laitman, 2009; Štěrba et al., 2000; Thewissen et al., 2006; Thewissen and Heyning, 2007). During dissections of adult dolphins, rudiments of the pelvic girdle and hind limbs may be found. Obviously, they are still needed as the site of origin for genital muscles. Very rarely, adult dolphins with one or two knobs near the genital region can be seen; these knobs can be regarded as kind of persisting outer limb buds (Thewissen et al., 2009).

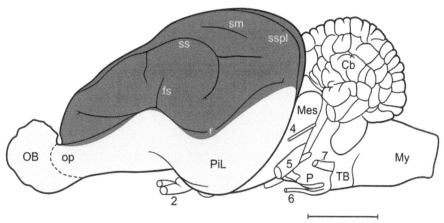

(a) Mouse deer (*Hyemoschus aquaticus*)

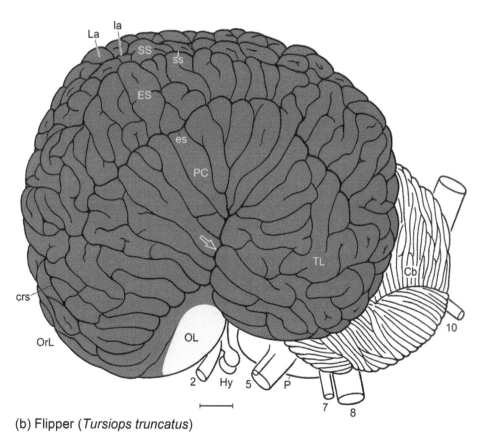

(b) Flipper (*Tursiops truncatus*)

FIGURE 10.2   **The phenomenon of brain growth in dolphins compared to that in a mouse deer (water chevrotain) (Scales = 1 cm).** The latter group of animals (family Tragulidae) is considered to be rather plesiomorphic among the living ruminant artiodactyls. The water chevrotain lives in West and Central Africa, is a good swimmer, can dive underwater and walk along the river bottom to elude predators. The body mass ranges from 7–16 kg and the brain mass from about 17–30 g (Sigmund, 1981). Above both brains have been brought to about the same length. There are profound differences between the two brains as to absolute and relative size, the relative size of the telencephalon and of the neocortex. The large absolute size of the telencephalic hemisphere in the dolphin is obviously correlated with a high degree of gyrification and thus with area enhancement in the cortical grey. Another correlation seems to exist between the size of the telencephalon and that of the cerebellum. Within the cerebral hemisphere, a striking difference between both brains is the size of the paleocortex and that of the neocortex. In the water chevrotain, the olfactory bulb is very large and the surface of the telencephalic hemisphere, to a considerable degree, is covered by paleocortex. The dolphin has lost the anterior part of its olfactory system (olfactory mucosa, olfactory bulb and tract; not all olfactory areas seen here) and the olfactory lobe is highly modified and presumably nonolfactory in function. As to the morphological nomenclature of the surface of the telencephalon, there is some discrepancy in the literature. Cb, cerebellum; crs, cruciate sulcus; es, ectosylvian sulcus; ES, ectosylvian gyrus; Hy, hypophysis; La, lateral gyrus; la, lateral sulcus; Mes, mesencephalon; My, myelencephalon; OB, olfactory bulb; OL, olfactory lobe; OrL, orbital lobe; op, olfactory peduncle; P, pons; PC, perisylvian cortex; PiL, piriform lobe; r, rhinal fissure; ss, suprasylvian sulcus; SS, suprasylvian gyrus; ssp, sulcus splenialis; TB, trapezoid body; TL, temporal lobe; 2–10, cranial nerves: 2, optic nerve; 4, trochlear nerve; 5, trigeminal nerve; 6, abducens nerve; 7, facial nerve; 8, vestibulocochlear nerve; 10, vagus nerve. Arrow pointing into sylvian cleft (sc). *(Artwork by Jutta Oelschläger.)*

TABLE 10.1 Head Structures of Extant Toothed Whales and Their Properties Related to Their Functional Implications

Area	Structure	Parts and Properties	Functional Implications, Remarks
Skull and soft parts	Skull roof (cranial vault)	Telescoping	Acoustic shield, impedance mismatch
Epicranial complex	Monkey lips, dorsal bursae, melon, nasal sacs, facial muscles	Synapomorphic cluster of specific features	Generator and transmitter of sonar (sound) signals; unique
	Melon, dorsal bursae	Acoustic fat	Unique
Temporal region	Temporomandibular joint	Simple hinge joint	Jaws grasping
	Lower jaw long but weak	Pan bone, acoustic window, mandibular canal, acoustic fat body	Reception and transmission of sound; unique among mammals
	Teeth	Teeth polyodont, homodont	No cutting or chewing
	Ear bones	Tympanoperiotic (TP)	Uncoupling from neighboring skull elements
	Tympanic air sacs	Acoustic isolation of TP	Underwater directional hearing
	Tympanic bulla	Pachyosteosclerosis	Underwater directional hearing
	Ossicles	Pachyosteosclerosis	Underwater directional hearing
	Periotic	Pachyosteosclerosis	Underwater directional hearing
	Cochlea	Comparatively large	Ultrasound adaptations
	Vestibular system	Very small	Reduced sensitivity
	Semicircular canals	Extremely reduced; unique	Reduced sensitivity
Cervical vertebrae		Fused	Limited moveability of head
Larynx	Large, duckbill-shaped	Sound generation	Air pressurization

In order to further increase the hydrodynamic performance of the dolphin body, the neck contour of toothed whales was smoothed out by the shortening of the cervical vertebral column and the strengthening of the neck musculature (see Chapter 3; Table 10.1). Concomitantly, the head was firmly integrated into the undulatory locomotion of the body. As a result, the toothed whales (dolphins) became spindle-shaped and, as warm-blooded mammals with a high metabolism, rather successful in chasing and catching agile and quick fish.

Many of these adaptational changes in the body shape and structure of dolphins have left their marks in the peripheral and central nervous system and can be seen today. They are most obvious in the structure of the sensory systems (olfaction, vision, audition, equilibrium, somatosensation) as well as in quantitative relationships of the relevant sensory centers within the brain which process the sensory input and feed the motor system.

Olfaction

This sensory system has been lost in fetal and postnatal dolphins (Table 10.2). Nevertheless it is mentioned here because

1. it is one of the leading senses or even the dominant sense in many terrestrial mammals,
2. it was certainly well developed in the terrestrial (artiodactyl) ancestors of the cetaceans,
3. the reduction of this system is seen during dolphin ontogenesis and the secondary cortical formations at the base of the telencephalon have been doubted as to a possible olfactory function, and
4. it was accompanied by a profound reconstruction of the nasal region with respect to sound emission.

In the toothed whale ancestors, the external nasal openings (nostrils) have migrated from the tip of the snout to the vertex of the head during evolution. At the same time, the nose has been modified into a highly sophisticated acoustic signal generator in the forehead (epicranial complex; see Chapter 5). With an ensemble of morphologically new structures around the upper respiratory tract, this nasal complex produces sound signals for communication between dolphins as well as ultrasound clicks for orientation in their environment and the detection of moving and hidden prey (biosonar). Such high-pressure changes for sound generation within the nasal tract to some extent may have been noncompatible with sensory function of the anterior olfactory system.

TABLE 10.2 Structures of the Peripheral Nervous System of the Head and Their Properties in Extant Toothed Whales

Area	Structure	Properties	Functional Implications, Remarks	Difference From the Presumed Ancestral Configuration
	Cranial nerves including nerve nuclei	Associated parts of the central nervous system		Progression (⬆) Regression (⬇)
Nose	I (olfactory nerve)	Paleocortex	Smell lost in embryos, echolocation (EC)	⬇⬇⬇
Eye	II, III, IV, VI	For example, superior colliculus, BSN	Significance as in ancestors	⬇?
Face	V (trigeminal)	Brainstem nuclei (BSN)	Rostrum, EC, vibrissal pressure receptors	➡
	VII (facial)	Facial motor nucleus	Nasofacial acoustic expression. Echolocation	⬆
Ear	VIIIc (cochlear)	Auditory pathway; for example, inferior colliculus, etc.	Dominant sensory system. Echolocation	⬆⬆
	VIIIv (vestibular)	Vestibular nuclei	Extremely reduced; unique in mammals	⬇⬇

Note: The number of arrows indicate the significance of progression and regression.

Vision

Dolphins have a fairly effective visual system that allows good sight above and below the water surface (Table 10.2). Their eyeballs are relatively large and all the ocular components are present and functional. However, their external eye muscles are reduced, as is the relevant innervation. The retina is about twice as thick as in terrestrial artiodactyl mammals of the same body dimension (cattle, horse). The density of ganglion cells within the inner layer of the retina is low, about one-fourth of that in the human. At the same time, the diameters of retinal somata are exceptionally large in dolphins (Chapter 5).

The optic nerve in dolphins is thicker than in terrestrial mammals of the same body dimension (Chapter 6). However, optic fiber density in the bottlenose dolphin is only about one-third to one-eighth of that found in the human. This correlates well with the situation in the retina (low density of very large ganglion cells). Also, the axons found here are thicker than in any other mammals investigated: compared to those in the human, their average diameter is 3–5 times higher, with a maximum of more than 15 μm (giant optic nerve fibers).

The exceptional thickness of the retina has been explained by glia elaboration and increased extraneural space. A similar situation may be seen in the optic nerve; here, more than 50% of the cross-section belongs to the nonneural tissue (glia) and increased extraneural space. It has been supposed that this feature may be associated with peculiarities in the tissue metabolism of diving animals. And it was taken into account that this phenomenon refers not only to the retina and optic nerve but also to other parts of the central nervous system (Dawson et al., 1983). This would be interesting for the assessment of how much volume of a brain is dedicated to neural function in the sense of computational power and how much is taken by the relevant vegetative (metabolic) support.

In dolphins, the structure of the retina, the distribution of its ganglion cells and the potential resolution power of the eye show adaptations to dim-light conditions under water. Here, the animals seem to rely more on sensitivity to detect the "contrast" of the motion of prey than on acuity (Murayama et al., 1995). The low density of thick and heavily myelinated axons may be related to an increase in conduction velocity and a relatively low resolution within the retina (Gao and Zhou, 1992). Peak conduction velocities in the dolphin optic nerve may be higher than in other mammals. An ethologically based argument for this could be the need for rapid visual target acquisition in turbid marine visual environments (Dawson et al., 1983).

In the case of massive worsening of light conditions in river systems with slow currents and permanently murky waters, such modifications mentioned previously for marine dolphins obviously can no longer help compensate for the loss of visual information. Here, some archaic river dolphins such as *Platanista gangetica*, that lives in the waters of the Ganges,

Brahmaputra, Karnaphuli, and Indus rivers and their tributaries (Smith and Braulik, 2009) have tiny eyes. There is no lens and the retina is very small. The optic nerve as well as the remaining visual pathway is also strongly reduced, including the visual cortical area which is small with respect to the situation in marine dolphins (Knopf et al. in press).

Audition

The hearing organ is the dominant sensory gateway in dolphins and other toothed whales (Tables 10.1 and 10.2). The external ear (pinna) is lost, the middle ear highly modified, the ear bones and ossicles extremely dense and voluminous (pachyostotic) and the ascending auditory pathway much hypertrophied with respect to the situation in most terrestrial mammals, including humans (see Chapters 5 and 6). This strong development of the sound/ultrasound receptor structures reflects both the importance of this sensory system for dolphins and the high energy level for aquatic sound transduction between the transmitter and receiver. With this high-performance sonar system dolphins can detect and discriminate/investigate objects over large distances. And within the spectrum of sensory modalities, hearing was presumably the major factor in the development of large to very large brains.

During cetacean evolution, the middle ear system was kept as a modulatory unit for preprocessing and transmission of incoming sound with a minimum of energy loss. Here, the mechanical properties of the medium water left deep traces in the organization of the tympanic cavity and its components (ossicles, etc.). At the same time, the middle ear retained its capacity to shield the inner ear against unfavorable or dangerous biological influences (protection against acoustic overload by, eg, the middle ear muscles). In correlation with the improvement of diving performance, the air-filled middle ear is adapted to changing hydrostatic pressures by surrounding secondary air sinuses. In order to have stable conditions for reliable audition within the uncoupled tympanoperiotic complex, the middle ear cavity is protected by the extremely dense and heavy ceramic-like periotic and tympanic bones which represent a solid mechanical framework for the hearing process.

The peripheral sound-conducting pathway of dolphins is unique among mammals. In different ways, selection pressure has profoundly changed the outer and middle ear as outposts for the collection, filtering, and mechanical preprocessing of sound and ultrasound from water to air (middle ear cavity) and again to water (endolymph). Due to different physical conditions for cetacean ancestors in the aquatic habitat, the access to the sound path had to be transferred from the outer ear to the lower jaw of dolphins, with its acoustic fat body guiding the sound waves to the middle ear. Although the mammalian outer ear has nearly been wiped out, the middle ear is obviously adapted to couple the incoming sound to the inner ear. The cochlea, however, seems to be remarkably conservative: it is coiled, contains three fluid-filled scalae and shows all essentials known from other mammals. Several structural peculiarities correlate with hearing range, behavior, and habitat, for example, the development of the bony spiral lamina in the dolphin cochlea (Chapter 5). Such correlations allow investigations on the hearing capabilities of extinct or rare living species that are inaccessible for behavioral investigations (Fleischer, 1976). Thus there are morphological specializations in dolphin ears that seem to correlate with different types of ultrasound spectrum used for orientation and navigation.

Major events in the evolution of dolphin hearing had to do with the uncoupling of the ear from the skull and the incorporation of the accessory bones (tympanic, periotic) into a new morphological and functional unit (tympanoperiotic complex). Within this new entity, sound reception and transduction was adapted to higher mechanical forces by the dedication of structures and components to new challenges: thus the tympanic membrane was changed into a ligament and replaced functionally by a very thin part of the tympanic bone (tympanic plate). The latter receives sound/ultrasound waves coming from the lower jaw and transfers them via the modified chain of middle ear ossicles to the vestibular foramen and thus the labyrinth (cochlea). (Chapter 5).

Sense of Balance

In contrast to the well-developed cochlea, the vestibular organ is only very small (Table 10.2), and in general (with one exception; see following sections and Chapter 6) this is also true for the central nuclei within the brainstem (Kern et al., 2009). Here, an unusual reduction has taken place which is unparalleled within the mammalia; it is maximal in the semicircular canals which are only minute. The vestibular ramus is rather thin in dolphins; less than 5% of the fibers in the whole vestibulocochlear nerve are devoted to vestibular function as compared to about 20–40% in terrestrial mammals including the human (Gao and Zhou, 1992). In dolphins, the vestigial vestibular system may act as a "vehicle-oriented accelerometer" that registers linear speed-ups (utricle) but less rotational inputs during turning maneuvers, thus avoiding "space-sickness" (Ketten, 1992).

In the evolution of dolphins, essential functional parameters influencing locomotion may have interfered with the sense of equilibrium (Chapter 6). In dolphins the vestibular system in the ear as well as the accessory brainstem nuclei are bound to their mode of locomotion: these animals have included the head into the powerful body stem and caudal fin (flukes) and move by vertical undulation (Chapter 2). Only the strong atlanto-occipital joints are left after the shortening and fusion of the cervical vertebrae and allow rotations of the head around a transverse axis through the two occipital condyles (Chapter 3). In contrast, the human head is balanced on the vertebral column in the atlanto-occipital and atlanto-axial joints. Relative stability in head position is largely maintained by complex and fine-tuned muscular activity. In dolphins this moveability, for the most part, has been abandoned in favor of high-power locomotion. As a drawback, sensory feedback loops involved in such complex orientation reactions are presumably restricted or lacking in dolphins. In terrestrial mammals, head position and movements are closely correlated with eye movements by means of reflexes and other types of sensorimotor connectivity (vestibuloocular reflex, optokinetic nystagmus, horizontal gaze holding, smooth pursuit). In comparison, dolphins seem to be rather different and may not have such a repertoire of eye movability. Also, they obviously do not have a wide field of binocular vision but only some visual overlap rostroventrally and dorsocaudally (Chapter 5). Despite these small areas of overlap, dolphin eye movements are not conjugate and the eyes seem to move independently. For all these reasons, a vestibuloocular or other sophisticated connectivity including the neck as seen in primates (humans) does not seem plausible for dolphins. Relevant specialized sensory neurons in terrestrial mammals like head direction cells, active within the "eye–head–neck" movement reference frame and located in the hippocampus and entorhinal area may also be lacking in dolphins. So far, however, it is not clear whether this assumption can help explain the extraordinarily small size of the hippocampus of dolphins (Chapter 6, Section "Limbic System").

As indicated previously, there is one exception as to the reduction of vestibular brainstem nuclei. This refers to the lateral vestibular (Deiters') nucleus which is very large in dolphins (Chapter 6). In contrast to the "genuine" vestibular nuclei which get their input from the sensory organs of the labyrinth and are intimately related to the visual system, Deiters' nucleus is obviously closely related to the cerebellum and projects into the spinal cord as part of the extrapyramidal system that is responsible for mass movements of the body stem.

Skin Innervation

Cutaneous sensitivity was reported to be fairly good in cetaceans (Chapters 2 and 5). However, the posterior white columns (gracilis and cuneate fascicles), the posterior horns of the gray matter as well as the posterior roots of the spinal nerves are small in whales and dolphins. This means that in these animals the representation of body sensitivity may be lower than in terrestrial mammals. This seems to be plausible with respect to a lower surface area of the body in cetaceans caused by the loss of the hind limbs and the modification of the forelimbs into steering organs. Therefore the medial lemniscus is rather thin in its caudal part, where it arises from the gracile and cuneate nuclei. The contribution of sensory fiber masses from the trigeminal system, however, is remarkable in dolphins (Chapter 6). These results are in line with the considerable diameter of the trigeminal nerve which, in this respect, is second to the vestibulocochlear nerve only. In the bottlenose dolphin, the total number of axons in the trigeminal nerve (a maximum within the dolphin cranial nerves) equals that in the human.

Confirming morphological data, physiological experiments reported head sensitivity in dolphins to be high around the blowhole, the eyes, and on the melon and lips (Chapter 5). The blowhole area (epicranial complex) is extremely important for dolphins with respect to sound and ultrasound generation and emission as the efferent part of their sonar system (Table 10.1). Efficient sonar orientation may need detailed somatosensory (trigeminal) input from the blowhole area for fine regulation of nasal (vocal) muscular activity by means of the well-developed facial nerve. The primary somatosensory area within the neocortex is moderately developed compared to that of the auditory cortex, as are those of the motor and visual cortices (Chapter 6). Unfortunately, nothing is known on the respective topographic projection maps within these neocortical areas.

BRANCHIOMOTOR NUCLEI

The facial nerve and nucleus as the motor unit of the so-called "epicranial complex" are both very well developed (Table 10.1). This secondary ensemble, consisting of the upper respiratory tract and potentially "new" structures such as the nasal air sacs and acoustic fat bodies, is involved in the generation and emission of (ultra)sound signals. The concentration of several sheets of musculature around the paired nasal air sacs, in addition, may allow dolphins to produce complicated and highly variable sound signals in the sense of "acoustic facial expressions." As in other mammals, the motor nucleus of the facial nerve consists of several subdivisions which can be distinguished cytologically and may have different functional implications in dolphin sound and ultrasound generation. In comparison (Figs. 6.48, 10.3 and 10.4), other motor nuclei (trigeminal, ambiguus) are rather inconspicuous. The trigeminal motor nucleus is comparatively small, in correlation with the moderately developed masticatory complex which can be regarded as reduced. The ambiguus nucleus is larger in baleen

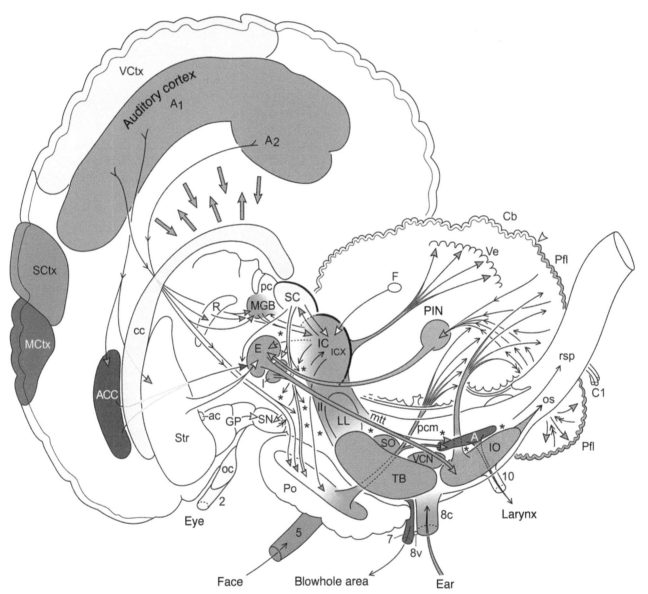

FIGURE 10.3 **Sagittal aspect of the bottlenose dolphin brain with selected functional systems.** The dominant auditory system (green) is put into the center of the circuitry which has already been described and discussed in Chapter 6. The auditory brainstem receives informations from two sides: afferent projections from the cochlear nerve (Ear) and efferent projections from the auditory cortex (A1, A2). Main relay stations outside the auditory pathway are the pontine nuclei (Po), elliptic nucleus and the inferior olivary complex. Motor system in red, somatosensory system blue, and visual system yellow. The facial nerve (7) innervates the blowhole muscles, the nucleus ambiguus (A) the larynx muscles. A, nucleus ambiguus; A1, primary auditory cortex; A2, secondary auditory cortex; ac, anterior (rostral) commissure; ACC, anterior cingulate cortex; C1, first cervical nerve; Cb, cerebellum; cc, corpus callosum; E, elliptic nucleus (Darkschewitsch); F, nucleus fastigii; GP, globus pallidus; I, interstitial nucleus of Cajal; IC, inferior colliculus; ICX, external cortex of IC; IO, inferior olive (medial accessory nucleus); ll, lateral lemniscus; LL, nucleus of the lateral lemniscus; MCtx, motor cortex; MGB, medial geniculate body; mtt, medial tegmental tract; oc, optic chiasm; os, olivospinal tract; pc, posterior commissure; pcm, pedunculus cerebellaris medius (auditory part); Pfl, paraflocculus (cortex in brown); PIN, posterior interposed nucleus; Po, pons with pontine nuclei; R, reticular thalamic nucleus; rsp, reticulospinal tract; SC, superior (rostral) colliculus; SCtx, somatosensory cortex; SN, substantia nigra; SO, superior olive; Str, striatum; TB, trapezoid body and nucleus; VCN, ventral cochlear nucleus; VCtx, visual cortex; Ve, vermis; 2, optic nerve; 5, trigeminal nerve; 8c, cochlear nerve; 8v, vestibular nerve; 10, vagus nerve; * * *, periaqueductal gray and reticular formation. *(Modified by Jutta Oelschläger after Oelschläger and Oelschläger, 2009. Brain. In: Perrin, W.F., Würsig, B., Thewissen, J.G.M. (Eds.), Encyclopedia of Marine Mammals. Academic Press, San Diego, CA, pp. 134–149.)*

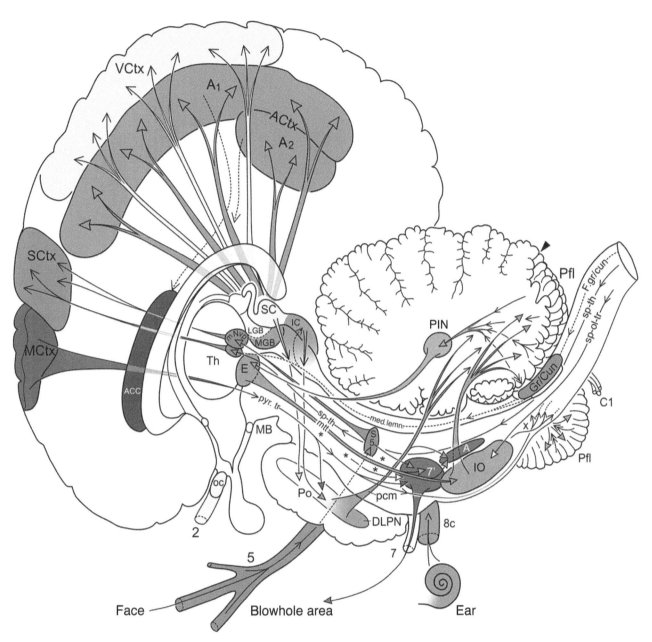

FIGURE 10.4   Sagittal view on the bottlenose dolphin brain with selected functional systems. This sketch shows the brain from the aspect of the long ascending and descending fiber systems (hence the auditory nuclei omission in the center of the brainstem). The somatomotor system (pyramidal tract; red) originates in the motor cortex (MCtx), gives off a branch to the elliptic nucleus (not shown) and to the pontine nuclei as well as the facial and ambiguus nuclei, passes below the inferior olive as a thin sheet, and expires immediately after the weak decussation of pyramids (X). The somatosensory system (blue) receives input from the face including the blowhole area. It is represented by the sensory root of the trigeminal nerve (5), principle sensory nucleus of trigeminal (S5), the medial nucleus ventralis posterior of thalamus (mNvp), and the somatosensory cortex (SCtx). In parallel, the somatosensory fibers from the spinal apparatus (Gr/Cun) ascend in the medial lemniscus (med.lemn; dotted line) and, relayed in the lateral nucleus ventralis posterior of thalamus (not labeled), terminate also in the somatosensory cortex. The visual system (yellow) is represented by the optic nerve (2), optic chiasm (oc), lateral geniculate body (LGB), superior colliculus (SC) and the visual cortex (VCtx). The auditory system is only present with the cochlea (Ear), cochlear nerve (8c) in the anterior myelencephalon, the inferior colliculus (IC) and the medial geniculate body (MGB) projecting to the auditory neocortex (ACtx, A1, A2). The auditory cortical field is underestimated here; it is located on the vertex of the hemisphere and therefore seen in the mediolateral projection. A, nucleus ambiguus; ACC, anterior cingulate cortex; A1, primary auditory cortex; A2, secondary auditory cortex; C1, first cervical nerve; Cun, cuneate nucleus; DLPN, dorsolateral pontine nuclei; E, elliptic nucleus (Darkschewitsch); F. gr/cun, gracile and cuneate fascicles; Gr, gracile nucleus; IC, inferior (caudal) colliculus; IO, inferior olive (medial accessory nucleus); MB, mamillary body; mtt, medial tegmental tract; pcm, pedunculus cerebellaris medius (auditory part); Pfl, paraflocculus; PIN, posterior interposed nucleus; Po, pons; pyr.tr, pyramidal tract; SC, superior (rostral) colliculus; sp-ol-tr, spino-olivary tract; sp-th, spinothalamic tract; Th, thalamus; 7, facial nerve; 7′, motor nucleus of facial nerve; X, decussation of pyramids; * * *, periaqueductal gray and reticular formation; arrowhead, border between the dorsal and ventral paraflocculus. *(Artwork by Jutta Oelschläger.)*

whales (mysticetes) than in toothed whales. In mammals, it innervates the muscles of the pharynx and larynx involved in swallowing and sound production. Whether its smaller size in odontocetes can be explained with a shift of the phonation process from the larynx to the epicranial complex, has to be shown in the future.

IMPORTANT FIBER SYSTEMS

In dolphins, certain fiber tracts are rather small whereas others are particularly well developed (Figs. 10.3 and 10.4). Such differences can be correlated either with the potential configuration in ancestral hoofed animals (pyramidal tract small; medial lemniscus well developed after the fusion with the trigeminal fiber masses; see Chapter 6). The crus cerebri, a strong fiber tract within the cerebral peduncle seems to be massive because of considerable somatomotor and auditory projections to the pontine nuclei. Here they are relayed to the cerebellar cortex for premotor processing (Fig. 10.5) and presumably very important for audiomotor navigation (Figs. 10.3 and 10.4). An obvious correlation can be found in the extraordinary degree of structural and quantitative development of the ascending auditory pathway, as a concession to excellent hearing with respect to echolocation by means of a high-capacity sonar system (for exceptions to this see Section "Auditory System" in this chapter). Within the extrapyramidal system on a larger scale, the medial tegmental tract is very well developed compared to terrestrial mammals like primates and their close relatives, the artiodactyls. This relatively large size has to be correlated with the outstanding development of the associated elliptic nucleus (Figs. 10.3 and 10.4) which is very small in humans and hoofed animals (also called nucleus of Darkschewitsch). In dolphins, its efferent fibers mainly run to the inferior olive via the medial tegmental tract, as part of the core loop of the cerebellum that projects back to the elliptic nucleus, thalamus, and cortex (Chapter 6). A parallel core loop is present in terrestrial mammals (primates) but originates in the red nucleus and uses other subnuclei, nuclei, and cortical areas in the inferior olivary complex and the cerebellum, respectively. One of the main sources of input into this core loop is the pyramidal tract, including a significant part of auditory projection. In dolphins, the loop itself can be regarded a flywheel for the integration of multisensory projections from the cortex and brainstem with respect to motor activity in phonation and locomotion. Here, the main input into this system should be of auditory origin (Chapter 6).

NEOCORTEX FUNCTION

As in many large-sized mammals, the neocortex is the dominant structure of the dolphin (odontocete) brain (Figs. 10.2 and 10.6) and it is characterized by a remarkable folding of the gray matter (Fig. 10.5; Table 10.3). Although the gyral organization of the cerebral cortex shows many specific patterns in dolphins, the sequence of the primary projection areas (motor, somatosensory, auditory, and visual fields) across the surface of the telencephalic hemisphere is similar to that in terrestrial mammals (Chapter 6). The auditory field of dolphins is always the largest primary projection area (Figs. 10.3 and 10.4) and comprises the largest cortical volume with the highest total number of neurons. The highest neuron densities in both layer III and V are found in the visual cortex of the common dolphin and the bottlenose dolphin. The neocortex of the human is distinctly thicker throughout all areas investigated.

In principle, the structure of the dolphin neocortex follows the mammalian bauplan, that is, it exhibits six cortical layers composed of different neuron populations with characteristic features as to size and shape of perikaryon as well as the arrangement and local distribution of their processes. In general the dolphin neocortex is dominated by large and medium-sized pyramidal neurons. Whereas in the human there is a layer IV in every sensory area (primary somatosensory, S1; primary auditory, A1; primary visual, V1), the situation in dolphins is inconsistent. It seems that in young specimens of the bottlenose dolphin, layer IV occurs in the visual field, while adult animals show only more or less discontinuous residues in a restricted area of V1 (Garey and Leuba, 1986; Morgane et al., 1988). Because layer IV is essential for the perception, processing, and projection of thalamocortical (sensory) input, the focus in studies on the dolphin neocortex was recently laid on the layer III–V complex (Kern et al., 2011). As to selective neuron density in layers III and V of the neocortical areas (primary motor, M1; S1, A1, V1), the human figures are sometimes higher and sometimes lower than in the dolphin, with the values rising markedly in the visual cortex (V1). Apart from that, the density values of the human sensory cortices are distinctly augmented by the extremely numerous but minute stellate cells in layer IV of the sensory cortices (see Chapter 6, Section "The Cortices").

SPECIFIC AND ALLOMETRIC PHENOMENA IN TOOTHED WHALE BRAINS

Comparison of the smallest and the largest brains among the toothed whales may help to understand morphological trends that seem to have occurred during evolution (Fig. 10.7). The two brains differ very much in size; the adult La

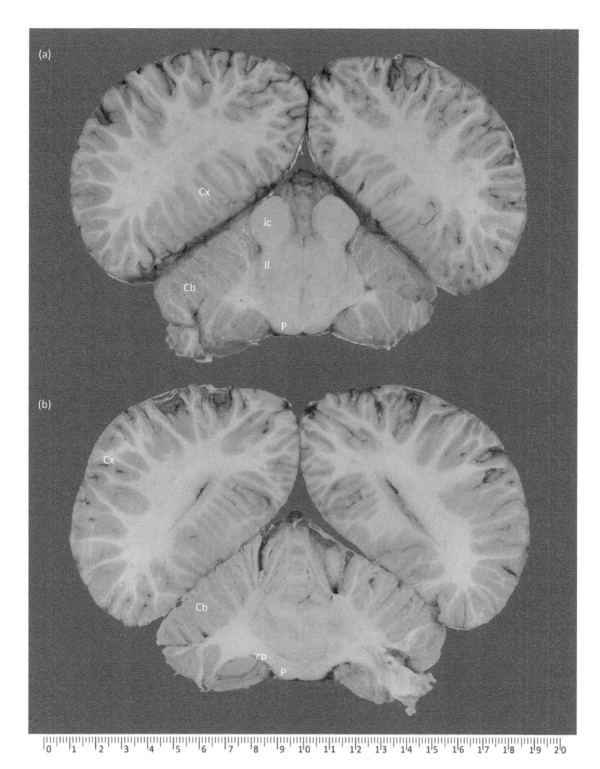

FIGURE 10.5 Transverse sections through a Risso's dolphin brain on level of the (a) inferior (caudal) colliculus (*ic*) and (b) the middle cerebellar peduncle (*cp*). *Cb*, cerebellum; *Cx*, cortex (cerebrum); *ll*, lateral lemniscus and its nuclei; *P*, Pons.

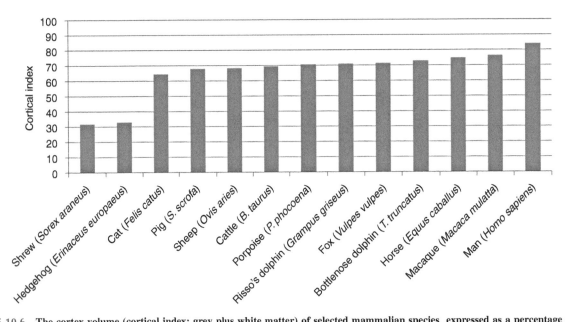

FIGURE 10.6 **The cortex volume (cortical index; grey plus white matter) of selected mammalian species expressed as a percentage of total brain volume.** The relative cortex volume of dolphins is in the range of hoofed mammals (cattle, horse) and carnivores (fox). Only humans have a distinctly larger relative cortex volume; insectivores (shrew, hedgehog) have significantly smaller cortex volumes. *(From Manger, 2006. An examination of cetacean brain structure with a novel hypothesis correlating thermogenesis to the evolution of a big brain. Biol. Rev. 81, 293–338.)*

TABLE 10.3 Structures of the Central Nervous System and Their Characteristics in Extant Toothed Whales

Area	Structure	Size	Functional Implications, Remarks	Difference From the Presumed Ancestral Configuration Progression (⬆) Regression (⬇)
Brain	Total brain	Very large	Strong encephalization	⬆⬆⬆
	Telencephalon	Very large	Strong telencephalization	⬆⬆⬆
	Cortex	Thin, extremely expanded and convoluted	Strong neocorticalization (parallel to pons, cerebellum)	⬆⬆⬆
	Hippocampus	Very small absolute and relative size	See main text	⬇⬇
	Brainstem	Large	Size of cranial nuclei: auditory, trigeminal systems, etc.	⬆
	Auditory system	Very large	Sonar	⬆⬆
	Pons/Cerebellum	Very large	Acousticomotor correlation, nasofacial expression, 3D locomotion	⬆⬆
	Medulla oblongata	Very large	Acousticomotor correlation, nasofacial expression, 3D locomotion	⬆⬆

Note: Number of arrows indicate the significance of progression and regression.

Plata dolphin (franciscana, *Pontoporia blainvillei*) has an average brain mass of a little more than 200 g, the largest sperm whales may attain nearly 10,000 g, a factor of 1: 50. However, the adult franciscana has a body mass of about 35 kg and the sperm whale even more than 35 t, a factor of 1: 1000. This comparison shows that the body mass grows much faster than brain mass. The biological implications of this phenomenon are not well understood. They may have to do with different strategies in the ontogenetic development of different species. But what we can see in this

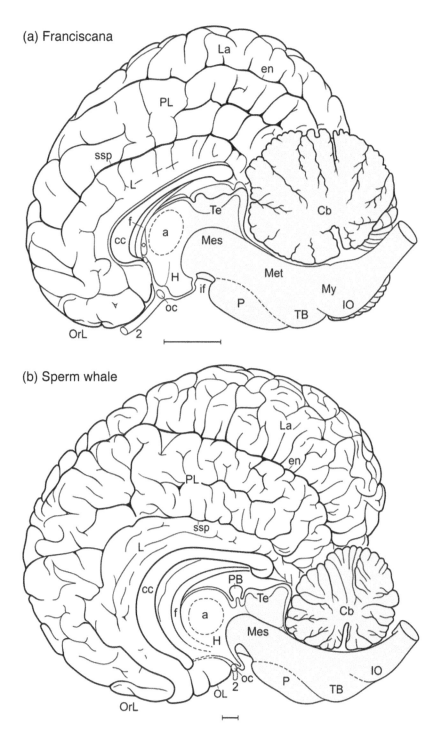

FIGURE 10.7 **Confrontation of the smallest and the largest extant toothed whale brains (Scales = 1 cm).** (a) franciscana (*Pontoporia blainvillei*), and (b) sperm whale (*Physeter microcephalus*) in mediosagittal section, brought to the same length. The size of typical dolphin brains (common dolphin, bottlenose dolphin) is between these two extremes but closer to the franciscana. The brain (a) shows the typical features known from other smaller toothed whales; the size of the brainstem in the franciscana (yellow; Mes, mesencephalon; Met, metencephalon; Cb, cerebellum; My, myelencephalon) is rather impressive compared to the rest of the brain. In contrast, the brainstem of the sperm whale is comparatively smaller in relation to the brain as a whole. As to the shape of the brain, that of the sperm whale is much more globular due to maximal cortex development. Also, the rotation of the cerebral hemisphere is much more advanced here, and the corpus callosum curved in a semicircle. In comparison with the situation in the franciscana (and genuine dolphins, delphinids), the ratio of the cerebellum volume in the whole brain volume is much lower in the sperm whale. Adult killer whales (*Orcinus orca*; the largest delphinids, not shown) and sperm whales have about the same brain mass (mean 6,728 g vs. 8,076 g) but very different body masses (3,000 kg vs. 35,000 kg); nevertheless, the cerebellum mass attains about 14% of the whole brain in the killer whale against only about 7% of the total brain mass in the sperm whale (Ridgway and Hanson, 2014). Within the brain, those parts of the cortex adjoining or near the corpus callosum (cc) seem to be conservative in the sperm whale as to relative size increase within to the brain as a whole; in contrast, the superiormost cortices (eg, lateral gyrus, La) may have extended the hemisphere dorsally due to massive growth of the auditory cortical fields on the vertex of the hemisphere. In dolphins, these auditory cortices are about as large as the other primary cortical areas together. As a result, the small franciscana brain obviously has a fully developed ascending auditory pathway in the brainstem as have larger delphinid brains which tend to have much more neocortex (cf. Fig. 6.7c). a, interthalamic adhesion (thalamus); Cb, cerebellum; cc, corpus callosum; en, entolateral sulcus; f, fornix; H, hypothalamus; if, interpeduncular fossa; IO, inferior olive; L, limbic lobe; La, lateral gyrus; oc, optic chiasm; OL, olfactory lobe; OrL, orbital lobe; P, pons; PB, pineal body; PL, paralimbic lobe; ssp, suprasplenial (limbic) sulcus; TB, trapezoid body; Te, Tectum; °, interventricular foramen; 2, optic nerve. (*Artwork by Jutta Oelschläger.*)

confrontation of the two brains is that brain growth is very much in favor of the cortex (neocortex) whereas the ratio of the brainstem falls back from the franciscana to the sperm whale. In this case, this is also true for the cerebellum which is relatively very small to total brain size in the sperm whale. Interestingly, this does not hold for the largest dolphin (delphinid), the killer whale; here the cerebellum is much larger in a brain that rivals that of the sperm whale in size (Ridgway and Hanson, 2014).

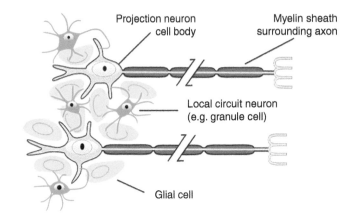

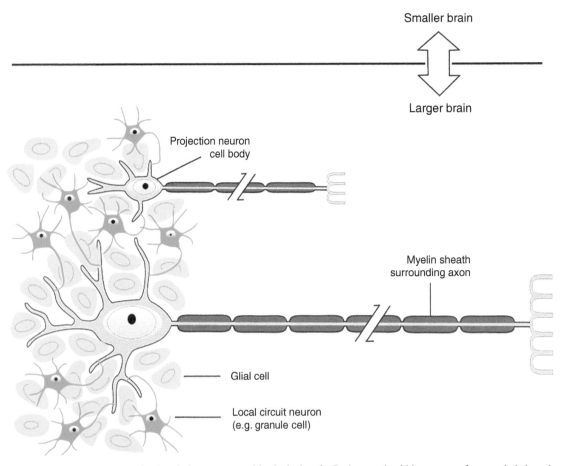

FIGURE 10.8 **Allometric phenomena in the whale cortex on a histological scale.** Brain growth within a group of mammals is bound to several morphological and quantitative cytoarchitectonic changes. In comparison with the smaller brain above, the projection neurons (layer V) in a larger brain are larger as to the size of their somata, the diameter of their axons and of their myelin sheaths. At the same time, the number of the projection neurons decreases and the ratio of the local neurons (interneurons, granule cells) and of the glia cells increases. Larger projection neurons have longer axons and higher transmission velocities to their periphery. Such allometric growth processes also mean that in a larger brain the size differences between largest and smallest neurons increase. *(Modified by Jutta Oelschläger from Deacon, 1990. Rethinking mammalian brain evolution. Am. Zool. 30, 629–705.)*

It is interesting to see what increase in the size of larger animals means in detail and with respect to the morphology of the neocortex as its largest component. A comparatively simple phenomenon is the size increase of the single neocortical neurons which may become very large (Fig. 10.8). At the same time, these neurons rarify; their density per volume unit of neocortex decreases allometrically (Poth et al., 2005). A significant difference to terrestrial mammals (primates) is cortex width; the latter is thinner in dolphins than in terrestrial artiodactyls of the same body dimension and much thinner than in the human. In both dolphins and hoofed animals, the molecular layer (layer I) is particularly thick and makes up about one-third of the whole cortex width. While adequate data on hoofed animals are not available, it was shown for the striped dolphin (*Stenella coeruleoalba*) and the pilot whale (*Globicephala melaena*) that 70% of all synapses in the neocortex of these animals are located in layers I and II. This is consistent with the argumentation of Deacon (1990) that in dolphins the principal part of input coming from the sensory organs via the thalamus ends in the molecular layer, and that this situation may have been derived from basic terrestrial mammals by means of the shift of recipient neurons from layer IV to layers I and II (Fig. 10.9). In primates the main termination site for incoming information is layer IV but this layer does not exist in many terrestrial mammals as, for example, hoofed animals or is incomplete and indistinct, respectively. Whether this specific situation in dolphins is related to a more vertical and quick mode of intracortical information processing is not clear to date. Also, for a better understanding of their situation, basic investigations on artiodactyl hoofed animals would be needed.

In this place, it may be worthwhile to summarize some of the most characteristic features of dolphins in order to understand their marvelous and successful evolution in an environment so problematic for mammals with terrestrial ancestors.

TOPOGRAPHY AND FUNCTIONAL IMPLICATIONS: A NEUROBIOLOGICAL SYNTHESIS

Apart from a plethora of other adaptive features across the body, dolphins are characterized by two major phenomena more or less strictly correlated with the physical parameters of the aquatic environment: their sensory equipment is partly restricted with the exception of the auditory system which is strongly hypertrophied, while their locomotory apparatus is highly derived. Figures 10.10 and 10.11 show the major topographic regions of the dolphin head in dorsoventral aspect in order to explain the most important details and characteristics of the evolutionary processes the dolphins have accomplished. A more formal representation of the dolphin chracteristics and adaptations is given in Tables 10.1–10.3: Table 10.1 gives a more general view on the topic (head), Table 10.2 focuses on the peripheral nervous system, and Table 10.3 on the central nervous system.

1. **The nose and the olfactory system** (Fig. 10.10, Table 10.1) were totally reorganized during evolution presumably because of problems with the readaptation of the chemoreceptor systems to the aquatic environment. At the same time, the nose was rededicated to a totally different function, that is, the production of sound by means of secondary "vocal cords" that are suspended in the upper respiratory tract (Chapter 5). The larynx, as the vocalization organ in terrestrial mammals, has obviously become a mechanical pump to produce air pressure needed for the generation of ultrasound for orientation and of sound for communication. Together with the high-capacity hearing organ, the epicranial complex forms a powerful sonar system which, in some respect, is comparable to that in bats. In dolphins, the lack of adequate olfactory stimuli in the aquatic environment and perhaps the mechanical stress in vocalization may have led to the reduction of the olfactory mucous membrane and, subsequently, to the loss of the anterior (rostral) part of the olfactory system. Secondary olfactory centers (Fig. 10.10: "OT") are still present in the dolphin brain but their functional implications are unclear. The mechanical apparatus of the acoustic nose became equipped with a periphery of auxiliary structures (epicranial or nasal complex) like facial muscles for operating the blowhole and upper respiratory tract and a set of acoustic fat bodies for the transmission of sound/ultrasound throughout the forehead and its emission into the surrounding water body.

    During the embryonal and early fetal period, the dolphin brain is similar to that in other mammals. Later on, it becomes shorter and more and more spherical, somehow reminiscent of a boxing glove (Fig. 10.2). This process parallels the so-called "telescoping effect" of the skull (see Chapter 5). In the adult, both brain and braincase are wider than long, a situation exceptional among mammals. In Figure 10.10, the brain surface and the anterior wall of the brain capsule are separated by a narrow gap; in the living dolphin this gap is occupied by the meninges including the subarachnoid space. Adjacent to the brain case rostrally is the nasal region with the bony nasal passages (in cross-section; arrows: air stream) which run perpendicular. In the area of the nasal tubes the braincase is equivalent to the former cribriform plate. With respect to terrestrial mammals however, the nasal tract of adult dolphins has been profoundly reconstructed during

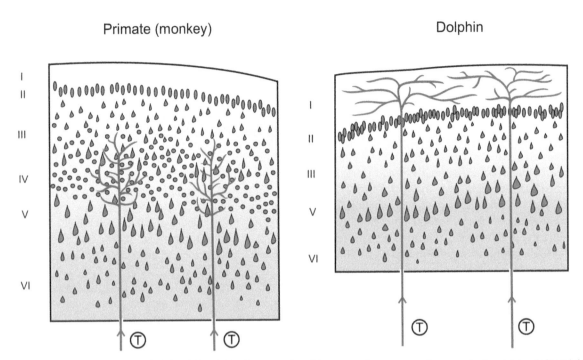

Primate (monkey)   Dolphin

FIGURE 10.9 Different projections of thalamic input into the neocortex of two exemplary groups of mammals. In primates (left) and dolphins (right) the main thalamic projections (T) ascend into different cortical layers. In primates, the terminations of the projections end in the wide granular and densely packed layer IV and on pyramidal cells of the inner zone of layer III. In the much thinner agranular pyramidal cortex of the dolphin, the thalamic fibers ascend up to the wide layer I which is poor in neurons. In this layer, the thalamic fibers widely ramify tangentially and preferentially contact dendrites of pyramidal neurons in layers II and V which send their axons into the periphery. *(Modified by Jutta Oelschläger from Deacon, 1990. Rethinking mammalian brain evolution. Am. Zool. 30, 629–705.)*

FIGURE 10.10 Dorsal aspect of the whole head of dolphin with the nasal sound generation apparatus (nasal complex), major sensory organs and accessory brain structures for an overview of principle functional connections and mechanisms. This presentation is deducted from our knowledge on the mammalian bauplan and the specifics of dolphin neurobiology; it is meant as a reference tool for functional and evolutionary aspects seen throughout the dolphin head and body. On the left, the core loop of the cerebellum (E and I, IO, Pfl, PIN, E) is seen from above. The contralateral projection of auditory input from the left pontine nuclei (DLPN) to the right cerebellar hemisphere is not shown here; this allows detailed insight into the transmission and dissemination of auditory information throughout the brainstem. Particularly interesting are the functional implications of the nuclear (PNO, PNC, Gi) and diffuse reticular formation (asterisks) as a superior relay and integration center for all kind of afferent systems (see detail in Fig. 10.11). By this, the projection of auditory information on motor target nuclei (5′, 7′, A) for dolphin vocalization is obvious. Other details (SC, 8v, red arrows) indicate functional implications of the vestibular system. As to audiomotor navigation, the superior (rostral) and inferior (caudal) colliculi seem to play an important role as integration centers. At the same time, the functional role of Deiters' nucleus (D), tegmental nuclei (IO, Gi, PNO, PNC), and of the rubrospinal tract in the innervation of the motor horns in the spinal cord is shown (locomotion and navigation after sound). The extremely small vestibular apparatus (v) is magnified (open arrow) to show the semicircular canals (sc) with the very thin vestibular nerve (8v). The potential output of the rudimentary vestibular nuclei to the small nuclei of the external eye muscles (3′, 4′) is indicated by broken red lines. Another small detail in this context is the innervation of the tensor tympani and stapedius muscles (tt, stap; protection of the inner ear against acoustic overload). The terminal nerve and ganglia (tn) connect the epithelium of the nasal passage with the area of the olfactory tubercle (OT). The ear bones (pe, t) are united in the tympanoperiotic complex; the latter is uncoupled from the skull (dotted line) except for the posterior process of tympanic (simple arrow), the only firm suspension of the ear bone complex. The neck region has been shortened; the cervical vertebrae are fused; the movability of the head is restricted to the transverse hinge- joint (atlantooccipital joint) between the occipital condyles (Occ) and the cervical vertebral complex (cvc); the atlantoaxial joint is lost. Motor and vestibular systems in red, somatosensory system blue, visual system light yellow, auditory system green, limbic system violet. The emitted sonar beam and the returning echoes are given as trains of green pulses (sound). Bones are shown in brown. The frame indicated here (broken line) delimits the magnified detail of the most complicated area in Figure 10.11. The colours and abbreviations largely follow those in preceding figures: A, nucleus ambiguus; A1, primary auditory cortex; A2, secondary auditory cortex; ACC, anterior cingulate cortex; Amy, amygdaloid complex; Co, cochlea; D, lateral vestibular nucleus of Deiters' (the other vestibular nuclei attached to it are rudimentary in dolphins); DLPN, dorsolateral pontine nuclei; E, elliptic nucleus (Darkschewitsch); eam, external auditory meatus (reduced); Fl, flocculus; Gi, gigantocellular reticular nucleus; I, interstitial nucleus of Cajal; IC, inferior (caudal) colliculus; IO, inferior olive (medial accessory); LSO, lateral superior olive (the medial superior olive (#) is rudimentary); MB, mamillary body; ml, medial lemniscus; mtt, medial tegmental tract; np, transverse section through nasal passages (arrows indicating the airstream); ns, nasal sac; o-sp, olivospinal tract; OT, olfactory tubercle; pcm, pedunculus cerebellaris medius (auditory part); pe, periotic bone; Pfl, paraflocculus; Ph, parahippocampal gyrus; PIN, posterior interposed nucleus; PNO, nucleus reticularis pontis oralis; PNC, nucleus reticularis pontis and caudalis (horizontal gaze center); Po, pons; py, pyramidal tract (fading out); rsp, reticulospinal tract; SC, superior (rostral) colliculus; spo, spinoolivary tract; St, striatum; stap, musculus stapedius; t, tympanic bone; tn, terminal nerve; tt, musculus tensor tympani; v, vestibular system; V, vermis; VCN, ventral cochlear nucleus; the dorsal cochlear nucleus (cross) is rudimentary; ve, ventricle; vsp, vestibulospinal tract; 2, optic nerve; 3′, nucleus of oculomotor nerve; 4′, nucleus of trochlear nerve; 5, trigeminal nerve with main sensory branches (5.1-5.3); 5′, motor nucleus of trigeminal; 7, facial nerve; 7′, motor nucleus of facial nerve innervating the blowhole musculature; 8c, cochlear nerve; 8v, vestibular nerve; * * *, periaqueductal gray and reticular formation. For more information on the ensemble of organ systems see the relevant chapters in this book. *(Artwork by Jutta Oelschläger.)*

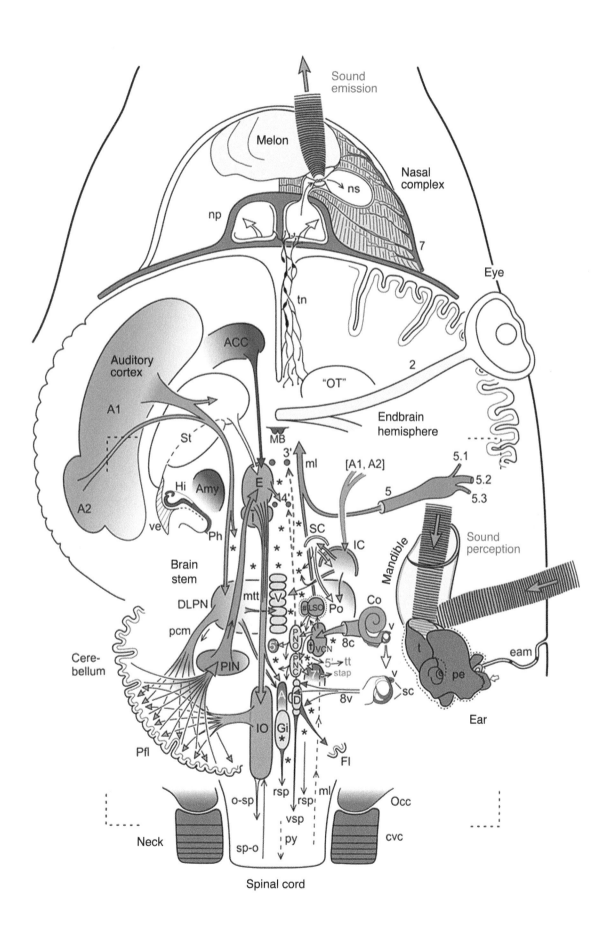

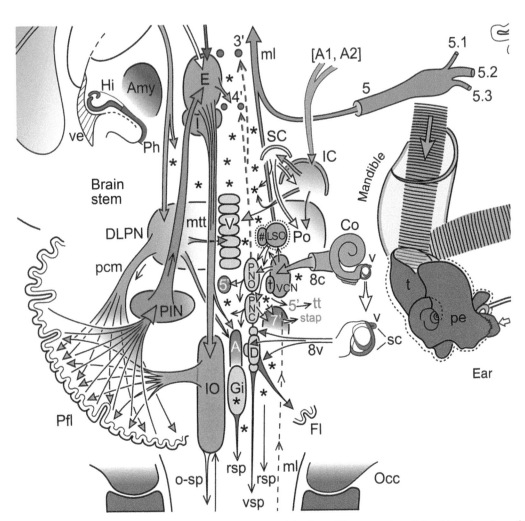

FIGURE 10.11 **Detail from Figure 10.10 with the center of brain connectivity.** Dotted lines around medial and lateral nucleus of superior olive (#, LSO) represent periolivary nuclei which project into the cochlea; dotted lines around the tympanoperiotic complex indicate its suspension and uncoupling respectively, by dense connective tissue. Arrow indicates the single direct bony contact of the tympanoperiotic complex (Ear) with the neighbouring bones. For abbreviations see Fig. 10.10. *(Artwork by Jutta Oelschläger.)*

ontogenesis: there are no olfactory foramina, no olfactory fiber bundles coming from the nasal mucous membrane, and no olfactory bulb. As a trace of all this can be regarded fiber bundles of the terminal nerve (tn) that arise from the olfactory placode of the embryo, in parallel to the anterior olfactory system (Chapter 6). In contrast to the anterior olfactory system, the terminal nerve persists until adulthood in dolphins and is maximally developed here among the Mammalia.

2. **The eye and its muscular armature** (Fig. 10.10, Table 10.2) was retained as an important sensory system but adapted to difficult light conditions under water and obviously to high-velocity information processing (see Chapters 5 and 6). Its retinal resolution is obviously moderate in favor of quick detection of objects in murky environments. As to visual acuity, the bottlenose dolphin is well behind the human, horse, dog, and cat; it equals the harbor seal and is followed by the goat, Indian elephant, and red deer.

3. **The somatosensory system** (Fig. 10.10, Table 10.2) is bipartite as to its functional importance; in the spinal cord, it is moderately developed. The medial lemniscus proper (ml), ascending to the brain and here to the thalamus and somatosensory cortex (Figs. 10.4, 10.10, and 10.11) is thin. In the tegmentum of the midbrain, however, it is distinctly reinforced by the strong somatosensory projection of the trigeminal nerve (trigeminal lemniscus). This is correlated with

a considerable thickness of this nerve which is the second largest cranial nerve in dolphins after the vestibulocochlear nerve and comprises a number of fibers similar to that in the human. As far as is known today, the trigeminal nerve which has a rather weak motor component, intensively innervates the face of the dolphin, particularly the blowhole, eye, and lip region. It may have a close functional association with the facial nerve that operates the blowhole musculature with respect to vocalization and respiration (Fig. 10.10).

4. **The auditory system** (Fig. 10.10, Tables 10.1 and 10.2) has been heavily modified, as hearing is extremely important for dolphin life. The outer ear is reduced obviously due to the relocation of the auditory gateway from the external pinna (lost) to a new sound channel in the lower jaw. The latter bone contains an acoustic fat body which may receive sound through the flat and thin outer wall of the mandible (panbone) and transmits it to the tympanic plate. This ultra-thin bony plate is part of the tympanoperiotic complex, a secondary formation consisting of the two large, extremely dense and heavy tympanic and periotic bones. The middle ear cavity with the chain of hyperostotic ossicles and their little muscles lies enclosed and protected between them. For optimal function as to the hearing process, the tympanoperiotic complex was uncoupled from the skull base, obviously in order to dampen or avoid bone conduction, with the result of better directional hearing. According to its dominant position in dolphins, the auditory system (Tables 10.2 and 10.3) is well represented in the brain and much hypertrophied with respect to other terrestrial mammals with the exception of bats.

5. **The vestibular system** (Figs. 10.10 and 10.11, Tables 10.1 and 10.2) is dramatically reduced, particularly the semicircular canals with their cupular organs. There is no other example for such a reduction of the vestibular system among the mammalia and even the vertebrates. This extraordinary situation in dolphins may have several reasons which are more or less substantiated by morphological criteria previously discussed. In Figs. 10.10 and 10.11 the absolute size of the cochlea and vestibular apparatus are obvious as is their size ratio to one another. This extreme difference in volume impressively shows how far apart these two systems are functionally.

   The morphological and functional situation of the cervical region may help to explain the small size of the vestibular system in dolphins. In the cetacean ancestors, the body had to become hydrodynamic for quick and enduring locomotion in the dense medium water; the neck was integrated in the incipient fusiform body contour by strengthening its musculature, and the head included in powerful sinusoid undulations. This required a compact transition from head to vertebral column, achieved by, first, the shortening and fusion of the cervical vertebrae and, second, the elimination of the atlanto-axial joints. Concomitantly, the remaining atlanto-occipital articulations were reinforced: they allow the movements necessary for such powerful undulations (namely simple rotations in a hinge-joint) with its axis running through both occipital condyles. The implementation of this restricted movability of the head was obviously a major cause for the reduction of the vestibular system as several feedback loops between head and neck became obsolete (see Chapter 6). The repetitive up-and-down movements of the head during locomotion may seem to be of limited informative value compared with orientation reactions in terrestrial mammals; they are, however, impressive when dolphins scan the environment before them with their sonar system. This mechanical simplification in the neck region probably affected also the visual reflexes and consequently the muscular equipment of the eye, confining the eye–head–neck movement reference frame.

6. **Locomotion** (Figs 10.3, 10.4, and 10.10). As in other mammals, the pyramidal tract should innervate (i) the brainstem nuclei (eg, those of the eye muscle nerves, trigeminal, facial, ambiguus, hypoglossal) and their accessory musculature, and (ii) the motor horns of the rostral cervical segments of the spinal cord for muscles of the neck, shoulder girdle and flipper. In dolphins, however, the spinal cord does not extend further caudally. Another part of the pyramidal tract (iii) leads via the pontine nuclei to the cerebellum which generally acts as a multisensory premotor integration center; the latter projection (cortico-ponto-cerebellar tract) is supposed to comprise the bulk of the cortical auditory output.

   Dolphins have nearly no spinal pyramidal tract but a strongly developed nonpyramidal system which may compensate for this deficit. Here, a special circuitry consisting of hypertrophic brainstem nuclei and the very large cerebellum is characterized by the considerable size of the nucleus of Darkschewitsch (elliptic nucleus). It is probable that inputs from the neocortex and spinal cord are processed in this central loop and then transferred to the reticular formation. Together with extrapyramidal projections into the spinal cord (reticulospinal, vestibulospinal, and olivospinal tracts) the particularly well developed reticular formation may be engaged in the premotor processing and projection of auditory and other sensory information and thus be important for audiomotor navigation. Perhaps this configuration (very large elliptic nucleus, cerebellum, and reticular formation) is advantageous for the aquatic life of dolphins which show an extreme type of mammalian locomotion, the enduring powerful sinusoidal undulation of the body.

REFERENCES

Boisserie, J.-R., Fisher, R.E., Lihoreau, F., Weston, E.M., 2011. Evolving between land and water: key questions on the emergence and history of the Hippopotamidae (Hippopotamoidea, Cetancodonta, Cetartiodactyla). Biol. Rev. 86, 601–625.

Dawson, W., Dawson, W.W., Hope, G.M., Ulshafer, R.J., Hawthorne, M.N., Jenkins, R.L., 1983. Contents of the optic nerve of a small cetacean. Aquat. Mamm. 10, 45–56.

Deacon, T.W., 1990. Rethinking mammalian brain evolution. Am. Zool. 30, 629–705.

Fleischer, G., 1976. Hearing in extinct cetaceans as determined by cochlear structure. J. Paleontol. 50, 133–152.

Gao, G., Zhou, K., 1992. Fiber analysis of the optic and cochlear nerves of small cetaceans. In: Thomas, J.A., Kastelein, R.A., Supin, A.Y. (Eds.), Marine Mammal Sensory Systems. Plenum Press, New York, pp. 39–52.

Garey, L.J., Leuba, G., 1986. A quantitative study of neuronal and glial numerical density in the visual cortex of the bottlenose dolphin: evidence for a specialized subarea and changes with age. J. Comp. Neurol. 247, 491–496.

Gatesy, J., Geisler, J.H., Chang, J., Buell, C., Berta, A., Meredith, R.W., Springer, M.S., McGowen, M.R., 2013. A phylogenetic blueprint for a modern whale. Mol. Phylogenet. Evol. 66, 479–506.

Kern, A., Seidel, K., Oelschläger, H.H.A., 2009. The central vestibular complex in dolphins and humans: functional implications of Deiters' nucleus. Brain Behav. Evol. 73, 102–110.

Kern, A., Siebert, U., Cozzi, B., Hof, P.R., Oelschläger, H.H.A., 2011. Stereology of the neocortex in Odontocetes: qualitative, quantitative, and functional implications. Brain Behav. Evol. 77, 79–90.

Ketten, D.R., 1992. The marine mammal ear: specializations for aquatic audition and echolocation. In: Webster, D.B., Popper, A.N., Fay, R.R. (Eds.), The Evolutionary Biology of Hearing. Springer, New York, NY, pp. 715–717.

Knopf, J.P., Hof, R., Oelschläger, H.H.A., (in press). The neocortex of Indian river dolphins (Genus Platanista): comparative qualitative and quantitative analysis.

LeDuc, R., 2009. Delphinids, overview. In: Perrin, W.F., Würsig, B., Thewissen, J.G.M. (Eds.), Encyclopedia of Marine Mammals. Academic Press, San Diego, CA, pp. 298–302.

Manger, P.R., 2006. An examination of cetacean brain structure with a novel hypothesis correlating thermogenesis to the evolution of a big brain. Biol. Rev. 81, 293–338.

Morgane, P.J., Glezer, I.I., Jacobs, M.S., 1988. Visual cortex of the dolphin: an image analysis study. J. Neurol. Comp. 273, 3–25.

Murayama, T., Somiya, H., Aoki, I., Ishii, T., 1995. Retinal ganglion cell size and distribution predict visual capabilities of Dall's porpoise. Mar. Mamm. Sci. 11, 136–149.

Oelschläger, H.H.A., 2008. The dolphin brain – a challenge for synthetic neurobiology. Brain Res Bull 75, 450–459.

Oelschläger, H.H.A., Oelschläger, J.S., 2009. Brain. In: Perrin, W.F., Würsig, B., Thewissen, J.G.M. (Eds.), Encyclopedia of Marine Mammals. Academic Press, San Diego, CA, pp. 134–149.

Poth, C., Fung, C., Güntürkün, O., Ridgway, S.H., Oelschläger, H.H.A., 2005. Neuron numbers in sensory cortices of five delphinids compared to a physeterid, the pygmy sperm whale. Brain Res. Bull. 66, 357–360.

Reidenberg, J.S., Laitman, J.T., 2009. Cetacean prenatal development. In: Perrin, W.F., Würsig, B., Thewissen, J.G.M. (Eds.), Encyclopedia of Marine Mammals. Academic Press, San Diego, CA, pp. 220–230.

Ridgway, S.H., Hanson, A.C., 2014. Sperm whales and killer whales with the largest brains of all toothed whales show extreme differences in cerebellum. Brain Behav. Evol. 83, 266–274.

Sigmund, L., 1981. Morphometrische Untersuchungen an Gehirnen der Wiederkäuer (Ruminantia, Artiodactyla, Mammalia). 1. Makromorphologie des Gehirns der Hirschferkel (Tragulidae). Vest. cs. Spolec. zool. 45, 144–156.

Smith, B.D., Braulik, G.T., 2009. Susu and Bhulan. Platanista gangetica gangetica and P. g. minor. In: Perrin, W.F., Thewissen, J.G.M. (Eds.), Encyclopedia of Marine Mammals, second edition Academic Press, San Diego, pp. 1135–1139.

Štěrba, O., Klima, M., Schildger, B., 2000. Embryology of dolphins—staging and ageing of embryos and fetuses of some cetaceans. Adv. Anat. Embryol. Cell Biol. 157, 1–133.

Thewissen, J.G.M., Cohn, M.J., Stevens, L.S., Bajpai, S., Heyning, J., Horton, W.E., 2006. Developmental basis for hind-limb loss in dolphins and origin of the cetacean bodyplan. Proc. Natl. Acad. Sci. 103, 8414–8418.

Thewissen, J.G.M., Cooper, L.N., George, J.C., Bajpai, S., 2009. From land to water: the origin of whales, dolphins, and porpoises. Evol. Educ. Outreach 2, 272–288.

Thewissen, J.G.M., Heyning, J.E., 2007. Embryogenesis and development in Stenella attenuata and other cetaceans. In: Miller, D.L. (Ed.), Reproductive Biology and Phylogeny of Cetacea—Whales, Dolphins and Porpoises, Reproductive Biology and Phylogeny. Science Publishers, Enfield, pp. 307–329.

Verhaart, W.J., 1970. Comparative anatomical aspects of the mammalian brain stem and the cord. Stud. Neuroanat. 2, 1–312.

Index